Reaction Rate Theory and Rare Events

Reaction Rate Theory and Rare Events

Baron Peters

elsevier.com

Elsevier
Radarweg 29, PO Box 211, 1000 AE Amsterdam, Netherlands
The Boulevard, Langford Lane, Kidlington, Oxford OX5 1GB, United Kingdom
50 Hampshire Street, 5th Floor, Cambridge, MA 02139, United States

Library of Congress Cataloging-in-Publication Data
A catalog record for this book is available from the Library of Congress

British Library Cataloguing-in-Publication Data
A catalogue record for this book is available from the British Library

ISBN: 978-0-444-56349-1

For information on all Elsevier publications
visit our website at https://www.elsevier.com

Publisher: John Fedor
Acquisition Editor: Kostas Marinakis
Editorial Project Manager: Sarah Jane Watson
Production Project Manager: Paul Prasad Chandramohan
Designer: Matthew Limbert
Cover Painting: Linden and Baron Peters

Typeset by VTeX

To my little illustrators

Aven Peters
age 5

Linden Peters
age 7

Contents

Preface

Chemists, engineers, physicists, and mathematicians have all made important contributions to the theory of reaction kinetics and rare events. But despite the cross-disciplinary nature of contemporary research, the *literature* on reaction rate theory and rare events remains compartmentalized into rather traditional subcategories. This book provides a uniquely broad and cross-disciplinary introduction to the most powerful and practical rate theories and rare events methods. Moreover, much of the content has not previously been included in any book. In presenting the latest developments, I have tried to use consistent notations throughout and to distill the often dense mathematics from the original literature into an accessible form. Of course, the mathematics cannot (and should not) be fully removed from the subject. My hope is that chemists, engineers, and physicists alike will come to embrace the necessary mathematics in reaction rate theory as they already do in statistical mechanics, transport phenomena, and quantum mechanics.

I am grateful to the Department of Chemical Engineering at UC Santa Barbara for allowing me to teach a graduate elective course from the nearly finalized book in Spring 2016. I am also grateful to the students of that course for vetting the book. A major challenge in my writing was the creation of homework exercises, especially for those chapters where there are no prior books to draw upon. As noted among the exercises, some of the students suggested their own original homework problems. I hope that instructors and students who use the book will also contribute exercises. With permission from (and credit to) the contributors, I will incorporate contributed exercises into future editions.

I thank Bryan Goldsmith, Geoffrey Poon, Mark Joswiak, Kartik Kamat, Nils Zimmermann, Attila Szabo, David Wales, Charlie Campbell, Jim Pfaendtner, Dima Makarov, and Ben Leimkuhler for helpful suggestions on the book. I thank the Theoretical Chemistry Division of the National Science Foundation for the CAREER award (0955502) that launched this book project and for continued support under CTMC award 1465289. I thank YiJing Yan and Jiang Jun for their warm hospitality during my sabbatical at USTC in Hefei, China (where most of this book was written).

I thank my mom and dad who sacrificed their own youth and permanently delayed their own educations to provide me (and my sister) with a loving home and a wonderful rural Missouri childhood. I am grateful to Moberly High School, especially Wendy Carter, Ed Miller, and Greg Klokkengae for strong foundations. I am grateful to Stephen Lombardo, Elias Saab, and many others at the University of Missouri for expanding my horizons. I thank my PhD and postdoctoral advisors: Arup Chakraborty, Alex Bell, Bernhardt Trout, and Berend Smit. I thank Andreas Heyden, Gregg Beckham, and Valeria Molinero for many discussions that have shaped my own interests and ideas. I thank Susannah Scott for advice, insight, generosity, and friendship. Finally, I thank Ban for her playful spirit, her matchless wit, her instinctive compassion, and her endless capacity to love and nurture three kids. I am forever grateful to her for all the joy in my life.

Baron Peters
Santa Barbara, CA

CHAPTER 1

Introduction

"In the years since 1940, only little cross-fertilization between physics and chemistry has taken place. ...books on physical chemistry and kinetics do not discuss Kramers results. Likewise, rarely does one find a book on kinetics or nonequilibrium statistical mechanics written by a physicist in which is discussed the important transition state theory..."

Hanggi et al. Rev. Mod. Phys. (1990)

1.1 Motivation for this book

This book provides a broad introduction to the most powerful theories and computational methods for understanding the kinetics and mechanisms of activated processes in chemistry and physics: chemical reactions, nucleation processes, non-adiabatic rate processes, protein folding, solid-state diffusion, etc. These topics are usually discussed in separate courses, in separate departments, to separate groups of students. Nearly three decades have passed since Hanggi et al. [1] noted the gulf between chemical kinetics and non-equilibrium statistical physics research in reaction rates and rare events, and still the gaps remain. However, a small but growing group of chemists, physicists, engineers, and applied mathematicians has been working to bridge these sub-branches of reaction rate theory and rare events. Their efforts have led to the discovery of several powerful new theories and to the development of entirely new types of rare events methods.

Figure 1.1.1 shows how this book bridges the gaps between chemistry [2], engineering [3], and chemical physics [4] oriented books on kinetics. The featured topics were selected to provide a *practical* foundation for theoretical and computational analyses. Chapters on chemical reaction equilibria, rate laws, and catalysis illustrate how rate constants and equilibrium constants enter phenomenological kinetic models. Several chapters focus on practical theoretical frameworks for predicting rate constants and kinetic trends, e.g. harmonic transition state theory [5], diffusion control theories [6], nonadiabatic reaction rate theories [7], and theories for overdamped barrier crossings [8]. Chapters on transmission coefficients address the effects of tunneling [9], and dynamical recrossing including the Kramers [10] and Grote-Hynes [11] theories and the reactive flux methods [12]. Several chapters focus on computational machinery including methods for finding saddle points [13], methods for computing free energy surfaces [14–16], stochastic simulation algorithms [17], transition path sampling methods [18,19], and the recently developed reaction coordinate identification methods [19]. The book concludes with a discussion of free energy relationships, powerful tools for discovering and explaining trends across series of similar reactions [20,21].

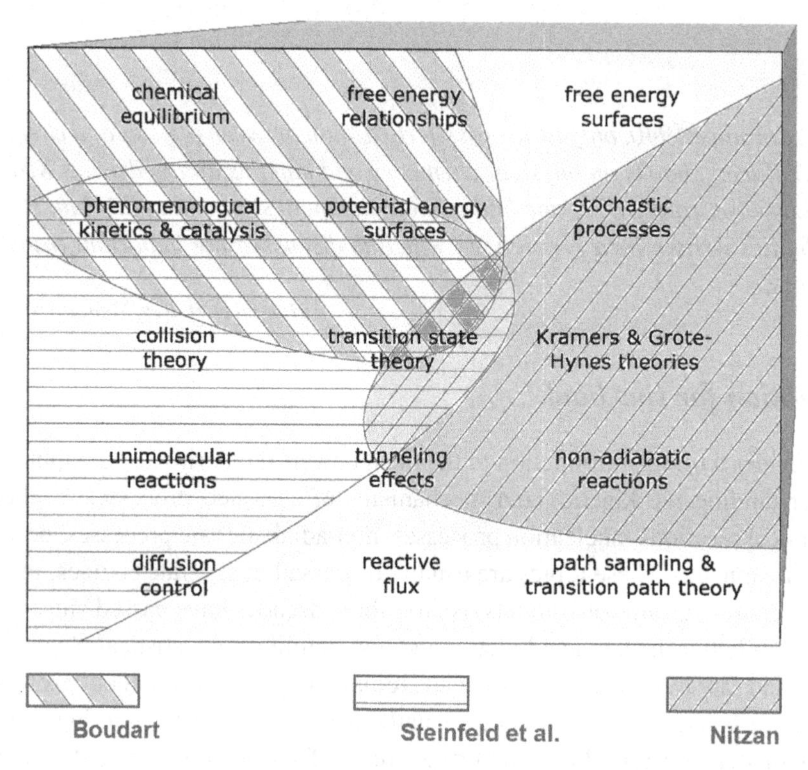

Figure 1.1.1: The shaded areas show included topics which are covered by some previous books. These include the Boudart (engineering) text "Kinetics of Chemical Processes," the Steinfeld et al. (chemistry) text "Chemical Kinetics and Dynamics," and the Nitzan (physics) text "Chemical Dynamics in Condensed Phases." This book spans longstanding gaps between these three branches of the literature on reaction rate theory and rare events.

Some preliminary topics that are included in many traditional books were omitted and/or abridged. For example, this book assumes a basic understanding of physical chemistry, equilibrium statistical mechanics, and some familiarity with non-equilibrium statistical mechanics. The book also assumes an understanding of basic computational chemistry ideas and basic simulation methods like molecular dynamics [22,23] and Monte Carlo methods [24,25]. Additionally, the book assumes some familiarity with basic mathematical techniques like Laplace transforms, eigenfunction expansions, and statistical regression procedures. All algorithms in the book are presented in language-independent pseudo-code. There are no computer codes or scripts in the book because prevalent programming styles and languages vary over time and between communities.

Several important topics were omitted or abridged because they are extensively discussed in other books. For example, there is no discussion of experimental methods for measuring rates of reaction, nucleation, folding, and electron transfer [26–28]. This book does not address molecular beam experiments [29], femtosecond spectroscopy [30], or quantum scattering theories [26]. A single introductory chapter combines acid base catalysis [31], enzyme catalysis [32], homogeneous catalysis [33,34], and heterogeneous catalysis [21,35,36]. Powerful tools for systematic reduction of complex reaction networks to simple rate laws are only briefly introduced [33,37,38]. We do not discuss classical electrostatic solvation models [39–42] nor the continuum solvation models for *ab initio* calculations [43–46]. Unfortunately, it was not possible to cover these and many other important topics.

1.2 Why are rare events important?

A colleague once heard me say that nucleation is a rare event, to which he objected: "Nucleation isn't a rare event. It happens all the time." Indeed, my colleague is correct. Many of the processes that theorists call rare events do happen all the time. Measured rates of nucleation range from 10^{-10}events$/cm^3/s$ (slow and rare) to about 10^{20}events$/cm^3/s$ (which seems quite fast). Measurable chemical reaction rates can be even faster. For example, a gas phase reaction with rate constant $10^{10}/s$ at standard conditions occurs at a blistering rate of 10^{30}events$/cm^3/s$. Similarly, rare events like electron transfer, protein folding, bimolecular reactions, etc. often exhibit high frequencies and/or extremely fast rates.

Why do we refer to these processes as rare events even when they are fast? The reason lies in their rates *relative to other relaxation processes.* $10^{10}/s$ is a fast reaction, but typical bond vibrations are much faster with frequencies on the order $10^{13}/s$. Thus before a molecule reacts, its chemical bonds will oscillate thousands of times, exchanging energy with each other and with the surroundings and entirely forgetting the initial conditions before a chance fluctuation funnels the required activation energy into the reaction coordinate. The story is similar for nucleation: thousands of nuclei will form and redissolve before a chance fluctuation creates a nucleus that is large enough to grow. In fact, across all rare events in chemical physics the mechanism is essentially the same: small and inconsequential excursions along the reaction coordinate happen all the time, but larger and more infrequent excursions along the reaction coordinate lead to new products and intermediates with qualitatively different structures and properties from the reactants.

Most experiments cannot directly see the reaction coordinate or track its fluctuations. Likewise, experiments cannot anticipate when or where a spontaneous reaction or nucleation event will occur, and thus they cannot directly observe the barrier crossing event. Rare events that occur at the molecular scale are only *indirectly* evident from macroscopic observations of new products, conformations, and/or phases. Because reaction coordinates, transition states, and free energy profiles elude direct observation, research in rare events and reaction rate theory has some unusual characteristics. In many other disciplines, simulation and computation are "third wheels" of the scientific method, but in kinetics they are vital components. To understand mechanisms of reactions and other rare events we rely heavily on a judicious combination of evidence from rate theories, experimental kinetics, *and* simulations. Indeed, first principles computation and simulations often provide the most direct evidence for or against molecular-level mechanistic hypotheses.

Note, however, that studying rare events with simulation is far more difficult than running a standard molecular dynamics simulation. Figure 1.2.1 shows the (approximate) time scales and length scales which are accessible by different simulation methods. These diagrams are common in the multiscale simulation literature, but they only show methods along the diagonal (the gray circles). Each method moving up the diagonal hierarchy sacrifices some detail and accuracy from the Hamiltonian to access longer time scales. However, the most prevalent methods in the coarse graining [47,48] hierarchy do not preserve the spectrum of time scales [49], especially for chemical reactions, electron transfer reactions, and nucleation and growth where the slowest time scales are associated with processes at very short length scales.

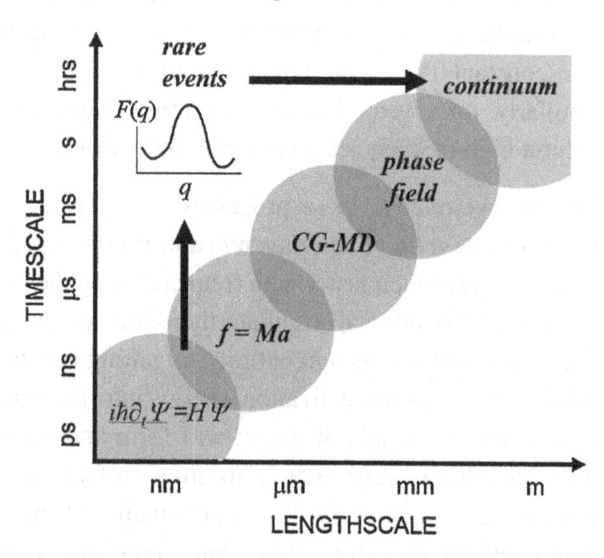

Figure 1.2.1: Multiscale simulation attempts to reach long time and length scales with a hierarchy of increasingly coarse grained methods. Rare events approaches obtain rates and rate laws without altering the Hamiltonian or the natural dynamics by taking advantage of natural time and length scale separations. The rate laws then become the species generation terms in continuum scale models.

Progress toward a molecular understanding of rare events began with simple rate theories and models that helped to understand observed rates and kinetic trends. Prominent examples are harmonic transition state theory [5], classical nucleation theory [50], and Marcus theory [7,51]. These do not always predict accurate rates, but they are extremely useful for understanding and correlating experimental data [52]. Computational rare events methods emerged later, enabled by advances in computational quantum chemistry [53,54], molecular simulation [22,24], and stochastic simulations [55–58]. As depicted in the upper left off-diagonal region of Figure 1.2.1, rare events methods exploit natural time scale separations to compute rates without simulating the waiting time between events. The rate constants and rate laws can then be used in species balance equations or population balance models to predict kinetics and dynamics at extremely long time scales. The rare events strategy thus bypasses the hierarchy of methods on the diagonal, and it does so with remarkable efficiency and without sacrificing *any* molecular-level resolution. Amazingly, these "too-good-to-be" claims are true. Of course, there are challenging aspects of these calculations. The predictions are only as good as the force fields and model chemistries, and often the most efficient rare events methods require an accurate reaction coordinate, i.e. an *a priori* mechanistic understanding. Fortunately, the past few decades of rare events research have led to powerful methods for computing free energy surfaces, for finding saddle points, for optimizing minimum free energy paths, for simulating reactive trajectories, for identifying reaction coordinates, and for computing accurate rates even without a complete mechanistic understanding.

Selecting the most appropriate rare events method or rate theory for a particular application can be a major pitfall even for veteran practitioners. This book outlines the strengths and limitations of the most practical theoretical frameworks and rare events methods – discussions which are often muted in the literature. In some cases the advantages of one theory over another are indisputable. For example, inertial rate theories like transition state theory are appropriate for chemical reactions, while overdamped theories are more appropriate for nucleation and protein folding. In other cases, the advantages of one method over another are more subtle issues of computational efficiency or ease of implementation. In general, two considerations should be weighed when choosing a computational method or theory:

1. *Correct and definitive answers.* The foremost merit criteria is a method's capacity to correctly answer the most important and interesting questions, and to do so with minimal uncertainty in the conclusions.
2. *Ease of implementation and computational efficiency.* These are also important (but secondary) criteria. Sometimes ease of implementation and efficiency considerations oppose each other, and in these cases their balance depends on available computational power and human expertise.

Computing power continues to grow and some supercomputers now boast petaflop/s performance. Additionally, enormous effort has been invested to develop accurate *ab initio* model chemistries and molecular force fields. Several software packages now implement state-of-the-art electronic structure methods and force fields so that users can bypass the difficult tasks of developing and implementing model chemistries. These developments have dramatically increased the system sizes, the time scales, and the breadth of different processes that can be studied in simulations. But despite advances in hardware and software, the time scales that are accessible to a straightforward simulation are still far too short for analysis of rare events. Let us now consider the types of questions which can be posed and answered with theoretical analyses, simulations, and rare events methods.

1.3 The role of computation and simulation

In the most literal sense, a simulation is a model that resembles a real process. Some computational studies of rare events basically report snapshots from long unbiased simulations and then anecdotally relate the observations. There are a few situations where long unbiased simulations are useful in studies of rare events. First, long unbiased trajectories are useful for characterizing the properties of reactant and product states. Second, a long unbiased trajectory sometimes discovers unexpected intermediates or new pathways, and these discoveries (while impossible to plan) can inspire important new areas of inquiry. Third, long time scale trajectories are useful for constructing Markov state models [59] and diffusion maps, but only when barriers are relatively small.

Rare events analyses more typically use simulations only as devices to compute averages, free energy barriers, and dynamical prefactors. Some rare events methods modify the dynamics to sample parts of phase space that would otherwise be rarely visited. Often we learn the most from simple models that explain trends and/or from analyses of static equilibrium properties like free energy landscapes. Those rare events methods which do require realistic dynamical simulations will only require extremely short trajectories.

Quantitative rate predictions

A 2010 US National Academy report estimated that a heat of formation measurement is 100 times more expensive than a heat of formation calculation of similar accuracy [60]. Because of these capabilities, computational thermochemistry is widely viewed as an inexpensive alternative to experimental measurements. However, the situation is different in kinetics. Rates are more difficult to compute for several reasons:

1. Rate calculations exponentially magnify small errors in activation barriers. *Ab initio* calculations and force fields are typically calibrated to match equilibrium properties, but precise transition state properties are rarely available for calibration.

2. Tunneling, spin crossings, and electron transfer processes require a correct description of quantum effects and excited states.
3. In systems where the dynamics deviate from those assumed by transition state theory, computing the kinetic prefactor can be highly non-trivial task.
4. The most practical rate theories require selection of a reaction coordinate, and identifying the reaction coordinate can be a major challenge.
5. Correctly predicting the observed kinetics often requires careful analysis of large reaction networks or catalytic cycles, e.g. using pseudo-steady states or other approximations. A computational error in just one elementary step may severely compromise the predicted kinetics.

In certain cases, state-of-the-art *ab initio* calculations and force fields can predict absolute rates, but for many systems accurate rate calculations are not yet possible. The fact that rates depend on so many factors makes it difficult to pinpoint the most important sources of error. Thus predicting rates continues to pose challenges, even for systems where predicting the thermodynamics has become straightforward.

Two types of kinetic trends, two different applications

In many applications we cannot accurately predict absolute rates, but we can often predict *trends* in the rate as a function of composition, temperature, or molecular characteristics. The reason that we can accurately predict trends even when we cannot predict rates is a rather unglamorous cancellation of errors. Fortunately, the trends are often more important than the absolute rates anyway. First, engineering better catalysts, drugs, crystal growth modifiers, etc. requires understanding kinetic trends as a function of composition, temperature, pressure, pH, solvent dielectric, etc. Second, the trends implied by computational analysis of a hypothesized mechanism can be used to refute or confirm the mechanism.

Note that there are actually two different types of trends and that they are used in different types of analyses. First, there are trends in the rate of one reaction across a family of related reactants and/or catalysts. In recent years, these trends have been widely pursued in computational studies of heterogeneous catalysts [21]. Many recent catalysis studies do not even attempt to compute rates, instead focusing on descriptor correlations that relate activation energies to adsorbate binding energies and other catalyst properties. Descriptor correlations are powerful ways to discover cheap and active catalyst materials from the vast possibilities in the periodic table. However, these trends describe the kinetics of one specific reaction, usually with a well-understood mechanism, occurring on different metal surfaces.

To test a new mechanistic hypothesis for a poorly understood reaction, we must instead cross-examine the observed and predicted activation parameters, rate laws, and reaction orders. Some would argue that trends are more important than quantitatively correct activation parameters or rate laws, but this viewpoint misses the fact that reaction orders and activation parameters **are** trends. Specifically, the activation parameters are trends in the rate vs. temperature, and the reaction orders are trends in the rate vs. concentration, partial pressure, etc. Reaction orders and activation parameters follow directly from the mechanism, and as such they are important fingerprints for identifying the mechanism. Where possible, one can also test predictions about abundant intermediates, spectra, molecular weights, etc. [61]. Ideally, one should cross-examine a comprehensive suite of properties and kinetic trends to provide many points of comparison to experiment. Each comparison is an opportunity to refute the hypothesized mechanism [62], and the more tests a hypothesis passes, the more confidently we can assert its veracity. Finally, note that the ideal starting point to identify a mechanism in computational work is a single well-defined catalyst and/or specific reactant(s). We should not *a priori* assume that the mechanism will be preserved across a family of catalysts, reactants, adsorbates, etc. Accordingly, we cannot build reliable descriptor correlations without first developing an understanding of the mechanism and establishing its validity for a family of catalysts and/or reactants.

In silico experiments

Simulations are often viewed as a branch of theory, but in some ways simulations are more like experiments. The role of simulations as idealized experiments was recognized at the dawn of molecular dynamics simulations by Fermi, Pasta, and Ulam (FPU) who described their study as an "experiment" to test a theory [60,63]. Theories are often oversimplified to facilitate analysis, whereas a simulation can retain all of the essential physics and sometimes reveal surprises. Experiments also retain all of the physics, but changing one variable in an experiment can inadvertently change other variables. The often messy nature of real experiments calls to mind the quote from Einstein: "A theory is something nobody believes, except the person who made it. An experiment is something everybody believes, except the person who made it." A simulation provides absolute and independent control over all parameters, even parameters that could never be changed in a real experiment. For example, we can turn off quantum effects, change the sizes of atoms, choose any masses we like, turn off dispersion interactions, add friction to the natural dynamics, etc. In this manner, simulations can be devised to perform proper controlled experiments that test hypotheses about these factors.

Mechanistic hypothesis testing

For many decades the principal way of testing a mechanistic hypothesis was to fit data to a phenomenological model. A sequence of elementary steps would be proposed along with assumptions about pseudo-steady-state behavior, quasi-equilibrium steps, rate determining steps, irreversible steps, etc. All rate constants in the resulting rate law would then be used as adjustable parameters in fitting the model to the experimental data. The rate law would then be accepted or refuted according to analysis of residual variance from the best fit. Classic hypothesis testing approaches are still valuable, but computational chemistry and molecular simulation have dramatically expanded the arsenal of hypothesis tests.

- In addition to checking for quality of a fit to experimental data, we can now ask whether the fitted rate constants are approximately consistent with independent estimates from *ab initio* computational chemistry. First principles calculations can also cross-examine the predicted and observed abundances of hypothesized intermediates.
- Classic phenomenological analyses like those of Hinshelwood, Hougen and Watson invoked abstract symbolic entities like "*" for an active site. Computational chemistry techniques now allow us to propose and test specific molecular models of the active site structure.

Hypothesis testing procedures based on first principles calculations are primarily useful for understanding chemical reactions and catalysis. In these applications, questions about mechanisms primarily concern the sequence of elementary steps and models of the active sites. Once transition states have been found for all elementary steps, the individual rates are easily computed, e.g. using harmonic transition state theory.

For processes like nucleation, self-assembly, and protein folding, simulations face many additional challenges. Are mechanistic conclusions robust to changes in the force field? Are the designated bath variables sufficiently fast relative to designated slow variables? Do dynamical trajectories really lose memory between the adopted milestones? Is the transmission coefficient small because of intrinsic friction in the dynamics or because of a poorly chosen dividing surface? The answers to these questions depend upon which reaction coordinate was selected for computing barriers and rates. There are now several methods for discovering reaction coordinates, for testing their accuracy, for using them to compute rates, and for constructing simple rate theories [19,52]. These new tools are conclusively answering mechanistic questions and pointing the way to simple theories for new types of reactions and rare events.

1.4 Polemics

This section addresses some of the occasional abuses, misuses, and misconceptions that appear in the rare events literature. It is not intended as a targeted critique of any specific work, but rather as a list of precepts and best-practices.

Units

Units are extremely important in scientific research, and yet we (engineers in particular) always try to eliminate them. Consistency of units is the foremost hurdle that must be passed by any successful theory or quantitatively meaningful statement. Nonsensical comparisons between quantities with incompatible units are all too common in the literature. Statements like "...the growth rate surpasses the nucleation rate..." or "...the reaction is faster than diffusion..." do not have a proper meaning. To compare quantities with different units, they must first be multiplied/divided by additional time scales, length scales, areas, volumes, concentrations, etc. to obtain an equivalent units. Proper comparisons and dimensional analyses often reveal important lengthscales and time scales apart from the intrinsic kinetics. As such, units are invaluable guides *in the early stages of model formulation.*

On the other hand, the units are a nuisance in mathematical manipulations and data analysis. Furthermore, dimensionless groups reduce the number of independent parameters that are needed to make quantitative predictions. Systems with different barriers, temperatures, frictions, concentrations, etc. might follow an equation with just one or two dimensionless parameters. In many cases, dimensional analyses identify certain factors in a model that can be entirely omitted. Finally, dimensionless equations and parameters suggest compact ways to correlate data across a diverse set of conditions.

Non-dimensionalization is ubiquitous in engineering, but less so in chemistry and physics. Scientists instead use different unit conventions in different areas, e.g. there are atomic units, cgs units, natural units, SI units, etc. [64]. Cussler quips [65] "Dimensionless numbers are weapons that engineers use to confuse scientists." but perhaps scientists have the more confusing system. For example, polymer scientists, quantum chemists, and non-equilibrium statistical mechanicians all solve similar Schrodinger equations, but their equations *look* different except for those cases where they are non-dimensionalized. This book discusses both formulation and analysis stages of kinetic modeling. Accordingly, some discussions involve quantities with units, and others are non-dimensionalized.

On the value of results that disagree with experiment

The reaction pathways and mechanisms that we investigate are often based on some preconceived ideas about the mechanism. Anyone can fail to anticipate the correct mechanism, but among those mechanisms that are investigated, computational results should be used and interpreted in an unbiased manner. Unfortunately, some investigators view *ab initio* calculations as a tool "to validate" their mechanistic hypotheses. Note the subtle bias in that phrase – we should embark on a calculation with the goal of performing an unbiased test, not to achieve a validation. The problem goes beyond semantics. A bias toward results that validate the hypotheses has led some investigators to excuse results that clearly and conclusively refute their mechanistic hypothesis as "expected errors from DFT".

If we believe that *ab initio* computational evidence can support a mechanistic hypothesis, then we must also believe that *ab initio* evidence can refute a mechanism [62]. Computational chemistry cannot accurately predict experimental results, but most computational methods are associated with some typical range of errors. When disagreements with experiment exceed the plausible range of computational errors, the theorist should revise his mechanistic ideas, reinterpret experimental results, consider new active site models, etc. Revisiting the mechanistic assumptions and hypotheses because of some discrepancy often leads to the most important discoveries: revised mechanisms, new interpretations for experiments, new insights about the active site, etc. In this sense, computational results that disagree with experiment are often more exciting than results which agree.

On the other hand, if implausibly large discrepancies are just brushed aside, then there really is no value in performing calculations at all. Unfortunately, studies that promote a mechanism despite strong computational counter-evidence are easy to find. In some cases, the computational counter-evidence is even presented as supporting evidence for a mechanism [61]. One scathing critique [66] concluded that *ab initio* calculations cannot predict anything and even described computational chemistry as a "dartboard". It should be noted that the most egregious errors cited in their critique resulted from (i) calculations that were done incorrectly (e.g. by ignoring entropy in association steps), or (ii) investigating the wrong mechanisms or active sites. It seems likely that many of the most egregious errors are not actually due to computational chemistry itself, but rather to human errors; i.e. to flawed analysis, applications, and interpretations of computational chemistry results.

Quests of questionable value

As computers and software for massively parallel simulations become more powerful, it is increasingly common to see fully atomistic simulations of processes that involve highly disparate length and time scales.[1] In some cases, large scale atomistic simulations have been used to model phenomena where there are very good continuum models. Some examples include atomistic simulations of solidification (where there are exquisite and accurate continuum scale models) [67,68], atomistic simulations of boundary layer transport and catalysis (despite rigorous continuum equations for coupling boundary layer resistances to surface kinetics) [69,70]

[1] These efforts are not to be confused with multiscale simulation methods, where phenomena at different scales are studied with different methods.

and sophisticated rare events analyses of water evaporation (which is accurately described by coupled heat and mass transfer equations) [65,71].

Each of the above examples involves gradients in temperature, composition, or fluid velocity coupled to molecular scale interfacial processes. Perhaps these seem like complicated situations which require multiscale simulation or (even less desirably) atomistic simulations with billions of atoms to span multiple scales. But before embarking on a massive simulation effort, one should ask whether the problem might be broken into separately soluble but coupled parts. We should also ask which molecular level details (if any) are needed to advance technology or fundamental understanding. When there is no need for molecular level detail, then an accurate continuum theory is superior to an atomistic simulation. This statement is not a matter of opinion or personal preference. Simple theories, by construction, omit thousands of molecular details, and when the details are irrelevant they should be omitted.

Consider, for example, the overall rate of a heterogeneous surface reaction. The intrinsic kinetics at the surface depend only on the concentrations of reactants immediately above the surface. Mass transfer from the bulk depends on concentrations in the bulk and concentrations at the surface. The problem can be rigorously broken into two parts: (1) determine the rate law as a function of concentration near the surface, and (2) self-consistently match the continuum mass transfer rate to the rate law. Matching reveals the concentrations at the surface and therefore the overall rate. The problem can be rigorously solved without simulating any part of the mass transfer process at atomistic resolution. The model that we obtain from the coupled analytic analysis is simpler than the results from a massive simulation. Importantly, the analytic model is also more useful. It immediately predicts the overall rates at many different bulk concentrations, whereas the massive simulation just gives the overall rate for the one concentration that was simulated.

On the proliferation and testing of new methods

There are hundreds of methods for optimizing transition states, hundreds of methods for computing free energy surfaces, and scores of methods for computing dynamically accurate rate constants. Some of these methods dramatically advanced rare events research through powerful theoretical principles and innovative computational strategies. Others (perhaps most) are minor variations on the major themes with little or no practical benefit.

In the long run, the merits of a method are easily judged by its impact in actual applications. But how can merit be assessed for new methods? Ideally, new methods should resolve some deficiency in the capabilities of existing methods, or else they should provide a great leap (e.g. $>2\times$) in efficiency. In principle, publications on methodology should include difficult example problems that demonstrate truly new capabilities, but modern funding structures require that we frequently document our progress. Thus, in practice, to require a challenging demonstration would stifle many creative, important, and ambitious new directions.

How do we balance the principled ideals against the practical realities? Ultimately, a method development effort should not be regarded as complete until its capabilities are established for rigorous and demanding test cases. There are subltle differences in the ways that new methods are tested and used in science, engineering, and mathematics. To a mathematician, assumptions have logical consequences (theorems) which can be studied apart from the properties of any real system. To the scientist, assumptions (hypotheses) have consequences that must be tested against observations of real systems. To an engineer, assumptions and models are measured by their broader impacts (technology).

The different priorities of scientists, engineers, and mathematicians are often evident in our respective contributions to the field.[2] Engineers tend to pursue important and ambitious applications, sometimes with insufficient rigor and before the available simulation methods are actually ready. Scientists tend to dwell on model systems like hard spheres and model peptides, sometimes with too little concern for discoveries with potentially practical impacts. Mathematicians tend to create test systems for which the requisite assumptions of a theorem are true by construction, and as such these tests are more properly regarded as illustrations. For example, overdamped dynamics on a low dimensional potential energy landscape is Markovian by construction, so all theorems and methods derived from that property must work perfectly well – on the system used in the illustration. After methods have been so illustrated, scientists and engineers sometimes use them on complex molecular systems with no further tests of the underlying assumptions. As Churchill has said "However beautiful the strategy, you should occasionally test the results." Clearly, scientists, engineers, and mathematicians can each learn valuable lessons from each other.

On corrections to transition state theory

Several chapters of this book focus on methods for obtaining accurate rate constants. Many readers will already be familiar with transition state theory (TST) and perhaps also with non-TST effects like tunneling and recrossing that require more elaborate calculations. There are two schools of thought on the importance of corrections for non-TST effects. One viewpoint maintains that errors in *ab initio* calculations are unavoidable and dominant sources of error, regardless of how carefully the non-TST corrections are estimated. Some proponents of this viewpoint omit non-TST corrections and focus on trends rather than absolute rates. Indeed,

[2] As an engineering professor, with a partial appointment in chemistry, and an undergraduate degree in mathematics, I feel less-underqualified-than-most to over-generalize.

there are many applications where non-TST effects can be ignored. There is no need to include non-TST effects in studies that examine one reaction across a series of catalysts with different adsorbate binding energies or d-band centers. Non-TST effects can also be ignored for examining trends in homogeneous reaction kinetics, e.g. trends in the rate vs. electron withdrawing characteristics of various substituent functional groups [72]. Non-TST effects can also be ignored for steps that do not control the overall rate.

On the other hand, corrections to TST can be extremely important when computational results are compared to experiment to test a hypothesized mechanism. For example, tunneling can alter the slope of Arrhenius and Eyring plots for many hydrogen and proton transfer reactions. Dynamical recrossing can lower the intercept on Arrhenius and Eyring plots. These effects cannot be "switched off" in the experiments, and accordingly they should not be switched off in computational rate predictions. The PES may well be the dominant source of error, but to omit a correction that influences the quantity being computed adds bias to an already imprecise calculation. In general, the potential impact of non-TST effects and the need for accurate rates should be assessed carefully on a case-by-case basis when justifying computational shortcuts.

Of course, one should not decide whether to add or omit terms solely to improve the agreement with experiment. Adding corrections because theoretical considerations predict they are significant is good practice. It is also good practice to add corrections when a discrepancy with experiment alerts us to unanticipated factors. But omitting a theoretically important correction because the data fits better without it is shameful. The model chemistry and the effects to be included are, in some sense, a part of the hypothesized theoretical model. In principle, the model chemistry, the approximations to be used, and the corrections which will be applied should be selected before the results are compared to experiment. When all potentially important corrections are included and the results agree with experiment, the agreement is significant. When all potentially important corrections are included and the results still disagree with experiment, the results are still useful. As elaborated above, carefully performed calculations that disagree with experiment indicate flaws in our mechanistic understanding and often lead us to new discoveries.

On science priority and impact metrics

Scientists (theorists in particular) are notoriously sensitive about credit and recognition (see Figure 1.4.1). Some investigators have even written articles which are chiefly about who used their ideas and methods without appropriate credit. Concerns over science priority can be understood from the many historical examples of misappropriated credit.[3] Journal editors cannot

[3] Stigler's law of mistaken eponymy says "No scientific theory is named after the person who actually discovered it." Amusingly, Stigler's law was first stated by R. Merton. There are many examples among the topics in this

check whether our citations are accurate and complete, and typical papers are read by only a few of our peers before publication. Therefore citations are largely an honor system. Most investigators feel naturally compelled to credit others, but even the most renowned scientists are occasionally slighted [73]. Proper citations are an important component of scholarly work for several reasons: they point readers to other important papers, they dispell the myth of a single genius working in isolation, and they set an example that scientific progress requires careful attention to the work of others.

Figure 1.4.1: Despite the stigma associated with self-promotion, all scientists, living and deceased, deserve credit for their original contributions. I have tried to give appropriate credit throughout, but I will undoubtedly forget and misdirect credit for certain important contributions. Please suggest references and topics for future editions.

Are issues of science priority important for the advancement of science itself? Yes, because professional survival for a scientist requires some degree of recognition. The perceived impact of our prior achievements influences our ability to secure funding, to attract talented students, and to stay employed. Young investigators in particular depend on the *short term impact* of their achievements – a fledgling research program can flounder if its most innovative contributions take a decade to discover.

To some degree, all research programs depend on short term impact, and unfortunately short term metrics like journal impact factors are imperfect. Of course, any conceivable metric for judging scientific contributions would have flaws, cf. Campbell's law: "The more any quantitative social indicator (or even some qualitative indicator) is used for social decision-making, the more subject it will be to corruption pressures and the more apt it will be to distort and corrupt the social processes it is intended to monitor." The widespread pressure to publish in high impact journals does have several undesireable side effects on the quality of our literature. First, the most prestigious journals tend to attract manuscripts with hyperbolic (and sometimes fraudulent) claims [74]. Second, it is increasingly common for editorial boards – rather than true experts – to judge the importance of submitted manuscripts. These preliminary editorial reviews typically prioritize work that is new and trendy over careful, comprehensive, and consequential work on longstanding challenges. Finally, the importance of short term impact forces

book: the Arrhenius law was put forth by van't Hoff, Fermi's Golden Rule was discovered by Dirac, Voronoi cells were first used by Descartes, the Bodenstein approximation was first used by Chapman, etc.

young investigators to pursue the hottest trends *en masse* – a feedback loop that exacerbates fads in scientific research and funding.

References

[1] P. Hanggi, P. Talkner, M. Borkovec, Rev. Mod. Phys. 62 (1990) 251–341.
[2] J.I. Steinfeld, J.S. Francisco, W.L. Hase, Chemical Kinetics and Dynamics, Prentice Hall, Englewood Cliffs, NJ, 1989.
[3] M. Boudart, Kinetics of Chemical Processes, Prentice Hall, Englewood Cliffs, NJ, 1968.
[4] A. Nitzan, Dynamics in Condensed Phases: Relaxation, Transfer, and Reactions in Condensed Molecular Systems, Oxford University Press, Oxford, 2006.
[5] E. Wigner, Trans. Faraday Soc. 34 (1938) 29–41.
[6] P. Debye, Trans. Electrochem. Soc. 82 (1942) 265–272.
[7] R.A. Marcus, Rev. Mod. Phys. 65 (1993) 599–610.
[8] L.S. Pontryagin, A.A. Andronov, A.A. Vitt, Zh. Eksp. Teor. Fiz. 3 (1933) 165–180.
[9] H.S. Johnston, Adv. Chem. Phys. 83 (1961) 1–9.
[10] H.A. Kramers, Physica A 7 (1940) 284–304.
[11] R.F. Grote, J.T. Hynes, J. Chem. Phys. 73 (1980) 2715–2732.
[12] D. Chandler, J. Chem. Phys. 68 (1978) 2959–2970.
[13] G. Henkelman, G. Johannesson, H. Jonsson, in: S.D. Schwartz (Ed.), Theoretical Methods in Condensed Phase Chemistry, Kluwer Academic, Dordrecht, 2000, pp. 269–300, chapter 10.
[14] G.M. Torrie, J.P. Valleau, J. Comput. Phys. 23 (1977) 187–199.
[15] N. Hansen, W.F.V. Gunsteren, J. Chem. Theory Comput. 10 (2014) 2632–2647.
[16] G. Bussi, D. Branduardi, Rev. Comput. Chem. 28 (2015) 1–49.
[17] D.T. Gillespie, Markov Processes: An Introduction for Physical Scientists, Academic Press, Boston, 1992.
[18] P.G. Bolhuis, C. Dellago, Rev. Comput. Chem. 27 (2009) 1–105.
[19] B. Peters, Mol. Simul. 36 (2010) 1265–1281.
[20] W.P. Jencks, Chem. Rev. 85 (1985) 511–527.
[21] J.K. Norskov, F. Studt, F. Abild-Pederson, T. Bligaard, Fundamental Concepts in Heterogeneous Catalysis, Wiley, Hoboken, NJ, 2014.
[22] M.P. Allen, D.J. Tildesley, Computer Simulation of Liquids, Clarendon Press, Oxford, UK, 1987.
[23] A. Leach, Molecular Modelling: Principles and Applications, Prentice Hall, Pearson Education, Englewood Cliffs, NJ, 2001.
[24] D. Frenkel, B. Smit, Understanding Molecular Simulation: From Algorithms to Applications, Academic Press, San Diego, 2002.
[25] D.P. Landau, K. Binder, A Guide to Monte Carlo Simulations in Statistical Physics, Cambridge University Press, Cambridge, UK, 2009.
[26] R.D. Levine, Molecular Reaction Dynamics, Cambridge University Press, Cambridge, UK, 2005.
[27] A.M. Kuznetsov, J. Ulstrup, Electron Transfer in Chemistry and Biology: An Introduction to the Theory, Wiley, New York, 1999.
[28] A. Lewis, M. Seckler, H. Kramer, G.M. van Rosmalen, Industrial Crystallization, Cambridge Press, 2015.
[29] Y.T. Lee, Science 236 (1987) 793–798.
[30] A.H. Zewail, J. Phys. Chem. 100 (1996) 12701–12724.
[31] J.H. Espenson, Chemical Kinetics and Reaction Mechanisms, McGraw-Hill, New York, 1995.
[32] P.C. Engel, in: M.I. Page (Ed.), The Chemistry of Enzyme Action, Elsevier, Amsterdam, 1984, pp. 73–110.
[33] F.G. Helfferich, Kinetics of Multistep Chemical Reactions, Elsevier, Amsterdam, 2004.
[34] P.W. van Leeuwen, Homogeneous Catalysis: Understanding the Art, Kluwer, Dordrecht, 2004.

[35] M. Boudart, G. Djega-Mariadassou, Kinetics of Heterogeneous Catalytic Reactions, Princeton University Press, Princeton, 1984.
[36] J.M. Thomas, W.J. Thomas, Principles and Practice of Heterogeneous Catalysis, 2nd ed., Wiley-VCH, Weinheim, 2015.
[37] J. Christiansen, Adv. Catal. 5 (1953) 311–353.
[38] G.B. Marin, G.S. Yablonsky, Kinetics of Chemical Reactions: Decoding Complexity, Wiley-VCH, Weinheim, 2011.
[39] K. Laidler, Reaction Kinetics: Reactions in Solution, vol. 2, Pergamon Press Press, Oxford, 1963.
[40] M. Born, Z. Phys. 1 (1920) 45.
[41] L. Onsager, J. Am. Chem. Soc. 58 (1938) 1486.
[42] J.G. Kirkwood, J. Chem. Phys. 7 (1939) 911.
[43] S. Miertus, E. Scrocco, J. Tomasi, Chem. Phys. 55 (1981) 117–129.
[44] V. Barone, M. Cossi, J. Phys. Chem. A 102 (1998) 1995–2001.
[45] M. Cossi, N. Rega, G. Scalmani, V. Barone, J. Comput. Chem. 24 (2003) 669–681.
[46] A. Klamt, Wiley Interdiscip. Rev. Comput. Mol. Sci. 1 (2011) 699–709.
[47] W.G. Noid, J. Chem. Phys. 139 (2013) 090901.
[48] M.S. Shell, in: Advances in Chemical Physics, 2016, in press.
[49] A. Davtyan, J.F. Dama, G.A. Voth, H.C. Andersen, J. Chem. Phys. 142 (2015) 154104.
[50] D.T. Wu, Solid State Phys. 50 (1997) 37–187.
[51] R.A. Marcus, J. Chem. Phys. 24 (1956) 966.
[52] B. Peters, J. Phys. Chem. B 119 (2015) 6349–6356.
[53] R.G. Parr, W. Yang, Density Functional Theory of Atoms and Molecules, Oxford Press, 1989.
[54] T. Helgaker, P. Jorgensen, J. Olsen, Molecular Electronic-Structure Theory, Wiley, Hoboken, NJ, 2013.
[55] A.B. Bortz, M.H. Kalos, J.L. Lebowitz, J. Comput. Phys. 18 (1975) 10–18.
[56] D.T. Gillespie, J. Comput. Phys. 22 (1976) 403–434.
[57] R. Erban, S.J. Chapman, Phys. Biol. 6 (2009) 46001.
[58] J. van Zon, P.R. ten Wolde, Phys. Rev. Lett. 94 (2005) 128103.
[59] G.R. Bowman, V.S. Pande, F. Noe, An Introduction to Markov State Models and Their Application to Long Timescale Molecular Simulation, Springer, Berlin, Heidelberg, 2013.
[60] J.R. Elliott, E. Maginn, Ind. Eng. Chem. Res. 49 (2010) 3059–3078.
[61] B. Peters, S.L. Scott, A. Fong, Y. Wang, A.E. Stiegman, Proc. Natl. Acad. Sci. USA 112 (2015) 4160–4161.
[62] K.R. Popper, Conjectures and Refutations, Routledge Classics, New York, 1963.
[63] J. Ford, Phys. Rep. 213 (1992) 271–310.
[64] B.N. Taylor, A. Thompson, The International System of Units (SI), National Institutes of Science and Technology, U.S. Department of Commerce, 2008.
[65] E.L. Cussler, Diffusion: Mass Transfer in Fluid Systems, 3rd ed., Cambridge Press, Cambridge, UK, 2009.
[66] R.E. Plata, D.A. Singleton, J. Am. Chem. Soc. 137 (2015) 3811–3826.
[67] K. Fisher, W. Kurz, Fundamentals of Solidification, Trans. Tech. Publications, 1986.
[68] J.A. Dantzig, M. Rappaz, Solidification, EPFL Press, Lausanne, 2012.
[69] C.N. Satterfield, Mass Transfer in Heterogeneous Catalysis, MIT Press, Cambridge, MA, 1970.
[70] J.J. Carberry, Chemical and Catalytic Reactor Engineering, Dover, Minneola, NY, 2001.
[71] R.B. Bird, W.E. Stewart, E.N. Lightfoot, Transport Phenomena, Wiley, New York, 1960.
[72] L.P. Hammett, J. Am. Chem. Soc. 59 (1937) 96.
[73] S.C.L. Kamerlin, J. Cao, E. Rosta, A. Warshel, J. Phys. Chem. B 113 (2009) 10905–10915.
[74] F.C. Fang, R.G. Steen, A. Casadevall, Proc. Natl. Acad. Sci. USA 109 (2012) 17028–17033.

CHAPTER 2

Chemical equilibrium

"...to calculate in advance the result of the action of the chemical forces for any conditions and substances whatsoever. ...we set forth the following two laws, namely the law of mass action and the law of volume action..."

Guldberg, Waage (1864); tr. Abrash, J. Chem. Educ. (1986)

Thermodynamics, while not the main focus of this book, is integral to nearly every aspect of kinetics. Indeed, many phenomena can be understood quite well without kinetics, using only equilibrium considerations. Where kinetics are important, we will find that, to a large extent, thermodynamics controls the populations of transition states and rates. Thus, elements of chemical reaction equilibria will appear intermittently throughout the book. The cursory introduction in this chapter bypasses most of thermodynamics and skips directly to the chemical potential, the mass action law, and equilibrium compositions of reaction mixtures.

The chemical potential μ_i is a partial molar Gibbs free energy with respect to species i. Specifically $\mu_i \equiv (\partial G/\partial n_i)_{T,P,n_j,j\neq i}$. The chemical potential is the thermodynamic driving force behind all processes that result in a change of composition, e.g. chemical reactions, electrochemical reactions, and some phase transitions. All of these transformations proceed toward an equilibrium that is determined by a balance of chemical potentials. The chemical potential balance that defines equilibrium may be between reactants and products, between redox agents and an electrical potential, or between two populations of the same species on two sides of a phase boundary. We begin with a discussion of chemical potentials, activities, and their dependence on composition in various types of mixtures.

2.1 Chemical potential and activity

Like most thermodynamic potentials, we define the chemical potential in relation to that of a reference system. Models of the chemical potential have the form [1]

$$\mu_i = \mu_i^o + k_B T \, ln \, a_i \tag{2.1.1}$$

where μ_i^o is a reference chemical potential and a_i is the activity of species i. The activity quantifies changes in the molar Gibbs free energy $(\mu_i - \mu_i^o)$ as one moves away from a reference condition *at constant temperature*.

In practice, equation (2.1.1) requires additional relationships between activities and specific composition variables like mole fractions, concentrations, partial pressures, molalities, etc. Usually, these relationships are first developed for idealized model systems, and then real systems are modeled by accounting for departures from the idealized model. There are many idealized models, each with corresponding expressions accounting for non-idealities. In practice, one must be careful to combine an idealized model with its appropriate non-ideal corrections. There is plenty of room for confusion, e.g. mixtures of ideal gases are not the same as ideal gas solutions. Here we present a short overview of the most commonly used models and conventions.

Gas mixtures

There are established conventions for modeling chemical equilibrium between gas phase species. The activity of each component is written as $a_i = f_i/f_i^o$ where f_i and f_i^o are the fugacity and reference fugacity for species i. The resulting chemical potentials are

$$\mu_i = \mu_i^o + k_B T \, ln \, f_i/f_i^o$$

The fugacity of each species depends on the temperature, pressure, and composition of the gas mixture.

1. For a **mixture of ideal gases** the fugacity of each species is its ideal partial pressure $f_i = x_i P$ where x_i is the mole fraction of species i. The standard fugacity is typically $f_i^o = P^o = 1$ bar or 1 atm. pressure. The chemical potential of species i becomes [1,2]

$$\begin{aligned} \mu_i &= \mu_i^o + k_B T \, ln P/P^o + k_B T \, ln \, x_i \\ &= \mu_i^* + k_B T \, ln \, x_i \qquad (Lewis - Randall\ rule) \end{aligned} \tag{2.1.2}$$

   where $\mu_i^* = \mu_i^o + k_B T \, ln \, P/P^o$. With these definitions, μ_i^o only depends on temperature, μ_i^* depends on temperature and pressure, and μ_i depends on temperature, pressure, and composition.
2. The fugacities in an **ideal gaseous solution** are $f_i = x_i f_i^{pure}$, where f_i^{pure} is the fugacity of pure species i at the temperature and pressure of the reaction mixture. The reference fugacities f_i^o are those of the pure substances at the reference temperature and pressure. The ideal gas solution model simplifies the treatment of fugacities to the well-known fugacities of pure gases [2,3]. Again a Lewis-Randall rule results from a regrouping to isolate the

composition dependence: $\mu_i = \mu_i^* + k_B T \ln x_i$ where now $\mu_i^* = \mu_i^o + k_B T \ln f_i^{pure}/f_i^o$ such that all composition dependence again appears in the $\ln x_i$ terms.

3. In the most general case of a **non-ideal gas solution**, the fugacity of species i is $f_i = \phi_i x_i P$. The quantity ϕ_i is a fugacity coefficient, and in general it depends on the temperature, pressure, and composition of the gas mixture. Again the reference fugacity is f_i^o, the fugacity of the pure substance i at 1 bar. Separating the composition dependent terms from purely pressure and temperature dependent terms now gives $\mu_i = \mu_i^* + k_B T \ln \phi_i x_i$.

The following brief outline of a typical fugacity coefficient calculation may be helpful in understanding their meaning. First, the critical temperature and the critical pressure for each pure gas are used to create a relative temperature and relative pressure scale. The relative pressure, relative temperature, and a molecular geometry "acentric" factor then determine the compressibility factor for pure component i: $z_i \equiv PV_i/k_B T$. Finally, the fugacity coefficient is obtained by integrating the compressibility factor, $\ln\phi_i = \int_0^P (z_i - 1)P^{-1}dP$. Fugacity coefficients for pure gases ($f_i^{pure} = \phi_i^{pure} P$) are extensively tabulated [3], but fugacity coefficients for non-ideal gas mixtures are not. Readers are referred elsewhere for more on fully non-ideal gas solutions [4].

Liquid solutions

There are several models and conventions in use for activities in liquids. Activities for salts dissolved in solutions are often described in terms of molalities. Mole fractions are used for components that would be liquids in their pure state at the temperature and pressure of the solution. The activities of dissolved gases and molecular solids are usually also described in terms of mole fractions. In many cases, it is useful to envision the solution in equilibrium with a hypothetical vapor at the same temperature and pressure. For example, if **A** is a volatile solute in aqueous solution, then the hypothetical equilibrium is

$$\mathbf{A}_{(aq)} \rightleftharpoons \mathbf{A}_{(g)}$$

Hypothetical equilibria are invoked to take advantage of Henry's law for dilute solutes and Raoult's law for nearly pure solvent components. Because the hypothetical vapor is in equilibrium with the solution, the chemical potentials of species in solution are equal to those in the gas phase.

Suppose that the hypothetical gas mixture behaves ideally. The partial pressure of a dilute and volatile solute in the hypothetical gas phase is given by Henry's law [1–3],

$$P_i = H_i x_i\,,$$

where x_i is very small ($x_i \approx 0$) and H_i is the Henry's constant. For a nearly pure solvent, the partial pressure in the hypothetical gas phase is given by Raoult's law [1–3],

$$P_i = P_i^{sat} x_i \,,$$

where x_i is near unity ($x_i \approx 1$) and where P_i^{sat} is the vapor pressure above a pure liquid of component i. Figure 2.1.1 illustrates the equilibrium gas phase partial pressure as a function of the liquid mole fraction. Figure 2.1.1 also shows Henry's law and Raoult's law models which are accurate in the dilute solute limit and pure-solvent limit, respectively. Note that many real solutes have a maximum solubility x_i^{sat} beyond which the solution becomes metastable.

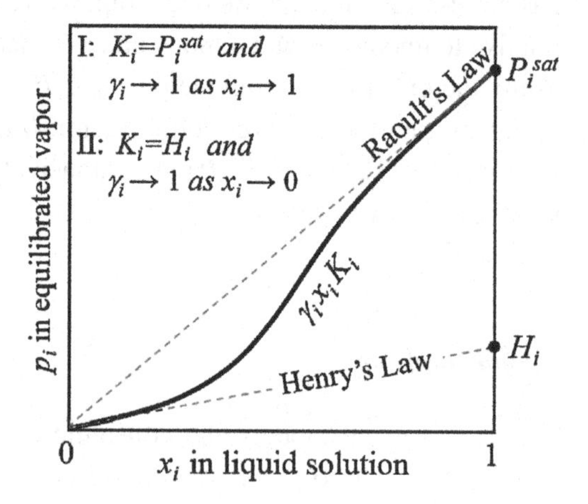

Figure 2.1.1: Vapor-liquid equilibria are useful in defining activities and chemical potentials of solvents and volatile solutes. When the solvent is nearly pure, Raoult's law predicts partial pressure (or fugacity) in the gas phase. When the solute is infinitely dilute, Henry's law predicts the gas phase partial pressures. The pure solvent and an (unphysical) extrapolation of Henry's law to unit mole fraction are used as reference conditions, both at the temperature and pressure of the solution. These two reference conditions are indicated as dots labeled P_i^{sat} and H_i, respectively.

In the hypothetical equilibrium, gas phase chemical potential models can be adapted to develop models of the solution phase chemical potentials. For a mixture of ideal gases, recall that $\mu_i = \mu_i^o + k_B T \, ln \, P_i / P^o$. Now using Henry's law and/or Raoult's law for P_i, the chemical potential becomes

$$\mu_i = \begin{cases} \mu_i^o + k_B T \, ln \, H_i x_i / P^o & x_i \approx 0 \\ \mu_i^o + k_B T \, ln \, P_i^{sat} x_i / P^o & x_i \approx 1 \end{cases} \tag{2.1.3}$$

Equation (2.1.3) can be further simplified by separating the $ln \, x_i$ terms from $ln \, H_i / P^o$ and $ln \, P_i^{sat} / P^o$ terms. H_i and P_i^{sat} depend on temperature, but not on composition. Of course, Henry's constants do strongly depend on the nature of the solvent. Typically, the temperature dependent reference potential μ_i^o is combined with $ln \, H_i / P^o$ or $ln \, P_i^{sat} / P^o$ to give solution phase reference chemical potentials μ_i^* and a Lewis-Randall rule for liquid solutions [1,2].

1. **Ideal solutions** obey a Lewis-Randall rule

$$\mu_i = \mu_i^* + k_B T \, ln \, x_i \tag{2.1.4}$$

Based on the considerations above, $\mu_i^* = \mu_i^o + k_B T \ln H_i/P$ for dilute volatile solutes ($x_i \approx 0$), and $\mu_i^* = \mu_i^o + k_B T \ln P_i^{sat}/P$ for nearly pure solvents ($x_i \approx 1$). The natural reference state for the solvent is the pure solvent at the solution temperature and pressure. The reference solute state extends Henry's law behavior all the way to the unattainable composition of a pure solute, i.e. to the point labeled H_i in Figure 2.1.1.

2. **Non-ideal solutions** obey a more general rule

$$\mu_i = \mu_i^* + k_B T \ln \gamma_i x_i \tag{2.1.5}$$

where γ_i is an activity coefficient. The activity coefficients are analogous to the fugacity coefficients (ϕ_i) for non-ideal gas mixtures. $\gamma_i > 1$ indicates an excess free energy driving solutes out of the solution, while $\gamma_i < 1$ indicates a solute with excess affinity for the solution environment. The reference conditions are those from the ideal solution analysis: $\gamma_i \to 1$ when component i is infinitely dilute, and $\gamma_i \to 1$ when component i becomes the pure solvent. Given the activity $a_i = \gamma_i x_i$ at a pressure P° the activity at another pressure P is given by [3,4]

$$a_i = \gamma_i x_i \, exp\left[\mathcal{V}_i (P - P^\circ)/k_B T\right] \tag{2.1.6}$$

where $\mathcal{V}_i$ is the volume per mole of the pure liquid i. For condensed phases, the exponential correction is usually unity except at very high pressures.

The pure solvent convention ($\gamma_i \to 1$ when $x_i \to 1$) is also used for all components of a liquid mixture that are liquids in their pure state at the temperature and pressure of solution. For example, the activities of ethanol and water in beer, wine, or vodka would be modeled using the pure solvent convention for ethanol and water, even though these beverages span a range of different compositions. The dilute solute convention based on mole fractions ($\gamma_i \to 1$ when $x_i \to 0$) is used for molecular solutes like sugar or urea in water. Dilute solute conventions using molalities (and sometimes concentrations) are used for dissolved salts, acids, or hydrates.

Electrolytes

The theory of electrolyte activities is among the greatest triumphs of statistical mechanics [5,6]. By practical convention, the chemical potential of a dissolved salt is often given in terms of molality. The molality m_i is the number of moles of dissolved solute (salt) per kg of solvent. The non-ideal chemical potential for the salt is again a Lewis-Randall type expression [2]:

$$\mu_i = \mu_i^\diamond + k_B T \ln(\gamma_i m_i/m^o) \tag{2.1.7}$$

The reference chemical potential $\mu_i^\diamond$ is that in a hypothetical solution in which there is only salt i with its ideal infinite dilution properties at a molality $m^o = 1$ mol salt/1 kg solvent. Again, the *solvent* chemical potential is $\mu_i = \mu_i^* + k_B T \ln \gamma_i x_i$ with $\gamma_i \rightarrow 1$ when $x_i \rightarrow 1$ such that the reference solvent is pure at the solution temperature and pressure.

Thus far we have only considered salt as a formula unit in solution. For purposes of compact tabulation and for solutions with multiple salts dissolved, it is convenient to write chemical potentials for the individual dissociated cations and anions. We cannot unambiguously define an absolute free energy for the separate ions, but given a reference free energy of formation for any one ion, the chemical potentials for all other solvated ions can be determined [2]. The hydrogen ion convention sets the free energy of formation of H^+ equal to zero at $m_{H^+} = 1$ mol/kg water [7]:

$$\frac{1}{2} H_{2(g,1atm)} \rightleftarrows H^+_{(aq)} + e^- \qquad \Delta G^o = 0$$

This hydrogen ion (or hydrogen electrode) standard is used at all temperatures so that, by the Gibbs-Helmholtz equation, the hydrogen electrode reaction also has $\Delta H^o = 0$ and also $\Delta S^o = 0$.

Now consider an ionic compound $\mathbf{A}_{\upsilon+}\mathbf{B}_{\upsilon-}$, which dissociates into solvated ions as

$$\nu_+ \mathbf{A}^{z+} + \nu_- \mathbf{B}^{z-}$$

where electroneutrality requires $\nu_+ z_+ + \nu_- z_- = 0$. Before considering equilibrium for $\mathbf{A}_{\upsilon+}\mathbf{B}_{\upsilon-}$ dissociation, let us establish some notational conventions. The total number of ions per dissociated formula unit is

$$\nu = \nu_+ + \nu_-$$

The (geometric) mean electrolyte molality including contributions from both cations and anions is

$$m_\pm = (m_+^{\nu+} m_-^{\nu-})^{1/\nu}$$

Note that $m_\pm$ equals the formula unit molality m only when there is complete dissociation and when complete dissociation creates an equal number of cations and anions. Dissociation is complete in the infinitely dilute limit, so we define the separate ion molalities in terms of the total formula unit molality m as [2]

$$m_+ = \nu_+ m \qquad \mathit{and} \qquad m_- = \nu_- m \tag{2.1.8}$$

The possibility of incomplete dissociation at higher molality is then accounted for within a mean electrolyte activity coefficient, $\gamma_\pm$. The mean activity coefficient including contributions from both cations and anions is

$$\gamma_\pm = (\gamma_+^{\nu+} \gamma_-^{\nu-})^{1/\nu}$$

Finally, the chemical potential including contributions from both anions and cations is

$$\mu_{\pm} = \nu_{+}\mu_{+}^{\diamond} + \nu_{-}\mu_{-}^{\diamond} + \nu k_B T \, ln \, \gamma_{\pm} m_{\pm}/m^o \tag{2.1.9}$$

Geometric means in the definitions of $\gamma_{\pm}$ and $m_{\pm}$ allow the $ln\gamma_{\pm}m_{\pm}$ term to be split into separate ion contributions.

$$\mu_{+} = \mu_{+}^{\diamond} + k_B T \, ln \, \gamma_{+} m_{+}/m^o$$

and

$$\mu_{-} = \mu_{-}^{\diamond} + k_B T \, ln \, \gamma_{-} m_{-}/m^o$$

For dilute electrolytes, the mean activity coefficients can be described by the Debye-Huckel limiting law [5]

$$ln \, \gamma_{\pm} = -\alpha |z_{+} z_{-}| \sqrt{I}$$

where $\alpha = 1.172$ and the ionic strength is $I = \frac{1}{2}\sum_i m_i z_i^2$ with the summation including all ions in solution. Figure 2.1.2 shows the Debye-Huckel limiting law and the actual activity coefficient for HCl in water at 25°C. As seen in the figure, the Debye-Huckel theory is only valid at very low ionic strengths. Additionally, Figure 2.1.2 shows that deviations from ideal behavior (and also from Debye-Huckel theory) can become quite large at high ionic strengths.

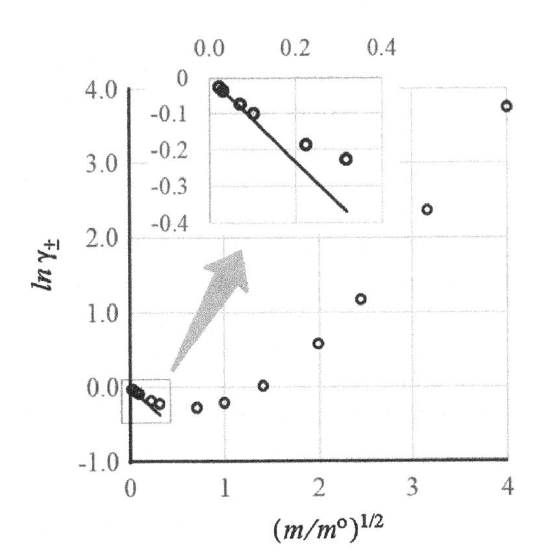

Figure 2.1.2: Activity coefficient vs. molality for HCl in aqueous solution at 25^oC. Predictions of the Debye-Huckel limiting law are shown as a line at low ionic strengths.

Pure solids and liquids

Many reactions consume or generate pure solids or liquids, e.g. the combustion of coal, the dissolution of solid NaCl in water, the formation of rust, or the formation of water at a fuel cell electrode. To correctly determine the ΔG_f^o for a pure solid or liquid, one must be careful to use values for the correct allotrope or polymorph. Apart from this pitfall, activities for pure solids or liquids are trivially easy to model. By definition the chemical potential of a pure solid or liquid is not composition dependent. Additionally, solids and liquids are usually impervious to changes in pressure. Therefore, we set the activity of pure solids and liquids to unity.

Within this section, we have been careful to use separate notations (μ_i^o, μ_i^*, and $\mu_i^\diamond$) for chemical potentials in different types of reference states. In later sections, and throughout the book the separate notations for different reference states will not be maintained. Of course the importance of using the appropriate activity expression for the available free energy of formation data will remain vital.

This chapter takes all free energies of formation from tabulated thermochemical data, but free energies of formation and reaction can also be computed using electronic structure theory and basic elements of statistical mechanics. *Ab initio* approaches are now fairly reliable for free energies in the gas phase. In the liquid phase, *ab initio* methods offer the exciting possibility of computing the free energy of any solute in any solvent. Unfortunately, *ab initio* calculations of free energies, chemical potentials, and activity coefficients in liquid solution are, as yet, not very reliable.

2.2 Equilibrium constants and compositions

Chemical reactions are represented at the most basic level by equations that give their stoichiometry. Consider the hypothetical reaction

$$a\mathbf{A} + b\mathbf{B} \rightleftarrows c\mathbf{C} + d\mathbf{D} \tag{2.2.1}$$

with stoichiometric coefficients $\nu_\mathbf{A} = -a$, $\nu_\mathbf{B} = -b$, $\nu_\mathbf{C} = c$, and $\nu_\mathbf{D} = d$. At constant temperature and pressure, the reaction proceeds to equilibrium when

$$dG = \mu_\mathbf{A} dN_\mathbf{A} + \mu_\mathbf{B} dN_\mathbf{B} + \mu_\mathbf{C} dN_\mathbf{C} + \mu_\mathbf{D} dN_\mathbf{D} = 0$$

The numbers of **A**, **B**, **C**, and **D** molecules do not change independently. Instead, they each change in proportion to a stoichiometric coefficient and the extent of reaction ξ,

$$d\xi = -\frac{1}{a} dN_\mathbf{A} = -\frac{1}{b} dN_\mathbf{B} = \frac{1}{c} dN_\mathbf{C} = \frac{1}{d} dN_\mathbf{D} \tag{2.2.2}$$

The extent of reaction ξ is the number of times the reaction occurs as written left to right according to its stoichiometry. Positive ξ indicates reaction progress to the right, and negative ξ indicates reaction progress to the left.

The equilibrium condition, because of equation (2.2.2), becomes

$$dG = \sum_i \nu_i \mu_i d\xi = 0$$

where ν_i are the stoichiometric coefficients. dG is zero for arbitrary $d\xi$ when

$$\sum_i \nu_i \mu_i = 0 \tag{2.2.3}$$

Inserting $\mu_i = \mu_i^o + k_B T \, ln \, a_i$ into equation (2.2.3) and rearranging gives

$$exp[-\beta \sum_i \nu_i \mu_i^o] = \prod_i a_i^{\nu_i} \tag{2.2.4}$$

where $\beta = 1/k_B T$. The argument in the exponential is the standard Gibbs free energy per mole of reaction at the reference conditions:

$$\Delta G^\circ_{rxn} = \sum_i \nu_i \mu_i^o \tag{2.2.5}$$

The Gibbs free energy of reaction can also be written in terms of standard free energies of formation: $\Delta G^\circ_{rxn} = \sum_i \nu_i \Delta G^\circ_{f,i}$. Combining equations (2.2.4) and (2.2.5) gives

$$exp[-\beta \Delta G^\circ_{rxn}] = \prod_i a_i^{\nu_i} \tag{2.2.6}$$

The individual activities depend on the temperature, the pressure, and the composition, but the left hand side of equation (2.2.6) only depends on the temperature and the chosen reference conditions. Accordingly, the left hand side is called an equilibrium constant

$$K_{eq} \equiv exp[-\beta \Delta G^\circ_{rxn}] \tag{2.2.7}$$

Combining all of the results above, gives the familiar mass action law [8,9] which relates the equilibrium constant and the stoichiometric coefficients to the thermodynamic activities at equilibrium

$$K_{eq} = \prod_i a_i^{\nu_i} \tag{2.2.8}$$

Equation (2.2.8) is the central result from the theory of chemical equilibrium. The left side depends on intrinsic properties of the reactants and products. The composition-dependent ratio of activities on the right side monotonically changes with extent of reaction. Equality is achieved at a unique point corresponding to thermodynamic equilibrium. Let us begin on the left side of the equation, by learning to compute equilibrium constants.

The equilibrium constant is a strong function of temperature, and frequently, one needs the equilibrium constant at a different temperature from that in the thermodynamic tables. Given the free energy of reaction at one temperature, the standard free energy of reaction at another temperature can be obtained from the Gibbs-Helmholtz equation [3,7].

$$\frac{\partial(\beta \Delta G^{\circ}_{rxn})}{\partial \beta} = \Delta H^{\circ}_{rxn} \tag{2.2.9}$$

The Gibbs-Helmholtz equation in this context is identical to the van't Hoff equation [2,7]

$$\frac{\partial ln K_{eq}}{\partial(1/k_B T)} = -\Delta H^{\circ}_{rxn} \tag{2.2.10}$$

When the temperature interval is small and when the reactants and products have the same heat capacity, the Gibbs-Helmholtz equation integrates to

$$\left.\frac{\Delta G^{\circ}_{rxn}}{k_B T}\right|_{T_2} = \left.\frac{\Delta G^{\circ}_{rxn}}{k_B T}\right|_{T_1} + \frac{\Delta H^{\circ}_{rxn}}{k_B}\left(\frac{1}{T_2} - \frac{1}{T_1}\right) \tag{2.2.11}$$

If the temperature interval is large then the dependence of ΔH°_{rxn} on temperature may become important. The temperature dependence in ΔH°_{rxn} can be accounted for by first integrating the heat capacities

$$\Delta H^{\circ}_{rxn}(T) = \Delta H^{\circ}_{rxn}(T_1) + \int_{T_1}^{T} \Delta C_P dT \tag{2.2.12}$$

where ΔC_P is the heat capacity of the products minus the heat capacity of the reactants. Expressions for the temperature dependence of heat capacity have been tabulated for many molecules.

From this point, the calculation of equilibrium compositions is a straightforward procedure. First, the equilibrium constant obtained from the van't Hoff equation is equated to the product of activities as in equation (2.2.8). The activities are then written in terms of mole fractions or molalities using the appropriate expressions from section 2.1. For example, if the reaction $a\mathbf{A} + b\mathbf{B} \rightleftarrows c\mathbf{C} + d\mathbf{D}$ occurs in the gas phase, then the individual species activities are $a_i = \phi_i x_i P/P^o$. Inserting the activities into equation (2.2.8) gives [10,11]

$$K_{eq} = \left(\frac{\phi^c_{\mathbf{C}}\phi^d_{\mathbf{D}}}{\phi^a_{\mathbf{A}}\phi^b_{\mathbf{B}}}\right)\left(\frac{x^c_{\mathbf{C}}x^d_{\mathbf{D}}}{x^a_{\mathbf{A}}x^b_{\mathbf{B}}}\right)\left(\frac{P}{P^\circ}\right)^\delta \tag{2.2.13}$$

where $\delta = c + d - a - b$. It is convenient to write this as $K_{eq} = K_\phi K_x (P/P^o)^\delta$ where $K_\phi = (\phi^c_{\mathbf{C}}\phi^d_{\mathbf{D}}/\phi^a_{\mathbf{A}}\phi^b_{\mathbf{B}})$ and $K_x = (x^c_{\mathbf{C}}x^d_{\mathbf{D}}/x^a_{\mathbf{A}}x^b_{\mathbf{B}})$. For low pressures and high temperatures the fugacity coefficients are usually close to unity, so that the factor K_ϕ may be omitted. It is also common to define $K = K_{eq}(P^\circ)^\delta$ so that K has units, and then $K_x P^\delta$ becomes a mass action ratio of the (ideal) partial pressures.

Example: Ammonia synthesis

In the Haber process, hydrogen and nitrogen react to make ammonia

$$\frac{3}{2}H_{2(g)} + \frac{1}{2}N_{2(g)} \rightleftarrows NH_{3(g)} \tag{2.2.14}$$

at high pressure and high temperature in the presence of a catalyst [10]. The standard (1 bar) properties of ammonia, nitrogen, and hydrogen at 298 K are given in the table below. The free energy and enthalpy are in units kJ/mol, and the parameters A_1, A_2, and A_3 give the heat capacity as a function of temperature (in Kelvin) for each species via the equation [3]

$$C_P/k_B = A_1 + 10^{-3}A_2 T + 10^5 A_3/T^2$$

The stoichiometric coefficients have been used to compute the change in each property upon reaction. Specifically, the entry in row "net" is $\Delta X_{rxn} = X_{NH_3} - 0.5X_{N_2} - 1.5X_{H_2}$ for each property X.

	$\Delta G^o_{f,298}$	$\Delta H^o_{f,298}$	A_1	A_2	A_3
NH_3	-16.45	-46.11	3.578	3.020	-0.186
N_2	0	0	3.280	0.593	0.040
H_2	0	0	3.249	0.422	0.083
net	-16.45	-46.11	-2.936	2.091	-0.331

The reaction is favorable at 298K, with an equilibrium constant $K_{eq} = exp[-\Delta G^\circ_{rxn}/k_B T] = 764.8$, but (even with a catalyst) high temperatures are needed to accelerate the reaction kinetics. The reaction is usually operated at temperatures from 700 to 750 K. The Gibbs free energy of reaction and the enthalpy of reaction can be used to obtain the entropy of reaction, $\Delta S^\circ_{rxn} = -99.53$ J/mol/K. Equations (2.2.12) and (2.2.11) can be numerically integrated to obtain the Gibbs free energy and equilibrium constant at high temperatures. The figure below shows the standard Gibbs free energy of reaction (1 bar) for all temperatures from 275 K to 725 K.

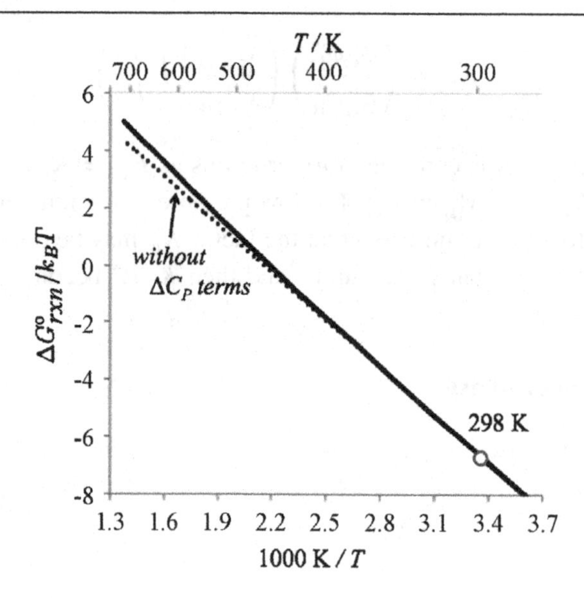

Even though the temperature interval in the figure is approximately 400K wide, the constant enthalpy approximation of equation (2.2.11) captures most of the temperature dependence. However, for quantitatively accurate equilibrium constants near 700K, the heat capacity contributions must be included.

Because of the negative reaction entropy, the reaction becomes thermodynamically unfavorable at high temperatures. By 723 K the Gibbs free energy has increased to 30.1 kJ/mol and the equilibrium constant has decreased to $K_{eq} = 0.0066$. Because of the small equilibrium constant, the Haber process requires operation at hundreds of atmospheres of pressure. Consider an initial mixture of n_0 moles of nitrogen gas and m_0 moles of hydrogen gas at a fixed pressure of 300 bar. Solving equation (2.2.13) for K_x gives $K_x = K_{eq} K_\phi^{-1} (P/P^\circ)^{-\delta}$. The fugacity coefficients at 723K and 300bar are $\phi_{N2} = 1.14$, $\phi_{H2} = 1.09$, and $\phi_{NH3} = 0.91$ [10]. Thus

$$K_\phi = \frac{\phi_{NH3}}{\phi_{N2}^{1/2} \phi_{H2}^{3/2}} = 0.75$$

Also using $\delta = -1$, $P/P^\circ = 300$, and $K_{eq} = 0.0066$,

$$K_x = \frac{x_{NH3}}{x_{N2}^{1/2} x_{H_2}^{3/2}} = 2.64 \tag{2.2.15}$$

The stoichiometric table below gives the amount of each reagent and its mole fraction in terms of the extent of reaction ξ.

Species	Initial	Equilibrium	Equilibrium mole fraction
N_2	n_0	$n_0 - \xi/2$	$(n_0 - \xi/2)/(n_0 + m_0 - \xi)$
H_2	m_0	$m_0 - 3\xi/2$	$(m_0 - 3\xi/2)/(n_0 + m_0 - \xi)$
NH_3	0	ξ	$\xi/(n_0 + m_0 - \xi)$
total	$n_0 + m_0$	$n_0 + m_0 - \xi$	

Now, for any initial values of n_0 and m_0, the equilibrium composition is obtained by solving for ξ_{eq} in the equation

$$\left.\frac{\xi/(n_0 + m_0 - \xi)}{(n_0 - \xi/2)^{1/2}(m_0 - 3\xi/2)^{3/2}/(n_0 + m_0 - \xi)^2}\right|_{\xi_{eq}} = 2.64 \qquad (2.2.16)$$

For example, with a stoichiometric composition $n_0 = 0.25$ mol N_2 and $m_0 = 0.75$ mol H_2, solving equation (2.2.16) gives an equilibrium extent of reaction $\xi_{eq} = 0.262$ mol [10]. Inserting ξ_{eq} into the mole fraction expressions shows that a stoichiometric mixture of nitrogen and hydrogen reactants gives an equilibrium composition of $x_{N_2} = 0.16$, $x_{H_2} = 0.48$, and $x_{NH_3} = 0.36$. Thus, even at 300 bar, the high temperature makes the equilibrium limitations severe.

Similar calculations can be performed to determine the equilibrium compositions for reactions in liquid solution. For reactions between solutes, the solvent components will often be in large excess so that the solvent activity is essentially constant over the course of a reaction. The appropriate activity convention for solutes depends on the nature of the solute. For a reaction $a\mathbf{A} + b\mathbf{B} \rightleftarrows c\mathbf{C} + d\mathbf{D}$ where all reagents have activities that depend on mole fractions, the equilibrium constant is

$$K_{eq} = \left(\frac{\gamma_{\mathbf{C}}^c \gamma_{\mathbf{D}}^d}{\gamma_{\mathbf{A}}^a \gamma_{\mathbf{B}}^b}\right)\left(\frac{x_{\mathbf{C}}^c x_{\mathbf{D}}^d}{x_{\mathbf{A}}^a x_{\mathbf{B}}^b}\right) exp\left[\beta \Delta \mathcal{V}_{rxn}(P - P^\circ)\right] \qquad (2.2.17)$$

which accounts for non-idealities and for changes in molar volume: $\Delta \mathcal{V}_{rxn} = c\mathcal{V}_{\mathbf{C}} + d\mathcal{V}_{\mathbf{D}} - a\mathcal{V}_{\mathbf{A}} - b\mathcal{V}_{\mathbf{B}}$. In liquid solution, the exponential term is usually very close to unity because $\beta \Delta \mathcal{V}_{rxn}(P - P^\circ)$ is small. Thus, except for extremely high pressures, we can usually approximate that

$$K_{eq} \approx \left(\frac{\gamma_{\mathbf{C}}^c \gamma_{\mathbf{D}}^d}{\gamma_{\mathbf{A}}^a \gamma_{\mathbf{B}}^b}\right)\left(\frac{x_{\mathbf{C}}^c x_{\mathbf{D}}^d}{x_{\mathbf{A}}^a x_{\mathbf{B}}^b}\right) \qquad (2.2.18)$$

For ideal solutions, the activity coefficients are unity giving $K_{eq} \approx \left(x_{\mathbf{C}}^c x_{\mathbf{D}}^d / x_{\mathbf{A}}^a x_{\mathbf{B}}^b\right)$ in accordance with the law of mass action. Solutions tend to be approximately ideal when they are comprised of chemically similar components, especially mixtures of isomeric alcohols or of similar olefins [2]. Otherwise, the ratio of activity coefficients $(\gamma_{\mathbf{C}}^c\gamma_{\mathbf{D}}^d/\gamma_{\mathbf{A}}^a\gamma_{\mathbf{B}}^b)$ is often far from unity leading to qualitative departures from the law of mass action. Non-idealities are particularly important for reactions involving ionic species which interact strongly with each other even at small concentrations. The following example uses the molality convention to predict the extent of adsorption of CO_2 into water, conversion to carbonic acid, and deprotonation to carbonate and bicarbonate ions [12].

Example: CO_2 speciation in water

Older thermodynamics books give the concentration of CO_2 in the atmosphere as 300 ppm. Current CO_2 levels are slightly over 400 ppm. One consequence of high CO_2 levels is the acidification of rainfall and ocean water due to the formation of carbonic acid.

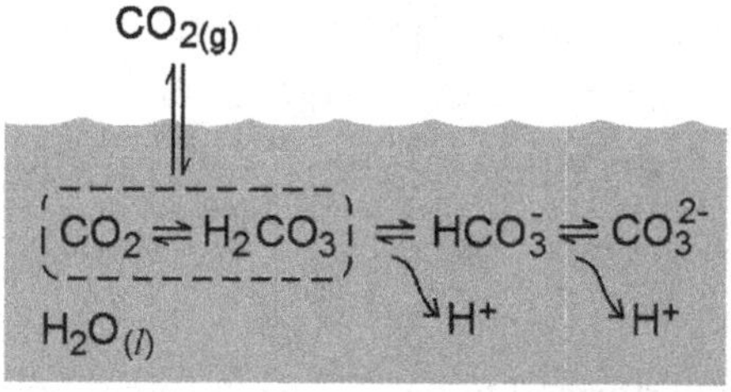

The concentrations of $CO_{2(aq)}$, $HCO_{3(aq)}^-$, and $CO_{3(aq)}^{2-}$ at 25°C can be computed using the following tabulated free energies of formation [12].

Species	ΔH_f^o	ΔG_f^o	Ref. state
$H_2O_{(l)}$	-285.8	-237.1	pure liq.
$H_{(aq)}^+$	0.0	0.0	1 mol/kg
$CO_{2(g)}$	-393.5	-394.4	pure gas
$[CO_{2(aq)}]$	-413.8	-386.0	1 mol/kg
$HCO_{3(aq)}^-$	-692.0	-586.9	1 mol/kg
$CO_{3(aq)}^{2-}$	-677.1	-527.9	1 mol/kg

Note that H_2CO_3 and CO_2 in aqueous solution are treated as one lumped species, denoted $[CO_{2(aq)}]$ in the table.

Henry's constant

The partitioning of molecular CO_2 between the gas and aqueous phases is governed by a Henry's constant. We can derive the Henry's constant by thinking of transfer across the phase boundary as a reaction.

$$CO_{2(g)} \rightleftharpoons CO_{2(aq)} \qquad \Delta G^o = +8.4kJ/mol$$

Using $K_{eq} = exp[-\beta \Delta G^o]$ with $\beta = 1/k_B T$ gives $K_{eq} = 0.0337$. Equating the equilibrium constant to the mass action ratio of activities gives

$$\begin{aligned} K_{eq} &= \frac{\gamma_{CO_2} m_{CO_2}/m^o}{\phi_{CO_2} y_{CO_2} P/f^o_{CO_2}} \\ &\approx \frac{1000\frac{g}{kg} \cdot x_{CO_2}(x_{H_2O} 18g/mol)^{-1}/(\frac{1mol}{kg})}{y_{CO_2} P/P^o} \end{aligned}$$

where the approximation assumes a mixture of ideal gases. At 1atm and 25°C, we have $P/P^o \approx 1$ and $x_{H_2O} \approx 1$ so that the equilibrium condition simplifies to

$$0.000607\, y_{CO_2} = x_{CO_2}$$

where y_{CO_2} is the mole fraction in the vapor and x_{CO_2} is the mole fraction of non-ionic CO_2 species dissolved in the liquid. Henry's law is $x_i H_i = y_i P$, so inserting $P = 1$atm identifies the Henry's constant as $H_{CO_2} = 1atm/0.000607$, or

$$H_{CO_2} = 1640atm$$

The calculated value of Henry's constant at 25°C is consistent with tabulated values of 1480atm at 20°C and 1940atm at 30°C [12].

Equilibrium composition

At 400ppm $CO_{2(g)}$, 25°C, and 1atm pressure the equilibrium molality of non-ionic dissolved CO_2/H_2CO_3 in the liquid is

$$m_{CO_2} \approx 0.0337 * 0.0004 * m^o = 1.35 \times 10^{-5} mol/kg$$

Now consider the equilibrium for the first deprotonation step

$$CO_{2(aq)} + H_2O_{(l)} \rightleftharpoons HCO^-_{3(aq)} + H^+_{(aq)} \qquad \Delta G^o = +36.2kJ/mol$$

giving $K_{eq} = 4.52 \times 10^{-7}$ and

$$4.52 \times 10^{-7} = \frac{\gamma_{H^+} \gamma_{HCO_3^-} m_{H^+} m_{HCO_3^-}}{\gamma_{H_2O} x_{H_2O} \gamma_{CO_2} m_{CO_2}}$$

$\gamma_{H_2 0} x_{H_2 O} \approx 1$ because CO_2 and dissolved ions are dilute. We have already shown that $m_{CO_2} = 1.35 \times 10^{-5} mol/kg$. The activity coefficient of a non-ionic and dilute species should be nearly unity, so $\gamma_{CO_2} \approx 1$. If $[H^+]$ is sufficiently, large then essentially all H^+ comes from dissociation of carbonic acid, i.e. $m_{H^+} \approx m_{HCO_3^-}$. For the moment, suppose that $\gamma_{H^+} \approx \gamma_{HCO_3^-} \approx 1$. Then

$$m_{H^+} \approx m_{HCO_3^-} \approx \sqrt{m_{CO_2} 4.52 \times 10^{-7}} \approx 2.47 \times 10^{-6} mol/kg$$

Indeed the proton concentration from dissolved CO_2 is significantly higher than the $10^{-7} M\ H^+$ concentration that results from water autodissociation, so the assumption that nearly all H^+ comes from carbonic acid is justified. We can also use Debye-Huckel theory [2] to check the assumption that $\gamma_{H^+} \approx \gamma_{HCO_3^-} \approx 1$. At ionic strength $I = 2.47 \times 10^{-6}$, $ln\, \gamma_{\pm} = -\alpha |z_+ z_-| \sqrt{I}$ gives $\gamma_{H^+} = \gamma_{HCO_3^-} = 0.998$. If the activities were farther from unity, one would need to iterate to self-consistency. In conclusion, the pH of water in equilibrium with CO_2 at 400ppm in the gas phase is approximately 5.6. The reader can check that the second deprotonation results in only trace carbonate and trace additional H^+ [12].

The conventions for solutes, solvents, gases, and electrolytes can be combined as needed to use available free energy of formation data. However, one must be extremely careful when pulling the associated activity coefficients from the literature. The numerical value of the activity coefficient depends on the convention for which it has been determined.

Multireaction equilibria

Equilibria in the presence of multiple reactions is a straightforward conceptual generalization, but there are complicated aspects of the calculations. Before embarking on calculations, it is important to ensure that the chemical reactions are independent [10,13]. Like vectors in linear algebra, a set of chemical reactions is independent if no reaction in the set can be written as a linear combination of the others. Once all redundant equations have been pared away, the equilibrium composition can be uniquely determined by introducing a separate extent of reaction variable for each independent reaction [10,13].

1. Each independent reaction introduces a degree of freedom and an associated equilibrium constant, i.e. a constraint on the equilibrium composition.
2. Even when there are many more compounds involved than reactions, the equilibrium is well-defined because separate compounds are not independent degrees of freedom. They are restrained by the stoichiometric coefficients of the available reactions.

These two considerations result in a set of coupled constraint equations involving as many extent of reaction variables to determine. Solving the coupled algebraic equations often requires numerical methods and/or approximations [10,13].

Exercises

1. Consider a reaction that involves a net generation of $\delta = \sum_i \nu_i$ molecules in a mixture of ideal gases. Does $\Delta G^{\circ}_{rxn} = \sum_i \nu_i \mu_i^o$ for all δ? Does $\Delta G^{\circ}_{rxn} = \sum_i \nu_i \mu_i^*$ for all δ?
2. Sometimes activity coefficients must be converted from one convention to activity coefficients for another convention.
   (a) Show that molality and mole fraction for a solute i are related by $m_i = 1000x_i / M_{solvent}x_{solvent}$ where $M_{solvent}$ is the molecular weight of the solvent.
   (b) In the limit of a pure solvent with only trace salt i, explain why the relationship between x_i and m_i implies the following relationship between activity coefficients for the molality and mole fraction conventions $\gamma_i^{(m)} = \gamma_i^{(x)} 1000x_i/(M_{solvent}m_i/m^o)$.
3. The catalytic hydrogenation of formaldehyde ("**A**" = CH_2O) to methanol ("**B**" = CH_3OH) has overall stoichiometry **A** + $H_2 \rightleftarrows$ **B**. The enthalpy and Gibbs free energy of the gas phase reactants and products at 300K and $f^o = 1bar$ are (approximately):

Species	ΔH_f^o(kJ/mol)	ΔG_f^o(kJ/mol)
B	-200	-162
H_2	0.0	0.0
A	-110	-102

   (a) Estimate the temperature at which $\Delta G^o_{rxn} = 0$. You may ignore non-idealities and you may assume that ΔH^o_{rxn} is independent of temperature.
   (b) For the temperature at which $\Delta G^o_{rxn} = 0$ and at $P = 1$bar, find the equilibrium extent of reaction starting from 1.0 mol of **A**, 0.0 mol of **B**, and a large excess of H_2.
4. [Hill and Root] [11] Consider the equilibrium between solid nickel, carbon monoxide, and nickel tetracarbonyl.

$$Ni_{(s)} + 4CO_{(g)} \rightleftarrows Ni(CO)_{4(g)}$$

For the reaction as written, the standard Gibbs free-energy change at 100^oC is 1292 cal/mol when the following standard states are used: $Ni_{(s)}$ = pure crystalline solid at 100^oC under its own vapor pressure, $CO_{(g)}$ = pure gas at 100^oC and 1atm, and $Ni(CO)_{4(g)}$ = pure gas at 100°C and 1atm.
   (a) If a vessel is initially charged with pure $Ni(CO)_4$ and maintained at a temperature of 100^oC and $2atm$. What fraction of the $Ni(CO)_4$ will decompose? The vapor pressure

of pure nickel at 100^oC is $1.23 \times 10^{-46} atm$. Assume that each component of the gas behaves like an ideal gas, and state other assumptions that you make.

(b) What pressure would cause 95% of the $Ni(CO)_4$ to decompose?

5. Carbonic acid dissolution and speciation in aqueous solution is dramatically influenced by pH. As a function of pH at 25°C, compute the fractions by mole of total carbonate species that are non-ionic carbonates ($[CO_{2(aq)}] = CO_2$ or H_2CO_3), bicarbonate HCO_3^-, and fully deprotonated carbonate CO_3^{2-}. Hint: the predominant protonation state will progressively change from neutral carbonate species, to bicarbonates, to fully deprotonated carbonate as pH increases.
6. CO_2 and NH_3 gases react to make aqueous urea according to the reaction

$$CO_{2(g)} + 2NH_{3(g)} \rightleftarrows CO(NH_2)_{2(aq)} + H_2O_{(l)}$$

At 25°C the reference states, free energies of formation, and corresponding chemical potentials are

Species	ν_i	Reference state	ΔG_f^{ref}	$\mu_i(T, P, x_1, x_2, ...)$
CO_2	-1	gas, 25°C, 1atm	-394.4	$\mu_{CO_2}^o + k_BT\, ln\,(f_{CO_2}/f_{CO_2}^o)$
NH_3	-2	gas, 25°C, 1atm	-16.6	$\mu_{NH_3}^o + k_BT\, ln\,(f_{NH_3}/f_{NH_3}^o)$
U	1	25°C, 1m aqueous	-237.2	$\mu_U^\diamond + k_BT\, ln\,(\gamma_U m_U/m^\diamond)$
H_2O	1	25°C, pure water	-203.8	$\mu_{H_2O}^* + k_BT\, ln\, \gamma_{H_2O} x_{H_2O}$

The Gibbs free energies of formation are given in kJ/mol, and U stands for urea, $CO(NH_2)_2$. Predict the equilibrium amount of dissolved aqueous urea in contact with a stoichiometric mixture of CO_2 and NH_3 at a total pressure of 1 bar in the gas phase. Does your answer depend on the amount of liquid water?

7. Use the data from problem 6 to predict the boiling point of one liter of aqueous solution containing 5 g of urea.

References

[1] G.N. Lewis, M. Randall, Thermodynamics, McGraw-Hill, New York, 1923.
[2] K.G. Denbigh, The Principles of Chemical Equilibrium, Cambridge University Press, Cambridge, UK, 1981.
[3] J.M. Smith, H.C. Van Ness, M.M. Abbott, Introduction to Chemical Engineering Thermodynamics, McGraw-Hill, New York, 1996.
[4] B.E. Poling, J.M. Prausnitz, J.P. O'Connell, The Properties of Liquids and Gases, McGraw-Hill, New York, 2001.
[5] P. Debye, E. Huckel, Z. Phys. 24 (1923) 185.
[6] E.A. Guggenheim, R.H. Stokes, Equilibrium Properties of Aqueous Solutions of Single Strong Electrolytes, Pergamon Press, London, 1969.
[7] D. Kondepudi, I. Prigogine, Modern Thermodynamics: From Heat Engines to Dissipative Structures, Wiley, Chichester, UK, 1998.
[8] C.M. Guldberg, P. Waage, Forhandlinger: Videnskabs, Selskabet i Christiania 35 (1864).

[9] H. Abrash, J. Chem. Educ. 63 (1986) 1044–1047.
[10] M.E. Davis, R.J. Davis, Fundamentals of Chemical Reaction Engineering, McGraw-Hill, New York, 2003.
[11] C.G. Hill, T.W. Root, Introduction to Chemical Engineering Kinetics and Reactor Design, Wiley, Hoboken, NJ, 2014.
[12] N. de Nevers, Physical and Chemical Equilibrium for Chemical Engineers, Wiley, New York, 2002.
[13] I. Tosun, The Thermodynamics of Phase and Reaction Equilibria, Elsevier, Amsterdam, 2012.

CHAPTER 3

Rate laws

> *"Almost every chemical reaction of practical interest consists of a network of elementary steps, each with its own contribution to kinetics. Single-step reactions are most often found in textbooks."*
>
> ***Helfferich, Kinetics of Multistep Reactions (2004)***

The overall kinetics of a chemical reaction depend on thermodynamic properties, transport properties, and intrinsic kinetics. All chemistry students learn that kinetics depend on concentrations and on temperature. Clearly, the kinetics of a chemical reaction also depend on the molecular characteristics of the reactants and catalysts. Indeed much of basic chemistry concerns differences in reactivity for molecules with different steric characteristics, different electrostatic characteristics, different frontier orbital characteristics, different acidities, etc. More subtle factors can also influence reactivity. For example, isotope effects can be approximately predicted and used to test hypotheses about rate limiting steps. The kinetics of reactions in solution also depend on solvent properties like viscosity, ionic strength, and hydrogen bonding propensity. The kinetics of catalytic reactions often depend on mass transfer and heat transfer limitations. The kinetics of gas phase reactions depend on vibrational coupling within molecules and sometimes on the partial pressure of inert collision partners. Some processes like electron transfer and protein folding can be driven by external potentials or forces which influence their kinetics. Each of these effects has been mechanistically informative and integral to the development of modern theories of kinetics.

The all-encompassing nature of kinetics makes the subject challenging and fun. In many cases, judicious analyses of time scales, length scales, and equilibrium properties can disentangle the effects of thermodynamics, transport, and kinetics. Later chapters will delve into molecular characteristics, potential energy surfaces, and dynamics, with a particular focus on the factors that influence *rate constants*. But first we must understand phenomenological rate laws, i.e. how *rates* depend on reactant concentrations, rate constants, and equilibrium constants. Theorists sometimes use the term "phenomenological" to disparage empiricisms, and indeed many phenomenological models are pure empiricisms. However, phenomenological rate laws can also be constructed based on specific mechanistic hypotheses, and these are essential in the theory of reaction kinetics. Mechanistically motivated rate laws can be derived from first principles and the law of large numbers (see Chapter 14), and thus they are actually far more than empiricisms. These special phenomenological rate laws yield accurate quantitative predictions across a wide range of reaction conditions (when the mechanism from which they are constructed is correct). Thus mechanistically motivated phenomenological rate laws are the quantitative links between experimental kinetics and theoretical predictions, and the subject could scarcely have made any progress without them.

3.1 Rates, mass balances, and reactors

The first step toward a useful phenomenological model is to determine whether the reaction occurs in the bulk, on a surface, at specific catalytic sites, etc. The nature of the domain in which the reaction occurs influences the entire structure of a kinetic model and even the units of the reaction rate. For a reaction that occurs in the bulk of a well-mixed fluid, the rate is the number of reaction events per unit volume per unit time. The rate of a surface reaction is the number of reaction events per unit area per unit time. Alternatively a surface reaction might occur at certain active sites, e.g. kink sites or isolated metal centers, so that the rate is best expressed as a turnover frequency, i.e. the number of reaction events per site per second. To see how these considerations influence the phenomenological model, consider a reaction catalyzed by supported metal nanoparticles. A single batch of the supported nanoparticle catalyst may have kinetics that are adequately summarized on a "rate per gram of catalyst" basis. However, different batches of the catalyst with different nanoparticle loadings would reveal that "rate per accessible nanoparticle surface area" leads to a single kinetic model that works for all batches of the catalyst.

The reaction rate is a property of the reaction mixture. It depends on the local composition and temperature, but does not directly depend on the reactor geometry or flow patterns into and out of the reaction vessel. A non-steady state balance on species i in the presence of diffusion, convection, and reaction gives the more general species balance equation [1,2]

$$\partial c_i/\partial t + \nabla \cdot (\mathbf{v}c_i) = D_i \nabla^2 c_i + r_i(\mathbf{c}) \qquad (3.1.1)$$

Here c_i is the concentration of species i, $\mathbf{v}$ is the convective velocity, and $r_i(\mathbf{c})$ is the concentration-dependent rate at which species i is generated by reaction(s). For reactors with imperfect mixing and temperature gradients, the species balance equation can be combined with energy balances and equations of fluid mechanics. These coupled equations can be numerically solved, but most reactors are designed to approximate idealized behavior for which the reaction kinetics are more easily modeled and controlled. This book is not about reaction engineering, but nonetheless the design equations for idealized reactors help to appreciate the real meaning of a reaction rate. Three common idealized reactors and their design equations are shown in Figure 3.1.1.

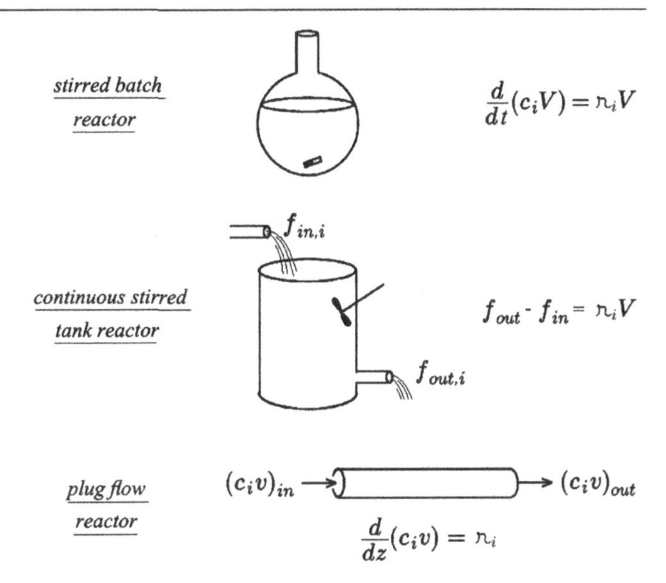

Figure 3.1.1: Idealized reactors and their design equations can be developed as special cases of the species balance equation. The variables $f_{out,i}$ and $f_{in,i}$ are the inlet flowrates of species i in moles per time. V is the volume of liquid in the continuous stirred tank reactor (CSTR). v is a flow velocity in the plug flow reactor. Note that the concentrations within the perfectly mixed CSTR are the same as those in the exiting stream.

When the volume of the reacting mixture is constant in a well-mixed batch reactor, then all terms involving spatial derivatives of the concentration field vanish and the species balance equations simplify to

$$dc_i/dt = r_i(\mathbf{c}) \tag{3.1.2}$$

Some chemistry books *define* the reaction rate as a time derivative of the concentration. This definition is only valid for the special case of a well-stirred constant volume batch reactor. It leads to serious contradictions when used elsewhere. For example, the concentrations in a steady-state flow reactor are not changing, and yet the reaction rate is clearly non-zero. Some batch reactor examples are solved in the following sections, but we should NOT think of the rate as a time derivative of the concentration.

It is important to remember that stoichiometry can make the reaction rate different from the rate at which products are made or reactants are consumed. This book uses the symbol r to represent the reaction rate, while the symbol r_i is the rate at which species i is generated by a reaction. For example, consider a reaction with stoichiometric equation

$$a\,\mathbf{A} + b\,\mathbf{B} \rightleftarrows c\,\mathbf{C} + d\,\mathbf{D}$$

If only this one reaction is occurring, then the reaction rate can equivalently be written in terms of the rate at which any species is being generated.

$$r = -\frac{1}{a} r_{\mathbf{A}} = -\frac{1}{b} r_{\mathbf{B}} = \frac{1}{c} r_{\mathbf{C}} = \frac{1}{d} r_{\mathbf{D}} \tag{3.1.3}$$

The reaction rate is the net number of moles of reaction events per volume per time. The reaction rate is positive if the net reaction rate is forward, i.e. from left to right as written in the stoichiometric equation. The reaction rate may also be negative, i.e. from right to left as written.

3.2 Reaction order and elementary reactions

Some basic terminology and definitions are needed for the discussion of reaction rate laws. Some of the earliest and still most practical rate law expressions are of power-law form. Power-laws can emerge from the mass action rate laws of elementary reactions, from empirical rate analyses, and also as limiting cases of derived rate laws for multistep reactions and catalytic cycles. A power-law rate expression separates the temperature and concentration dependences of the rate into a rate constant $k(T)$ and a product of power-law concentration dependences,

$$\Re = k(T) \prod_i c_i^{\alpha_i}.$$

The product of concentrations runs over different reagents i, including reactants, products, and possibly other species. The exponents α_i are the reaction orders for each species, and these need not be the stoichiometric coefficients. Reaction orders also need not be integers. For example, the rate law $\Re = k[\mathbf{B}][\mathbf{A}_2]^{1/2}$ arises when **B** reacts irreversibly with the product of $\mathbf{A}_2$ dissociation. For empirical or derived rate laws, we define the reaction order as

$$\alpha_i = \left. \frac{\partial \ln \Re(\mathbf{c})}{\partial \ln c_i} \right|_{[j],\ j \neq i} \tag{3.2.1}$$

Note that the concentration of a specific species i is frequently denoted $[i]$ instead of c_i throughout the book.[1] For catalytic reactions, multistep reactions, and reactions with various alternative mechanisms, the reaction orders may change as the concentrations change.

Elementary reactions

Reactions that occur directly, as written in the stoichiometric equation with no stable intermediates, are called elementary reactions. The stoichiometric coefficients in an elementary reaction specifically indicate the numbers of molecules of each type that are involved in the transition state complex along the reaction pathway. Most elementary reactions involve one, two, and sometimes three reactant entities. We refer to these as unimolecular, bimolecular, and trimolecular reactions, respectively. Elementary reactions can occur in fluid phases or at sites on

[1] Why two different notations? It is easier to write a vector list of concentration variables as **c**, and easier to specify the chemical species with the notation $[i]$ than placing chemical formulas in the subscript of c_i.

surfaces. For surface reactions, one of the reactant entities might be a catalyst site or a surface intermediate. In this book, surface sites are denoted as * and the symbol $\mathbf{A}*$ denotes the species or surface intermediate $\mathbf{A}$ adsorbed at surface site *. $\theta_{\mathbf{A}}$ denotes the fraction of surface sites which have an adsorbed $\mathbf{A}*$ species.

The rate of an elementary reaction is a constant multiplied by the concentrations (or surface coverages) of each reactant raised to its stoichiometric coefficient. The table below provides a partial list of some types of elementary reactions and their rate laws.

Reaction type	(Forward) reaction	(Forward) rate law
unimolecular	$\mathbf{A} \rightarrow$	$r_f = k_f[\mathbf{A}]$
bimolecular	$\mathbf{A}+\mathbf{B} \rightarrow$	$r_f = k_f[\mathbf{A}][\mathbf{B}]$
trimolecular	$\mathbf{A}+\mathbf{B}+\mathbf{C} \rightarrow$	$r_f = k_f[\mathbf{A}][\mathbf{B}][\mathbf{C}]$
surface reaction	$\mathbf{A}^* \rightarrow$	$r_f = k_f \sigma \theta_{\mathbf{A}}$
surface + fluid species	$\mathbf{A}^* + \mathbf{B} \rightarrow$	$r_f = k_f \sigma \theta_{\mathbf{A}}[\mathbf{B}]$
surface + surface species	$\mathbf{A}^* + \mathbf{B}^* \rightarrow$	$r_f = k_f \sigma \theta_{\mathbf{A}} \theta_{\mathbf{B}}$

The factor of σ in the surface reaction rate expressions is the total concentration or surface density of active sites. Examples of unimolecular reactions include dissociation reactions, certain isomerization reactions, and conformational transitions. Examples of bimolecular reactions include atom transfer reactions, bimolecular association reactions, electron transfer reactions, and adsorption steps in catalysis. Trimolecular elementary reactions are rare but they do occur. For example, low pressure gas phase association reactions occur by the elementary mechanism

$$\mathbf{A}+\mathbf{B}+\mathbf{X} \longrightarrow \mathbf{AB}+\mathbf{X} \tag{3.2.2}$$

The third partner $\mathbf{X}$ may seem irrelevant, but at low pressures it provides an essential energy transfer sink. Its role is to 'quench' the association product $\mathbf{AB}$, which would otherwise remain excited and immediately re-dissociate. For more details see Chapter 9. Trimolecular elementary reactions also occur when a water molecule facilitates proton transfer between a donor and acceptor [3–7]. Even tetramolecular elementary reactions, where two water molecules facilitate proton transfer, have been proposed [8,9]. Examples of these proton transfer reactions are shown in Figure 3.2.1.

In each of the reactions described above, the proportionality constant k_f is a rate constant. Like equilibrium constants, rate constants have only weak dependence on concentration.[2] The units of a unimolecular rate constant are $k_f \; [=] \; s^{-1}$, the units of a bimolecular rate constant are $k_f \; [=] \; L\,mol^{-1}\,s^{-1}$, and the units of a trimolecular reaction are $k_f \; [=] \; L^2\,mol^{-2}\,s^{-1}$. Surface reactions and their rate constants are discussed in Chapter 4.

[2] They would have no dependence on concentration if not for non-idealities.

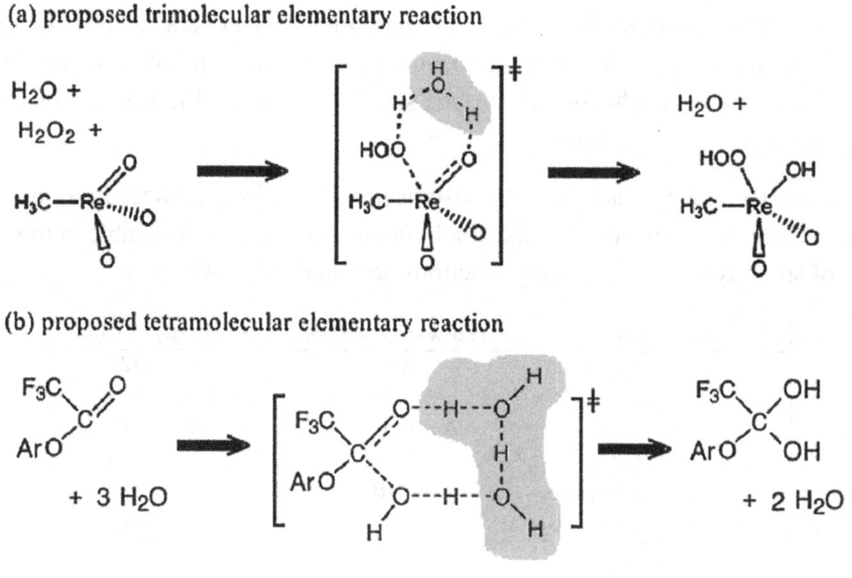

(a) proposed trimolecular elementary reaction

(b) proposed tetramolecular elementary reaction

Figure 3.2.1: Tri- and tetra-molecular elementary reactions are rare, but they have been proposed for proton transfer reactions in aqueous solution. In both of the examples shown, water molecules help bridge the donor and acceptor sites with a proton transfer 'wire'. These reactions are only possible because water molecules are present at high concentration in the solvent. [See Goldsmith et al. J. Am. Chem. Soc. 137, 9604-16 (2015) and Frasson et al. J. Phys. Org. Chem. 19, 143-47 (2006).]

For reversible reactions, the overall rate is a difference of forward and reverse rates. Consider again the generic reversible reaction $a\,\mathbf{A} + b\,\mathbf{B} \rightleftarrows c\,\mathbf{C} + d\,\mathbf{D}$. If this is an elementary reaction, then its rate law is

$$r = k_f[\mathbf{A}]^a[\mathbf{B}]^b - k_b[\mathbf{C}]^c[\mathbf{D}]^d \tag{3.2.3}$$

Note that the backward reaction rate is constructed according to the mass action law in the same way as the forward reaction rate. Also note that the forward and backward rate laws must be consistent with chemical equilibrium by construction. Quantitatively, the *net* rate of a reversible reaction vanishes when its reactants and products are at equilibrium, i.e. $r = 0$ when

$$K_{eq} = \frac{k_f}{k_b} = \frac{[C]^c[D]^d}{[A]^a[B]^b} \tag{3.2.4}$$

The forward and backward reaction events are still occurring at equilibrium, but with rates that balance giving no net fluxes.

Equation (3.2.4) is the condition of detailed balance, that forward and backward rate constants between each pair of states are such that all fluxes cancel at equilibrium. The closely related principle of microscopic reversibility says that, at conditions near equilibrium, reactions in the forward and reverse directions occur via analogous mechanisms and transition states. The principles of microscopic reversibility and detailed balance have long histories. The idea first appears in work by Maxwell [10], who arrived at the principle of detailed balance because it is "impossible to assign a reason" otherwise. These principles entered reaction rate theory with the work of La Mer, whose early thermodynamic rate theory bore many similarities to the later transition state theory [11]. For very large departures from equilibrium, reactions become irreversible and then microscopic reversibility is moot.

As noted in the epigraph by Helfferich, most processes do not occur by a single elementary reaction. Instead, they are comprised of complex networks involving many coupled elementary reactions. These complex reaction processes can give rise to overall rate laws with reaction orders that are not simple integers and which may even be negative. Nevertheless, complex reactions may *appear* in certain limits to have first or second order dependence on reactant concentrations. An elementary reaction always implies a simple reaction order, but simple reaction orders do not imply elementary mechanisms. One should not assume that reactions with simple first or second order rate laws are elementary.

3.3 Initial rates and integrated rate laws

This section describes two traditional techniques for determining reaction orders and rate laws from experimental data: the method of initial rates and methods based on integrated rate laws. In practice, the traditional approaches are rather tedious compared to emerging techniques for using data to rapidly discriminate between trial rate expressions [12]. However, the traditional approaches are broadly applicable and adequately instructive for our introductory discussion. Consider a reaction of unknown mechanism and reaction order which converts reactants **A** and **B** to products. To determine the reaction order with respect to [**A**], the rate can be measured at different **A** concentrations, with [**B**] held constant. One then lumps the unknown [**B**]-dependence into an effective rate constant:

$$\mathfrak{r} = k_{eff}([\mathbf{B}])[\mathbf{A}]^{\nu} \tag{3.3.1}$$

The ratio of rates at the two different **A** concentrations, after taking logarithms, is

$$ln\frac{\mathfrak{r}_1}{\mathfrak{r}_2} = \nu\, ln\frac{[\mathbf{A}]_1}{[\mathbf{A}]_2} \tag{3.3.2}$$

which is easily solved for the reaction order ν, but using just two data points to find ν is not advised. More typically one prepares a van't Hoff plot showing the natural logarithm of the initial rates vs. the log of the concentrations for a range of initial concentrations. If the rate law is indeed of the form assumed in equation (3.3.1), then the data will all fall on a line with the slope ν being the reaction order. After the reaction order with respect to reactant **A** has been determined, a similar procedure can determine the reaction order with respect to **B**. As shown in Figure 3.3.1, van't Hoff plots are often constructed from initial rates starting with pure reactant mixtures to eliminate the effects of the reverse reaction.

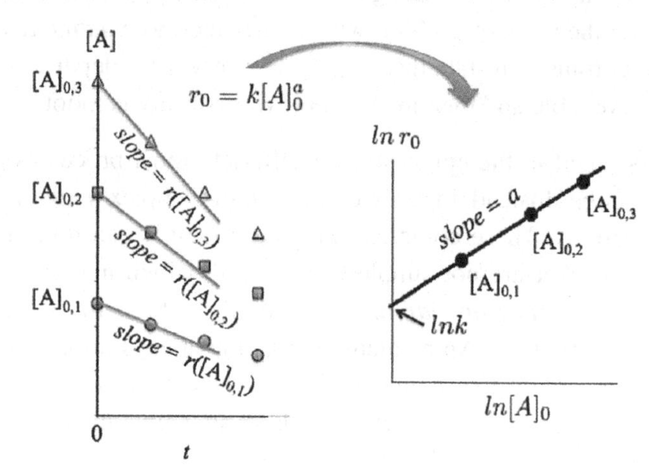

Figure 3.3.1: The method of initial rates varies the initial concentration and uses the resulting initial rates to identify the reaction order and rate constant with respect to a particular reactant.

Some common sense is required in the analysis of initial rates. The data must truly be representative of the *initial* rate and not from times where the reaction is well on its way to completion. Also note that the method of initial rates *assumes* a power law concentration dependence for all species. Obtaining a reasonable value for the reaction order ν does not ensure that equation (3.3.1) is actually a correct model. The data may appear to fit a power law form over relatively narrow concentration ranges, but use of the rate law beyond the fitted range will then yield inaccurate results. For example, the Michaelis Menten rate law $r = k[E][S]/(K + [S])$ predicts a reaction order that changes from $\nu = 1$ to $\nu = 0$ at ca. $[S] = K$.

For sufficiently simple rate laws, the differential equation for concentrations in a constant volume batch reactor, $dc_i/dt = r_i(\mathbf{c})$, can be exactly integrated. Integrated rate laws are instructive for developing an intuitive understanding of rate laws and for analyses of experimental data. For analyses that include volume expansion at constant pressure, readers are referred to the chemical engineering literature [13–15]. Some simple cases with constant volume are integrated in the discussions below.

Example: First order irreversible reaction

A first order irreversible reaction $\mathbf{A} \rightarrow (products)$ in a constant volume batch reactor gives the differential equation

$$-\frac{d[\mathbf{A}]}{dt} = k[\mathbf{A}] \tag{3.3.3}$$

Integrating the rate law gives

$$[\mathbf{A}](t) = [\mathbf{A}]_0 exp[-kt] \tag{3.3.4}$$

Similar exponential decay laws govern many first order rate processes beyond chemical reactions: radioactive decay, survival probabilities in nucleation, autocorrelation functions for protein folding/unfolding, etc. Each case retains the basic $exp[-kt]$ portion of the integrated first order rate law with $[\mathbf{A}](t)/[\mathbf{A}]_0$ replaced by an appropriate ratio of quantities.

The integrated rate law can be used as a regression model to find the rate constant k. For convenience, let $y_i = [\mathbf{A}](t_i)/[\mathbf{A}]_0$ be the ith data point, and let $y(t_i) = exp[-kt_i]$ be the predicted value at the ith ordinate. There are two common ways to infer the rate parameter k from data:

1. **Nonlinear regression**. Directly minimize the squared residuals by varying k in

$$S_{NL}(k) = \sum_i (y_i - y(t_i))^2$$

2. **Linear regression**. Linearize the integrated rate law to $ln\ y(t) = -kt$ and then vary k to minimize

$$S_L(k) = \sum_i (ln\ y_i + k\,t_i)^2$$

Both types of regression are widely used to extract rate constants and to determine the reaction order by comparing fits from alternative rate laws.

Most textbooks (including this one!) primarily emphasize linearized regression procedures, but direct nonlinear regression often provides better parameter estimates. The traditional emphasis on linear regression has been widely criticized [13,16]. Linear regression became the standard procedure because it was the only viable option in the era of slide rules and graph paper. But even with modern computers, there are some good reasons to use/teach linear regression. For starters, linear regression is beautifully modular and it guarantees a unique best fit line. Any model that can be arranged into linear slope-intercept form can be fitted using the same formula and the same computer program. There are no lines of code to write, no objective functions to differentiate, no iterative optimizations, and no possibility of getting trapped in local minima.

But how badly do linearized regression procedures distort the importance of certain errors/residuals in the data? Square error minimizations, whether in linear or nonlinear regression, assume that the error is identically distributed at every data point. The algebraic manipulations that transform a nonlinear model into linear form distort the errors unevenly and compromise parameter estimates by exaggerating the importance of errors at one end of the data. Critics of linearization say these error distortions make direct nonlinear fitting the better procedure. However, their argument rests on an assumption that errors *in the raw data* are distributed identically at every data point. Kittrell wrote that this assumption "is believed to be more justifiable from an experimental point of view" [16]. But there are some exceptions. Depending on the nature of the errors and the quantity being modeled, the most appropriate regression procedure may be a linearized regression, a direct nonlinear fit, or neither of the above.

Consider regression to find the rate constant in a first order rate law. When errors are uniformly superimposed on the real value at each data point, the optimal procedure for fitting a first order rate law is nonlinear regression. When errors are proportional to the correct value at each data point, the optimal procedure is the linearized regression. These cases are illustrated in Figure 3.3.2.

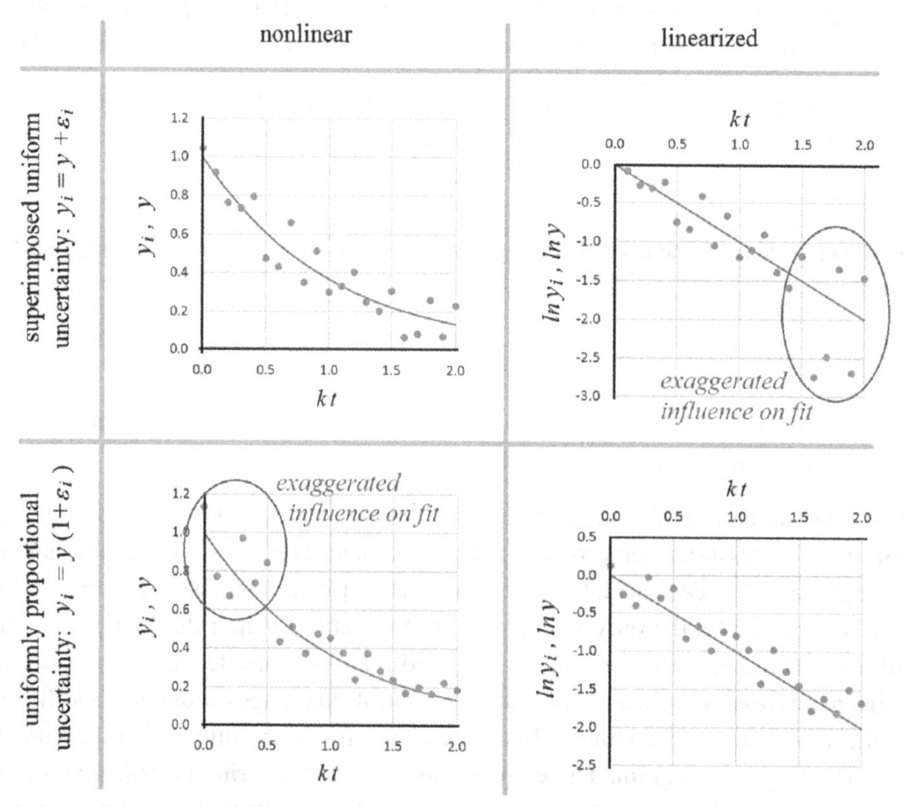

Figure 3.3.2: The examples show the effects of linearization for an exponential decay (solid line) with added Gaussian noise (dots). In the upper panels, the random noise ε_i with zero mean and *uniform* variance is superimposed on each $y(t_i)$. In the lower panels, a *proportional* noise is superimposed on each $y(t_i)$. Linearization distorts these errors in the sum of square regression residuals.

Rates are essentially Poisson parameters for the number of events per unit time. They tell us how many ticks per minute are expected from the Geiger counter, how many microfluidic droplets per second will host a nucleation event, how many molecules per second are expected to become products, etc. For a Poisson process the expected (mean) number of events per unit time is equal to the variance in the number of events per unit time. For a chemical reaction, the number of reaction events per unit time is on the order of 10^{23}. Therefore the standard deviation relative to the mean is $10^{23/2}/10^{23} = 10^{-11.5}$. Clearly, the experimental kineticist can ignore intrinsic fluctuations and focus entirely on instrumentation error sources. Assuming that errors from instruments are uniformly distributed, then Kittrell would be correct.

But now consider Geiger counter ticks, the first nucleation event in a droplet, and the number of reacting molecules in a $100(nm)^3$ simulation box. In each of these cases, there is no ambiguity about the measurement – all uncertainty comes from intrinsically stochastic waiting times. The events being counted are far less numerous, and now Kittrell's assumption is in doubt.

Example: Geiger counter ticks for ^{137}Ba

The ^{137}Ba half-life is on the order of minutes. The rate of Geiger counter ticking η per unit time is proportional to the rate of ^{137}Ba decay. The rate of ^{137}Ba decay is proportional to the amount of remaining ^{137}Ba in the sample, so that $\eta = \eta_0 exp[-kt]$ where η_0 depends on the distance from the sample to the Geiger counter and the size of the sample. To estimate k, we must count ticks during short intervals, e.g. ten seconds $\ll$ minutes half-life, to safely ignore changes in the decay rate during the short observation interval. How will the number of observed ticks in each ten second interval (y_i) be distributed around the true expected number, i.e. around $y(t_i) = \eta(t_i)\delta t$? Based on Poisson statistics, we should expect

$$y_i \approx y(t_i) + y(t_i)^{1/2}\varepsilon_i$$

where ε_i is a random variable with zero mean and unit variance. We have ignored measurement/instrumentation error because there is no uncertainty in detecting a Geiger counter tick. Now let us compare linear and nonlinear regression procedures to estimate k. For direct nonlinear regression:

$$\begin{aligned} S_{NL} &= \sum_i (y(t_i) + y(t_i)^{1/2}\varepsilon_i - y(t_i))^2 \\ &= \sum_i y(t_i)\varepsilon_i^2 \end{aligned}$$

For the linearized least squares fit using $ln\ y$ vs. t,

$$S_L = \sum_i (ln[y(t_i) + y(t_i)^{1/2}\varepsilon_i] - ln[y(t_i)])^2$$
$$= \sum_i ln^2[1 + \varepsilon_i / y(t_i)^{1/2}]$$

The degrees of error distortion in the linear and nonlinear formulae are similar, with the distortions in S_L being very slightly weaker. Thus for first order rate laws with Poisson distributed data, linearization of the regression model is not deleterious and perhaps it is even beneficial. Later chapters will show that Poisson distributed data are very common in rare events simulations.

In general, the optimal regression procedure always depends on the distribution of errors as a function of the quantity being estimated. A fitted model should be allowed to deviate widely from those data points where the uncertainty is large, and should very nearly pass through those data points where the uncertainty is small. Thus each data point in a least squares fit should ideally be weighted by the inverse of its statistical uncertainty,

$$S_W = \sum_i (y_i - y(t_i))^2 / 2\sigma_i^2$$

where y_i is the ith data point, $y(t_i)$ is the model prediction at the ith ordinate, and σ_i is the uncertainty in the ith data point. In a few cases the σ_i which define the weights are known, e.g. in a Poisson process $\sigma_i = y(t_i)^{1/2}$. More typically the σ_i are not known, so the best choice between linear, nonlinear, or weighted regression is not entirely clear. However, linearized regression always has a few practical advantages over nonlinear fitting procedures. (i) Our eyes are good (we think...) at spotting severe error distortions in the data and systematic deviations from a linearized model. (ii) Manipulating the rate expressions to obtain a linear model helps to understand them. (iii) Linear regression provides initial guesses and sanity checks for nonlinear regression. (iv) If there is a large difference between parameter estimates from linearized and nonlinear regression, then it seems likely that the parameter estimates from both procedures are rather unreliable. In a sense, the last reason (iv) is a softened version of Rutherford's maxim: "If you need statistics, then you did the wrong experiment."

Before moving on to more complex rate laws, note that even the most carefully analyzed data can be misinterpreted. Nearly any model can fit a sufficiently narrow range of data [17]. Reactions under an unsteady-state diffusion limitation can appear to be first order [18]. Rate constants for a catalytic reaction may only reflect a small portion of accessible catalyst material [13]. Data that appears to fit a homogeneous nucleation model might actually be from heterogeneous nucleation [19]. These errors of interpretation, often called kinetic falsifications [20], can go entirely undiscovered until someone attempts to scale-up to larger equipment or to operate at different conditions. Thus being cognizant of potential differences between what is measured and what is modeled is often more important than the differences between parameter estimation protocols. Now let us return to the integration of elementary rate laws.

Example: Second order irreversible reaction

A second order reaction $2\mathbf{A} \rightarrow (products)$ with rate law $r = k[\mathbf{A}]^2$ gives the differential equation

$$-\frac{1}{2}\frac{d[\mathbf{A}]}{dt} = k[\mathbf{A}]^2 \tag{3.3.5}$$

Integration of the rate law gives

$$\frac{1}{[\mathbf{A}](t)} = \frac{1}{[\mathbf{A}]_0} + 2kt \tag{3.3.6}$$

Figure 3.3.3 shows how time-dependent concentration data can be plotted to test hypotheses about irreversible rate laws and to obtain rate constants. A procedure known as Essen's method [21] creates a plot from rate laws of each order and selects the reaction order as that which gives the most linear plot. Note that Essen's method is entirely based on linearized versions of the integrated rate laws. As shown in the exercises, linearization usually lowers the accuracy of rate parameter estimates for a second order rate law [21].

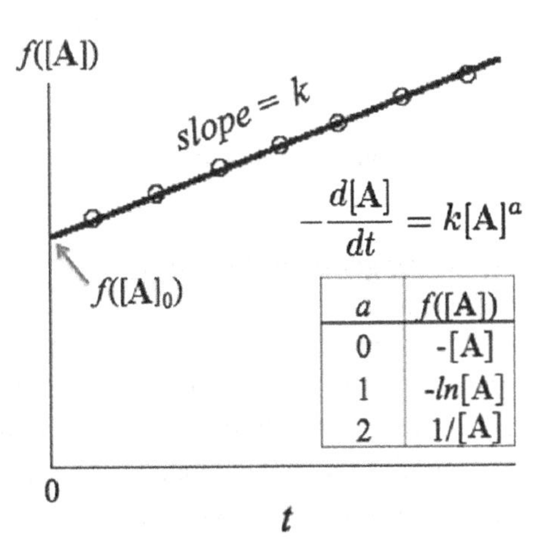

a	f([A])
0	-[A]
1	-ln[A]
2	1/[A]

Figure 3.3.3: (Schematic) Integration of a hypothesized rate law suggests a function of concentration to linearize the time-dependent concentration data. These plots can be used to test hypothesized rate laws and to extract rate constants. Note that error distortions due to linearization become more severe as the reaction order increases.

The above examples integrated [**A**] directly because **A** was the only reactant. When there are multiple reactants, we can integrate the rate law using extent of reaction or conversion variables [13].

Example: Bimolecular reaction with two reactants

Consider the bimolecular irreversible reaction $\mathbf{A} + \mathbf{B} \rightarrow (products)$ and rate law $r = k[\mathbf{A}][\mathbf{B}]$. Both **A** and **B** are consumed by the reaction, so we introduce an extent of reaction per volume $\hat{\xi}$. Then, in terms of $\hat{\xi}$ we have

$$\begin{aligned} [\mathbf{A}] &= [\mathbf{A}]_0 - \hat{\xi} \\ [\mathbf{B}] &= [\mathbf{B}]_0 - \hat{\xi} \end{aligned} \tag{3.3.7}$$

From the definition of $\hat{\xi}$ we have the rate $r = d\hat{\xi}/dt$. Therefore the differential equation governing $\hat{\xi}$ as a function of time is

$$d\hat{\xi}/dt = k([\mathbf{A}]_0 - \hat{\xi})([\mathbf{B}]_0 - \hat{\xi}) \tag{3.3.8}$$

Integration gives

$$ln\left(\frac{[\mathbf{B}]_0([\mathbf{A}]_0 - \hat{\xi})}{[\mathbf{A}]_0([\mathbf{B}]_0 - \hat{\xi})}\right) = ([\mathbf{A}]_0 - [\mathbf{B}]_0)kt \tag{3.3.9}$$

Equation (3.3.9) is useful for second order reactions in which the starting composition is not a stoichiometric mixture. In special limiting cases, it can be simplified considerably. When the starting mixture is stoichiometric, i.e. when $[\mathbf{A}]_0 = [\mathbf{B}]_0$, equation (3.3.8) simplifies to $d\hat{\xi}/dt = k([\mathbf{A}]_0 - \hat{\xi})^2$ before integration. Integration of this special stoichiometric case yields a behavior identical to that for the dimerization reaction $2\mathbf{A} \rightarrow (products)$. Additionally, when there is a limiting reagent such that $[\mathbf{A}]_0 \ll [\mathbf{B}]_0$ or $[\mathbf{B}]_0 \ll [\mathbf{A}]_0$, the behavior reverts to first order consumption of the limiting reagent.

Excess reagents and flooding

In the flooding technique, one intentionally prepares conditions where all reactants are in large excess relative to one limiting reactant [22]. Under these conditions, the concentrations of all species except the limiting reactant can be treated as constants during the reaction. Flooding thus isolates the dependence of the kinetics on the limiting reactant. Start with the bimolecular reaction $\mathbf{A} + \mathbf{B} \rightarrow (products)$ that we have just studied, and suppose that reactant **B** is in large excess. The extent of reaction is then limited, for all times, such that $0 \le \hat{\xi} < [\mathbf{A}]_0$. Now because $[\mathbf{A}]_0 \ll [\mathbf{B}]_0$, we always have $\hat{\xi} \ll [\mathbf{B}]_0$, and the differential equation (3.3.8) is approximately

$$\frac{d\hat{\xi}}{dt} = k[\mathbf{B}]_0([\mathbf{A}]_0 - \hat{\xi}) \qquad (3.3.10)$$

We can then treat $k[\mathbf{B}]_0$ as an effective first order rate constant to obtain the solution $\hat{\xi}(t) = [\mathbf{A}]_0\{1 - exp(-k[\mathbf{B}]_0 t)\}$, or equivalently

$$[\mathbf{A}](t) = [\mathbf{A}]_0 exp(-k[\mathbf{B}]_0 t). \qquad (3.3.11)$$

$k[\mathbf{B}]_0$ is then easily extracted from a plot of $ln[\mathbf{A}]$ $vs.$ t.

Example: Excess and limiting reagents

The reaction $CH_3O + Br \rightarrow CH_2O + HBr$ was studied in a batch reactor with an excess of bromine atoms. The following chart gives data from Aranda et al. [21,23]

CH_3O conversion at ~		const. [Br] ($*10^{10}$M)	
t/ms	[Br] = 3.69	[Br] = 7.44	[Br] = 10.8
0.00	0.00	0.00	0.00
0.50	0.08	0.15	0.21
1.00	0.15	0.28	0.38
1.50	0.21	0.39	0.51
2.00	0.27	0.48	0.61
2.50	0.33	0.56	0.70
3.00	0.38	0.63	0.76
5.00	0.55	0.81	0.91

To test the hypothesis that the rate law is $r = k[CH_3O][Br]$, we can check whether the kinetics under excess bromine follow the equation

$$d\hat{\xi}/dt \approx k[Br]_0([CH_3O]_0 - \hat{\xi}) \qquad (3.3.12)$$

The data is given in terms of conversion X.

$$X = \hat{\xi}/[CH_3O]_0$$

Note that we define conversion with respect to the limiting reagent. Divide both sides of equation (3.3.12) by $[CH_3O]_0$ to obtain

$$dX/dt \approx k[Br]_0(1 - X)$$

and integrate

$$-ln(1-X) = k[Br]_0 t.$$

If the data is plotted as $-ln(1-X)$ vs. t and if the hypothesis that $r = k[CH_3O][Br]$ is correct, we should obtain three lines, each with a different slope determined by the initial $[Br]_0$ concentration. If we further divide both sides by $[Br]_0$, then all three lines should collapse to one line with common slope k,

$$-[Br]_0^{-1} ln(1-X) = kt$$

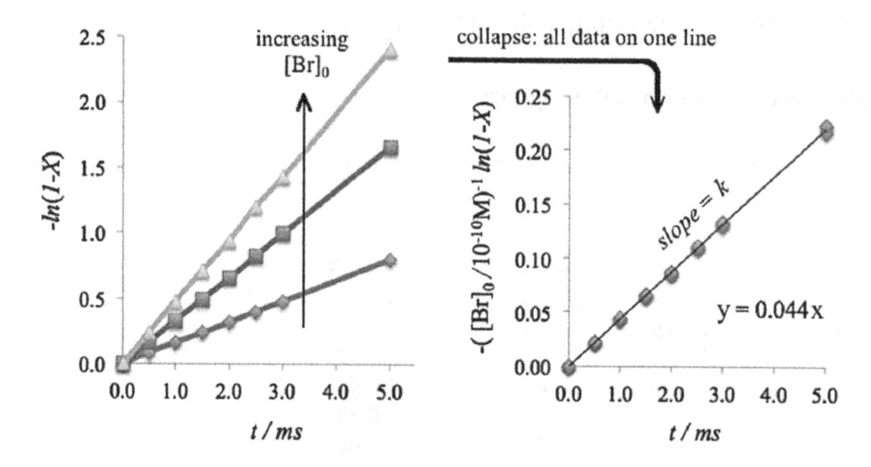

The best fit slope gives $k = 4.4 \cdot 10^{11} L/(mol \cdot s)$.

When fitted to data over a wide range of conversion, integrated rate laws enable practical tests for assumptions about reaction orders and reversibility. However, for all but the simplest rate expressions, the integration must be done numerically. The lack of a closed form integrated rate law often makes analysis of data with integrated rate laws arduous. Here, initial rate analyses have the advantage of simplicity, even if they are experimentally tedious.

3.4 Reversible reactions

Within the subject of kinetics, the words reversible and irreversible are used in two important contexts with very different meanings.

- In the context of dynamics, a reversible process proceeds slowly, traversing a manifold of equilibrium states from start to finish. Irreversible processes proceed more rapidly such that "extra" work is lost through dissipative non-equilibrium processes. Reversibility in this context is particularly important in understanding single molecule pulling experiments and in understanding the differences between adiabatic and non-adiabatic reactions.

- In the present context of phenomenological kinetic models, a reversible reaction has an equilibrium which is neither completely to the right nor completely to the left. Irreversible reactions are those for which the reaction conditions cause the reaction to consume essentially all of the available reactants.

Theoretical work on chemical kinetics requires simultaneous use of both terminologies. The intended meaning should usually be clear from the context.

The previous sections of this chapter considered only elementary irreversible reactions. This section considers reversible reactions, again integrating the rate equations for constant volume batch reactors as examples. Reversibility tends to complicate the analysis, but analytic solutions for the composition as a function of time illustrate the approach to an equilibrium composition.

Example: Reversible unimolecular isomerization

Consider the reversible unimolecular isomerization reaction $\mathbf{A} \rightleftarrows \mathbf{B}$ where both forward and reverse reactions follow simple rate laws with rates $k_f[\mathbf{A}]$ and $k_b[\mathbf{B}]$, respectively. Because the reaction is reversible, the rate law includes both forward and backward rates

$$d[\mathbf{A}]/dt = -k_f[\mathbf{A}] + k_b[\mathbf{B}] \tag{3.4.1}$$

The equation for $[\mathbf{B}]$ is trivially related to that for $[\mathbf{A}]$ because of stoichiometry, $d[\mathbf{A}]/dt = -d[\mathbf{B}]/dt$. In terms of an extent of reaction per volume

Species	Initial	After $\hat{\xi}$
$\mathbf{A}$	$[\mathbf{A}]_0$	$[\mathbf{A}]_0 - \hat{\xi}$
$\mathbf{B}$	$[\mathbf{B}]_0$	$[\mathbf{B}]_0 + \hat{\xi}$

At equilibrium, $k_f([\mathbf{A}]_0 - \hat{\xi}_{eq}) = k_b([\mathbf{B}]_0 + \hat{\xi}_{eq})$ which is consistent with the usual relation between the equilibrium constant and the rate constants, $K_{eq} = k_f/k_b$. Solving for the equilibrium extent of reaction gives

$$\hat{\xi}_{eq} = (k_f[\mathbf{A}]_0 - k_b[\mathbf{B}]_0)/(k_f + k_b)$$

Now equation (3.4.1) can be written in terms of the equilibrium extent of reaction as

$$d\hat{\xi}/dt = -(k_f + k_b)(\hat{\xi} - \hat{\xi}_{eq}) \tag{3.4.2}$$

Equation (3.4.2) shows that the rate at all times is proportional to the remaining deviation from equilibrium. Moreover, it shows that the concentrations approach equilibrium with a characteristic reaction time

$$\tau_{rxn} = (k_f + k_b)^{-1} \tag{3.4.3}$$

After solving equation (3.4.2), the solution can be written as

$$\begin{aligned} [\mathbf{A}](t) &= [\mathbf{A}]_{eq} + ([\mathbf{A}]_0 - [\mathbf{A}]_{eq}) exp[-t/\tau_{rxn}] \\ [\mathbf{B}](t) &= [\mathbf{B}]_{eq} + ([\mathbf{B}]_0 - [\mathbf{B}]_{eq}) exp[-t/\tau_{rxn}] \end{aligned} \tag{3.4.4}$$

with $[\mathbf{A}]_{eq} = [\mathbf{A}]_0 - \hat{\xi}_{eq}$ and $[\mathbf{B}]_{eq} = [\mathbf{B}]_0 + \hat{\xi}_{eq}$. It is trivially seen from the solution that equilibrium is recovered when $t/\tau_{rxn} \gg 1$.

Example: Reversible dimerization

Now consider the elementary reversible dimerization reaction $2\mathbf{A} \rightleftarrows \mathbf{B}$ with forward and backward rate constants k_f and k_b, respectively. Again it is useful to introduce an extent of reaction per volume $\hat{\xi}$. The stoichiometric coefficients relate $\hat{\xi}$ to the concentrations and initial concentrations as

$$\begin{aligned} [\mathbf{A}] &= [\mathbf{A}]_0 - 2\hat{\xi} \\ [\mathbf{B}] &= [\mathbf{B}]_0 + \hat{\xi} \end{aligned} \tag{3.4.5}$$

The rate equation is

$$d\hat{\xi}/dt = k_f([\mathbf{A}]_0 - 2\hat{\xi})^2 - k_b([\mathbf{B}]_0 + \hat{\xi}) \tag{3.4.6}$$

Direct integration leads to an ugly result, but note that the right hand side of equation (3.4.6) is a second order polynomial in $\hat{\xi}$. Moreover, the right hand side of equation (3.4.6) must vanish at equilibrium

$$0 = k_f([\mathbf{A}]_0 - 2\hat{\xi}_{eq})^2 - k_b([\mathbf{B}]_0 + \hat{\xi}_{eq}) \tag{3.4.7}$$

We can exactly write equation (3.4.6) as

$$d\hat{\xi}/dt = (\hat{\xi} - \hat{\xi}_{eq})\{4k_f(\hat{\xi} + \hat{\xi}_{eq}) - 4[\mathbf{A}]_0 k_f - k_b\} \tag{3.4.8}$$

After some non-trivial integration and simplification,

$$\hat{\xi}/\hat{\xi}_{eq} = \frac{1 - e^{-t/\tau_{rxn}}}{1 - \alpha e^{-t/\tau_{rxn}}} \tag{3.4.9}$$

where $\tau_{rxn}^{-1} = 4[\mathbf{A}]_0 k_f + k_b - 4k_f \hat{\xi}_{eq}$ and $\alpha = 4k_f \hat{\xi}_{eq}/(4[\mathbf{A}]_0 k_f + k_b)$. α is a fixed dimensionless group characteristic of the reaction equilibria and kinetics. The figure below shows $\hat{\xi}/\hat{\xi}_{eq}$ as a function of t/τ_{rxn} for several values of α.

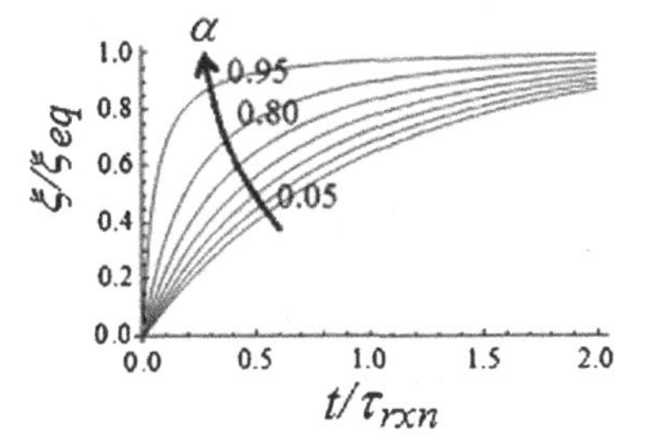

The extent of reaction evolves in time moving rightward along curves of constant α. Note that this result describes both cases where the reaction runs forward ($\hat{\xi}_{eq} > 0$) and backward ($\hat{\xi}_{eq} < 0$).

We now state two results related to the previous example without derivation. First, consider the reversible elementary reaction

$$\mathbf{A} \rightleftarrows \mathbf{B} + \mathbf{C}$$

with forward and backward rate constants k_f and k_b, respectively. When $[\mathbf{A}](0) = [\mathbf{A}]_0$ and $[\mathbf{B}](0) = [\mathbf{C}](0) = 0$, the fractional conversion as a function of time is [24]

$$\hat{\xi}(t)/[\mathbf{A}]_0 = K\left(K/2 + \left(\hat{\xi}_{eq} + K/2\right)\coth\left[k_b\left(\hat{\xi}_{eq} + K/2\right)t\right]\right)^{-1} \tag{3.4.10}$$

where $K = k_f/k_b$ and $\hat{\xi}_{eq} = K\left[([\mathbf{A}]_0/K + 1/4)^{1/2} - 1/2\right]$. For the same reaction but with $[\mathbf{A}](0) = 0$, $[\mathbf{B}](0) = [\mathbf{B}]_0$, and $[\mathbf{C}](0) = [\mathbf{C}]_0$, the concentration of $\mathbf{A}$ as a function of time is

$$[\mathbf{A}](t) = [\mathbf{B}]_0[\mathbf{C}]_0\left[\frac{1}{2}([\mathbf{B}]_0 + [\mathbf{C}]_0 + K) + \Phi\coth(k_b\Phi t)\right]^{-1} \tag{3.4.11}$$

where $\Phi^2 = ([\mathbf{B}]_0 + [\mathbf{C}]_0 + K)^2/4 - [\mathbf{B}]_0[\mathbf{C}]_0$.

As shown in both examples of this section, the solutions are most easily obtained by first solving for the equilibrium composition and then writing the rate equations in terms of deviations from equilibrium. When there is only a single reversible reaction, all components of the solution will monotonically evolve toward the equilibrium concentration as time increases. When there are multiple reactions occurring, the nature of the solutions can become far more complicated. As the next section shows, analyses of multistep reaction networks often require approximations or numerical techniques.

3.5 Multistep reactions

Reactions that involve one or several intermediates are common across all branches of chemistry. The discussion in this section centers on reversible and irreversible versions of the $\mathbf{A} \leftrightarrow \mathbf{B} \leftrightarrow \mathbf{C}$ reaction sequence. The example reaction may seem to describe a very limited set of processes, e.g. isomerizations in series. However, there are actually many complex reaction systems that behave like the $\mathbf{A} \leftrightarrow \mathbf{B} \leftrightarrow \mathbf{C}$ reaction sequence. For example, the oxidation of methanol in excess O_2 might be described in stages: methanol to formaldehyde followed by formaldehyde to CO_2. The fact that both reactions involve an O_2 co-reactant is completely irrelevant because under excess O_2, the reactions are effectively first order in methanol or formaldehyde [14].

Example: Irreversible biphasic kinetics

Consider two irreversible reactions which occur in sequence to create product $\mathbf{C}$ from $\mathbf{A}$ via intermediate $\mathbf{B}$.

$$\mathbf{A} \xrightarrow{k_1} \mathbf{B} \xrightarrow{k_2} \mathbf{C} \tag{3.5.1}$$

The system of rate equations corresponding to equation (3.5.1) is

$$\begin{aligned} d[\mathbf{A}]/dt &= -k_1[\mathbf{A}] \\ d[\mathbf{B}]/dt &= -k_2[\mathbf{B}] + k_1[\mathbf{A}] \\ d[\mathbf{C}]/dt &= k_2[\mathbf{B}] \end{aligned} \tag{3.5.2}$$

Linear systems of equations like these can be solved by several methods: sequential integration and substitution, Laplace transforms, or matrix diagonalization. For example, the first equation clearly gives $[\mathbf{A}] = [\mathbf{A}]_0 exp[-k_1 t]$. The solution for $[\mathbf{A}]$ can be substituted in the second equation to obtain an equation for $[\mathbf{B}]$: $d[\mathbf{B}]/dt + k_2[\mathbf{B}] = k_1[\mathbf{A}]_0 exp[-k_1 t]$. Taking a Laplace transform ($[\mathbf{B}](t) \rightarrow [\hat{\mathbf{B}}](s)$) and solving for $[\hat{\mathbf{B}}](s)$,

$$[\hat{\mathbf{B}}](s) = \frac{s[\mathbf{B}]_0 + k_1([\mathbf{A}]_0 + [\mathbf{B}]_0)}{(s+k_1)(s+k_2)} \tag{3.5.3}$$

Inversion of the Laplace transform gives

$$[\mathbf{B}](t) = \left(\frac{[\mathbf{A}]_0 k_1}{k_1 - k_2} + [\mathbf{B}]_0\right) e^{-k_2 t} - \frac{[\mathbf{A}]_0 k_1}{k_1 - k_2} e^{-k_1 t} \tag{3.5.4}$$

Clearly, $[\mathbf{C}](t) + [\mathbf{B}](t) + [\mathbf{A}](t) = [\mathbf{A}]_0 + [\mathbf{B}]_0 + [\mathbf{C}]_0$, so $[\mathbf{C}](t)$ can be obtained from $[\mathbf{A}](t)$ and $[\mathbf{B}](t)$.

Approximations for analyzing more complicated multistep reactions will be discussed in later sections. The remainder of this section describes reversible biphasic reactions for which the equations can again be solved exactly. We also take this opportunity to illustrate eigenvectors and eigenvalues as another method of solution.

Example: Reversible biphasic kinetics

Now consider the series of reversible reactions

$$\mathbf{A} \underset{k_{-1}}{\overset{k_1}{\rightleftarrows}} \mathbf{B} \underset{k_{-2}}{\overset{k_2}{\rightleftarrows}} \mathbf{C}$$

In a constant volume batch reactor, the concentrations evolve according to the equations

$$\begin{aligned} d[\mathbf{A}]/dt &= -k_1[\mathbf{A}] + k_{-1}[\mathbf{B}] \\ d[\mathbf{B}]/dt &= -(k_{-1} + k_2)[\mathbf{B}] + k_1[\mathbf{A}] + k_{-2}[\mathbf{C}] \\ d[\mathbf{C}]/dt &= -k_{-2}[\mathbf{C}] + k_2[\mathbf{B}] \end{aligned} \tag{3.5.5}$$

which is again a linear system of equations. To simplify the notation, introduce the concentration vector, $\mathbf{x}(t) = ([\mathbf{A}](t), [\mathbf{B}](t), [\mathbf{C}](t))$, and the initial condition vector, $\mathbf{x}_0 = ([\mathbf{A}]_0, [\mathbf{B}]_0, [\mathbf{C}]_0)$. Then the system of equations becomes

$$d\mathbf{x}/dt = -\mathbf{K}\mathbf{x} \tag{3.5.6}$$

with

$$\mathbf{K} = \begin{pmatrix} k_1 & -k_{-1} & 0 \\ -k_1 & (k_{-1} + k_2) & -k_{-2} \\ 0 & -k_2 & k_{-2} \end{pmatrix} \tag{3.5.7}$$

These equations can be solved by diagonalizing $\mathbf{K}$, i.e.

$$\mathbf{S}^{-1}\mathbf{K}\mathbf{S} = \mathbf{\Lambda} \tag{3.5.8}$$

where $\mathbf{\Lambda}$ is a diagonal matrix of eigenvalues and where the columns $\mathbf{s}_1$, $\mathbf{s}_2$, and $\mathbf{s}_3$ of matrix $\mathbf{S}$ are the eigenvectors of matrix $\mathbf{K}$. One of the three eigenvalues of $\mathbf{K}$ is zero ($\lambda_{11} = 0$) and its eigenvector $\mathbf{s}_1$ will correspond to the equilibrium state. The other two nonzero eigenvalues correspond to fast and slow relaxation rates in this little reaction network. Specifically, $\lambda_{22} = k_{slow}$ and $\lambda_{33} = k_{fast}$, such that $\mathbf{K}\mathbf{s}_2 = k_{slow}\mathbf{s}_2$, and $\mathbf{K}\mathbf{s}_3 = k_{fast}\mathbf{s}_3$. The slow and fast rates are

$$\begin{aligned} k_{slow} &= \tfrac{1}{2}\{\varphi - \sqrt{\varphi^2 - 4\psi}\} \\ k_{fast} &= \tfrac{1}{2}\{\varphi + \sqrt{\varphi^2 - 4\psi}\} \end{aligned} \tag{3.5.9}$$

where $\varphi = k_1k_2 + k_1k_{-2} + k_{-1}k_{-2}$ and $\psi = k_1 + k_2 + k_{-1} + k_{-2}$. Note that k_{fast} and k_{slow} also satisfy $k_{fast}k_{slow} = \psi$ and $k_{fast} + k_{slow} = \varphi$. The rate constants are real and positive because $\varphi^2 > \varphi^2 - 4\psi = (k_1 - k_2 + k_{-1} - k_{-2})^2 > 0$. The eigenvector of equilibrium concentrations is

$$\mathbf{s}_1 = (k_{-1}k_{-2},\ k_1k_{-2},\ k_1k_2).$$

The equilibrium concentrations are found by scaling $\mathbf{s}_1$ so that the sum of its elements add to the total initial concentration: $[tot] = [\mathbf{A}]_0 + [\mathbf{B}]_0 + [\mathbf{C}]_0$. The resulting equilibrium concentrations are

$$\begin{aligned} [\mathbf{A}]_{eq} &= [tot]k_{-1}k_{-2}/(k_{fast}k_{slow}) \\ [\mathbf{B}]_{eq} &= [tot]k_1k_{-2}/(k_{fast}k_{slow}) \\ [\mathbf{C}]_{eq} &= [tot]k_1k_2/(k_{fast}k_{slow}) \end{aligned} \tag{3.5.10}$$

The two equilibrium constants k_1/k_{-1} and k_2/k_{-2} in addition to k_{slow} and k_{fast} comprise four equations and four unknowns by which k_1, k_{-1}, k_2, and k_{-2} can be determined.

The original system of equations was $d\mathbf{x}/dt = -\mathbf{K}\mathbf{x}$. Because $\mathbf{S}^{-1}\mathbf{K}\mathbf{S} = \mathbf{\Lambda}$, we can insert $\mathbf{K} = \mathbf{S}\mathbf{\Lambda}\mathbf{S}^{-1}$ into equation (3.5.6)

$$d\mathbf{x}/dt = -\mathbf{S}\mathbf{\Lambda}\mathbf{S}^{-1}\mathbf{x} \tag{3.5.11}$$

Now we multiply on the left by $\mathbf{S}^{-1}$ to obtain

$$d\chi/dt = -\mathbf{\Lambda}\chi \tag{3.5.12}$$

where the new variable $\chi = \mathbf{S}^{-1}\mathbf{x}$. The equations are no longer coupled because $\mathbf{\Lambda}$ is diagonal. Each element of vector χ follows a simple exponential dependence on time $\chi_k(t) = exp[-\mathbf{\Lambda}_{kk}t]\chi_k(0)$. The initial vector $\chi(0)$ is obtained from the initial condition as $\chi(0) = \mathbf{S}^{-1}\mathbf{x}(0)$. Finally, the full solution is obtained by mapping $\chi(t)$ back to $\mathbf{x}(t)$ using $\mathbf{S}\chi = \mathbf{x}$.

$$\mathbf{x}(t) = \mathbf{S}e^{-\mathbf{\Lambda}t}\mathbf{S}^{-1}\mathbf{x}(0) \tag{3.5.13}$$

Formally, equation (3.5.13) is equivalent to $\mathbf{x}(t) = e^{-\mathbf{K}t}\mathbf{x}(0)$ with $e^{-\mathbf{K}t} = \mathbf{1} - \mathbf{K}t/1! + (\mathbf{K}t)^2/2! - \ldots$ For the special case where $\mathbf{x}(0) = ([\mathbf{A}]_0,\ 0,\ 0)$, the procedure outlined above gives

$$
\begin{aligned}
\frac{[\mathbf{A}](t)}{[\mathbf{A}]_0} &= \frac{k_{-1}k_{-2}}{k_{fast}k_{slow}} + \frac{k_{-1}(k_{fast}-k_2-k_{-2})}{k_{fast}(k_{fast}-k_{slow})}e^{-k_{fast}t} \\
&\quad - \frac{k_{-1}(k_{slow}-k_2-k_{-2})}{k_{slow}(k_{fast}-k_{slow})}e^{-k_{slow}t} \\
\frac{[\mathbf{B}](t)}{[\mathbf{B}]_0} &= \frac{k_1k_{-2}}{k_{fast}k_{slow}} + \frac{k_1(k_{-2}-k_{fast})}{k_{fast}(k_{fast}-k_{slow})}e^{-k_{fast}t} \\
&\quad - \frac{k_1(k_{-2}-k_{slow})}{k_{slow}(k_{fast}-k_{slow})}e^{-k_{slow}t} \\
\frac{[\mathbf{C}](t)}{[\mathbf{C}]_0} &= \frac{k_1k_2}{k_{fast}k_{slow}} + \frac{k_1k_2}{k_{fast}(k_{fast}-k_{slow})}e^{-k_{fast}t} \\
&\quad - \frac{k_1k_2}{k_{slow}(k_{fast}-k_{slow})}e^{-k_{slow}t}
\end{aligned}
\tag{3.5.14}
$$

The time dependent concentrations look terribly complicated, but actually each one contains just a constant term (equilibrium), a slowly decaying term ($\propto exp[-k_{slow}t]$), and a rapidly decaying term ($\propto exp[-k_{fast}t]$).

If the UV-vis extinction coefficients are known for species **A**, **B**, and **C** at two different wavelengths, then time-dependent UV-vis spectra at the two wavelengths can be used to determine k_{fast} and k_{slow} [22]. The absorbances as a function of time at any wavelength are a linear combination of the concentrations [**A**], [**B**], and [**C**]. Specifically, the absorbance is given by the Beer-Lambert law:

$$a_\lambda(t) = \ell\varepsilon_{\mathbf{A}}(\lambda)[\mathbf{A}] + \ell\varepsilon_{\mathbf{B}}(\lambda)[\mathbf{B}] + \ell\varepsilon_{\mathbf{C}}(\lambda)[\mathbf{C}]$$

where each term includes the path length ℓ and an extinction coefficient $\varepsilon_i(\lambda)$. Note that the absorbance $a_\lambda(t)$ is just a combination of three terms: one that is constant in time, one that is proportional to $exp[-k_{slow}t]$, and one that is proportional to $exp[-k_{fast}t]$. The shape of $a_\lambda(t)$ depends on λ because of the $\varepsilon_i(\lambda)$. But in principle any value of λ where the extinction coefficients are not degenerate can be used to extract the time constants k_{fast} and k_{slow}. Thus even when the optical extinction coefficients are not known, UV-vis spectra can still be used to extract k_{fast} and k_{slow}. In practice, it is useful to examine $a_\lambda(t)$ for multiple wavelengths, especially those wavelengths which show the most pronounced changes in the spectra. Figure 3.5.1 shows UV-vis results at two different wavelengths for a reaction of type $\mathbf{A} \rightleftarrows \mathbf{B} \rightleftarrows \mathbf{C}$. Note that the fast time scale may correspond to the quasi-equilibration of **A** with **B** or **B** with **C**.

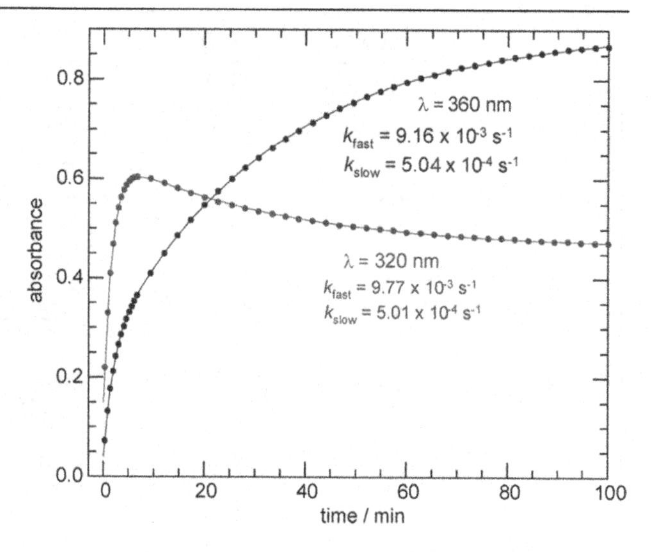

Figure 3.5.1: Experimental results (UV-vis spectra as a function of time) that suggest sequential reactions with two very different rate constants. [From Hwang et al. Inorg. Chem. 52, 13904-17 (2013).]

For systems involving more steps and more components, the rate equations can only be solved in certain special cases [25]. Alternatively, one can always resort to numerical solutions. Numerical solvers are widely available in programs like MatLab, but numerical solutions have two disadvantages. First, the differential equations are often terribly stiff, i.e. they involve vastly different time scales [26]. For example, when the fastest and slowest time scales differ by a factor 10^{10}, the timesteps must be smaller than the fastest time scale. Therefore a solution that reaches the longest time scale will require at least 10^{10} time steps. Second, the analytic solution naturally identifies groups of rate parameters that control the kinetic behavior, e.g. k_{fast} and k_{slow} in the example above. Numerical solutions require a separate investigation for each choice of the rate parameters, and for large networks with many rate parameters a complete analysis is not feasible. Thus analytic solutions to the rate equations, when they can be obtained, have many advantages. Fortunately, there are systematic and accurate approximations for deriving closed form rate laws even for complex reaction networks.

3.6 The pseudo-steady-state approximation

As explained at the beginning of this chapter, phenomenological rate laws are extremely important to both experimental and theoretical research in kinetics. However, we have also seen that, even for relatively simple reaction networks, the equations and their exact solutions can become extremely complicated. Because small differences in barriers lead to large differences in rate constants, processes involving multiple reactions tend to involve vastly different time scales. Giant differences in time scales are an onerous challenge for numerical methods, but time scale gaps actually facilitate the development of simple and highly accurate phenomenological rate laws.

The book by Helfferich contains several chapters on approximations for translating complex reaction networks into simple analytic rate laws. He develops solutions for catalytic reactions involving side paths, coupled cycles, and cycles coupled to external reactions, etc. Despite their long history, rigor, and straightforward nature [27–29], many modern investigators are only loosely familiar with standard techniques for analysis of complex reaction networks. Chapters 3 and 4 of this book present only the most basic approximations, but readers are encouraged to see other books which chiefly focus on complexity reduction techniques [28,29].

The least severe approximation in the hierarchy of complexity reduction tools is a pseudo-steady-state approximation (PSSA), sometimes called a Bodenstein approximation. The PSSA is a *pseudo*-steady-state approximation because, it neither implies nor requires an actual steady-state condition. The PSSA eliminates the differential equations associated with trace reactive intermediates that are often difficult to observe or isolate. To see how the PSSA emerges, let us reconsider the series of reactions

$$\mathbf{A} \xrightarrow{k_1} \mathbf{B} \xrightarrow{k_2} \mathbf{C}$$

which was exactly solved in section 3.5. For initial conditions $[\mathbf{A}] = [\mathbf{A}]_0$ and $[\mathbf{B}]_0 = [\mathbf{C}]_0 = 0$, the (exact) time-dependent concentrations of $[\mathbf{A}]$ and $[\mathbf{B}]$ were

$$\begin{aligned} [\mathbf{A}] &= [\mathbf{A}]_0 e^{-k_1 t} \\ [\mathbf{B}] &= [\mathbf{A}]_0 e^{-k_1 t} \frac{k_1}{k_2} (1 - k_1/k_2)^{-1} \{1 - e^{-k_2(1-k_1/k_2)t}\} \end{aligned} \tag{3.6.1}$$

with $[\mathbf{C}]$ determined by the mass balance $[\mathbf{A}]+[\mathbf{B}]+[\mathbf{C}] = [\mathbf{A}]_0$. The solution for $[\mathbf{B}]$ has been written in a form that emphasizes the important role of the small parameter k_1/k_2. If $k_1/k_2 \ll 1$ then shortly after **B** molecules are created, they are converted to **C** molecules. The form of equation (3.6.1) shows that $[\mathbf{B}] \ll [\mathbf{A}]$ for all times. The concentration of the reactive intermediate **B** never accumulates, and for nearly all times it is also small relative to the concentration of **C**.

As the parameter k_1/k_2 becomes very small, equation (3.6.1) becomes [30]

$$\begin{aligned} [\mathbf{A}] &= [\mathbf{A}]_0 e^{-k_1 t} \\ [\mathbf{B}] &= [\mathbf{A}]_0 e^{-k_1 t} \frac{k_1}{k_2} \{1 - e^{-k_2 t}\} \end{aligned} \tag{3.6.2}$$

After a time k_2^{-1} which is short compared to the time on which $[\mathbf{A}]$ changes (k_1^{-1}), we recover the PSSA for the concentration of species **B**:

$$[\mathbf{B}] \approx \frac{k_1}{k_2} [\mathbf{A}] \tag{3.6.3}$$

The exact solution for $[\mathbf{B}](t)$ and the PSSA approximation to $[\mathbf{B}](t)$ are shown in Figure 3.6.1.

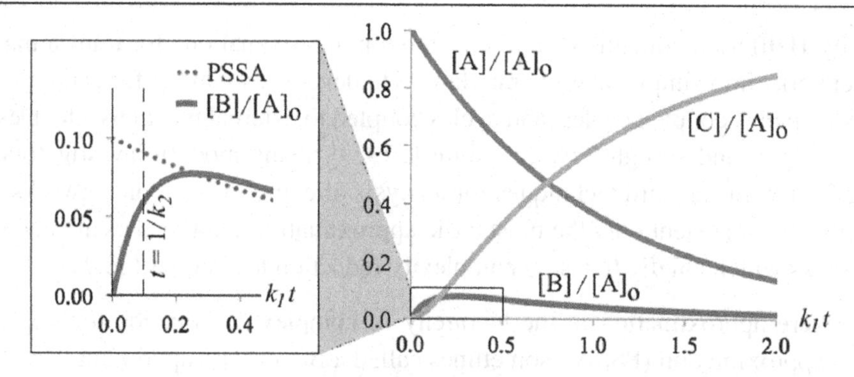

Figure 3.6.1: Time dependent concentration profiles for the sequential reactions A → B → C when the rate constant k_2 for B → C is ten times the rate constant k_1 for A → B. The inset at left shows the exact early time behavior of [B] when $k_2 = 10k_1$ along with the time dependence of [B] that emerges from the pseudo-steady-state approximation (PSSA). The approximation has $o(k_1/k_2)$ accuracy except for times near k_2^{-1} and shorter.

Readers familiar with techniques of perturbation theory will recognize the PSSA as a zeroth order singular perturbation result. However, equation (3.6.3) is also exactly the expression for [**B**] that would emerge from a steady-state approximation on species **B**:

$$d[\mathbf{B}]/dt = -k_2[\mathbf{B}] + k_1[\mathbf{A}] \approx 0 \tag{3.6.4}$$

$d[\mathbf{B}]/dt \approx 0$ does not actually imply a constant [**B**] concentration because the concentrations of the other species (in this case, [**A**] and [**C**]) are still changing. A true steady state does not emerge unless *all* species are at steady state, and hence the term *pseudo*-steady-state approximation (PSSA). Instead, the PSSA relates the reactive intermediate concentration [**B**] to the concentrations of the reactants, products, and abundant intermediates. Note that simplified rate laws from the PSSA can be plugged into the species balance (equation (3.1.1)) and used to model systems at steady state or to model systems which are not at steady state. Thus instead of writing the PSSA on [**B**] as $d[\mathbf{B}]/dt \approx 0$, it is more correct to write

$$\mathcal{R}_{\mathbf{B}} \approx 0 \tag{3.6.5}$$

which does not specify the steady-state or time-varying nature of the overall process.

Operationally, the PSSA always replaces an ordinary differential equation with an algebraic equation. The algebraic equation relates the unknown intermediate concentration to concentrations of other species. These may be time varying or static concentrations. The PSSA applies not only to simple sequences like **A** → **B** → **C**. In general, when the *separate rates* at which a reactive intermediate is consumed and produced are both be much larger than its *net rate* of generation, then the intermediate is only transiently visited with negligible accumulation. The

PSSA can often be justified based on the equilibrium properties, e.g. from the free energies of the intermediates. Let us define adjoining species to an intermediate as those which create or consume the intermediate via a single elementary reaction. If an intermediate has high free energy relative to all adjoining species, then the PSSA is justified for that intermediate. This rule is useful in the derivation of rate laws from *ab initio* computational results.

The PSSA can be applied to intermediates that are involved in reversible or irreversible steps of any order. Moreover, the PSSA can be simultaneously applied to many intermediates in a reaction network. These features make the PSSA a versatile and powerful tool for extracting simple rate laws from complex multistep reaction mechanisms. Helfferich has given three useful rules for applications of the PSSA:

1. The algebraic equation $\mathcal{r}_B = 0$ that results from the PSSA for intermediate **B** should be used to eliminate the intermediate **B** from the overall rate equations. The equation $\mathcal{r}_{\mathbf{B}} = 0$ should not be used to eliminate a separate intermediate **X**.
2. When PSSA is applied to a series of intermediates along a simple reaction path, the algebra is simplified by starting from the last intermediate in the sequence.
3. The overall rate laws that emerge from PSSA can be converted to "one-plus" form (see examples below) to minimize the number of independent parameters in the rate law.

The PSSA can be used to obtain rate laws for arbitrarily complex networks of first order (or pseudo-first order) steps. If the transient intermediates in the network do not react with each other, then a closed form solution for the rate law is assured [31–33].

Example: Aromatic nitration

This example illustrates how a complex series of reactions can be systematically reduced to a simple rate law. The example concerns the Gillespie-Ingold mechanism for nitration of aromatics [28,34] in the presence of a sulfuric acid solution. The reaction involves several intermediates (labeled $\mathbf{X}_i$ for convenience) as shown in the figure below. $H\mathbf{B}$ represents sulfuric acid and $\mathbf{B}^-$ represents the HSO_4^- ion. Double-headed arrows indicate reversible steps, and each step indicates an elementary reaction. Rate constants are not shown, but let k_{ij} be the rate constant from $\mathbf{X}_i$ to $\mathbf{X}_j$.

$$X_0 \underset{B^-}{\overset{HB}{\rightleftharpoons}} X_1 \underset{H_2O}{\rightleftharpoons} X_2 \xrightarrow{ArH} X_3 \underset{HB}{\overset{B^-}{\longrightarrow}} X_4$$

$X_0 = HNO_3$ stoichiometric reactant

$X_1 = H_2NO_3^+$ } reactive intermediates

$X_2 = NO_2^+$

$X_3 = ArNO_2H^+$

$X_4 = ArNO_2$ stoichiometric product

net reaction: $ArH + HNO_3 \longrightarrow ArNO_2 + H_2O$

A simple rate law can be obtained by sequential applications of the PSSA. The only assumption is irreversibility of the last two steps as indicated in the figure. We will start at the last step and work back to the beginning as suggested by Helfferich. The overall rate is

$$\mathcal{R}_{X4} = k_{34}[\mathbf{X}_3][\mathbf{B}^-]$$

Applying the PSSA to $\mathbf{X}_3$,

$$\mathcal{R}_{X3} = k_{23}[\mathbf{X}_2][ArH] - k_{34}[\mathbf{X}_3][\mathbf{B}^-] \approx 0$$

and solving for $[\mathbf{X}_3]$ gives

$$[\mathbf{X}_3] = k_{23}[\mathbf{X}_2][ArH]/k_{34}[\mathbf{B}^-]$$

Rewrite the rate law without the reactive intermediate $[\mathbf{X}_3]$:

$$\mathcal{R}_{X4} = k_{23}[\mathbf{X}_2][ArH]$$

Note that the PSSA already gave an interesting result. Namely, the rate expression $k_{23}[\mathbf{X}_2][ArH]$ is that which we would have written by assuming that the first irreversible step $\mathbf{X}_2 + ArH \rightarrow \mathbf{X}_3$ is rate determining.

The rate law still involves the reactive intermediate $[\mathbf{X}_2]$, so apply PSSA again

$$\mathcal{R}_{X2} = k_{12}[\mathbf{X}_1] - k_{21}[\mathbf{X}_2][H_2O] - k_{23}[\mathbf{X}_2][ArH] \approx 0$$

Now solve for $[\mathbf{X}_2]$:

$$[\mathbf{X}_2] = k_{12}[\mathbf{X}_1]/(k_{23}[ArH] + k_{21}[H_2O])$$

Rewrite the rate law without the reactive intermediate $[\mathbf{X}_2]$:

$$r_{\mathbf{X}4} = \frac{k_{12}k_{23}[\mathbf{X}_1][ArH]}{k_{23}[ArH] + k_{21}[H_2O]}$$

Now there is only $[\mathbf{X}_1]$ to be eliminated with one last application of the PSSA.

$$r_{\mathbf{X}1} = k_{01}[\mathbf{X}_0][H\mathbf{B}] + k_{21}[\mathbf{X}_2][H_2O] - k_{12}[\mathbf{X}_1] - k_{10}[\mathbf{B}^-][\mathbf{X}_1] \approx 0$$

Solving for $[\mathbf{X}_1]$ gives

$$[\mathbf{X}_1] = \frac{k_{01}[\mathbf{X}_0][H\mathbf{B}] + k_{21}[\mathbf{X}_2][H_2O]}{k_{12} + k_{10}[\mathbf{B}^-]}$$

which reintroduces $[\mathbf{X}_2]$. This is not a serious problem. We can reuse the expression from PSSA on $[\mathbf{X}_2]$ which already showed that $[\mathbf{X}_2]$ is proportional to $[\mathbf{X}_1]$. The resulting expression is

$$[\mathbf{X}_1] = \frac{k_{01}[\mathbf{X}_0][H\mathbf{B}](k_{23}[ArH] + k_{21}[H_2O])}{k_{12}k_{23}[ArH] + k_{10}k_{23}[ArH][\mathbf{B}^-] + k_{10}k_{21}[\mathbf{B}^-][H_2O]}$$

Finally, after applying the PSSA to all reactive intermediates, the rate law is

$$r_{ArNO2} = \frac{k_{01}k_{12}k_{23}[HNO_3][H\mathbf{B}][ArH]}{k_{12}k_{23}[ArH] + k_{10}k_{23}[\mathbf{B}^-][ArH] + k_{10}k_{21}[\mathbf{B}^-][H_2O]}$$

Now all of the reactive intermediates have been eliminated, and the only remaining step is to write the rate law in one-plus form. These last simplifications are facilitated by some assumptions. First, if water is the solvent then its concentration is a constant. Second, acid-base protonation is fast and assumed to be in equilibrium throughout the course of the reaction. Therefore, $[H^+][\mathbf{B}^-]/[H\mathbf{B}] = K_{H\mathbf{B}}$. Let $[\mathbf{B}]_{tot} = [\mathbf{B}^-] + [H\mathbf{B}]$ be the total amount of acid before considering its deprotonated forms. Also define the equilibrium constants $K_{01} = k_{01}/k_{10}$ and $K_{12} = k_{12}/k_{21}$. Notice that these ratios appear only for the reversible steps. The simplified expression is [28]

$$r_{ArNO2} = \frac{k_a[HNO_3][H^+][ArH]}{1 + k_b[ArH](1 + k_c\{1 + [H^+]/K_{HB}\}/[\mathbf{B}]_{tot})}$$

with

$$k_a = \frac{K_{01}K_{12}k_{23}}{K_{HB}[H_2O]} \qquad k_b = \frac{k_{23}}{k_{21}[H_2O]} \qquad k_c = \frac{k_{12}}{k_{10}}.$$

The Gillespie-Ingold mechanism thus yields a rate law (from PSSA) which (correctly) predicts that

1. the rate is proportional to undissociated HNO_3.
2. the rate is also proportional to $[H^+]$ in the numerator, but can also be inhibited at low pH because of $[H^+]$ dependence in the denominator.
3. the reaction order in $[ArH]$ can be one or zero with the latter prevailing at sufficiently high $[ArH]$.

Note that the concentration of HNO_3 decreases with increasing $[H^+]$. One can expect that the pH dependence of this reaction will be quite complex.

It is remarkable that the structure of the reaction network is sufficient to predict so many aspects of the overall kinetics. In the aromatic nitration example, one might have qualitatively anticipated effects of $[ArH]$ and $[HNO_3]$, but the dependence on pH dependence and effects of $K_{H\mathbf{B}}$ are certainly less obvious. Furthermore, the PSSA rate law has identified the grouped kinetic parameters k_a, k_b, and k_c which control the overall kinetics. There are some important limitations that should be remembered when using the PSSA:

1. Rate laws obtained using PSSA do not accurately describe the behavior at short times while the intermediate populations are being established.
2. When reactive intermediates are generated through branching steps, they can build-up rapidly to very high levels. For example, two branching steps in H_2 combustion are

$$\begin{aligned} H + O_2 &\rightleftarrows OH + O \\ O + H_2 &\rightleftarrows OH + H \end{aligned}$$

   Each of these branching steps creates two reactive radicals from one and therefore the radical population is not small – it will be of the same order as the extent of reaction. Readers are directed to more specialized books for more on explosions and combustion chemistry [35], but in general the PSSA should not be used for intermediates that are involved in branching steps.

It is also important to remember that the rate law is merely one consequence of the proposed mechanism. The rate law must be tested in the laboratory, but even if it is confirmed the underlying mechanism might still be incorrect. Consider, for example, the reversible reaction

$$H_2 + I_2 \rightleftarrows 2HI \tag{3.6.6}$$

Bodenstein studied this reaction and found a rate law corresponding to the stoichiometric equation. For the next 70 years following his work, this reaction was the standard textbook example of an elementary reaction [36]. There was no evidence to suggest otherwise. Then in 1967, Sullivan studied the reaction in the presence of visible light at a frequency that dissociates I_2 [37].

The rate law

$$\mathfrak{r} = k_f[H_2][I_2] - k_b[HI]^2$$

corresponding to the single elementary step mechanism would predict no significant effect if a small fraction of the I_2 is dissociated during the reaction. The surprise was that the rate is dramatically enhanced by irradiation. I_2 also has a strong tendency to thermally homolyze into I atoms, so the irradiated experiments suggested an important mechanistic role for dissociated I atoms in the thermally activated mechanism as well. The currently accepted mechanism involves an initial dissociation step $I_2 \rightleftarrows 2I$ followed by a trimolecular step $H_2 + 2I \rightleftarrows 2HI$. An alternative to the trimolecular step involves two bimolecular steps, $H_2 + I \rightleftarrows H_2I$ followed by $H_2I + I \rightleftarrows 2HI$. The alternatives are both consistent with the observed kinetics [37].

When such ambiguities arise, theory provides an important partner to experiment. For starters, the Woodward-Hoffmann orbital symmetry rules [38] forbid the elementary reaction (3.6.6) via the four-membered-ring transition state that had been proposed. In other cases, there may be no qualitative theoretical reasons to favor one mechanism over another, but still computational analyses can help. At the present time, *ab initio* calculations cannot reliably predict accurate rate constants, but *in silico* tests can often rule out mechanisms that are completely implausible. *Ab initio* calculations can also provide internal consistency checks. For example, if the proposed mechanism involves an irreversible step, then the Gibbs free energy for that step at the process conditions should be decidedly downhill. If the mechanism invokes two states that rapidly interconvert, as in the quasi-equilibrium approximation, then their interconversion should have a small barrier.

3.7 Rate determining steps and quasi-equilibrated steps

Many reaction networks involve reversible steps with rapid forward and backward rates. If the forward and reverse rates are much faster than the rates of other steps in the network, then it is often useful to treat the fast reaction as a quasi-equilibrated (QE) step. The QE approximation is useful in many contexts, especially physisorption steps in catalysis and protonation steps in aqueous phase reactions involving acids and bases. Many examples of the QE approximation can be found in Chapter 4. Here we discuss conditions which justify a QE approximation. Consider a fast reversible elementary reaction which is embedded within a larger reaction network as shown below.

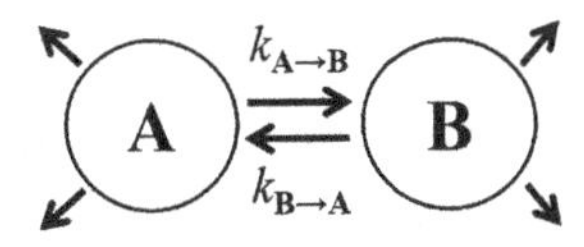

The QE approximation effectively merges the **A** and **B** species into one lumped state. Within the equilibrated **A**, **B** pair, a fraction

$$x_{\mathbf{A}|\mathbf{A},\mathbf{B}} = \frac{k_{\mathbf{B}\to\mathbf{A}}}{k_{\mathbf{A}\to\mathbf{B}} + k_{\mathbf{B}\to\mathbf{A}}}$$

will be in state **A**, and a fraction

$$x_{\mathbf{B}|\mathbf{A},\mathbf{B}} = \frac{k_{\mathbf{A}\to\mathbf{B}}}{k_{\mathbf{A}\to\mathbf{B}} + k_{\mathbf{B}\to\mathbf{A}}}$$

will be in state **B**. Note that the $x_{\mathbf{A}|\mathbf{A},\mathbf{B}}$ and $x_{\mathbf{B}|\mathbf{A},\mathbf{B}}$ are not regular mole fractions. Instead they are conditional mole fractions within the subsystem including **A** and **B**. How does one justify the QE approximation? The time scale for equilibration of $\mathbf{A} \rightleftarrows \mathbf{B}$ should be much shorter than the time for escape from the equilibrated $\mathbf{A} \rightleftharpoons \mathbf{B}$ state. Let $\overline{\mathbf{A},\mathbf{B}}$ refer to the set of all states beyond **A** and **B**. Then the criteria for QE is

$$(k_{\mathbf{A}\to\mathbf{B}} + k_{\mathbf{B}\to\mathbf{A}})^{-1} \ll (x_{\mathbf{A}|\mathbf{A},\mathbf{B}} \sum k_{\mathbf{A}\to\overline{\mathbf{A},\mathbf{B}}} + x_{\mathbf{B}|\mathbf{A},\mathbf{B}} \sum k_{\mathbf{B}\to\overline{\mathbf{A},\mathbf{B}}})^{-1} \tag{3.7.1}$$

where the left side of the inequality is the $\mathbf{A},\mathbf{B}$ equilibration time and the right side is the $\mathbf{A},\mathbf{B}$ escape time. Criterion (3.7.1) can identify quasi-equilibrated pairs of states within a network of first order reactions. The requirements are more complicated for bimolecular steps. To justify the quasi-equilibrium assumption in networks with biomolecular steps, we usually make the assumption, use it to obtain concentrations of the intermediates, and then check whether the rates within the putative QE steps are appropriately faster than the rates of other steps. When this requirement is fulfilled, a QE approximation effectively combines the reactants and products of the fast step with a single "lumped" intermediate species [28]. For an example of lumping QE states, see the model of ethylene polymerization by the Phillips catalyst in Chapter 4 and tools for simplifying Markov State Models in Chapter 14 [41,42]. Note that lumping also refers to a method for modeling reactions in mixtures that contain species with similar rate coefficients [13,43, 44].

A linear series of steps often has a rate determining step (RDS) that controls the overall rate. We frequently describe the RDS as "slow" even in steady-state situations where all steps must have the same rate. More properly the RDS is a dynamical bottleneck that limits the overall rate. Reactants accumulate ahead of the RDS, much like traffic jams accumulate ahead of road construction. The overall rate is particularly sensitive to the forward and backward rate constants for the RDS and nearly independent of rate constants for other steps in the network.[3]

[3] Strictly speaking, statements about overall rates and sensitivity to rate constants are justified only when forward and backward rate constants are varied together so as to maintain the equilibria for elementary steps [39,40].

The figure below illustrates the situation where there is a clear RDS, i.e. a step where the forward and backward rates are each slower than all other rates in the series. When forward and backward rates are separately considered, the RDS can properly be identified as slow.

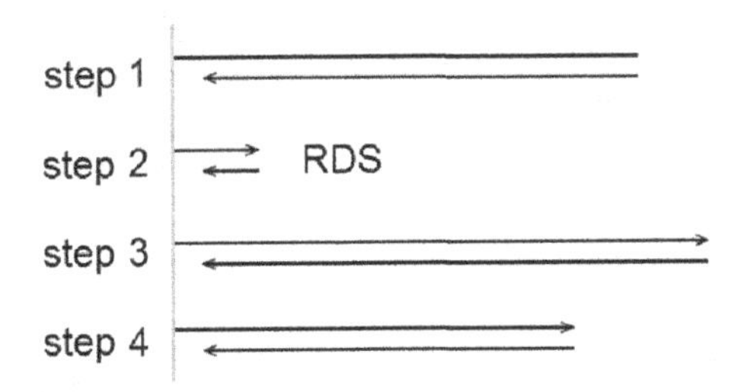

When there are multiple parallel pathways each having multiple steps, the bottleneck is the slow step along the fastest path. The existence of a single RDS should be assumed with some trepidation for two key reasons.

- A change in temperature can change the RDS. This tends to occur when the RDS and some comparably slow step have very different activation energies. Steps with the higher activation energies are more sensitive to temperature.
- A change in concentrations can change the RDS. This occurs when the RDS and a comparably slow step have different reaction orders with respect to some reagent. Steps with high reaction orders are highly sensitive to concentration.

Both of the points above emphasize that the existence of an RDS depends on relative *rates*, not on relative *rate constants*. When a RDS is assumed, one should check whether the implied assumptions about relative rates are true *at the operating conditions*. It is not correct to infer the RDS from standard free energies or standard activation free energies unless the reaction conditions of interest are those of the standard state.

When there is a clear RDS, many simplifications follow. Steps which come after the RDS will have no impact on the overall rate [32]. Steps which come before the RDS will have sufficient time to (approximately) reach equilibrium [32]. For these steps we can apply the QE approximation and dramatically simplify the rate law. Figure 3.7.1 shows a useful fluid flow analogy for the RDS and QE assumptions in a sequential multistep reaction mechanism. The QE intermediate pools that form ahead of the RDS have a capacity that increases exponentially with their depth according to Boltzmann statistics. Thus nearly all trajectories will be trapped in a deep pool ahead of the RDS with plenty of time to equilibrate. Note that the RDS and QE assumptions are always special limiting cases of the PSSA rate law as depicted in Figure 3.7.2. This is easiest to see by examining the rate laws that emerge from the PSSA.

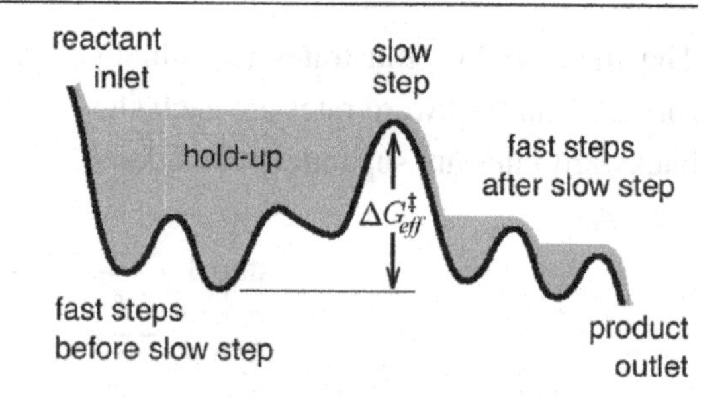

Figure 3.7.1: A fluid mechanics model showing how the slow step makes subsequent steps irrelevant. At steady state, the slow step causes intermediates to accumulate until the slow step proceeds at the same rate as all other steps.

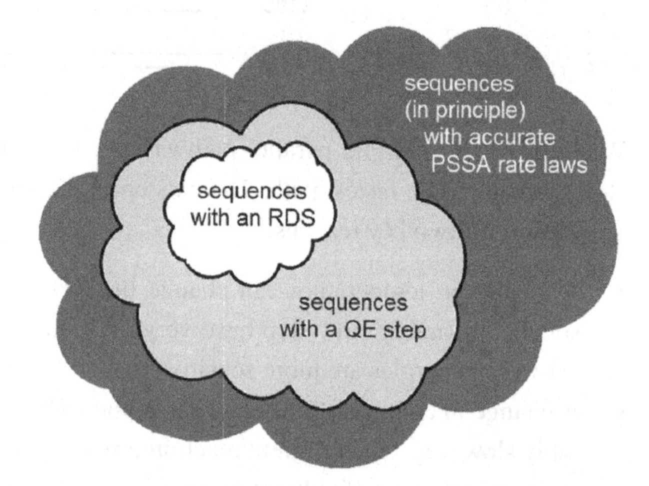

Figure 3.7.2: The most general approximation for a sequence including several intermediate steps is the PSSA, but if the sequence is not simple the PSSA may not yield an analytic rate law. Quasi-equilibrium (QE) can be assumed for steps whose forward and backward rates are much faster than those which produce and consume its reactants and products. Invoking an RDS assumes there is only one slow step.

The less general rate laws that emerge from RDS and QE assumptions are still used because they are easier to derive and they contain fewer parameters. Moreover, reaction networks in which intermediates react with each other usually yield analytic rate laws only after QE and RDS assumptions. However, the RDS and/or QE assumptions are more drastic than a PSSA assumption, so one should appeal to data or to *ab initio* calculations for justification. This section gives no in-depth examples on deriving rate laws with RDS or QE assumptions, but examples can be found in other parts of the book (especially in the chapter on catalysis).

In addition to linear sequences of reactions, complex reaction networks and catalytic cycles can be treated with RDS and QE assumptions. In catalysis, the Michaelis-Menten (and Langmuir-Hinshelwood) rate laws are well-known examples. However, distinctions between steps that come 'before' and 'after' the RDS will need to be amended for catalytic cycles. See Chapter 4 for more on RDS and QE assumptions in catalysis.

For complex networks, it is useful to think of the network as a graph [28,29]. The network is first broken into piecewise linear sequences that link the branching points in the network. Separate rate laws are derived for these sequences in terms of the species at the branching vertices. These separate rate laws may invoke the PSSA, QE, RDS, and/or lumping approximations as justified.

Then the separate rate expressions are merged by balancing fluxes at each branching vertex of the graph. The procedure is identical to the way one might approach a circuit analysis with Kirchoff's laws. Along individual edges of the network one may invoke the PSSA, QE, RDS, and/or lumping approximations as justified *en route* to a rate law. Over the past few decades, this process has been reduced to a formulaic algorithm [28,29,32,33]. However, justifying the RDS and QE steps for the edges of a complex network is more difficult than for reactions considered in isolation. The criteria for justifying RDS and QE assumptions along the separate edges of the network depend on relative rates, and the rates in turn depend on reactant and product concentrations. The rate law must be derived for each edge in the network, flux-matched to adjoining edges, and then checked as concentrations evolve according to the overall rate law to justify the individual QE and/or RDS approximations for each edge.

Exercises

1. [S.L. Scott] In a closed batch reactor, show that a first order irreversible reaction with rate constant k_1 approaches completion faster than a first order reversible reaction with (forward) rate constant k_1 approaches equilibrium. Explain the physical reason for this result.
2. Derive equations (1) and (2) in Kaern et al. Nature Reviews: Genetics, 6, 451 (2005). Hint: remember that an mRNA molecule can be used to create multiple proteins.
3. [B.L. Trout] Protein aggregation is related to certain diseases as well as problems in manufacturing and stability. Aggregation is thought to occur via the mechanism $\mathbf{P} \rightarrow \mathbf{I}$ and $2\mathbf{I} \rightarrow \mathbf{P}_2$ where $\mathbf{P}$ is the isolated protein, $\mathbf{I}$ is the partially unfolded intermediate, and $\mathbf{P}_2$ is a dimer aggregate. What reaction orders could be observed for this process? Describe experiment(s) that could determine the true overall reaction order.
4. For the reversible reaction $a\mathbf{A}+b\mathbf{B} \rightleftarrows c\mathbf{C}+d\mathbf{D}$, a power law type rate expression can be constructed as

$$r = k_f(T)\prod_i c_i^{\alpha_i} - k_b(T)\prod_i c_i^{\beta_i}$$

   where the species index i includes **A**, **B**, **C**, **D**, and possibly additional species. For the sake of illustration, let the additional species be **X**. What conditions must be satisfied by $k_f(T)$, $k_b(T)$, $\{\alpha_i\}_{i=\mathbf{A},\mathbf{B},\mathbf{C},\mathbf{D},\mathbf{X}}$, and $\{\beta_i\}_{i=\mathbf{A},\mathbf{B},\mathbf{C},\mathbf{D},\mathbf{X}}$?
5. Some nanoparticle syntheses exhibit a long induction (nucleation) period followed by an autocatalytic stage (growth). To model the kinetics of nucleation and growth, Watzky and Finke, *J. Am. Chem. Soc.* 119, 10382 (1997), proposed a two-step mechanism:

$$\mathbf{A} \xrightarrow{k_1} \mathbf{B} \qquad \textit{and} \qquad \mathbf{A}+\mathbf{B} \xrightarrow{k_2} 2\mathbf{B}$$

where **A** and **B** are abstract entities as opposed to specific species. The two-step model accurately fits experimental data on the fraction of remaining precursor as a function of time, but comment on the degree of phenomenology in this model. Specifically, discuss (i) the units of k_1 and k_2 vs. those for nucleation and growth by surface attachment, and (ii) whether the data demonstrates that k_1 and k_2 are true concentration independent rate constants. Should the two-step model be regarded as an empirical fit or as a mechanism?

6. [A. Bhan] Consider the yield of a product **B** from conversion of reactant **A**. If **B** is a primary product that decomposes in a subsequent step to a secondary product, would the yield of **B** vs. conversion of **A** exhibit upward or downward curvature? If **B** is a secondary product that does not decompose further, would the yield of **B** vs. conversion of **A** exhibit upward or downward curvature?
7. Autocatalytic reactions are catalyzed by their own products. Examples are found in metal nanoparticle synthesis and in acid catalyzed ester hydrolyses. An autocatalytic rate law for the reaction $\mathbf{A} \to \mathbf{B}$ is $\mathscr{r} = k[\mathbf{A}][\mathbf{B}]$. Find the conversion of $[\mathbf{A}]$ and $[\mathbf{B}]$ as functions of time in a constant volume batch reactor. Start with non-zero initial concentrations $[\mathbf{A}]_0$ and $[\mathbf{B}]_0$.
8. Derive the extent of the reaction $\mathbf{A} \rightleftarrows \mathbf{B} + \mathbf{C}$ as a function of time for a constant volume batch reactor and for general initial conditions. Verify the expressions given in equations (3.4.10) and (3.4.11) when there are only reactants or products, respectively, at the initial time.
9. Consider a batch reaction $2\mathbf{A} \to \textit{products}$ with rate law $\mathscr{r} = k\,[\mathbf{A}]^2$. Assume errors in concentration measurements that are uniformly superimposed on the actual value of $[\mathbf{A}]$, i.e. $[\mathbf{A}]_i = [\mathbf{A}](t_i) + \varepsilon_i$ where ε_i has the same distribution (with zero mean) at each time t_i.
   (a) Show that nonlinear regression with $[\mathbf{A}] = [\mathbf{A}]_0/(1+2[\mathbf{A}]_0 k\,t)$ gives $S_{NL} = \sum_i \varepsilon_i^2$ to leading order in the ε_i. Also show that linearized regression with $1/[\mathbf{A}] = 1/[\mathbf{A}]_0 + 2kt$ gives $S_L = \sum_i \varepsilon_i^2/[\mathbf{A}]_i^4$ to leading order in the ε_i.
   (b) Rather than expanding in the ε_i show that the nature of $S_L \approx \sum_i \varepsilon_i^2/[\mathbf{A}]_i^4$ can be anticipated from $|d(1/[\mathbf{A}])| = |d[\mathbf{A}]/[\mathbf{A}]^2|$.
   (c) Based on the results in (a), what can you say about linearization vs. nonlinear regression for second order rate laws? How should the data points be weighted in linear regression to obtain a more balanced fit?
10. The reaction between methyltrioxorhenium (CH_3ReO_3, **MTO**) and hydrogen peroxide creates species **A** ($CH_3ReO_2(\eta_2 - O_2)$), a potent and selective agent for epoxidation. The **MTO** activation steps are

$$\mathbf{MTO} + H_2O_2 \underset{k_{-1}}{\overset{k_1}{\rightleftarrows}} CH_3ReO(OH)(OOH) \underset{k_{-2}}{\overset{k_2}{\rightleftarrows}} \mathbf{A} + H_2O$$

The reaction occurs in an aqueous solvent where $[H_2O] \gg [Re]_{total}$. The intermediate is not observed.

(a) Derive a rate expression for the reversible formation of **A** from **MTO** and H_2O_2.

(b) Hwang et al. *Inorg. Chem.* (2013) showed that the **MTO** activation reaction is accelerated by water. What does that tell you about the mechanism and rate constants which are given above?

11. [Adapted from Missen, Mims, Saville] Houser and Lee, *J. Phys. Chem.* 1967 proposed a free radical chain reaction mechanism for the ethyl nitrate ($C_2H_5ONO_2$, **A**) pyrolysis. The main products are formaldehyde (CH_2O, **B**) and methyl nitrate (CH_3NO_2, **D**). The proposed mechanism is

$$\begin{aligned} \mathbf{A} &\overset{k_1}{\rightarrow} C_2H_5O^\bullet + NO_2 \\ C_2H_5O^\bullet &\overset{k_2}{\rightarrow} CH_3^\bullet + \mathbf{B} \\ CH_3^\bullet + \mathbf{A} &\overset{k_3}{\rightarrow} C_2H_5O^\bullet + \mathbf{D} \\ 2C_2H_5O^\bullet &\overset{k_4}{\rightarrow} CH_3CHO + C_2H_5OH \end{aligned}$$

where one expects acetaldehyde and ethanol to be minor products.

(a) Derive a rate law for the formation of **B** and **D**. Label the initiation, propagation, and termination steps for this chain reaction.

(b) The initial rate has been measured as $r_0 = 0.0230$ mM/s at 250°C and $[\mathbf{A}]_0 =$ 0.271 mM. Suppose the rate law from part (a) is correct and use it to predict [**A**] at later times.

(c) The initial rates have been measured at 250°C and at a series of concentrations of the reactant **A**:

[A]/mM	**.0713**	**.0759**	**.0975**	**.235**	**.271**
r_0/(mM s^{-1})	.0121	.0122	.0134	.0209	.0230

Are these results consistent with the rate law from part (a)?

(d) The apparent activation energy is $E_a = -d\ln r/d(1/k_BT)$. How does E_a relate to the individual activation free energies for steps 1, 2, and 4?

12. Let **M** be a stable monomer, let $\mathbf{A}_n$ be an active polymer chain of length n, and let $\mathbf{P}_n$ be a stable (not growing) polymer chain of length n. Suppose that polymerization occurs by the mechanism

$$\mathbf{M} \overset{k_1}{\rightarrow} \mathbf{A}_1 \qquad \textit{initiation}$$

$$\mathbf{A}_n + \mathbf{M} \overset{k_P}{\rightarrow} \mathbf{A}_{n+1} \qquad \textit{propagation}$$

$$\mathbf{A}_n \overset{k_T}{\rightarrow} \mathbf{P}_n \qquad \textit{termination}$$

where the propagation and termination steps can occur for all chain lengths $n = 1, 2, 3, \ldots$. Derive a rate law for the consumption of **M**. Also derive the (arithmetic) average polymer chain length. Note that another definition of the average chain length weights each polymer by its molecular weight. See Chapter 4 for details.

References

[1] R.B. Bird, W.E. Stewart, E.N. Lightfoot, Transport Phenomena, Wiley, New York, 1960.
[2] E.L. Cussler, Diffusion: Mass Transfer in Fluid Systems, 3rd ed., Cambridge Press, Cambridge, UK, 2009.
[3] K. Morokuma, C. Muguruma, J. Am. Chem. Soc. 116 (1994) 10316–10317.
[4] T. Loerting, K.R. Liedl, Proc. Natl. Acad. Sci. USA 97 (2000) 8874–8878.
[5] E. Vohringer-Martinez, B. Hansmann, H. Hernandez, J.S. Francisco, J. Troe, B. Abel, Science 315 (2007) 497–502.
[6] D.M. Koch, C. Toubin, G.H. Peslherbe, J.T. Hynes, J. Phys. Chem. C 112 (2008) 2972–2980.
[7] T. Hwang, B.R. Goldsmith, B. Peters, S.L. Scott, Inorg. Chem. 52 (2013) 13904–13917.
[8] C.M.L. Frasson, T.A.S. Brandao, C. Zucco, F. Nome, J. Phys. Org. Chem. 19 (2006) 143–147.
[9] B.R. Goldsmith, T. Hwang, S. Seritan, B. Peters, S.L. Scott, J. Am. Chem. Soc. 137 (2015) 9604–9616.
[10] J.C. Maxwell, Philos. Trans. R. Soc. Lond. 157 (1867) 49–88.
[11] V.K. La Mer, J. Chem. Phys. 1 (1933) 289.
[12] D.H. Coller, B.C. Vicente, S.L. Scott, Chem. Eng. Sci. 303 (2016) 182–193.
[13] G.F. Froment, K.B. Bischoff, Chemical Reactor Analysis and Design, Wiley, Hoboken, NJ, 1990.
[14] M.E. Davis, R.J. Davis, Fundamentals of Chemical Reaction Engineering, McGraw-Hill, New York, 2003.
[15] C.G. Hill, T.W. Root, Introduction to Chemical Engineering Kinetics and Reactor Design, Wiley, Hoboken, NJ, 2014.
[16] J.R. Kittrell, W.G. Hunter, C.C. Watson, AIChE J. 11 (1965) 1051–1057.
[17] K.J. Laidler, J. Chem. Educ. 49 (1972) 343.
[18] B. Peters, Chem. Eng. Sci. 71 (2012) 367–374.
[19] A. Lewis, M. Seckler, H. Kramer, G.M. van Rosmalen, Industrial Crystallization, Cambridge Press, 2015.
[20] D.E. Rosner, AIChE J. 9 (1963) 321–331.
[21] R.I. Masel, Chemical Kinetics and Catalysis, Wiley, New York, 2001.
[22] J.H. Espenson, Chemical Kinetics and Reaction Mechanisms, McGraw-Hill, New York, 1995.
[23] A. Aranda, D. Daele, G. Lebras, G. Poulet, Int. J. Chem. Kinet. 30 (1998) 249.
[24] E.A. Moelwyn-Hughes, The Chemical Statics and Kinetics of Solutions, Academic Press, London, 1971.
[25] J.I. Steinfeld, J.S. Francisco, W.L. Hase, Chemical Kinetics and Dynamics, Prentice Hall, Englewood Cliffs, NJ, 1989.
[26] R.L. Burden, J.D. Faires, Numerical Analysis, Prindle, Weber, and Schmidt, Boston, 1993.
[27] J. Christiansen, Adv. Catal. 5 (1953) 311–353.
[28] F.G. Helfferich, Kinetics of Multistep Chemical Reactions, Elsevier, Amsterdam, 2004.
[29] G.B. Marin, G.S. Yablonsky, Kinetics of Chemical Reactions: Decoding Complexity, Wiley-VCH, Weinheim, 2011.
[30] R. Aris, Am. Sci. 58 (1970) 420–428.
[31] J. Wei, C.D. Prater, AIChE J. 9 (1963) 77–81.
[32] M. Boudart, Kinetics of Chemical Processes, Prentice Hall, Englewood Cliffs, NJ, 1968.
[33] M. Boudart, G. Djega-Mariadassou, Kinetics of Heterogeneous Catalytic Reactions, Princeton University Press, Princeton, 1984.
[34] D.T. Gillespie, E.D. Hughes, C.K. Ingold, D.J. Millen, Nature 163 (1949) 599.
[35] B. Lewis, G. von Elbe, Combustion, Flames, and Explosions of Gases, Academic Press, London, 1987.
[36] L.S. Kassel, Kinetics of Homogeneous Gas Reactions, Reinhold, New York, 1932.

[37] J.H. Sullivan, J. Chem. Phys. 46 (1967) 73.
[38] F. Jensen, Introduction to Computational Chemistry, Wiley, West Sussex, 1999.
[39] C. Stegelmann, A. Andreasen, C.T. Campbell, J. Am. Chem. Soc. 131 (2009) 8077–8082.
[40] S. Kozuch, S. Shaik, Acc. Chem. Res. 44 (2011) 101–110.
[41] S.A. Trygubenko, D.J. Wales, J. Chem. Phys. 124 (2006) 234110.
[42] N.-V. Buchete, G. Hummer, J. Phys. Chem. B 112 (2008) 6057.
[43] P.G. Coxson, K.B. Bischoff, Ind. Eng. Chem. Res. 26 (1987) 1239–1248.
[44] N. Lempesis, D.G. Tsalikis, G.C. Boulougouris, D.N. Theodorou, J. Chem. Phys. 135 (2011) 204507.

Catalysis

"Catalysis is the acceleration of a chemical reaction, which proceeds slowly, by a foreign substance. ... after the end of the reaction the foreign body can again be separated from the field of the reaction..."

Ostwald, Z. Phys. Chem. (1894)

A catalyst accelerates a reaction without being consumed. Catalysts take many forms: water, acids, bases, ions, radicals, organometallic complexes, enzymes, metals, oxides, and sometimes even light can be a catalyst. Catalysts are integral to energy conversion and fuel production, to the chemicals industry, to waste and pollution mitigation, and in the production of fertilizers that feed the world's population (now an alarming 10^{10} people). Approximately 90% of all chemical processes use catalysts, and the resulting products account for 20% of the GDP in the United States of America [1]. Much of the research in computational chemistry has focused on understanding, discovering, and improving catalysts. As depicted in Figure 4.0.1, catalysts accelerate reactions by stabilizing transition states and/or by creating entirely new pathways and intermediates [2]. This chapter presents some basic principles and prototypical examples of acid/base catalysis, enzyme catalysis, and heterogeneous catalysis. The focus is primarily on phenomenological rate laws because first-principles techniques for computing rate constants are outlined in later chapters. However, we do attempt in places to show how first-principles evidence can help cross-examine mechanistic hypotheses.

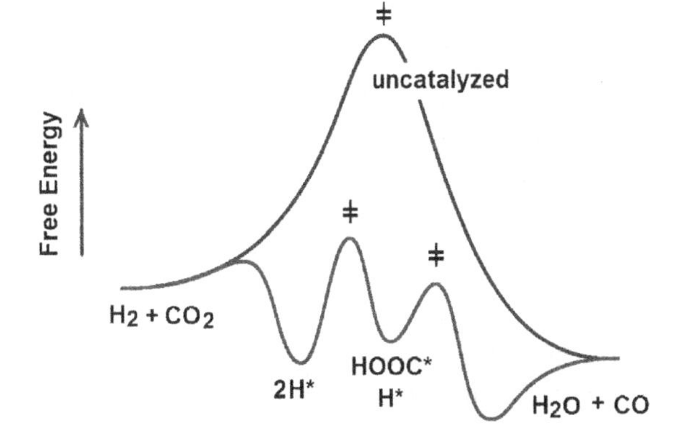

Figure 4.0.1: Catalysts lower activation barriers, usually by creating stable intermediates along the uncatalyzed reaction pathway or by enabling entirely new pathways. The figure depicts barriers and pathways for the water-gas shift reaction: $H_2 + CO_2 \rightarrow H_2O + CO$.

4.1 Acid-base catalysis

Many reactions in aqueous solution have rates that depend strongly on pH. Both Bronsted and Lewis acids can be potent catalysts, but usually acid-base catalysis refers to catalysis by Bronsted acids and bases, e.g. H^+, OH^-, a deprotonated molecule A^-, or a protonated molecule AH^+. The populations of these four reagents can be separately modulated by controlling the pH and by the addition of specific conjugate acids and bases. By measuring the rates carefully at different conditions, the contributions of each species to the reaction rate can be separately determined. Thus it is customary to write the rate of an acid or base catalyzed reaction as [3]

$$r = (k_0 + k_{H^+}[H^+] + k_{OH^-}[OH^-] + k_{HA}[HA] + k_{A^-}[A^-])[S] \tag{4.1.1}$$

where HA is the acid, S is the substrate/reactant, and where

k_0 = uncatalyzed rate constant
k_{H^+} = rate constant for specific acid catalysis (by H^+)
k_{OH^-} = rate constant for specific base catalysis (by OH^-)
k_{HA} = rate constant for general acid catalysis (by HA)
k_{A^-} = rate constant for general base catalysis (by A^-)

Analyses of acid/base catalyzed reactions will rely on the relationships between pK_a, pH, and concentrations of various species, e.g. $pK_a = -\Delta G^0_{HA \to A^- + H^+}/(2.303 k_B T)$ and the Henderson-Hasselbalch equation: $pH = pK_a + log_{10}([A^-]/[HA])$.[1] Models of acid/base catalysis treat protons, hydroxyls, and general acids or bases in almost the same way as other reagents. However, the protonation states rapidly interconvert, so protonation and deprotonation can usually be treated as quasi-equilibrium (QE) steps. The QE assumption leads to effective pseudo-rate constants which contain (and accurately describe) the pH dependence [3,5].

Acid/base pre-equilibrium

Protonation equilibria, particularly QE steps before the rate determining step are a common reason for pH-dependent rates. Let us consider first a substrate S which is first protonated to SH^+ and then reacts to form P

$$\begin{aligned} SH^+ &\overset{K_a}{\rightleftharpoons} S + H^+ \\ SH^+ &\xrightarrow{k} P + H^+ \end{aligned} \tag{4.1.2}$$

[1] Here we make no distinction between protons H^+, hydronium H_3O^+ ions, Zundel $H_5O_2^+$ ions, or Eigen $H_9O_4^+$ ion concentrations, but the distinction is important for *ab initio* calculations of $\Delta G^0_{HA \to A^- + H^+}$ [4].

The equilibrium between SH^+ and $S + H^+$ gives

$$K_a = \frac{[S][H^+]}{[SH^+]} \tag{4.1.3}$$

The rate is $k[SH^+]$ which can be written using equations (4.1.3) and $[S]_{tot} = [S] + [SH^+]$ as

$$r = \frac{k[H^+]}{[H^+] + K_a}[S]_{tot} \tag{4.1.4}$$

Equation (4.1.4) provides a simple example of pH dependence in the rate law. The reaction will be acid catalyzed for $[H^+] < K_a$, and it will become pH-independent for $[H^+] > K_a$.

Now consider a similar pre-equilibrium, but let the *deprotonated* species react to form the product P.

$$\begin{gathered} SH^+ \overset{K_a}{\rightleftharpoons} S + H^+ \\ S \xrightarrow{k} P \end{gathered} \tag{4.1.5}$$

Again equation (4.1.3) applies, but $r = k[S]$ which can be written as

$$r = \frac{kK_a}{[H^+] + K_a}[S]_{tot} \tag{4.1.6}$$

Equations (4.1.4) and (4.1.6) predict characteristics of the r vs. pH curve for reactions that follow an equilibrated protonation or deprotonation step. If the reaction consumes the protonated substrate, then the rate will be acid catalyzed for $[H^+] < K_a$ and pH independent for $[H^+] > K_a$. If the reaction consumes the deprotonated substrate, the rate will be base catalyzed for $[H^+] > K_a$, and approximately pH-independent for $[H^+] < K_a$. Note that in both cases the trend in r vs. pH shows a downward bend at the transition between dominant protonation states.

Obviously, increasing portions of the r vs. pH curve are base catalyzed with $r \sim [H^+]^{-1}$. They indicate a rate limiting transition state that is deprotonated relative to the predominant substrate. Decreasing portions of the r vs. pH curve are acid catalyzed with $r \sim [H^+]$. Flat portions of the r vs. pH curve are uncatalyzed and correspond to a transition state with the same protonation state as the predominant substrate. With some practice, even more insight can be extracted from features of the r vs. pH curve. Downward turns in the pH dependence correspond to a change in the dominant protonation state or to a change in the bottleneck along the sequence of reactions. Downward turns in r vs. pH cannot correspond to an overall change in the reaction mechanism. Loudon [5] gave the following helpful guidelines for deducing acid/base catalyzed mechanisms from the r vs. pH profile.

1. Cut the r vs. pH profile into regions separated by *upward* bends in the rate. Regions on opposite sides of an upward turn correspond to entirely different sequences of intermediates.

2. For each *downward* bend, examine whether the substrate can be protonated or deprotonated. If not, then seek a change in rate determining step at the pH corresponding to the downward bend.
3. As shown in the k vs. pH profiles of Figure 4.1.1, protonation steps in the mechanism correspond to downward turns that fall off after a flat portion, while deprotonation steps correspond to downward turns that level off after a rising portion.

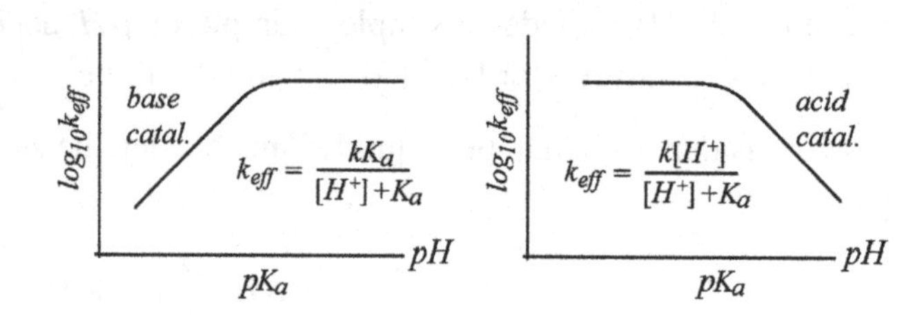

Figure 4.1.1: Typical pH-dependent rate constant that emerges when the reaction follows a pre-equilibrium interconversion between protonated and unprotonated forms of a reactant. Base catalysis emerges when the unprotonated reactant is consumed. Acid catalysis emerges when the protonated reactant is consumed.

The following detailed example shows how the rules of Loudon, in combination with *ab initio* calculations, can be used to understand pH-dependent kinetics.

Example: Asparagine deamidation

Asparagine deamidation exemplifies many of the key concepts in acid-base catalysis, and the reaction has several important consequences [6]. For example, asparagine deamidation is important in cataract formation, in the apoptosis response to chemotherapy, and in the degradation of therapeutic antibodies. As shown in Figure 4.1.2, the literature describes two pathways of asparagine deamidation to aspartate and iso-aspartate residues [7,8]. One pathway converts asparagine (Asn) directly to aspartate (Asp^-) at low pH. The pathway at intermediate to high pH converts asparagine to aspartate and iso-aspartate ($isoAsp^-$) via a succinimide (Suc) intermediate.

Figure 4.1.2: Observed reactants, intermediates, and products in asparagine deamidation. Of course, observed intermediates are valuable mechanistic clues. [From Peters and Trout, Biochemistry, 45, 5384-92 (2006).]

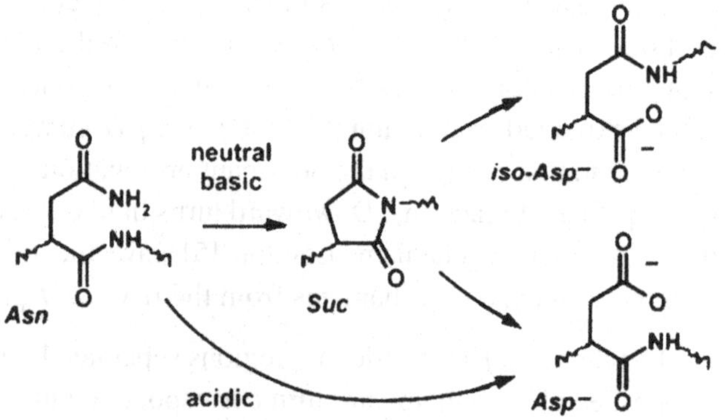

Figure 4.1.3 combines data from Patel and Borchardt [7] at low pH with data from Capasso et al. [8] at high pH. The pH dependence of the deamidation rate suggests a complicated mechanism with intermediates in multiple protonation states. Figure 4.1.3 contains two downward turns (at $pH = 7$ and $pH \approx 11$) and two upward turns (at $pH = 3.5$ and 8).

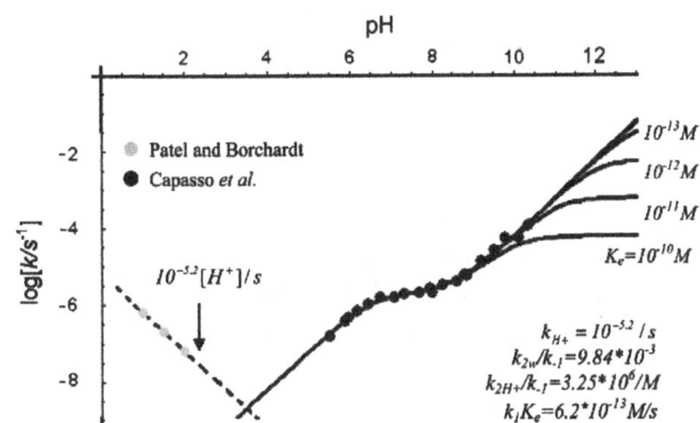

Figure 4.1.3: pH**-dependence of the deamidation rate for a model pentapeptide. Note acid catalyzed, base catalyzed, and** pH**-independent regions. [From Peters and Trout, Biochemistry 45, 5384-92 (2006).]**

Based on Loudon's rules and computed reaction pathways, we can infer that the reaction occurs via three different pathways depending on the pH. Figure 4.1.4 shows a network of likely intermediates and pathways arranged with protonated intermediates on the left (low pH structures) and deprotonated intermediates on the right (high pH structures). As pH increases, protonated species should become scarce causing pathways involving neutral and deprotonated pathways to dominate the kinetics. For $pH < 3.5$ the reaction mechanism follows the acid catalyzed pathway on the left side of Figure 4.1.4. From $3.5 < pH < 6$, the mechanism follows the path $Asn \to Asn^- \to \ddagger \to Tet^- \to Tet \to Suc$ where $\ddagger$ indicates the bottleneck location. From $6 < pH < 8$, the sequence of intermediates is the same, but the bottleneck shifts to a later (and neutral) transition state: $Asn \to Asn^- \to Tet^- \to Tet \to \ddagger \to Suc$. The shift in bottleneck location is driven by increasing stability of the $Asn^- \to \ddagger \to Tet^-$ transition state as pH increases. Beyond $pH = 8$, the mechanism follows the rightmost pathways in the network: $Tet^- \to SucOH^- \to Suc$ rather than $Tet^- \to Tet \to Suc$. Hence the upward turn in the *rate* vs. pH curve at $pH = 8$. Finally, beyond $pH = -logK_e \approx 11$, both $[Tet^-]$ and $[Asn^-]$ are dominant protonation states and therefore the rate is no longer base catalyzed.

How do we test whether these are viable mechanistic hypotheses? First, we can derive a rate law in which a single pseudo-rate constant absorbs all of the pH-dependence. If the mechanism is correct, the resulting rate law should provide a reasonable fit to each and every section of the measured r vs. pH curve. To derive the overall rate law, we assume QE between the neutral (Asn) and deprotonated (Asn^-) forms which gives $K_e = [Asn^-][H^+]/[Asn]$. Also using $[Asn]_{tot} = [Asn] + [Asn^-]$ gives

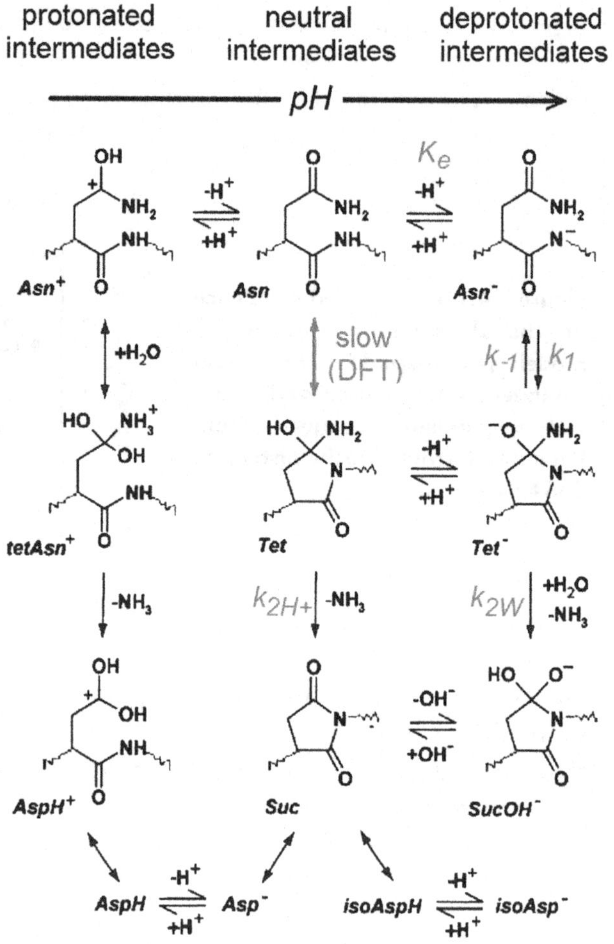

Figure 4.1.4: Protein deamidation intermediates arranged by prevalence as a function of pH. [From Peters and Trout, Biochemistry, 45, 5384-92 (2006).]

$$[Asn] = [Asn]_{tot}[H+]/(K_e + [H^+])$$

and

$$[Asn^-] = [Asn]_{tot} K_e/(K_e + [H^+])$$

Asn^- converts directly to Tet^-, a deprotonated form of the tetrahedral intermediate. The deprotonated form can take a proton from solution to become the neutral form again if the pH is sufficiently low. Again assuming fast quasi-equilibrium interconversion between Tet and Tet^-, we write $K_{Tet} = [Tet^-][H^+]/[Tet]$ and $[Tet]_{tot} = [Tet] + [Tet^-]$ to obtain

$$[Tet] = [Tet]_{tot}[H^+]/(K_{Tet} + [H^+])$$

and

$$[Tet^-] = [Tet]_{tot} K_{Tet}/(K_{Tet} + [H^+])$$

Now applying the PSSA to the unobserved tetrahedral intermediates gives

$$0 \approx -k_{2H^+}[Tet] - k_{2W}[Tet^-] - k_{-1}[Tet^-] + k_1[Asn^-] \tag{4.1.7}$$

This can be rewritten as

$$\{k_{2H^+}[H^+] + k_{2W}K_{Tet} + k_{-1}K_{Tet}\} \frac{[Tet]_{tot}}{K_{Tet} + [H^+]} = \frac{k_1 K_e}{K_e + [H^+]}[Asn]_{tot} \tag{4.1.8}$$

At low concentrations of ammonia [6], the reaction from Tet to Suc is irreversible. Furthermore, succinimide (Suc) is far more stable than the hydroxylated succinimide species ($SucOH^-$), so $SucOH^-$ should quickly and irreversibly convert to a neutral succinimide residue. Now the rate expression can be reduced entirely to a function of $[Asn]_{tot}$ and $[H^+]$:

$$\begin{aligned} \mathcal{r} &= k_{2W}[Tet^-] + k_{2H^+}[Tet] \\ &= [Tet]_{tot} \left\{ \frac{k_{2W}K_{Tet}}{K_{Tet} + [H^+]} + \frac{k_{2H^+}[H^+]}{K_{Tet} + [H^+]} \right\} \\ &= \frac{k_1 K_e [Asn]_{tot}}{K_e + [H^+]} \cdot \frac{k_{2W}K_{Tet} + k_{2H^+}[H^+]}{k_{2W}K_{Tet} + k_{2H^+}[H^+] + k_{-1}K_{Tet}} \\ &= \frac{k_1 K_e}{K_e + [H^+]} \left\{ 1 - \frac{1}{1 + (k_{2W} + k_{2H^+}[H^+]/K_{Tet})/k_{-1}} \right\} [Asn]_{tot} \end{aligned}$$

The last expression contains four independent equilibrium and rate parameters because K_{Tet} can be absorbed into k_{2H^+}. According to this mechanism, asparagine deamidation is first order in total asparagine, but has a complicated dependence on pH. The pH dependence can be absorbed into an effective rate constant

$$k_{eff} = \frac{k_1 K_e}{K_e + [H^+]} \left\{ 1 - \frac{1}{1 + (k_{2W} + k_{2H^+}[H^+]/K_{Tet})/k_{-1}} \right\} \tag{4.1.9}$$

As shown in Figure 4.1.3, expression (4.1.9) fits the data well. However, several fit parameters were used: $k_1 K_e$, K_e, k_{2W}/k_{-1}, and k_{2H^+}/k_{-1} and there could be other mechanisms that equally fit the $\mathcal{r}$ vs. pH data.

How do we know if the agreement between the data and the derived-and-fitted rate law is fortuitous? *Ab initio* calculations provide additional independent tests of the proposed mechanism. For example, which steps are limiting at each pH? Are the assumed pseudo-steady-state intermediates sufficiently rare according to free energy calculations? Do the *predicted* pH values where changes in the mechanism occur adequately match the locations and types of observed bends in n vs. pH? *Ab initio* calculations can answer each of these questions, and they also provide transition state structures for the proposed mechanism. Transition states yield additional insights and testable predictions. For example, Peters and Trout found that the rate determining transition state at some pH values involves explicit water, while the rate determining transition state at other pH values does not. Steric effects in the transition states involving explicit water explained why sequence design rules for preventing deamidation in therapeutic antibodies had been ineffective at the pH values of therapeutic formulations [6] (see Figure 4.1.5).

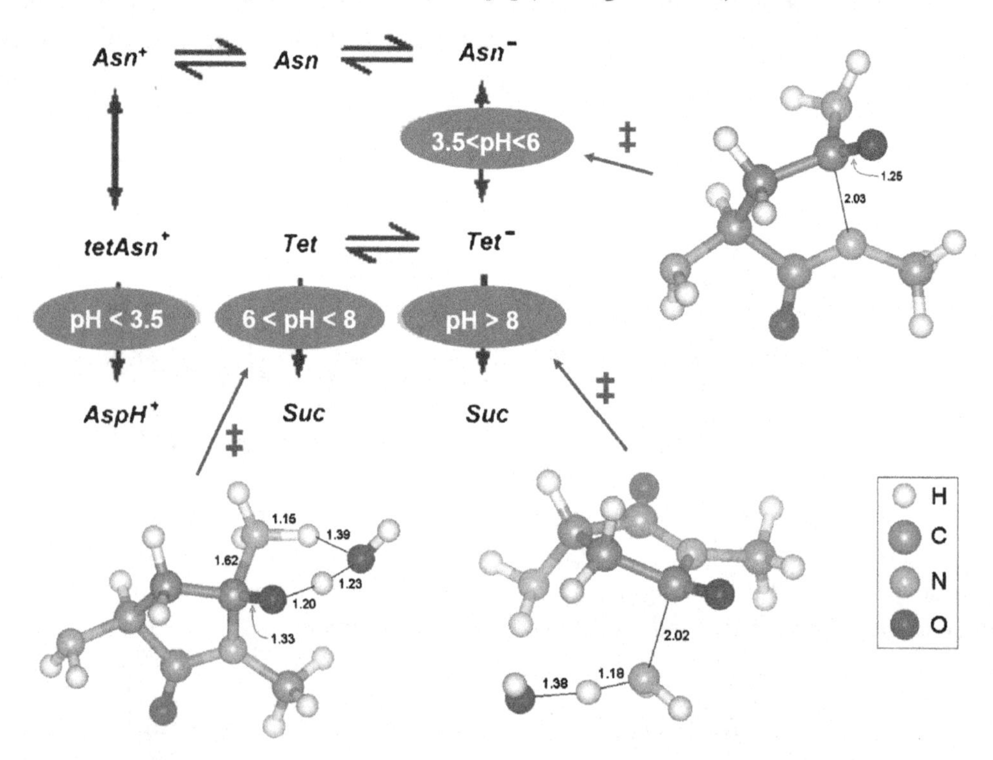

Figure 4.1.5: A simplified reaction network showing only key intermediates and the rate limiting transition state along the dominant path in each pH range. [From Peters and Trout, Biochemistry 45, 5384-92 (2006).]

As seen in the deamidation example, many reactions in aqueous solution are only feasible because of protonation and deprotonation steps. In this sense, H^+ or OH^- ions are catalysts because they create new pathways with lower barriers, without being consumed in the process. The above example also shows how experimental data, phenomenological modeling, and *ab initio* calculations can be combined to elucidate reaction mechanisms. Later sections will elaborate on computational methods that can (1) find structures of intermediates and transition states, (2) estimate equilibrium constants and rate constants, and (3) identify the rate limiting transition states at each condition.

4.2 Enzymes

Enzymes are the catalysts which make life possible. They can accelerate some reactions by as much as 10^{17}-fold [9], often with remarkably high selectivity [10]. They operate at relatively mild conditions. They distinguish between enantiomers. Many enzymes can "intelligently" switch on or off in the presence of signaling molecules or in a specific pH range. Enzymes demonstrate that catalysis with cheap metals like iron, manganese, and copper is feasible. Most importantly, enzymes are agents of biological control – organisms control their metabolism by regulating the amount of each enzyme through transcription and translation. Finally, enzymes are a multibillion dollar industry [11,12]. For all of these reasons, enzymes have inspired decades of work in catalysis and in theoretical chemistry.

Enzymes are proteins[2] that fold to create an active site comprised of a few precisely arranged amino acid residues. Some enzymes are entirely made from amino acids, but often a metal ion and its ligands reside in the active site. For example, catalase contains iron within a heme group, methane monooxygenase contains a dinuclear copper center, and carbonic anhydrase contains a zinc ion [14].

In the simplest possible case, an enzyme reaction involves a substrate (reactant) binding step, a chemical step, and a product release step. Fischer proposed that the substrate is like a key that fits precisely into the active site presented by the enzyme [15]. The lock-and-key picture explains substrate specificity, but not activity. As an alternative to the lock-and-key explanation, Koshland proposed the induced fit hypothesis in which the enzyme and substrate are modified together upon binding [16]. The active site binds to, exerts forces upon, and polarizes or protonates the substrate in ways that distort the substrate at least part of the way towards its transition state. As these distortions activate the substrate, they also alter the enzyme structure, at least near the binding site.

Enzymes employ elegant tricks of physical chemistry to lure substrates into their active sites. For example, nonpolar substrates with only weak van der Waals non-bonded interactions require enzymes with hydrophobic residues near the active sites [17]. Other enzymes have active

---

[2] Some enzymes are comprised entirely or partly from RNA [13]. Many enzymes also incorporate sugars and other organic molecules.

sites that bind one or several water molecules when the substrate is not present. When the substrate binds, water molecules are displaced by the substrate. The positive entropy of water release helps to offset the entropic losses upon substrate binding [18].

Enzymes must somehow lower activation barriers like other catalysts. Many enzymes effectively stabilize the structure of the uncatalyzed transition state [19–21]. This simple idea has recently guided the computational design of working enzymes for reactions that have no naturally occurring enzymes [22,23]. Note, however, that enzymes may also lower activation barriers by enabling an entirely different mechanism from that of the uncatalyzed reaction [24].

Enzymes can have amazing *specificity*, but their *versatility* can be just as remarkable. The combined requirements of substrate binding and substrate activation explain why some enzymes can act upon large families of similar substrates while others very specifically act upon one substrate [25]. For example, a substrate impostor might be sufficiently similar to the ideal substrate to bind in the active site, but the impostor may not induce correctly aligned interactions to stabilize its transition state. In this case the enzyme would only work on the specific substrate and not on an impostor even when the differences are only in peripheral functionalities. On the other hand, some enzymes react with a specific functional group regardless of the rest of the molecule. For example, there are endonucleases which cut DNA indiscriminately and there are also restriction endonucleases which cut DNA only at specific sequences.

Phenomenological enzyme catalysis models

A rate law can be developed for the enzyme mechanism shown in Figure 4.2.1. To obtain the most general rate law possible we should assume that all steps are reversible. The resulting rate law would contain one forward term and one backward term, with both being rather messy expressions. If the overall process $\mathbf{A} \rightarrow \mathbf{B}$ is irreversible, then at least one step must be irreversible. Taking only the product expulsion step to be irreversible:

$$\mathbf{E} + \mathbf{A} \underset{k_{-1}}{\overset{k_1}{\rightleftarrows}} \mathbf{EA} \underset{k_{-2}}{\overset{k_2}{\rightleftarrows}} \mathbf{EB} \overset{k_3}{\rightarrow} \mathbf{E} + \mathbf{B} \tag{4.2.1}$$

The PSSA can be applied to each state of the enzyme cycle. In this case, the PSSA is justified not by the standard free energies of the enzyme complexes, but instead by the relative scarcity of the enzyme. The substrate is usually at much higher concentration than the enzyme, so that each enzyme "turns over" many times before the substrate or product concentrations change appreciably. Thus the substrate and product concentrations can be treated as fixed parameters while deriving the rate law for the enzymatic reaction. Applying the PSSA to [**E**] and [**EA**]

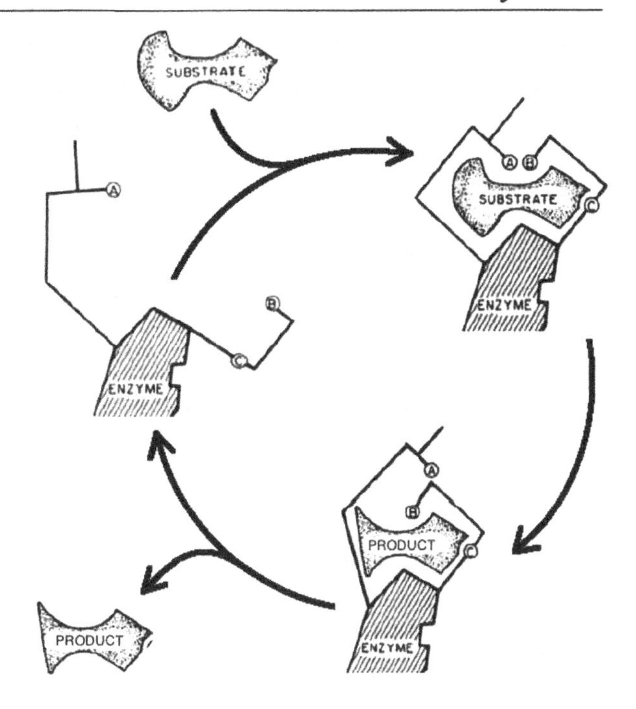

Figure 4.2.1: A simple enzymatic reaction cycle illustrates that both the enzyme and the substrate are modified upon binding. Groups A, B, and C represent enzyme residues that interact with the substrate upon binding. Product expulsion also involves conformational changes that restore the naked enzyme configuration. [Adapted from Koshland, Angew. Chem. Int. Ed. 33, 2375-78 (1994).]

gives

$$\mathcal{R}_{\mathbf{E}} = -k_1[\mathbf{E}][\mathbf{A}] + k_{-1}[\mathbf{EA}] + k_3[\mathbf{EB}] \approx 0 \tag{4.2.2}$$

and

$$\mathcal{R}_{\mathbf{EA}} = k_1[\mathbf{E}][\mathbf{A}] - k_{-1}[\mathbf{EA}] - k_2[\mathbf{EA}] + k_{-2}[\mathbf{EB}] \approx 0. \tag{4.2.3}$$

A third pseudo-steady-state equation for $[\mathbf{EB}]$ could be written in the same manner. However, the equation $\mathcal{R}_{\mathbf{EB}} \approx 0$ would be linearly dependent with the equations $\mathcal{R}_{\mathbf{EA}} \approx 0$ and $\mathcal{R}_{\mathbf{E}} \approx 0$. Even without doing the algebra, we know that the three pseudo-steady-state equations must be linearly independent. Conservation of enzymes – a requirement for calling the enzyme a catalyst – implies that $\mathcal{R}_{\mathbf{E}} + \mathcal{R}_{\mathbf{EA}} + \mathcal{R}_{\mathbf{EB}} = 0$. The enzyme conservation equation gives

$$[\mathbf{E}]_{tot} = [\mathbf{E}] + [\mathbf{EA}] + [\mathbf{EB}] \tag{4.2.4}$$

with $[\mathbf{E}]_{tot}$ being a constant.

After some algebra, we can use equations (4.2.2)–(4.2.4) to write the rate $\mathcal{R} = k_3[\mathbf{EB}]$ in terms of $[\mathbf{E}]_{tot}$, $[\mathbf{A}]$, and elementary rate constants. The result is

$$\mathcal{R} = \frac{k_2 k_3 (k_2 + k_{-2} + k_3)^{-1} [\mathbf{E}]_{tot} [\mathbf{A}]}{[\mathbf{A}] + \frac{k_{-1}k_{-2} + k_{-1}k_3 + k_2 k_3}{k_1(k_2 + k_{-2} + k_3)}}. \tag{4.2.5}$$

That appears to be a complicated rate law, but actually it is a remarkable simplification. Equation (4.2.1) had five elementary rate parameters, but the overall rate law contains just two. In the numerator, there is the maximum rate parameter

$$\nu_M = k_2 k_3 (k_2 + k_{-2} + k_3)^{-1} \quad (4.2.6)$$

and in the denominator there is the Michaelis constant

$$K_M = \frac{k_{-1}k_{-2} + k_{-1}k_3 + k_2 k_3}{k_1(k_2 + k_{-2} + k_3)} \quad (4.2.7)$$

so that the overall rate expressed as a turnover frequency, i.e. $\nu = \mathcal{r}/[\mathbf{E}]_{tot}$, is

$$\nu = \frac{\nu_M [\mathbf{A}]}{[\mathbf{A}] + K_M}. \quad (4.2.8)$$

Equations (4.2.6) and (4.2.7) are the most general definitions of ν_M and K_M in terms of elementary rate parameters for the sequence of steps in mechanism (4.2.1). Contrary to the results of more simplified derivations, ν_M and K_M do not necessarily correspond to specific rate constants or equilibrium constants for any one step [10,26,27].

The turnover frequency switches from first order in [**A**] when $[\mathbf{A}] \ll K_M$ to zeroth order in [**A**] when $[\mathbf{A}] \gg K_M$. ν_M and K_M can be obtained from a slope-intercept analysis of rate data. Equation (4.2.8) can be written as

$$\frac{1}{\nu} = \frac{1}{\nu_M} + \frac{K_M}{\nu_M [\mathbf{A}]} \quad (4.2.9)$$

which motivates the Lineweaver-Burk plot, or as done in Eadie-Hofstee plots

$$\nu = \nu_M - K_M \nu / [\mathbf{A}] \quad (4.2.10)$$

Schematic examples of these plots are shown in Figure 4.2.2. Note that the $1/[\mathbf{A}]$ axis in a Lineweaver-Burk plot distorts the spacing of data which has been collected at regular [**A**] intervals. Additionally, the $1/\nu$ axis in a Lineweaver-Burk plot exaggerates errors in the rate where the rate is small. The Eadie-Hofstee analysis tends to provide more accurate estimates of K_M and ν_M.

Some simplifications lead to well known enzymatic rate laws (and also to some loss of generality). For example, the Briggs-Haldane mechanism

$$\mathbf{E} + \mathbf{A} \underset{k_{-1}}{\overset{k_1}{\rightleftarrows}} \mathbf{EA} \overset{k_2}{\rightarrow} \mathbf{E} + \mathbf{B} \quad (4.2.11)$$

Figure 4.2.2: (Left) A Lineweaver-Burk plot. (Right) An Eadie-Hofstee plot.

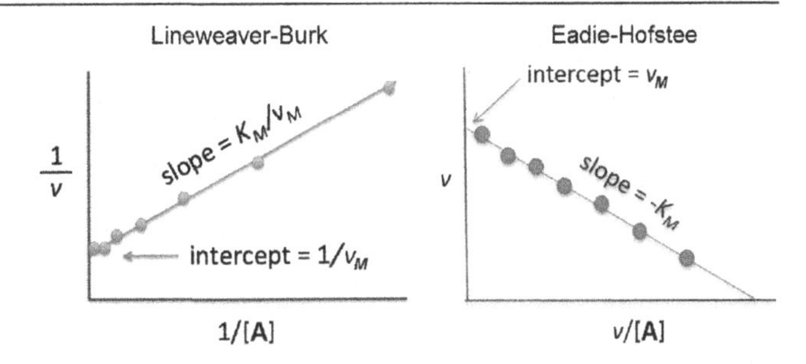

is recovered when $k_3 \gg k_2$ and $k_3 \gg k_{-2}$, i.e. when the product P is expelled as soon as the chemical step $\mathbf{EA} \rightarrow \mathbf{EB}$ occurs, with no chance for back reaction [28]. In this case, the rate law is again $v = v_M[\mathbf{A}]/([\mathbf{A}] + K_M)$ but the constant K_M simplifies to

$$K_M \approx \frac{k_{-1} + k_2}{k_1} \tag{4.2.12}$$

and

$$v_M \approx k_2 \tag{4.2.13}$$

If we further assume that the chemical step is much slower than substrate binding and unbinding, i.e. that $k_2 \ll k_1[\mathbf{A}]$ and $k_2 \ll k_{-1}$, then the general mechanism (4.2.1) simplifies to the Michaelis-Menten mechanism [29].

$$\mathbf{E} + \mathbf{A} \rightleftharpoons \mathbf{EA} \xrightarrow{k_2} \mathbf{E} + \mathbf{B} \tag{4.2.14}$$

The rate law, yet again has the form $v = v_M[\mathbf{A}]/([\mathbf{A}] + K_M)$. However, the Michaelis constant K_M becomes the equilibrium substrate (un)binding constant k_{-1}/k_1. The maximum turnover frequency is again $v_M \approx k_2$. The Michaelis-Menten mechanism is widely known, but it makes rather narrow assumptions that may not be true even when the resulting rate law appears to fit kinetic data [26]. Note that all three variations of the enzyme mechanisms considered above give a rate law with the same overall form. They can only be distinguished if one experimentally or computationally obtains the relative sizes of k_3, k_2, k_{-2}, k_1, and k_{-1}.

Enzymatic reactions cannot exceed equilibrium limitations, but biochemical inhibition mechanisms can limit product concentrations to values well below equilibrium. For example, a product that reaches a critical concentration may bind to the enzyme and turn off its activity. Inhibition can also be affected by molecules other than products. In one common mode of inhibition, an inhibitor competes with the substrate in binding to the active site. The competitive inhibition mechanism is shown in Figure 4.2.3.

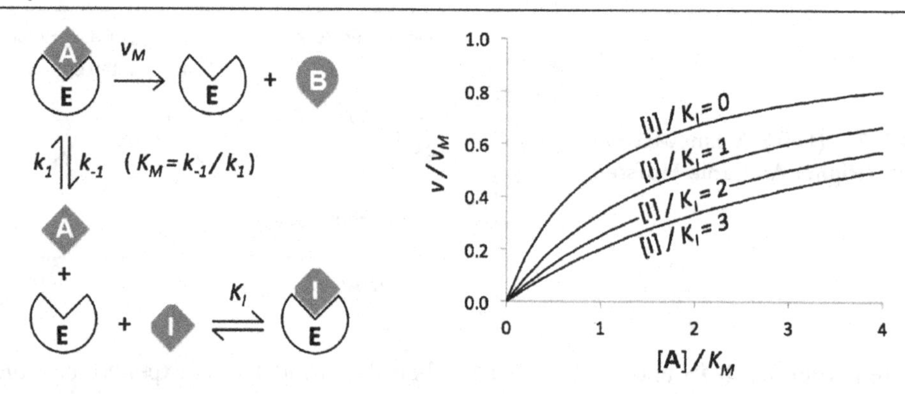

Figure 4.2.3: (Left) In competitive inhibition, an inhibitor binds to the active site and thereby blocks an enzymatic reaction. The rate law (given as a turnover frequency) in the text corresponds to the binding equilibria and rate constants shown. (Right) As the dimensionless inhibitor concentration $[\mathrm{I}]/K_{\mathrm{I}}$ increases, it takes higher and higher $[\mathrm{A}]/K_M$ to achieve the same turnover frequency.

The rate law for competitive inhibition is [30]

$$\nu = \frac{\nu_M[\mathbf{A}]}{[\mathbf{A}] + K_M(1 + [\mathbf{I}]/K_{\mathbf{I}})} \tag{4.2.15}$$

When the inhibitor is absent, the usual Michaelis-Menten expression is recovered. The approach to the maximum turnover frequency as a function of $[\mathbf{A}]/K_M$ is shown at different inhibitor concentrations in Figure 4.2.3. Note that the inhibitor **I** may be the product itself, a feedback strategy that biology exploits to control metabolite concentrations.

How large are the rate constants invoked by these mechanisms? A detailed discussion will require free energies of transition states, stability of intermediates, and sometimes an understanding of conformational dynamics [31] and tunneling [32]. However, analyses of rate data suggests that most enzymes have Michaelis constants in the range $10^{-7}M \leq K_M \leq 10^{-1}M$ and maximum turnover frequencies in the range $1 \leq \nu_M \leq 10^5$. There are some "perfect enzymes" for which ν_M is so fast that the enzyme operates in the diffusion controlled limit [9]. For a small molecule substrate and an enzyme of typical size in water, the diffusion controlled rate constant is on the order $10^8 M^{-1}s^{-1}$. Examples of enzymes that operate near the diffusion control limit include catalase which decomposes hydrogen peroxide with $\nu_M/K_M = 4.0 \cdot 10^8 M^{-1}$ and fumarase which hydrates fumarate with $\nu_M/K_M = 1.6 \cdot 10^8 M^{-1}$ [30].

4.3 Heterogeneous catalysis

Before discussing heterogeneous catalysis, note that we have only discussed a few special homogeneous catalysts: acids, bases, and enzymes. Homogeneous catalysts also include organometallic complexes with important applications in polymerization, olefin metathesis, hydroformylation, hydrogenation, and carbonylation [33]. These catalysts are homogeneous in the sense that they are co-dissolved in solution with the reactants. Homogeneous catalysts have a number of important advantages: (1) the intimate mixture of catalyst with reactants eliminates mass transfer limitations, (2) all of the catalyst sites are the same, so they are relatively easy to model and characterize, and (3) the catalysts are often active under mild conditions. However, recovering and recycling a homogeneous catalyst from the product mixture is often difficult or impossible. This is a serious problem for those homogeneous catalysts that are based on expensive metals like Ir, Rh, Pt, or Re [33].

Heterogeneous catalysts are solids that facilitate reactions between reactants in a fluid phase. Heterogeneous catalysts are readily incorporated in continuous flow reactors without being lost among the fluid mixture of reactants and products. This practical advantage has made them important for petroleum refining, chemical manufacture, exhaust treatment, and many other processes that support our modern lifestyles. Peeking at one of these catalysts after operation might leave some doubts about whether it is a true catalyst. They slowly become coked and fouled during operation, but nevertheless good heterogeneous catalysts remain active for hours or even years during which they might facilitate millions of reactions per active site.

Catalysis research seeks materials with improved activity per atom, improved selectivity, reduced reliance on rare metals, high surface area, easy regeneration, and low susceptibility to deactivation even under harsh conditions with many impurities and potential poisons in the feed stream [11,34–36]. The main considerations in catalyst design are

1. **turnover frequency** (TOF): – the number of reactions per site per unit time. Our discussion focuses on the TOF during steady-state operation.
2. **turnover number** (TON): – the typical number of reactions per site before the catalyst deactivates, i.e. the TOF multiplied by the lifetime. Studies of catalyst robustness should consider the TON under *realistic* operating conditions [37].
3. **recovery, regeneration, and replacement costs**: Catalysts can be a significant portion of capital and operating costs if they are made from rare metals, especially if they are difficult to recover and regenerate.

All three considerations are intertwined with reactor design and process design considerations. Typical catalysts have several components: the support, the catalytic agent, and sometimes promoters. The catalyst material can take various forms:

1. Metal surfaces, e.g. the surface of an electrode [38,39] or nanoparticles dispersed on a support [40].
2. Metal oxides and sulfides, e.g. TiO_2 photocatalysts or MoS_2 for desulfurization [41,43].

3. Shape selective catalysts, e.g. the zeolites that are used in hydrocarbon cracking [44].
4. Isolated metal atoms on amorphous silica, silica-alumina, etc. [45].

Much of catalysis research concerns the synthesis and properties of metal nanoparticles. These provide a large surface area per metal volume – an important factor because metal nanoparticle catalysts often inorporate rare metals. Another reason for creating metal nanoparticles is that some typically inert metals like gold can become highly active when prepared as nanoparticles or as clusters of a few atoms [46,47]. In this regard, catalysis is one of the foremost applications of nanotechnology.

The support is usually a cheap and thermally robust material like silica (SiO_2) or alumina (Al_2O_3). Catalytic reactions occur at the fluid-solid interface, so the support usually has a porous morphology to provide a large surface area for contact between the reactants and the catalyst. The support also immobilizes the catalyst and protects it from sintering and deactivation during operation [48,49]. In some cases, the support may also function as a solid acid, or help capture adsorbates, or help activate the catalyst through charge transfer processes [50]. The support can indirectly influence catalysis through its effect on the structure of catalyst particles, e.g. the exposed surfaces of metal nanoparticles depends on the degree of wetting between the support and the different surfaces of the metal. Research on heterogeneous catalysis often uses the techniques and idealized materials of surface science, but real catalysts are tremendously complicated.

Phenomena at the reactor [51,52], pellet [52–54], and pore [54] scale are important for the overall performance of a catalytic reactor. Figure 4.3.1 depicts some of the processes occurring at each scale. Heat transfer limitations can cause heat generated by exothermic reactions to build up in the core of a reactor or catalyst pellet. The elevated temperatures can dramatically *accelerate* reaction rates [55,57]. Mass transfer limitations can deprive the core of a catalyst pellet or entire parts of a reactor of reactants, and thereby dramatically *lower* reaction rates [55, 56,58]. This book focuses on intrinsic kinetics, but the importance of transport considerations in catalysis cannot be overstated. Before using *ab initio* calculations to model the intrinsic reactivity of a catalyst, one should always ensure that the experimentally reported rates reflect the actual reactivity and not transport coefficients.

While transport does influence reaction rates, we rarely (if ever) need to explicitly model heterogeneous reactions and boundary layer transport in one molecular simulation or in a multiscale simulation. Reactions involve very short length scales ($\sim$ 1nm) while the concentrations in a boundary layer change over a length scale of ℓ $\sim$μm-mm depending on flow characteristics. Thus a surface reaction sees an apparently homogeneous concentration at the solid-fluid surface. Similarly, reactant molecules out in the fluid see only the local gradient driving their diffusion. The diffusion in the boundary layer and surface reactions are only coupled to each other through a diffusion-flux-equals-rate-per-area boundary condition. First, we compute the

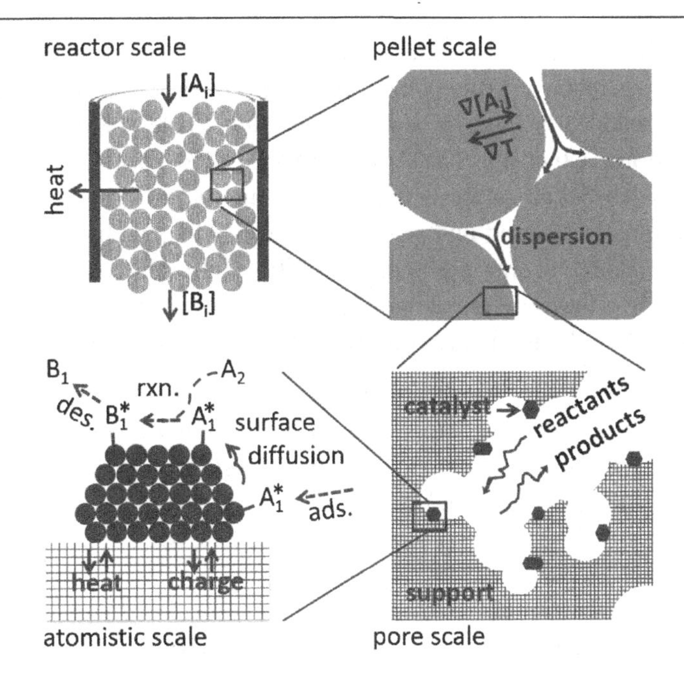

Figure 4.3.1: Heterogeneous catalysis involves phenomena at many different scales. At the reactor scale, heat and mass transfer are accelerated by fluid streamlines that split and recombine as they flow around pellets. For an exothermic reaction, the core of a catalytic reactor develops a high temperature zone where conversion advances quickly. At the pellet scale, the overall rate may become vanishingly small if the reactants are consumed before reaching sites deep in the pellet interior. At the pore scale, reactants must diffuse from the pore mouth to the active sites. The pore diameter has a strong effect on the rate and mechanism of diffusion processes. This book primarily concerns rate constants associated with phenomena at the atomistic scale: adsorption, surface reaction, and desorption.

rate law for the surface reaction as a function of the local surface concentration. Second, we equate the rate of reaction with the rate of transport from the bulk via a reaction-diffusion boundary condition for the surface concentration. This self-consistent procedure determines the surface concentrations and the overall rate, and it accounts for both the intrinsic kinetics and the transport resistances.

Modeling heterogeneous catalysts

Chemical engineers have developed excellent phenomenological models that account for transport to and from a heterogeneous catalyst [52,56,59]. Therefore, modern computational studies mostly focus on the "atomistic scale" processes in Figure 4.3.1. *Ab initio* computational frameworks (and also most phenomenological theories) in catalysis rely upon many simplifying assumptions about the nature of the catalyst. The foremost assumption – made in nearly all theoretical analyses – is that every site is the same [60]. This assumption is approximately true for some materials, and clearly false for some others. Nevertheless, the elegant simplifications that follow from uniform active sites have inspired a fruitful, decades-long quest to synthesize, characterize, and model special catalysts that do have highly uniform sites. For example, academic research has largely focused on crystalline materials like zeolites, highly ordered metal oxides, and well-defined metal surfaces under carefully controlled conditions. Studies of these materials have been prioritized deliberately to gain mechanistic understanding, not necessarily to create good catalysts under industrial operating conditions.

Given a model of the active site, remarkable progress can be made with microkinetic models, *ab initio* calculations, and fairly straightforward applications of harmonic transition state theory (see Chapter 10). *Ab initio* computational models of catalysis primarily use periodic slab models or cluster models as shown in Figure 4.3.2.

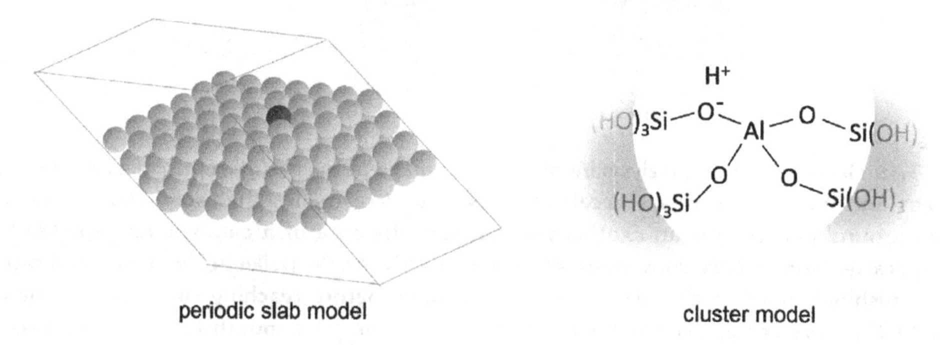

Figure 4.3.2: (Left) Periodic slabs are used for modeling crystalline catalysts with very small unit cells and for metallic or semiconducting materials. The dark atom depicts a dopant. The cell must be large enough that dopants and surface species do not interact with their periodic images. (Right) Cluster models are used for insulating catalyst materials with large unit cells. The example depicts a small cluster model of the Bronsted acid site in a zeolite. OH groups in the shaded peripheral region are typically fixed in space to mimic the approximately rigid connections to the surrounding crystal.

Periodic slab models are useful for crystalline materials and they are essential for conducting materials and semiconductors. Periodic models also avoid the need to create artificial capping atoms and peripheral atom constraints. On the other hand, periodic models must be large enough that adsorbates do not interact with their own periodic images, and thick enough to resemble the bulk material in lattice spacing and electronic structure. Other resources provide recommendations on the ideal slab thickness, periodic box dimensions, basis sets, and functionals for periodic density functional theory calculations [46,61].

Cluster models are useful for atomically dispersed metal atoms or small clusters on insulating supports like silica, zeolites, or silica-alumina. Cluster models often have only 10–100 atoms, so calculations with highly accurate model chemistries are affordable. Moreover, the lack of periodic boundary conditions allows the use of atom-centered Gaussian bases which describe the electron density more compactly than plane wave bases. Other resources provide recommendations for cluster modeling procedures: capping atoms, peripheral constraints, embedding effects, and model chemistries [62–64].

Looking to the future, we must use the lessons from idealized materials and models to cross the "pressure-gap" (from mTorr to high pressure conditions) and the "materials-gap" (from single crystal surfaces to imperfect materials and supports) [65–68]. High temperatures and various adsorbates can induce significant changes in surface structure and surface chemistry. Figure 4.3.3 shows temperature induced changes in nanoparticles of rhodium supported on amorphous silica (for NO_x reduction in catalytic converters). Adsorbates can accelerate ripening [69], induce reconstruction [70], oxidize metal surfaces [71], facilitate sintering [72], and promote surface-segregation of alloy components [73,74].

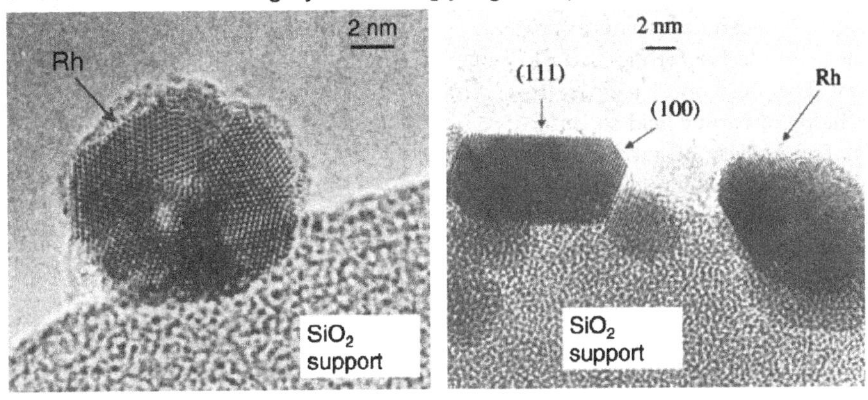

Figure 4.3.3: Rhodium nanoparticles after oxidation show massive defects including internal twinning, grain boundaries, and rough surfaces. At high temperature reducing conditions the faceted and internally annealed nanoparticles are recovered. Such changes in surface structure and chemistry are among the foremost challenges in modeling catalysis. [From Logan et al., J. Phys. Chem. 95, 5568-74 (1991) and Datye, Topics in Catalysis 13, 131-38 (2000).]

Sabatier's principle

Sabatier's principle says that bonds formed between the catalyst and its reactant should be just right, not too strong and not too weak. Reactants which bind too strongly to the catalyst will form highly stable non-reactive adsorbates/complexes and poison the catalyst. Reactants that bind too weakly will leave the surface devoid of reactants, again resulting in an inactive catalyst.

In addition to modulating coverage, the bonds that form between a reactant and the catalyst directly "activate" the reactant by weakening its other bonds. These weakened bonds are more easily broken through surface reactions with other adsorbates. In many cases the weakened bonds can be directly observed by techniques like *in situ* infrared (IR) spectroscopy [75].

One of the clearest consequences of Sabatier's principle is the volcano plot. A volcano plot for a reaction shows catalyst activity (for one catalytic reaction) vs. heat of reactant adsorption across a series of different catalysts [76,77]. The volcano plot typically has a sharp peak corresponding to the catalysts whose interactions with the reactant are just right. Figure 4.3.4 shows the volcano plot for formic acid decomposition on a variety of metals. More recent analyses suggest that the activities of the catalysts represented in the plot are affected by C and O adsorbates – a reminder that the actual surface chemistry responsible for the volcano plot may be more complex than that envisaged for a clean metal surface.

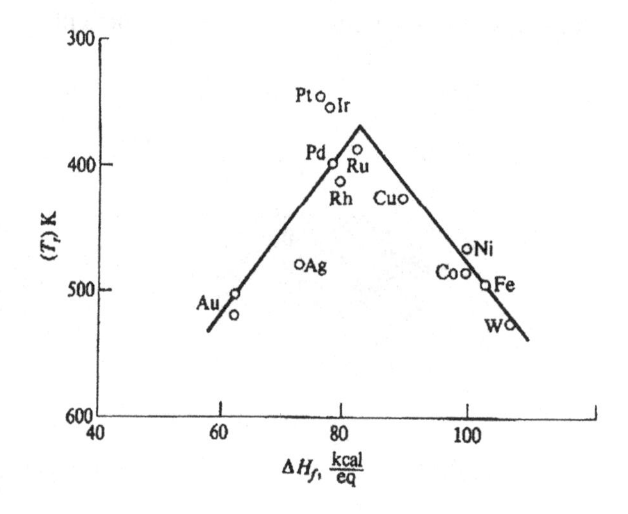

Figure 4.3.4: A volcano plot illustrating the Sabatier principle for formic acid decomposition. The horizontal axis is the binding enthalpy of formic acid to different metals. The activity of each catalyst is represented by its isokinetic temperature, i.e. the temperature at which formic acid decomposition occurs at reference rate. [Boudart, Chem. Eng. Prog. 57, 33 (1961)]

Why does Sabatier's principle work for so many different reactions and catalysts? The reason is rooted in two principles.

1. When the catalyst is changed such that the free energy of reaction for one step changes, the corresponding barrier for the same step (usually) changes in the same direction. In other words, if step i is made more favorable then the barrier for step i also becomes lower. See Chapter 22 on free energy relationships and Bronsted-Evans-Polanyi relations for justification [46,47].
2. The overall free energy change associated with the reaction is a catalyst-independent quantity. Therefore, if a particular catalyst makes one step more favorable (faster) then it must make at least one other step less favorable (slower). The fastest catalyst is that for which the two opposing effects balance so that no one step is rate limiting.

In practice, there are often multiple sites with slightly different optimal reactant binding strengths and different relationships between the barrier heights and free energy differences. These all contribute to the observed activity so that, in many cases, the top of the volcano curve is less sharp than might be expected from idealized models.

Sabatier's principle can be generalized to include binding energies of multiple reactants and intermediates. Sabatier's principle can also be generalized by predicting activities from the binding energies of convenient adsorbates (often single atoms) instead of actual reactants and intermediates [47,78–80]. These generalizations of the Sabatier principle result in two-dimensional volcano plots, where turnover frequency (height of the volcano) is plotted against the binding energies for two adsorbates. The two-dimensional volcano plot in Figure 4.3.5 is for methanation ($CO + 3H_2 \rightarrow CH_4 + H_2O$), a reaction that involves CO bond breaking and desorption of carbon and oxygen containing adsorbates. The 2D volcano plot shows that the energies of $O*$ and $C*$, i.e. adsorbed oxygen and carbon atoms, are enough to identify the "Goldilocks catalyst" among many possible metals and alloys.

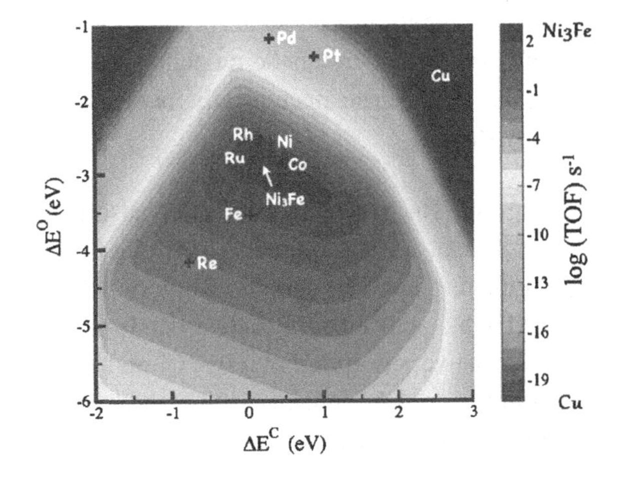

Figure 4.3.5: An *ab initio* volcano plot for methanation with two descriptors, the adsorption energies of atomic oxygen and carbon. CO bond breaking controls the rate for inactive catalysts like Ni and Pt. For faster catalysts, the rate is controlled by desorption from a surface that is fully covered by oxygen and carbon containing fragments. The optimal catalyst (Ru and the *de novo* computational prediction Ni_3Fe) have balanced rates of desorption, adsorption, and CO bond breaking. [From Norskov et al. Proc. Nat. Acad. Sci. USA 108, 937-43 (2012)]

Ab initio computational analyses based on Sabatier's principle and volcano plots have resulted in the discovery of new catalysts including near surface alloys [81] and highly active alloys with remarkably inexpensive constituent metals [82]. In many cases, the catalytically active alloys are "good children from bad parents", i.e. active alloys with inactive constituent metals drawn from opposite sides of the volcano curve [82–84]. Note that volcano plots are usually constructed without accounting for adsorbate-adsorbate interactions, and these can influence the binding enthalpy and the catalyst activity [85]. Interactions between adsorbates can also cause patches of surface coverage, e.g. in the case of $CO*$ and $O*$ patches on RuO_2 during CO oxidation [86]. Finally, note that Sabatier's principle can break down in certain situations, e.g. when the free energies of elementary steps are poorly correlated to their barriers, or when bare sites are not part of the catalytic cycle so that binding free energies are not related to the free energy of any particular step.

4.4 Microkinetic models

A microkinetic model is a sequence of hypothesized elementary steps and rate constants that can be translated into a rate law. The procedure resembles that by which rate laws were derived for multistep homogeneous reactions in Chapter 3. Microkinetic models describe only the chemical processes of adsorption, surface reaction, and desorption, i.e. the intrinsic kinetics for fixed local reactant and product concentrations near a site. Microkinetic models yield rate laws which themselves do not account for transport limitations, but they can be coupled to continuum models of transport processes [55–57].

A microkinetic model begins with a mechanistic hypothesis about the elementary steps by which the catalytic reaction occurs. Hinshelwood [87,88] developed the first microkinetic models from Langmuir models of adsorption, desorption, and a surface reaction step. Hougen and Watson extended microkinetic modeling techniques to more complex catalytic processes [89, 90]. Catalytic mechanisms typically involve at least two or three of the following types of elementary steps:

Adsorption The reactants must first adsorb onto an active site. Catalysis often requires chemisorption, i.e. the formation of strong bonds to the active site that tend to weaken bonds within the reactant. The destabilization of bonds within the reactants upon adsorption is often referred to as "activation". Reactant bonds may also completely break upon adsorption, e.g. in the dissociative adsorption of H_2, N_2, and O_2 on certain transition metal surfaces. As illustrated in Figure 4.4.1, chemisorption can have a significant activation barrier. In this chapter, the notation **A**$*$ indicates a species **A** adsorbed to an active site $*$.

Adsorbate diffusion On some catalysts, the surface intermediates can migrate across the catalyst surface by hopping from one active site to another. The activated hops affect diffusion at a rate given by the frequency of hopping multiplied by the squared distance between adjacent sites. Diffusion of adsorbates in narrow pores, e.g. in a zeolite, is a similarly activated process. Adsorbate diffusion is mechanistically important when intermediates derived from different reactants must encounter one another on the surface in order to react. Figure 14.1.2 in section 14.4 depicts H-atom diffusion on the surface of $Ni[100]$.

Unimolecular surface reactions Surface intermediates can undergo various types of unimolecular surface reactions. For example, a surface intermediate might isomerize or decompose to

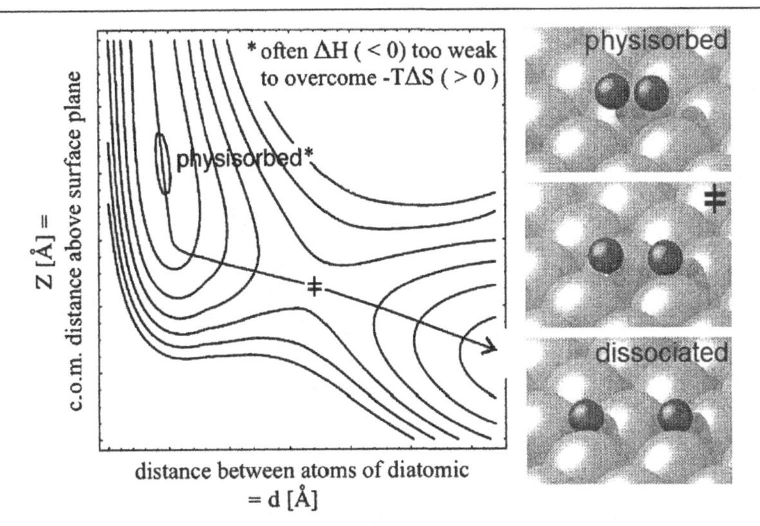

Figure 4.4.1: Many transition metal surfaces dissociate diatomic molecules like H_2, O_2, and N_2 into $H*$, $O*$, and $N*$ adsorbates. Depending on the metal and the conditions, these chemisorption processes may be thermodynamically favorable ($\Delta G < 0$), or unfavorable ($\Delta G > 0$). Dissociative chemisorption may be barrierless (e.g. H_2 on $Pt[111]$) or activated (e.g. N_2 on $Ru[1111]$). The figure shows a potential energy surface and minimum energy dissociation path for N_2 on $Ru[1111]$. [Adapted from Hammer and Norskov, Adv. Catalysis 45, 71–129 (2000).]

give two surface intermediates. Alternatively, a surface intermediate might decompose to release a product and leave a new intermediate on the surface. For example, the β-elimination reaction that terminates some polymerization processes is $*CH_2CH_2R \to *H + H_2C = CHR$.

Eley-Rideal surface reactions Some surface intermediates can directly react with species in the gas phase as shown in Figure 4.4.2. The rate of an Eley-Rideal step is proportional to the local gas phase reactant concentration and to the coverage of the surface intermediate. Note that gas phase reactants almost always have at least a weak physisorption minima on the *potential* energy landscape. If adsorption is sufficiently weak that the gas-phase co-reactant has an uphill *free* energy of adsorption, then the reaction will exhibit Eley-Rideal type kinetics.

Langmuir-Hinshelwood surface reactions These are reactions between two surface intermediates as shown in Figure 4.4.3. Prototypical examples include recombination of adsorbed O atoms in the decomposition of N_2O, and reaction of adsorbed H-atoms with adsorbed ethylene in hydrogenation.

Product desorption Desorption is often associated with an activation barrier. Strongly bound products that cannot desorb will occupy all of the sites and thereby poison the surface.

Figure 4.4.2: Scheffler and coworkers investigated an Eley-Rideal mechanism for CO oxidation on a Ruthenium catalyst. The CO molecule approaches the $O*$ species from the gas phase. Even in the transition state (right), the CO fragment is not bonded to the surface. [Adapted from Stampfl and Scheffler, Phys. Rev. Lett. 78, 1500-3 (1997).]

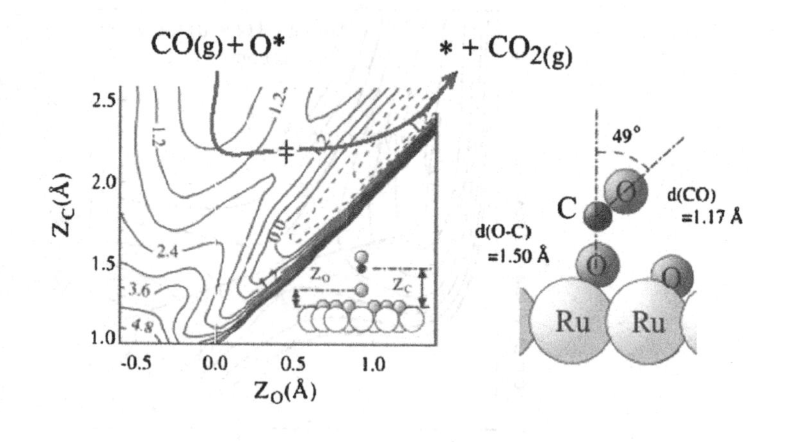

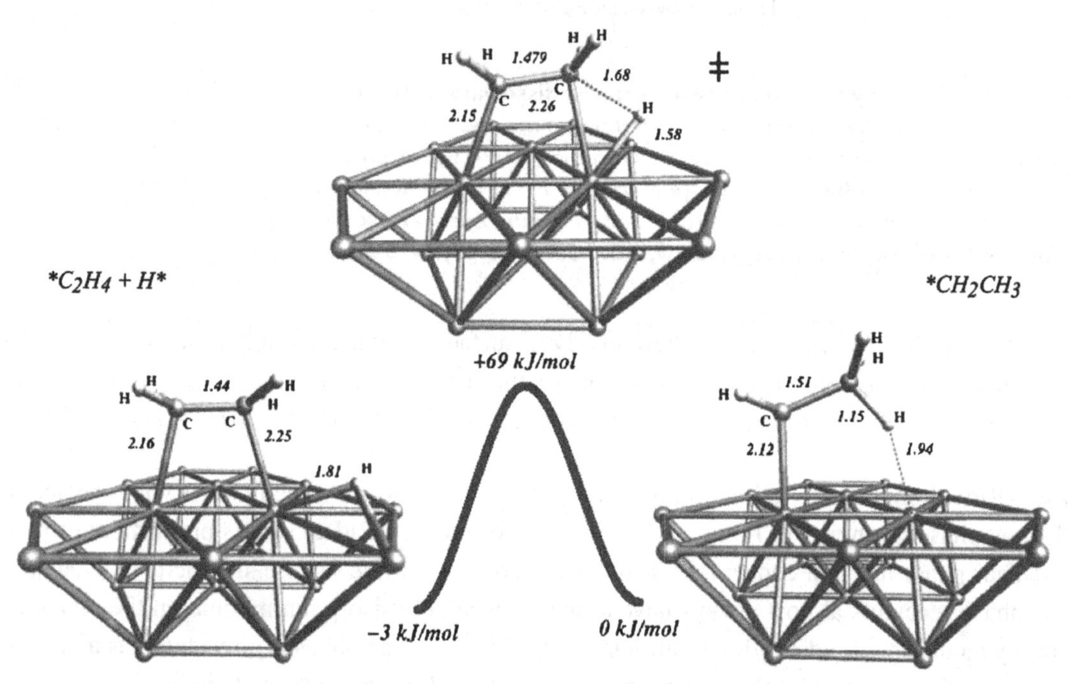

Figure 4.4.3: In ethylene hydrogenation, $H*$ species on a $Pd[111]$ surface react with chemisorbed ethylene to create monodentate ethyl adsorbates. The ethyl fragments react with additional $H*$ species to create an ethane molecule. [From M. Neurock and R. A. van Santen, J. Phys. Chem. B. 104, 11127 (2000).]

Complex catalytic mechanisms can be constructed by combining these elementary steps. Then, overall rate laws can be derived from the rates of the elementary steps much like we did for multistep chemical reactions. In the most general formulations, each step is reversible, with mass action rate expressions for the forward and backward rates. The species involved in the elementary steps may include free sites, surface intermediates, and/or gas phase species. The elementary rate expressions for surface reactions depend on surface population densities in the same way that rate laws for elementary reactions in the bulk depend on concentrations. Surface population densities are typically quantified in terms of coverages. The coverage of a reactant, product, or intermediate is the fraction of sites (assuming they are identical) that are occupied by that species. The coverage of species **A** is denoted $\theta_{\mathbf{A}}$. The coverage of bare sites is denoted θ_*.

The hypothesized sequence of elementary steps and their rate parameters are the basic ingredients for a microkinetic model. In some cases (not all) the overall rate law can be derived by eliminating all reactive intermediates with the pseudo-steady-state approximation (PSSA). In Chapter 3 the PSSA was justified by a small ratio of rate parameters. In contrast, the PSSA for catalytic intermediates is more generally justified any time the ratio of catalyst sites to reactant molecules in the bulk fluid is very small [91]. The rationale is simple: the catalyst can turn through many cycles without appreciably changing the reactant and product concentrations. Thus, applying the PSSA to all surface intermediates yields catalytic rate laws that depend only on approximately constant concentrations of species in the bulk fluid: reactants, products, and sometimes desorbed intermediates. The validity of the PSSA is only guaranteed for the surface intermediates. Desorbed intermediates may accumulate during catalysis, and the PSSA should not be applied to these species without additional justification [92].

Catalytic rate laws obtained from the PSSA with no further approximations often contain too many rate parameters to determine experimentally. Thus further assumptions are often made about which steps are rate determining or quasi-equilibrated. The hypothesized mechanism and assumptions are then tested by comparing experimental kinetics to the predicted rate law. If a microkinetic model has been constructed from the correct mechanism and assumptions, then it should "fit" experimental data over a range of concentrations for each species.

When the rate law that emerges from a microkinetic model fits the available kinetic data one must still be cautious. A successful data fit may be fortuitous – even empirical power-law models often fit as well as mechanistically motivated microkinetic models. Boudart [60], Kiperman et al. [93] and others vigorously debated whether the mechanistic steps in microkinetic models were "fortuitous placebos" or "the real thing" in the 1980s. Their debates focused on what criteria should be met before a microkinetic model is accepted as reality. Since that time, *ab initio* calculations have emerged as a powerful and independent way of testing mechanistic hypotheses. If their debate were continued today, then validation of rate parameters by *ab initio* calculations would undoubtedly be added to the criteria for accepting a microkinetic model as reality. Kinetic parameters from *ab initio* calculations are not quantitatively accurate, but with uncertainty estimates from *ab initio* benchmarking studies [94,95] they can reject some mechanisms or active site models as totally implausible. *Ab initio* calculations can additionally predict spectroscopic signatures and intermediate abundances to further cross examine the predictions of a hypothesized mechanism. Examples of these additional *ab initio* analyses can be seen in section 4.1 (the deamidation case study) and in later parts of this section.

One-site mechanisms

In a one-site mechanism, all of the steps take place at a single catalytic site.[3] One-site mechanisms always yield a rate law after only minimal PSSA approximations. Remarkably, this is true even for highly complex one-site mechanisms like coupled cycles [92] and one-site models with an infinite number of states as in polymerization [96]. Formulaic approaches to derive the one-site rate laws have been given by Christiansen [97], Temkin [98], Helfferich [92], and by Kozuch and Shaik [99] using an energy-span representation. Below we make the additional simplifying approximations that adsorption steps are fast and equilibrated.

Example: One-site model for reaction $\mathbf{A} \rightleftarrows \mathbf{B}$

Consider the surface catalyzed reaction $\mathbf{A} \rightleftarrows \mathbf{B}$ in which reactant adsorption and product desorption are QE steps and the surface reaction is a single elementary RDS. The elementary steps and their corresponding rate equations or equilibrium relations are

$$\begin{array}{llll}
\mathbf{A} + * & \overset{K_A}{\rightleftarrows} & \mathbf{A}* & \theta_A = K_A[\mathbf{A}]\theta_* \\
\mathbf{A}* & \underset{k_{-2}}{\overset{k_2}{\rightleftarrows}} & \mathbf{B}* & r_2 = k_2\sigma\theta_A - k_{-2}\sigma\theta_B \\
\mathbf{B}* & \underset{K_B}{\rightleftarrows} & \mathbf{B} + * & \theta_B = K_B[\mathbf{B}]\theta_* \\
\hline
\mathbf{A} & \rightarrow & \mathbf{B} &
\end{array} \tag{4.4.1}$$

Here σ is the density of sites per area or per mass of catalyst. To complete the rate law, θ_A and θ_B must be expressed entirely in terms of $[\mathbf{A}]$ and $[\mathbf{B}]$, i.e. in terms of variables that can be controlled and/or measured. A convenient starting point is to find an expression for θ_*. The total number of active sites is conserved and therefore the fractional coverages

[3] A multi-site mechanism includes some elementary steps in which surface intermediates on different sites react with each other. For multi-site mechanisms, the PSSA approximations yield non-linear equations that (usually) cannot be solved without further approximations and assumptions about the RDS and QE steps.

must add to unity

$$\begin{aligned} 1 &= \theta_* + \theta_{\mathbf{A}} + \theta_{\mathbf{B}} \\ &= \theta_*(1 + K_{\mathbf{A}}[\mathbf{A}] + K_{\mathbf{B}}[\mathbf{B}]) \end{aligned} \tag{4.4.2}$$

Solving for θ_* (the fraction of unoccupied sites) and then using $\theta_{\mathbf{A}} = K_{\mathbf{A}}[\mathbf{A}]\theta_*$ and $\theta_{\mathbf{B}} = K_{\mathbf{B}}[\mathbf{B}]\theta_*$ in the expression for $\mathcal{r}$ gives

$$\mathcal{r} = \frac{k_2 K_{\mathbf{A}} \sigma \{[\mathbf{A}] - [\mathbf{B}]/K\}}{1 + K_{\mathbf{A}}[\mathbf{A}] + K_{\mathbf{B}}[\mathbf{B}]} \tag{4.4.3}$$

where $K = K_{\mathbf{A}} K_2 / K_{\mathbf{B}}$ is the equilibrium constant for the uncatalyzed reaction. Note that the rate law involves a thermodynamic driving force term, a modified rate constant, a density of sites, and a denominator that accounts for the availability of sites. Limiting regimes of the microkinetic model can help confirm or reject the mechanistic hypothesis. For example, the model predicts $\mathcal{r} = 0$ when $K = [\mathbf{B}]/[\mathbf{A}]$, $\mathbf{B}$-inhibition when $K_{\mathbf{B}}[\mathbf{B}] \gg 1 + K_{\mathbf{A}}[\mathbf{A}]$, and a first order dependence on $[\mathbf{A}]$ when $K_{\mathbf{A}}[\mathbf{A}] \ll 1 + K_{\mathbf{B}}[\mathbf{B}]$.

Note that the relative sizes of 1, $K_{\mathbf{A}}[\mathbf{A}]$, and $K_{\mathbf{B}}[\mathbf{B}]$ correspond to θ_*, $\theta_{\mathbf{A}}$, and $\theta_{\mathbf{B}}$, i.e. to the relative abundances of different surface adsorbates at steady state. Because adsorption constants vary over many orders of magnitude, one term among 1, $K_{\mathbf{A}}[\mathbf{A}]$, and $K_{\mathbf{B}}[\mathbf{B}]$ is often several orders of magnitude larger than the others. For example, suppose that $K_{\mathbf{A}}[\mathbf{A}] \gg K_{\mathbf{B}}[\mathbf{B}]$ and $K_{\mathbf{A}}[\mathbf{A}] \gg 1$. Then we would have $\theta_{\mathbf{A}} \approx 1$ while $\theta_{\mathbf{B}} \approx \theta_* \approx 0$, and accordingly species $\mathbf{A}*$ would be the most abundant reactive intermediate (MARI). Sometimes the most abundant surface species is a non-reactive adsorbate, e.g. a poison. In these cases, it is useful to distinguish the most abundant *surface* intermediate (MASI) from the most abundant *reactive* intermediates (MARI).

With some practice, one can directly jump from a mechanism and RDS assumption to the overall rate law. A variety of simple catalytic cycles can be solved in this manner, but the following polymerization example is more difficult. The polymerization example also illustrates how QE steps can be treated as lumped species within more complicated kinetic networks. Finally, the polymerization example illustrates how microkinetic modeling in combination with *ab initio* calculations can be used to test a mechanistic hypothesis.

Example: Testing a one-site polymerization mechanism

The Phillips catalyst converts ethylene into approximately half of the world's polyethylene [100]. It was discovered in the 1950's but its mechanism has remained elusive, especially the initiation step. In 2014, an explanation for the mysterious initiation step was put forth based on DFT calculations and IR spectroscopy evidence [101]. The new mechanism invoked dormant (**D**) sites and a series of active Si(OH)Cr-R sites that can be numbered according to the length of the R chain as shown in Figure 4.4.4.

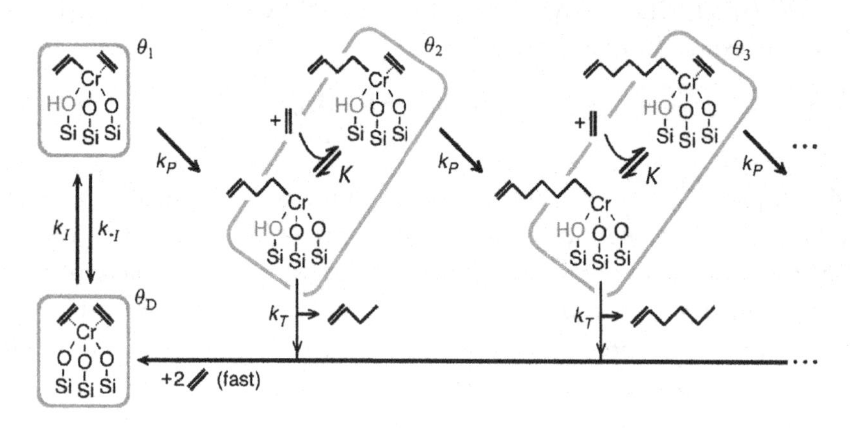

Figure 4.4.4: Initiation (I), propagation (P), and termination (T) steps in the mechanism of Delley et al. [101]. The authors reported standard free energy changes and barriers for the elementary steps from *ab initio* calculations. Adsorption processes, i.e. the steps within rounded boxes, are assumed to be fast and equilibrated. Figure from Peters et al. Proc. Nat. Acad. Sci. USA, 112, E4160-61 (2015).

A microkinetic model for the mechanism of Figure 4.4.4 yields expressions for the polymerization rate, the molecular weight of the polymers, and the abundance of various catalytic intermediates. The lumped species introduce a slight change to the way that the rate equations are constructed. First, applying the PSSA to species 1 [102]:

$$r_1 = -k_P\sigma\theta_1 - k_{-I}\sigma\theta_1 + k_I\sigma\theta_D \approx 0$$

Applying PSSA to each lumped state i with $i > 2$ gives

$$r_i = \frac{k_P K[||]\sigma}{1+K[||]}\theta_{i-1} - \frac{k_P K[||]\sigma}{1+K[||]}\theta_i - \frac{k_T\sigma}{1+K[||]}\theta_i \approx 0 \tag{4.4.4}$$

Note that the adsorption equilibria which create π-bound ethylene complexes are built into equation (4.4.4). For example, the factor $(1+K[||])^{-1}$ is the fraction of a sites in lumped state i *without* a π-bound ethylene and $K[||](1+K[||])^{-1}$ is the fraction of sites in lumped state i *with* a π-bound ethylene. Applying the PSSA to the dormant surface species **D**:

$$\mathcal{R}_{\mathbf{D}} = k_T \sigma \sum_{i=2}^{\infty} \frac{\theta_i}{1 + K[||]} - k_I \sigma \theta_{\mathbf{D}} + k_{-I} \sigma \theta_1 \approx 0$$

We leave the special case $i = 2$ to the reader. Each PSSA equation relates the abundance of one surface species to a linear combination of the other surface species. To obtain independent equations, an additional site balance is needed to normalize the total population:

$$1 = \theta_{\mathbf{D}} + \sum_{i=1}^{\infty} \theta_i$$

These equations are somewhat complicated, but they can be solved exactly because the sum $\theta_2 + \theta_3 + \ldots$ is a geometric series owing to equation (4.4.4).

The solution yields all surface abundances in terms of k_P, k_T, k_I, k_{-I}, K, and $[||]$. From these one can obtain essentially any kinetic property of interest. The overall polymerization rate, expressed as the rate of consumption of ethylene per site, is [102]

$$\mathcal{R}_{||} = \frac{k_P K[||]\sigma}{1 + K[||]} \sum_{i=1}^{\infty} \theta_i$$

The weight-averaged molecular weight is

$$W_W = \sum_{i=1}^{\infty} i^2 \theta_i / \sum_{i=1}^{\infty} i \theta_i$$

The arithmetic averaged molecular weight is

$$W_A = \sum_{i=1}^{\infty} i \theta_i / \sum_{i=1}^{\infty} \theta_i$$

Density functional theory (DFT) calculations by Delley et al. [101] can be used to predict the polymerization rates and molecular weights. Relative to the experimental measurements, the predicted polymerization rate is a few orders of magnitude too small and the molecular weight is a few orders of magnitude too high [102,103]. Typical errors in free energies from DFT calculations (ca. $20kJ/mol$) translate to factors of $exp[(20kJ/mol)/k_B T] \approx 10^3$ errors in rates, selectivities, and weights. Thus, the rate and molecular weight predictions could be viewed as acceptable departures from the experimental observations [102].

The intermediate abundances are important for interpreting the spectroscopic signal intensities in *in situ* IR experiments. Delley et al. proposed that the bridging hydroxyl

$Si(OH)Cr - R$ sites were responsible for IR peaks at 3640 and 3605cm^{-1} that appear during catalyst operation. To be IR-visible, a surface intermediate should have coverage comparable to unity. According to the proposed mechanism, the total abundance of these species is [102]

$$\theta_{Si(OH)Cr-R} = \theta_1 + \theta_2 + \theta_3 + \ldots = \frac{k_T + (1 + K[|||])k_P}{k_T + (1 + K[|||])k_P + k_T(k_{-I} + k_P)/k_I}$$

Using transition state theory and the DFT calculations by Delley et al. [101] to estimate the individual rate parameters yields a surface coverage $\theta_{Si(OH)Cr-R} \approx 10^{-9}$ [102]. Converting to a free energy difference via $k_B T\, ln[10^9]$ suggests we would need to invoke unusually large (ca. 70 kJ/mol) errors in the *ab initio* calculations to rationalize the discrepancy. What does such a large discrepancy mean? There is always a chance that errors in the DFT calculations are unusually large, giving errors far beyond the mean absolute deviations from benchmark studies. Alternatively, the new IR peaks may have been misinterpreted. IR bands at 3605 and 3640 are present even after catalyst removal, and others have assigned them to intrinsic combination bands of polyethylene [102]. These findings illustrate the synergistic power of *ab initio* calculations, microkinetic modeling, and experimental characterization tools for testing mechanistic hypotheses.

We are accustomed to thinking about the assumed sequence of elementary reactions as a mechanistic hypothesis, but assumptions about the active site are also a type of mechanistic hypothesis. Specifically, assumptions about the active sites propagate through all subsequent calculations into the predicted rates, spectra, activation energies, molecular weights, etc. Thus when predictions from an *ab initio* study disagree with experimental observations, it may be that the proposed elementary steps are wrong, that the active site model is wrong, or both.

As *ab initio* calculations become more reliable, cross examinations of computational results can become more and more exacting. Scrutiny and cross-examination are central to the scientific method [104], so these new capabilities should be welcomed by all investigators. However, some contend that *ab initio* calculations cannot make quantitative predictions and therefore they are better used for studying more qualitative trends. Indeed, recent studies that examine trends with BEP relations and volcano plots have had tremendous impact on the selection of alloy compositions and dopants. Given the limitations of *ab initio* calculations, what is the best way to use them? The answer depends on how much we already know about the mechanism of the reaction or catalyst in question.

- When a mechanism is well understood, it tells us which sites, steps, and trends to examine. Then, as demonstrated by many recent studies, BEP relations, descriptors, and volcano plots can be used to quickly screen the periodic table for optimal materials. Precise rate calculations and highly accurate quantum chemistries are usually not necessary for these analyses.

- When a mechanism is unknown, we must start by quantitatively computing rates, spectra, and other observables to identify the most plausible mechanism, rate limiting step, most abundant intermediate, and active site structure. Investigating trends *before* understanding the reaction might only reveal rather useless insights about kinetically irrelevant steps and sites.

Mechanistic hypothesis testing requires highly accurate model chemistries and accurate rate calculations. It seems likely that quantitative *ab initio* tests of hypothesized mechanisms and active site structures will become more common as model chemistries improve.

We have seen that one of the gentlest approximations for deriving rate laws is the PSSA. Many *ab initio* studies illustrate that one can now compute the free energy barriers and rate constants for every elementary step, but still the overall rate law is rarely examined. Christiansen pioneered a graph theoretic approach for constructing catalytic rate laws [97]. His results were extended and applied by King and Altman [105], and by Temkin [106]. For one-site catalytic cycles, the Christiansen mathematics provides exact rate laws within the PSSA. The book by Helfferich develops rate laws for coupled cycles, competing cycles, connected cycles, and cycles with off-pathway equilibrium branches [92]. Here we give only the result for a single simple catalytic cycle. The cycle can include an arbitrary number of steps, co-reactants, and by-products, but each forward and backward step should involve only one catalyst site.

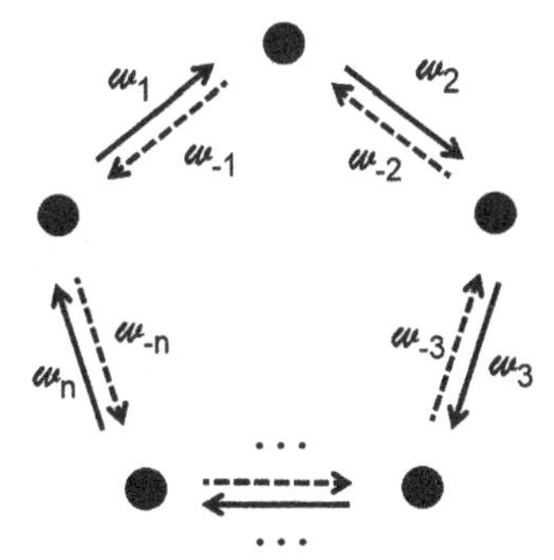

Figure 4.4.5: A single-site catalytic cycle showing the most general case of entirely reversible steps. Reactants, co-reactants, products, by-products, etc. are not shown, but they do enter the state-to-state transition frequency expressions as described in the text.

Consider for example, the catalytic cycle shown in Figure 4.4.5. For each step, define forward and backward transition frequencies w_i and w_{-i} as the corresponding forward and backward rates without the corresponding site coverages. For example if step i is $\mathbf{A} + * \rightarrow \mathbf{A}*$ then $w_i = k_{*\rightarrow \mathbf{A}*}[\mathbf{A}]$ without the factor of θ_*. The turnover frequency is

$$\mathfrak{r} = \mathbb{C}^{-1}\left(\prod_{i=1}^{n} \mathfrak{w}_i - \prod_{i=1}^{n} \mathfrak{w}_{-i}\right)$$

where $\mathbb{C}$ is the Christiansen denominator. The Christiansen denominator is constructed as a sum over all terms in the Chrisiansen matrix

$$\begin{array}{ccccc}
w_2 w_3 \cdots w_n & w_{-1} w_3 \cdots w_n & w_{-1} w_{-2} w_4 \cdots w_n & \cdots & w_{-1} w_{-2} \cdots w_{-(n-1)} \\
w_3 w_4 \cdots w_1 & w_{-2} w_4 \cdots w_1 & w_{-2} w_{-3} w_5 \cdots w_1 & \cdots & w_{-2} w_{-3} \cdots w_{-n} \\
\vdots & \vdots & \vdots & & \vdots \\
w_1 w_2 \cdots w_{n-1} & w_{-n} w_2 \cdots w_{n-1} & w_{-n} w_{-1} w_3 \cdots w_{n-1} & \cdots & w_{-n} w_{-1} \cdots w_{-(n-2)}
\end{array}$$

Figure 4.4.6 shows a graphical representation of the Christiansen matrix.

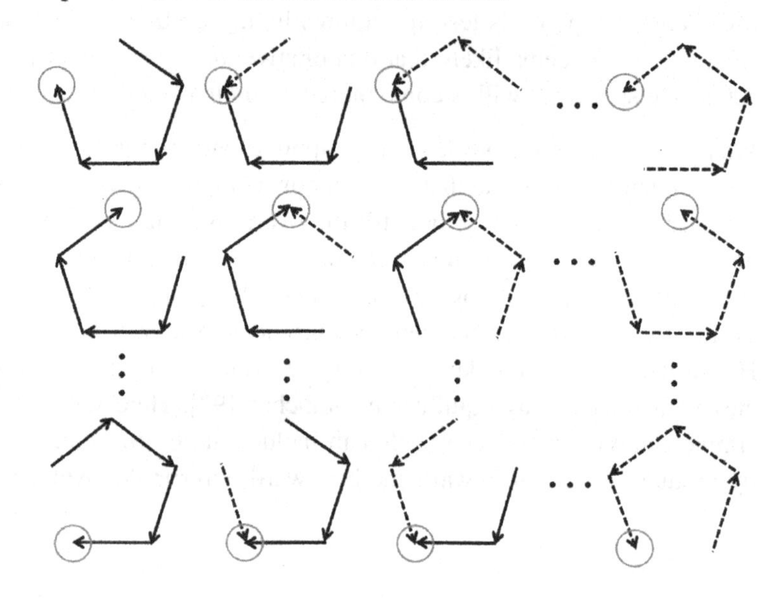

Figure 4.4.6: The Christiansen denominator is a sum of products of transition frequencies. Terms of the sum correspond to diagrams in the figure with forward edges corresponding to a factor w_i and backward edges corresponding to a factor w_{-i}. Note that each cycle leaves one edge (one transition frequency) out. The circles emphasize that arrows in each row end at the same vertex in the graph.

Fractional coverages in the PSSA can also be obtained from the Christiansen matrix. Specifically, adding elements in the ith row gives quantity $\mathbb{C}_i$. The fractional coverage of site i, i.e. the circled site in row i of Figure 4.4.6, is

$$\theta_i = \mathbb{C}_i / \mathbb{C}$$

Example: Dock-lock mechanism in amyloid fibril growth

The Christiansen formulas are not limited in scope to catalysis applications. For example, amyloid fibrils are thought to grow by a dock-lock mechanism [107], a two-stage cycle as depicted in the figure below. The figure below has been constructed using images with permission from Schor et al., *Biophysical J.* (2012) [108].

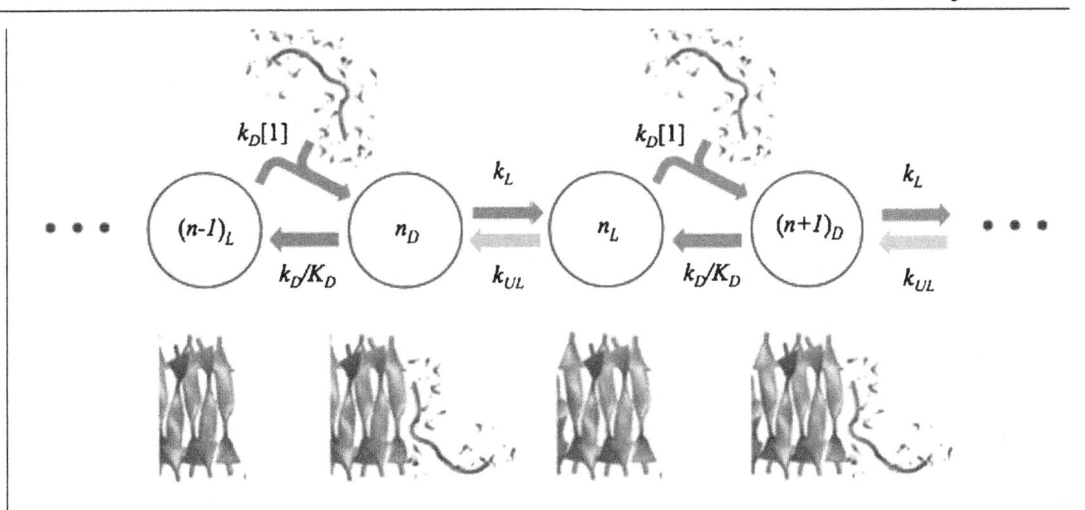

In the first stage, a peptide loosely associates with the fibril end via non-native interactions. In the second stage, the peptide locks into a configuration where its residues form appropriate hydrogen bonds and hydrophobic interactions with the fibril. At each point in time the end of the fibril is either in the docked (D) state or the locked (L) state, and the number of turnovers is recorded in the length of the fibril. Let the dock step be diffusion controlled so that the docking rate is $k_D[1]$ where [1] is the concentration of individual peptides. Let the lock step have first order rate constant k_L. Their reverse rates are k_{UL} and k_D/K_D where K_D is the equilibrium constant for the docking step. Based on these elementary rates, the Christiansen formula yields the net growth rate as

$$\mathfrak{r} = \frac{k_L k_D[1] - k_D K_D^{-1} k_{UL}}{k_L + k_D/K_D + k_{UL} + k_D[1]} \tag{4.4.5}$$

Many valuable insights are buried within this kinetic model. For example, equation (4.4.5) reveals that the overall growth rate becomes k_L when the monomer concentration is sufficiently large. The overall growth rate becomes proportional to monomer concentration if the lock step is irreversible and limiting. Embedded within the rate constants are other insights. For example, the docking rate constant k_D is a diffusion control rate parameter with a weak temperature dependence and a well-known viscosity dependence [107]. The docking association constant K_D also plays an important role: if docking and locking are sufficiently irreversible (i.e. if K_D and k_L/k_{UL} are both extremely large) then the growth law reduces to a partial diffusion control expression $\mathfrak{r} = k_L k_D[1]/(k_L + k_D[1])$. By starting from mechanistic models and rate laws like that in equation (4.4.5), we can use rare events simulations (or experimental data) to check the microscopic interpretations and obtain microscopically accurate mechanistic models.

If the dock-lock hypothesis is correct, then $\mathcal{R}$ corresponds to the overall growth rate that has been directly measured in some experiments [109,110] and used as a fit parameter in others. For example, $\mathcal{R}$ would correspond to the parameter k_+ in the master equation of Knowles et al. [111]. The advantage of this formulation is that all components of $\mathcal{R}$ can be estimated from diffusion control theories, peptide size, and non-native binding propensities. Moreover, the rate parameters within $\mathcal{R}$ introduce well-understood dependences on temperature, viscosity, and other properties that can be manipulated in experiments.

To facilitate connections between experiment and computational analyses, Kozuch and Shaik assumed that each step in the cycle is described by transition state theory. Combining this assumption with the Christiansen formula yields a rate expression that is entirely based on free energies and activation free energies [99,112]. Their energy span expression for the turnover frequency is

$$\mathcal{R} = \frac{k_B T}{h} \frac{\Delta}{\Omega} \tag{4.4.6}$$

where Δ quantifies the thermodynamic driving force and Ω is an overall kinetic resistance. The quantity Δ is

$$\Delta = exp[-\beta \Delta G_{rxn}] - 1 \tag{4.4.7}$$

which is positive in the direction with favorable thermodynamics, i.e. $\Delta \geq 0$ when $\Delta G_{rxn} \leq 0$. The resistance is defined as a double sum over all intermediates [99]

$$\Omega = \sum_{i,j} exp[\beta(G_i^{\ddagger} - G_j - \delta G_{i,j})] \tag{4.4.8}$$

with

$$\delta G_{i,j} = \begin{cases} \Delta G_{rxn} & i > j \\ 0 & i \leq j \end{cases} \tag{4.4.9}$$

Note that the intermediates and transition states along the catalytic cycle must be numbered in sequence and the double sum includes $i = j$ terms. Also note that there are no concentrations in the formulas because each ΔG is evaluated *at the actual concentrations*, not at reference concentrations.

Multi-site mechanisms

Thus far we have considered only one-site mechanisms, but some of the most important catalytic reactions involve elementary steps of Langmuir-Hinshelwood type. For example, Figure 4.4.7 shows the hypothesized mechanism for N_2O decomposition which is important for mitigating pollution from automobile exhaust.

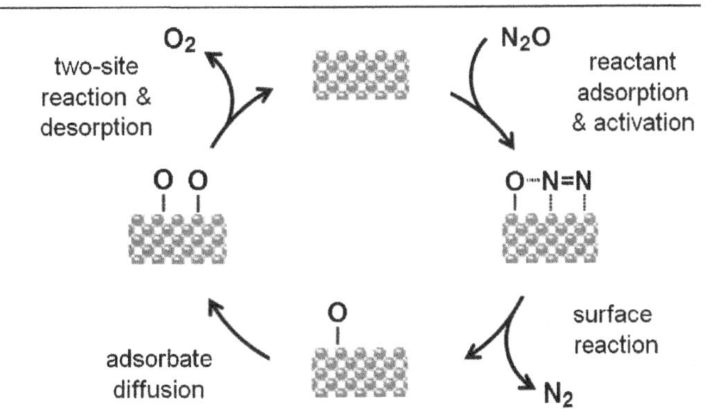

Figure 4.4.7: N_2O **decomposition on a** Ru **catalyst involves four steps: adsorption and activation of** N_2O**, a unimolecular surface reaction to release** N_2**, diffusion of the remaining** $O*$ **species, and a Langmuir-Hinshelwood type recombination of** $O*$ **at neighboring sites to release** O_2**.**

The following example illustrates the analysis of a two-site catalyst mechanism. Obtaining rate laws for two-site mechanisms typically requires assumptions about the RDS and QE steps. Hougen and Watson developed formulaic procedures for transforming these mechanisms and assumptions into rate laws [89]. The formulaic approach is fast, but doing the derivation forces one to think carefully about each proposed step and each assumption. That time is often well-spent.

Example: A two-site mechanism for $\mathbf{A}+\mathbf{B}\rightarrow\mathbf{C}+\mathbf{D}$

Consider a reaction in which reactants $\mathbf{A}$ and $\mathbf{B}$ adsorb on separate sites, then the surface intermediates $\mathbf{A}*$ and $\mathbf{B}*$ react on the surface to make $\mathbf{C}*$ and $\mathbf{D}*$, and finally $\mathbf{C}$ and $\mathbf{D}$ desorb to restore the free catalyst sites. The steps, equilibria, and rates are shown below.

$$\begin{array}{cccl}
\mathbf{A}+* & \overset{K_\mathbf{A}}{\rightleftharpoons} & \mathbf{A}* & \theta_\mathbf{A}=K_\mathbf{A}[\mathbf{A}]\theta_* \\
\mathbf{B}+* & \overset{K_\mathbf{B}}{\rightleftharpoons} & \mathbf{B}* & \theta_\mathbf{B}=K_\mathbf{B}[\mathbf{B}]\theta_* \\
\mathbf{A}*+\mathbf{B}* & \underset{k_b}{\overset{k_f}{\rightleftharpoons}} & \mathbf{C}*+\mathbf{D}* & r_3=k_f\sigma^2\theta_\mathbf{A}\theta_\mathbf{B}-k_b\sigma^2\theta_\mathbf{C}\theta_\mathbf{D} \\
\mathbf{C}* & \underset{K_\mathbf{C}}{\rightleftharpoons} & \mathbf{C}+* & \theta_\mathbf{C}=K_\mathbf{C}[\mathbf{C}]\theta_* \\
\mathbf{D}* & \underset{K_\mathbf{D}}{\rightleftharpoons} & \mathbf{D}+* & \theta_\mathbf{D}=K_\mathbf{D}[\mathbf{D}]\theta_* \\
\hline
\mathbf{A}+\mathbf{B} & \rightleftharpoons & \mathbf{C}+\mathbf{D} &
\end{array}$$

The fractional coverages are five unknowns: θ_*, $\theta_\mathbf{A}$, $\theta_\mathbf{B}$, $\theta_\mathbf{C}$, and $\theta_\mathbf{D}$. The adsorption equilibria for $\mathbf{A}$, $\mathbf{B}$, $\mathbf{C}$, and $\mathbf{D}$ are only four equations. The fifth equation requires that the sum of fractional coverages add to unity

$$1 = \theta_* + \theta_\mathrm{A} + \theta_\mathrm{B} + \theta_\mathrm{C} + \theta_\mathrm{D}$$

Using the expressions for θ_A, θ_B, θ_C, and θ_D in the rate law to obtain an expression for θ_* provides all terms in the rate expression.

$$\mathcal{R} = \frac{k_f \sigma^2 K_\mathrm{A} K_\mathrm{B} \{[\mathrm{A}][\mathrm{B}] - [\mathrm{C}][\mathrm{D}]/K\}}{\{1 + K_\mathrm{A}[\mathrm{A}] + K_\mathrm{B}[\mathrm{B}] + K_\mathrm{C}[\mathrm{C}] + K_\mathrm{D}[\mathrm{D}]\}^2} \quad (4.4.10)$$

where $K = K_A K_B k_f/(K_\mathrm{D} K_\mathrm{C} k_b)$ is to the equilibrium constant of the overall (uncatalyzed) reaction. The rate law is a thermodynamic driving force term multipled by $k_f \sigma^2$ and two fractional coverages:

$$\theta_\mathrm{A} = K_\mathrm{A}[\mathrm{A}]/\{1 + K_\mathrm{A}[\mathrm{A}] + K_\mathrm{B}[\mathrm{B}] + K_\mathrm{C}[\mathrm{C}] + K_\mathrm{D}[\mathrm{D}]\}$$

and

$$\theta_\mathrm{B} = K_\mathrm{B}[\mathrm{B}]/\{1 + K_\mathrm{A}[\mathrm{A}] + K_\mathrm{B}[\mathrm{B}] + K_\mathrm{C}[\mathrm{C}] + K_\mathrm{D}[\mathrm{D}]\}$$

The denominator is squared because the RDS involves reactions between two surface species. For the same reason, the overall rate is proportional to the square of the density of sites per unit area. Finally, note that a single RDS was assumed and all other steps are assumed to be QE.

Analysis by microkinetic models was, for many decades, the standard approach for testing mechanistic hypotheses. Now, the same *ab initio* rate constants and equilibrium constants that are used to construct a microkinetic model can instead be used in a (stochastic) kinetic Monte Carlo (kMC) simulation. Rates are obtained by averaging over many kMC simulations. (See Chapter 14 for a discussion of kMC.) kMC simulations are increasingly common alternatives to microkinetic modeling in state-of-the-art analyses. There are cases where the average over stochastic kMC simulations is different from (and superior to) microkinetic model predictions [71,86,113,114]. The kMC approach is particularly valuable when there are *unstable* fluctuations around mean field/deterministic behavior, or when the system of interest is so small that typical fluctuations around mean field behavior are significant.

More typically, there are many sites and a large reservoir of reactants. In these cases, more may be lost than gained by using kMC instead of a microkinetic model. Simple rate laws that describe the kinetics over a range of concentrations are easily obtained from the microkinetic modeling approach. By comparison, it is difficult to discover concentration dependences and temperature dependences from a mountain of kMC simulation data.

4.5 Degree-of-rate-control

We saw in the previous section that catalytic rate laws are dramatically simplified if one can identify a single RDS. Simplified models are not just shortcuts for lazy theorists. Rather, these simplifications eliminate the need to estimate many parameters and monitor trace intermediates when comparing to experiments. A rate determining step also helps to identify those changes in the catalyst, solvent, or reaction conditions that can be exploited to modify the overall rate.

Ab initio calculations are increasingly used to predict transition states, intermediate abundances, and rates. Murdoch [115], Dumesic [116,117], Campbell [118], and Kozuch and Shaik [99, 112], have developed formalisms for identifying the RDS from the computed rate parameters in a multistep mechanism. Campbell's degree-of-rate-control is a particularly intuitive formulation. If $\mathcal{R}$ is the overall rate and k_i the rate constant for step i in the reaction network, then the degree-of-rate-control for step i is

$$\chi_i \equiv \left(\frac{\partial ln \mathcal{R}}{\partial ln k_i}\right)_{K_i, k_j \ for \ j \neq i} \tag{4.5.1}$$

In equation (4.5.1), the equilibrium constant for step i and rate constants for all steps (forward and backward) other than for step i are being held constant. Thus, as the rate constant for step i is increased, its reverse rate constant must also increase to maintain K_i. The coupled differential changes in k_i and k_{-i} could be affected by changing the free energy of the transition state for step i, while holding the free energies of all other transition states and intermediates constant. Thus it is not surprising that the degree-of-rate-control can also be written in terms of derivatives with respect to the free energies of individual transition states [99,112,119]. If the rate of each step in a reaction network is presumed to follow transition state theory, then (at constant temperature and concentration) the degree-of-rate-control can also be written as[4]

$$\chi_i \equiv -\left(\frac{\partial ln \mathcal{R}}{\partial \beta \Delta G_i^{\ddagger,\circ}}\right)_{\Delta G_i^{\circ}, \Delta G_j^{\ddagger,\circ} \ for \ j \neq i} \tag{4.5.2}$$

Clearly, degree-of-rate-control is a valuable concept for catalyst design. We must first understand which steps limit the overall rate before attempting to engineer a better catalyst.

Dumesic has shown that the degrees-of-rate-control for any reaction network leading to a single overall reaction sum to unity, i.e. that [117]

$$\sum_i \chi_i = 1 \tag{4.5.3}$$

[4] Expressions like equation (4.5.1) can also be used to quantify sensitivity of the overall rate to the stability of surface intermediates [119].

with each step in a catalytic cycle having a degree-of-rate-control between zero and one. Depending on the system and the reaction conditions, one step may have $\chi_i \approx 1$ with all other steps having $\chi_i \approx 0$. In these cases, one can properly designate one step as rate limiting and dramatically simplify the rate law.

For complex reaction networks and coupled catalytic cycles that make multiple products, the degree-of-rate-control can be generalized to a matrix degree-of-rate-control for each product (columns) and each step (rows). Less is known about formal degree-of-rate-control properties for complex reaction networks with multiple products, but numerical results show that the individual degrees-of-rate-control for some reactions and products can exceed unity. In fact, χ can be very large for reactions that govern the selectivity of minor products [42,120].

For simple multistep reactions and catalytic cycles, Murdoch [115] and Kozuch and Shaik [99] suggested graphical procedures to identify the RDS and the MARI. The graphical procedure is outlined below and illustrated in Figure 4.5.1.

Algorithm: Graphical RDS and MARI identification

For a single simple catalytic cycle containing any number of steps, the RDS and MARI can be determined from *ab initio* free energy calculations as follows:

1. Replicate the free energy profile for the catalytic cycle as shown in Figure 4.5.1.
2. From each barrier, extend a horizontal line leftward until it crosses the free energy profile.
3. The RDS is the transition state whose leftward extending line passes the highest above a free energy minimum.
4. The deepest free energy minimum to the left of the RDS is the MARI.

Steps on the (forward) path from the MARI to the surface species that is consumed in the RDS are QE steps. Steps on the (forward) path between the products of the RDS onward to the MARI may be irreversible, reversible, or QE. In Figure 4.5.1, the products of the RDS are not part of the basin created around the MARI. Therefore, we should not apply the QE approximation to the step leading from the RDS products to the MARI. If the reversibility [121], $exp[\Delta G/k_B T]$, for this step is zero then it can be treated as irreversible.

The graphical procedure results in a diagram with a series of pools and spillways, much like the depth profile of a trout stream. However, the capacity (site abundance) of the pool ahead of each barrier depends exponentially on the pool's depth. Thus a pool designated as the MARI may be only a few $k_B T$ deeper than the other pools, but it will still accumulate an overwhelming

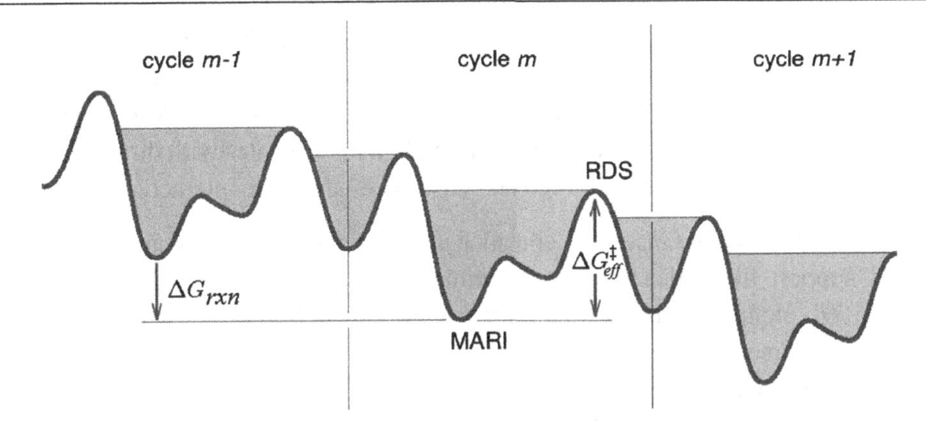

Figure 4.5.1: A graphical procedure that identifies the rate determining step (RDS) and the most abundant reactive intermediate (MARI). These diagrams are constructed *from Gibbs free energies at the reaction conditions*, not from standard Gibbs free energies.

majority of the sites during steady-state operation. Relative depths of these pools determine the degree of confidence in *ab initio* RDS and MARI predictions. If the pool corresponding to the MARI is many $k_B T$ deeper than all other pools, we can be very confident of the RDS/MARI. If there are several transition states with comparably deep pools on their left, then the RDS and MARI predictions will be sensitive to changes in concentration, to changes in temperature, and to uncertainties in the *ab initio* calculations.

4.6 Catalysts with non-uniform sites

The previous sections focused on structural models and microkinetic models for catalysts with uniformly active sites. These are instructive, but they are not always the reality. Rideal, Taylor, and others suggested that real catalysts have an ensemble of active sites with different structures and activities [122]. The overall activity from an ensemble of sites will be dominated by the most active sites [90,123–125]. Non-uniform sites pose a serious problem for both computational and experimental studies. Computational models may not be representative of the most active sites, and standard characterization techniques like EXAFS, NMR, etc. probe the most common sites, even though these may be inactive [100,126]. Even progressive poisoning and other active site counting techniques that provide an approximate distribution of site activities rely on rather uncertain assumptions [127].

How can we investigate active sites when the site population is dominated by inactive sites? First, let us consider the two potential sources of variance in the ensemble of active sites: equilibrium fluctuations and quenched disorder. For an example of *equilibrium* site non-uniformity, consider the reactivity of an alloy. The reactivity at each site on the surface is influenced by the local composition and configuration of the metal atoms. If the temperature is high enough, surfaces may undergo reconstructions and changes in surface chemistry. At realistic operating pressures, even an idealized metal surface like Pt(111) will have sites with different reactivities because of variations in the local coverage and adsorbate-adsorbate interactions [128].

For an example of *quenched disorder*, consider an atomically dispersed catalyst on an amorphous solid support like silica or silica-alumina. These catalysts are used in olefin polymerization [100], olefin metathesis [129,130], olefin epoxidation [131], and partial oxidation [132]. Each single metal atom is isolated from the others [45], but they are each grafted to a unique local environment on the amorphous support. The amorphous solid matrix around each site remains static during catalysis apart from modest degrees of strain. Thus the disorder is "quenched" into a permanent non-equilibrium amorphous distribution [133]. Materials with quenched disorder are notoriously difficult to model. Without the Boltzmann distribution, where does one even start?

Several different computational approaches have been pursued to model catalysts on amorphous supports. Some investigators have used silsesquioxanes and other small clusters to mimic the support [103,134–136]. Others have used fragments of β-crystobalite or zeolite crystals as cluster models [137,138]. Tielens [139], Ugliengo [140], Johnson [141], and others have prepared large slabs of amorphous silica by quenching molten SiO_2, carving out a surface, and then hydroxylating and/or annealing the dangling bonds. These *in silico* protocols are quite different from experimental preparations, but some yield materials with realistic hydroxylation densities and energetics [141]. Large scale *in silico* models are just beginning to be used in studies of catalysis and catalyst preparation [50,142,143].

Goldsmith et al. [144,145] used sequential quadratic programming to create a family of simple cluster models with rigid peripheral atom constraints and a range of different activities. Their procedure attempts to approximate the distribution of non-uniform sites by generating low energy (and hopefully representative) model sites at each activation energy (see Figure 4.6.1). The method of Goldsmith et al. only crudely approximates the distribution of site activities, e.g. by identifying the range of possible activation energies [144,145].

Kinetics with quenched disorder

For more quantitative predictions, several challenges must be overcome. Ideally, *in silico* support models should be prepared via annealing, aging, and hydroxylation protocols that mimic their experimental counterparts. It may eventually become possible to directly construct support models from atomistically resolved images of amorphous support films [146]. Recent efforts suggest that accurate models of amorphous supports are on the horizon, and that future studies

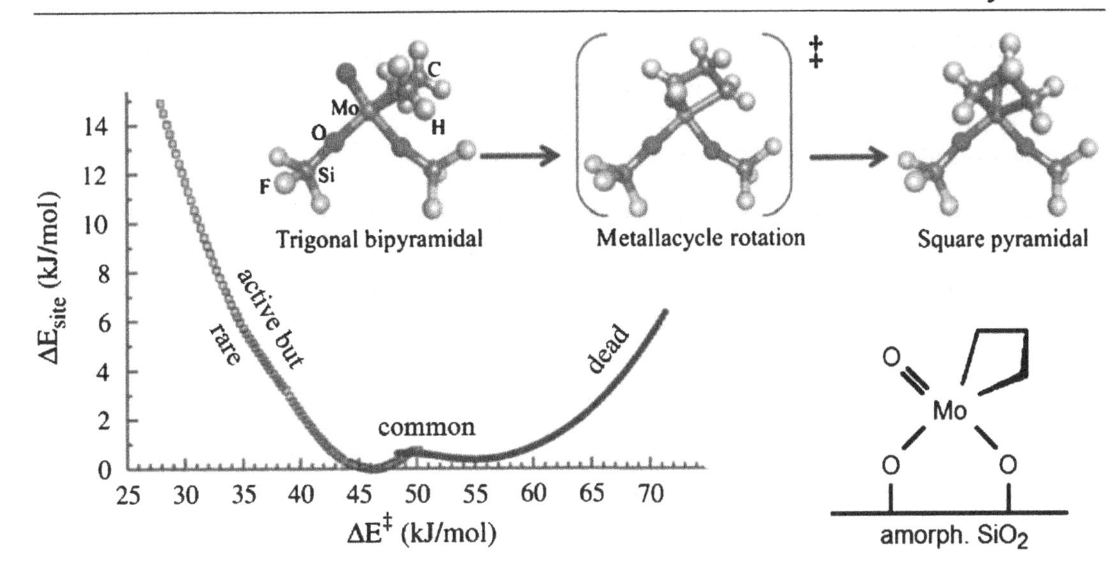

Figure 4.6.1: The diagram at right depicts a Mo-catalyst grafted to amorphous silica. To understand how different local support environments influence structure and reactivity, Goldsmith et al. developed an algorithm that systematically generates amorphous cluster models with the lowest possible site formation energy at each value of the activation energy. Sites with the lowest formation energies are assumed to be common. Highly active sites have higher site formation energies and therefore are presumably less common in the site distribution. Figure from Goldsmith et al. J. Chem. Phys. 138, 204105 (2013).

will need a statistical framework for modeling catalysts with quenched distributions of non-uniform active sites. Looking ahead to more accurate amorphous support models, we present early developments toward such a statistical framework.

Suppose that each site on a catalyst has the same RDS and the same most abundant intermediate, but that the sites have a distribution of apparent activation parameters. To avoid complications with complex multistep rate laws, assume the kinetics on each site approximately follows a power law $r(\mathbf{x}) = k(\mathbf{x}) \prod_i c_i^{\alpha_i}$ where $k(\mathbf{x})$ is a rate constant with Arrhenius temperature dependence. Here the variable $\mathbf{x}$ represents the local site geometry and the rate constant on a site with geometry $\mathbf{x}$ is [133]

$$k(\mathbf{x}) = A(\mathbf{x})exp[-\beta E_a(\mathbf{x})] \tag{4.6.1}$$

with $\beta = 1/(k_B T)$. We further assume that the prefactor $A(\mathbf{x})$ depends on geometry but not on temperature. Then the overall rate is

$$\langle k \rangle = \int k(\mathbf{x})\rho(\mathbf{x})d\mathbf{x} \tag{4.6.2}$$

where $\rho(\mathbf{x})$ is the probability density for sites of specific geometry $\mathbf{x}$. Equation (4.6.2) also serves to define the quenched average indicated by the brackets $\langle\cdot\rangle$. During catalyst operation, even over a range of different temperatures, we assume that $\rho(\mathbf{x})$ is a static distribution that was permanently quenched-in during the catalyst synthesis protocol.

The distribution $\rho(\mathbf{x})$ can be projected onto the activation energy E_a to obtain $\rho(E_a) \equiv \int d\mathbf{x}\,\rho(\mathbf{x})\delta[E_a - E_a(\mathbf{x})]$. $\rho(E_a)$ is a density of sites as a function of the activation energy, analogous to a density of states at energy E in equilibrium statistical mechanics. Also define

$$k(E_a) \equiv \rho(E_a)^{-1} \int d\mathbf{x}\,\rho(\mathbf{x})k(\mathbf{x})\delta[E_a - E_a(\mathbf{x})]$$

so that

$$\langle k\rangle = \int k(E_a)\rho(E_a)dE_a$$

Using these formulas, the slope of the Arrhenius plot is [133]

$$\partial ln\,\langle k\rangle\,/\partial\beta = -\,\langle E_a\rangle_k \tag{4.6.3}$$

where

$$\langle E_a\rangle_k = \langle k\rangle^{-1} \int k(E_a)\,\rho(E_a)\,E_a\,dE_a$$

is a k-weighted average of the activation energy. The k-weighted activation energy is always smaller than the average activation energy in the unweighted distribution $\rho(E_a)$. The following example considers the Arrhenius slope for a distribution $\rho(E_a)$ that is Gaussian.

Example: Arrhenius parameters from Gaussian $\rho(E_a)$

Consider a catalyst for which the distribution of activation energies is

$$\rho(E_a) = \frac{1}{\sigma_{E_a}\sqrt{2\pi}} exp[-\frac{(E_a - E_{a,0})^2}{2\sigma_{E_a}^2}]$$

with mean $E_{a,0}$ and standard deviation σ_{E_a}. If the prefactor A is $\mathbf{x}$-independent, then equation (4.6.2) for $\langle k\rangle$ gives

$$\langle k\rangle = A\,exp[-\beta E_{a,0} + \beta^2\sigma_{Ea}^2/2]$$

Therefore, the apparent rate constant is *faster than* the rate constant at the average activation energy. Likewise, the apparent activation energy is smaller than $E_{a,0}$:

$$-\partial ln\,\langle k\rangle\,/\partial\beta = E_{a,0} - \beta\sigma_{Ea}^2 \tag{4.6.4}$$

Equation (4.6.4) shows that [133]

1. The apparent Arrhenius slope depends on the temperature during catalyst operation ($\beta = 1/k_B T$) because high temperatures allow sites with higher activation energies to contribute.
2. The catalyst synthesis protocol influences the mean and width of the distribution $\rho(E_a)$ which in turn influences the activation energy.

The Gaussian density of sites model is a useful illustration, but the actual activation energy distribution may soon be available from *in silico* models of catalysts on amorphous supports. Note that the statistical framework which is outlined above can also account for activation energy distributions that emerge from an equilibrium ensemble of active sites [128].

Exercises

1. Derive the rate law for competitive inhibition, equation (4.2.15).
2. [S.L. Scott] For an enzyme with Michaelis-Menten kinetics in a well-stirred batch reactor, the substrate concentration follows the equation

$$\frac{d[\mathbf{A}]}{dt} = -\frac{V_{max}[\mathbf{A}]}{[\mathbf{A}] + K_M}$$

   Show that

$$\frac{[\mathbf{A}]}{K_M} = W\left(\frac{[\mathbf{A}]_0}{K_M} exp\left[\frac{[\mathbf{A}]_0 - V_{max}t}{K_M}\right]\right)$$

   where W is the Lambert-W function such that $ye^y = x \Leftrightarrow y = W(x)$ for all $y > 0$. Plot $[\mathbf{A}]/K_M$ as a function of $V_{max}t/K_M$. Note that the same result applies for Briggs-Haldane and Langmuir-Hinshelwood kinetics.
3. Consider a tetrameric enzyme that has two forms R and T. The R form binds $\mathbf{A}$ at active sites, while the T form binds $\mathbf{A}$ at inhibitor sites.
   (a) Suppose that the binding of $\mathbf{A}$ to each active site and each inhibitor site is independent with equilibrium constants

$$K_R = \frac{[*R][\mathbf{A}]}{[\mathbf{A} * R]} \quad and \quad K_T = \frac{[*T][\mathbf{A}]}{[\mathbf{A} * T]}$$

   Now define states of the enzyme according to the R, T-form and the number of $\mathbf{A}$ substrates bound. The states of the enzyme interconvert through the series of reactions shown below.

$$\begin{array}{ccccccccc} \mathbf{A}_4R & \overset{K_{R4}}{\rightleftarrows} & \mathbf{A}_3R & \overset{K_{R3}}{\rightleftarrows} & \mathbf{A}_2R & \overset{K_{R2}}{\rightleftarrows} & \mathbf{A}_1R & \overset{K_{R1}}{\rightleftarrows} & R \\ & & & & & & & & \updownarrow K \\ \mathbf{A}_4T & \overset{K_{T4}}{\rightleftarrows} & \mathbf{A}_3T & \overset{K_{T3}}{\rightleftarrows} & \mathbf{A}_2T & \overset{K_{T2}}{\rightleftarrows} & \mathbf{A}_1T & \overset{K_{T1}}{\rightleftarrows} & T \end{array}$$

Give an expression for each equilibrium constant K_{Ri} and K_{Ti} that accounts for the number of available binding sites.

(b) In the absence of reaction, show that the fraction Y of sites that are occupied by **A** *and* in the R state is

$$Y = \frac{y(1+y)^{n-1}}{(1+y)^n + K(1+cy)^n}$$

where $y = [\mathbf{A}]/K_R$, $c = K_R/K_T$, K, and $n = 4$ for a tetrameric enzyme. Hint: see Monod, Wyman, and Changeux, *J. Mol. Biol.* 12, 88–118 (1965).

(c) Plot Y vs. $log_{10}y$ for the case $K = 0.1$ and $c = 10$. Also plot Y vs. $log_{10}y$ for the case $K = 1$ and $c = 1$. Explain the metabolic significance of these two cases.

4. For certain applications, e.g. in computing reactor volumes in engineering [52], we might *want* the best possible model for $1/r$. Suppose the reactor has a catalyst with a Michaelis-Menten (or Langmuir-Hinshelwood) rate law. If the goal is to accurately predict $1/r$ at all concentrations, are the error distortions in a Lineweaver-Burk plot deleterious or helpful?

5. The catalytic hydrogenation of formaldehyde ("**A**" $= CH_2O$) to methanol ("**B**" $= CH_3OH$) has overall stoichiometry $\mathbf{A} + H_2 \rightleftarrows \mathbf{B}$.

(a) Initially, when $[\mathbf{B}] \approx 0$ the reverse reaction is negligible. The initial rate has been measured for a known mass of catalyst as a function of gas phase $[H_2]$ and $[\mathbf{A}]$ concentrations.

$[H_2]$/M	$[\mathbf{A}]$/M	r/[mol/(g cat s)]
1.0	1.0	3.1
1.0	4.0	12.4
10.0	1.0	10.0

Determine the reaction orders with respect to H_2 and **A**. Additionally, estimate the rate constant k for the empirical power law expression $r = k[H_2]^x[\mathbf{A}]^y$.

(b) The proposed Eley-Rideal type mechanism involves a partially hydrogenated surface intermediate ("$\mathbf{I}*$" $= *CH_2OH$) with elementary steps:

$$H_{2(g)} + 2* \rightleftharpoons 2H* \qquad K_{H2} = \theta_H^2/\theta_*^2[H_2]$$

$$\mathbf{A}_{(g)} + H* \underset{k_{-1}}{\overset{k_1}{\rightleftarrows}} \mathbf{I}* \qquad \textit{reversible RDS}$$

$$\mathbf{I}* + H* \rightleftharpoons \mathbf{B}_{(g)} + 2* \qquad K_B = \theta_I\theta_H/\theta_*^2[\mathbf{B}]$$

Note that all equilibrium constants are written for the adsorption process, regardless of whether the overall reaction requires adsorption or desorption. Derive the rate law for this mechanism.

6. [Helfferich] A proposed mechanism for base catalyzed aldol condensation of aldehyde **A** is shown below.

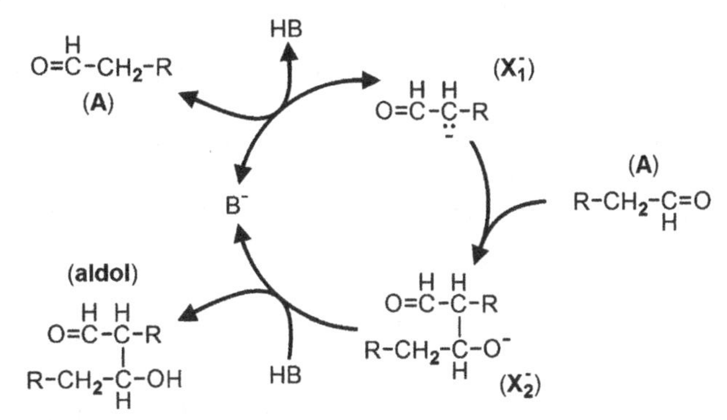

(a) Derive a rate law for production of the aldol. Assume double-headed and single-headed arrows correspond to reversible and irreversible steps, respectively.

(b) Sketch the predicted dependence of the rate on pH.

(c) Suppose the solution contains two aldehydes, one with group R and the other with group R'. What is the rate of production of the aldol containing two R groups?

7. [Helfferich] Consider the effects of different assumptions about the aldol condensation process in problem (6). Suppose the step $\mathbf{A} + \mathbf{B}- \rightleftarrows \mathbf{X}_1^- + H\mathbf{B}$ is quasi-equilibrated, the step $\mathbf{X}_1^- + \mathbf{A} \rightleftarrows \mathbf{X}_2^-$ is reversible, and (as before) the step $H\mathbf{B} + \mathbf{X}_2^- \rightarrow$ **aldol** is irreversible.

(a) Show that the rate law becomes

$$r_{aldol} = \frac{k[\mathbf{A}]^2[\mathbf{B}^-]}{1 + K[H\mathbf{B}]}$$

(b) Is the rate law in part (2a) kinetically distinguishable from that in part (1)?

(c) Find k and K using the following (illustrative) data. Compare the parameter estimates from a Lineweaver-Burke type linearization and from Eadie-Hofstee type linearization.

$[\mathbf{B}^-]/M$	$[H\mathbf{B}]/M$	$(r_{aldol}/[\mathbf{A}]^2)/(M^{-1}min^{-1})$
.099	0.011	0.0237
.099	0.033	0.0173
.099	0.068	0.0139
.099	0.090	0.0119
.030	0.030	0.0065
.059	0.059	0.0093
.171	0.171	0.0157

8. Consider the sequence of elementary steps

$$\begin{array}{cccl}
\mathbf{A}+* & \underset{}{\overset{K_{\mathbf{A}}}{\rightleftharpoons}} & \mathbf{A}* & \theta_{\mathbf{A}} = K_{\mathbf{A}}[\mathbf{A}]\theta_* \\
\mathbf{A}*+\mathbf{B} & \underset{k_{-1}}{\overset{k_1}{\rightleftharpoons}} & \mathbf{AB}* & r = k_1\sigma\theta_{\mathbf{A}}[\mathbf{B}] - k_{-1}\sigma\theta_{\mathbf{AB}} \\
\mathbf{AB}* & \underset{k_{-2}}{\overset{k_2}{\rightleftharpoons}} & \mathbf{C}*+\mathbf{D} & r = k_2\sigma\theta_{\mathbf{AB}} - k_b\sigma\theta_{\mathbf{C}}[\mathbf{D}] \\
\mathbf{C}* & \underset{K_{\mathbf{C}}}{\rightleftharpoons} & \mathbf{C}+* & \theta_{\mathbf{C}} = K_{\mathbf{C}}[\mathbf{C}]\theta_* \\
\hline
\mathbf{A}+\mathbf{B} & \rightleftharpoons & \mathbf{C}+\mathbf{D} &
\end{array}$$

(a) Derive the rate law assuming that step $\mathbf{A}*+\mathbf{B} \rightleftharpoons \mathbf{AB}*$ is the RDS and that all other steps are QE.
(b) Derive the rate law assuming that step $\mathbf{AB}* \rightleftharpoons \mathbf{C}*+\mathbf{D}$ is the RDS and that all other steps are QE.
(c) Derive the rate law without assuming an RDS, using only the PSSA and QE assumptions as indicated for the adsorption/desorption steps.
(d) Compare $r_{(c)}$ vs. $(1/r_{(a)} + 1/r_{(b)})^{-1}$.

9. Use transition state theory and the (exact) Christiansen formula to derive the energy span formula of Kozuch and Shaik (equation (4.4.6)). Then read the paper by Kozuch and Martin, *ACS Catalysis*, 2, 2787 (2012) as well as the critique by Lente, *ACS Catalysis*, 3, 381, (2013). Respond to each of the nine points raised by Lente.
10. Temperature programmed desorption experiments can provide the distribution of desorption energies for non-uniform sites [147–150]. Likewise, temperature programmed decompositions of preadsorbed molecules can determine the activation energy distribution for a specific decomposition step [151,152]. Why is it fundamentally more difficult to probe the activation energy distribution for the overall catalytic rate law?
11. By analogy to the Gibbs-Feynman-Bogulioubov-Jensen inequality, argue that the TOF for a site at the mean activation energy in a non-uniform active site distribution is always smaller than (or equal to) the site-averaged TOF.
12. Suppose that a fraction f of the sites on a catalyst perform a reaction with activation energy E_a and a fraction $(1-f)$ perform the same reaction with activation energy $E_a + \Delta E_a$ and with the same prefactor. Find the apparent activation energy in terms of f, E_a, ΔE_a, and the temperature.
13. Suppose an activation energy distribution that is a power law starting at the absolute lowest possible activation energy $E_{a,min}$:

$$\rho(E_a) \sim \gamma \cdot (E_a - E_{a,min})^\alpha$$

for $E_{a,min} \leq E_a \lesssim E_{a,min} + 2(\alpha + 1)k_B T$. Beyond this leftmost tail region, the shape of the distribution is largely irrelevant except that its normalization serves to fix γ. Find the apparent activation energy that results from the power law distribution.

References

[1] National Academies Press, Catalysis Looks to the Future, National Research Council, Washington DC, 1992.
[2] M. Polanyi, Z. Electrochem. 27 (1921) 142.
[3] J.H. Espenson, Chemical Kinetics and Reaction Mechanisms, McGraw-Hill, New York, 1995.
[4] D. Marx, M.E. Tuckerman, J. Hutter, M. Parrinello, Nature 397 (1999) 601–604.
[5] G.M. Loudon, J. Chem. Educ. 68 (1991) 973.
[6] B. Peters, B.L. Trout, Biochemistry 45 (2006) 5384–5392.
[7] K. Patel, R.T. Borchardt, Pharm. Res. 7 (1990) 703–711.
[8] S. Capasso, S. Mazarella, J. Chem. Soc. Perkin Trans. (1993).
[9] A. Radzicka, R. Wolfenden, Science 267 (1995) 90–93.
[10] O. Moe, R. Cornelius, J. Chem. Educ. 65 (1988) 137–141.
[11] S.T. Oyama, G.A. Somorjai, J. Chem. Educ. 65 (1990) 765–769.
[12] B. Erickson, J.E. Nelson, P. Winters, Biotechnol. J. 7 (2012) 176–185.
[13] G.M. Mccorkle, S. Anman, M. Scovell, J. Chem. Educ. 64 (1987) 221–226.
[14] P.D. Boyer, The Enzymes, 3rd ed., Academic Press, New York, 1976.
[15] E. Fischer, Ber. Dtsch. Chem. Ges. 27 (1894) 2985–2993.
[16] D.E. Koshland, Angew. Chem., Int. Ed. Engl. (1994) 2375–2378.
[17] W. Kauzmann, Adv. Protein Chem. 14 (1959) 1–63.
[18] M.J. Snider, R. Wolfenden, Biochemistry 40 (2001) 11364–11371.
[19] L. Pauling, Chem. Eng. News 24 (1946) 1375–1377.
[20] R. Wolfenden, Nature 223 (1969) 704–705.
[21] S.J. Benkovic, S. Hammes-Schiffer, Science 301 (2003) 1196–1202.
[22] L. Jiang, E.A. Althoff, F.R. Clemente, L. Doyle, D. Röthlisberger, A. Zanghellini, J.L. Gallaher, J.L. Betker, F. Tanaka, C.F. Barbas, D. Hilvert, K.N. Houk, B.L. Stoddard, D. Baker, Science 319 (2008) 1387–1391.
[23] D. Rothlisberger, O. Khersonsky, A.M. Wollacott, L. Jiang, J. DeChancie, J. Betker, J.L. Gallaher, E.A. Althoff, A. Zanghellini, O. Dym, S. Albeck, K.N. Houk, D.S. Tawfik, D. Baker, Nature 453 (2008) 190–195.
[24] R. Wolfenden, Biophys. Chem. 105 (2003) 559–572.
[25] M.I. Page, in: M.I. Page (Ed.), The Chemistry of Enzyme Action, Elsevier, Amsterdam, 1984, pp. 1–54.
[26] P.C. Engel, in: M.I. Page (Ed.), The Chemistry of Enzyme Action, Elsevier, Amsterdam, 1984, pp. 73–110.
[27] D.B. Northrop, J. Chem. Educ. 75 (1998) 1153–1158.
[28] G.E. Briggs, J.B.S. Haldane, Biochem. J. 19 (1925) 338–339.
[29] L. Michaelis, M.L. Menten, Biochem. Z. 49 (1913) 333–369.
[30] R. Chang, Physical Chemistry for the Biosciences, University Science Books, Sausalito, CA, 2005.
[31] W.W. Cleland, Acc. Chem. Res. 8 (1974) 145–151.
[32] M. Garcia-Viloca, J. Gao, M. Karplus, D.G. Truhlar, Science 303 (2004) 186–195.
[33] P.W. van Leeuwen, Homogeneous Catalysis: Understanding the Art, Kluwer, Dordrecht, 2004.
[34] M. Beller, A. Renken, R.A. van Santen, Catalysis: From Principles to Applications, Wiley-VCH, Weinheim, 2012.
[35] G.C. Bond, Heterogeneous Catalysis: Principles and Applications, Clarendon, Oxford, 1987.
[36] R.A. van Santen, J.W. Niemansverdriet, Chemical Kinetics and Catalysis, Plenum Press, New York, 1995.
[37] C.W. Jones, Top. Catal. 53 (2010) 942–952.
[38] E. Skulason, G.S. Karlberg, J. Rossmeisl, T. Bligaard, J. Greeley, H. Jonsson, J.K. Norskov, Phys. Chem. Chem. Phys. 9 (2007) 3241–3250.

[39] K.-Y. Yeh, M.J. Janik, J. Comput. Chem. 32 (2011) 3399–3408.
[40] A.K. Datye, Top. Catal. 13 (2000) 131.
[41] U.I. Gaya, A.H. Abdullah, J. Photochem. Photobiol., C, Photochem. Rev. 9 (2008) 1–12.
[42] C. Stegelmann, N.C. Schiodt, C.T. Campbell, P. Stoltze, J. Catal. 221 (2004) 630–649.
[43] M. Daage, R.R. Chianelli, J. Catal. 149 (1994) 414–427.
[44] J. Cejka, A. Corma, S. Zones, Zeolites and Catalysis, Wiley-VCH, Weinheim, 2010.
[45] J.M. Thomas, Proc. R. Soc. A, Math. Phys. Eng. Sci. 468 (2012) 1884.
[46] R.A. van Santen, M. Neurock, Molecular Heterogeneous Catalysis: A Conceptual and Computational Approach, Wiley-VCH, Weinheim, 2006.
[47] J.K. Norskov, F. Studt, F. Abild-Pederson, T. Bligaard, Fundamental Concepts in Heterogeneous Catalysis, Wiley, Hoboken, NJ, 2014.
[48] C.H. Bartholomew, Appl. Catal. A, Gen. 212 (2001) 17–60.
[49] J.A. Moulijn, A.E. Van Diepen, F. Kapteijn, Appl. Catal. A, Gen. 212 (2001) 3–16.
[50] C.S. Ewing, M.J. Hartmann, K.R. Martin, A.M. Musto, S.J. Padinjarekutt, E.M. Weiss, G. Veser, J.J. McCarthy, J.K. Johnson, D.S. Lambrecht, J. Phys. Chem. C 119 (2015) 2503–2512.
[51] B. Rosendall, B. Finlayson, Comput. Chem. Eng. 19 (1995) 1207–1218.
[52] G.F. Froment, K.B. Bischoff, Chemical Reactor Analysis and Design, Wiley, Hoboken, NJ, 1990.
[53] R. Aris, Chem. Eng. Sci. 6 (1957) 262–268.
[54] M.E. Davis, R.J. Davis, Fundamentals of Chemical Reaction Engineering, McGraw-Hill, New York, 2003.
[55] P.B. Weisz, J.S. Hicks, Chem. Eng. Sci. 17 (1962) 265.
[56] C.N. Satterfield, Mass Transfer in Heterogeneous Catalysis, MIT Press, Cambridge, MA, 1970.
[57] J.J. Carberry, Chemical and Catalytic Reactor Engineering, Dover, Minneola, NY, 2001.
[58] B. Peters, Chem. Eng. Sci. 71 (2012) 367–374.
[59] E.L. Cussler, Diffusion: Mass Transfer in Fluid Systems, 3rd ed., Cambridge Press, Cambridge, UK, 2009.
[60] M. Boudart, Ind. Eng. Chem. Res. 25 (1986) 656–658.
[61] D.S. Sholl, J.A. Steckel, Density Functional Theory: A Practical Introduction, Wiley, Hoboken, NJ, 2009.
[62] J. Sauer, Chem. Rev. 89 (1989) 199–255.
[63] R.A. van Santen, G.J. Kramer, Chem. Rev. 95 (1995) 637–660.
[64] B.L. Trout, A.K. Chakraborty, A.T. Bell, J. Phys. Chem. (ISSN 0022-3654) 100 (1996) 17582–17592.
[65] D. Vlachos, L. Schmidt, R. Aris, J. Chem. Phys. 93 (1990) 8306–8313.
[66] D.W. Goodman, Chem. Rev. 95 (1995) 523–536.
[67] G. Ertl, H.-J. Freund, Phys. Today 52 (1999) 32.
[68] J.R. Morris, J.N. Russell, C.J. Karwacki, J. Phys. Chem. Lett. 6 (2015) 4923–4926.
[69] R. Ouyang, J.X. Liu, W.X. Li, J. Am. Chem. Soc. 135 (2013) 1760–1771.
[70] L.H. Germer, A.U. MacRae, Proc. Natl. Acad. Sci. USA 48 (1962) 997–1000.
[71] J. Rogal, K. Reuter, M. Scheffler, Phys. Rev. Lett. (2007).
[72] J. Sehested, J.A.P. Gelten, I.N. Remediakis, H. Bengaard, J.K. Norskov, J. Catal. 223 (2004) 432–443.
[73] A.V. Ruban, H.L. Skriver, J.K. Norskov, Phys. Rev. B 59 (1999) 15990.
[74] S.A. Tupy, A.M. Karim, C. Bagia, W. Deng, Y. Huang, D.G. Vlachos, J.G. Chen, ACS Catal. 2 (2012) 2290–2296.
[75] M. Hunger, J. Weitkamp, Angew. Chem., Int. Ed. Engl. 40 (2001) 2954–2971.
[76] W.M.H. Sachtler, J. Fahrenfort, Proc. Int. Congr. Catal. 2nd, Paris 1 (1961) 831.
[77] M. Boudart, Kinetics of Chemical Processes, Prentice Hall, Englewood Cliffs, NJ, 1968.
[78] E. Shustorovich, Surf. Sci. Rep. 31 (1998) 1–119.
[79] B. Hammer, J.K. Norskov, Adv. Catal. 45 (2000).
[80] C.H. Christensen, J.K. Norskov, J. Chem. Phys. 128 (2008) 182503.
[81] J. Greeley, M. Mavrikakis, Nat. Mater. 3 (2004) 810–815.
[82] C.J.H. Jacobsen, S. Dahl, B.G.S. Clausen, S. Bahn, A. Logadottir, J.K. Norskov, J. Am. Chem. Soc. 123 (2001) 8404–8405.

[83] D.A. Hansgen, D.G. Vlachos, J.G. Chen, Nat. Chem. 2 (2010) 484–489.
[84] J.K. Norskov, F. Abild-Pedersen, F. Studt, T. Bligaard, Proc. Natl. Acad. Sci. USA 108 (2011) 937–943.
[85] Z. Ulissi, V. Prasad, D.G. Vlachos, J. Catal. 281 (2011) 339–344.
[86] K. Reuter, D. Frenkel, M. Scheffler, Phys. Rev. Lett. 93 (2004) 116105.
[87] C.N. Hinshelwood, The Kinetics of Chemical Change, Clarendon, Oxford, 1940.
[88] C. Hinshelwood, Ann. Rep. Prog. Chem. 24 (1927) 314–317.
[89] O.A. Hougen, K.M. Watson, Chemical Process Principles, vol. III, Wiley, New York, 1943.
[90] M. Boudart, G. Djega-Mariadassou, Kinetics of Heterogeneous Catalytic Reactions, Princeton University Press, Princeton, 1984.
[91] R. Aris, Am. Sci. 58 (1970) 420–428.
[92] F.G. Helfferich, Elsevier, Amsterdam, 2004.
[93] S. Kiperman, K. Kumbilieva, L.A. Petrov, Ind. Eng. Chem. Res. 28 (1989) 379–380.
[94] N. Mardirossian, J.A. Parkhill, M. Head-Gordon, Phys. Chem. Chem. Phys. 13 (2011) 19325–19337.
[95] Y. Zhao, D.G. Truhlar, J. Chem. Theory Comput. 7 (2011) 669–676.
[96] T. Keii, Heterogeneous Kinetics: Theory of Ziegler-Natta-Kaminsky Polymerization, Springer, New York, 2004.
[97] J. Christiansen, Adv. Catal. 5 (1953) 311–353.
[98] M. Temkin, Int. Chem. Eng. 11 (1971) 709.
[99] S. Kozuch, S. Shaik, Acc. Chem. Res. 44 (2011) 101–110.
[100] M.P. McDaniel, Adv. Catal. 53 (2010) 123–606.
[101] M.F. Delley, F. Nunez-Zarur, M.P. Conley, A. Comas-Vives, G. Siddiqi, S. Norsic, V. Monteil, O.V. Safonova, C. Coperet, Proc. Natl. Acad. Sci. USA 111 (2014) 11624–11629.
[102] B. Peters, S.L. Scott, A. Fong, Y. Wang, A.E. Stiegman, Proc. Natl. Acad. Sci. USA 112 (2015) 4160–4161.
[103] A. Fong, Y. Yuan, S.L. Scott, B. Peters, ACS Catal. 5 (2015) 3360–3374.
[104] K.R. Popper, Conjectures and Refutations, Routledge Classics, New York, 1963.
[105] E.L. King, C. Altman, J. Phys. Chem. 60 (1956) 1375–1378.
[106] M. Temkin, Adv. Catal. 28 (1979) 173–291.
[107] J.E. Straub, D. Thirumalai, Ann. Rev. Phys. Chem. 62 (2011) 437–463.
[108] M. Schor, J. Vreede, P.G. Bolhuis, Biophys. J. 103 (2012) 1296–1304.
[109] T. Ban, M. Hoshino, S. Takahashi, D. Hamada, K. Hasegawa, H. Naiki, Y. Goto, J. Mol. Biol. 344 (2004) 747–767.
[110] K. Garai, C. Frieden, Proc. Natl. Acad. Sci. USA 110 (2013) 3321–3326.
[111] T.P.J. Knowles, C. Waudby, G.L. Devlin, S. Cohen, A. Aguzzi, M. Vendruscolo, E.M. Terentjev, M.E. Welland, C.M. Dobson, Science 326 (2009) 1533–1537.
[112] S. Kozuch, S. Shaik, J. Phys. Chem. A 112 (2008) 6032–6041.
[113] K. Reuter, M. Scheffler, Phys. Rev. B 73 (2006) 45433.
[114] J. Rogal, K. Reuter, M. Scheffler, Phys. Rev. B 77 (2008) 155410.
[115] J.R. Murdoch, J. Chem. Educ. 58 (1987) 32–36.
[116] J.A. Dumesic, D.F. Rudd, L. Aparicio, J.E. Rekoske, A.A. Trevino, The Microkinetics of Heterogeneous Catalysis, American Chemical Society, Washington, 1993.
[117] J.A. Dumesic, J. Catal. 204 (2001) 525–529.
[118] C. Campbell, J. Catal. 204 (2001) 520–524.
[119] C. Stegelmann, A. Andreasen, C.T. Campbell, J. Am. Chem. Soc. 131 (2009) 8077–8082.
[120] S.T. Dix, J.K. Scott, R.B. Getman, C.T. Campbell, Faraday Discuss. 188 (2016) 21–38.
[121] R.D. Cortright, J.A. Dumesic, Adv. Catal. 46 (2002) 161–264.
[122] E. Rideal, W.A. Bone, G. Ingle-Finch, E.C.C. Baly, W.C.M. Lewis, E. Edser, E.F. Armstrong, T.P. Halditch, H.E. Holtorp, I. Langmuir, S. Arrhenius, Trans. Faraday Soc. 17 (1922) 655–675.
[123] H.S. Taylor, Proc. R. Soc. A 108 (1925) 105–111.
[124] F.H. Constable, Proc. R. Soc. A, Math. Phys. Eng. Sci. 108 (1925) 355–378.

[125] D. Halsey, J. Chem. Phys. 17 (1949) 758–761.
[126] P.A. Lee, P.H. Citrin, P. Eisenberger, B.M. Kincaid, Rev. Mod. Phys. 53 (1981) 769–806.
[127] R. Cipullo, S. Mellino, V. Busico, Macromol. Chem. Phys. 215 (2014) 1728–1734.
[128] C. Wu, D.J. Schmidt, C. Wolverton, W.F. Schneider, J. Catal. 286 (2012) 88–94.
[129] J.C. Mol, Catal. Today 51 (1999) 289.
[130] A.W. Moses, C. Raab, R.C. Nelson, H.D. Leifeste, N.A. Ramsahye, S. Chattopadhyay, J. Eckert, B.F. Chmelka, S.L. Scott, J. Am. Chem. Soc. 129 (2007) 8912–8920.
[131] R.A. Sheldon, M.C.A. van Vliet, in: R.A. Sheldon, H. van Bekkum (Eds.), Fine Chemicals Through Heterogeneous Catalysis, Wiley-VCH, Weinheim, 2001, pp. 473–490.
[132] W.C. Vining, J. Strunk, A.T. Bell, J. Catal. 281 (2011) 115.
[133] B. Peters, S.L. Scott, J. Chem. Phys. 142 (2015) 104708.
[134] J. Sauer, P. Ugliengo, E. Garrone, V.R. Saunders, Chem. Rev. 94 (1994) 2095–2160.
[135] O. Espelid, K.J. Borve, J. Catal. 205 (2002) 366–374.
[136] A.T. Bell, M. Head-Gordon, Ann. Rev. Chem. Biomol. Eng. 2 (2011) 453–477.
[137] J. Handzlik, J. Phys. Chem. C 111 (2007) 9337–9348.
[138] J. Handzlik, J. Ogonowski, J. Phys. Chem. C 116 (2012) 5571–5584.
[139] F. Tielens, C. Gervais, J. Franc, F. Mauri, D. Costa, Chem. Mater. 3 (2008) 3336–3344.
[140] P. Ugliengo, M. Sodupe, F. Musso, I.J. Bush, R. Orlando, R. Dovesi, Adv. Mater. 20 (2008) 4579–4583.
[141] C.S. Ewing, S. Bhavsar, G. Veser, J.J. McCarthy, J.K. Johnson, Langmuir 30 (2014) 5133–5141.
[142] H. Guesmi, R. Grybos, J. Handzlik, F. Tielens, Phys. Chem. Chem. Phys. 16 (2014) 18253–18260.
[143] D.C. Tranca, A. Wojtaszek-Gurdak, M. Ziolek, F. Tielens, Phys. Chem. Chem. Phys. 17 (2015) 22402–22411.
[144] B.R. Goldsmith, A.M. Fong, B. Peters, in: K. Han (Ed.), Reaction Rate Computations: Theories and Applications, RSC Publishing, 2013, pp. 213–221.
[145] B.R. Goldsmith, E.D. Sanderson, D. Bean, B. Peters, J. Chem. Phys. 138 (2013) 204105.
[146] J.A. Boscoboinik, X. Yu, B. Yang, F.D. Fischer, R. Wlodarczyk, M. Sierka, S. Shaikhutdinov, J. Sauer, H.-J. Freund, Angew. Chem., Int. Ed. 51 (2012) 6005–6008.
[147] M. Temkin, V. Levich, Russ. J. Phys. Chem. 20 (1946) 1441.
[148] U. Landman, E.W. Montroll, J. Chem. Phys. 64 (1976) 1762–1767.
[149] Z. Du, A.F. Sarofim, J.P. Longwell, Energy Fuels 4 (1990) 296–302.
[150] M. Arai, Y. Nishiyama, T. Masuda, K. Hashimoto, Appl. Surf. Sci. 89 (1995) 11–19.
[151] J.J.G. van Bokhoven, A.E.T. Kuiper, J. Medema, J. Catal. 43 (1976) 181–191.
[152] J.J.G. van Bokhoven, A.E.T. Kuiper, J. Medema, J. Catal. 180 (1976) 168–180.

CHAPTER 5

Diffusion control

Rapid coagulation and chemical reaction processes are opposing limits in chemical kinetics. The former is a pure diffusional phenomenon, while the latter involves yet unknown valence phenomena that cause only a minimal fraction of collision encounters to result in chemical transformation.

Smoluchowski, transl., Z. Phys. Chem. (1917)

Chapter 4 briefly discussed the transport limitations that are encountered in heterogeneous catalysis. Those were associated with specific length scales: the reactor diameter, the pellet diameter, the pore diameter, etc. This chapter discusses a different type of diffusion control – one which sets a fundamental speed limit on bimolecular reactions in a liquid solution. The only length scale involved is the range of interactions between the reactants and solvent molecules. The earliest version of the theory, by Smoluchowski in 1917 [1], is simple and yet surprisingly accurate for many real reactions. This chapter also presents a few of the many theories [2] for partial diffusion control and diffusion control under the influence of long range interactions. Finally, we present the method of Northrup, Allison, and McCammon for computing the rates of diffusion controlled reactions between species with non-spherically symmetric interactions [3,4].

5.1 Complete diffusion control

Consider hard spheres **A** and **B** that react instantaneously upon contact. Let the radii of **A** and **B** be $R_{\mathbf{A}}$ and $R_{\mathbf{B}}$, respectively. Initially, imagine that the spheres of type **A** are stationary targets and that the **B** spheres are diffusing. Because **B** reacts with **A** upon contact, the concentration of **B** at a distance $r = R_{\mathbf{A}} + R_{\mathbf{B}}$ from the **A**-center must be zero, as shown in Figure 5.1.1. The rate of transport of **B** to the surface around **A** by diffusion is

$$\overline{N}_{\mathbf{B}} = 4\pi r^2 D_{\mathbf{B}} \frac{dc_{\mathbf{B}}}{dr} \tag{5.1.1}$$

where $\overline{N}_{\mathbf{B}}$ is a constant representing the rate of reaction per molecule **A** in mol/s. The sign convention is that $\overline{N}_{\mathbf{B}} > 0$ means transport to **A**, not away from **A**. Note that $c_{\mathbf{B}}(r)$ is not a true concentration [5]. Indeed to speak of a true concentration at the length scale $R_\mathbf{A} + R_\mathbf{B}$ is nonsense. Rather $c_\mathbf{B}(r)$ is a radial distribution function under the quasi-stationary rescue and replace conditions: each time the molecules reach $R_\mathbf{A} + R_\mathbf{B}$, we 'rescue' them and replace them at random locations in the bulk liquid. The rate at which molecules reach $R_\mathbf{A} + R_\mathbf{B}$ then corresponds to the purely diffusion controlled reaction rate per **A** molecule at conditions of steady bulk **B** concentration.

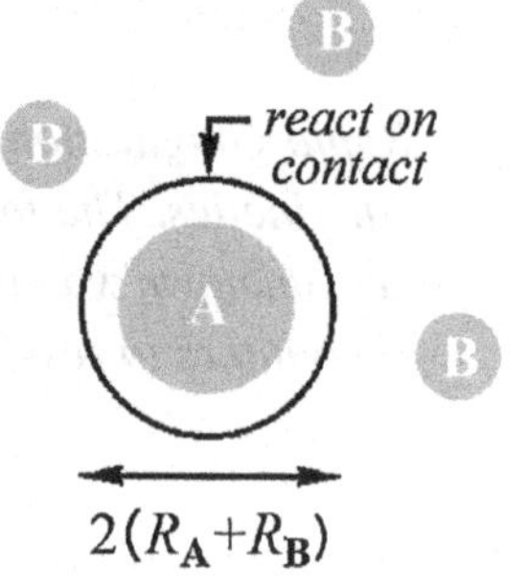

Figure 5.1.1: Showing the distance at which spherical reactants A and B react and its relation to the radii of A and B.

Now separate variables in equation (5.1.1) and integrate

$$\int_0^{c_\mathbf{B}(r)} dc_\mathbf{B} = \frac{\overline{N}_\mathbf{B}}{4\pi D_\mathbf{B}} \int_{R_\mathbf{A}+R_\mathbf{B}}^{r} \frac{dr}{r^2}$$

to obtain the concentration as a function of r,

$$c_\mathbf{B}(r) = \frac{\overline{N}_\mathbf{B}}{4\pi D_\mathbf{B}} \left(\frac{1}{R_\mathbf{A} + R_\mathbf{B}} - \frac{1}{r} \right) \tag{5.1.2}$$

Let $r \to \infty$ and use $c_B(\infty) = [\mathbf{B}]$ to obtain

$$\overline{N}_B = 4\pi D_\mathbf{B}(R_\mathbf{A} + R_\mathbf{B})[\mathbf{B}]$$

Remember that $N_\mathbf{B}$ is the rate of consumption of **B** per molecule **A**. Therefore the total rate of reaction per unit volume of liquid is given by $N_\mathbf{B}$ times the number of **A** molecules per unit volume, i.e.

$$\mathscr{r} = 4\pi D_\mathbf{B}(R_\mathbf{A} + R_\mathbf{B})[\mathbf{B}][\mathbf{A}]$$

Thus the rate constant between **A** and **B** under complete diffusion control with stationary **A** molecules is $k_D = 4\pi D_\mathbf{B}(R_\mathbf{A} + R_\mathbf{B})$.

Our assumption of static **A** molecules is perfectly acceptable when **A** is much larger than **B** or when **A** is actually a small active site on a large surface. However, for reactions between molecules in solution, both the **A** and **B** molecules are moving about by diffusion. The effective diffusivity for the distance between freely diffusing molecules **A** and **B** is the sum of their diffusivities.

$$D = (D_{\mathbf{A}} + D_{\mathbf{B}}) \tag{5.1.3}$$

This can be shown from a joint Smoluchowski (diffusion) equation for the positions of **A** and **B** which is then projected onto their center-to-center distance. More simply, note that the mean square displacement is a variance, and variances of independent random variables are additive: $\langle[\Delta(\mathbf{x_A} - \mathbf{x_B})]^2\rangle = \langle\Delta\mathbf{x}_\mathbf{A}^2\rangle + \langle\Delta\mathbf{x}_\mathbf{B}^2\rangle = 6D_\mathbf{A}t + 6D_\mathbf{B}t$. For diffusion in a liquid, the Stokes-Einstein equation [6] predicts $D_i = k_B T/6\pi\eta R_i$ where η is the fluid viscosity and R_i is a molecular (hydrodynamic) radius. Using Stokes-Einstein for both $D_\mathbf{A}$ and $D_\mathbf{B}$ gives

$$D = \frac{k_B T}{6\pi\eta}\left(\frac{1}{R_\mathbf{A}} + \frac{1}{R_\mathbf{B}}\right) \tag{5.1.4}$$

Thus the rate constant for complete diffusion control with both **A** and **B** molecules diffusing is $k_D = 4\pi(D_\mathbf{A} + D_\mathbf{B})(R_\mathbf{A} + R_\mathbf{B})$, or

$$k_D = \frac{2k_B T}{3\eta}\frac{(R_\mathbf{A} + R_\mathbf{B})^2}{R_\mathbf{A} R_\mathbf{B}} \tag{5.1.5}$$

Note that this rate constant will not show an Arrhenius temperature dependence, but the solvent viscosity in the denominator does tend to increase with temperature. Combining the typical temperature dependence of the viscosity with the explicit factor of T in the numerator suggests an overall temperature dependence like $k_D \sim T^{1.5-1.8}$. This relatively weak temperature dependence (much weaker than Arrhenius) is one hallmark of a diffusion controlled reaction.

We can obtain a typical diffusion control limit for reaction rates in solution by assuming $R_\mathbf{A} \approx R_\mathbf{B}$. In that case the diffusion controlled rate constant reduces to

$$k_D = \frac{8k_B T}{3\eta} \tag{5.1.6}$$

Amazingly, the diffusion controlled limit depends only weakly on the size of the molecules as long as they are of comparable sizes. The physical reason for this invariance is that the slower diffusion is exactly compensated by a larger area for reaction. For **A** and **B** of very different sizes k_D will be larger because the diffusion rate will be dominated by the smaller molecule and the area for reaction will be dominated by the larger molecule. For water at 25°C, $k_D = 7 \times 10^9 L/mol.s$. Thus no reaction in aqueous solution for molecules with short range interactions can occur faster than this k_D speed limit.

Example: Diffusion control in the gas phase?

Suppose first that every collision in a gas of hard spheres leads to a reaction. Then from collision theory (Chapter 6) we have $k_{coll} = \pi d^2 \langle |v| \rangle$. Now from the kinetic theory of gases, the diffusion constant for hard spheres is $D = \lambda \langle |v| \rangle /3$ where $\langle |v| \rangle$ is the mean absolute velocity and λ is the mean free path

$$\langle |v| \rangle = \sqrt{8k_BT/(\pi\mu)} \quad and \quad \lambda = k_BT/(\pi d^2 P\sqrt{2})$$

where P is pressure, μ is reduced mass for two collision partners, and d is the collision diameter. Using D we can estimate $k_D = 4\pi Dd$. Clearly, the diffusion controlled gas reaction rate and the collision theory rate are not equivalent. Their ratio is

$$\frac{k_D}{k_{coll}} = \frac{4\lambda}{3d}$$

At room temperature and moderate pressures for a nearly ideal gas, the mean free path is much longer than the collision diameter, so the diffusion control theory overestimates the rate. Why? Equation (5.1.5) requires that the mean free path be much smaller than the collision diameter. The reason can be seen in equation (5.1.2): a depleted zone in the diffusing concentration field around a reactant is nonsense when the mean free path exceeds the thickness of the depleted zone.

This section has considered the diffusion controlled rates of association reactions, but unimolecular dissociation reactions also face a diffusion limit. Detailed detailed balance requires the association (forward) and dissociation (backward) rates to have prefactors with the same dependence on viscosity.

5.2 Partial diffusion control

When the resistance to reaction from diffusion (k_D^{-1}) is comparable to the resistance to reaction from the intrinsic kinetics (k_{rxn}^{-1}), the overall rate constant can be obtained by adding the resistances in series. Then the overall rate constant is

$$\frac{1}{k} = \frac{1}{k_D} + \frac{1}{k_{rxn}} \tag{5.2.1}$$

In many situations, one term on the right-hand-side turns out to be vanishingly small compared to the other, so that the slowest process dominates the observed kinetics. (See Figure 5.2.1.)

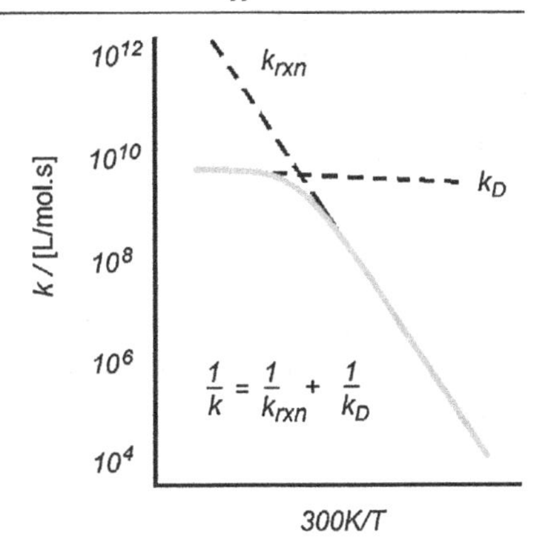

Figure 5.2.1: The weak temperature dependence of k_D and the strong Arrhenius-like temperature dependence of k_{rxn} lead to a crossover temperature at which the dominant resistance changes from reaction to diffusion. The gray curve shows how the partial diffusion control formula bridges these two regimes.

Assuming transition state theory, $k_{rxn} = k_B T h^{-1} V_o exp[-\beta \Delta G^{\ddagger}]$, we can ask what free energy of activation makes k_{rxn} equal to k_D ($7 \times 10^9 L/mol.s$). Using a reference concentration of $1M$, a temperature of $25°C$, and the viscosity of water, the crossover free energy barrier is $\Delta G^{\ddagger} = 4kcal/mol$. Thus diffusion resistances can influence the rates of reactions with barriers as high as $7k_BT$. The importance of diffusion control is perhaps especially important for biomolecular association processes, where reported barriers are often lower than $7k_BT$.

5.3 Diffusion control with long range interactions

The theory of Smoluchowski assumes that reactants are non-interacting up to a critical distance at which they instantly react. Collins and Kimball [5] generalized the Smoluchoswki treatment for cases with partially absorbing boundary conditions, but there are two problems with the Smoluchowski and Collins-Kimball models.

1. Real reactants interact with each other over a range of distances, especially for ions. These interactions influence the probability of close (and potentially reactive) encounters.
2. In some cases, e.g. in electron transfer, the reaction event can also occur with the reactants separated by a range of distances rather than at one specific distance.

This section shows how both of these simplified aspects of the Smoluchowski and Collins-Kimball models can be improved. We begin by following Debye[7] who retained a critical distance R^* at which the reaction is instantaneous, but who included the effects of a potentially long-ranged pair potential. Let $U(r)$ be the pair potential (potential of mean force) between two ions. Adding the chemical and electrostatic forces on ions as they approach now gives two contributions to the flux.

$$j = -D\left\{\frac{d\beta U(r)}{dr} + \frac{d}{dr}\right\} c_{\mathbf{B}}(r) \tag{5.3.1}$$

Again consider steady-state rescue and replace conditions. From the continuity equation $[\partial c_{\mathbf{B}}/\partial t = -\nabla \cdot j\,]$ at steady state in spherical coordinates and with no dependence on angular terms,

$$0 = \frac{1}{r^2}\frac{d}{dr}(r^2 j)$$

i.e. the divergence of j must be zero. From $d(r^2 j)/dr = 0$ we find that $r^2 j$ is constant, i.e. that the integrated flux through any spherical surface is a constant. Following our previous convention, define the constant $\overline{N}_{\mathbf{B}}$ to be the rate of reaction at a single molecule **A**

$$\overline{N}_{\mathbf{B}} = 4\pi r^2 D\left\{\frac{d\beta U(r)}{dr} + \frac{d}{dr}\right\} c_{\mathbf{B}}(r) \tag{5.3.2}$$

Now change variables using $c_{\mathbf{B}}(r)/[\mathbf{B}] = \xi(r)exp[-\beta U(r)]$ where $[\mathbf{B}]$ is the steady-state bulk **B** concentration. Equation (5.3.2) becomes

$$\frac{d\xi}{dr} = \frac{\overline{N}_{\mathbf{B}} e^{\beta U(r)}}{4\pi D[\mathbf{B}]r^2}$$

Integrating from $R*$ to ∞ gives

$$\xi(\infty) - \xi(R*) = \frac{\overline{N}_{\mathbf{B}}/[\mathbf{B}]}{4\pi D\lambda} \tag{5.3.3}$$

where λ is *defined* as the effective capture radius:

$$\lambda^{-1} \equiv \int_{R*}^{\infty} \frac{1}{r^2} e^{\beta U(r)} dr \tag{5.3.4}$$

Boundary conditions on $\xi(r)$ correspond to boundary conditions on $c_{\mathbf{B}}(r)$ as shown in the table.

	$\exp[\beta U(r)]$	$c_{\mathbf{B}}(r)$	$\xi(r)$
$r \to R*$	$\exp[\beta U(R*)]$	0	0
$r \to \infty$	1	$[\mathbf{B}]$	1

Inserting the transformed boundary conditions into equation (5.3.3) and solving for $\overline{N}_{\mathbf{B}}$ gives

$$\overline{N}_{\mathbf{B}} = 4\pi D\lambda[\mathbf{B}]$$

$\overline{N}_{\mathbf{B}}$ is the rate per **A** molecule-ion, so the rate in a unit volume is $\mathcal{R} = 4\pi D\lambda[\mathbf{B}][\mathbf{A}]$. Thus the diffusion controlled rate constant for reactions between strongly interacting ions or molecules is [7]

$$k_D = 4\pi D\lambda \tag{5.3.5}$$

with the capture radius λ defined in equation (5.3.4). The capture radius is a functional that decreases with increasing $\beta U(r)$ at all locations. For reactions between an anion and a cation, λ and k_D are large. For reactions between two cations or between two anions, λ and k_D are small.

Example: Diffusion control for reactions between ions

For Coulomb interactions, the pair potential in units of $k_B T$ can be expressed in terms of the Bjerrum length, i.e. the distance at which the electrostatic potential between two monovalent ions is $k_B T$ [7].

$$\beta U(r) = \frac{z_\mathbf{A} z_\mathbf{B}}{r/\ell_B} \quad with \quad \ell_B = \frac{e^2}{4\pi \varepsilon_0 \varepsilon k_B T}$$

For pure water at $300K$, $\varepsilon = 80$ and the Bjerrum length is approximately $\ell_B = 7\text{Å}$. The capture radius is

$$\lambda^{-1} = \int_{R*}^{\infty} \frac{exp[z_\mathbf{A} z_\mathbf{B} \ell_B / r]}{r^2} dr$$

which simplifies to

$$\lambda / R* = \frac{z_\mathbf{A} z_\mathbf{B} \ell_B / R*}{exp[z_\mathbf{A} z_\mathbf{B} \ell_B / R*] - 1}$$

Figure 5.3.1 shows $\lambda/R*$ as a function of $z_\mathbf{A} z_\mathbf{B} \ell_B / R*$. $\lambda/R*$ can be viewed as correction factor to the Smoluchowski hard-sphere result, which is recovered when $z_\mathbf{A} z_\mathbf{B} \ell_B / R* = 0$. Accordingly, reactions between like-charged ions are slower and reactions between oppositely-charged ions are faster than the hard sphere result.

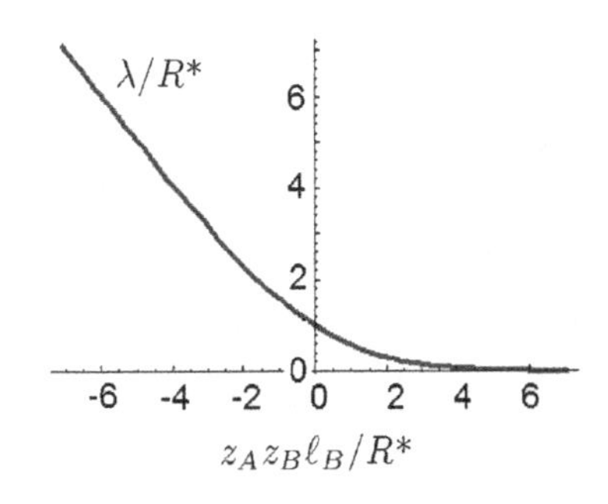

Figure 5.3.1: $\lambda/R*$ **as a function of** $z_\mathbf{A} z_\mathbf{B} \ell_B / R*$.

For reactions that are known to be diffusion limited and for which the reactant diffusivities are known, the experimental rate constants in combination with equation (5.3.5) provide an estimate of $R*$. For most reactions between small ions $R*$ is a few Angstroms. One exception is the reaction between H^+ and OH^- for which this analysis predicts $R* = 8.5\text{Å}$. This is thought to be due to proton wire transfer through one or more water molecules by the Grotthus mechanism as shown in Figure 5.3.2.

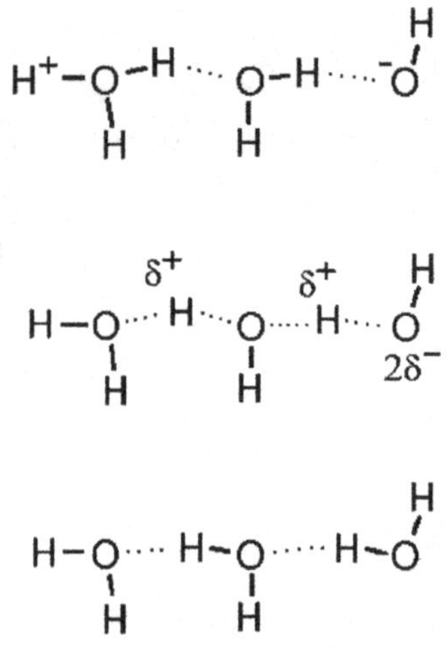

Figure 5.3.2: In the Grotthus mechanism of proton transport, protons are transferred across water molecules by the breaking of one O-H bond and the formation of another O-H bond. The figure depicts how the Grotthus mechanism can facilitate the annihilation of hydronium and hydroxide.

The Debye model still assumes that the reaction happens instantly (and only) at a specific distance $R*$. Electron transfer and some other types of reactions can occur at a range of distances, so a preferable model would include the rate as a function of r in the species balance equation. For a given distance r between **A** and **B**, the electron transfer between **A** and **B** occurs with first order kinetics. The rate of consumption of **B** per molecule of **A** at each r is $k_{ET}(r)c_{\mathbf{B}}(r)$ per unit volume. Thus the steady-state species balance equation is

$$0 = \frac{1}{r^2}\frac{d}{dr}(r^2\frac{dc_{\mathbf{B}}}{dr}) - k_{ET}(r)c_{\mathbf{B}}(r) \tag{5.3.6}$$

Again we can use the diffusion flux with a non-constant potential of mean force, equation (5.3.1), to obtain a differential equation for $c_{\mathbf{B}}(r)$. In general cases, one could solve equation (5.3.6) using Green's functions and integral equations, or numerical techniques. The following example shows a special case in which a closed-form solution of equation (5.3.6) can be obtained.

Example: Diffusion control from a reaction-diffusion equation

If the reaction is an electron transfer involving a neutral ion, then to a good approximation $\beta U(r) = 0$ and $k_{ET}(r) = k_0 exp[-r/L]$. In this case the differential equation for $c_{\mathbf{B}}(r)$ is

$$e^{-\bar{r}} c_{\mathbf{B}}(\bar{r}) = \frac{D}{k_0 L^2} \frac{1}{\bar{r}^2} \frac{d}{d\bar{r}} \left(\bar{r}^2 \frac{dc_{\mathbf{B}}}{d\bar{r}} \right) \tag{5.3.7}$$

where r has been rescaled according to $\bar{r} = r/L$. One boundary condition is $c_B \to [\mathbf{B}]$ as $\bar{r} \to \infty$. The other condition requires more careful analysis, but it can be shown (see exercise 5) that for c_B to remain bounded we must impose $dc_B/d\bar{r} \to 0$ at $\bar{r} \to 0$. The solution to equation (5.3.7) for these conditions involves modified Bessel functions:

$$c_B(r) = [\mathbf{B}] \frac{2L}{r} K_0(2s\, e^{-r/2L}) - [\mathbf{B}] \frac{2L\, K_0(2s)}{r\, I_0(2s)} I_0(2s\, e^{-r/2L})$$

where $s = \sqrt{k_0 L^2/D}$ is a Damkohler number [8]. Using $L = 0.75$Å, $k_0 = 10^{13}/sec$, and $D = 10^{-9} m^2/sec$, suggests that a typical value for s is on the order of 0.1. The figure below shows $c_B(r)$ for cases $s = 0.1$, 1.0, and 10.0 corresponding to reaction control, partial diffusion control, and diffusion controlled regimes.

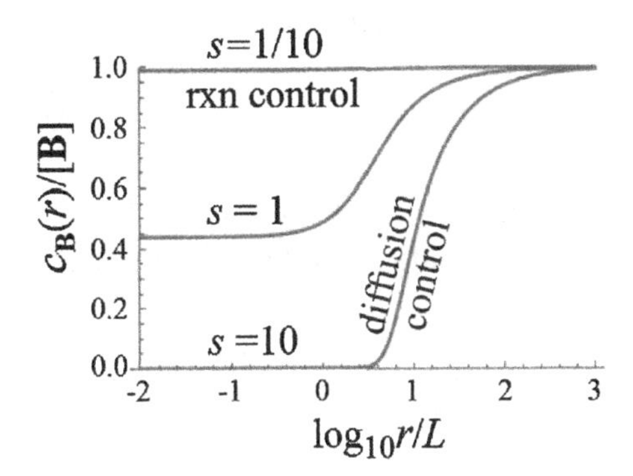

This model could be further improved by including a non-zero potential of mean force between the donor and acceptor ions. I thank S. Cray for his contributions to this example.

5.4 Diffusion control for irregularly shaped reactants

For many reactions of interest, the interaction potential between the reactants is not spherically symmetric. Diffusion controlled reaction rates can be computed for spheres with symmetrically arranged reactive patches [9,10]. Apart from certain special symmetries, diffusion controlled reaction rates can only be computed using simulation methods. For example, the reaction of a substrate with the active site of an enzyme takes place in a complicated and irregular electrostatic environment. Far from the active site the net Coulomb interaction will dominate all dipole and short ranged interactions. Figure 5.4.1 from McCammon et al. [11,12] shows the electrostatic potential around a superoxide dismutase enzyme in an electrolyte solution.

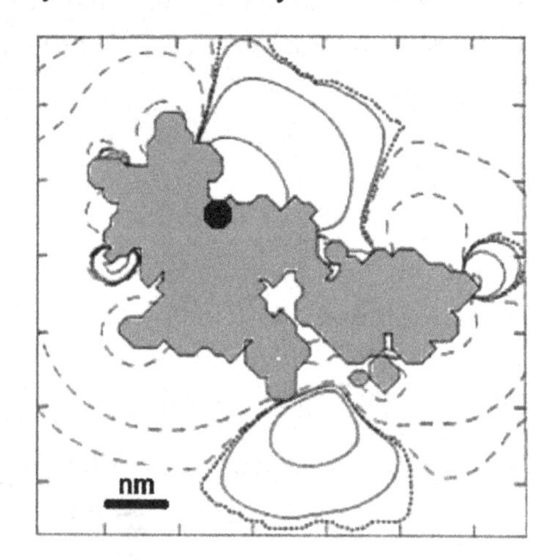

Figure 5.4.1: The electrostatic interaction between superoxide dismutase and O_2^- in solution. The contours, moving away from the enzyme, are at $\pm 1.0 k_B T$, $\pm 0.1 k_B T$, and $\pm 0.01 k_B T$. Solid contours show attractive interactions and dashed contours show repulsive interactions. The dotted lines are contours where the electrostatic interaction potential is zero. The enzyme and its substrate O_2^- each have a net negative charge. Figure adapted from McCammon et al. Faraday Disc. 213, 213–222 (1987).

For complicated electrostatic environments like that above, NAM gave the diffusion control rate constant in terms of the position-dependent net flux $j(\mathbf{x})$ through a closed surface S around the reaction site.

$$\mathcal{R} = k_{NAM} c_{\mathbf{B}} = \oiint_S n \cdot j \, da \tag{5.4.1}$$

where the rate is on a per enzyme basis. The challenge is to computationally and efficiently evaluate the surface integral in equation (5.4.1). NAM related the net flux through surface element da to two other quantities. First, they defined a local first passage flux $j_0(\mathbf{x})$ as that portion of the flux through S at location $\mathbf{x}$ which is reaching the surface for the first time. Second, they defined a probability $p_B(\mathbf{x})$ that a substrate at point $\mathbf{x}$ on surface S will react before diffusing to an infinite distance from the enzyme. Readers familiar with transition path sampling and transition path theory will recognize $p_B(\mathbf{x})$ as a committor and/or a splitting probability. The committor is denoted by $p_B(\mathbf{x})$ in later chapters, so we have adopted the modern notation here for consistency. From the definitions of $p_B(\mathbf{x})$ and $j_0(\mathbf{x})$, the net flux through S at point $\mathbf{x}$ is

$$j(\mathbf{x}) = j_0(\mathbf{x}) p_B(\mathbf{x}) \tag{5.4.2}$$

The obvious problem with equation (5.4.2) is that it is impossible to estimate $p_B(\mathbf{x})$ by following trajectories to infinity. At any finite distance from the enzyme, every trajectory has a small but nonzero chance of returning. To circumvent this limitation, NAM choose a distance b_1 at which all but the spherically symmetric Coulomb interactions are negligible. NAM then choose a second distance $b_2 > b_1$ as shown in Figure 5.4.2.

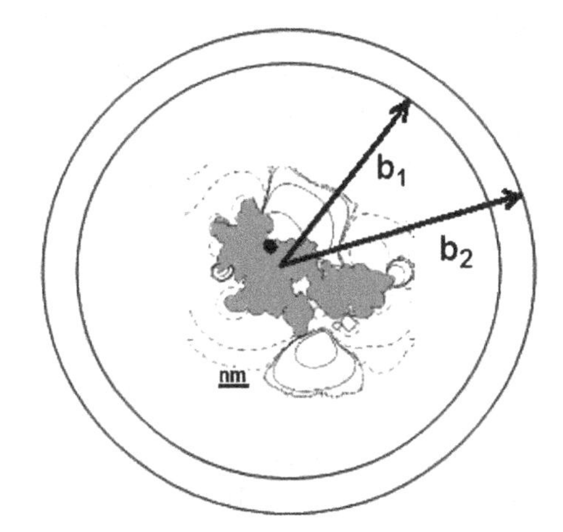

Figure 5.4.2: The diffusion controlled rate between irregularly shaped reactants can be computed in stages by choosing two radii b_1 and b_2 both of which are large enough to ignore all but spherically symmetric terms in the interaction potential. Details of the calculation are given in the text.

For each of the two large spheres b_1 and b_2, the diffusion controlled rate constant can be computed using Debye's equation for spherically symmetric interactions (5.3.5).

$$k_D(b_1) = 4\pi D \Bigg/ \int_{b_1}^{\infty} r^{-2} exp[\beta U(r)]dr$$

and similarly for $k_D(b_2)$. NAM define a parameter ϕ as the probability that a trajectory initiated at b_1 will react before diffusing to b_2. In contrast to the value of p_B for points on the surface b_1, the parameter ϕ can be computed using short Brownian dynamics trajectories. It should be relatively easy to compute ϕ accurately using transition interface sampling [13] (see Chapter 20). NAM prove that ϕ is related to p_B at the distance b_1 by

$$p_B = \frac{\phi}{1 - (1 - \phi) k_D(b_1)/k_D(b_2)}$$

These equations assume that b_1 is sufficiently large to ignore all but spherically symmetric interactions. They also assume fast rotational diffusion of both reactants (the enzyme and the substrate) so that substrates diffusing from each point on the sphere b_1 have the same values of ϕ. If these assumptions are met then the diffusion controlled rate constant for the complicated electrostatic environment is simply

$$k_{NAM} = k_D(b_1) p_B \tag{5.4.3}$$

All steric and electrostatic effects are accounted for in the calculation of p_B. Figure 5.4.3 shows how the reaction rate depends on the ionic strength for the superoxide dismutase enzyme. As ionic strength increases the electrostatics are more screened and the reaction rate *decreases*. This is remarkable because both the O_2^- substrate ion and the enzyme are negatively charged. NAM interpreted this to mean that the local electrostatic field around the enzyme steers the substrate into the active site. Getzoff et al. confirmed the electrostatic steering hypothesis by designing mutant superoxide dismutase enzymes to enhance the electrostatic steering effect [14]. Note that the work of Getzoff et al. demonstrates that a diffusion controlled reaction can be accelerated!

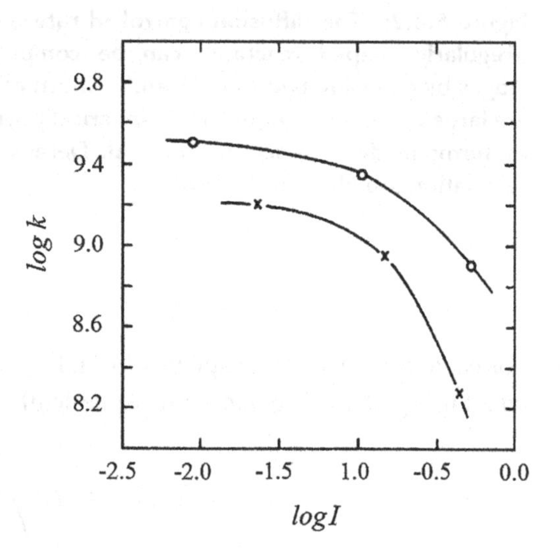

Figure 5.4.3: The bimolecular diffusion limited rate constant k (in $M^{-1}s^{-1}$) for reaction between superoxide dismutase and O_2^- as a function of ionic strength. The circles are computed results and crosses are experimental data. Figure from McCammon et al. Faraday Disc. 83, 213-222 (1987).

The NAM approach has tremendous potential to resolve some contemporary simulation challenges. For example, most simulations of amyloid fibril nucleation and growth use closed simulation boxes that correspond to extremely high peptide concentrations. The resulting rates reveal little about the true concentration dependent kinetics of fibril formation. The NAM approach (if one could only convince the NIH) could predict proper second order rate constants for the elementary peptide attachment steps during amyloid fibril nucleation and growth. The NAM approach may similarly facilitate studies of attachment kinetics during crystal nucleation and growth.

Exercises

1. Plot $ln[8k_BT/3\eta]$ vs $1/T$ for CO_2, water, and glycerol to see that temperature dependent viscosity can masquerade as an activation energy for a reaction that is actually diffusion controlled.

2. It is often suggested, e.g. on Wikipedia as of 2016, that stirring or agitation can accelerate a diffusion controlled reaction. Consider the length scales and diffusivities involved in a diffusion controlled reaction and comment on the shear rates that would be required to accelerate the rate. Hint: the Peclet number [6] should be of order unity before shear rates become important.
3. Bimolecular reactions in solution can be viewed as a series of two steps. First, reactants **A** and **B** create **AB** encounter complexes at a diffusion controlled rate. Then, the encounter complex can either dissociate or react.

$$\mathbf{A}+\mathbf{B} \underset{k_{-D}}{\overset{k_D}{\rightleftarrows}} \mathbf{AB} \xrightarrow{k_{rxn}} \mathbf{P}$$

   (a) Use this scheme to derive equation (5.2.1).
   (b) If $k_{rxn}=0$, then k_D/k_{-D} is the association equilibrium constant for the **AB** encounter complex. Derive the association equilibrium constant and k_{-D} in terms of the potential of mean force between **A** and **B**.
4. Derive the elongation (growth) rate of a fibril with diameter $2R$ and hemispherical cap that grows by diffusion controlled attachment of monomers to the fibril tip. Assume that monomers have radius R_A and that the monomers pack perfectly into the fibril structure with no excess packing volume. State any additional approximations that you need to compute the rate of attachment to the fibril tip.
5. Compute the functional derivative of the capture radius with respect to $\beta U(r)$, i.e. $\delta\lambda/\delta\beta U(r)$. What does your answer say about the sensitivity of the diffusion controlled reaction rate to changes in $U(r)$ at different distances.
6. An example in section 5.3 solved a reaction-diffusion equation without fully explaining the origin of the boundary condition at $r=0$.
   (a) Explain the following balance equation for a small sphere around the origin

$$4\pi r^2 D\frac{dc}{dr}\bigg|_R = \int_0^R dr\cdot 4\pi r^2 k_0 c\, e^{-r/L}$$

   (b) Taylor expand $c(r)$ around $r=0$ in the integrand of the balance equation, and then do the integrals of leading order in R. Finally, take the limit as $R\to 0$ to show that a bounded solution at the origin requires $dc/dr=0$ at $r=0$.
7. Consider a circular catalyst patch of radius R on a flat surface. Reactant **A** diffuses from the bulk and reacts instantaneously upon contact with any point in the area of the circle.

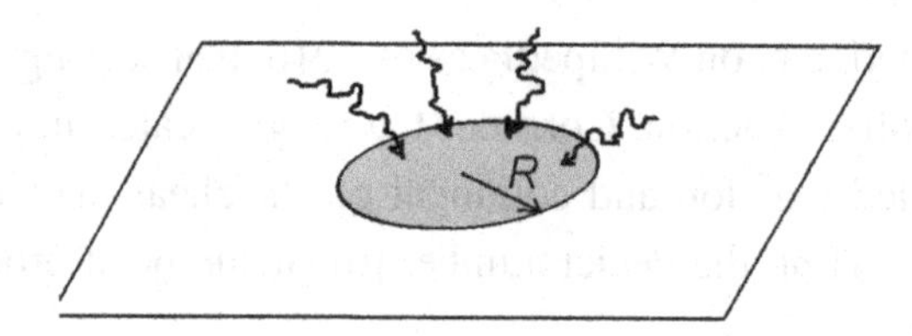

(a) Show that the reaction frequency at the site is $\mathcal{R} = 4DR[\mathbf{A}]$ where D is the bulk diffusivity of **A**.
(b) Compare the frequency of reactions per area on the circle to the frequency of diffusion controlled reactions with a stationary sphere of the same radius in 3D. Also compare frequencies if the stationary sphere in 3D has the same area.
(c) Read Alberty and Hammes [15] and Chou et al. [16] to see how this model has been used and improved to describe diffusion controlled enzyme reactions.

8. Diffusion control in one-dimension is important for transcriptase enzymes that diffuse along a DNA strand until they encounter an open reading frame codon, and for adatoms that diffuse along terrace edges of a crystal until they are incorporated at a kink site. In both cases, the diffusing moieties (enzymes or adatoms) can reversibly bind and unbind with rates k_{on} and k_{off}, respectively. The reversible binding establishes an equilibrium adsorbate density on the one-dimensional line at locations far from the sink.
   (a) In the absence of a sink, develop a kinetic model of adsorption in which k_{on}, k_{off}, and the bulk adsorbate concentration result in a specific adsorbate density per unit length on the one-dimensional "line" at equilibrium.
   (b) Augment the adsorption model with a model of lateral diffusion along the line and a one dimensional point-sink. Solve for the steady-state concentration profile near the sink and the rate at which adsorbates reach the sink.
9. Again consider a circular patch of catalyst with radius R on a flat support surface, but this time assume that only the perimeter of the circle is active. Reactant **A** adsorbs onto, desorbs from, and diffuses on the support surface with rates $k_{on}[\mathbf{A}](1-\theta_{\mathbf{A}})$, $k_{off}\theta_{\mathbf{A}}$, and D, respectively. Reactants that reach the edge of the circular patch react instantaneously. To simplify the analysis, assume there is no adsorption onto the middle of the catalyst patch and that concentration gradients in the fluid above the surface are negligible. Under these conditions, the surface coverage $\theta_{\mathbf{A}}$ satisfies

$$\frac{\partial \theta_{\mathbf{A}}}{\partial t} = D\frac{1}{r}\frac{\partial}{\partial r}\left[r\frac{\partial \theta_{\mathbf{A}}}{\partial r}\right] + k_{on}[\mathbf{A}](1-\theta_{\mathbf{A}}) - k_{off}\theta_{\mathbf{A}} \qquad for \quad r > R$$

Far away from the catalyst patch ($r \to \infty$), $\theta_{\mathbf{A}}$ satisfies $k_{on}[\mathbf{A}](1-\theta_{\mathbf{A},\infty}) = k_{off}\theta_{\mathbf{A},\infty}$. Adsorbed **A** instantly reacts at $r = R$ so that $\theta_{\mathbf{A}}(R) = 0$. Let $\theta_{\mathbf{A}}(r) = \theta_{\mathbf{A},\infty} + \Delta(r)$, non-dimensionalize, and solve for $\Delta(r)$ at steady state. Obtain the diffusion controlled reaction rate for the catalyst patch. Hint: the solution involves a modified Bessel function.

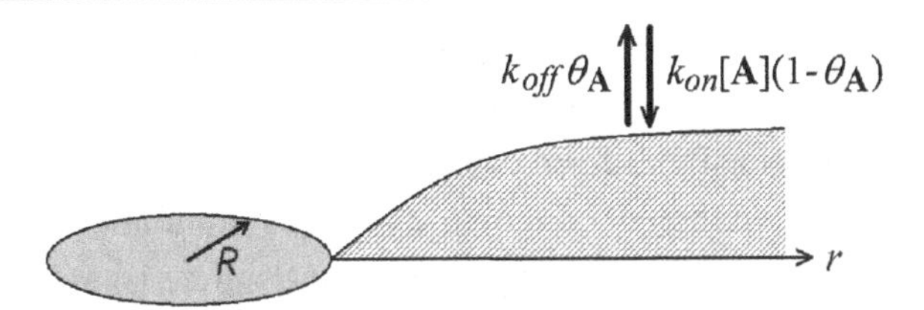

10. This problem concerns the fate of a point particle that diffuses from an initial position at a distance $b_1 = R + \Delta R$ from a spherically symmetric sink of radius R. The particles' position as a function of time is given by Fick's second law

$$\partial u/\partial t = \nabla^2 u$$

i.e. the Smoluchowski equation with no potential of mean force. The initial and boundary conditions are

$$u(r, 0) = \delta[r - r_0] \quad and \quad u(1, \tau) = u(\infty, \tau) = 0$$

where r is distance in units of R and $r_0 = (R + \Delta R)/R$. Time has been rescaled to $\tau = Dt/R^2$.

(a) Explain why the initial condition can be smeared over the sphere of radius r_0, rather than located at a single point.

(b) Show that the spatiotemporal probability distribution is

$$u(r, \tau) = \frac{r_0}{r\sqrt{4\pi\tau}}(e^{-(r_0-r)^2/4\tau} - e^{-(r-2+r_0)^2/4\tau})$$

Hint: Laplace transform, solve for $\hat{u}(r, s)$ using Green's function techniques, and then invert the Laplace transform.

11. Using the solution to problem (10), the total amount of solute material surrounding the nucleus at the initial time is $4\pi r_0^2$.

(a) Obtain the fraction which has not been absorbed by the nucleus at later times.

(b) In your solution to problem (a), take the limit as $t \to \infty$ and show that the fraction of diffusion trajectories that permanently escape without ever touching the sink is

$$p_{esc} = \frac{r_0 - 1}{r_0} = \frac{\Delta R}{R + \Delta R} \tag{5.4.4}$$

(c) How does p_{esc} relate to the parameter ϕ in the theory of Northrup, Allison, and McCammon [3]?

12. According to the flycasting mechanism [17], the binding of proteins can be accelerated by protein disorder because it increases the capture radius. The flycasting mechanism has been supported by some studies and refuted by others.

(a) Outline the arguments for and against the flycasting mechanism. What factors other than the capture radius in the diffusion controlled rate expression might be influenced by a propensity to form disordered rather than tightly folded structures?

(b) Develop an overdamped Langevin dynamics simulation of a heavy point particle (mass M_L) attached by a spring to a lighter sphere (mass M_S). The heavy particle should react on contact with a circular target of radius R_T on a flat surface. Let the small sphere interact with the center of the target via a Lennard-Jones potential. To investigate the flycasting mechanism, design springs between the small and large spheres so that one type of spring occasionally exposes a "sticky tippet". For example, compare the case where the small sphere is tethered to the heavy mass by a soft spring to the case where they are tethered by a stiff spring, both of equilibrium length zero.

(c) Compare the rates as predicted by the Debye formula, and as predicted from forward flux sampling simulations (see Chapter 19).

13. Read M. Tachiya, *J. Chem. Phys.* 69, 1375 (1978) [18]. Explain how Tachiya's route to the escape probability relates to that of the computational NAM approach, to Debye's capture radius formula, and to many concepts in Chapter 18. Check your answers against those in Shoup and Szabo, *Biophys. J.* 40, 33–39 (1982).

14. Gopich and Szabo, *Proc. Nat. Acad. Sci. USA*, 110, 19784-89 (2013) examined diffusion controlled reactions where one reactant has multiple reaction sites, e.g. the reaction between an enzyme that can phosphorylate two sites on one substrate. Naively, the rate equations under conditions of diffusion control would seem to involve two sequential diffusion controlled reactions:

$$\mathbf{S}_0 + \mathbf{E} \xrightarrow{k_{rxn} p_{esc}} \mathbf{S}_1 + \mathbf{E} \xrightarrow{k_{rxn} p_{esc}} \mathbf{S}_2 + \mathbf{E}$$

where $\mathbf{E}$ is the enzyme, $\mathbf{S}_i$ is the substrate with i phosphorylated sites, k_{rxn} is the rate constant for the chemical step with no diffusion limitations, and $p_{esc} = k_D/(k_D + k_{rxn})$ is the probability that the substrate with escape without phosphorylation upon contact. Gopich and Szabo show that, as diffusion becomes slow, it becomes increasingly likely for the enzymes to phosphorylate both sites in procession before diffusing away from the substrate.

$$\mathbf{S}_0 + \mathbf{E} \xrightarrow{k_{rxn} p_{esc}^2} \mathbf{S}_1 + \mathbf{E} \xrightarrow{k_{rxn} p_{esc}} \mathbf{S}_2 + \mathbf{E}, \qquad \mathbf{S}_0 + \mathbf{E} \xrightarrow{k_{rxn} p_{esc}(1 - p_{esc})} \mathbf{S}_2 + \mathbf{E}$$

The derivation of these results by Gopich and Szabo includes additional transient effects. Derive the results above with an entirely steady-state version of their analysis.

References

[1] M. von Smoluchowski, Z. Phys. Chem. 92 (1917) 129.
[2] G.H. Weiss, J. Stat. Phys. 42 (1986) 3–36.
[3] S.H. Northrup, S.A. Allison, J.A. McCammon, J. Chem. Phys. 80 (1984) 1517–1524.
[4] J.A. McCammon, S.H. Northrup, S.A. Allison, J. Phys. Chem. 90 (1986) 3901–3905.
[5] F.C. Collins, G.E. Kimball, J. Colloid Sci. 4 (1949) 425–437.
[6] E.L. Cussler, Diffusion: Mass Transfer in Fluid Systems, 3rd ed., Cambridge Press, Cambridge, UK, 2009.
[7] P. Debye, Trans. Electrochem. Soc. 82 (1942) 265–272.
[8] G.F. Froment, K.B. Bischoff, Chemical Reactor Analysis and Design, Wiley, Hoboken, NJ, 1990.
[9] K. Solc, W.H. Stockmayer, J. Chem. Phys. 54 (1971) 2981.
[10] D. Shoup, G. Lipari, A. Szabo, Biophys. J. 36 (1981) 697–714.
[11] J.A. McCammon, R.J. Bacquet, S.A. Allison, S.H. Northrup, Faraday Discuss. 213 (1987) 213–222.
[12] J.A. McCammon, Science 238 (1987) 486–491.
[13] T.S. van Erp, P.G. Bolhuis, J. Comput. Phys. 205 (2005) 157–181.
[14] E.D. Getzoff, D.E. Cabelli, C.L. Fisher, H.E. Parge, M.S. Viezzoli, L. Banci, R.A. Hallewell, Nature 358 (1992) 347–351.
[15] R.A. Alberty, G.G. Hammes, J. Phys. Chem. 62 (1958) 152–159.
[16] K.C. Chou, G.P. Zhou, J. Am. Chem. Soc. 104 (1982) 1409–1413.
[17] B.A. Shoemaker, J.J. Portman, P.G. Wolynes, Proc. Natl. Acad. Sci. USA 97 (2000) 8868–8873.
[18] M. Tachiya, J. Chem. Phys. 69 (1978) 2375.

CHAPTER 6

Collision theory

> *"If all these molecules were flying in the same direction, they would constitute a wind blowing at the rate of seventeen miles a minute, and the only wind which approaches this velocity is that which proceeds from the mouth of a cannon. How, then, are you and I able to stand here? Only because the molecules happen to be flying in different directions, so that those which strike against our backs enable us to support the storm which is beating against our faces."*
>
> ***Maxwell, Nature (1873)***

Early work on reaction rate theory proceeded along two separate lines, both seeking an interpretation of the highly successful, but empirical, Arrhenius law. Marcelin initiated work toward a statistical mechanical theory for reaction rates. Marcelin's effort was impaired by his early demise [1], but it would eventually become the transition state theory. The other direction, pioneered by Trautz [2] and Lewis [3,4], built upon the kinetic theory of gases to explain the prefactor and the Boltzmann factor in terms of collision frequencies and energies. Collision theory led to the theory of reaction cross sections which is highly successful, and in principle exact, for low pressure gas phase reactions.

Chapter 5 has already emphasized that molecules must approach each other to react. The simplest encounters between molecules are those of a rarified gas where "collisions", whether reactive or non-reactive, are pairwise events with no complications from third party molecules. In principle, all of gas phase chemistry could be taught entirely from the vantage of collision theory and quantum scattering cross sections. Indeed, a large community of scientists would equate the term "molecular dynamics" not with molecular simulations of $\boldsymbol{f}=\boldsymbol{Ma}$, but rather with experiments and theories of scattering processes. The objective of this chapter is to introduce the most basic elements of collision theory. Our discussion is limited to hard spheres and simple atom exchange reactions.

6.1 Hard spheres: Trautz and Lewis

Collisions occur because real molecules have finite sizes, or more properly because they interact over non-zero distances. Repulsive forces at short distances and (sometimes) attractive forces at longer distances can bend and deflect the otherwise linear trajectories that would be taken by the non-interacting particles. The simplest model that captures some of these effects is the hard sphere gas. Hard spheres occupy an important historical place in statistical mechanics and molecular simulation. For example, surprisingly accurate gas phase diffusion constants, thermal conductivities, and viscosities can be derived by considering collisions of hard spheres [5,6].

Here we follow Trautz [2] and Lewis [3,4] in considering a gas mixture with two types of hard spheres, **A** and **B**. Suppose the concentrations of **A** and **B** particles are [**A**] and [**B**], their sizes are $R_{\mathbf{A}}$ and $R_{\mathbf{B}}$, and their masses are $m_{\mathbf{A}}$ and $m_{\mathbf{B}}$. Collisions between the **A** and **B** particles occur each time they approach within a distance $d_{\mathbf{AB}} = R_{\mathbf{A}} + R_{\mathbf{B}}$. Each pair of **A** and **B** molecules moves with a relative velocity $\mathbf{v}_{\mathbf{AB}}$ as shown in Figure 6.1.1.

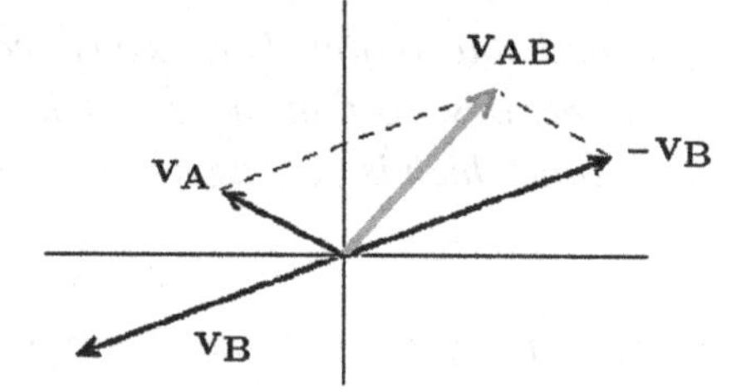

Figure 6.1.1: Depicting the individual velocities of A and B molecules and their relative velocity.

$\mathbf{v}_{\mathbf{AB}}$ is the velocity of **A** as seen from **B**. To obtain $\mathbf{v}_{\mathbf{AB}}$, the individual (laboratory) velocities of collision partners are transformed to relative and center-of-mass (COM) velocities by the equations

$$\mathbf{v}_{\mathbf{AB}} = \mathbf{v}_{\mathbf{A}} - \mathbf{v}_{\mathbf{B}} \tag{6.1.1}$$

$$\mathbf{v}_{COM} = (m_{\mathbf{A}}\mathbf{v}_{\mathbf{A}} + m_{\mathbf{B}}\mathbf{v}_{\mathbf{B}})/(m_{\mathbf{A}} + m_{\mathbf{B}}) \tag{6.1.2}$$

The kinetic energies before (left) and after (right) the change of velocity representation are

$$\frac{1}{2}m_{\mathbf{A}}v_{\mathbf{A}}^2 + \frac{1}{2}m_{\mathbf{B}}v_{\mathbf{B}}^2 = \frac{1}{2}\mu_{\mathbf{AB}}v_{\mathbf{AB}}^2 + \frac{1}{2}(m_{\mathbf{A}} + m_{\mathbf{B}})v_{COM}^2$$

The velocity transformation identifies the reduced mass for a collision pair as

$$\mu_{\mathbf{AB}} = \frac{m_{\mathbf{A}}m_{\mathbf{B}}}{m_{\mathbf{A}} + m_{\mathbf{B}}} \tag{6.1.3}$$

The root mean square relative velocity between **A** and **B** pairs is

$$\overline{v}_{\mathbf{AB}} = \left(\frac{8k_BT}{\pi\mu_{\mathbf{AB}}}\right)^{1/2} \tag{6.1.4}$$

The root mean square velocity for atoms **A** and **B** with a reduced mass $\mu_{\mathbf{AB}} = 1amu$ at 300K is $\overline{v}_{\mathbf{AB}} = 2520m/s$. Of course, $\mu_{\mathbf{AB}} = 1amu$ only occurs for two deuterium atoms, but $\overline{v}_{\mathbf{AB}} = 2520m/s$ and $\mu_{\mathbf{AB}}$ are easily scaled to the appropriate values for other temperatures and masses.

The collision cross section between **A** and **B** hard spheres is the area

$$\sigma = \pi d_{\mathbf{AB}}^2 \tag{6.1.5}$$

as depicted in Figure 6.1.2. The relative motion of the **A**, **B** pair sweeps a volume $\pi d_{\mathbf{AB}}^2 \overline{v}_{\mathbf{AB}}$ per unit time. Note that **A**, **A** collisions and **B**, **B** collisions are happening too. These are important for maintaining the Maxwell-Boltzmann distribution, but from the standpoint of reactions between **A** and **B** they are unimportant. Figure 6.1.2 shows the swept collision volume for a molecule **B** moving at velocity $\overline{v}_{\mathbf{AB}}$ through **A** molecules.

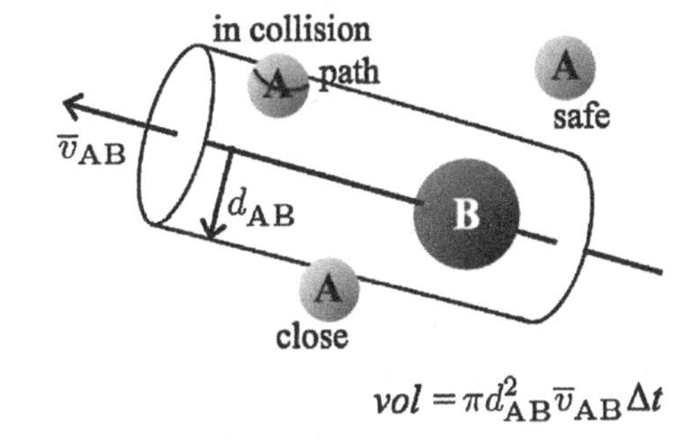

Figure 6.1.2: The cross-sectional area of the cylinder is the collision cross section. The collision volume per unit time and the concentrations of collision partners gives the number of collisions per unit time.

Multiplying the collision volume per time by the concentrations gives the number of collisions per unit volume per time.

$$\nu_{\mathbf{AB}} = \pi d_{\mathbf{AB}}^2 \overline{v}_{\mathbf{AB}} [\mathbf{A}][\mathbf{B}] \tag{6.1.6}$$

Reactions cannot occur more frequently than collisions, so $\nu_{\mathbf{AB}}$ is an upper bound on the reaction rate in the gas phase. We have not directly used the mean free path between collisions, $\lambda_{MFP} = 4k_B T/\pi d_{\mathbf{AB}}^2 P$, but it is often useful to compare the mean free path length to pore diameters or other lengthscales.

Collision theory for atom exchange reactions

The schematic in Figure 6.1.3 shows reactive and non-reactive collisions between an atom **A** and a diatomic molecule **BC**. The basic ideas of collision theory can be illustrated with the following simple hard sphere model.

The frequency of collisions between **A** and **BC** per unit volume is $\nu_{\mathbf{A-BC}} = \pi d_{\mathbf{A-BC}}^2 \times \overline{v}_{\mathbf{A-BC}}[\mathbf{A}][\mathbf{BC}]$. Where $d_{\mathbf{A-BC}}$ is an effective collision diameter for **A** and a non-spherical **BC** molecule. Trautz and Lewis recommended setting $d_{\mathbf{A-BC}}$ as the sum of the van der Waals radii for **A** and **BC**. We cannot choose forward and backward collision diameters independently if we wish to obtain rates that obey microscopic reversibility. In other words, the forward and reverse collision frequencies must balanced: $\nu_{\mathbf{AB-C}} = \nu_{\mathbf{A-BC}}$. Clearly, the relative velocities

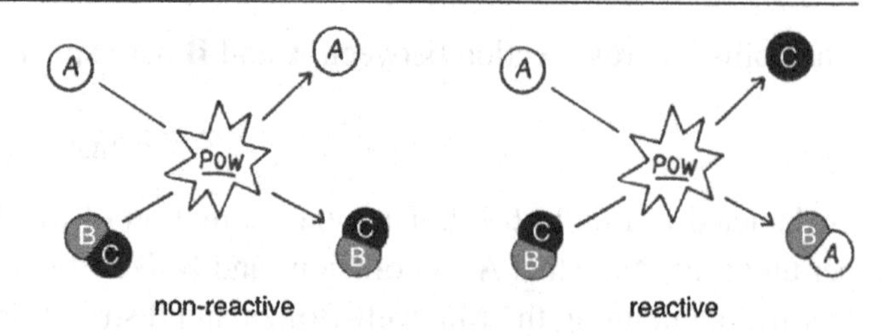

Figure 6.1.3: (Left) A non-reactive collision and (right) a reactive collision resulting in an atom exchange reaction $\mathbf{A}+\mathbf{BC}\rightarrow\mathbf{AB}+\mathbf{C}$.

$\overline{v}_{\mathbf{A-BC}}$, and $\overline{v}_{\mathbf{AB-C}}$ are set by the atomic masses and the temperature. Therefore, once the collision diameter for the reactants is chosen, the collision diameter for the products is also effectively specified. In terms of the equilibrium constant for the reaction $\mathbf{A}+\mathbf{BC}\rightarrow\mathbf{AB}+\mathbf{C}$, the collision diameters must obey

$$d^2_{\mathbf{AB-C}}=\frac{d^2_{\mathbf{A-BC}}\overline{v}_{\mathbf{A-BC}}}{K_{eq}\overline{v}_{\mathbf{AB-C}}} \tag{6.1.7}$$

Trautz and Lewis used collision frequencies and diameters to estimate prefactors, i.e. per volume collision frequencies. They then equated the reaction rate to a product of the collision frequency per volume and a factor $exp[-E_a/k_BT]$. Combining the collision frequency with the factor of $exp[-E_a/k_BT]$ gives the second order rate law

$$r = \{\pi d^2_{\mathbf{A-BC}}\overline{v}_{\mathbf{A-BC}}\cdot exp[-E_a/k_BT]\}[\mathbf{A}][\mathbf{BC}] \tag{6.1.8}$$

The term in curly brackets is the collision theory rate constant

$$k_{coll}=\pi d^2_{\mathbf{A-BC}}\overline{v}_{\mathbf{A-BC}}\cdot exp[-E_a/k_BT] \tag{6.1.9}$$

Readers will recognize an obvious oversimplification in the derivation of k_{coll}. The use of independent collision frequency and Arrhenius factors implies that the probability of reaction is independent of the collision energy. Clearly, the collision energy and $v_{\mathbf{A-BC}}$ are not independent. Indeed, the activation energy is only attained by the small fraction of violent collisions that result from a very large $v_{\mathbf{A-BC}}$. The correct Arrhenius factor and collision frequency must emerge from an average over all possible collision parameters. However, Trautz and Lewis believed that the activation energy of Arrhenius was separately supplied by radiation. In fact, Trautz was the first proponent of the radiation hypothesis [7].

Despite their oversimplifications, equations (6.1.8) and (6.1.9) have several satisfying properties. First, the rate law is second order, consistent with kinetics of gas phase reactions except for association reactions at very low pressures. Second, the prefactor emerges from a sensible and quasi-first-principles argument. Third, the k_{coll} rate constant has an Arrhenius-like temperature dependence and is appropriately independent of reactant concentrations. Fourth, collision theory makes a specific quantitative prediction about the prefactor of the rate constant. Thus collision theory was a definite improvement over Arrhenius' equation in which both the prefactor and the activation energy were entirely empirical parameters.

Table 6.1: Prefactors (in 10^{10}L/mol/s) for atom exchange reactions according to collision theory and experiment. [Data from Masel, Chemical Kinetics and Catalysis]

Reaction	$\pi d^2_{\mathbf{A-BC}} \overline{v}_{\mathbf{A-BC}}$	Experiment
$H + C_2H_6 \rightarrow C_2H_5 + H_2$	37.3	9.6
$H + CH \rightarrow H_2 + C$	24.1	0.1
$O + C_2H_6 \rightarrow OH + C_2H_5$	11.4	1.5
$2OH \rightarrow H_2O + O$	7.5	0.6

Table 6.1 compares prefactors from experiments (Arrhenius law fits) to prefactors from collision theory with van der Waals radii. Overall, the collision theory is remarkably successful given that its only inputs are van der Waals radii and masses. The prefactors from collision theory are typically about an order of magnitude larger than the experimental prefactors. Let us examine some of the possible reasons for its systematic overestimation errors.

1. Trautz and Lewis assume that collisions with the molecules at any relative orientation to each other can be reactive. In reality, molecules have reactive zones that must be favorably oriented during a collision to facilitate the reaction. Note that each of the reactions in Table 6.1 involves a non-spherically symmetric and therefore orientation-sensitive reactant. Hinshelwood introduced steric factors to represent the fraction of collisions that occur with orientations that enable a reaction [8].
2. All of the reactions in Table 6.1 are hydrogen transfers. When compared to classical rate constants, tunneling results in smaller prefactors and smaller activation energies [9]. (See Chapter 12.)
3. Trautz and Lewis assume that the collision frequency and the threshold energy for a reactive collision can be obtained independently and then multiplied. Likewise, their collision cross section, $\pi d^2_{\mathbf{A-BC}}$, is independent of the collision energy and reaction probability.

We have already discussed how the third deficiency stems from the belief that activation energy comes from radiation and not from the collision itself. Later chapters will address the first and second deficiencies of this simple collision theory. The next section discusses more realistic models that can appropriately account for the small frequency of high energy reactive collisions.

6.2 Cross sections and rate constants

This section presents another simple illustrative model of cross sections. However, the model does begin to resemble the modern framework of scattering theory which can, in principle, yield exact cross sections and rate constants. Again consider the collisions of two spherical molecules **A** and **B**, but now suppose that they only react when the kinetic energy E_T associated with translation along their center-to-center axis exceeds a threshold energy E_0 at the moment of collision. The threshold energy requirement accounts for the fact that a direct hit at low velocity (or a glancing blow at high velocity) may fail to induce a reaction. Figure 6.2.1 illustrates a collision in the center of mass coordinate system for the collision partners. Note that all collisions in this coordinate system occur with the velocities in parallel but opposing directions. This concept is critical for obtaining a thermal rate constant from the analysis here and from crossed molecular beam experiments. For hard spheres, the collision occurs at the moment when the center-to-center distance is $d_{\mathbf{AB}} = R_{\mathbf{A}} + R_{\mathbf{B}}$. The relative velocity $v_{\mathbf{AB}} = v_{\mathbf{A}} - v_{\mathbf{B}}$ can be decomposed into a component parallel to the center-to-center axis at the moment of collision, and other orthogonal components. The center-to-center component determines whether the reaction occurs. The orthogonal components are experimentally important in extracting impact parameters from scattered products and reactants.

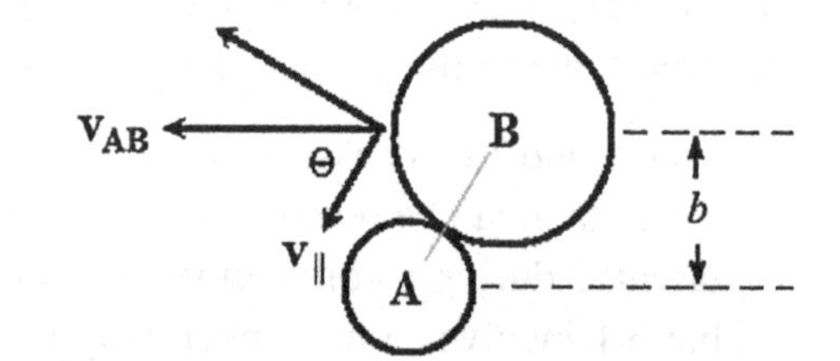

Figure 6.2.1: For a collision between A and B in their center of mass coordinate system, the velocities are in opposite, parallel directions with a collision parameter *b*.

From the geometry at the moment of collision, the angle θ satisfies $cos\theta = \sqrt{1 - b^2/d_{\mathbf{AB}}^2}$. Clearly, collisions can only occur if the impact parameter b satisfies $b < d_{\mathbf{AB}}$. The center-to-center velocity at the moment of collision is $v_{||} = v_{\mathbf{AB}} cos\theta$. The kinetic energy associated with the center-to-center motion is $\mu_{\mathbf{AB}} v_{||}^2/2$. If the reaction occurs when the center-to-center motion involves a threshold of energy E_0, then the requirement for reaction is [10,11]

$$E_T \left\{1 - b^2/d_{\mathbf{AB}}^2\right\} > E_0 \tag{6.2.1}$$

where $E_T = \mu_{\mathbf{AB}} v_{\mathbf{AB}}^2/2$. For any total relative kinetic energy E_T between **A** and **B**, the reaction occurs if $b < b_0$ where

$$b_0 < d_{\mathbf{AB}} \sqrt{\frac{E_T - E_0}{E_T}} \tag{6.2.2}$$

The reaction probability $p(b)$ that a collision will result in a reaction can now be expressed in terms of the critical impact parameter b_0 as

$$p(b) = \begin{cases} 1 & if\ b < b_0 \\ 0 & otherwise \end{cases} \tag{6.2.3}$$

The probability $p(b)$ is called an opacity function [12]. Note that $p(b)$ does depend on E_T through the definition of b_0. For collisions between real molecules $p(b)$ would also depend on many other details including relative orientations, angular momentum states, and internal quantum states of the collision partners.

The differential cross section is defined as the effective differential area that leads to a collision when the impact parameter is between b and $b + db$.

$$d\sigma = p(b)\, 2\pi b\, db \tag{6.2.4}$$

Integrating the differential cross section over all impact parameters b gives the total reaction cross section for collisions at translational energy E_T. The integration in our example is trivial, $\sigma = \int_0^\infty p(b) 2\pi b\, db$ gives

$$\sigma = \begin{cases} \pi d_{\mathbf{AB}}^2 (E_T - E_0)/E_T & if\ E_T > E_0 \\ 0 & otherwise \end{cases} \tag{6.2.5}$$

The cross section now appropriately depends on the kinetic energy $E_T = \mu_{\mathbf{AB}} v_{\mathbf{AB}}^2/2$. (See Figure 6.2.2.)

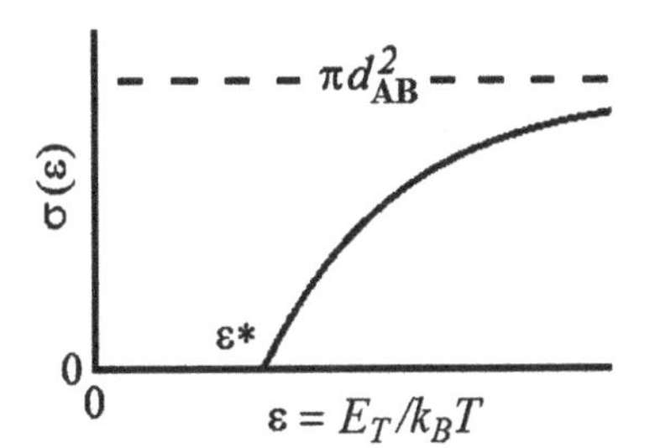

Figure 6.2.2: The reactive cross section as a function of the dimensionless kinetic energy ϵ for hard sphere collisions. At high kinetic energies, the simple cross section model used by Trautz and Lewis becomes correct.

Collision theory for atom exchange reactions

The cross section in equation (6.2.5) depends on the collision energy (and therefore on relative velocity), so now we cannot simply multiply a velocity averaged collision frequency by a Boltzmann factor as Trautz and Lewis did. Instead, the rate constant is a Boltzmann weighted sum over contributions from collisions at all velocities. The calculation must be done carefully to account for our reduction from two separate velocities to one relative velocity.

The derivation leading to equation (6.2.5) was done entirely within a center-of-mass coordinate system and considering only relative velocities. All collisions in the center of mass coordinate system occur with collinear velocities and thus we were able to work with scalar quantities.

Obtaining a rate constant from the cross section requires a return to the laboratory frame and a sum over all separate vector velocities. The inverse velocity transformations that we need are

$$\mathbf{v}_\mathbf{A} = \mathbf{v}_{COM} - m_\mathbf{B}\mathbf{v}_\mathbf{AB}/(m_\mathbf{A} + m_\mathbf{B})$$

and

$$\mathbf{v}_\mathbf{B} = \mathbf{v}_{COM} - m_\mathbf{A}\mathbf{v}_\mathbf{AB}/(m_\mathbf{A} + m_\mathbf{B})$$

The Jacobian for the velocity transformation is unity

$$d\mathbf{v}_\mathbf{A} d\mathbf{v}_\mathbf{B} = d\mathbf{v}_{COM} d\mathbf{v}_\mathbf{AB} \tag{6.2.6}$$

Now the rate constant is an integral over the relative velocity times the relative velocity dependent cross section, i.e.

$$k(T) = \int d\mathbf{v}_\mathbf{A} d\mathbf{v}_\mathbf{B}\, f(\mathbf{v}_\mathbf{A})\, f(\mathbf{v}_\mathbf{B}) v_\mathbf{AB} \sigma(v_\mathbf{AB}). \tag{6.2.7}$$

Here $f(\mathbf{v})$ is the Maxwell-Boltzmann distribution,

$$f(\mathbf{v}) = \left(\frac{m}{2\pi k_B T}\right)^{3/2} exp\left[-\frac{mv^2}{2k_B T}\right], \tag{6.2.8}$$

for a particle of mass m. The product $f(\mathbf{v}_A) f(\mathbf{v}_B)$ becomes

$$\left(\frac{\mu_\mathbf{AB}}{2\pi k_B T}\right)^{3/2} \left(\frac{m_\mathbf{A} + m_\mathbf{B}}{2\pi k_B T}\right)^{3/2} exp\left[-\frac{\mu_\mathbf{AB} v_\mathbf{AB}^2}{2k_B T}\right] exp\left[-\frac{(m_\mathbf{A} + m_\mathbf{B}) v_{COM}^2}{2k_B T}\right].$$

The change to center-of-mass and relative velocity separates integral (6.2.7) into a product of two integrals. The first integral over the center-of-mass velocity integrates to a factor of 1. The second integral contains the $v_\mathbf{AB}$-dependent cross section and $v_\mathbf{AB}$ factors. The rate constant is thus obtained from

$$k(T) = \left(\frac{\mu_\mathbf{AB}}{2\pi k_B T}\right)^{3/2} \int_0^\infty v_\mathbf{AB} \sigma(v_\mathbf{AB}) exp\left[-\frac{\mu_\mathbf{AB} v_\mathbf{AB}^2}{2k_B T}\right] 4\pi v_\mathbf{AB}^2 dv_\mathbf{AB} \tag{6.2.9}$$

Alternatively, the rate constant can be written as an integral over the center-to-center kinetic energy

$$k(T) = \frac{1}{k_B T} \left(\frac{8}{\pi \mu_\mathbf{AB} k_B T}\right)^{1/2} \int_0^\infty E_T \sigma(E_T) e^{-E_T/k_B T} dE_T \tag{6.2.10}$$

These hard sphere models for molecules are not directly useful for chemistry, but they illustrate the importance of molecular beam experiments that can precisely control collision velocities and detect reaction products [13,14]. These experiments provide direct measurements of reaction cross sections, and therefore a way to obtain accurate rate constants for specific (gas phase) reactions. On the theoretical side, classical and quantum scattering calculations can predict cross sections and rate constants for collisions involving small molecules [12,15]. Modern analyses can predict the effects of collision velocity, excited vibrational states, and non-zero angular momentum on cross sections and rate constants. A few additional results involving cross sections and rate constants are found in the following chapter, but for advanced treatments of molecular beam experiments and scattering theories we direct the reader to other sources [10, 12,15].

Exercises

1. For each of the reactions in Table 6.1, verify the results by doing your own calculations with van der Waals radii. Then recompute the prefactors using covalent bond radii instead. Do the answers improve or worsen relative to experiment?
2. Write a computer code to simulate the parallel collisions between Lennard-Jones atoms. Your code should output the distance of closest approach between the atoms as a function of the relative velocity and the impact parameter b. Scale all energies by the well-depth parameter ϵ and all distances by the Lennard-Jones distance parameter σ.
3. The text presented a model with parallel velocities and an impact parameter b. Instead, we could have used a one-dimensional framework in which the only variable is the distance r between particles in a collision where the energy and angular momentum of the collision pair are conserved [16].
   (a) Show that the angular momentum is $L = \mu b v_{\mathbf{AB}}$ where $v_{\mathbf{AB}}$ is the relative velocity at infinite separation and μ is the reduced mass of the collision pair.
   (b) The total energy of the collision pair is $E = \frac{1}{2}\mu\dot{r}^2 + \frac{1}{2}\mu r^2\dot{\theta}^2 + V(r)$. Show that the angular kinetic energy term $\frac{1}{2}\mu r^2\dot{\theta}^2$ can be written as $L^2/(2\mu r^2)$.
   (c) Use the result in part (b) to define an effective pair potential $V_{eff}(r)$ for motion along the r coordinate and to derive equations of motion for $r(t)$.
   (d) Comment on the dimensionality reduction affected by this transformation and its value for simulations with complicated spherical interaction potentials.
4. Use the results of problem (3) to predict the distance of closest approach between Lennard-Jones atoms using the r-coordinate system. As in problem (2), use the well-depth parameter ϵ and distance parameter σ to scale all energies and distances.
5. For each collision parameter, velocity, and pair potential, one can define a deflection angle between the incoming and outgoing velocities after a collision. Deflection angles are measured in scattering experiments to learn about the collision process [17]. Read Greene and Kuppermann, *J. Chem. Educ.* 45, 361-69 (1968) and derive their equation (14) for the deflection angle.

6. When averaged over all orientations, a dipole interacts with a point charge with a potential $V(r) = -z^2 a e^2/(2r^4)$, where a is related to the dipole moment and the free space permittivity.
   (a) Show that the ion-dipole collision cross section is $\sigma(v_{\mathbf{AB}}) = 2\pi z e \sqrt{a/(\mu v_{\mathbf{AB}}^2)}$.
   (b) Use the collision cross section to obtain the collision rate constant, $k = 2\pi z e \sqrt{a/\mu}$. Comment on quantities that appear in your answer and on quantities that have vanished.

References

[1] K.J. Mysels, J. Chem. Educ. 63 (1985) 740.
[2] M. Trautz, Z. Phys. Chem. 96 (1) (1916).
[3] W.C.M. Lewis, J. Chem. Soc. 113 (1918) 471.
[4] W.C.M. Lewis, J. Chem. Soc. 113 (1918) 47.
[5] E.H. Kennard, Kinetic Theory of Gases, McGraw-Hill, New York, 1938.
[6] S. Chapman, T.G. Cowling, The Mathematical Theory of Non-Uniform Gases, 3rd ed., Cambridge Press, Cambridge, UK, 1970.
[7] K.J. Laidler, M.C. King, J. Phys. Chem. 87 (1983) 2657–2664.
[8] C.N. Hinshelwood, The Kinetics of Chemical Change, Clarendon, Oxford, 1940.
[9] H.S. Johnston, Adv. Chem. Phys. 83 (1961) 1–9.
[10] J.I. Steinfeld, J.S. Francisco, W.L. Hase, Chemical Kinetics and Dynamics, Prentice Hall, Englewood Cliffs, NJ, 1989.
[11] R.I. Masel, Chemical Kinetics and Catalysis, Wiley, New York, 2001.
[12] R.D. Levine, Molecular Reaction Dynamics, Cambridge University Press, Cambridge, UK, 2005.
[13] Y.T. Lee, Science 236 (1987) 793–798.
[14] D.M. Neumark, M. Wodtke, G.N. Robinson, T.P. Schafer, Y.T. Lee, J. Chem. Phys. 82 (1985) 3067–3077.
[15] G.C. Schatz, M.A. Ratner, Quantum Mechanics in Chemistry, Dover, Mineola, NY, 2002.
[16] R.E. Weston, H.A. Schwarz, Chemical Kinetics, Prentice-Hall, Englewood Cliffs, NJ, 1972.
[17] E.F. Greene, A. Kuppermann, J. Chem. Educ. 45 (1968) 361–369.

CHAPTER 7

Potential energy surfaces and dynamics

"...the whole burden of philosophy seems to consist in this – from the phenomena of motions to investigate the forces of nature, and then from these forces to demonstrate the other phenomena..."

Newton, Principia (1686)

Most of the rare events methods and reaction rate theories which are discussed in this book assume that nuclei evolve according to classical mechanics on a ground state potential energy surface. Accordingly, this chapter primarily describes the important features of ground state potential energy surfaces. However, it is also useful to understand when electronically excited states may be important. Obviously photochemistry (not covered in this book) involves excited electronic states, but electronic excitations can also be important for reactions that happen in the dark. Examples include reactions that involve spin crossings, reactions that occur at extremely high temperatures, and reactions that occur on the surface of a metal. The following section briefly describes the equations which define ground state potential energy surfaces and the conditions that determine whether excited electronic states can be ignored.

7.1 Molecular potential energy surfaces

At a first-principles level, the energy of a molecule is determined by a wavefunction and a quantum mechanical Hamiltonian involving electron-electron, electron-nucleus, and nucleus-nucleus interactions [1,2]. The energy and its dependence on the nuclear coordinates can, in principle, be obtained by solving the Schrodinger equation. In practice, a number of approximations are made in obtaining energy landscapes from electronic structure theory [2,3]. First, adiabatic and Born-Oppenheimer approximations are invoked to separate equations for the nuclear and electronic degrees of freedom. Second, approximations are invoked in solving the electronic structure problem itself. Third, approximations are invoked in modeling the dynamics of the nuclei subject to forces emerging, in part, from the electronic structure.

The details of these approximations are only briefly mentioned here, starting with the effects of moving nuclei. As the nuclei move, the ground state electronic configuration feels a continuous perturbation that tries to excite it into higher energy levels. The description of reaction dynamics as motion of nuclei on a ground state electronic energy surface implicitly invokes the adiabatic approximation: that moving nuclei do not induce electronic excitations [1,2]. According to the adiabatic approximation, the ground state continually (or "adiabatically") evolves as the nuclei move. The adiabatic approximation breaks down when excited energy levels are narrowly separated from the ground state, when nuclei move at extremely high speeds, or when energy levels for diabatic states cross [4,5]. The adiabatic approximation is also violated for photochemical reactions that involve an initial excitation followed in some cases by a cascade of electronic and vibrational transitions.

The Born-Oppenheimer approximation [6] additionally assumes that along a dynamical trajectory, the electrons continually adopt the ground state that would be obtained *with the nuclei fixed in their instantaneous positions*. Formally, the Born-Oppenheimer approximation ignores the coupling between the nuclear kinetic energy and the electronic wavefunction with fixed nuclei [2]. The usual justification is that nuclei are at least 2000 times heavier than electrons and therefore nuclei will move slowly relative to the rate at which the electronic state changes. Technically, the small electron-to-nucleus mass ratio alone is not a sufficient justification. Because the Born-Oppenheimer approximation is an extension of the adiabatic approximation, there must also be an adequate separation between energy levels. This latter requirement is important for the dynamics of reactions on metal surfaces [5,7] where adsorbates see a continuum of energy levels. Motion of the adsorbates on the surface excites electrons in the metal which dissipates energy into electronic modes and exerts a friction on the adsorbate motion [4,5]. The importance of electronic friction is easily demonstrated by rolling a cylindrical magnet on a flat surface. If the surface is a non-metal the magnet will roll freely. If the surface is a metal, the magnet will quickly come to a stop because of electronic friction.

Formally, using the Born-Oppenheimer approximation in electronic structure theory involves solving two separate problems. First, the electronic equation for the ground state energy landscape is [2]

$$\hat{H}\Psi_0(\mathbf{r_e}|\mathbf{x}) = E_0(\mathbf{x})\Psi_0(\mathbf{r_e}|\mathbf{x}) \tag{7.1.1}$$

where $\mathbf{r_e}$ are electronic degrees of freedom, $\mathbf{x}$ are coordinates of the nuclei, and

$$\hat{H} = \hat{T}_e + \hat{V}_{ne} + \hat{V}_{ee}$$

The terms within $\hat{H}$ are the electron kinetic energy, the nucleus-electron interactions, and the electron-electron interactions, respectively [2,3]. Notice that $\hat{T}_n$ is omitted. The electrons exert forces on the nuclei because the ground state electronic energy $E_0(\mathbf{x})$ depends on the coordinates of the nuclei. Additionally, the nuclei exert forces on each other through repulsive

Coulomb interactions. The combination of $E_0(\mathbf{x})$ and all nucleus-nucleus repulsions $V_{nn}(\mathbf{x})$ is the Born-Oppenheimer potential energy surface (PES)

$$V(\mathbf{x}) = V_{nn}(\mathbf{x}) + E_0(\mathbf{x}) \tag{7.1.2}$$

The Born-Oppenheimer potential is purely a function of $\mathbf{x}$. The nuclei evolve on the Born-Oppenheimer energy landscape according to

$$i\hbar\frac{\partial}{\partial t}\Phi(\mathbf{x}) = (\hat{T}_n + V(\mathbf{x}))\Phi(\mathbf{x}) \tag{7.1.3}$$

where $\hat{T}_n$ is the nuclear kinetic energy operator. Note that we have already made the Born-Oppenheimer approximation, and the nuclei are still quantum particles. In most applications, we further assume that the nuclei obey classical mechanics with forces emerging from the Born-Oppenheimer $V(\mathbf{x})$.

Much of electronic structure theory is devoted to solving equation (7.1.1) for the electronic ground state potential energy. Excellent introductions to electronic structure theory can be found in other books [2,3,8,9]. There are also a number of commercial electronic structure codes. A proper review of electronic structure theory is far beyond the scope of this book, but here we attempt to outline some contributions to the electronic Hamiltonian and some approximations used in prominent electronic structure methods [10]:

- Coulomb interaction – nucleus-electron coulomb interactions are the primary stabilizing forces for molecules, but these are partially offset by repulsive nucleus-nucleus and electron-electron coulomb interactions.
- Exchange interaction – the electronic wavefunction must change sign when the identity of electrons are swapped. This antisymmetry property forces wavefunctions to take the form of determinants and leads to important properties like the hard repulsive cores of atoms and the orbital symmetry rules that make some reactions forbidden. The exchange energy also stabilizes the pairing of electrons to form covalent bonds.
- Correlation energy – this is actually an artifact of the self-consistent field (SCF) approach for computing the wavefunction. The correlation energy is an attempt to regain a proper treatment of the electron-electron coulomb repulsion.

The most widely used methods for electronic structure calculations are Hartree-Fock [3], Moller-Plesset perturbation theory [11], coupled cluster methods [3], multi-reference methods [3], and density functional theory [12]. SCF methods use basis functions [13–16] to construct models of the wave function or the electron density. Depending on the application, the basis functions may be periodic plane waves, atom-centered Gaussians, Slater type orbitals, muffin-tin functions, etc. Each strategy for describing the Born-Oppenheimer PES is associated with a suite of computational methods and further approximations. Some details are purely related to efficiency, memory usage, and accuracy. Other details like relativistic treatments for core electrons of heavy nuclei [17], polarization by an implicit solvent medium [18–22], and spin polarization [8] have qualitative chemical consequences.

Bond breaking/making reactions in liquids or in enzymes require a quantum mechanical description, but the surrounding environment is usually too large to model quantum mechanically and too important to ignore. These reactions typically require a quantum mechanics/molecular mechanics (QM/MM) model, i.e. a quantum mechanical model of the breaking/forming bonds combined with a classical force field model of the environment [23–26]. The QM and MM models must be linked across the interface of the QM and MM regions, and in certain applications the linkages pose major difficulties [27].

Later sections of the book will discuss activated processes that do not break or make chemical bonds. For example, simulations of nucleation, self-assembly, protein folding, and fluid-phase equilibria need only describe the reshuffling of "non-bonded" interactions. Purely classical force fields are often sufficient for these applications. A classical force field typically includes dispersion interactions and electrostatic interactions between molecules. Within molecules, the "bonded" interactions include harmonic springs between bonded atoms, spring constants for bond angle distortions, and torsional potentials for dihedral angles [2,9,28]. Each of these terms is designed to mimic the forces that naturally emerge from the Born-Oppenheimer PES.

7.2 Atom-exchange reactions

For a system with N atoms, there are $3N$ degrees of freedom. Even for reactions between small molecules the dimensionality of the PES quickly becomes too large for a complete map. For example, the Cartesian coordinate $\mathbf{x}$ for an atom exchange reaction $\mathbf{A} + \mathbf{BC} \rightleftarrows \mathbf{AB} + \mathbf{C}$ is already nine-dimensional. Computing the Born-Oppenheimer PES at each point of a nine-dimensional grid with 10 points along each axis would already require 10^9 calculations.

Fortunately, there is no need to compute the PES in Cartesian coordinates. Three of the nine degrees of freedom are conserved center-of-mass translations which can be entirely ignored. Except for analyses of rovibrational coupling, the rotational degrees of freedom can also be ignored. After conversion to internal coordinates, the atom exchange reaction $\mathbf{A} + \mathbf{BC} \rightleftarrows \mathbf{AB} + \mathbf{C}$ has only three internal degrees of freedom (from $3N - 6 = 3$). Internal coordinates like bond lengths, angles, and dihedrals generally help to provide a low dimensional representation of the PES.

Early analyses of atom exchange reactions made one further approximation that the atoms are collinear at all stages of the reaction. This simplification leaves only two collinear internal degrees of freedom, and dramatically reduces the cost of computing the full PES.

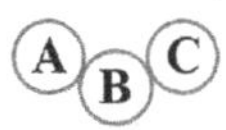

collinear
- three center-of-mass translations
- two rotations
- two (degenerate) bending modes
- two bond stretch modes

bent
- three center-of-mass translations
- three rotations
- one bond angle vibration
- two bond stretch modes

By symmetry, when $\theta_{\mathbf{ABC}} = 180°$ there are no forces that couple to the bending motion. Therefore, a collision that is initiated in collinear fashion will remain collinear throughout and the only important degrees of freedom are the two bond stretching modes.

If we consider only the collinear case, then the PES is just $V(r_{\mathbf{AB}}, r_{\mathbf{BC}})$. When **A** is far from **BC**, the **BC** molecule will vibrate around its equilibrium bondlength. Similarly, when **C** is far from the product **AB** molecule, **AB** will vibrate around its equilibrium bondlength. As **A** approaches **BC**, the **BC** bond begins to break and the **AB** bond begins to form. $V(r_{\mathbf{AB}}, r_{\mathbf{BC}})$ has a saddle point where both $r_{\mathbf{AB}}$ and $r_{\mathbf{BC}}$ are slightly elongated [29], and two channels running along the $r_{\mathbf{AB}}$ and $r_{\mathbf{BC}}$ axes corresponding to the stable **BC** and **AB** molecules, respectively. Figure 7.2.1 shows the PES for such a collinear reaction.

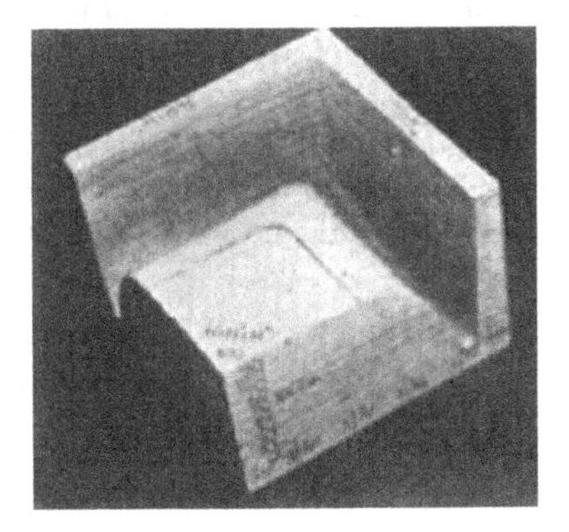

Figure 7.2.1: Collinear $H + H_2$ potential energy surface constructed by C.F. Goodeye. [Photograph from Wigner and Eyring, *The Scientific Monthly*, 6, 564 (1937).] The collinear $H + H_2$ potential depends on the two bondlengths to the central hydrogen at four different bond angles.

Figure 7.2.1 shows the third energy axis, but more typically the same information is presented as a 2D contour plot. See for example, Figure 7.2.2 which shows the energy landscape for the same atom-exchange reaction as a contour plot.

Even when the saddle geometry is linear, most reactive trajectories will actually cross the dividing surface in bent configurations. The reason is that the collinear arrangement has a vanishingly small probability[1] relative to bent configurations. At any non-zero temperature, entropy (and zero-point motion) favors complexes that are at least slightly bent. Figure 7.2.2 shows two PESs from Karplus et al. [30] for the $H + H_2$ reaction, one collinear, and one bent. Note that the collinear reaction pathway has the smallest energy barrier. Collinear reactions were favorite test systems for early developments in molecular reaction dynamics because: (1) the saddle point and all configurations on the minimum energy path are often collinear, (2) the collinear configurations are described by just two coordinates, and (3) many atom exchange reactions show qualitatively similar features even when the transition state and minimum energy path are not collinear. For example, the $H + CH_3OH \rightleftarrows H_2 + CH_2OH$ reaction has a saddle point which is slightly bent, but otherwise involves a similar potential energy profile along the minimum energy path.

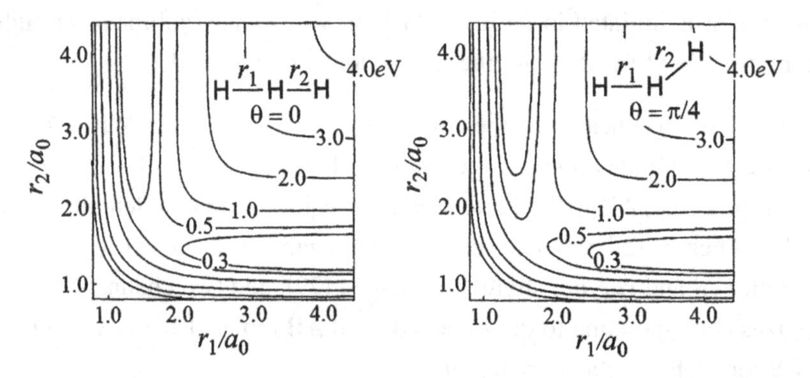

Figure 7.2.2: The potential for the $H + H_2$ reaction as a function of the two distances to the transferring H atom. Energies are measured relative to an infinitely separated H atom and H_2 molecule. Distances are given in Bohrs (a_0). The PES on the left is for collinear $H - H - H$ configurations. The PES on the right is for 45° bent $H - H - H$ configurations. The collinear transition state has a lower energy than the bent transition states. [Modified from R.N. Porter and M. Karplus, J. Chem. Phys., 40, 1105, (1964).]

Constructing model PESs for atom exchange reactions and using them in trajectory based rate calculations occupied many theorists from 1935–1970. The dynamics may, at first glance, seem simple for these atom exchange reactions with three atoms. It is often said that the dynamics can be envisioned as "a marble rolling on the potential energy landscape," but actually there are some flaws in this imagery. The most obvious problem is that marbles have rotational inertia, but even for an idealized point particle rolling on the PES there are problems with the analogy.

The first problem is that rotation and vibration are coupled motions in real molecules. Centrifugal forces from rotation tend to *expand* a rotating molecule around its center of mass, and

[1] Collinear geometries are a measure zero part of configuration space. Because of zero point vibrations, this remains true even at T = 0K.

rotational energy from a collision can be transferred into internal vibrations. In short, the dynamics on the PES are complicated when rovibrational coupling is included. Karplus, Porter, and Sharma developed algorithms for studying atom exchange reactions with rotation in internal coordinates [30]. Figure 7.2.3 shows two of their classical trajectories for the $H + H_2$ reaction.

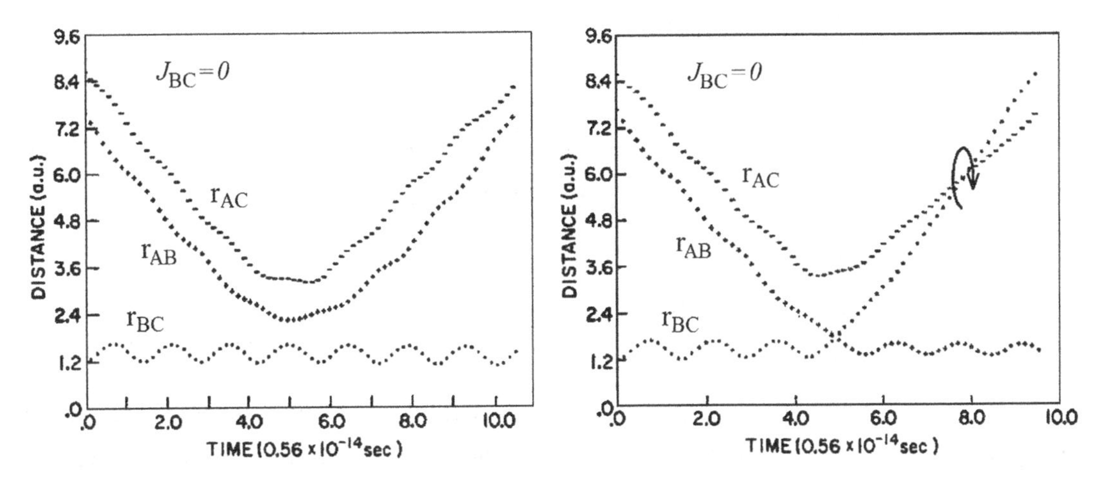

Figure 7.2.3: Non-collinear $H + H_2$ collisions with reactants in their rotational and vibrational ground states ($J = 0$ and $v = 0$) can inelastically induce rotation in the products. The trajectory in the left panel is non-reactive, and the trajectory in the right panel is reactive. From Karplus, Porter, and Sharma, *J. Chem. Phys.* 43, 3259 (1965).

A second problem with the marble analogy is that coordinates which simplify the PES often come at the price of an extremely complicated kinetic energy operator. Even for the collinear and non-rotating ($J = 0$) atom exchange reaction $\mathbf{A} + \mathbf{BC} \rightleftarrows \mathbf{AB} + \mathbf{C}$, the $J = 0$ Hamiltonian in the center-of-mass coordinate system has the rather complicated form:

$$H = \frac{1}{2} m_{\mathbf{A-B}} \left(\frac{dr_{\mathbf{AB}}}{dt} \right)^2 + \frac{1}{2} m_{\mathbf{AB-C}} \left(\frac{dr_{\mathbf{BC}}}{dt} + \frac{m_{\mathbf{A}}}{m_{\mathbf{A}} + m_{\mathbf{B}}} \frac{dr_{\mathbf{AB}}}{dt} \right)^2 + V(r_{\mathbf{AB}}, r_{\mathbf{BC}})$$

As indicated by the associated masses, the two terms in the kinetic energy are for the motion of **A** relative to **B** and for the motion of the **AB** 'molecule' relative to **C**. The second part of the kinetic energy term makes the dynamics in this space complicated. The two momenta are coupled by a $(dr_{\mathbf{BC}}/dt)(dr_{\mathbf{AB}}/dt)$ term which must be included in trajectories like those in Figure 7.2.3. Because of the coupling term, trajectories plotted directly on the $V(r_{\mathbf{AB}}, r_{\mathbf{BC}})$ surface do not follow an intuitive '$f_i = m_i a_i$' equation in which each degree of freedom separably accelerates in response to its separate force.

F.T. Smith and others [31,32] have shown how mass weighted *internal* coordinates can decouple the kinetic energy term. These are special examples of a general procedure known as a

Jacobi coordinate transformation [33,34]. For collinear atom exchange reactions, the transformation is

$$q_1 = r_{\mathbf{AB}} + \gamma r_{\mathbf{BC}} \tag{7.2.1}$$

$$q_2 = \varepsilon r_{\mathbf{BC}} \tag{7.2.2}$$

with parameters ε and γ defined as

$$\gamma = m_{\mathbf{C}}/(m_{\mathbf{B}} + m_{\mathbf{C}}) \tag{7.2.3}$$

$$\varepsilon = \gamma\sqrt{m_{\mathbf{B}} m_{\mathbf{ABC}}/(m_{\mathbf{A}} m_{\mathbf{C}})} \tag{7.2.4}$$

$$m_{\mathbf{ABC}} = m_{\mathbf{A}} + m_{\mathbf{B}} + m_{\mathbf{C}} \tag{7.2.5}$$

The $J = 0$ Hamiltonian in Jacobi (q_1, q_2) coordinates is

$$H = \frac{m_{\mathbf{A}}(m_{\mathbf{B}} + m_{\mathbf{C}})}{2m_{\mathbf{ABC}}}\left[\left(\frac{dq_1}{dt}\right)^2 + \left(\frac{dq_2}{dt}\right)^2\right] + V(r_{\mathbf{AB}}(q_1, q_2), r_{\mathbf{BC}}(q_1, q_2))$$

The advantage of these coordinates is that kinetic energy is now "diagonal," i.e. the kinetic energy term contains no coupling between dq_1/dt and dq_2/dt. Moreover, the two degrees of freedom q_1 and q_2 are now associated with a single effective mass

$$\mu_{eff} = \frac{m_{\mathbf{A}}(m_{\mathbf{B}} + m_{\mathbf{C}})}{m_{\mathbf{ABC}}} \tag{7.2.6}$$

Thus in the (q_1, q_2) coordinates, the dynamics *are* consistent with motion of a point particle on the PES. Mahan writes, "...the motion of three particles in two dimensions (collinear r_1, r_2) has been reduced to something which can be more easily visualized: the motion of one particle in two dimensions." Mahan has also shown that the resulting (q_1, q_2) coordinates are just rescaled versions of coordinates $r_{\mathbf{A-BC}}$ and $r_{\mathbf{BC}}$ as depicted in Figure 7.2.4 [32].

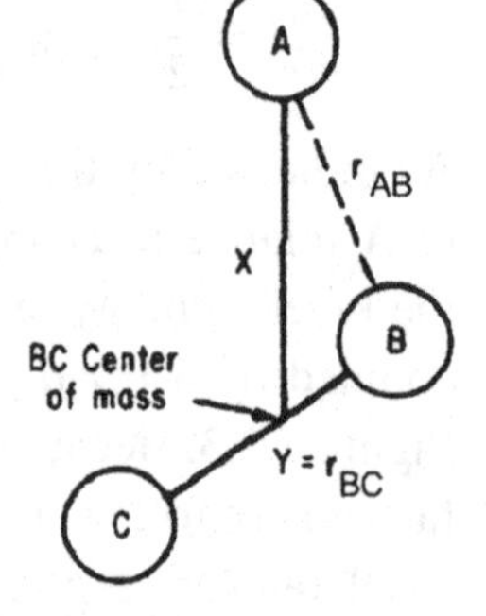

Figure 7.2.4: The (q_1, q_2) coordinates are rescaled versions of coordinates $r_{\mathbf{BC}}$ and $r_{\mathbf{A-BC}}$. Figure adapted from Mahan, *J. Chem. Educ.* 51, 308 (1974).

In the (q_1, q_2) coordinate system, lines of constant $r_{\mathbf{AB}}$ are no longer perpendicular to lines of constant $r_{\mathbf{BC}}$. This is easily seen from the inverse coordinate transformations: $r_{\mathbf{BC}} = q_2/\varepsilon$

and $r_{\mathbf{AB}} = q_1 - (\gamma/\varepsilon)q_2$. Lines of constant q_2 correspond to lines of constant $r_{\mathbf{BC}}$, but lines of constant $r_{\mathbf{AB}}$ deviate from constant q_2 by a skew angle φ:

$$\varphi = tan^{-1}\sqrt{\frac{m_{\mathbf{B}} m_{\mathbf{ABC}}}{m_{\mathbf{A}} m_{\mathbf{C}}}} \tag{7.2.7}$$

φ reflects the mass of the central **B** atom being transferred relative to the masses of the **A** and **C** atoms. A key consequence of the mass weighted internal coordinate representation is that the angle φ becomes very small when a light atom **B** is being transferred between two heavy atoms **A** and **C** [35–38]. The table below shows a list of atom transfer reactions and their skew angles.

Reaction	Skew angle
$H + Cl_2 \rightleftarrows HCl + Cl$	83.2°
$HD + H \rightleftarrows H + DH$	70.5°
$H_2 + H \rightleftarrows H + H_2$	60.0°
$DH + H \rightleftarrows D + H_2$	54.7°
$ClH + Cl \rightleftarrows Cl + HCl$	13.5°

The small skew angle for 'heavy-light-heavy' atom-exchange reactions causes non-adiabatic effects in which reactive trajectories do not follow the minimum energy path over the saddle [36]. Figure 7.2.5 shows a (hand-drawn) reactive classical trajectory for the heavy-light-heavy re-

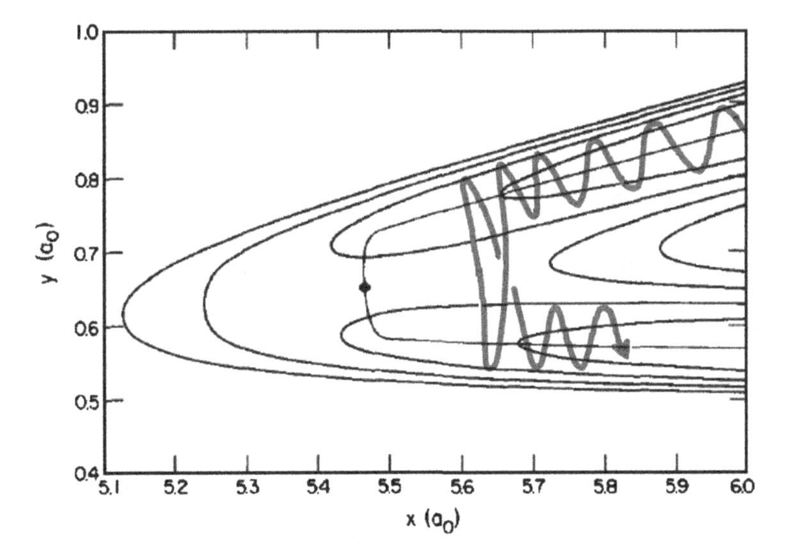

Figure 7.2.5: A schematic hand-drawn classical trajectory on the PES for the heavy-light-heavy reaction where H is transferred between Cl atoms in skewed (q_1, q_2) coordinates. The trajectory crosses the plane of symmetry three times, and each crossing occurs before the trajectory has reached the saddle point. The sharp curvature near the bottleneck leads to the shown non-adiabatic behavior, and also to tunneling between the narrowly separated channels when quantum effects are considered. The PES is from Bondi et al. *J. Chem. Phys.* 78, 5981-9 (1983).

action $ClH \cdots Cl \rightleftarrows Cl \cdots HCl$ reaction [38]. Additionally, the small skew angle enables tunneling by a corner-cutting mechanism (see Chapter 12). The effects of a small skew angle on tunneling and nonadiabatic characteristics in the dynamics can be generalized to proton transfer between heavy donors and acceptors even when the complex fragments are not linear [39]. The atom-exchange reaction exemplifies the generalizable mechanistic insights that can be gained from (1) studying simple model systems, and (2) choosing coordinates that make dynamics consistent with features of the energy or free-energy landscape. The following section introduces a more general coordinate system for describing those key features of a PES that influence the equilibrium properties, dynamics, and reactivity.

7.3 Mass weighted coordinates and normal modes

In many contexts, the chosen coordinate system can have a dramatic effect on the difficulty of a calculation. Many coordinate systems have been used in chemical kinetics, each with specific advantages and disadvantages for specific tasks. For example, geometry optimizations are frequently facilitated by internal coordinates [40–43]. On the other hand, dynamical simulations and vibrational frequency analyses are facilitated by mass weighted Cartesian coordinates [44,45]. Let the coordinate vector $\mathbf{x}$ be the full list of Cartesian coordinates $(x_1, y_1, z_1, x_2, y_2, \ldots, x_N, y_N, z_N)$ where N is the number of atoms. In the Cartesian coordinate system the Hamiltonian is

$$H = \frac{1}{2}\mathbf{p}^{\mathbf{T}}\mathbf{M}^{-1}\mathbf{p} + V(\mathbf{x}) \tag{7.3.1}$$

where the diagonal matrix $\mathbf{M}$ contains the atomic masses,

$$\mathbf{M} = diag[m_1, m_1, m_1, \cdots, m_N, m_N, m_N]. \tag{7.3.2}$$

Newton's equation of motion, $\boldsymbol{f} = \mathbf{M}\boldsymbol{a}$, for the nuclei in the Cartesian coordinate system is

$$-\frac{\partial V}{\partial \mathbf{x}} = \mathbf{M}\frac{d^2\mathbf{x}}{dt^2} \tag{7.3.3}$$

The matrix $\mathbf{M}$ can be removed from the dynamical equations by rescaling the original Cartesian variables to obtain mass weighted Cartesian (MW) coordinates

$$\mathbf{x}_{MW} = \frac{1}{\sqrt{m_0}}\mathbf{M}^{1/2}\mathbf{x} \tag{7.3.4}$$

and

$$\mathbf{p}_{MW} = \sqrt{m_0}\,\mathbf{M}^{-1/2}\mathbf{p} \tag{7.3.5}$$

where the constant scale factor m_0 is usually chosen for convenience as $m_0 = 1amu$ [46]. With this choice, the Hamiltonian becomes

$$H = \frac{1}{2m_0}\mathbf{p}_{MW}^2 + V(\mathbf{x}_{MW}) \tag{7.3.6}$$

so that all degrees of freedom are associated with a $1amu$ mass. To maintain a less cumbersome notation, we use the new scaled variables as arguments of the original potential function V. Mathematicians and programmers are asked to forgive this notational shortcut.

In mass weighted coordinates, $f = Ma$ becomes

$$-\frac{\partial V}{\partial \mathbf{x}_{MW}} = m_0 \frac{d^2}{dt^2}\mathbf{x}_{MW} \tag{7.3.7}$$

so that the different masses vanish from the dynamical equations. To compute the vibrational frequencies and modes, expand around a molecular conformation that minimizes the potential energy. Taylor expanding to second order gives the local quadratic approximation to the PES,

$$V(\mathbf{x}_{MW}^{min} + \Delta\mathbf{x}_{MW}) = V(\mathbf{x}_{MW}^{min}) + \frac{1}{2}\Delta\mathbf{x}_{MW}^{\dagger}\mathbf{H}_{MW}\Delta\mathbf{x}_{MW}.$$

$\mathbf{H}_{MW}$ is the mass weighted Hessian, a matrix of second derivatives of the potential energy with respect to the mass weighted coordinates. $\mathbf{H}_{MW}$ is easily computed from the second derivative matrix in regular Cartesian coordinates via the transformation $\mathbf{H}_{MW} = m_0\,\mathbf{M}^{-1/2}\mathbf{H}\mathbf{M}^{-1/2}$. The gradient is similarly transformed by the change of coordinates: $\mathbf{g}_{MW} = \sqrt{m_0}\mathbf{M}^{-1/2}\mathbf{g}$, but the gradient is zero at a potential energy minimum. Inserting the second order expansion into the equation of motion gives a set of coupled second order linear differential equations

$$m_0 \frac{d^2}{dt^2}\mathbf{\Delta x}_{MW} = -\mathbf{H}_{MW}\mathbf{\Delta}x_{MW} \tag{7.3.8}$$

These can be decoupled by diagonalizing the matrix $\mathbf{H}_{MW}$

$$\mathbf{U}^{\dagger}\mathbf{H}_{MW}\mathbf{U} = m_0\mathbf{\Omega}^2$$

where the columns of matrix $\mathbf{U}$ are the eigenvectors (normal modes) and the $\mathbf{\Omega}^2$ matrix is diagonal with non-negative eigenvalues. From the spectral theorem, $\mathbf{U}^{\dagger}\mathbf{U} = \mathbf{U}\mathbf{U}^{\dagger} = \mathbf{1}$, i.e. $\mathbf{U}$ is unitary. Writing the displacement vector in the eigenvector basis

$$\xi = \mathbf{U}^{\dagger}\mathbf{\Delta x}_{MW}$$

gives the decoupled equations of motion

$$\frac{d^2}{dt^2}\xi = -\mathbf{\Omega}^2\xi. \tag{7.3.9}$$

The columns of matrix **U** are the normal vibrational modes and the diagonal elements of $\mathbf{\Omega}$ are the corresponding angular vibrational frequencies. Many electronic structure packages can find minima on the PES and compute/diagonalize the Hessian matrix. Computationally predicted vibrational frequencies and modes are important links to experimental IR and Raman spectroscopy results. The spectroscopic predictions and observations can be used to improve force fields, to interpret/assign spectroscopic peaks, and to test hypothesized mechanisms as we have already seen in Chapter 4. The normal modes also provide a coordinate system that we will use frequently in later chapters.

7.4 Features of molecular potential energy surfaces

At a minimum energy configuration, the gradient of the potential is zero and the second derivative matrix (Hessian) is positive definite. The structure of stable molecules and intermediates can be predicted by locating energy minima, and their vibrational frequencies (as shown in the previous section) section can be predicted by diagonalizing the mass-weighted Hessian. In addition to minima, we will define basins to compute free energies of stable species and intermediates, identify saddle points to characterize transition states, use the minimum energy path to measure dynamical progress, and define a dividing surface to compute fluxes. Figure 7.4.1 shows the potential energy minima, a dividing surface, a saddle point, the unstable mode, and the MEP on the model PES of Muller and Brown [51].

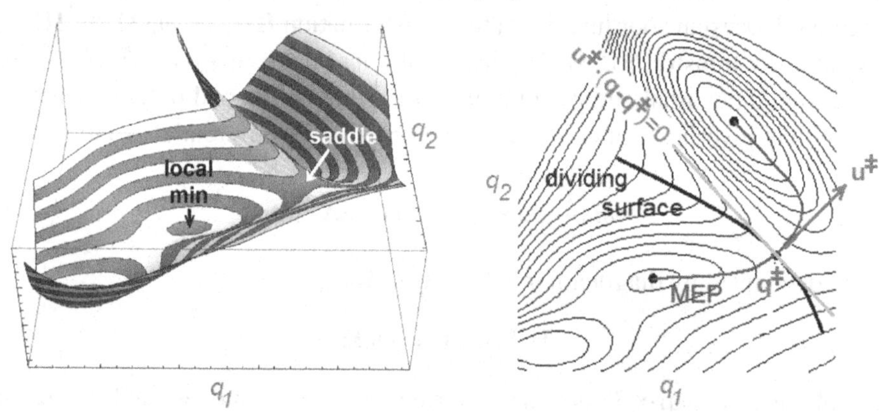

Figure 7.4.1: (Left) A three-dimensional plot of the Muller-Brown PES with constant energy contours on the surface. (Right) A contour plot of the Muller-Brown PES showing only the contours in light black. The plot on the right also shows a dividing surface (the heavy black curve), a minimum energy path (MEP, light gray curve), a saddle point (black dot at the intersection of the dividing surface and MEP), and two minima (black dots at the ends of the MEP).

The basin around an energy minimum refers to the set of configurations that relax to the minimum upon steepest descent potential energy minimization. Two basins of attraction on a PES meet at a dividing surface between the stable basins. As Marcelin described in his PhD thesis, (1914) [47,48], the dividing surface defines the boundary for computing the flux from one basin to another. Saddle points are special low energy configurations within the dividing surface. At a saddle point the potential energy gradient vanishes and the second derivative matrix has exactly one imaginary frequency. The unstable eigenmode corresponding to the imaginary frequency points along the downhill directions from the saddle point to the reactants and products. For many elementary reactions, the saddle point corresponds to a dynamical bottleneck in the sense that nearly all reactive trajectories cross the dividing surface in the vicinity of the saddle point [29]. A hyperplane perpendicular to the unstable mode in mass weighted coordinates at the saddle point defines a convenient dividing surface between the reactants and products.

A steepest descent calculation started from the saddle point forward and backward along the unstable mode on the PES will reach the minima on either side of the dividing surface. The steepest descent path computed in this way with mass weighted coordinates is known as the minimum energy path (MEP) [49,50].

$$\frac{d\boldsymbol{a}}{ds} = -\frac{\mathbf{g}_{MW}}{\|\mathbf{g}_{MW}\|} \tag{7.4.1}$$

Here $\mathbf{g}_{MW}$ is the gradient of the potential with respect to the mass weighted coordinates, and $\boldsymbol{a}(s)$ is the MEP in mass weighted Cartesian coordinates. In equation (7.4.1) the normalized gradient is computed at the local position along the MEP, $\boldsymbol{a}(s)$.

Equation (7.4.1) describes a curve $\boldsymbol{a}(s)$ that is parameterized by its own arclength in mass weighted coordinates. By convention the MEP is usually reparameterized such that point $\boldsymbol{a}(0)$ corresponds to the saddle point with $\boldsymbol{a}(s)$ for $s > 0$ corresponding to the product side of the path and $\boldsymbol{a}(s)$ for $s < 0$ corresponding to the reactant side of the path. This reparameterization only requires changing the sign of s on the backward (saddle to reactant) section of the path after solving (7.4.1).

The MEP is useful for visualizing the potential energy along the reaction pathway, for defining a reaction path Hamiltonian (section 7.5), and for computing rates especially when tunneling is significant (Chapter 12). Contrary to some descriptions in the literature [52–54], the MEP does not minimize the classical action and therefore it is not a dynamical path. As depicted in Figure 7.2.5, dynamical paths tend to oscillate around the MEP as they move along the reaction pathway.

Starting from the saddle point, the MEP can (in principle) be followed downward to the neighboring minima using the steepest descent algorithm of Euler. However, the Euler algorithm requires extremely small stepsizes. Preferable alternatives include a stabilized Euler method [55], and algorithms based on local quadratic approximations (LQAs) and Hessian updates [56–58].

The Gonzalez-Schlegel algorithm [56] and the Hratchian-Schlegel algorithm [58] with Bofill updates [59] are accurate and easily implemented ways to generate the MEP.

In some cases, one wants to construct the reaction path Hamiltonian in addition to computing the MEP. The LQA algorithm of Page and McIver does these calculations simultaneously. The Page-McIver algorithm gives a smooth and accurate approximation to the MEP [60], but the calculation is non-trivial because it requires *ab initio* Hessians at each point along the MEP. [2]

Algorithm: Page-McIver algorithm

Page and McIver [60] introduced an efficient and accurate method for approximating the MEP as series of points $\{\boldsymbol{a}(s_1), \boldsymbol{a}(s_2), \ldots, \boldsymbol{a}(s_n)\}$ by walking down from the saddle point, $\boldsymbol{a}(s_0)$. At each point $\boldsymbol{a}(s_k)$ a local quadratic approximation provides the local gradient at points displaced by a small amount $\Delta\boldsymbol{a}$ from $\boldsymbol{a}(s_k)$:

$$\mathbf{g}_{MW}(\boldsymbol{a}(s_k) + \boldsymbol{\Delta a}) = \mathbf{g}_{MW}(\boldsymbol{a}(s_k)) + \mathbf{H}_{MW}(\boldsymbol{a}(s_k))\boldsymbol{\Delta a}$$

Here $\mathbf{H}_{MW}(\boldsymbol{a})$ is the mass weighted second derivative matrix of the potential at $\boldsymbol{a}$ and $g(\boldsymbol{a})$ is the gradient of the potential with respect to the mass weighted coordinates at $\boldsymbol{a}$. These should both be computed at each point $\boldsymbol{a}(s_k)$. After diagonalization of $\mathbf{H}_{MW}$ to obtain $\mathbf{U}^\dagger \mathbf{H}_{MW} \mathbf{U} = m_0 \boldsymbol{\Omega}^2$ and $\gamma = \mathbf{U}^\dagger \mathbf{g}_{MW}$, a steepest descent trajectory on the local quadratic PES follows the curve of displacements given by

$$\boldsymbol{\Delta a}(t) = -\mathbf{U} m_0^{-1} \boldsymbol{\Omega}^{-2} (1 - exp[-m_0 \boldsymbol{\Omega}^2 t])\gamma$$

Here t is *not* time. It is a non-physical variable used for parameterization of the local curve. $\Delta\boldsymbol{a}(t)$ initially follows the steepest descent direction and then bends to follow softer modes as t advances. Recall that $\boldsymbol{a}(s)$ should be parameterized by its own arclength, so the relationship between t and arclength s along the displacement trajectory is needed. As measured from point $\boldsymbol{a}(s_k)$, the differential arclength ds and dt are related by [60]

$$ds = \left\{\gamma^\dagger e^{-2m_0 \boldsymbol{\Omega}^2 \mathbf{t}} \gamma\right\}^{1/2} dt \tag{7.4.2}$$

For any desired stepsize $s_{k+1} - s_k = \Delta s$ in arclength along the MEP, equation (7.4.2) can be numerically integrated from $t = 0$ to find the appropriate value of t for the chosen stepsize. Steps of fixed size Δs are generated in this manner until a value smaller than Δs is obtained from equation (7.4.2) as $t \to \infty$. The figure below shows steepest descent

[2] These are especially costly calculations with plane-wave DFT codes where the Hessians at each point along the MEP must be finite-differenced from analytic gradients.

paths obtained from the Euler method and from the Page-McIver method with the same step sizes.

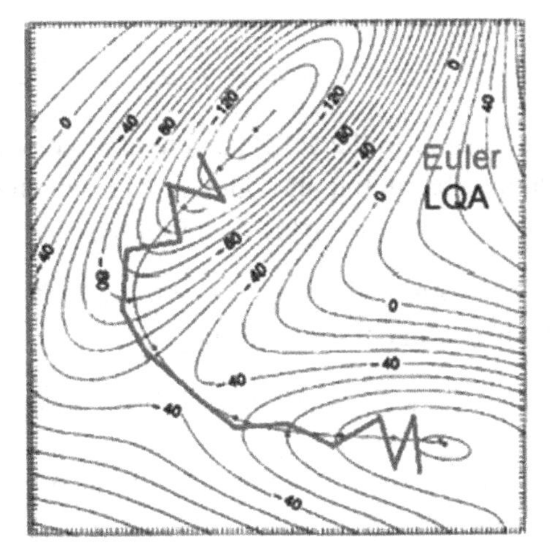

The Page-McIver (LQA) path closely follows the exact MEP, while the Euler path zig-zags from one side of the MEP to the other. The instability in the Euler method is most severe where there are large forces that oppose a displacement perpendicular to the MEP.

Each of the above methods for computing the MEP and the reaction path Hamiltonian requires *a priori* knowledge of the saddle point on the PES. On a low dimensional model PES, the saddle point can be identified by visual inspection. However, real molecules can have hundreds of degrees of freedom, so finding saddle points can be a major challenge. Algorithms that find saddle points are discussed in Chapter 8.

7.5 Reaction path Hamiltonian

The Page-McIver method for computing the MEP provides second derivative information that can be used to model the PES as a curved valley with harmonic walls [61]. Miller, Handy, and Adams developed a reaction path Hamiltonian (RPH) to describe the dynamics in a curved harmonic valley. The RPH is similar in spirit to Marcus' natural collision coordinates for atom exchange reactions [62] but the RPH applies to more complex types of reactions.

The arclength along the MEP provides the reaction coordinate in the RPH. Recall that the MEP is parameterized by its signed arclength s with the saddle point at $s = 0$. Using the notation from the previous section, the MEP is a vector function $\boldsymbol{a}(s)$, and the saddle point is $\boldsymbol{a}(0)$. This definition is adequate for points on the MEP, but it remains to define the reaction coordinate for points displaced from the MEP. If the MEP has been computed in mass weighted coordinates, then the path tangent $\mathbf{t}(s) = d\boldsymbol{a}(s)/ds$ defines a hyperplane normal to the MEP at each point s. All points $\mathbf{x}$ such that

$$\mathbf{t}(s)^{\dagger}(\mathbf{x} - \boldsymbol{a}(s)) = 0 \tag{7.5.1}$$

lie on the hyperplane perpendicular to the MEP tangent at s. For each point $\mathbf{x}$, the reaction coordinate $s(\mathbf{x})$ can be determined by finding the value of s for which equation (7.5.1) is satisfied, or equivalently by finding s such that $|\mathbf{x} - \boldsymbol{a}(s)|$ is minimized.

A coordinate system for the remaining degrees of freedom is obtained from the second derivative matrix at $\boldsymbol{a}(s)$ with the path tangent $\boldsymbol{t}(s)$ projected out.

$$\mathbf{K}(s) \equiv (\mathbf{1} - \mathbf{t}(s) \circ \mathbf{t}(s)^{\dagger})\mathbf{H}_{MW}(s)(\mathbf{1} - \mathbf{t}(s) \circ \mathbf{t}(s)^{\dagger})$$

Diagonalization of $\mathbf{K}(s)$ gives transverse frequencies and transverse eigenvectors that span the hyperplane perpendicular to the MEP.

$$m_0 \breve{\boldsymbol{\Omega}}^2(s) = \mathbf{L}(s)^{\dagger}\mathbf{K}(s)\mathbf{L}(s) \tag{7.5.2}$$

The eigenvectors of $\mathbf{K}(s)$ are the columns of $\mathbf{L}(s)$ and the associated eigenvalues are the squared transverse angular frequencies $\breve{\boldsymbol{\Omega}}^2(s)$. Each transverse coordinate $\breve{x}_k$ is the displacement from the path along the transverse eigenvector $\mathbf{L}_k(s)$:

$$\breve{\mathbf{x}} = \mathbf{L}(s)^{\dagger}(\mathbf{x} - \boldsymbol{a}(s)). \tag{7.5.3}$$

By construction $\breve{\mathbf{x}}$ is the vector of displacements perpendicular to the MEP. Motion along the MEP in regions where the MEP is curved causes excitation of vibrational modes perpendicular to the MEP tangent. The RPH includes coupling between the reaction coordinate and the bath modes through the path curvature coupling coefficients [61],

$$\mathbf{b}(s) = \mathbf{L}(s)^{\dagger}\mathbf{t}'(s)$$

where $\mathbf{t}'(s) = d\mathbf{t}/ds$. The harmonic valley potential is exact along the MEP. For points displaced from the MEP, errors in the PES grow as the third power of distance from the MEP. Figure 7.5.1 shows the true Muller potential along the reaction pathway alongside the harmonic valley approximation.

For systems with more than two internal degrees of freedom, e.g. s, $\breve{x}_1$, and $\breve{x}_2$, the transverse eigenvectors in $\mathbf{L}(s)$ can also twist about the steepest descent path leading to Coriolis coupling coefficients [61].

$$\mathbf{B}(s) = \mathbf{L}(s)^{\dagger}\mathbf{L}'(s) \tag{7.5.4}$$

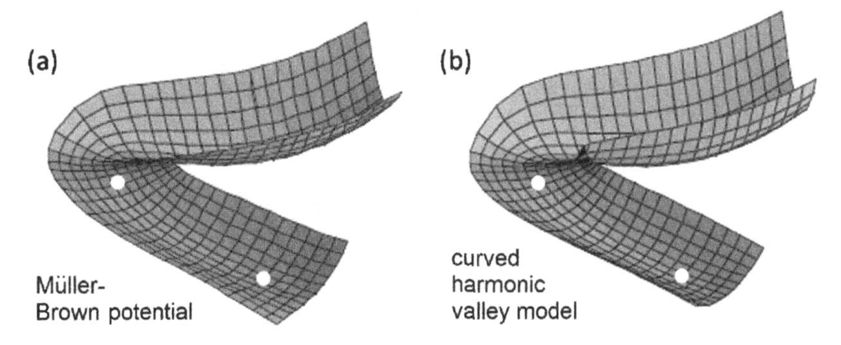

Figure 7.5.1: (a) The reaction pathway region on the exact Muller-Brown PES with the $(s, \breve{x})$ coordinate grid shown. (b) The approximate curved harmonic valley model that is used in the RPH. The surfaces match exactly for points on the MEP, and they match to within one percent of barrier height over most of the region shown. [Figure from PhD Thesis, B. Peters, Berkeley (2004).]

A nonzero Coriolis coupling matrix $\mathbf{B}(s)$ allows the transfer of energy amongst the bath modes. Let the conjugate momenta to $(s, \breve{\mathbf{x}})$ be $(p_s, \breve{\mathbf{p}})$. Then the RPH for a system with zero total angular momentum is [61]

$$H = \frac{(p_s + \breve{\mathbf{x}}^\dagger \mathbf{B}(s)\breve{\mathbf{x}})^2}{2m_0(1 + \breve{\mathbf{x}}^\dagger \mathbf{b}(s))^2} + \frac{1}{2m_0}\breve{\mathbf{p}}^2 + V_0(s) + \frac{m_0}{2}\breve{\mathbf{x}}^\dagger \breve{\boldsymbol{\Omega}}^2(s)\breve{\mathbf{x}} \tag{7.5.5}$$

where $V_0(s) = V(\boldsymbol{a}(s))$ is the potential energy profile along the MEP. To a first approximation, rotational kinetic energy can be added to the RPH as a rigid rotation of the molecule in configuration $(s, \breve{\mathbf{x}})$. It is also possible to extend the RPH to include rovibrational coupling, i.e. coupling between rotation and internal modes of the molecule [61].

In practice, the RPH must be computed from a collection of discrete points along the path $\boldsymbol{a}(s)$. The diagonalization of $\mathbf{K}(s)$ at a point s_k on the MEP yields an unordered transverse eigensystem that must be matched to the transverse eigensystem at the neighboring points on the MEP, i.e. to the eigensystems of $\mathbf{K}(s_{k+1})$ and $\mathbf{K}(s_{k-1})$. The construction requires a *diabatic matching* so that the transverse eigenvectors change continuously as a function of s and give smoothly varying coupling coefficients, $\mathbf{B}(s)$ and $\mathbf{b}(s)$ [63]. Figure 7.5.2 illustrates the difference between diabatic and adiabatic procedures for matching the eigenmodes.

The RPH provides a remarkably simple way to capture nearly all features of the reaction dynamics with only a modest number of electronic structure calculations. Why then, has the RPH not been more widely used? Let us consider its predictions about reaction rates and reaction dynamics. The natural reaction coordinate is the arclength s, which evolves according to Hamilton's equation $ds/dt = \partial H/\partial p_s$. Expressions for the time derivatives of s as well as p_s, $\breve{\mathbf{p}}$,

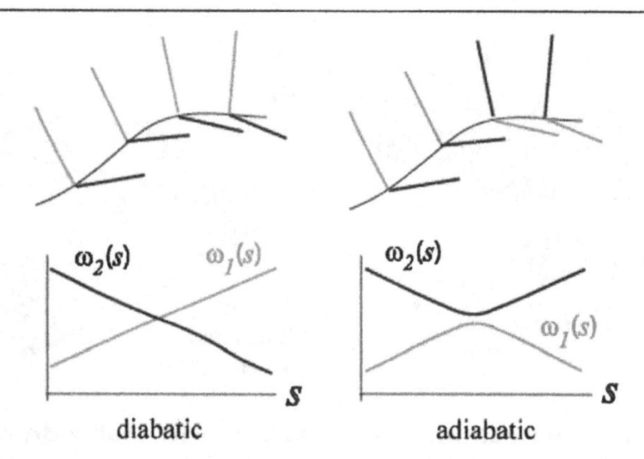

Figure 7.5.2: The top figures depict two transverse eigenvectors (black and gray) along the MEP. The lower figures depict the corresponding transverse frequencies as functions of the reaction coordinate. The diabatic construction leads to transverse eigenmodes that vary continuously with motion along s. The adiabatic construction (based on ordering of the transverse frequencies) leads to transverse eigenmodes that abruptly change with motion along s. [From PhD Thesis, B. Peters, Berkeley (2004).]

and $\check{\mathbf{x}}$ can be found in ref. [63]. Hu et al. used a simplified RPH with only select curvature coupling terms and no Coriolis coupling to compute transmission coefficients for Diels-Alder and Claisen condensation reactions [64]. Hu et al. found transmission coefficients of nearly 90%, so a vastly simpler harmonic transition state theory calculation would have been adequate for all practical applications. In a similar calculation for cyclohexane interconversion, Peters et al. [63] found transmission coefficients of approximately 50%. Again the reasonably large transmission coefficient suggests that dynamical effects of curvature coupling and Coriolis coupling are of secondary importance. These results do not mean that transmission coefficients are always near unity, but when the bottleneck corresponds to a saddle point on the PES, harmonic transition state theory (perhaps with a tunneling correction) is usually adequate. When the bottleneck *does not* correspond to a saddle point, then neither harmonic transition state theory nor the RPH framework are applicable. Instead, the latter situations require entirely different rare events methods based on collective variables, free energies, and/or path sampling methods.

7.6 Empirical valence bond models

The empirical valence bond (EVB) framework is a relatively simple way to construct a ground state PES that is approximately correct near its potential energy minima and saddle points. As we have already noted in section 7.5, these are the most critical regions of the PES for computing accurate rates. EVB [65,66] uses separate diabatic models of the reactant and product states to create a model for the adiabatic ground state PES. The reactant and product models can be arbitrarily elaborate, and the EVB approach has the important advantage that experimental structure, thermodynamics, and even kinetic information can be incorporated in the parameterization.

In the simplest EVB implementation, one begins with two force fields that accurately describe the reactant and product states. Let $E_{\mathbf{A}}(\mathbf{x})$ and $E_{\mathbf{B}}(\mathbf{x})$ be diabatic models of the energy landscape within states **A** and **B**.[3] For a given set of nuclear coordinates **x**, the functions $E_{\mathbf{A}}(\mathbf{x})$ and $E_{\mathbf{B}}(\mathbf{x})$ correspond to energy landscapes for two different electronic configurations. For example, at any given bond distance r the covalent complex **XY** could also exist as an ionic complex $\mathbf{X}^{-}\mathbf{Y}^{+}$. These two electronic states would correspond to the diabatic energy functions $V_{\mathbf{XY}}(r)$ and $V_{\mathbf{X-Y+}}(r)$. The diabatic potential energy surfaces ("diabats") in combination with an empirical coupling term $V_{\mathbf{AB}}$ define a matrix Hamiltonian [65]:

$$\mathbf{V}(\mathbf{x}) = \begin{bmatrix} E_{\mathbf{A}}(\mathbf{x}) & V_{\mathbf{AB}} \\ V_{\mathbf{AB}} & E_{\mathbf{B}}(\mathbf{x}) \end{bmatrix} \tag{7.6.1}$$

In the simplest cases $V_{\mathbf{AB}}$ is a constant, but EVB models with coordinate dependent coupling are also common [67]. Adiabatic energy landscapes emerge as roots of the secular equation $det[\mathbf{V}(\mathbf{x}) - V_i(\mathbf{x})\mathbf{I}] = 0$. The roots are interpreted as a ground state $(V_0(\mathbf{x}))$ and an excited state $(V_1(\mathbf{x}))$. These are

$$V_0(\mathbf{x}) = \frac{1}{2}(E_{\mathbf{A}}(\mathbf{x}) + E_{\mathbf{B}}(\mathbf{x})) - \frac{1}{2}\left\{\Delta E(\mathbf{x})^2 + 4V_{\mathbf{AB}}^2\right\}^{1/2} \tag{7.6.2}$$

and

$$V_1(\mathbf{x}) = \frac{1}{2}(E_{\mathbf{A}}(\mathbf{x}) + E_{\mathbf{B}}(\mathbf{x})) + \frac{1}{2}\left\{\Delta E(\mathbf{x})^2 + 4V_{\mathbf{AB}}^2\right\}^{1/2} \tag{7.6.3}$$

Here $\Delta E(\mathbf{x})$ is the vertical energy gap

$$\Delta E(\mathbf{x}) = E_A(\mathbf{x}) - E_B(\mathbf{x}).$$

A schematic illustration of the adiabatic EVB potential from two diabatic landscapes is shown in Figure 7.6.1.

The parameterization of the diabatic states can be tuned to give the appropriate relative free energy of reaction. Additionally, the parameter $V_{\mathbf{AB}}$ can be tuned so that the barrier matches that from experiment. Thus, the tremendous advantage of EVB is that an entire reactive PES can be constructed from the spectroscopic, structural, thermodynamic, and kinetic properties of the

[3] This book denotes abstract molecular entities with bold capital letters, e.g. **A**, **B**, **C**, etc. Later chapters will use A and B to denote the regions of configuration space that correspond to reactants and products, respectively. Regardless of the configuration **x**, the diabatic states of an EVB model correspond to reactant and product entities, so we denote the diabatic states as **A** and **B**.

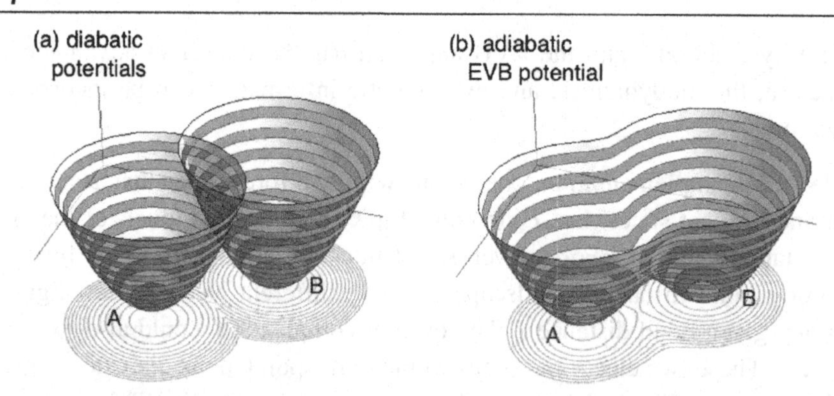

Figure 7.6.1: Diabatic models of the local PESs in states A and B (shown in (a)) can be converted into a model of the adiabatic ground state PES $V_0(\mathbf{x})$ using the EVB construction. As the coupling vanishes, i.e. as $V_{\mathbf{AB}} \to 0$, the EVB PES $V_0(\mathbf{x})$ approaches $min[E_{\mathbf{A}}(\mathbf{x}), E_{\mathbf{B}}(\mathbf{x})]$.

reactants and products [68–70]. The diabatic state models can also incorporate solvent-solute and solvent-solvent interactions. Because of these capabilities, EVB provides a straightforward route to accurate models of reactions in complex solution and enzyme environments. Finally, the energy gap provides a natural collective variable to use as a reaction coordinate. The dividing surface in EVB is the locus of points where $\Delta E(\mathbf{x})$ is zero, i.e. the set of configurations $\mathbf{x}$ for which $E_{\mathbf{B}}(\mathbf{x}) = E_{\mathbf{A}}(\mathbf{x})$. The surface $\Delta E(\mathbf{x}) = 0$ is not a hyperplane. It naturally curves through the ridge between basins **A** and **B**, following the surface of nearest avoided crossing between $V_0(\mathbf{x})$ and $V_1(\mathbf{x})$. The dividing surface defined by $\Delta E(\mathbf{x}) = 0$ is particularly useful for enzyme reactions and for reactions in solution, where the hyperplanar dividing surface of harmonic transition state theory is not sufficiently accurate. Several applications of the EVB approach to enzyme reactions have found transmission coefficients of order unity [71,72]. These calculations confirm the accuracy of the energy gap dividing surface $\Delta E(\mathbf{x}) = 0$ within an EVB Hamiltonian. Of course, it must be remembered that the dividing surface in EVB is not defined by the true adiabatic ground state potential, but rather by the intersection of two diabatic ground state models.

The discussion here covers only the absolute simplest elements of EVB. More elaborate EVB models have been developed for modeling water autodissociation [69,70], for proton transfer in water [73], and for several enzymatic reactions [70–72]. These EVB models use fully detailed atomistic force fields for the diabatic state models as well as more complicated off-diagonal coupling elements. See Warshel, Truhlar, and coworkers for additional details on the construction of EVB potentials and related models [65,74].

7.7 Disconnectivity graphs

Thus far we have focused on complete descriptions of the 3N-dimensional PES as a continuous manifold in the molecular configuration space. A continuous PES is essential for molecular dynamics simulations, but for systems with multitudes of local minima and saddles, a concise visual summary of the minima and the saddles between them is useful. There are two strategies for summarizing the features of a high dimensional PES. One strategy projects the high dimensional PES onto a few collective variables, order parameters, or reaction coordinates to obtain a lower dimensional free energy surface. This strategy will command much of our focus in later chapters, and its principle advantage is that – when the correct reaction coordinate is used – the kinetics are easily estimated and understood from the free energy surface and the dynamics of the reaction coordinate.

The second strategy is to create a disconnectivity graph. Disconnectivity graphs are powerful ways to visualize the diversity of local minima and the lowest energy barriers between them. The tip of each root in a disconnectivity graph corresponds to a distinct local energy minimum. Moving upward from a pair of root tips, their separate roots merge at the energy maximum along the lowest energy path between the structures at their tips. Assuming that rates are always dominated by the lowest energy pathways, branches of the graph that separate at higher energies connect groups of states that interconvert on longer time scales. Figure 7.7.1 shows how the disconnectivity graph is created from a simple potential energy surface with several minima and saddles. Figure 7.7.1 also shows the first disconnectivity graph from the work of Czerminski and Elber [75].

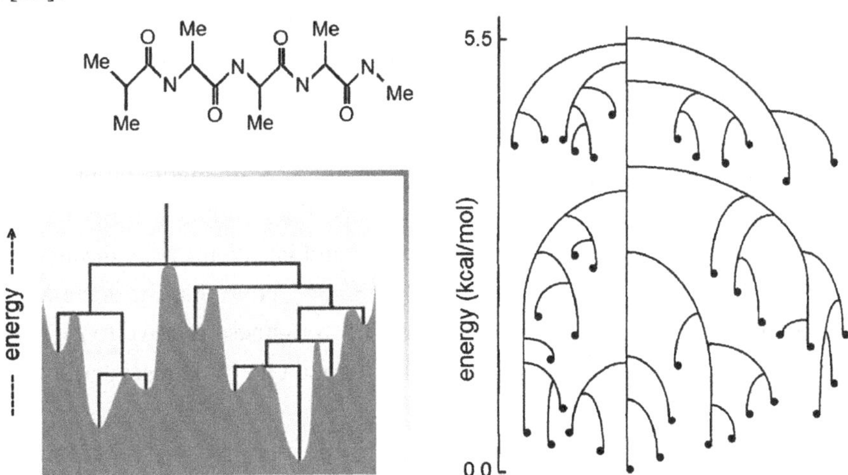

Figure 7.7.1: (Lower left) Illustrating how a disconnectivity graph is created from the minima and saddle points on a potential energy landscape. (Upper left and right) The disconnectivity graph for an isobutyryl-(ala)3-NH-methyl (IAN) tetrapeptide. [From Czerminski and Elber, J. Phys. Chem. 92, 5580 (1990).]

Disconnectivity graphs become extremely dense even for relatively small systems, but still they are vastly easier to visualize than high dimensional energy landscapes. Wales, Krivov, and coworkers have constructed disconnectivity graphs for many systems [76–79]. Wales and coworkers in particular have made several improvements to the algorithms for identifying the complete set of minima and saddle points [80–82]. Their results show how disconnectivity graph structure reveals characteristics of the time scale spectrum in complex systems [77,79]. The dynamics of relaxation from one state on the graph to equilibrium depend, in a rough sense, on the branching structure of the root system. If all minima are connected via small downhill barriers to the main taproot, then relaxation to the global minimum will be fast. For example, such disconnectivity graphs have been extensively studied in connection with models that envision the protein folding landscape as a funnel centered on the folded structure [77, 79]. These studies have contributed to our understanding of native and non-native interactions in protein folding kinetics. Alternatively, if the main roots branch at high energies and then separately diverge into thousands of sub-branches containing low energy states, then relaxation will be slower. For example, the disconnectivity graph of a glassy system has thousands of minima with no clear depth stratification, nor a clear stratification of energies in the branching junctions between minima [83,84].

Note that the disconnectivity graph only shows the *lowest* energy bottleneck between two distinct minima. Many pathways may pass through alternative bottlenecks that are just narrowly higher than the saddle shown in the disconnectivity graph. The ensemble of higher energy pathways can be vastly more important in the kinetics than the one lowest energy path. Thus disconnectivity graphs alone cannot be used to quantitatively predict the kinetics. However, the procedures for constructing a disconnectivity graph involve locating a multitude of stationary points, and the complete collection of saddles and minima *can* be used to estimate transition rates. Discrete path sampling approaches use the complete set of saddle points and minima to estimate the overall transition rate over rugged barriers. Each saddle corresponds to one transition in the discrete master equation, and the forward and backward rates through each saddle are obtained from harmonic transition state theory [85,86]. Note that the accuracy of the harmonic approximation depends on the scale of the PES roughness relative to $k_B T$. Alternative methods to compute rates in complex systems with rugged energy landscapes are described in later chapters.

Exercises

1. The figure below, modified from Karplus et al. [30], shows the computed cross section for the $H_2 + H$ reaction as a function of impact parameter b. The reaction probability $P(b)$ and

b were defined in Chapter 6. The data are from non-collinear classical trajectories on the Born-Oppenheimer surface with reactants in their rotational and vibrational ground states.

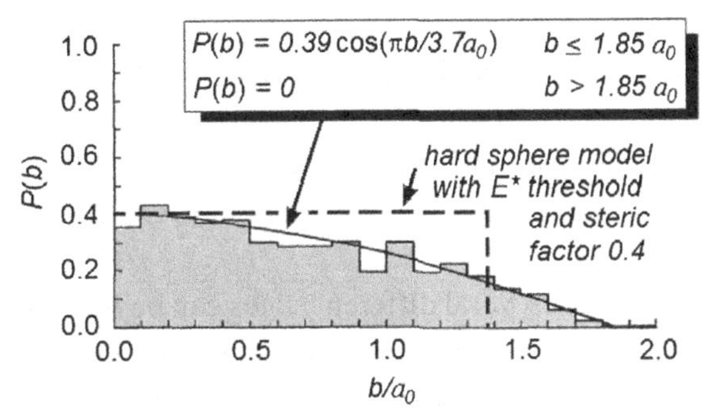

Set up equations to find the special energy threshold and steric factor which yield the same rate constant and reaction probability for the reactive hard spheres model. Use van der Waals radii for H and H_2.

2. Consider three atoms **A, B,** and **C** which are constrained to the x-axis. Suppose that their potential energy in the vicinity of a stationary point is given by

$$V(r_{\mathbf{AB}}, r_{\mathbf{BC}}) = V_0 + \frac{1}{2}\kappa_1 \Delta r_{\mathbf{AB}}^2 + \frac{1}{2}\kappa_2 \Delta r_{\mathbf{BC}}^2 + \kappa_{12} \Delta r_{\mathbf{AB}} \Delta r_{\mathbf{BC}}$$

where $\Delta r_{\mathbf{AB}} = x_{\mathbf{B}} - x_{\mathbf{A}} - \ell_{\mathbf{AB}}$, $\Delta r_{\mathbf{BC}} = x_{\mathbf{C}} - x_{\mathbf{B}} - \ell_{\mathbf{BC}}$, and V_0 is the potential energy at the stationary point.

(a) Construct the mass-weighted Hessian matrix from $\mathbf{H} = \left[\partial^2 V / \partial \mathbf{x} \partial \mathbf{x}\right]$ and $\mathbf{M} = diag[m_A, m_B, m_C]$. Note that these are 3×3 matrices because x_A, x_B, and x_C are positions along the x-axis.

(b) Diagonalize the mass weighted Hessian. Show that one frequency is zero and that the other two frequencies can be written in the form $\omega_{\pm}^2 = \omega_0^2 \pm \sqrt{\omega_0^4 + \chi}$.

(c) For what values of ω_0^2 and χ are both of the internal mode frequencies real, e.g. like those of the CO_2 molecule? For what values of ω_0^2 and χ does one of the internal mode frequencies become imaginary, e.g. like the $H_2 + H$ transition state?

3. Derive expression $\mathbf{b}(s) = \mathbf{L}(s)^{\dagger}\mathbf{t}'(s)$ for the coupling between a reaction coordinate s and the bath modes $\mathbf{q}$ in the reaction path Hamiltonian (RPH).

4. See Chapter 17 for a discussion of BCHO models, i.e. models in which the reaction coordinate is bilinearly coupled to a harmonic oscillator bath. BCHO models and the RPH both have potential energy surfaces that resemble a harmonic valley, but they also have important differences. Does the MEP tangent vector in a BCHO model overlap with its bath modes? How do these differences influence the equations of motion from the two models.

5. Professor X builds an EVB model to describe enzyme kinetics by starting from an established force field and known structures for the reactants and products. The empirical

coupling parameter in the EVB model is tuned so that the computed reaction rates match those from experiment. What predictions of the model can be tested by comparison to experiment?

6. Create an EVB model for the left and right protonation states of malonaldehyde, cf. Bicerano et al. *J. Am. Chem. Soc.* 105, 2550-3 (1983). Use ωB97X-D/TZVP calculations to describe vibrational modes and molecular geometries. Use CCSD(T) calculations to obtain accurate energies at the minima and saddle points. Choose V_{AB} to match the barrier between the two states.
7. In principle, diabatic PESs for several different states can be incorporated into a single EVB model, but solving the secular equation would become a complicated numerical exercise. Instead, convince yourself that alternative empirical models with the form

$$V(\mathbf{x}) = -a^{-1} ln \left\{ e^{-a_A V_A(\mathbf{x})} + e^{-a_B V_B(\mathbf{x})} + e^{-a_C V_C(\mathbf{x})} + \ldots \right\}$$

can incorporate an arbitrary number of "diabatic states" while maintaining a closed form (albeit entirely heuristic) expression for the ground state PES. As an example, construct a one-dimensional model of proton transport along a water wire as shown in Figure 5.3.2. Use a series of harmonic wells with centers separated by the diameter (ca. 3Å) of a water molecule for the diabatic states. Tune the spring constants and the coupling to give a $15kJ/mol$ barrier between each well.
8. In a real protein, all residues interact with all other residues to some degree. The complete set of interactions stabilizes non-native intermediates en route from unfolded to folded protein structures. A Go-model of protein folding only includes interactions between residues which contact each other in the natively folded state. Make a qualitative sketch of the disconnectivity graph for a Go model and the corresponding model with all interactions turned on. Which model more perfectly conforms to the "folding funnel" picture of protein folding?

References

[1] P. Atkins, R. Friedman, Molecular Quantum Mechanics, 4th ed., Oxford Press, New York, 2005.
[2] F. Jensen, Introduction to Computational Chemistry, Wiley, West Sussex, 1999.
[3] T. Helgaker, P. Jorgensen, J. Olsen, Molecular Electronic-Structure Theory, Wiley, Hoboken, NJ, 2013.
[4] J.C. Tully, J. Chem. Phys. 137 (2012) 22A301.
[5] M. Head-Gordon, J.C. Tully, J. Chem. Phys. (ISSN 0021-9606) 103 (1995) 10137.
[6] M. Born, J.R. Oppenheimer, Ann. Phys. 84 (1927) 457.
[7] Z. Zhang, K. Haug, H. Metiu, J. Chem. Phys. 93 (1990) 31–34.
[8] R.G. Parr, W. Yang, Density Functional Theory of Atoms and Molecules, Oxford Press, 1989.
[9] C.J. Cramer, Essentials of Computational Chemistry: Theories and Models, Wiley, Hoboken, NJ, 2013.
[10] M. Head-Gordon, J. Phys. Chem. 100 (1996) 13213–13225.
[11] C. Moller, M.S. Plesset, Phys. Rev. 46 (1934) 618.

[12] W. Kohn, A.D. Becke, R.G. Parr, J. Phys. Chem. 100 (1996) 12974–12980.
[13] S.F. Boys, Proc. R. Soc. Lond. A 200 (1950) 542.
[14] S. Huzinaga, J. Chem. Phys. 42 (1965) 1293–1302.
[15] P.C. Hariharan, J.A. Pople, Theor. Chim. Acta (ISSN 0040-5744) 28 (1973) 213–222.
[16] D.S. Sholl, J.A. Steckel, Density Functional Theory: A Practical Introduction, Wiley, Hoboken, NJ, 2009.
[17] G. Kresse, J. Hafner, J. Phys. Condens. Matter 6 (1994) 8245.
[18] S. Miertus, E. Scrocco, J. Tomasi, Chem. Phys. 55 (1981) 117–129.
[19] V. Barone, M. Cossi, J. Phys. Chem. A 102 (1998) 1995–2001.
[20] M. Cossi, N. Rega, G. Scalmani, V. Barone, J. Comput. Chem. 24 (2003) 669–681.
[21] J. Ho, A. Klamt, M.L. Coote, J. Phys. Chem. A 114 (2010) 13442–13444.
[22] A. Klamt, Wiley Interdiscip. Rev. Comput. Mol. Sci. 1 (2011) 699–709.
[23] A. Warshel, M. Levitt, J. Mol. Biol. 103 (1976) 227–249.
[24] H. Lin, D.G. Truhlar, Theor. Chem. Acc. 117 (2006) 185–199.
[25] H. Hu, W. Yang, Annu. Rev. Phys. Chem. 59 (2008) 573–601.
[26] H.M. Senn, W. Thiel, Angew. Chem., Int. Ed. Engl. 48 (2009) 1198–1229.
[27] A. Heyden, D.G. Truhlar, J. Chem. Theory Comput. 4 (2008) 217–221.
[28] A. Leach, Molecular Modelling: Principles and Applications, Prentice Hall, Pearson Education, Englewood Cliffs, NJ, 2001.
[29] E. Wigner, H. Eyring, Sci. Mon. 44 (1937) 564–567.
[30] M. Karplus, R.N. Porter, R.D. Sharma, J. Chem. Phys. 43 (1965) 3259.
[31] F.T. Smith, J. Chem. Phys. 31 (1959) 1352.
[32] B.H. Mahan, J. Chem. Educ. 51 (1974) 308.
[33] W.H. Miller, Adv. Chem. Phys. 25 (1974) 69–177.
[34] N.E. Henriksen, F.Y. Hansen, Theories of Molecular Reaction Dynamics, Oxford Press, 2008.
[35] M. Baer, J. Chem. Phys. 62 (1975) 305–306.
[36] V.K. Babamov, R.A. Marcus, J. Chem. Phys. 74 (1981) 1790.
[37] B.C. Garrett, D.G. Truhlar, A.F. Wagner, T.H. Dunning, J. Chem. Phys. 78 (1983) 4400.
[38] D.K. Bondi, J.N.L. Connor, B.C. Garrett, D.G. Truhlar, J. Chem. Phys. 78 (1983) 5981.
[39] J. Gao, D.G. Truhlar, Annu. Rev. Phys. Chem. 53 (2002) 467–505.
[40] P. Pulay, G. Fogarasi, J. Chem. Phys. 96 (1992) 2856–2860.
[41] C. Peng, P.Y. Ayala, H.B. Schlegel, M.J. Frisch, J. Comput. Chem. 17 (1995) 49–56.
[42] J. Baker, F. Chan, J. Comput. Chem. 17 (1996) 888–904.
[43] V. Bakken, T. Helgaker, J. Chem. Phys. 117 (2002) 9160.
[44] E.B. Wilson, J.C. Decius, P.C. Cross, Molecular Vibrations: The Theory of Infrared and Raman Vibrational Spectra, Dover, New York, 1980.
[45] D.J. Wales, J. Chem. Phys. 113 (2000) 3926–3927.
[46] D.G. Truhlar, B.C. Garrett, J. Phys. Chem. A 107 (2003) 4006.
[47] K.J. Laidler, M.C. King, J. Phys. Chem. 87 (1983) 2657–2664.
[48] K.J. Mysels, J. Chem. Educ. 63 (1985) 740.
[49] R.A. Marcus, J. Chem. Phys. 45 (1966) 4493–4499.
[50] R.A. Marcus, J. Phys. Chem. 72 (1968) 891–899.
[51] K. Muller, L.D. Brown, Theor. Chim. Acta 53 (1979) 75–93.
[52] R.I. Masel, Chemical Kinetics and Catalysis, Wiley, New York, 2001.
[53] E.G. Lewars, Computational Chemistry: Introduction to the Theory and Applications of Molecular and Quantum Mechanics, 2nd ed., Springer, Dordrecht, 2011.
[54] J.K. Norskov, F. Studt, F. Abild-Pederson, T. Bligaard, Fundamental Concepts in Heterogeneous Catalysis, Wiley, Hoboken, NJ, 2014.
[55] K. Ishida, K. Morokuma, A. Komornicki, J. Chem. Phys. 66 (1977) 2153–2156.
[56] C. Gonzalez, H.B. Schlegel, J. Chem. Phys. 90 (1989) 2154–2161.

[57] K. Ruedenberg, J.Q. Sun, J. Chem. Phys. 99 (1993) 5257–5268.
[58] H.P. Hratchian, H.B. Schlegel, J. Chem. Theory Comput. 1 (2005) 61–69.
[59] J.M. Bofill, J. Comput. Chem. 15 (1994) 1–11.
[60] M. Page, J.W. McIver, J. Chem. Phys. 88 (1987) 922–935.
[61] W.H. Miller, N.C. Handy, J.E. Adams, J. Chem. Phys. 72 (1980) 99–112.
[62] R.A. Marcus, J. Chem. Phys. 49 (1968) 2610–2616.
[63] B. Peters, A.T. Bell, A. Chakraborty, J. Chem. Phys. 121 (2004) 4453–4460.
[64] H. Hu, M.N. Kobrak, C. Xu, S. Hammes-Schiffer, J. Phys. Chem. A 104 (2000) 8058–8066.
[65] A. Warshel, R.M. Weiss, J. Am. Chem. Soc. 102 (1980) 6218–6226.
[66] S.C.L. Kamerlin, A. Warshel, Wiley Interdiscip. Rev. Comput. Mol. Sci. 1 (2011) 30–45.
[67] A. Warshel, Computer Modeling of Chemical Reactions in Enzymes and Solutions, Wiley, New York, 1991.
[68] Y.-T. Chang, W.H. Miller, J. Phys. Chem. 7 (1990) 5884–5888.
[69] J. Aqvist, A. Warshel, Chem. Rev. 93 (1993) 2523–2544.
[70] J. Aqvist, T. Hansson, J. Phys. Chem. 100 (1996) 9512–9521.
[71] P.K. Agarwal, S.R. Billeter, S. Hammes-Schiffer, J. Phys. Chem. B 106 (2002) 3283–3293.
[72] J.B. Watney, P.K. Agarwal, S. Hammes-Schiffer, J. Am. Chem. Soc. 125 (2003) 3745–3750.
[73] U. Schmitt, G.A. Voth, J. Phys. Chem. B 201 (1998) 5547–5551.
[74] Y. Kim, J.C. Corchado, J. Villa, J. Xing, D.G. Truhlar, J. Chem. Phys. 112 (2000) 2718–2735.
[75] R. Czerminski, R. Elber, J. Chem. Phys. 92 (1990) 5580–5601.
[76] J.P.K. Doye, M.A. Miller, D.J. Wales, J. Phys. Chem. 110 (1999) 6896–6906.
[77] D. Wales, J.P.K. Doye, M.A. Miller, P.K. Mortenson, T.R. Walsh, Adv. Chem. Phys. 115 (2000) 1–112.
[78] S.V. Krivov, M. Karplus, Proc. Natl. Acad. Sci. USA 101 (2004) 14766–14770.
[79] D.J. Wales, T.V. Bogdan, J. Phys. Chem. B 110 (2006) 20765–20776.
[80] L. Munro, D. Wales, Phys. Rev. B 59 (1999) 3969–3980.
[81] D.J. Wales, J.P.K. Doye, J. Phys. Chem. A 101 (1997) 5111–5116.
[82] J.M. Carr, S.A. Trygubenko, D.J. Wales, J. Chem. Phys. 122 (2005) 234903.
[83] F. Calvo, T.V. Bogdan, V.K. de Souza, D.J. Wales, J. Chem. Phys. 127 (2007) 044508.
[84] V.K. de Souza, D. Wales, J. Chem. Phys. 129 (2008) 164507.
[85] D.J. Wales, Mol. Phys. 100 (2002) 3285.
[86] D.J. Wales, Mol. Phys. 102 (2004) 891–908.

CHAPTER 8

Saddles on the energy landscape

Thus the whole reaction rate depends only on the nature of the landscape in the immediate neighborhood of the pass, the nature of the valley floor and the temperature. It has practically nothing to do, however, with the intervening country.

Wigner and Eyring, Scientific Monthly (1937)

Many of the rate theories that will be discussed in later chapters require information about the potential energy surface (PES) near a saddle point. Finding saddle points on an *ab initio* PES can be extremely difficult. Many saddle point search algorithms have been developed, each with its own advantages and limitations. Some early methods maneuvered ridge-straddling pairs of configurations down the ridge and toward each other as the pair converged to the saddle point [1–3]. Another group of methods found minimum energy configurations on a series of planes between the reactant and product minima [4,5]. Some of these early methods became obsolete because other methods proved more efficient [6], and others were just never incorporated into commercial software packages.

The methods that are widely used today can be grouped into a few main categories. First, chemical intuition can often provide a reasonable guess for the geometry of the transition state. Second, when the reaction coordinate can be anticipated, coordinate driving methods can push the system over the saddle region by using a series of constrained minimizations. Third, Nudged Elastic Band and similar methods can interpolate and optimize the positions of each atom along a path that connects the reactant and product states. Fourth, there are methods that systematically navigate upwards to the headwaters of a valley in search of a saddle point. Finally, the emerging reduced energy landscape algorithms have enormous potential to bring new capabilities.

8.1 Newton-Raphson

The Newton-Raphson algorithm [7] is primarily used to refine approximate saddle points. It is based on a local quadratic approximation (LQA) to the PES at a current position $\mathbf{x}$,

$$V(\mathbf{x} + \Delta\mathbf{x}) \approx V(\mathbf{x}) + \Delta\mathbf{x}^{\dagger}\,\mathbf{g}|_{\mathbf{x}} + \frac{1}{2}\Delta\mathbf{x}^{\dagger}\,\mathbf{H}|_{\mathbf{x}}\Delta\mathbf{x}$$

where

$$\mathbf{g}|_\mathbf{x} = \left.\frac{\partial V}{\partial \mathbf{x}}\right|_\mathbf{x} \quad \textit{and} \quad \mathbf{H}|_\mathbf{x} = \left.\frac{\partial^2 V}{\partial \mathbf{x} \partial \mathbf{x}}\right|_\mathbf{x}.$$

The Newton-Raphson algorithm chooses the step $\Delta\mathbf{x}$ to make the gradient at $\mathbf{x}+\Delta\mathbf{x}$ zero. According to the LQA, the gradient at $\mathbf{x}+\Delta\mathbf{x}$ is

$$\mathbf{g}|_{\mathbf{x}+\Delta\mathbf{x}} = \mathbf{g}|_\mathbf{x} + \mathbf{H}|_\mathbf{x}\,\Delta\mathbf{x}. \tag{8.1.1}$$

Let $\mathbf{U}$ be the unitary matrix of column eigenvectors that diagonalizes $\mathbf{H}$ and let $\mathbf{\Lambda}$ be the diagonal matrix of eigenvalues

$$\mathbf{U}^\dagger\,\mathbf{H}|_\mathbf{x}\mathbf{U} = \mathbf{\Lambda}$$

Also define the gradient and displacement in the eigenvector basis,

$$\boldsymbol{\gamma} = \mathbf{U}^\dagger\,\mathbf{g}|_\mathbf{x} \quad \text{and} \quad \boldsymbol{\xi} = \mathbf{U}^\dagger\,\Delta\mathbf{x},$$

respectively. Multiply both sides of $\mathbf{g}|_{\mathbf{x}+\Delta\mathbf{x}} = \mathbf{0}$ by $\mathbf{U}^\dagger$ to obtain the decoupled equations

$$\boldsymbol{\gamma} + \mathbf{\Lambda}\boldsymbol{\xi} = \mathbf{0}$$

These would have solution $\boldsymbol{\xi} = -\mathbf{\Lambda}^{-1}\boldsymbol{\gamma}$, except that both γ_i and λ_i are zero for rotational and translational modes. While the matrix $\mathbf{\Lambda}$ cannot actually be inverted, a generalized inverse [8] of $\mathbf{\Lambda}$ can be constructed by inverting only the nonzero eigenvalues of $\mathbf{\Lambda}$. This results in the step

$$\xi_i = -\gamma_i/\lambda_i \qquad \textit{for} \quad \lambda_i \neq 0$$

and

$$\xi_i = 0 \qquad \textit{for} \quad \lambda_i = 0$$

Finally, the Newton-Raphson step is

$$\Delta\mathbf{x} = \mathbf{U}\,\boldsymbol{\xi} \tag{8.1.2}$$

If the starting point $\mathbf{x}$ is near a minimum, then all (non-zero) eigenvalues λ_k in $\mathbf{\Lambda}$ will be positive and the step in direction $\Delta\mathbf{x}$ will be downhill in energy towards a minimum. For a saddle point, $\mathbf{\Lambda}$ should have only one negative eigenvalue (the unstable mode) and the rest of the (non-zero) eigenvalues should be positive. The step $\Delta\mathbf{x}$ will then point uphill from $\mathbf{x}$ along the unstable mode and downhill along all of the other modes. After updating the molecular geometry to $\mathbf{x}+\Delta\mathbf{x}$, the procedure is repeated starting with calculation of the gradient and Hessian at the new configuration.

Instead of using generalized inverses, difficulties with rotational and translational degrees of freedom can also be avoided by using internal coordinates. Redundant internal coordinates complicate the optimization algorithms but they have significant efficiency advantages for most problems [8–10]. The remainder of this chapter makes no further distinction between inverses, generalized inverses, and coordinate systems. Also note that mass-weighted coordinates are not used in this chapter because they are not necessary for finding saddle points.

There are several issues to address during the optimization of transition state geometries. For example, the magnitude of the Newton-Raphson step may exceed the trust radius for the quadratic Taylor expansion. In these cases, the step should be rescaled to remain within a reasonable trust radius [11] around $\mathbf{x}$. The need to take small steps and to compute the Hessian at each step makes Newton-Raphson optimizations costly. Additionally, the starting configuration for Newton-Raphson optimization must be sufficiently close to the saddle point to ensure that the Hessian has just one negative eigenvalue [9,12], i.e. one imaginary frequency corresponding to the reaction coordinate direction. Because of this restriction, Newton-Raphson is only used to refine approximate transition state structures which have been obtained via other methods.

Algorithm: Newton-Raphson

The starting point $\mathbf{x}$ should be an approximate transition state structure having one imaginary frequency. Set the trust radius R_{trust}, the stepsize convergence criterion R_{conv}, and the gradient tolerance g_{conv}.

1. Compute the gradient $\mathbf{g}$ and Hessian $\mathbf{H}$.
2. Diagonalize the Hessian: $\mathbf{U}^{\dagger}\mathbf{H}\mathbf{U} = \mathbf{\Lambda}$.
3. Convert the gradient to the eigenvector basis $\gamma = \mathbf{U}^{\dagger}\mathbf{g}$.
4. Compute $\Delta\mathbf{x} = -\mathbf{U}\mathbf{\Lambda}^{-1}\gamma$ from a generalized inverse.
5. If $||\Delta\mathbf{x}|| > R_{trust}$, rescale step: $\Delta\mathbf{x} \leftarrow R_{trust}\Delta\mathbf{x}/||\Delta\mathbf{x}||$.
6. Update geometry: $\mathbf{x} \leftarrow \mathbf{x} + \Delta\mathbf{x}$.
7. If $||\Delta\mathbf{x}|| < R_{conv}$ and $||\mathbf{g}|| < g_{conv}$, then stop. Else return to step (1) and repeat.

For large systems it is difficult to guess a configuration with just one unstable mode for refinement by the Newton-Raphson algorithm. The guessed configuration often possesses several unstable modes, and sometimes none of the imaginary modes resemble motion along the reaction coordinate of interest. Additionally, many of the saddle points on a complicated PES are decoys that correspond to chemically uninteresting isomerizations, e.g. 120° rotations of $-CH_3$ groups. Furthermore, Newton-Raphson is not guaranteed to find a saddle point from all geometries having exactly one imaginary frequency. Figure 8.1.1 shows the regions of the Muller-Brown PES that have one imaginary frequency and the much smaller region of the

Muller-Brown PES for which the Newton-Raphson algorithm converges to the high energy saddle point. The zone of convergence in the right panel of Figure 8.1.1 is for Newton-Raphson with an infinite trust radius. The zone of convergence to the saddle increases for a small trust radius but it does not fully cover the gray region at left.

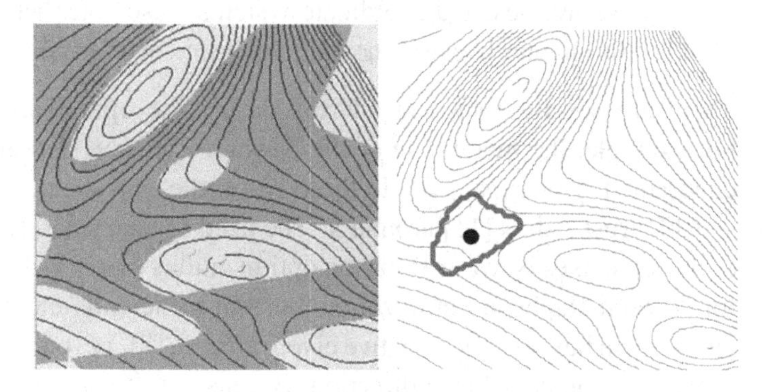

Figure 8.1.1: (Left) regions with a single imaginary frequency (dark) and regions with zero or two imaginary frequencies (light) on the Muller-Brown PES. (Right) the small zone around a saddle where Newton-Raphson iterations converge to the saddle. [From B. Peters, PhD thesis (2004).]

Although Newton-Raphson requires an accurate guess, several alternative algorithms can find saddle points even from initial guesses with no imaginary frequencies or with many imaginary frequencies. The table below lists some important saddle point search algorithms along with their key advantages and disadvantages.

8.2 Cerjan-Miller algorithm

Poppinger [20] developed an early method to find saddles by maximizing the energy along the reaction coordinate while minimizing the energy along other degrees of freedom. Poppinger's suggestion inspired the Cerjan-Miller (CM) algorithm [12] and many others [9,13,14,21,22]. The key innovation of Cerjan and Miller was a strategy that can "discover" saddles without being guided by chemical intuition or *a priori* chosen reaction coordinates. The Cerjan-Miller

Algorithm	Features
Newton-Raphson	Requires Hessian; quadratic convergence to saddle within small zone around saddle point; initial guess must have a single unstable mode;
Cerjan-Miller	Local quadratic approximation (LQA) to PES with one shift parameter λ for all modes; can initiate search from a stable basin [12];
Partitioned-RFO	Rational function approximation to local PES; secant updates for the Hessian matrix [9]; separate shift parameters for bath and reaction coordinate modes; quadratic convergence to saddle point; eigenvector-following version tracks the reaction coordinate mode between steps [13,14];
Reduced-PES	Reduces energy landscape to key internal coordinates to find saddle points that break and make selected bonds; secant updates; mitigates rediscovery [15,16];
Dimer	Eliminates need for Hessian; rotates and maneuvers a dimer on the PES [17–19];

algorithm seeks saddle points by climbing upward on the PES along the vibrational mode with the smallest frequency. While the algorithm is often successful, there is no guarantee that directions corresponding to small frequencies will lead to chemically interesting saddle points. The heuristic justification is that a steepest descent path from any point on the PES quickly optimizes the high frequency degrees of freedom and then follows a long winding valley downward to a minimum. The Cerjan-Miller algorithm (and many others) attempt to reverse the steepest descent path by climbing upward along long valleys that correspond to low frequency modes [13–15,22,23].

A Cerjan-Miller search trajectory begins near the minimum and zig-zags up the PES in the direction corresponding to the smallest frequency as illustrated in Figure 8.2.1. Let the sequence of configurations generated by the CM algorithm be numbered from the starting point $\mathbf{x}^{(0)}$, until the step $\mathbf{x}^{(n)}$ which first crosses the inflection surface, i.e. the edge of the region around the minimum where one eigenvalue of the second derivative matrix becomes negative. The configuration $\mathbf{x}^{(n)}$ is then used as a starting geometry for a Newton-Raphson search. The crossover from Cerjan-Miller to Newton Raphson steps is also illustrated in Figure 8.2.1.

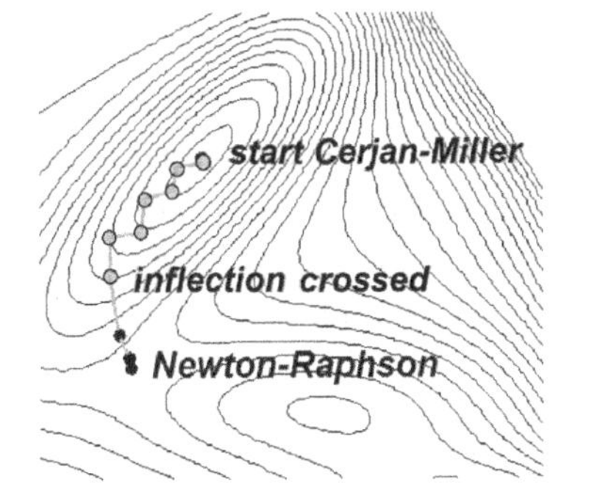

Figure 8.2.1: Cerjan-Miller transition state search on the Muller-Brown PES.

Each Cerjan-Miller step depends on the local gradient and second derivative matrix. If $\mathbf{x}^{(k)}$ is the kth point in the search trajectory, then $\mathbf{x}^{(k+1)} = \mathbf{x}^{(k)} + \Delta\mathbf{x}$ is a stationary point on the local quadratic PES subject to a constraint of constant stepsize Δ from $\mathbf{x}^{(k)}$. The Lagrangian used to find the step $\Delta\mathbf{x}$ is

$$L(\Delta\mathbf{x}, \lambda) = \Delta\mathbf{x}^{\dagger}\mathbf{g} + \frac{1}{2}\Delta\mathbf{x}^{\dagger}\mathbf{H}\Delta\mathbf{x} - \frac{\lambda}{2}(\|\Delta\mathbf{x}\|^{2} - \Delta^{2}) \quad (8.2.1)$$

The step can be expressed in terms of the Lagrange multiplier λ, the matrix of (column) eigenvectors $\mathbf{U}$, the eigenvalues of the Hessian matrix $\mathbf{\Lambda} = diag(\lambda_1, \cdots, \lambda_m)$, and the gradient in the eigenvector basis $\boldsymbol{\gamma} = \mathbf{U}^{\dagger}\mathbf{g}$. The step and quadratic approximation to the change in energy for this step are

$$\Delta\mathbf{x}(\lambda) = \mathbf{U}(\lambda\mathbf{1} - \mathbf{\Lambda})^{-1}\boldsymbol{\gamma} \quad (8.2.2)$$

and

$$\Delta V(\lambda) = \boldsymbol{\gamma}^{\dagger} \frac{(\lambda \mathbf{1} - \boldsymbol{\Lambda}/2)}{(\lambda \mathbf{1} - \boldsymbol{\Lambda})^2} \boldsymbol{\gamma} \tag{8.2.3}$$

A system with m internal degrees of freedom has m stationary points, each corresponding to different values of λ. The stationary point corresponding to $\lambda = 0$ gives the Newton-Raphson step which leads back to the minimum as long as all frequencies remain real. The Cerjan-Miller algorithm instead selects the shift parameter λ to minimize ΔV between the smallest and second smallest frequencies. Figure 8.2.2 shows the function $\Delta V(\lambda)$ and the value λ_0 that is selected by the Cerjan-Miller algorithm.

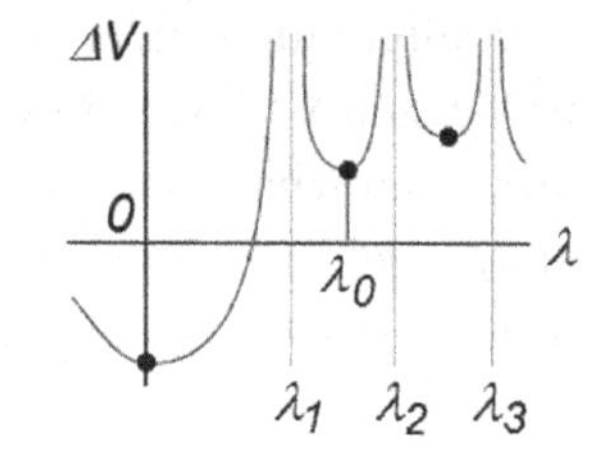

Figure 8.2.2: The search direction in the Cerjan-Miller algorithm is determined by minimizing ΔV in equation (8.2.3) between the smallest and second smallest eigenvalue in Λ. The resulting step is always uphill and directed primarily along the two eigenmodes with the smallest frequencies.

The Cerjan-Miller chooses λ_0 to ensure a step that is uphill along the eigenvector $\mathbf{u}_1$ corresponding to the eigenmode with the smallest positive eigenvalue, and downhill along all other eigenmodes for a net uphill step. The decreasing eigenvalue λ_1 eventually changes from positive to negative as the inflection surface is crossed in the uphill direction. From this point, Newton-Raphson can be used to find the exact saddle point.

8.3 Partitioned-Rational Function Optimization

The Partitioned-Rational Function Optimization (P-RFO) algorithm [13] and the related Eigenvector Following algorithm are among the most efficient and reliable methods for finding transition states [14,19,24]. This section describes the P-RFO algorithm with a list of its advantages over the Cerjan-Miller algorithm. First, P-RFO is based on a rational function approximation to the PES instead of a local quadratic approximation (LQA) [13]. The rational function approximation has the form

$$V(\mathbf{x} + \Delta\mathbf{x}) \approx V(\mathbf{x}) + \frac{\Delta\mathbf{x}^{\dagger}\mathbf{g} + \frac{1}{2}\Delta\mathbf{x}^{\dagger}\mathbf{H}\Delta\mathbf{x}}{1 + \Delta\mathbf{x}^{\dagger}\Delta\mathbf{x}}$$

The rational function approximation matches the LQA near $\Delta\mathbf{x} = \mathbf{0}$, but in contrast to an LQA the rational function remains bounded for large displacements. Hence the rational function approximation to $V(\mathbf{x} + \Delta\mathbf{x})$ is a better model for the energy of bond-dissociation at large bond extensions.

A second advantage of P-RFO is its ability to find the neighborhood of a transition state and also converge upon the precise transition state structure. In contrast, recall that the Cerjan-Miller algorithm required a switch from Cerjan-Miller to Newton-Raphson in the final stages of optimization. Moreover, P-RFO has quadratic convergence properties like the Newton-Raphson algorithm. Thus P-RFO can efficiently and seamlessly complete the entire saddle point search and optimization procedure.

A third advantage of P-RFO is that it uses two separate shift parameters, λ_{RC} and λ_{BATH}. The algorithm determines λ_{RC} such that each step attempts to maximize the energy along the designated reaction coordinate mode. The other shift parameter, λ_{BATH}, is determined so that each step attempts to minimize the energy along the set of bath modes. By using two shift parameters, P-RFO can find transition states even when initiated at configurations with multiple imaginary frequencies, and even when none of the initially unstable modes correspond to the reaction coordinate mode.

To identify the two shift parameters, a particular eigenvector corresponding to the reaction coordinate must be chosen from the set of vibrational modes. The special mode can be identified based on

1. visual inspection of animated vibrational modes,
2. overlap with the path tangent from a partially optimized Nudged Elastic Band,
3. overlap with the gradient of an *a priori* specified reaction coordinate, or
4. overlap with the mode that was followed in the previous iteration (see notes below on Eigenvector Following).

Let the selected reaction coordinate mode be the kth eigenvector of the Hessian $\mathbf{u}_k$ with eigenvalue λ_k. The two shift parameters are defined in a manner similar to the shift parameter in the Cerjan-Miller algorithm. Using the notation from section 8.1 for the gradient in the eigenvector basis and for the Hessian eigenvalues, the P-RFO shift parameters are roots of the equations [13]

$$\lambda_{BATH} = \sum_{j \neq k} \frac{\gamma_j^2}{\lambda_{BATH} - \lambda_j} \tag{8.3.1}$$

and

$$\lambda_{RC} = \frac{\gamma_k^2}{\lambda_{RC} - \lambda_k} \tag{8.3.2}$$

The P-RFO algorithm chooses λ_{BATH} as the smallest root of equation (8.3.1) and λ_{RC} as the largest root of equation (8.3.2). The step is then constructed in the eigenvector basis as

$$\xi_j = \begin{cases} -\gamma_j/(\lambda_j - \lambda_{BATH}) & j \neq k \\ -\gamma_k/(\lambda_k - \lambda_{RC}) & j = k \end{cases} \tag{8.3.3}$$

In the original basis, the step is $\Delta \mathbf{x} = \mathbf{U}\xi$. By construction, the P-RFO step is downhill along all modes except for the kth mode which was identified as the reaction coordinate direction. Figure 8.3.1 shows graphically how the two shift parameters are identified.

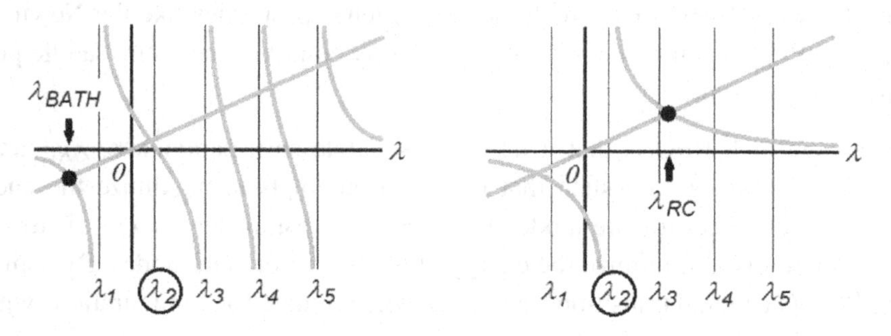

Figure 8.3.1: The P-RFO shift parameters from equations (8.3.1) and (8.3.2) in the case where the second mode is the reaction coordinate direction. The P-RFO algorithm uses one frequency shift parameter to direct the step upwards along the reaction coordinate (RC) and a second frequency shift parameter to direct the steps downward along all other bath modes.

Schlegel [9] began using secant updates to avoid recomputing the Hessian matrix at each step during a saddle search. The secant update innovation has been adopted by many later saddle search algorithms including P-RFO. The central idea of a secant update is that the difference in gradients between two points on the PES provides a directional second derivative along the vector between the two points. Secant updates use the directional second derivative information to revise the approximate Hessian after each step. The secant equation requires that [11]

$$\Delta \mathbf{g} = \mathbf{H}^{(n)} \Delta \mathbf{x} \tag{8.3.4}$$

where $\Delta \mathbf{g} = \mathbf{g}^{(n)} - \mathbf{g}^{(n-1)}$ and $\Delta \mathbf{x} = \mathbf{x}^{(n)} - \mathbf{x}^{(n-1)}$, as shown in Figure 8.3.2. The superscripts (n) and $(n-1)$ indicate gradients and positions at the new and previous locations, respectively. To satisfy the secant equation, the Hessian at the old point is modified to obtain a Hessian at the new point:

$$\mathbf{H}^{(\mathrm{n})} = \mathbf{H}^{(\mathrm{n}-1)} + \Delta \mathbf{H} \tag{8.3.5}$$

Combining equations (8.3.4) and (8.3.5) gives an underdetermined equation for the Hessian update $\Delta \mathbf{H}$ [11].

$$\Delta \mathbf{H} \Delta \mathbf{x} = \Delta \mathbf{g} - \mathbf{H}^{(\mathrm{n}-1)} \Delta \mathbf{x} \tag{8.3.6}$$

Because equation (8.3.6) is underdetermined, the Hessian update $\Delta \mathbf{H}$ can be constructed in many ways. The BFGS update preserves positive definite Hessian structure, a property that makes BFGS updates the favorite for minimization [25]. Choosing an appropriate update for optimizing transition states is more difficult [26–29]. Most early transition state algorithms used the symmetric rank-one (Murtagh-Sargeant) updates [25] or Powell updates [30]. Recent works [31] suggest that the Bofill update [28] and the exponential update (also developed by Bofill) [29] are the most effective update formulas for saddle search algorithms.

Figure 8.3.2: Illustrating how gradients at previous $(n-1)$ and new (n) points on the PES provide information about the directional second derivative along the line between the points. This information can be used to correct the Hessian matrix between steps without a complete re-evaluation.

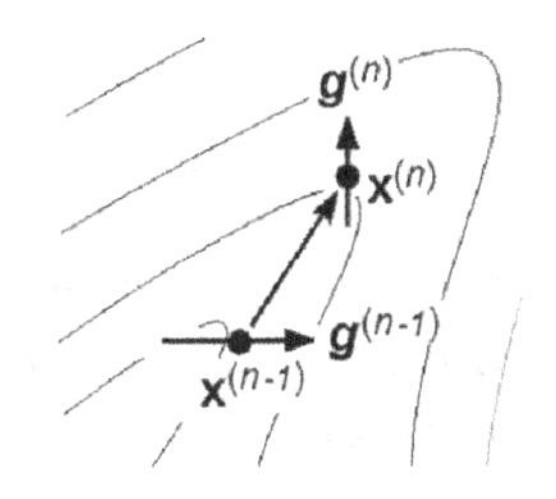

Algorithm: P-RFO

Start from point $\mathbf{x}$ with initial Hessian $\mathbf{H}$ and gradient $\mathbf{g}$. Set the trust radius R_{trust}, the force tolerance g_{conv}, and the stepsize convergence criterion R_{conv}.

1. Diagonalize the Hessian: $\mathbf{U}^{\dagger}\mathbf{H}\mathbf{U} = \mathbf{\Lambda}$.
2. Identify the eigenmode to be followed, i.e. that corresponding to the reaction coordinate.
3. Convert the gradient to the eigenvector basis $\gamma = \mathbf{U}^{\dagger}\mathbf{g}$.
4. Compute λ_{BATH} and λ_{RC} from equations (8.3.1) and (8.3.2).
5. Compute the step $\Delta\mathbf{x} = \mathbf{U}\xi$ with ξ from equation (8.3.3).
6. If $||\Delta\mathbf{x}|| > R_{trust}$, rescale step: $\Delta\mathbf{x} \leftarrow R_{trust}\Delta\mathbf{x}/||\Delta\mathbf{x}||$.
7. Update geometry: $\mathbf{x} \leftarrow \mathbf{x} + \Delta\mathbf{x}$.
8. Compute the new gradient $\mathbf{g}^{(n)}$ and $\mathbf{\Delta g} = \mathbf{g}^{(n)} - \mathbf{g}$.
9. Compute Hessian update $\Delta\mathbf{H}$ from $\Delta\mathbf{g}$, $\Delta\mathbf{x}$, and $\mathbf{H}$ using Bofill formula.
10. Reset the Hessian: $\mathbf{H} \leftarrow \mathbf{H} + \Delta\mathbf{H}$. Reset the gradient: $\mathbf{g} \leftarrow \mathbf{g}^{(n)}$.
11. If $||\Delta\mathbf{x}|| < R_{conv}$ and $||\mathbf{g}|| < g_{conv}$ then stop. Else return to step (1) and repeat.

In the P-RFO algorithm, the mode corresponding to the reaction coordinate must be identified at each step. The re-identification is necessary because the mode of interest can shift to a new position in the spectrum as its eigenvalue changes. Eigenvector Following [30] extensions of the P-RFO algorithm "track" the reaction coordinate mode from one step to the next. After each step, the Hessian is updated and re-diagonalized. Then its eigenmodes are checked for overlap with the mode that was previously being followed. The new reaction coordinate mode is the eigenmode having the largest overlap (positive or negative) with the mode that was followed on the previous step.

Inaccurate Hessian updates can scramble the updated eigenmodes causing Eigenvector Following to lose track of the mode being followed. To mitigate this problem, Peters et al. [32] followed the mode having maximum overlap with an average mode constructed from a history of several recently followed modes.[1] Alternatively, in step (2) of the P-RFO algorithm, one can automatically select a mode having maximal overlap with the gradient of some internal reaction coordinate, e.g. the bond length being broken minus the bond length being formed. Finally, note that many P-RFO or Eigenvector Following searches can be launched from one configuration by choosing different modes or different internal coordinates to follow. By comparison, Cerjan-Miller (and also Dimer searches – see below) require different starting locations for different search outcomes.

8.4 The dimer method

Like the Cerjan-Miller, P-RFO, and Eigenvector Following methods, the Dimer method can find transition states without *a priori* reference to specific products or reactants. As the name suggests, the Dimer method maneuvers a pair of configurations on the energy landscape to locate saddle points. The Dimer method exploits a variational property of quadratic forms to avoid actually computing the Hessian. The Hessian **H** is symmetric, and therefore the quadratic form $\mathbf{u}^\dagger\mathbf{H}\mathbf{u}$ for any unit vector **u** satisfies the inequality

$$\lambda_{min} \leq \mathbf{u}^\dagger\mathbf{H}\mathbf{u} \leq \lambda_{max} \tag{8.4.1}$$

where λ_{min} and λ_{max} are the smallest and largest eigenvalues of **H**. Voter [33] noted that equation (8.4.1) could help identify the unstable mode even when **H** is unavailable. $\mathbf{u}^\dagger\mathbf{H}\mathbf{u}$ attains the value λ_{min} of the smallest eigenvalue only when **u** is the softest eigenmode of the Hessian **H**. To identify the softest eigenmode without knowing **H**, the dimer method optimizes the orientation of a dimer, i.e. a pair of points displaced from each other by a small fixed distance. For configurations $\mathbf{x}_1$ and $\mathbf{x}_2$ subject to $||\mathbf{x}_1 - \mathbf{x}_2|| = \Delta$, the dimer energy can be approximated from a quadratic Taylor expansion of the PES around the dimer midpoint $\mathbf{x_{mid}} = (\mathbf{x}_1 + \mathbf{x}_2)/2$.

$$\begin{aligned} V_{dimer}(\mathbf{x}_1, \mathbf{x}_2) \approx\ & 2V(\mathbf{x_{mid}}) + (\mathbf{x}_2 - \mathbf{x_{mid}})^\dagger\mathbf{g} + (\mathbf{x}_1 - \mathbf{x_{mid}})^\dagger\mathbf{g} \\ & + \tfrac{1}{2}(\mathbf{x}_2 - \mathbf{x_{mid}})^\dagger\mathbf{H}(\mathbf{x}_2 - \mathbf{x_{mid}}) + \tfrac{1}{2}(\mathbf{x}_1 - \mathbf{x_{mid}})^\dagger\mathbf{H}(\mathbf{x}_1 - \mathbf{x_{mid}}) \end{aligned} \tag{8.4.2}$$

The two gradient terms exactly cancel because $(\mathbf{x}_2 - \mathbf{x_{mid}}) = -(\mathbf{x}_1 - \mathbf{x_{mid}})$. We can write the displacements $(\mathbf{x}_2 - \mathbf{x_{mid}})$ and $(\mathbf{x}_1 - \mathbf{x_{mid}})$ in terms of the unit vector $\mathbf{u} = (\mathbf{x}_2 - \mathbf{x}_1)/||\mathbf{x}_2 - \mathbf{x}_1||$ for the dimer orientation and the dimer separation Δ.

$$\mathbf{x}_2 - \mathbf{x_{mid}} = \mathbf{u}\Delta/2 \quad and \quad \mathbf{x}_1 - \mathbf{x_{mid}} = -\mathbf{u}\Delta/2$$

[1] When adding a new eigenmode to the list of recently followed modes, care must be taken to reverse the sign as needed so that the eigenmodes all point in approximately the same direction. Otherwise, eigenvectors representing the same mode, but with opposite directions can nullify each other in the average.

With this substitution it also becomes clear that two terms involving the Hessian are actually identical to each other. The dimer energy simplifies to

$$V_{dimer}(\mathbf{x_1}, \mathbf{x_2}) \approx 2V(\mathbf{x_{mid}}) + \frac{\Delta^2}{4}\mathbf{u}^\dagger \mathbf{H}\mathbf{u} \tag{8.4.3}$$

Thus, up to second order in Δ, the dimer energy depends only on the dimer orientation $\mathbf{u}$ with the lowest dimer energy corresponding to a dimer oriented along the most unstable mode of $\mathbf{H}$. This clever idea forms the basis of some important transition state search algorithms that do not require the Hessian [17–19,34].

The original Dimer methods [17,18] optimized the dimer orientation numerically and iteratively. The optimized dimer orientation was then used to modify the force on the dimer midpoint into a force pointing uphill along the least stable mode and downhill along all other modes. Specifically, the forces are reversed along the dimer orientation vector by the operation:

$$\mathbf{f}' = (\mathbf{1} - \mathbf{2u} \circ \mathbf{u}^\dagger)\mathbf{f} \tag{8.4.4}$$

Figure 8.4.1 shows how the orientation and force determine the search direction in the Dimer algorithm. The modified force then guides a conjugate gradient algorithm [11] to the saddle point. Dimer searches can be initiated from any point on the PES.

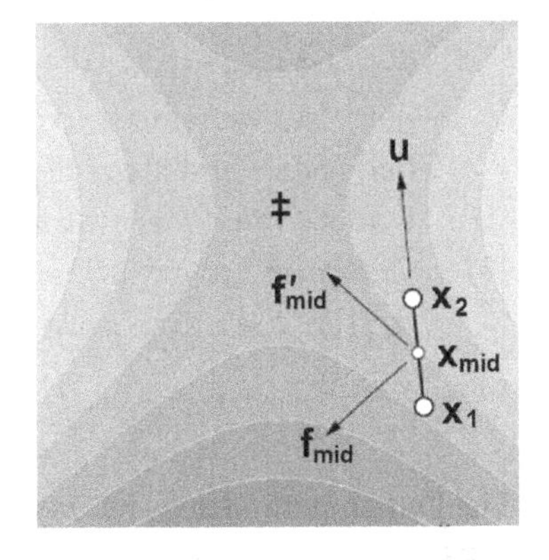

Figure 8.4.1: Minimizing the energy of the dimer aligns its orientation with the direction of the least stable eigenmode. The component of the force parallel to the dimer is then reversed to give the search direction. The Dimer method has the important advantage that no Hessians are required.

Heyden et al. [19] showed how an approximately optimal dimer orientation can be identified with just three gradients per iteration. The Heyden reorientation procedure is not exact because each orientational adjustment is restricted to the space spanned by the initial dimer orientation and the instantaneous torque. However, the restricted optimizations are adequate because, given sufficiently small steps on the PES, changes in the optimal dimer orientation from one step to the next are small. The Dimer method is especially useful in plane wave (periodic) DFT studies where Hessian matrix evaluations are prohibitively expensive.

Algorithm: Dimer method (with rotations of Heyden et al.)

This version of the Dimer method uses conjugate gradient steps guided by the forces with the components along the optimized dimer orientation reversed. Each step begins with raw unmodified forces, $\mathbf{f} = -\partial V/\partial \mathbf{x}$. For notational convenience, the algorithm below stores the dimer configuration as its midpoint, its orientation, and its distance: $(\mathbf{x_{mid}}, \mathbf{u}, \Delta)$. Configurations $\mathbf{x_2}$ and $\mathbf{x_1}$ are at locations $\mathbf{x_{mid}} + \mathbf{u}\Delta/2$ and $\mathbf{x_{mid}} - \mathbf{u}\Delta/2$, respectively.

1. Set dimer distance Δ, trust radius R_{trust}, stepsize convergence tolerance R_{conv}, and force convergence tolerance $||\mathbf{f}||_{conv}$. Select a starting dimer configuration. Compute the force on the dimer midpoint: $\mathbf{f_{mid}} = \mathbf{f}(\mathbf{x_{mid}})$.
2. Compute the force $\mathbf{f_2} = \mathbf{f}(\mathbf{x_2})$. Estimate $\mathbf{f_1} = 2\mathbf{f_{mid}} - \mathbf{f_2}$.
3. Compute the direction of the rotational force on the dimer (net force without components along the dimer orientation): $\mathbf{v} = \mathbf{w}/||\mathbf{w}||$ with $\mathbf{w} = (\mathbf{f_2} - \mathbf{f_1}) - \mathbf{u}^T(\mathbf{f_2} - \mathbf{f_1})$.
4. Rotate in $span(\mathbf{u}, \mathbf{v})$: $\mathbf{x_4} = \mathbf{x_{mid}} + \Delta \cdot \{\mathbf{u}\,cos(\pi/4) + \mathbf{v}\,sin(\pi/4)\}$.
5. Compute $\mathbf{f_4} = \mathbf{f}(\mathbf{x_4})$ and estimate $\mathbf{f_3} = 2\mathbf{f_{mid}} - \mathbf{f_4}$.
6. Define the rotational forces $\tau(\pi/4) = \Delta^{-1}(\mathbf{f_4}\text{-}\mathbf{f_3}) \cdot \mathbf{v}$ and $\tau(0) = \Delta^{-1}(\mathbf{f_2}\text{-}\mathbf{f_1}) \cdot \mathbf{v}$.
7. Compute the optimal rotation angle: $\theta_{opt} = \frac{1}{2}tan^{-1}(-\tau(0)/\tau(\pi/4))$.
8. Reset $\mathbf{u} \leftarrow \mathbf{u}\,cos(\theta_{opt}) + \mathbf{v}\,sin(\theta_{opt})$.
9. Compute $\mathbf{f}'_{\mathbf{mid}} = (1 - 2\mathbf{u} \circ \mathbf{u}^T)\mathbf{f_{mid}}$. Take conjugate gradient step $\Delta\mathbf{x_{mid}}$ based on the previous steps and modified forces. If $||\Delta\mathbf{x_{mid}}|| > R_{trust}$, rescale step: $\Delta\mathbf{x_{mid}} \leftarrow R_{trust}\Delta\mathbf{x_{mid}}/||\Delta\mathbf{x_{mid}}||$. The conjugate gradient step calculation is not elaborated here, but remember to add the $\mathbf{f}'_{\mathbf{mid}}$ and the final step $\Delta\mathbf{x_{mid}}$ to list of previous modified forces and steps after each iteration.
10. Update the dimer position: $\mathbf{x_{mid}} \leftarrow \mathbf{x_{mid}} + \Delta\mathbf{x_{mid}}$.
11. Compute $\mathbf{f_{mid}} = \mathbf{f}(\mathbf{x_{mid}})$. If $||\Delta\mathbf{x_{mid}}|| > R_{conv}$ and $||\mathbf{f_{mid}}|| < ||\mathbf{f}||_{conv}$, stop. Else return to step 2 with the present dimer $(\mathbf{x_{mid}}, \mathbf{u}, \Delta)$.

The Dimer method does have two disadvantages. First, in contrast to Eigenvector Following and P-RFO methods, the Dimer method can only pursue the mode which is locally most unstable – a direction that may not correspond to the reaction coordinate of interest. Thus while dimer searches can be initiated at any point on the PES, it is useful to carefully initiate searches so that they pursue promising directions.

The second disadvantage of the Dimer method is also a disadvantage of the P-RFO, Eigenvector Following, and Cerjan-Miller algorithms. Search trajectories from all of these methods tend to rediscover known saddle points. Recall that these algorithms were chiefly designed to discover transition states independent of chemical intuition. Thus, by construction, they also tend to rediscover what we already know and to discover some saddles that are not chemically interesting. Rediscoveries and chemically uninteresting discoveries become more and more problematic for large systems with many local minima and saddles, e.g. enzyme reactions, clusters of atoms, etc. Certain methods like discrete path sampling [35] and adaptive kinetic Monte Carlo [36] require the ability to find all saddle points that connect to a given minimum on the PES. To appreciate how many saddle points and minima are needed for a complete description, see works by Stillinger [37] and Wales and Doye [38]. As shown in Figure 8.4.2 [6] up to 90% of search trajectories may rediscover previously discovered transition states. Therefore, an algorithm that eliminates rediscovery could dramatically accelerate the discovery of *new* transition states.

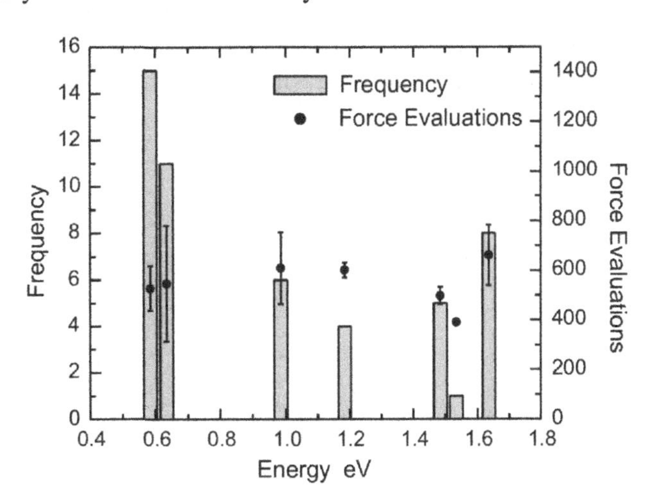

Figure 8.4.2: Several saddle point search algorithms tend to rediscover the same saddles again and again. These results (from Dimer searches) show that approximately 90 percent of the computational effort may be spent on rediscovery. [From Henkelman et al. "Methods for Finding Saddle Points and Minimum Energy Paths", p. 269–300 in *Theoretical Methods in Condensed Phase Chemistry*, S.D. Schwartz, Ed., Kluwer Academic (2000).]

There is a systematic learning-bias algorithm that prevents rediscovery of known saddles when searching for new saddles [39]. However, it is computationally expensive, difficult to implement, and requires that searches be completed in series. Algorithms that use reduced potential energy surfaces offer a more promising solution for the rediscovery problem [40].

8.5 Reduced landscape algorithms

To construct a reduced potential energy surface (RPES), one first selects appropriate "reaction coordinates" $\mathbf{q_r}$ and "bath coordinates" $\mathbf{q_b}$. Constrained optimizations then reduce $V(\mathbf{q_r}, \mathbf{q_b})$ to the lower dimensional RPES $V_R(\mathbf{q_r})$ by minimization over $\mathbf{q_b}$ at fixed $\mathbf{q_r}$, i.e.

$$V_R(\mathbf{q_r}) = \min_{\mathbf{q_b}} V(\mathbf{q_r}, \mathbf{q_b}). \tag{8.5.1}$$

The selected reaction coordinates can be bond lengths, bond angles, torsion angles, or even energy gaps. The main requirement is that $\mathbf{q_r}$ and $\mathbf{q_b}$ should both be differentiable functions of the atom positions to facilitate the constrained energy minimizations. If the reaction coordinates are suitably chosen, then minima and saddle points on the RPES directly correspond to minima and saddle points in the full space [15].

RPESs have a long history in computational studies of chemical reaction rates and mechanisms. The earliest examples were two-dimensional PESs for collinear $H_2 + H$ reactions where the $\mathbf{q_r}$ coordinates were H-H bond lengths and $\mathbf{q_b}$ was the H-H-H angle. The constrained optimization of equation (8.5.1) was not necessary in the early $H_2 + H$ studies because the H-H-H angle is naturally collinear. RPES calculations for molecules with many internal degrees of freedom appeared in 1972 [41,42], and some commercial electronic structure programs now include automated optimization routines for generating the RPES as a function of specified $\mathbf{q_r}$ variables. Figure 8.5.1 shows an RPES as used in a More-O'Ferrell-Jencks analysis [43,44] to determine which steps occur early and which steps are late along a reaction pathway.

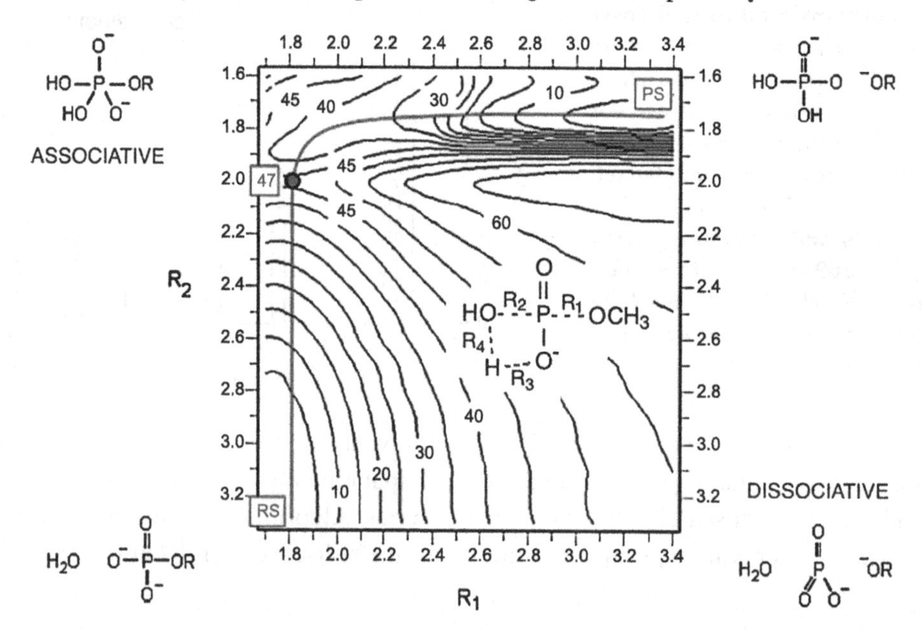

Figure 8.5.1: More-O'Ferrall-Jencks analysis showing the RPES retaining the P-OR and HO-P bondlengths. This analysis was used to determine whether associative or dissociative pathways are preferred for phosphate ester hydrolysis. Note that the actual process involves four critically important bondlengths, but the authors treat the potentially important R_3 and R_4 degrees of freedom as part of the bath. [Adapted from Klahn et al., *J. Am. Chem. Soc.* 128, 15310–15323 (2006).]

In principle, one only needs to specify the reaction coordinate and then the coordinate driving energy profile (the 1D version of an RPES) would suffice, but the correct reaction coordinate is difficult to anticipate. It is far easier to anticipate which degrees of freedom are important in the reaction coordinate. Note that the calculation performed in Figure 8.5.1 has computed a two dimensional landscape to analyze a reaction in which four bonds are changed. The authors have *a priori* decided how four important degrees of freedom can be reduced to two. Such *a priori* assumptions are necessary because the cost of generating an RPES increases dramatically with the number of reaction coordinates that are retained. The ideal four dimensional RPES is far too expensive for *ab initio* calculations, and usually even a three-dimensional RPES calculation is prohibitively expensive.

Bofill and Anglada introduced an interesting alternative to the exhaustive RPES calculation. The Bofill-Anglada algorithm efficiently navigates to saddle points on the RPES *without generating the entire RPES*. The Bofill-Anglada strategy can retain the bondlengths of all bonds that are being broken or formed in a reaction while relegating all other degrees of freedom to the bath. From equation (8.5.1), it follows that the forces along $\mathbf{q_b}$ are zero.

$$\mathbf{g_b} = \frac{\partial V(\mathbf{q_r},\mathbf{q_b})}{\partial \mathbf{q_b}} = \mathbf{0_b} \tag{8.5.2}$$

To understand how the RPES changes with $\mathbf{q_r}$, we first expand the full PES $V(\mathbf{q_r},\mathbf{q_b})$ around location $(\mathbf{q_r}, \mathbf{q_b})$.

$$\begin{aligned} V(\mathbf{q_r} + \Delta\mathbf{q_r}, \mathbf{q_b} + \Delta\mathbf{q_b}) \approx\; & V(\mathbf{q_r}, \mathbf{q_b}) + \begin{bmatrix} \Delta\mathbf{q_r^\dagger} & \Delta\mathbf{q_b^\dagger} \end{bmatrix} \begin{bmatrix} \mathbf{g_r} \\ \mathbf{g_b} \end{bmatrix} \\ & + \frac{1}{2} \begin{bmatrix} \Delta\mathbf{q_r^\dagger} & \Delta\mathbf{q_b^\dagger} \end{bmatrix} \begin{bmatrix} \mathbf{H_{rr}} & \mathbf{H_{rb}} \\ \mathbf{H_{br}} & \mathbf{H_{bb}} \end{bmatrix} \begin{bmatrix} \Delta\mathbf{q_r} \\ \Delta\mathbf{q_b} \end{bmatrix} \end{aligned} \tag{8.5.3}$$

In equation (8.5.3), $\Delta\mathbf{q_r}$ is a reduced coordinate displacement, $\Delta\mathbf{q_b}$ is a bath coordinate displacement, $\mathbf{g_r}$ is the gradient of V with respect to the reduced coordinates, and $\mathbf{g_b}$ is the gradient of V with respect to the bath coordinates at $(\mathbf{q_r}, \mathbf{q_b})$. The Hessian is partitioned into four block matrices: the reaction coordinates block $(\mathbf{H_{rr}})$, the bath coordinates block $(\mathbf{H_{bb}})$, and the coupling blocks $(\mathbf{H_{br}}$ and $\mathbf{H_{rb}})$. Equation (8.5.4) predicts how a change in the reaction coordinates will change the optimal bath coordinates:

$$\Delta\mathbf{q_b} = -\mathbf{H_{bb}^{-1}}(\mathbf{g_b} + \mathbf{H_{br}}\Delta\mathbf{q_r}) \tag{8.5.4}$$

The remainder of this discussion omits the bath gradient which is zero $\mathbf{g_b} = \mathbf{0_b}$ if the bath coordinates are always properly optimized. In that case, equation (8.5.4) simplifies to $\Delta\mathbf{q_b} = -\mathbf{H_{bb}^{-1}}\mathbf{H_{br}}\Delta\mathbf{q_r}$. Figure 8.5.2 illustrates the re-optimization of the bath following a displacement in the reaction coordinates.

Assuming that the bath coordinate displacements are always optimal given the reaction coordinate displacements, then the Taylor expansion on the full PES becomes a Taylor expansion in

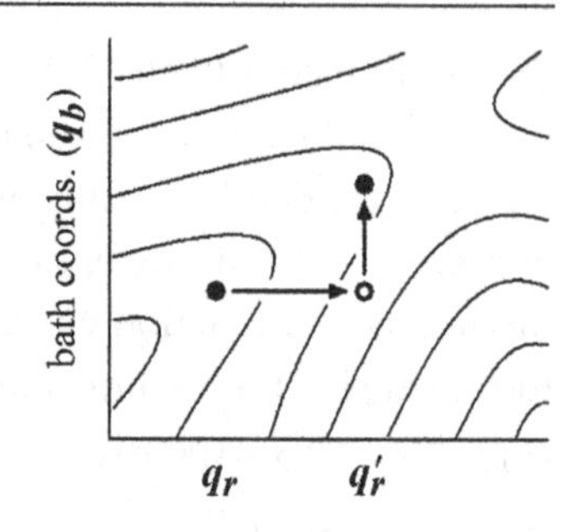

Figure 8.5.2: A displacement in the reaction coordinate space must be accompanied by a displacement in the bath space to maintain an adapted equilibrium in the bath.

$\Delta\mathbf{q_r}$ on the RPES:

$$V_R(\mathbf{q_r} + \Delta\mathbf{q_r}) \approx V_R(\mathbf{q_r}) + \Delta\mathbf{q_r^\dagger g_r} + \frac{1}{2}\Delta\mathbf{q_r^\dagger H_r}\Delta\mathbf{q_r} \tag{8.5.5}$$

where the reduced Hessian is "softened" by Schur complement terms:

$$\mathbf{H_r} = \mathbf{H_{rr}} - \mathbf{H_{rb}H_{bb}^{-1}H_{br}}$$

Equation (8.5.5) can be used to implement transition state search algorithms like P-RFO or the Dimer method on the RPES. The Bofill-Anglada algorithm is a Cerjan-Miller algorithm in the RPES. Ayers and co-workers developed an RPES-based saddle search algorithm that does not require the Hessian [45]. Further advances in this area may lead to (1) efficient methods for finding transition states in QM/MM models of reactions in solution, (2) methods that prevent rediscovery by simply choosing different sets of $\mathbf{q_r}$ coordinates for each search [40], (3) new free energy methods which ensemble average the bath degrees of freedom rather than minimize them, and (4) methods that work in synergy with new methods to automatically generate the list of reaction pathways (and therefore the $\mathbf{q_r}$ coordinates) to be investigated [46,47].

8.6 Coordinate driving

In the coordinate driving approach, a particular degree of freedom is chosen as the reaction coordinate. Constrained energy minimizations are then performed at a sequence of fixed reaction coordinate values starting at the reactant state and ending at the product state [48–50]. Coordinate driving is an RPES calculation in the limit where just one reaction coordinate is retained in $\mathbf{q_r}$. If the selected reaction coordinate is sufficiently accurate, then coordinate driving generates a chain of structures and a continuous energy profile between the reactant and product states.

Algorithm: Coordinate Driving

Start with minimum energy configuration $\mathbf{x}$ in reactant state. *Carefully* select a reaction coordinate $q(\mathbf{x})$. Minimize the energy in the reactant state (A) to find $\mathbf{x}_A$ and set $q_A = q(\mathbf{x}_A)$. Similarly minimize the energy in the product state (B) to find $q_B = q(\mathbf{x}_B)$. Start from q_A and incrementally advance the reaction coordinate value: $q_k = q_A + k\Delta q$ where $\Delta q \ll q_B - q_A$. For each q_k between q_A and q_B, minimize $V(\mathbf{x})$ subject to $q(\mathbf{x}) = q_k$ to find the minimum energy geometry $\mathbf{x}_{opt}(q_k)$. The constrained minimum with the highest energy is an approximate saddle point geometry.

The selected degree of freedom for coordinate driving may be a bond length, a torsional angle, a difference of two bond lengths, etc. Some coordinate driving methods avoid choosing one degree of freedom by using geometric distance coordinates. For example, the activation relaxation technique (ART) [51,52] drives configurations away from the reactant minimum while minimizing on hyperplanes perpendicular to the displacement from the minimum. Coordinate driving and related molecular dynamics [53] approaches require a carefully selected reaction coordinate to avoid hysteresis artifacts [23,42]. Figure 8.6.1 depicts the hysteresis that emerges from use of an inaccurate reaction coordinate.

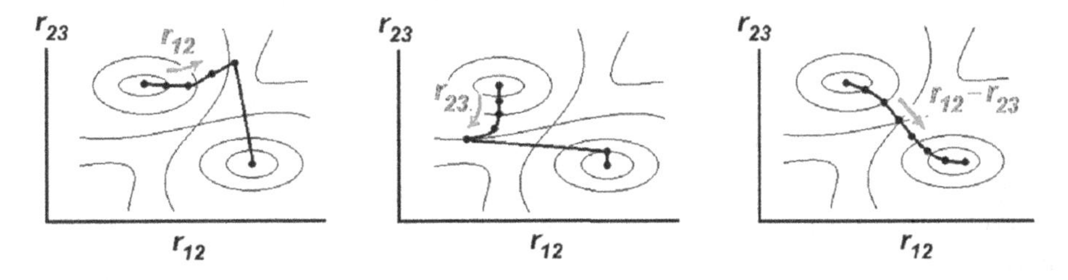

Figure 8.6.1: Coordinate driving generates a series of constrained minima at increasing values of an assumed reaction coordinate. If the coordinate is chosen poorly as shown in the first two panels, hysteresis causes coordinate driving to miss the transition state. The last panel depicts how coordinate driving works when the assumed coordinate is accurate.

8.7 Nudged elastic band

Nudged elastic band (NEB) was the culmination of several path optimization strategies to obtain saddle points and the minimum energy paths (MEPs). Let us begin with a brief discussion of linear synchronous transit (LST) [54]. LST interpolates molecular geometries between the reactant and product states. LST can be viewed as a crude version of coordinate driving in which there are no constrained optimizations. Let $\mathbf{x}^{(0)}$ be the energy minimum in the reactant state and let $\mathbf{x}^{(n)}$ be the energy minimum in the product state. The discretized LST path is the sequence

$$\mathbf{x}^{(k)} = \mathbf{x}^{(0)} + \frac{k}{n}\{\mathbf{x}^{(n)} - \mathbf{x}^{(0)}\} \tag{8.7.1}$$

where $0 \leq k \leq n$ so that the path contains $(n+1)$ configurations including its endpoints. The discrete states are often called "images" because they are like frames in an animated movie of the transition. The image with maximum energy along the LST path provides a (very crude) estimate of the saddle point. Note that the geometries along 'a straight line' depend on the coordinate system being used. For example, a straight line path in Cartesian coordinates from linear HCN to linear HNC brings the proton through the nitrogen and carbon cores. However, a straight line using internal coordinates r_{HC}, r_{CN}, and the angle θ_{HCN} approximates the MEP reasonably well. Thus, as with many optimization techniques and rare events methods, the coordinate system matters. Figure 8.7.1 shows the LST path on the Muller-Brown PES [55,56].

LST inspired many related methods including synchronous transit methods with optimization on conjugate planes [57], conjugate peak refinement (CPR) [5], and a synchronous transit method for periodic systems [58]. Here we only discuss Nudged elastic band (NEB) [59,60] and related methods that aim to directly optimize the minimum energy path (MEP). In addition to the LST path, Figure 8.7.1 shows the MEP which visits two saddle points and an intermediate minimum *en route* from A to B.

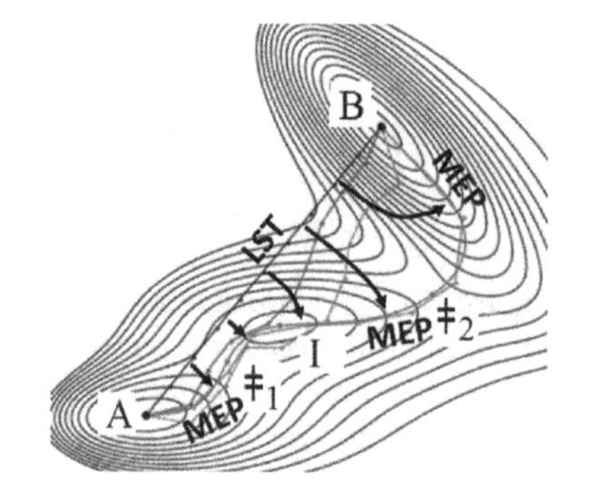

Figure 8.7.1: LST finds the point of maximum energy along a linearly interpolated path between the reactant and product minima, A and B. NEB optimizes a path between the reactant and product minima to obtain the MEP. [Figure modified from Schlegel, *Wiley Interdisc. Reviews,* 1, 790–809 (2011).]

The first step toward the NEB method was the path optimization method of Elber and Karplus. Their method introduced springs to maintain uniform distance between images during the path optimization. The objective function they minimized is [61]

$$S[\{\mathbf{x}^{(k)}\}_{0\leq k\leq n}] = \frac{1}{L}\sum_{j=1}^{n} \Delta\ell_k V(\mathbf{x}^{(k)}) + \lambda \sum_{k=1}^{n} (\Delta\ell_k - \Delta\ell)^2$$

where $\Delta\ell_k = ||\mathbf{x}^{(k)} - \mathbf{x}^{(k-1)}||$, $L = \sum_k \Delta\ell_k$, and

$$\Delta\ell = \left[\frac{1}{n}\sum_{j=1}^{n}(\Delta\ell_j)^2\right]^{1/2}.$$

The first term penalizes paths that traverse long and/or high energy regions. The second term includes a spring constant λ that penalizes deviations from a uniform root mean square distance $\Delta\ell$ between the discrete configurations on the path. At high energy curves in the MEP, the path of Elber and Karplus climbs the inside walls of the reaction channel. Because of this corner cutting artifact, the method of Elber and Karplus does not actually converge to the MEP [62]. At each stage of the path optimization, the Elber-Karplus springs have a non-zero equilibrium length that increases with the overall chain length. Thus the corner cutting problem does *not* emerge from the spring forces, but rather from the first term in the objective function which favors short paths through the barrier region. Olender and Elber [62] fixed the corner cutting problem, but their fix required (at least approximate) second derivatives of the potential energy.

NEB [59,63] resolved the corner cutting problem while maintaining uniform node spacing and using only gradients of the PES. The images in NEB are connected by springs of zero equilibrium length. If not for the clever "nudging" procedure of Mills et al., springs of zero-equilibrium length would exert forces in the path curvature direction and severely exacerbate the corner cutting problem. The nudging procedure is motivated by two observations:

1. At curves in the MEP, the artificial (spring) forces in NEB try to pull images in the direction of path curvature which would cause severe corner cutting.
2. On steep portions of the MEP true (non-spring) forces try to pull images downward along the path tangent to give a non-uniform image spacing.

The nudging procedure applies the spring and real forces on each image to the tangential and perpendicular displacements, respectively. Thus in the NEB algorithm, the PES only exerts lateral forces on the path, and the springs only exert tangential forces on the path. The separation of forces at each image into tangential ($\mathbf{f}_{||}$) and perpendicular ($\mathbf{f}_{\perp}$) components is illustrated in Figure 8.7.2.

The tangent vector at each image provides a way to resolve tangential components of spring forces and perpendicular components of real forces. The original NEB implementations [59,63] obtained tangents from a center difference approximation $\mathbf{t}^{(k)} = (\mathbf{x}^{(k+1)} - \mathbf{x}^{(k-1)})/||\mathbf{x}^{(k+1)} - \mathbf{x}^{(k-1)}||$ or from an average of the forward

$$\mathbf{t}_{+}^{(k)} = (\mathbf{x}^{(k+1)} - \mathbf{x}^{(k)})/\left\|\mathbf{x}^{(k+1)} - \mathbf{x}^{(k)}\right\|$$

and backward

$$\mathbf{t}_{-}^{(k)} = (\mathbf{x}^{(k)} - \mathbf{x}^{(k-1)})/\left\|\mathbf{x}^{(k)} - \mathbf{x}^{(k-1)}\right\|$$

differences. Henkelman and Jonsson [60,63] discovered that the original tangent definitions can cause instabilities, kinked NEB paths, and failed convergence. The instabilities occur when the transverse restoring forces are small compared to the tangential forces along the string.

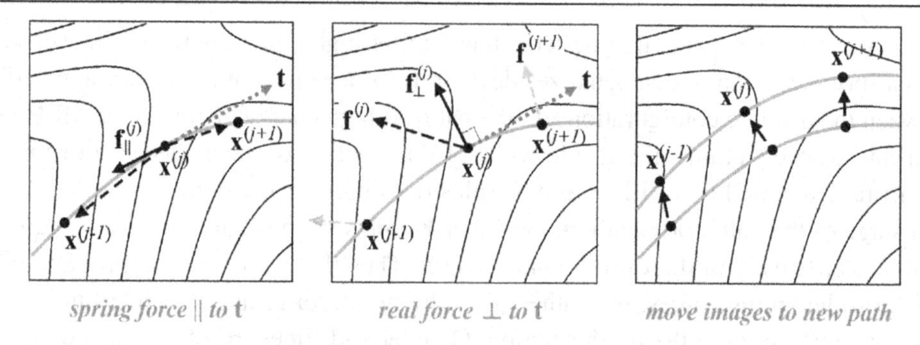

Figure 8.7.2: The NEB algorithm optimizes a chain of states (images) connected by springs to find the MEP between specified reactant and product states. Only the tangential component of the spring forces and the perpendicular component of the physical forces are used in the optimization. This "nudging" strategy fixes the corner cutting problem and maintains a uniform distance between images.

Displacement of one image changes the tangent for the next image in the uphill direction. The distorted tangent at the uphill image causes real forces along the MEP (which should be removed as the path converges) to be counted as perpendicular forces. The stability problem was resolved by adopting an "upward difference" definition of the path tangents [60]:

$$\mathbf{t}^{(k)} = \begin{cases} \mathbf{t}_{+}^{(k)} & if \quad V(\mathbf{x}^{(k+1)}) > V(\mathbf{x}^{(k)}) > V(\mathbf{x}^{(k-1)}) \\ \mathbf{t}_{-}^{(k)} & if \quad V(\mathbf{x}^{(k+1)}) < V(\mathbf{x}^{(k)}) < V(\mathbf{x}^{(k-1)}) \end{cases} \tag{8.7.2}$$

At images that are local energy maxima or minima along the path the upward tangent is not well-defined. For those images, Henkelman and Jonsson recommend an energy-weighted tangent that smoothly switches between the forward and backward tangents [60]. Specifically, the weighted tangent is the normalized vector in direction

$$\mathbf{t}^{(k)} = \begin{cases} \mathbf{t}_{+}^{(k)} \Delta V_{max}^{(k)} + \mathbf{t}_{-}^{(k)} \Delta V_{min}^{(k)} & if \quad V(\mathbf{x}^{(k+1)}) > V(\mathbf{x}^{(k-1)}) \\ \mathbf{t}_{+}^{(j)} \Delta V_{min}^{(j)} + \mathbf{t}_{-}^{(j)} \Delta V_{max}^{(j)} & if \quad V(\mathbf{x}^{(k+1)}) < V(\mathbf{x}^{(k-1)}) \end{cases} \tag{8.7.3}$$

where

$$\Delta V_{max}^{(k)} = max\left\{ \left| V(\mathbf{x}^{(k+1)}) - V(\mathbf{x}^{(k)}) \right|, \left| V(\mathbf{x}^{(k-1)}) - V(\mathbf{x}^{(k)}) \right| \right\} \tag{8.7.4}$$

and

$$\Delta V_{min}^{(k)} = min\left\{ \left| V(\mathbf{x}^{(k+1)}) - V(\mathbf{x}^{(k)}) \right|, \left| V(\mathbf{x}^{(k-1)}) - V(\mathbf{x}^{(k)}) \right| \right\} \tag{8.7.5}$$

Once the tangent vectors have been defined, the force at each image for the NEB optimization is computed as

$$\mathbf{f}_{NEB}^{(k)} = \mathbf{f}_{||}^{(k)} + \mathbf{f}_{\perp}^{(k)}.$$

Here, $\mathbf{f}_{||}^{(k)} = \left\langle \mathbf{f}_{spring}^{(k)} | \mathbf{t}^{(k)} \right\rangle \mathbf{t}^{(k)}$ is the tangential component of all spring forces exerted on image k, and $\mathbf{f}_{\perp}^{(k)} = (\mathbf{1} - \mathbf{t}^{(k)} \circ \mathbf{t}^{(k)})\mathbf{f}^{(k)}$ is the perpendicular component of the real forces on image k. The forces on the n images are concatenated to obtain a single force vector of dimension $3nN$ and then optimization is done with a conjugate gradient or BFGS minimizer [59,60].

Algorithm: Nudged Elastic Band

To start the algorithm, one needs geometries of reactant and product endpoints, $\mathbf{x}^{(0)}$ and $\mathbf{x}^{(n)}$. Place intermediate images at $\mathbf{x}^{(k)} = \mathbf{x}^{(0)} + (\mathbf{x}^{(n)} - \mathbf{x}^{(0)}) \cdot k/n$ with $1 \leq k \leq n-1$. Choose a spring constant λ that is comparable to the vibrational force constants between the atoms.

1. Compute tangents $\mathbf{t}^{(1)}, \mathbf{t}^{(2)}, \ldots, \mathbf{t}^{(n-1)}$ using the upward scheme, equations (8.7.2) and (8.7.3).
2. Compute actual forces at each image $\mathbf{f}^{(1)}, \mathbf{f}^{(2)}, \ldots, \mathbf{f}^{(n-1)}$.
3. Compute spring forces at each image $\mathbf{f}_{spring}^{(1)}, \mathbf{f}_{spring}^{(2)}, \ldots, \mathbf{f}_{spring}^{(n-1)}$. The (raw) spring forces are $\mathbf{f}_{spring}^{(k)} = \lambda(\mathbf{x}^{(k+1)} - \mathbf{x}^{(k)}) + \lambda(\mathbf{x}^{(k-1)} - \mathbf{x}^{(k)})$.
4. Remove tangential components of the real forces to get $\mathbf{f}_{\perp}^{(1)}, \mathbf{f}_{\perp}^{(2)}, \ldots,$ and $\mathbf{f}_{\perp}^{(n-1)}$.
5. Remove perpendicular components of the spring forces to get $\mathbf{f}_{||}^{(1)}, \mathbf{f}_{||}^{(2)}, \ldots,$ and $\mathbf{f}_{||}^{(n-1)}$.
6. Add the tangential spring forces and normal real forces: $\mathbf{f}_{NEB}^{(k)} = \mathbf{f}_{||}^{(k)} + \mathbf{f}_{\perp}^{(k)}$.
7. Create a $3nN$-dimensional vector from the NEB forces.
8. Modify the discretized path using a BFGS or conjugate gradient step.
9. Stop if the path is converged, i.e. if the forces and image displacements are sufficiently small. Else return to step (1) and repeat.

The force calculations in step (2) are easily parallelized. Convergence criteria can be loose when NEB is being used to generate an approximate saddle for refinement by other methods.

Most applications of NEB only require the saddle point, e.g. to apply harmonic transition state theory. The MEP itself is rarely needed, except to confirm that a transition state connects the intended reactants and products. When the MEP is needed, it should be noted that NEB (and similar path optimization methods) converge rather slowly because of the high

$3nN$-dimensionality of the discretized path space. It is usually easier to find the saddle point and then compute the MEP with an efficient algorithm for following steepest descent paths [31,64–66]. Note that NEB can also identify minimum energy seam crossing points if, at each image, the real forces are computed for the lower energy surface.

The chief advantage of path optimization methods is that they are more robust than constrained optimization strategies and high dimensional search algorithms. However NEB (and the other path optimization algorithms described below) also require more *a priori* understanding of the mechanism. Specifically, NEB requires the reactant and product endpoints of the path and the atom-by-atom correspondence between the reactants and products. In other words, at the outset of an NEB calculation, we must know enough about the mechanism to predict the outcome of every possible radio-labeling experiment. Thus, unlike the algorithms in sections 8.2–8.4, path optimization methods cannot discover entirely unexpected mechanisms or products.

In practice, NEB is mainly used to obtain an approximate transition state structure which can then be refined using faster P-RFO, Eigenvector Following, or Dimer optimizations [19,32, 67,68]. To accelerate the acquisition of a suitable approximation to the saddle point, Ayala and Schlegel [69] and Henkelman et al. [67] introduced the "climbing image" methods. The climbing image nudged elastic band (CI-NEB) method reverses the tangential forces on one special climbing image

$$\mathbf{f}'_{NEB} = (1 - 2\mathbf{t} \circ \mathbf{t}^T)\mathbf{f}_{NEB}.$$

The modified forces guide the climbing image toward the energy maximum along the interpolated path. The climbing image modification is easily made at step (6) in the algorithm above. The approximate unstable mode at the saddle point is also provided by the path tangent at the climbing image. CI-NEB avoids the need to interpolate for the location of maximum energy, so CI-NEB calculations can be performed with only 5–7 images. Thus CI-NEB is particularly efficient at generating approximate saddle point geometries for optimization by other methods [67].

Several other variations on the NEB algorithm have also been proposed. For example, Trygubenko and Wales proposed the efficient limited memory-BFGS method and an alternative double-nudging procedure for stabilizing the NEB path [68]. Maragakis et al. proposed an adaptive NEB approach that increases the resolution of the MEP near the saddle point [70]. Crehuet and Field introduced a finite temperature version of NEB [71].

Intuitively, NEB seems like a discretized approximation to a continuous path functional optimization, but NEB was entirely formulated in terms of discrete configurations along a chain of states. In this regard, the continuous String method formulations of Vanden-Eijnden and coworkers provided valuable insight about the NEB method [72–74]. In the zero temperature string method (ZTS) [72] the discretized string moves in the direction of the normal force at each image, as in the NEB method. However, instead of including tangential spring forces to maintain image spacing, the images are redistributed uniformly along the interpolated string after each iteration. The string method also incorporated many of the innovations that had improved NEB, e.g. an upward difference scheme for estimating string tangents [75] a climbing image scheme for improving transition state approximations [76], and a finite temperature version of the string method [77] in which the images are average positions from NVT simulations restrained to constant arclength isosurfaces along the path.

Peters et al. [32] used the reparametrization framework of the string method to eliminate images from the interior of the path during early stages of string optimization. The growing string method (GSM) begins with two short strings that are focused in the reactant and product minima. These are initially connected to each other only by a splined and image-less path between them. As forces on the end segments converge, images are added to splined portions of the string interior [32]. Once the separate halves of the string merge, GSM becomes the zero temperature string method [72]. Goodrow et al., and Zimmerman et al. improved the GSM by including superlinear optimization methods [78], internal coordinates [78,79], and climbing image modifications [79,80]. GSM is faster than NEB and ZTS in cases where the initial (often linearly interpolated) path is very different from the MEP, but otherwise GSM performance is similar to the NEB and ZTS methods [81].

Where multiple pathways might exist, remember that NEB and the other path optimization methods discussed in this section are deterministic minimization schemes. Path optimization approaches in general will only find the pathway which is immediately downhill from the initial pathway guess. As the number of pathways, local minima, and saddles increase, NEB (and saddle search algorithms in general) become less and less useful. NEB is not appropriate for processes with rugged energy landscapes: biomolecular processes like protein folding, chemical reactions in solution, nucleation processes, etc. See Chapter 11 for more suitable methods based on collective variables and free energy landscapes.

Exercises

1) Exercise 7.2 describes $V(r_{\mathbf{AB}}, r_{\mathbf{BC}})$, a locally quadratic model of the PES near saddle points for collinear atom-exchange reactions. Suppose that the optimal bond lengths and the three force constants in V are unknown.

   a) Start from a "wrong" geometry and implement the dimer method with analytic (Heyden) rotations to find the saddle point.

   b) For some model chemistries, e.g. FCI, CCSD(T), etc., we cannot compute analytic gradients of the local PES. Devise a "trimer" algorithm to find the saddle point based on the energies at three configurations: the current configuration, a symmetrically stretched configuration, and an asymmetrically stretched configuration. Use the anticipated quadratic nature of the potential at the saddle point to guide your search. Hint: the three configurations provide forces along the two internal modes. Implement your algorithm, again starting from a "wrong" geometry.

2) Implement a transition state search algorithm of your choice to find saddle points on the PES for a two-dimensional LJ-7 cluster.

   a) Construct a discrete master equation to describe the complete set of states and transitions in the LJ-7 cluster.

   b) Compress the state space without losing any kinetic resolution by lumping degenerate symmetry equivalent states.

3) Implement the RPES transition state search algorithm of Bofill and Anglada for the two-dimensional LJ-7 cluster.

   a) Select a low dimensional coordinate system to efficiently find saddle points on the PES.

   b) Attempt to eliminate rediscoveries by choosing different active space degrees of freedom for each search.

4) NEB has been used to study solid-solid nucleation processes by allowing the lattice parameters to vary along the solid transformation pathway [82–84].

   a) Discuss these calculations as ways of computing the relative stability of the crystal polymorphs at the endpoint images.

   b) Discuss these calculations as ways of studying nucleation and growth mechanisms. Specifically comment on periodic boundary conditions, variable lattice parameters, and the diversity of distinct nucleation pathways.

   c) Compare the solid-state NEB approach to path sampling analyses of solid-solid nucleation [85–87]? Comment on the relative advantages, disadvantages, and scope of these approaches for understanding polymorph stability and the transformation kinetics.

   d) [Project problem] Discuss the role of elastic stresses on solid-solid phase transformations in real systems and in periodic boundary conditions. Formulate and implement an "Eshelby" elastic stress-balanced boundary condition [88] to improve computational methods for solid-to-solid nucleation.

References

[1] M.J.S. Dewar, E.F. Healy, J.J.P. Stewart, J. Chem. Soc. Faraday Trans. 2 (80) (1984) 227.
[2] I.V. Ionova, E.A. Carter, J. Chem. Phys. 98 (1993) 6377.
[3] R. Miron, K. Fichthorn, J. Chem. Phys. 115 (2001) 8742.
[4] C. Choi, R. Elber, J. Chem. Phys. 94 (1991) 751.
[5] S. Fischer, M. Karplus, Chem. Phys. Lett. 194 (1992) 252–261.

[6] G. Henkelman, G. Johannesson, H. Jonsson, in: S.D. Schwartz (Ed.), Theoretical Methods in Condensed Phase Chemistry, Kluwer Academic, Dordrecht, 2000, pp. 269–300, chapter 10.
[7] R.L. Burden, J.D. Faires, Numerical Analysis, Prindle, Weber, and Schmidt, Boston, 1993.
[8] P. Pulay, G. Fogarasi, J. Chem. Phys. 96 (1992) 2856–2860.
[9] H.B. Schlegel, J. Comput. Chem. 3 (1982) 214–218.
[10] C. Peng, P.Y. Ayala, H.B. Schlegel, M.J. Frisch, J. Comput. Chem. 17 (1995) 49–56.
[11] R. Fletcher, Practical Methods of Optimization, Wiley, Chichester, 1987.
[12] C.J. Cerjan, W.H. Miller, J. Chem. Phys. 75 (1981) 2800–2806.
[13] A. Banerjee, N. Adams, J. Simons, R. Shepard, J. Phys. Chem. 89 (1985) 52–57.
[14] J. Baker, F. Chan, J. Comput. Chem. 17 (1996) 888–904.
[15] J.M. Bofill, J.M. Anglada, Theor. Chem. Acc. 105 (2001) 463–472.
[16] X. Prat-Resina, M. Garcia-Viloca, G. Monard, A. Gonzalez-Lafont, J.M. Lluch, J.M. Bofill, J.M. Anglada, Theor. Chem. Acc. 107 (2002) 147–153.
[17] G. Henkelman, H. Jonsson, J. Chem. Phys. 111 (1999) 7010–7022.
[18] L. Munro, D. Wales, Phys. Rev. B 59 (1999) 3969–3980.
[19] A. Heyden, A.T. Bell, F.J. Keil, J. Chem. Phys. 123 (2005) 224101.
[20] D. Poppinger, Chem. Phys. Lett. 35 (1975) 550–554.
[21] S. Bell, R. Fletcher, J.S. Crighton, Chem. Phys. Lett. 82 (1981) 122–126.
[22] J. Simons, P. Jorgensen, H. Taylor, J. Ozment, J. Phys. Chem. 87 (1983) 2745–2753.
[23] M. Hirsch, W. Quapp, J. Comput. Chem. 23 (2002) 887–894.
[24] R.A. Olsen, G.J. Kroes, G. Henkelman, A. Arnaldsson, H. Jonsson, J. Chem. Phys. 121 (2004) 9776–9792.
[25] J. Nocedal, S. Wright, Numerical Optimization, Springer, Berlin, 2006.
[26] J.M. Bofill, J. Comput. Chem. 15 (1994) 1–11.
[27] J.M. Bofill, M. Comajuan, J. Comput. Chem. 16 (1995) 1326–1338.
[28] J.M. Bofill, Chem. Phys. Lett. 260 (1996) 359–364.
[29] J.M. Anglada, J.M. Bofill, J. Comput. Chem. 19 (1998) 349–362.
[30] J. Baker, J. Comput. Chem. 7 (1986) 385–395.
[31] H.B. Schlegel, J. Comput. Chem. 24 (2003) 1514–1527.
[32] B. Peters, A. Heyden, A.T. Bell, A. Chakraborty, J. Chem. Phys. 120 (2004) 7877–7886.
[33] A.F. Voter, Phys. Rev. Lett. 78 (1997) 3908–3911.
[34] C. Shang, Z.-P. Liu, J. Chem. Theory Comput. 6 (2010) 1136–1144.
[35] D.J. Wales, Mol. Phys. 102 (2004) 891–908.
[36] G. Henkelman, H. Jonsson, J. Chem. Phys. 115 (2001) 9657.
[37] F.H. Stillinger, T. Weber, Science 225 (1984) 983–989.
[38] D.J. Wales, J.P.K. Doye, J. Chem. Phys. 119 (2003) 12409.
[39] B. Peters, W.-Z. Liang, A.T. Bell, A. Chakraborty, J. Chem. Phys. 118 (2003) 9533.
[40] B.R. Goldsmith, A.M. Fong, B. Peters, in: K. Han (Ed.), Reaction Rate Computations: Theories and Applications, RSC Publishing, 2013, pp. 213–221.
[41] A. Rastelli, A.S. Pozzoli, G. Del Re, J. Chem. Soc. Perkin Trans. II (1972) 1571–1575.
[42] J.W. McIver, A. Komornicki, J. Am. Chem. Soc. 5516 (1972) 2625–2633.
[43] W.P. Jencks, Chem. Rev. 85 (1985) 511–527.
[44] M. Klahn, E. Rosta, A. Warshel, J. Am. Chem. Soc. 128 (2006) 15310–15323.
[45] S.K. Burger, P.W. Ayers, J. Chem. Theory Comput. (ISSN 1549-9618) 6 (May 2010) 1490–1497.
[46] R. Vinu, L.J. Broadbelt, Annu. Rev. Chem. Biomol. Eng. 3 (2012) 29–54.
[47] P.M. Zimmerman, Mol. Simul. 41 (2014) 43–54.
[48] M.J. Rothman, L.L. Lohr, Chem. Phys. Lett. 70 (1980) 405–409.
[49] P. Scharfenberg, Chem. Phys. Lett. 79 (1981) 115–117.
[50] P. Scharfenberg, J. Comput. Chem. 3 (1982) 277–282.
[51] N. Mousseau, G.T. Barkema, Phys. Rev. E 57 (1972) 2419–2424.

[52] L.K. Beland, P. Brommer, F. El-Mellouhi, J.-F. Joly, N. Mousseau, Phys. Rev. E 84 (2011) 46704.
[53] B. Isralewitz, S. Izrailev, K. Schulten, Biophys. J. 73 (1997) 2972–2979.
[54] T.A. Halgren, W.N. Lipscohlb, Chem. Phys. Lett. 49 (1977) 1977.
[55] K. Muller, L.D. Brown, Theor. Chim. Acta 53 (1979) 75–93.
[56] H.B. Schlegel, Wiley Interdiscip. Rev. Comput. Mol. Sci. 1 (2011) 790–809.
[57] S. Bell, J.S. Crighton, J. Chem. Phys. 80 (1984) 2464.
[58] N. Govind, M. Petersen, G. Fitzgerald, D. King-Smith, J. Andzelm, Comput. Mater. Sci. 28 (2003) 250–258.
[59] G. Mills, H. Jonsson, G.K. Schenter, Surf. Sci. 324 (1995) 305–337.
[60] G. Henkelman, H. Jonsson, J. Chem. Phys. 113 (2000) 9978–9985.
[61] R. Elber, M. Karplus, Chem. Phys. Lett. 139 (1987) 375–380.
[62] R. Olender, R. Elber, J. Mol. Struct., Theochem 398–399 (1997) 63–71.
[63] H. Jonsson, G. Mills, K.W. Jacobsen, in: B.J. Berne, G. Ciccotti, D.F. Coker (Eds.), Classical and Quantum Dynamics in Condensed Phase Simulations, World Scientific, Singapore, 1998, pp. 385–404.
[64] M. Page, J.W. McIver, J. Chem. Phys. 88 (1987) 922–935.
[65] C. Gonzalez, H.B. Schlegel, J. Chem. Phys. 90 (1989) 2154–2161.
[66] H.P. Hratchian, H.B. Schlegel, J. Chem. Theory Comput. 1 (2005) 61–69.
[67] G. Henkelman, B.P. Uberuaga, H. Jonsson, J. Chem. Phys. 113 (2000) 9901.
[68] S. Trygubenko, D.J. Wales, J. Chem. Phys. 120 (2004) 2082–2094.
[69] P.Y. Ayala, H.B. Schlegel, J. Chem. Phys. 107 (1997) 375.
[70] P. Maragakis, S.A. Andreev, Y. Brumer, D.R. Reichman, E. Kaxiras, J. Chem. Phys. 117 (2002) 4651.
[71] R. Crehuet, M.J. Field, J. Chem. Phys. 118 (2003) 9563.
[72] W. E, W. Ren, E. Vanden-Eijnden, Phys. Rev. B 66 (2002) 52301.
[73] E. Vanden-Eijnden, in: M. Ferrario, G. Ciccotti, K. Binder (Eds.), Computer Simulations in Condensed Matter: From Materials to Chemical Biology, in: Lecture Notes in Physics, vol. 703, Springer, Berlin, Heidelberg, 2006, pp. 439–478.
[74] W. E, E. Vanden-Eijnden, Annu. Rev. Phys. Chem. 61 (2010) 391–420.
[75] W. Ren, Commun. Math. Sci. 1 (2003) 377–384.
[76] W. Ren, E. Vanden-Eijnden, J. Chem. Phys. 138 (2013) 134105.
[77] W. E, W. Ren, E. Vanden-Eijnden, J. Phys. Chem. B 109 (2005) 6688–6693.
[78] A. Goodrow, A.T. Bell, M. Head-Gordon, J. Chem. Phys. 129 (2008) 174109.
[79] P.M. Zimmerman, J. Chem. Phys. 138 (2013) 184102.
[80] P.M. Zimmerman, J. Chem. Theory Comput. 9 (2013) 3043–3050.
[81] E.F. Koslover, D.J. Wales, J. Chem. Phys. 127 (2007) 134102.
[82] A.R. Oganov, R. Martonak, A. Laio, P. Raiteri, M. Parrinello, Nature 438 (2005) 1142–1144.
[83] K. Casperson, E.A. Carter, Proc. Natl. Acad. Sci. USA 102 (2005) 6738.
[84] D. Sheppard, P. Xiao, W. Chemelewski, D.D. Johnson, G. Henkelman, J. Chem. Phys. 136 (2012) 074103.
[85] G.T. Beckham, B. Peters, C. Starbuck, N. Variankaval, B.L. Trout, J. Am. Chem. Soc. 129 (2007) 4714–4723.
[86] G.T. Beckham, B. Peters, B.L. Trout, J. Phys. Chem. B 112 (2008) 7460–7466.
[87] M. Grunwald, C. Dellago, Nano Lett. 9 (2009) 2099–2102.
[88] R. Phillips, Crystals, Defects, and Microstructures: Modeling Across Scales, Cambridge Press, Cambridge, UK, 2001.

CHAPTER 9

Unimolecular reactions

> *"[Perrin] assumes that he may extrapolate to such low concentrations that practically no collisions occur and that the reaction velocity will then still be constant. He concludes, therefore, that reaction velocity cannot be due to collisions but can only be due to some outside agency, which must obviously be radiation. The argument is convincing ... but surely our evidence of this is quite inadequate at low pressures."*
>
> ***Lindemann, Trans. Faraday Soc. (1922)***

Unimolecular reaction rate theory describes the isomerization, dissociation, or decomposition of a single reactant molecule or complex in the gas phase. Early work on unimolecular reactions was hampered by experimental difficulties and theoretical misconceptions. Some purportedly unimolecular reactions turned out to be multistep reactions, chain reactions, or reactions catalyzed by reactor walls. For truly unimolecular reactions, it was not initially clear why the rate should scale with the first power of concentration while the frequency of collisions scales with the square of concentration. Perrin and many others interpreted this observation as evidence for activation by radiation rather than by collisions. What proponents of the radiation hypothesis did not anticipate is that unimolecular reaction rates *do* scale with the square of concentration at sufficiently low pressures. Later theories by Lindemann and others would explain how collisions and intramolecular energy transfer processes yield complex rate laws that change from first to second order as the pressure decreases. The experimental validation of these predictions marked the end of the radiation hypothesis and the beginning of an exciting effort to develop quantitative reaction rate theories. See articles by Harned [1] and King and Laidler [2] for engaging reviews on these early theoretical and experimental developments.

This chapter outlines the key developments up through the RRKM theory. Starting from a threshold activation energy, vibrational frequencies, and rotational moments of inertia – quantities that are now readily computed with electronic structure theory – the RRKM theory can predict absolute rates, activation parameters, isotope effects, etc. The ability to compute these kinetic properties from *ab initio* calculations has made RRKM theory essential in applications like atmospheric chemistry, combustion chemistry, and photochemistry.

9.1 Lindemann-Christiansen mechanism

Modern unimolecular reaction rate theories began from a model that was put forth independently by Lindemann [3] and Christiansen [4]. The Lindemann-Christiansen mechanism says that a reactant molecule **A** gains sufficient energy to react by collision with another molecule **X**. **X** can be another **A** molecule or an inert collision partner. The activated molecule **A**$*$ may transfer its energy in another collision or it may react. These processes correspond to the elementary steps

$$\begin{gathered}\mathbf{X}+\mathbf{A} \underset{k_{-1}}{\overset{k_1}{\rightleftarrows}} \mathbf{X}+\mathbf{A}* \\ \mathbf{A}* \xrightarrow{k_2} \mathbf{P}\end{gathered} \tag{9.1.1}$$

If the total concentration of the collision partner [**X**] is sufficiently high, the rate at which **A**$*$ is quenched will surpass the rate at which it can react to form **P**. If [**X**] is sufficiently low, then each **A**$*$ that is created from a collision will react before being quenched by another collision. For each mole of **A**$*$ that is created, a fraction $k_2[\mathbf{A}*]/(k_{-1}[\mathbf{A}*][\mathbf{X}]+k_2[\mathbf{A}*])$ is converted to **P**. Of course **A**$*$ cannot react to form **P** faster than it is created by the forward rate $k_1[\mathbf{A}][\mathbf{X}]$, so the collisions that create **A**$*$ must limit the overall rate at very low concentrations. The rate law from the Lindemann-Christiansen mechanism is the rate at which **A**$*$ is formed multiplied by the fraction of **A**$*$ that converts to **P**.

$$\mathfrak{r} = \frac{k_1 k_2 [\mathbf{A}][\mathbf{X}]}{k_{-1}[\mathbf{X}] + k_2} \tag{9.1.2}$$

Equation (9.1.2) predicts a rate that changes from first order to second order in the reactant concentration below some threshold concentration. The Lindemann-Christiansen mechanism also predicts that the rate depends on the concentration of inert collision partners when $k_{-1}[\mathbf{X}] \ll k_2$. If written as a pseudo-first order rate law $\mathfrak{r} = k_{eff}[\mathbf{A}]$, the pseudo-first order rate constant is

$$k_{eff} = \frac{k_2 K_{eq}}{1 + k_2/(k_{-1}[\mathbf{X}])} \tag{9.1.3}$$

Here, $K_{eq} = k_1/k_{-1}$ is the equilibrium constant for the formation of **A**$*$. k_{eff} asymptotically approaches the constant value $k_2 K_{eq}$ as [**X**] increases, i.e. in the high pressure limit. Figure 9.1.1, from Schneider and Rabinovitch [5], confirms the qualitative behavior predicted by the Lindemann-Christiansen mechanism. The results in Figure 9.1.1 have been normalized by the apparent rate constant $k_\infty \equiv k_2 K_{eq}$ in the high pressure limit.

Early investigators interpreted the activation energy in the high pressure limit as coming entirely from the energy to create **A**$*$, i.e. they assumed $K_{eq} = exp[-E_a/k_B T]$. This assumption

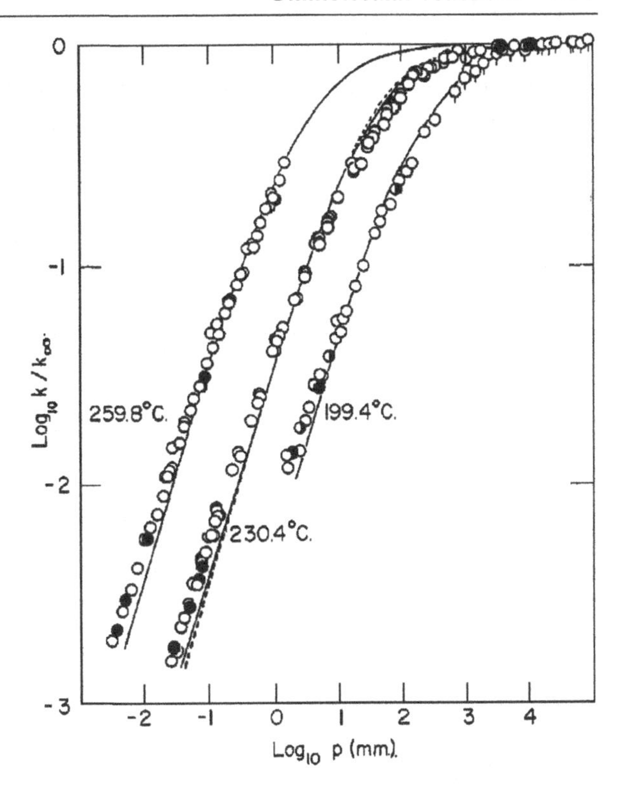

Figure 9.1.1: $log(k_{eff}/k_{\infty})$ **as a function of pressure for gas phase isomerization of pure** CH_3NC **to** CH_3CN**. Here** k_{∞} **is the observed rate constant in the limit of high pressure where the reaction appears to be first order in** CH_3NC**. The data qualitatively confirms the Lindemann-Christiansen model. The curves are actually from a more elaborate RRKM theory. [From Schneider and Rabinovitch, J. Am. Chem. Soc. 84, 4215-30 (1962).]**

allowed them to extract specific and separate values of K_{eq} and k_2 from data. Values of k_{-1} were independently estimated from collision theory. From the independent estimates of k_2 and k_{-1}, the crossover concentration at which the rate becomes 1/2 of its high pressure value should then be $k_2/k_{-1} = [\mathbf{X}]_{1/2}$. These theoretical estimates of $[\mathbf{X}]_{1/2}$ did not agree with observations [6]. In hindsight, the discrepancies surely stem from oversimplifications in the collision theory and the assumption that only K_{eq} depends on temperature.

9.2 Hinshelwood and RRK theories

In the same way that unimolecular reactions can appear to have second order rate laws at low concentrations, a bimolecular association reaction can appear to have a third order rate law at low pressures. Association reactions always leave the product molecule with enough energy to escape over the barrier by which they were created. The time for back reaction after association is given by the inverse of a microcanonical rate constant. If this time is less than the time between collisions of the excited complex with other molecules, the molecule will re-dissociate. Thus at very low concentrations, the concentration of collision partners becomes part of the rate law so that association appears to be third order.

Marcelin [7] and Rodebush [8] noted that molecules typically have several vibrating modes and proposed that a sufficient amount of their energy must be transferred to the reaction coordinate to promote the reaction. Building upon these considerations, Hinshelwood [9] developed a statistical mechanical model for the activation energy in the first step of the Lindemann mechanism in terms of the temperature and molecular size. Hinshelwood proposed that a molecule **A**$*$ needed at least a threshold E_0 of energy to enable the reaction **A**$* \rightarrow$ **P**. Hinshelwood then counted the number of ways that the total energy in a molecule could be partitioned among n vibrational modes. If each vibrational mode has the same frequency, and $\varepsilon = E/k_B T$, then the probability density to find an energy between ε and $\varepsilon + d\varepsilon$ for a molecule with n-oscillators is

$$\rho(\varepsilon)d\varepsilon = \frac{1}{(n-1)!}\varepsilon^{n-1} exp[-\varepsilon]d\varepsilon \tag{9.2.1}$$

Note that $\rho(\varepsilon)d\varepsilon$ is also $k_{-1}^{-1}k_1(\varepsilon)d\varepsilon$ which involves the differential excitation rate to energies between ε and $\varepsilon + d\varepsilon$ [10]. Hinshelwood then computed $\int_{\varepsilon_0}^{\infty} \rho(\varepsilon)d\varepsilon$ with $\varepsilon_0 = E_0/k_B T$ to obtain the fraction of molecules at equilibrium having at least the threshold activation energy. When $E_0 \gg k_B T$ the integration gives

$$f(\varepsilon_0) \approx \frac{\varepsilon_0^{n-1} exp[-\varepsilon_0]}{(n-1)!} \tag{9.2.2}$$

where terms of higher order in the small parameter $k_B T/E_0$ have been omitted. Finally, Hinshelwood gave the rate constant k_1 as the collision frequency times the fraction of molecules $f(\varepsilon_0)$ with energy beyond the threshold:

$$k_1 = \pi d_{AB}^2 \bar{v}_{AB} \frac{\varepsilon_0^{n-1} exp[-\varepsilon_0]}{(n-1)!} \tag{9.2.3}$$

Inserting Hinshelwood's k_1 expression into equation (9.1.2) gives an expression for the overall rate. Hinshelwood's analysis began an important series of efforts to derive rate parameters from statistical mechanics. In contrast to the purely phenomenological model of Lindemann and Christiansen, the Hinshelwood theory began to predict relationships between rates and molecular characteristics, e.g. size and vibrational frequencies. Additionally, the crossover concentrations from the Hinshelwood model more closely agree with experiment than those from the Lindemann-Christiansen model [10]. However, several shortcomings remained in the Hinshelwood theory:

1. Fits to experimental data required artificially small numbers of oscillator modes.
2. Plots of the pseudo-first order rate constants as $1/k_{eff}$ vs. $1/[A]$ are not always linear as predicted by equation (9.1.3).
3. k_2 is a constant in the Hinshelwood theory, but experiments suggest that k_2 decreases with increasing molecular size.

4. The factor $(E_0/k_BT)^{n-1}$ predicts a strongly temperature dependent prefactor in equation (9.2.3), but experiments do not confirm this prediction.

Rice-Ramsperger-Kassels theory

The Hinshelwood model assumed that any $\mathbf{A}*$ molecule activated beyond the energy barrier threshold would convert to products with the same rate constant k_2. Rice, Ramsperger [11], and Kassel [12–14] (RRK) proposed that molecules excited to different energy levels above the threshold would react at different rates. Rice and Ramsperger developed a classical theory while Kassel used quantized energy levels. Importantly, RRK distinguished energized molecules $\mathbf{A}*$ from activated molecules $\mathbf{A}^{\ddagger}$. According to RRK, activated molecules are special energized molecules whose energy is become sufficiently concentrated into the reaction coordinate to cross over the barrier.

$$\begin{gathered}\mathbf{X}+\mathbf{A} \underset{k_{-1}}{\overset{k_1}{\rightleftarrows}} \mathbf{X}+\mathbf{A}* \\ \mathbf{A}* \xrightarrow{k_{RRK}} \mathbf{A}^{\ddagger} \longrightarrow \mathbf{P}\end{gathered} \tag{9.2.4}$$

The process $\mathbf{A}* \rightarrow \mathbf{A}^{\ddagger}$ has a finite rate because energy transfer from one mode to another may take many vibrational periods. Additionally, RRK explained how k_{RRK} depends on the energy of $\mathbf{A}*$. When $\mathbf{A}*$ only has slightly more energy than the threshold, nearly all of its vibrational energy must be concentrated into the reaction coordinate as depicted in Figure 9.2.1. This unlikely partitioning of the energy quanta poses a substantial entropic barrier.

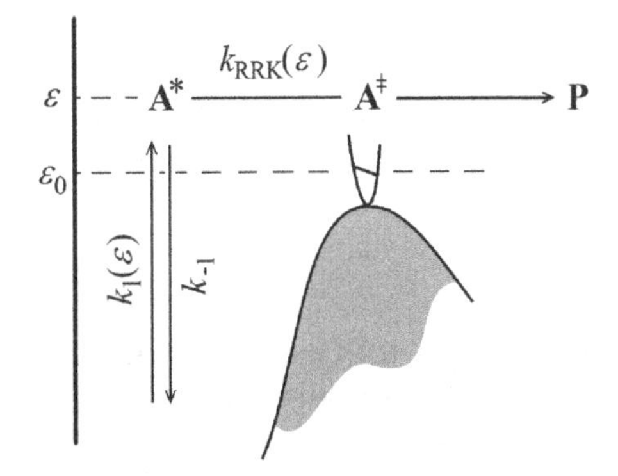

Figure 9.2.1: RRK theory differs from the Lindemann-Christiansen and Hinshelwood theories by accounting for the internal microcanonical energy transfer time between an energizing collision and concentration of that energy into the reaction coordinate.

RRK assumed (like Hinshelwood) that all vibrations have the same frequency. They also assumed that the vibrational energy quanta are randomly re-distributed among the modes after each vibrational period. If the collision process creates an energized molecule $\mathbf{A}*$ with j quanta of energy, $E = j\hbar\omega$, then the number of ways the quanta can be partitioned among the n modes (see exercises) is

$$\frac{(j+n-1)!}{j!(n-1)!} \tag{9.2.5}$$

Then RRK derived the number of ways to partition j energy quanta among n modes subject to a requirement that the special reaction coordinate mode gets at least j_0 quanta:

$$\frac{(j-j_0+n-1)!}{(j-j_0)!(n-1)!} \tag{9.2.6}$$

The ratio of expressions (9.2.6) and (9.2.5) gives the probability that at least j_0 energy quanta will be concentrated into the reaction coordinate mode when excited by j energy quanta:

$$p(j, n, \geq j_0) = \frac{(j-j_0+n-1)!j!}{(j-j_0)!(j+n-1)!}$$

Stirling's approximation applied to the case $j - j_0 \gg n-1$ gives $p(j, n, \geq j_0) \approx [(j-j_0)/j]^{n-1}$. Multiplying the numerator and denominator by $(n-1)$ factors of the dimensionless energy quanta $\hbar\omega/k_B T$ gives

$$p(j, n, \geq j_0) \approx \left(\frac{\varepsilon - \varepsilon_0}{\varepsilon}\right)^{n-1} \tag{9.2.7}$$

If the energy is randomly repartitioned after each oscillation, the rate constant k_{RRK} is

$$k_{RRK} = k_{\star\to\ddagger}\left(\frac{\varepsilon - \varepsilon_0}{\varepsilon}\right)^{n-1} \tag{9.2.8}$$

where $k_{\star\to\ddagger}$ is an empirical parameter. As shown in Figure 9.2.2, the RRK theory predicts the effects of energy and molecular size on the rate of the unimolecular activation step.

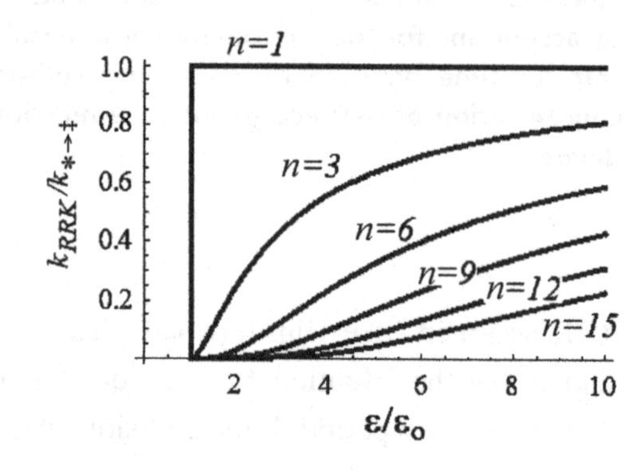

Figure 9.2.2: RRK theory predicts that the rate of reaction after excitation to energy ε beyond ε_0 depends on the relative energy $\varepsilon/\varepsilon_0$ and the number of vibrational modes n. Internal energy transfer to the reaction coordinate takes longer for larger molecules.

As in the Hinshelwood theory, the overall rate requires integration over the probability of excitations to different energy levels, but the integration in RRK theory must also account for the energy-dependence in $k_{RRK}(\varepsilon)$. In the Lindemann theory the probability of reaction after excitation was independent of ε. In the RRK theory, the probability to react after excitation is

$$k_{RRK}(\varepsilon)/(k_{RRK}(\varepsilon)+k_{-1}[\mathbf{X}])$$

Using the same arguments as Lindemann, Christiansen, and Hinshelwood, but now integrating over the rates at different energy levels,

$$\mathcal{R}=\int_{\varepsilon_0}^{\infty}\frac{k_1(\varepsilon)k_{RRK}(\varepsilon)[\mathbf{A}][\mathbf{X}]}{k_{-1}[\mathbf{X}]+k_{RRK}(\varepsilon)}d\varepsilon$$

Divide the top and bottom by $k_{-1}[\mathbf{X}]$ and recall that $\rho(\varepsilon)d\varepsilon=k_{-1}^{-1}k_1(\varepsilon)d\varepsilon$. Insert the harmonic oscillator model for $\rho(\varepsilon)d\varepsilon$ (equation (9.2.1)) and the microcanonical energy transfer rate constant $k_{RRK}(\varepsilon)$ (equation (9.2.8)). Finally, simplify the integral and divide by [**A**] to obtain

$$k_{eff}=\frac{k_{\star\to\ddagger}exp[-\varepsilon_0]}{(n-1)!}\int_0^{\infty}\frac{u^{n-1}e^{-u}du}{1+(k_{\star\to\ddagger}/k_{-1}[\mathbf{X}])(u/(u+b))^{n-1}} \tag{9.2.9}$$

where $b=\varepsilon_0/k_BT$ and where the dummy variable u came from $(\varepsilon-\varepsilon_0)/k_BT$.

When $k_{\star\to\ddagger}/k_{-1}$, ε_0/k_BT, and n are regarded as adjustable parameters and when $k_{\star\to\ddagger}$ is effectively removed from the analysis by plotting $k_{eff}/lim_{[\mathbf{X}]\to\infty}k_{eff}$, the RRK theory closely fits experimental rate data. Moreover, an Arrhenius expression is recovered from the RRK theory under a high pressure of collision partners, i.e.

$$\lim_{[\mathbf{X}]\to\infty}k_{eff}=k_{\star\to\ddagger}exp[-E_0/k_BT]$$

For directly predicting rates, n is the number of vibrational modes and $\varepsilon_0=E_0/k_BT$ can (in principle) be obtained from *ab initio* calculations. However, there is no definitive procedure for computing the one common vibrational frequency [15,16]. Later extensions by Marcus lead to a more powerful theory for predicting absolute rates.

9.3 RRKM theory

Marcus and Rice [17,18] reformulated the RRK equations to explicitly count vibrational states for the reactant and transition state by using their actual vibrational frequencies. They also properly counted the quantum states for translation along the reaction coordinate to eliminate the adjustable parameter $k_{\star\to\ddagger}$ from the RRK theory. The resulting RRKM theory is more complicated than the RRK theory, but it is truly predictive because all of its input parameters are readily obtained from *ab initio* calculations.

Again in RRKM theory, the threshold activation energy E_0 is the energy at the saddle point on the potential energy surface with the additional zero point vibrational energy of all bound vibrational modes. The excess energy $E - E_0$ could be entirely focused into translational motion along the reaction coordinate, but the excess energy is more typically partitioned between the reaction coordinate and other bound degrees of freedom. It is convenient to write the following energy balance equation:

$$E_{\cap} + E_{\cup} = E - E_0 \tag{9.3.1}$$

where $E_{\cap}$ is translational energy in the reaction coordinate at the transition state and $E_{\cup}$ is the energy in bound vibrations (bath modes) at the transition state. The right hand side is the total energy beyond the reaction threshold (see Figure 9.3.1). At the transition state, the value of $E_{\cap}$ identifies the momentum along the reaction coordinate,

$$p = (2mE_{\cap})^{1/2} \tag{9.3.2}$$

where m is the effective mass of the reaction coordinate (q). In practice vibrational modes are identified in mass weighted coordinates where all masses are unity, but regardless the effective mass m will vanish from the final RRKM result.

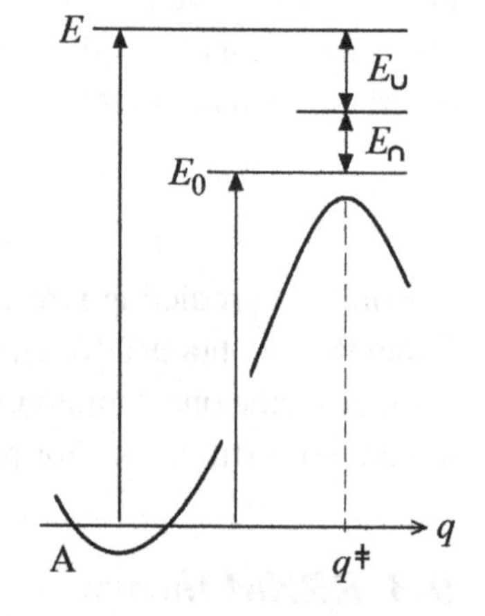

Figure 9.3.1: Relationships between energy levels in the RRKM theory. E_0 is the zero point energy relative to the zero point energy of the reactants.

For each energy level $E_{\cap}$, the rate at which systems cross the transition state is the product of $v = p/m$ and the density of states per length along q. For a one-dimensional system, the number of states per differential phase space volume element is [19] $dq\,dp/h$, so the number of states per unit length along the reaction coordinate is dp/h. Therefore, multiplying v and

dp/h gives the differential contribution to the rate (in one dimension) for each differential range of momenta

$$d\mathcal{R} = v\,dp/h$$

From the relationships between $E_\cap$, p, and v, we have $dE_\cap = v\,dp$. Therefore [20],

$$d\mathcal{R} = dE_\cap/h \tag{9.3.3}$$

Pause to think about this weird and wonderful result. Regardless of $E_\cup$ or other characteristics of the energized molecule, the differential contribution to the rate is $1/h$ per differential unit of translational energy in the reaction coordinate.

If the reaction coordinate does not exchange energy with the bath modes during the barrier crossing (i.e. if the motions are locally separable), then $E_\cup$ will be constant during the barrier crossing.[1] From $E = E_\cap + E_\cup + E_0$ with constant $E_\cup$ we have $dE_\cap = dE$. Thus *for each bath state in a multidimensional system*, the energy interval from E to $E + dE$ must contribute a differential amount dE/h to the rate. We can further simplify the counting of bath states by assuming that the barrier crossings are vibrationally adiabatic [21], i.e. that in addition to a constant $E_\cup$ the bath modes do not exchange energy with each other during the barrier crossing process. Now the total contribution to the rate from energies between E and $E + dE$ is

$$d\mathcal{R} = N^{\ddagger}(E - E_0)dE/h \tag{9.3.4}$$

where $N^{\ddagger}(E - E_0)$ is the number of bath states at the dividing surface with a total internal energy in the range $0 < E_\cup < (E - E_0)$. Equation (9.3.4) gives the contributions of all states with energies between E and $E + dE$ to the rate. We count from E_0 because no state on the dividing surface can have $E < E_0$. Also note that the *absolute* translational energy $E_\cap$ in the reaction coordinate does not matter. Differential contributions to the rate from energy in the reaction coordinate have already been counted in the dE/h term.

Now to obtain a rate constant in the microcanonical ensemble *at energy* E, divide $d\mathcal{R}$ by $\rho(E)dE = N(E + dE) - N(E)$, i.e. by the number of reactant states in the same energy range. The result is

$$k_{RRKM}(E) = \frac{N^{\ddagger}(E - E_0)}{h\rho(E)}. \tag{9.3.5}$$

$N^{\ddagger}(E - E_0)$ and $\rho(E)$ can be computed by assuming harmonic vibrational modes at the reactant minimum and at the saddle point. The reaction coordinate mode corresponds to the

[1] Dynamical separability in the neighborhood of the transition state is tantamount to the TST assumption of no recrossing. To reverse the direction of motion along the reaction coordinate while still near the barrier top, kinetic energy must at least temporarily be transferred from the reaction coordinate to the bath modes.

imaginary vibrational frequency at the saddle point. Only the real frequencies are considered in counting $N^{\ddagger}(E-E_0)$. Modes with real frequencies at the saddle point have discrete energy levels that lead to stepwise increases in $N^{\ddagger}(E-E_0)$ as a function of energy. As shown in Figure 9.3.2, some experiments have directly observed the predicted stepwise increases in $k_{RRKM}(E)$ [22].

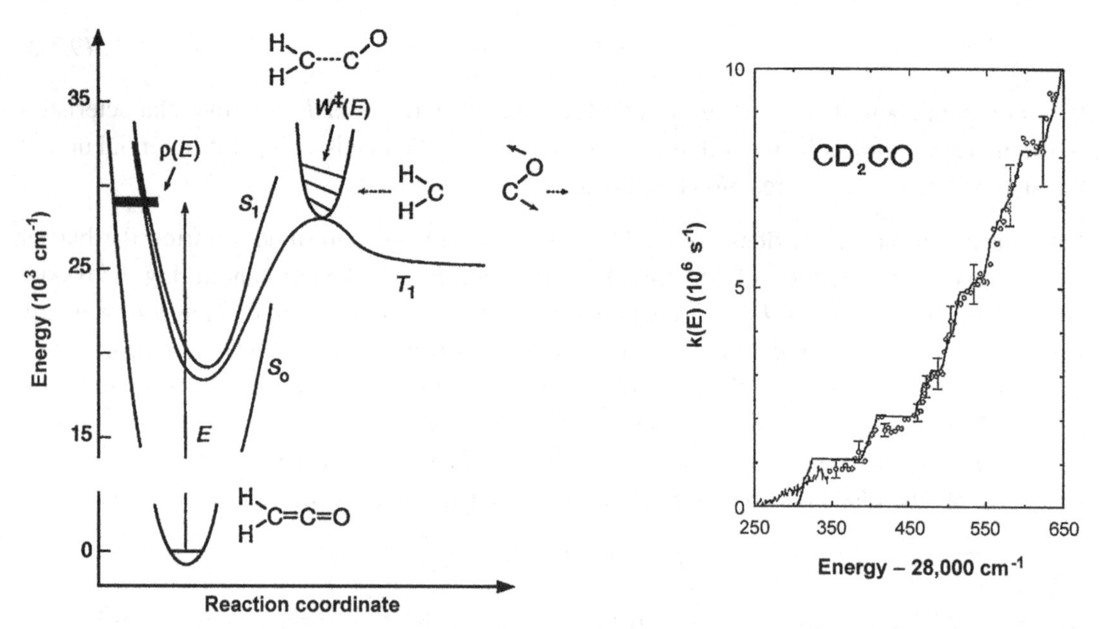

Figure 9.3.2: (Left) Dissociation mechanism of triplet ketene. Energy surfaces represent the singlet ground state (S_0), the first excited singlet (S_1), and first triplet (T_1) electronic states as functions of the $C-C$ distance. [From Lovejoy et al. *Science* 256, 1541-44 (1992).] (Right) Deuterated ketene was photochemically excited from S_0 to energies above 28000cm^{-1} where the dissociation rate is measured and compared to predictions from RRKM theory. [From Kim et al. *J. Chem. Phys.* 102, 3202-19 (1995).]

The (modified) Beyer-Swinehart algorithm can be used to directly count the energy levels for small molecules [23]. Explicit state counting becomes infeasible for large molecules, so approximate state counting schemes by Marcus and Rice [17] and by Whitten and Rabinovitch [24] are needed. In classical statistical mechanics the number of states with a total energy less than or equal to energy E is

$$N(E)=\frac{1}{h^n}\int_{H(\mathbf{q},\mathbf{p})\leq E} dq_1\cdots dq_n dp_1\cdots dp_n \tag{9.3.6}$$

Approximations to $N^{\ddagger}(E-E_0)$, $N(E)$ and $\rho(E)dE$ can be developed by treating all of the modes as uncoupled harmonic oscillators. In mass weighted coordinates where all masses are

$m_0 = 1amu$, the Hamiltonian is

$$H(\mathbf{q}, \mathbf{p}) = \frac{1}{2} \sum_k \left(p_k^2 / m_0 + m_0 \omega_k^2 q_k^2 \right). \tag{9.3.7}$$

Because the oscillators are uncoupled, the energy in each mode can be separately identified. If E_k is the energy in the kth mode, then the oscillator traverses an ellipse with equation

$$\frac{p_k^2}{2m_0 E_k} + \frac{m_0 \omega_k^2 q_k^2}{2E_k} = 1 \tag{9.3.8}$$

For the kth oscillator, the number of states up to energy E_k is the area of the ellipse divided by h, i.e. $(\pi/h)(2m_0 E_k)^{1/2}(2E_k/m_0\omega_k^2)^{1/2}$ which simplifies to $E_k/(\hbar\omega_k)$. Now consider n oscillators with different frequencies and total energy E. $N(E)$ can be determined from the volume of an n-dimensional wedge with a boundary defined by $E_1 + E_2 + \cdots + E_n \leq E$. Figure 9.3.3 shows the wedge that defines the number of states $N(E)$ for three oscillators with different frequencies.

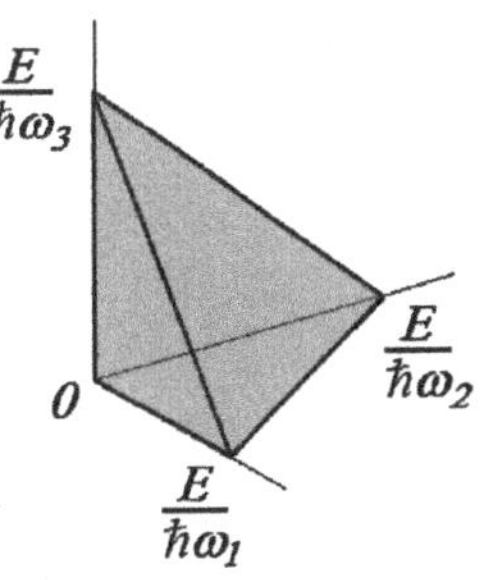

Figure 9.3.3: The number of states $N(E)$ for three oscillators with different frequencies depicted as a volume.

The hypervolume of the wedge is

$$N(E) = \frac{E^n}{n! \prod_{k=1}^n \sum \hbar\omega_k} \tag{9.3.9}$$

To improve the accuracy of the estimate, the ground state and its zero point energy can be included.

$$N(E) = 1 + \frac{(E - E_{zp})^n}{n! \prod_{k=1}^n \sum \hbar\omega_k} \tag{9.3.10}$$

with $E_{zp} = \sum_k \hbar\omega_k/2$. The density of states $\rho(E)$ can then be obtained from $\rho(E)dE = N(E + dE) - N(E)$.

$$\rho(E) = \frac{(E - E_{zp})^{n-1}}{(n-1)! \prod_{k=1}^n \hbar\omega_k} \tag{9.3.11}$$

Equations (9.3.9) or (9.3.10) can be directly used to count internal vibrational states at the dividing surface.

Using these (approximate) expressions in the RRKM rate constant and remembering that the sum over transition states involves one fewer vibrational mode than the sum for reactant states gives

$$k_{RRKM}(E) \approx \frac{\omega_n}{2\pi}\left(\prod_{k=1}^{n-1}\frac{\omega_k}{\omega_{\ddagger,k}}\right)\left(\frac{E-E_0}{E}\right)^{n-1} \tag{9.3.12}$$

Equation (9.3.12) replaces the discrete jumps in $k_{RRKM}(E)$ with a smooth curve. Although the discrete jumps have actually been observed in some experiments, the more easily computed continuous approximation in equation (9.3.12) is adequate for many applications.

An important prediction is that *the RRKM rate constant is a decreasing power law function in the number of vibrational modes*. k_{RRKM} also depends on the distribution of frequencies as seen in equation (9.3.10), but for certain reactions we can change n while only minimally changing the distribution of frequencies. Figure 9.3.4 shows a validation of the power-law dependence of k_{RRKM} on n.

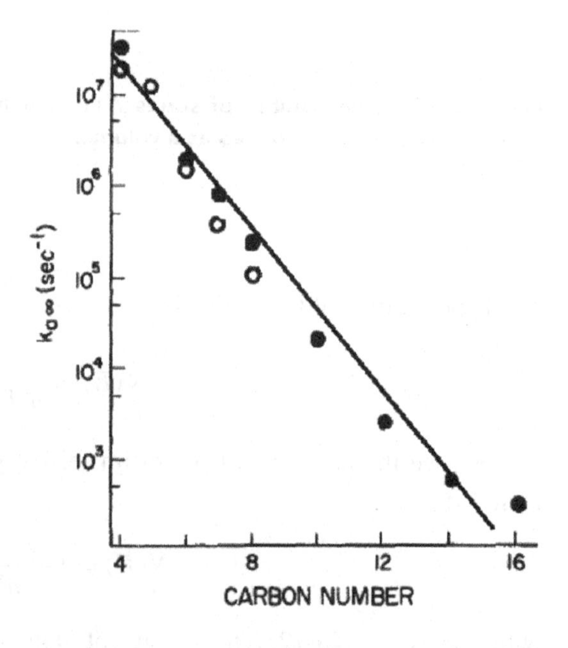

Figure 9.3.4: Experiments of Rabinovitch et al. and Pearson et al. for the dissociation rates of linear alkanes with a reactive radical head group. Conditions of the experiments are similar except for carbon chain length which is linearly related to the number of modes n. Compare these results to the RRKM rate constant expression (9.3.12). [Figure from R. Levine, Molecular Reaction Dynamics (2006).]

This RRKM derivation has separated the detailed characteristics of states in the reactant well from the detailed characteristics of the transition states by assuming that every internal state at the dividing surface is equally likely, but we have not ensured a conservation of angular momentum throughout the transition. Several versions of RRKM theory account for the rotational states [20,25–27]. The central result is that the rate constant depends not only on the energy, but also on the angular momentum J.

$$k_{RRKM}(E, J) = \frac{N^{\ddagger}(E - E_0, J)}{h\rho(E, J)} \tag{9.3.13}$$

The vibrational density of states calculation can still be used with non-zero angular momentum if the energy associated with rotation is removed in the vibrational state counting

$$k_{RRKM}(E, J) = \frac{N^{\ddagger}(E - E_0 - E^{\ddagger}_{rot})}{h\rho(E - E_{rot})} \tag{9.3.14}$$

For dissociation reactions, a simple model for the rotational energy is

$$E_{rot} = J(J+1)B_e(r_e/r)^2$$

where J is the angular momentum quantum number, $B_e = \hbar^2/(2\mu r_e^2)$, μ is the reduced mass, r is the reaction coordinate (the distance between the two dissociating fragments), and r_e is the equilibrium distance. The transition state stretches to longer values of $r^{\ddagger}$ which tends to lower the effective dissociation barrier. Forst has outlined strategies for including rotational effects for additional types of reactions [26].

Applications of RRKM theory go far beyond thermally activated unimolecular reactions [6,28, 29].

1. RRKM applies to any situation where a molecule has been activated to an energy where it can dissociate or isomerize at constant energy. Thus RRKM theory is often used to understand photochemical reactions [30,31].
2. For recombination (association) reactions, RRKM theory can predict whether the newly formed product complex survives long enough to be quenched by collisions with other gas molecules [28,32].
3. RRKM theory has been extensively used to understand the overall rates and selectivities of reactions that proceed via transient association complexes [33].

Predictions from RRKM theory have been verified by many experiments [6,10], but there have also been some failures [10,26,28]. The assumption of rapid intramolecular vibrational energy redistribution is often responsible for errors. When the vibrational energy redistribution is not sufficiently rapid, the kinetics behave like those of a molecule with fewer vibrational modes [34]. Extensions of RRKM theory include (i) models that separate modes that are active and inactive for energy exchange with the reaction coordinate [34], (ii) random matrix models of intramolecular energy transfer [35], and (iii) models that include the effects of tunneling [36]. See the excellent texts by Forst [26] and by Robinson and Holbrook [6] for a wealth of additional theoretical and experimental results on unimolecular reaction rate theory.

9.4 Transition state theory from RRKM theory

RRKM theory predicts the rate of unimolecular isomerization, dissociation, or decomposition by microcanonical dynamics at constant energy: $k_{RRKM}(E) = h^{-1}N^{\ddagger}(E - E_0)/\rho(E)$. It is essentially a microcanonical version of transition state theory (TST). Technically, $k_{RRKM}(E)$ only accounts for the rate of product generation from a pre-energed $\mathbf{A}*$ molecule, i.e.

$$\mathbf{A}* \overset{k_{RRKM}}{\longrightarrow} \mathbf{A}^{\ddagger} \longrightarrow \mathbf{P} \tag{9.4.1}$$

Like we did for RRK theory, the overall unimolecular rate according to RRKM theory can be obtained by integrating the rate at which collisions create $\mathbf{A}*$ at different energy levels and the probability that they become products before deactivation. At sufficiently high pressure, frequent collisions will maintain an approximate thermal equilibrium population of the $\mathbf{A}*$ species. Here we confirm that a thermal average of the RRKM rate reduces to the canonical TST rate.

At equilibrium, the probability of finding the reactants between energy E and $E + dE$ is $\rho(E)e^{-\beta E}dE$. Therefore, the thermally averaged rate constant is

$$k(T) = Q_R^{-1} \int_0^{\infty} k_{RRKM}(E)\rho(E)e^{-\beta E}dE \tag{9.4.2}$$

where $Q_R = \int \rho(E)exp[-\beta E]dE$ is the partition sum over *reactant* energy levels and where $E = 0$ corresponds to the reactants in their ground state. Inserting equation (9.3.5) for $k_{RRKM}(E)$ gives

$$k(T) = \frac{1}{hQ_R} \int_0^{\infty} N^{\ddagger}(E - E_0)e^{-\beta E}dE \tag{9.4.3}$$

Integration by parts yields

$$k(T) = -\frac{k_BT}{hQ_R}e^{-\beta E}N^{\ddagger}(E - E_0)\Big|_0^{\infty} + \frac{k_BT}{hQ_R}\int_0^{\infty} e^{-\beta E}\frac{d}{dE}N^{\ddagger}(E - E_0)dE \tag{9.4.4}$$

The boundary term vanishes because of the limiting behavior of $exp[-E/k_BT]$ as $E \to \infty$ and because $N^{\ddagger}(-E_0)$ is zero, i.e. because $E = 0$ is below the threshold energy where there are no classically allowed states. Our analysis of $N(E - E_0)$ in the previous section confirms that the exponential decay with energy is sufficiently fast to make the boundary term vanish in the limit as $E \to \infty$. Changing variables to $E' = E - E_0$ and using $dE' = dE$ gives

$$k(T) \quad = \quad \frac{k_BT}{hQ_R}e^{-\beta E_0}\int_0^{\infty} e^{-\beta E'}\rho_{\ddagger}(E')dE'$$

$$= \frac{k_B T}{h} \frac{Q_\ddagger}{Q_R} e^{-\beta E_0} \tag{9.4.5}$$

which is the canonical TST rate constant. Here we have used $\rho_\ddagger(E)dE$ for the number of transition states with energy between E and $E + dE$ above the threshold energy. $Q^\ddagger$ is a partition function over all transition states with their energy measured relative to the threshold E_0. If we properly account for different rotational energy levels, then the thermal rate constant is [37]

$$k(T) = Q_R^{-1} \sum_J (2J+1) \int_0^\infty k(E,\ J)\rho(E,\ J)e^{-\beta E} dE \tag{9.4.6}$$

Note that RRKM theory, like canonical TST, says that the forward rate constant from **A** to products has nothing to do with the concentration of products. According to RRKM theory, the forward reaction occurs in the same way and at the same rate whether the overall reaction mixture is at equilibrium or far from equilibrium. This assumption is justified on the grounds that the reactants become activated and pass through the transition state region independently of trajectories that might be moving in the opposite direction.

Exercises

1. Derive equations (9.2.5) and (9.2.6). Hint: consider the number of distinguishable sequences

   $$\bullet\bullet \mid \bullet\bullet\bullet\bullet \mid \bullet \mid \bullet \ldots \bullet \mid \bullet\bullet\bullet\bullet \mid \bullet\bullet$$

   where the dots represent energy quanta and the vertical dashes separate different vibrational modes.
2. Derive equation (9.2.1).
3. Show that k_{RRKM} should exhibit steps of size $1/h\rho(E)$ because of the transition state quantization. What phenomenon can cause rounding of the discrete steps in k_{RRKM}?
4. Show that $k_{RRKM}(E)$ obeys detailed balance in the microcanonical ensemble.
5. Equation (9.4.5) shows that a thermal equilibrium average of the RRKM rate constant gives the TST rate. Show that the Lindemann-Christiansen rate law with k_{RRKM} substituted for k_2 is consistent with a thermal equilibrium distribution of **A*** at high pressure.
6. [Adapted from Henriksen and Hansen [38].] The isomerization reaction $HCN \rightarrow HNC$ has a saddle point energy that is 1.51eV above the minimum of energy in the reactant state. The vibrational frequencies of HCN are 3300cm^{-1}, 2100cm^{-1}, 713cm^{-1}, and 713cm^{-1}. The (real) vibrational frequencies at the transition state are 3000cm^{-1} and 2000cm^{-1}. Compute $k_{RRKM}(E)$ starting from the appropriate value of E_0.

7. [Adapted from Forst [26].] HeI_2 is a T-shaped complex with the $He \cdots I_2$ bond being orthogonal to the $I - I$ bond. The complex vibrates with frequencies $\omega_{I-I} = 128\text{cm}^{-1}$ and $\omega_{He-I} = 26\text{cm}^{-1}$. The complex dissociates to give He and I_2 via a small dissociation barrier of 18cm^{-1}. The $I - I$ bond is excited by injection of a 1280cm^{-1} photon. Compute the RRKM theory rate and see Gray et al. [39] for the experimental rate. Explain the unusually large discrepancy. In your answer, discuss the relative vibrational frequencies and the extent of coupling between the $I - I$ and $He \cdots I_2$ modes.
8. Many unimolecular reactions that break strong bonds have unusually large prefactors. For example, thermal homolysis of ethane ($C_2H_6 \rightarrow 2CH_3$) has an Arrhenius prefactor of $A = 10^{17.4}s^{-1}$. Explain this observation.
9. Ozone can be formed by the reaction $O + O_2 + X \rightarrow O_3 + X$ where X is a collision partner that quenches the nascent O_3 molecule. Estimate $k_{RRKM}(E)$ for a "hot" O_3 molecule in rotational state $J = 0$.
   (a) Obtain the required energies, frequencies, and geometries from the literature or from electronic structure calculations.
   (b) Compare the results from exact counting of vibrational energy levels at the transition state to the approximation in equation (9.3.10).
   (c) Convert the $k_{RRKM}(E)$ to an estimated lifetime of the "hot" O_3 molecule. Plot lifetime as a function of E and add to your figure the typical time between collisions between O_3 and molecules in the air. For what energies could one use transition state theory?
10. Create an RRKM theory for a system with $J = 0$ and with all harmonic oscillators replaced by square well potentials.

References

[1] H.S. Harned, J. Franklin Inst. 196 (1923) 181–202.
[2] M.C. King, K.J. Laidler, Arch. Hist. Exact Sci. 30 (1984) 45–86.
[3] F.A. Lindemann, Trans. Faraday Soc. 17 (1922) 598.
[4] J.A. Christiansen, Z. Phys. Chem. 103 (1922) 91–98.
[5] F.W. Schneider, B.S. Rabinovitch, J. Am. Chem. Soc. 84 (1962) 4215–4230.
[6] P.J. Robinson, K.A. Holbrook, Unimolecular Reactions, Wiley, New York, 1972.
[7] R. Marcelin, Ann. Phys. 3 (1915) 120.
[8] W.H. Rodebush, J. Am. Chem. Soc. 45 (1923) 606–614.
[9] C.N. Hinshelwood, Proc. R. Soc. A 113 (1927) 230.
[10] J.I. Steinfeld, J.S. Francisco, W.L. Hase, Chemical Kinetics and Dynamics, Prentice Hall, Englewood Cliffs, NJ, 1989.
[11] O.K. Rice, H.C. Ramsperger, J. Am. Chem. Soc. 49 (1927) 1617.
[12] L.S. Kassel, Kinetics of Homogeneous Gas Reactions, Reinhold, New York, 1932.
[13] L.S. Kassel, J. Phys. Chem. 32 (1928) 1065.
[14] L.S. Kassel, J. Phys. Chem. 32 (1928) 225.
[15] E. Thiele, I. Stone, M.F. Goodman, Chem. Phys. Lett. 76 (1980) 579.

[16] J.R. Barker, J. Chem. Phys. 72 (1980) 3686.
[17] R.A. Marcus, O.K. Rice, J. Phys. Colloid Chem. 55 (1951) 894–908.
[18] R.A. Marcus, J. Chem. Phys. 20 (1952) 359.
[19] R.D. Levine, Molecular Reaction Dynamics, Cambridge University Press, Cambridge, UK, 2005.
[20] R.A. Marcus, J. Chem. Phys. 43 (1965) 2658.
[21] R.A. Marcus, J. Chem. Phys. 45 (1966) 4493–4499.
[22] E.R. Lovejoy, S.K. Kim, C.B. Moore, Science 256 (1992) 1541–1544.
[23] D.C. Astholz, J. Troe, W. Wieters, J. Chem. Phys. 70 (1979) 5107.
[24] G.Z. Whitten, B.S. Rabinovitch, J. Chem. Phys. 41 (1964) 1883.
[25] G.M. Wieder, R.A. Marcus, J. Chem. Phys. 37 (1962) 1835.
[26] W. Forst, Unimolecular Reactions, Cambridge University Press, Cambridge, UK, 2003.
[27] D.M. Wardlaw, R.A. Marcus, Adv. Chem. Phys. 231 (1988) 70.
[28] J. Troe, J. Phys. Chem. 83 (1979) 114–126.
[29] W.L. Hase, Acc. Chem. Res. 31 (1998) 659–665.
[30] A.H. Zewail, J. Phys. Chem. 100 (1996) 12701–12724.
[31] E.W. Diau, J.L. Herek, Z.H. Kim, A. Zewail, Science 279 (1998) 847–851.
[32] J. Troe, Annu. Rev. Phys. Chem. 29 (1978) 223–250.
[33] W.L. Hase, Science 266 (1994) 998–1002.
[34] D.L. Bunker, W.L. Hase, J. Chem. Phys. 59 (1973) 4621.
[35] W.H. Miller, R. Hernandez, C.B. Moore, W.F. Polik, J. Chem. Phys. 93 (1990) 5657.
[36] W.H. Miller, J. Am. Chem. Soc. 101 (1979) 6810.
[37] G.C. Schatz, M.A. Ratner, Quantum Mechanics in Chemistry, Dover, Mineola, NY, 2002.
[38] N.E. Henriksen, F.Y. Hansen, Theories of Molecular Reaction Dynamics, Oxford Press, 2008.
[39] S.K. Gray, S.A. Rice, M.J. Davis, J. Chem. Phys. 90 (1986) 3470.

CHAPTER 10

Transition state theory

> *... for a molecule to leave this region, we require ... for it to reach a certain region of state space, for its speed to exceed a certain limit, for its internal structure to correspond to an unstable configuration, etc.; in other words, it will have to cross a certain surface S_1, in state space, which we will call the critical surface.*
>
> ***Marcelin, (tr. D.C. Strub) Ann. Phys. (1915)***

Transition state theory (TST) is the cornerstone of chemical reaction kinetics. TST does have limitations, but overall it is remarkably successful at predicting and interpreting experimental rates. The successes are all the more remarkable when one considers its simplicity. As Wigner and Eyring noted [1], TST requires only minimal detail about the potential energy surface (PES) in the immediate vicinity of transition states and reactants. Theories that go beyond TST to account for recrossing, tunneling, and non-adiabatic effects require a far more detailed description of the PES, and in many cases they only marginally improve upon the accuracy of a simple TST rate calculation.

The route to TST began with Arrhenius' 1889 work on the kinetics of sugar isomerizations [2]. Arrhenius reported rate constants with a $k = \upsilon_0 exp[-E_a/k_BT]$ temperature dependence and postulated the importance of activated molecules to explain the temperature dependence. Following Arrhenius, several investigators made important contributions.

- W.C.M. Lewis (1918) suggested the rate is a product of the collision frequency and the fraction of activated molecules expressed as a Boltzmann factor [3].
- Rodebush (1923) recognized that internal vibrations (in addition to collisions) can provide the energy to activate molecules [4].
- Eyring and Polanyi (1931) computed PESs for reactions like $H_2 + H$ that focused on the saddle region corresponding to activated complexes [5].
- Pelzer and Wigner (1932) computed a TST rate constant for parahydrogen decay via the $H_2 + H$ atom-exchange reaction [6].
- La Mer (1933) suggested a rate equation with separate activation enthalpy and activation entropy parameters [7].
- Evans and Polanyi (1938) introduced volumes of activation to explain pressure effects [8].

Eyring [9] and Laidler [10] have given more extensive summaries of the contributions to TST. The study by Pelzer and Wigner is particularly remarkable. They depict nuclei moving on the Born-Oppenheimer surface, they invoke a thermal equilibrium population at the saddle point, and they estimate the equilibrium one-directional flux crossing a dividing surface. They even note that their calculation is approximate because it ignores recrossing [6]. Pelzer and Wigner are not widely credited for developing TST because their paper narrowly focuses on specific details and procedures for the $H_2 + H$ reaction.

Eyring's formulation relates molecular properties of the transition state [11] to partition functions and thermodynamic activation parameters [12]. The easily used formalism of Eyring was applied to many different reactions. Furthermore, it enabled the *interpretation* of experimental rates in terms of molecular characteristics of the activated complex. TST gained some early support and also some fierce criticism [10]. Transition states are intrinsically ephemeral species, so some opponents argued that they are ill-defined species – not properly described by equilibrium statistical mechanics. Concerns over the equilibrium assumption in TST have merit, but some of the residual objections that persist today are less justified. For example, some say that TST cannot describe processes "where dynamics are important", "where transition states are fluxional", or "where an ensemble of transition states" is involved. For each of these concerns, there are counterexamples where TST provides an excellent description. Taking these concerns one by one:

1. Dynamics are important in all reaction rate theories, TST included. The most common "harmonic" version of TST *appears* to be a static description because velocities do not explicitly appear in the expression. Velocities are in fact critical to the derivation, but they do not explicitly appear because they have been thermally averaged [11]. Some works claim that dynamical effects can "promote" or "drive" reactions in ways that cannot be described by TST [13]. In fact, all deviations from the direct dynamics of TST lead to rates which are *slower* than TST rate estimates.
2. All formulations of TST invoke an ensemble of transition states, not a single transition state configuration. The saddle point is often colloquially called "the transition state", but this terminology should not be taken literally. Even in harmonic TST where saddle points are of central importance, the saddle point is merely a special member of the transition state ensemble. In some cases, the dividing surface can be constructed with no reference to saddle points at all [14–16].
3. Fluxional transition states are not inconsistent with TST. Reactants and/or transition states that convert between forms, symmetric or otherwise, are the norm for reactions in polar solvents and enzymes. In many cases, these reactions are well-described by TST [17], albeit not by harmonic TST. Also note that the route from reactant(s) to the transition state is largely irrelevant. It can be direct or arbitrarily circuitous, involving one degree of freedom or many, as long as the basic assumptions of TST are valid.

This chapter aims to introduce the theory, to clarify that seemingly different TST expressions emerge from one common foundation, and to provide some illustrative examples of TST calculations.

10.1 Foundations

Whether a reaction continues to occur once the reactants and products reach equilibrium depends on how we define reaction events. One might argue that a reaction is the net conversion of reactants to products to establish chemical equilibrium. From this perspective, the reaction should stop at equilibrium. On the other hand, one can microscopically define a reaction as the breaking of old bonds and creation of new bonds. From this perspective the reaction continues at equilibrium, but with forward and reverse rates that exactly balance. TST estimates the forward (and reverse) rate at conditions where the reactants, products, and transition states are at equilibrium. The key assumptions *in the derivation of* TST are [18,19]:

1. Classical dynamics on a single Born-Oppenheimer PES.
2. A dividing surface that separates the reactants and products.
3. States on the dividing surface are populated as though at equilibrium with the reactants.
4. Trajectories that cross the dividing surface lead directly to products with no recrossing events.

These assumptions are illustrated in Figure 10.1.1. The equilibrium assumption is a particularly unique and interesting aspect of TST. Even though TST rate expressions are *derived* at equilibrium, they are regularly applied to reactions that are far from equilibrium. Additionally, even for perfectly equilibrated conditions, the TST rate is an approximation because of its dynamical assumptions. Dividing surfaces with absolutely no recrossing are difficult (or even impossible [20]) to construct.

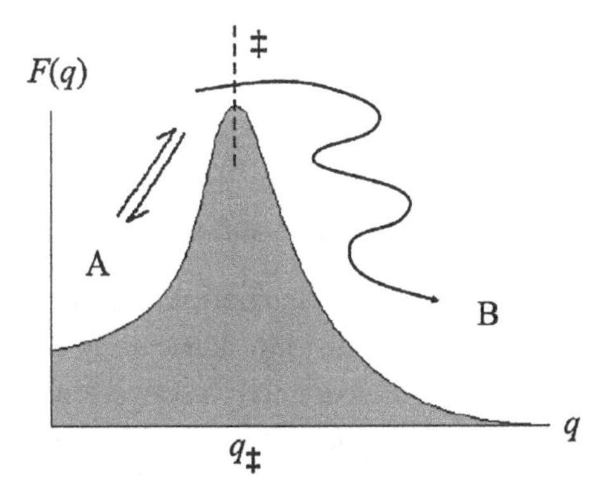

Figure 10.1.1: Assumptions in the derivation of transition state theory (TST) include a dividing surface between reactants and products, equilibrium between reactants and transition states, and no-recrossing of the dividing surface. TST also assumes classical adiabatic dynamics and a clear separation between the reaction time scale and all molecular relaxation time scales.

Chandler [21] emphasized the importance of time scale separation for properly defining a rate constant. Specifically, the waiting time between reactions (τ_{rxn}) should be much longer than the brief molecular relaxation time (τ_{mol}) within the stable basins. If $\tau_{mol} \ll \tau_{rxn}$, then the system will forget the route by which it arrived in the reactant state long before the next reaction occurs. Likewise, different vibrational and rotational states within the reactant population can be lumped together if these differences relax more quickly than the reaction time. Thus it is time scale separation that justifies the exponential distribution of escape times, the associated forgetfulness property, and phenomenological rate laws that simply multiply the entire population of a state by a single characteristic escape rate.

In a system with a clear time scale separation, the reaction coordinate as function of time makes infrequent but rapid transitions between the reactant and product states. In most cases, we cannot directly observe the dynamics of individual molecules, but there are a few exceptions as shown in Figure 10.1.2. When there is no clear time scale separation, there is no rate constant because different initial conditions within the reactant state lead to discernible differences in reactant lifetimes. As a useful rule of thumb, the free energy barrier should be about $5k_BT$ or more to properly define a rate constant.

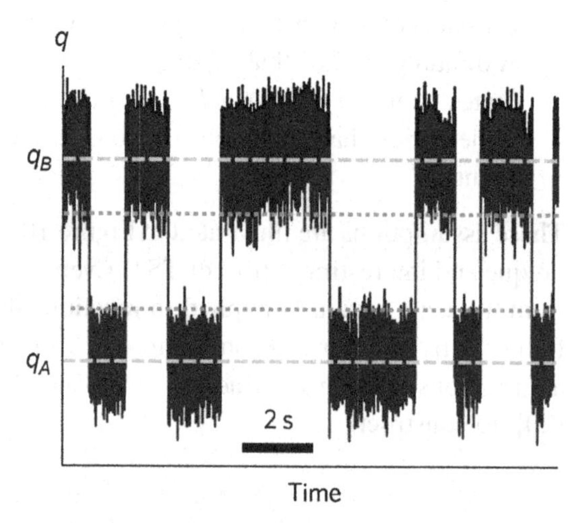

Figure 10.1.2: Extension vs. time data for a single protein in leaser tweezers provides a direct illustration of time scale separation. The trajectory spends long times in the folded and unfolded states, punctuated by infrequent and fast transitions. [Figure adapted from Neupane et al. Nature Physics, Mar (2016)]

Time scale separation also gives the reactants time to sample the equilibrium ensemble of reactant configurations between transitions. Thus the separation of time scales also helps to justify the use of equilibrium statistical mechanics for the reactant states.

Finally, time scale separation arguments provide insight on the errors that result from the assumption that transition states are in equilibrium with reactants. During the time τ_{rxn}, reactants will acquire *almost* enough activation energy to escape many many times, only to then have the excitation dissipate. Thus equilibrium must hold from the reactant minimum *up to a few* k_BT *before* the dividing surface. TST extrapolates just a little bit farther by assuming that equilibrium holds even for transition states *at* the dividing surface. Near the precipice, some theoretical frameworks explicitly include deviations from equilibrium, while others assume equilibrium at the transition state and include dynamical effects. There are interesting connections between nonequilibrium effects and dynamical effects that will be discussed further in Chapter 13. For now, we ignore both issues and proceed to develop TST.

From this point forward, in keeping with earlier chapters, bold **A** and **B** will refer to generic molecules or particles while unbolded A and B will indicate "reactant" and "product" regions of phase space. Sometimes we will describe stable basins A and B with the terminology of molecular properties, e.g. there are vibrational frequencies associated with motion along the reaction coordinate in states A and B. These are not my own machinations. They are historical notations and terminologies used by the two communities which this book straddles. Readers are asked to discern the difference from context and from the bold vs. regular typeface.

10.2 Statistical mechanics for chemical equilibria

Let us briefly review the statistical mechanics of chemical equilibrium. In Chapter 2 we learned that equilibrium constants are related to the Gibbs free energy of reaction at a standard reference condition

$$K_{eq} = exp[-\beta \Delta G^{\circ}_{rxn}]$$

where $\beta = 1/k_B T$. The equilibrium composition of a reacting mixture can be obtained by relating the equilibrium constant to a mass action ratio of activities.

$$K_{eq} = \prod_i a_i^{\nu_i}$$

Chapter 2, as per convention, presented activities in terms of mole fractions ($a_i = \gamma_i x_i$) or molalities ($a_i = \gamma_i m_i / m^{\circ}$). The activities are always defined relative to a standard state, and ΔG°_{rxn} must be computed at that same standard reference state.

Instead of mole fractions or molalities, one can equally define activities and activity coefficients relative to a reference concentration, e.g. $a_i = \gamma_i c_i / c^{\circ}$. For example, consider the reaction $a\mathbf{A} + b\mathbf{B} \rightleftarrows c\mathbf{C} + d\mathbf{D}$. The equilibrium constant is

$$\widetilde{K}_{eq} = \frac{(Q_{\mathbf{C}}/\mathcal{V})^c (Q_{\mathbf{D}}/\mathcal{V})^d}{(Q_{\mathbf{A}}/\mathcal{V})^a (Q_{\mathbf{B}}/\mathcal{V})^b} = \frac{[\mathbf{C}]^c \, [\mathbf{D}]^d}{[\mathbf{A}]^a \, [\mathbf{B}]^b} \tag{10.2.1}$$

where $Q_{\mathbf{X}}$ is the canonical partition function for species **X**. The twiddle over $\widetilde{K}_{eq}$ reminds that this equilibrium constant may still have concentration units because it is not expressed in terms of proper activities. However, each factor of volume $\mathcal{V}$ eliminates a factor of $\mathcal{V}$ in the translational partition function, so the quantity in equation (10.2.1) is a pure function of the temperature. Both sides of equation (10.2.1) can be multiplied by $\mathcal{V}_o^{c+d-a-b}$ which is equivalent to adopting a standard reference concentration of $c^o = \mathcal{V}_o^{-1}$. With this convention, we obtain a dimensionless equilibrium constant

$$K_{eq} = \frac{Q_{\mathbf{C}}{}^c Q_{\mathbf{D}}{}^d}{Q_{\mathbf{A}}{}^a Q_{\mathbf{B}}{}^b} = \frac{a_{\mathbf{C}}^c a_{\mathbf{D}}^d}{a_{\mathbf{A}}^a a_{\mathbf{B}}^b}$$

where the partition functions are computed with a reference volume $\mathcal{V}_o$ and where the activities are $a_i = \gamma_i c_i \mathcal{V}_o$. The concentrations can easily be converted to other bases, e.g. to write the activities in terms of pressure one would replace $c_i \to P_i/k_B T$, $\mathcal{V}_o^{-1} \to P^o/k_B T$, and $\gamma_i \to$ fugacity coefficient. Most of the discussion that follows assumes ideality, i.e. activity coefficients and fugacity coefficients are assumed to have values of unity.

Partition functions are discussed in many excellent statistical mechanics textbooks [23,24], so here we give only a highly condensed summary. For each species **X**, the canonical partition function is

$$Q_{\mathbf{X}} = Q_{trans}^{\mathbf{X}} Q_{rot}^{\mathbf{X}} Q_{el}^{\mathbf{X}} Q_{vib}^{\mathbf{X}} exp\left[-\beta V_{zp}^{\mathbf{X}}\right] \tag{10.2.2}$$

i.e. a product of translational, rotational, electronic, and vibrational partition functions. The zero point energy is $V_{zp} = V + \frac{1}{2}\sum_i \hbar\omega_i$ where V is the Born-Oppenheimer potential at the stationary point and the ω_i are vibrational angular frequencies. The electronic partition function is determined by the electronic degeneracy of the ground state, except where there is a low lying excited state. The translational, rotational, and vibrational partition functions are given in Table 10.1. The translational and rotational factors have been written in terms of thermal de Broglie wavelengths (λ_T) and thermal angles (φ_T), respectively.

Table 10.1: Canonical partition functions.

Type	Partition function
translation (3D)	$Q_{trans} = \mathcal{V}_o/\lambda_T^3$
rotation (nonlinear molec.)	$Q_{rot} = 8\pi^2/\{\sigma_{sym}\prod_{k=1}^{3}\varphi_T(I_k)\}$
rotation (linear molecule)	$Q_{rot} = 4\pi/\{\sigma_{sym}\varphi_T(I)^2\}$
internal free rotor	$Q_{rotor} = 2\sqrt{\pi}/\varphi_T(I)$
vibration (real frequencies)	$Q_{vib} = \prod_i(1 - e^{-\beta\hbar\omega_i})^{-1}$

Table 10.2 gives formulas for the de Broglie wavelengths, de Broglie angles, and other quantities which are needed to compute the partition functions. Each quantity has also been evaluated at a specific reference temperature, mass, frequency, and inertia. The reference values enable easy scaling-based calculations of each quantity at other temperatures, masses, frequencies, and inertias.

Some molecules also have rotors like $-CH_3$ or $-phenyl$ which are not truly free, but instead are better characterized as hindered rotors. See Pfaendtner et al. for a review on the statistical

Table 10.2: Useful combinations and reference values.

Quantity	Reference condition	Reference value
k_BT/h	$T = 300K$	$6.25 \cdot 10^{12}/s$
k_BT	$T = 300K$	$2.4945\,kJ/mol$
$\hbar\omega$	$\omega = 1000cm^{-1}$	$11.95\,kJ/mol$
$\lambda_T = h/\sqrt{2\pi m k_B T}$	$T = 300K$ $m = 1.0amu$	1.004Å
$\varphi_T = h/\sqrt{2\pi I k_B T}$	$T = 300K$ $I = 1.0(\text{Å})^2 amu$	1.004
$\beta\hbar\omega$	$T = 300K$ $\omega = 1000cm^{-1}$	4.79

mechanics of hindered rotor models [22]. Symmetry numbers for each molecular point group are given in Table 10.3. There are some important exceptions to the usual handling of symmetry numbers and these will be illustrated in the TST rate calculations that follow.

Table 10.3: Symmetry numbers for molecular point groups.

Point group	σ_{sym}	Point group	σ_{sym}
$Atom,\ C_1,\ C_i,\ C_s,\ C_{\infty v}$	1	C_7	7
$C_2,\ C_{2v},\ C_{2h},\ D_{\infty h},\ S_4$	2	$C_8,\ D_4,\ D_{4d},\ D_{4h}$	8
$C_3,\ C_{3v},\ C_{3h},\ S_6$	3	$D_5,\ D_{5d},\ D_{5h}$	10
$C_4,\ C_{4v},\ C_{4h}$	4	$T,\ T_d,\ T_h,\ D_6,\ D_{6d},\ D_{6h}$	12
$D_2,\ D_{2d},\ D_{2h},\ S_8$	4	D_{8h}	16
$C_5,\ C_{5v},\ C_{5h}$	5	$O,\ O_h$	24
$C_6,\ C_{6v},\ C_{6h},\ D_3,\ D_{3d},\ D_{3h}$	6	I_h	60

10.3 Harmonic transition state theory

Consider a second order reaction of the type $\mathbf{A} + \mathbf{B} \to products$, and suppose that all of the reactive trajectories pass through a high col on the PES. Eyring equated the reaction rate to a product of the equilibrium concentration of transition states and the frequency $\nu_\ddagger$ at which transition states are converted to products. Effectively, this corresponds to the mechanism:

$$\mathbf{A} + \mathbf{B} \rightleftharpoons \ddagger \to products \tag{10.3.1}$$

where $\ddagger$ denotes a transition state. The reaction rate is then

$$\mathcal{R}_{TST} = \nu_\ddagger[\ddagger] \tag{10.3.2}$$

where $\nu_\ddagger$ is the rate at which transition states become products, and $[\ddagger]$ is the concentration of transition state species. Transition states lie within a narrow interval of width δq centered at $q_\ddagger$ along the reaction coordinate $q(\mathbf{x})$, i.e. the transition state ensemble includes all configurations $\mathbf{x}$ such that $q_\ddagger - \delta q/2 < q(\mathbf{x}) < q_\ddagger + \delta q/2$.[1] The frequency $\nu_\ddagger$ with which transition states exit

[1] The precise width δq is immaterial – it ultimately cancels in the product of $\nu_\ddagger$ and $[\ddagger]$.

the transition state region is the average absolute velocity along q at the transition state divided by the width δq. At equilibrium only half of these trajectories are moving toward the product state. Therefore,

$$\nu_{\ddagger} = \frac{1}{2\delta q} \langle |\dot{q}| \rangle_{\ddagger} \tag{10.3.3}$$

The concentration of transition states, assuming ideality and equilibrium with the reactants, is

$$[\ddagger] = \widetilde{K}_{\ddagger}[\mathbf{A}][\mathbf{B}] \tag{10.3.4}$$

where

$$\widetilde{K}_{\ddagger} = \frac{\delta q}{\lambda_{T,q}} \cdot \frac{Q_{\ddagger}/\mathcal{V}_o}{(Q_{\mathbf{A}}/\mathcal{V}_o)(Q_{\mathbf{B}}/\mathcal{V}_o)} \tag{10.3.5}$$

The transition state partition function has been separated into factors $Q_{\ddagger}$ and $\delta q/\lambda_{T,q}$. $Q_{\ddagger}$ excludes the reaction coordinate degree of freedom, and $\delta q/\lambda_{T,q}$ treats the q degree of freedom as a free translation in one dimension. Here

$$\lambda_{T,q} = \frac{h}{\sqrt{2\pi \mu_q k_B T}} \tag{10.3.6}$$

is the de Broglie wavelength and μ_q is the reduced mass for motion along the reaction coordinate. Also note that the volume $\mathcal{V}_o$ (from the reference concentration $\mathcal{V}_o^{-1}$) was used to compute and to normalize the partition functions of equation (10.3.5). Any volume could be used in place of $\mathcal{V}_o$, but the standard reference volume will facilitate the later formulations of TST in terms of thermodynamic reference potentials. Inserting all components into the rate equation and dividing by [**A**][**B**] gives $k_{TST} = \frac{1}{2} \langle |\dot{q}| \rangle \lambda_{T,q}^{-1} \cdot \mathcal{V}_o^{\nu-1} Q_{\ddagger}/Q_{\mathbf{A}} Q_{\mathbf{B}}$. For our example, $\nu = 2$ because the reaction is second order.

The prefactor $\frac{1}{2} \langle |\dot{q}| \rangle \lambda_{T,q}^{-1}$ can be simplified considerably. First, let us compute the average absolute velocity along q:

$$\langle |\dot{q}| \rangle_{\ddagger} = \frac{\int_{-\infty}^{\infty} d\dot{q}\, exp[-\beta\mu_q \dot{q}^2/2] \cdot |\dot{q}|}{\int_{-\infty}^{\infty} d\dot{q}\, exp[-\beta\mu_q \dot{q}^2/2]} = \sqrt{\frac{2k_B T}{\pi \mu_q}} \tag{10.3.7}$$

Now the product $\frac{1}{2} \langle |\dot{q}| \rangle \lambda_{T,q}^{-1}$ simplifies to the familiar $k_B T/h$. Finally, the harmonic TST rate constant is

$$k_{TST} = \frac{k_B T}{h} \cdot \mathcal{V}_o^{\nu-1} \frac{Q_{\ddagger}}{Q_{\mathbf{A}} Q_{\mathbf{B}}} \tag{10.3.8}$$

In general the denominator in equation (10.3.8) contains one partition function for each reactant molecule, and ν is the overall reaction order.

Notice that the reduced mass for coordinate q has vanished. Additionally, the final expression for k_{TST} contains no remnants of the arbitrary interval δq. Finally, note that Planck's constant h appears in the TST prefactor only because the reaction coordinate partition function $\delta q\, \lambda_{T,\ddagger}^{-1}$ was conveniently separated from the rest of the transition state partition function $Q_\ddagger$. The factor of h originates entirely from within $\tilde{K}_\ddagger$. If action was counted by some unit other than h, a different prefactor would emerge from the product $\frac{1}{2}\langle|\dot{q}|\rangle\, \lambda_{T,\ddagger}^{-1}$ and the reactant partition functions would change in compensatory fashion. Thus $k_B T/h$ is a special prefactor associated with the unstable mode reaction coordinate and with the convention for counting states in classical statistical mechanics. See section 10.5 for a more general TST expression in which the prefactor is *not* $k_B T/h$.

The simplicity of equation (10.3.8) makes harmonic TST the single most important theory for analyses of chemical reactions. It has been used to understand and model *thousands* of activated processes: chemical reactions in vapor and liquid phases, activated diffusion in solids, heterogeneous catalysis, etc. Harmonic TST is applicable whenever the dynamical bottleneck is a high saddle on the molecular (Born-Oppenheimer) PES. In these cases there is little need for path optimizations or dynamical calculations because harmonic TST with an accurate model chemistry, a transition state search algorithm, and basic equilibrium statistical mechanics usually provides accurate rate constants. The following example illustrates the TST calculation for a simple hard sphere model.

Example: TST for hard spheres

Consider hard spheres $\mathbf{A}$ and $\mathbf{B}$ that react irreversibly when the collision energy exceeds some threshold $E_\ddagger$. There is only the center-to-center distance between the hard spheres and there are no bath modes to absorb the activation energy, so true hard spheres would immediately violate the no-recrossing dynamical assumption of TST. However, we can imagine the hard spheres as real molecules with internal modes. Using the center-to-center distance r as a reaction coordinate and $r_\ddagger = (d_\mathbf{A} + d_\mathbf{B})/2$ as a dividing surface, the TST rate constant is

$$
\begin{aligned}
k_{TST} &= \frac{k_B T}{h} V_o \frac{Q_\ddagger}{Q_\mathbf{A} Q_\mathbf{B}} \\
&= \frac{k_B T}{h} V_o \frac{Q_{trans}^\ddagger}{Q_{trans}^\mathbf{A} Q_{trans}^\mathbf{B}} Q_{rot}^\ddagger e^{-\beta E_\ddagger} \\
&= \frac{k_B T}{h} \cdot \frac{h^3 (2\pi m_\ddagger k_B T)^{3/2}}{(2\pi m_\mathbf{A} k_B T)^{3/2} (2\pi m_\mathbf{B} k_B T)^{3/2}} \cdot \frac{4\pi (2\pi I_\ddagger k_B T)}{\sigma_{sym}^\ddagger h^2} e^{-\beta E_\ddagger} \\
&= \frac{\pi r_\ddagger^2}{\sigma_{sym}} \left(\frac{8 k_B T}{\pi \mu_\mathbf{AB}} \right)^{1/2} e^{-\beta E_\ddagger} \quad (using\ I_\ddagger = \mu_\mathbf{AB} d_\mathbf{AB}^2)
\end{aligned}
$$

Note that TST gives the same answer that we obtained from integrating the reactive cross section in Chapter 6.

Now consider what would happen if we could magically take the **A**, **B** mixture and make **A** = **B**. Then we would obtain $\sigma_{sym} = 2$ for the transition state, and the rate constant would become half as large. However, both terms in the product [**A**][**B**] would double. The net effect would be to double the rate. This makes sense because now **AA**, **AB**, **BA**, and **BB** collisions (twice as many as before) can all lead to reactions.

Symmetry numbers in the rotational partition functions account for the number of equivalent reaction channels by which the reaction can proceed. For example, consider the reaction $F + H_2 \rightarrow HF + H$. The F atom can attack either of the two H atoms. When the rotational partition functions are computed, the symmetric reactant H_2 with $\sigma_{sym} = 2$ appropriately doubles the rate constant. The paper by Pollak and Pechukas [25] offers some useful recommendations and interpretations for the symmetry numbers in TST. In many cases the symmetry correction can be intuited by counting degenerate reaction pathways, but a formulaic approach using symmetry numbers (usually) avoids errors. There is one exception to remember. Some transition states are special points of high symmetry along the reaction coordinate. When an infinitesimal displacement along the reaction coordinate in either direction from the saddle point breaks symmetry, the lower (broken) symmetry factor should be used [25].

We have said very little about the electronic factors in the partition function. For a radical (doublet) species, the electronic symmetry factor is $Q_{el} = 2$ because of the degenerate spin-up and spin-down states. Of course, the electronic factors should be included, but in most cases radical transition states come from reactants that include a radical. The effect is that the electronic factors cancel. When there are crossing or low-lying electronic energy surfaces, caution should be exercised in using the adiabatic framework of TST. See Chapter 21 for a discussion of non-adiabatic reaction rate theories.

$Q_{\ddagger}$ excludes the reaction coordinate

Most discussions of TST just assert that the reaction coordinate contribution to $Q_{\ddagger}$ should be excluded by omitting the imaginary frequency. Why is excluding the imaginary frequency the correct recipe? Let us take a moment to appreciate the elegant justification for this step. At the outset of the harmonic TST discussion, we restricted our analysis to reactions whose trajectories pass through the vicinity of a high saddle point on the PES. By invoking vibrational partition functions, we have also assumed a locally quadratic energy landscape at the saddle point and at the equilibrium reactant geometries. Recall from Chapter 7, that transforming $\mathbf{f} = \mathbf{Ma}$ to mass weighted coordinates gave $-\mathbf{H}_{MW}\Delta\mathbf{x}_{MW} = m_0 d^2\Delta\mathbf{x}_{MW}/dt^2$ where m_0 =

1amu, $\mathbf{H}_{MW} = m_0 \mathbf{M}^{-1/2} \mathbf{H} \mathbf{M}^{-1/2}$ is the mass weighted Hessian matrix, $\mathbf{M}$ is the diagonal matrix of atomic masses, and $\Delta \mathbf{x}_{MW} = m_0^{-1/2} \mathbf{M}^{1/2} \Delta \mathbf{x}$ are mass weighted displacements from the saddle point. Diagonalization of $\mathbf{H}_{MW}$ yielded the full set of normal vibrational modes: $\mathbf{H}_{MW} \mathbf{u}_i = m_0 \omega_i^2 \mathbf{u}_i$.

At a saddle point, there is one *unstable* eigenmode with an imaginary vibrational frequency. Let the imaginary frequency and its unstable eigenmode be denoted $\omega_\ddagger^2$ and $\mathbf{u}_\ddagger$, respectively. The reaction coordinate in harmonic TST is

$$q = \Delta \mathbf{x}_{MW}^T \mathbf{u}_\ddagger \tag{10.3.9}$$

which identifies the extent of progress along the unstable mass-weighted eigenvector, as depicted in Figure 10.3.1. The dividing surface in harmonic TST is a hyperplane perpendicular to the unstable mode at the saddle point. Thus the transition state ensemble includes all displacements from the saddle such that $q = \Delta \mathbf{x}_{MW}^T \mathbf{u}_\ddagger = 0$. The perpendicular hyperplane is exactly spanned by the other 3N-7 eigenvectors (the modes with real frequencies) because $\mathbf{H}_{MW}$ is symmetric. Moreover, the dynamics of the normal mode coordinates are locally separable in the vicinity of the saddle.[2] Within the quadratic approximation, it is rather easy to show that $q = 0$ is a perfect dividing surface with no recrossing [26]. Dynamical tests suggest that the harmonic TST approximation is also quite accurate for "real" PESs of gas phase chemical reactions [27] and even for some reactions in polar solvents and enzymes [15,16].

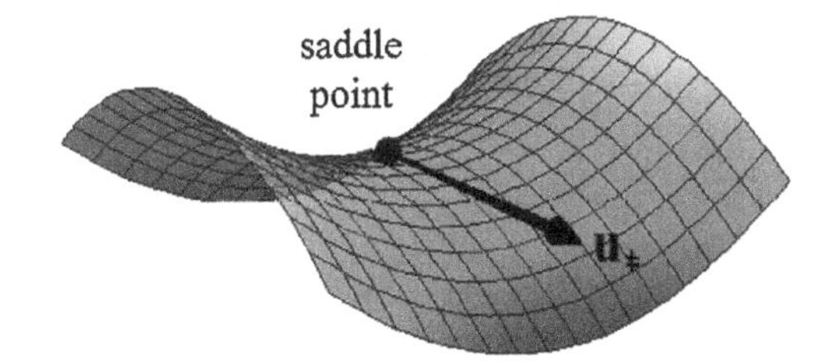

Figure 10.3.1: Diagonalization of the mass weighted second derivative matrix at the saddle point identifies an unstable mode reaction coordinate that is dynamically separable from other coordinates in the neighborhood of the saddle.

To foreshadow later parts of this book, note the strategy behind harmonic TST. The dynamics are carefully analyzed to identify a reaction coordinate *before* the calculation of free energies and rates. It is this reaction-coordinate-first strategy which makes harmonic TST robust for estimating rates, for identifying the properties of transition states, and for understanding the precise role of each degree of freedom in the mechanism. In contrast, many modern rare events studies compute free energies and rates *without* investigating the actual dynamics to identify the reaction coordinate. Transition path sampling (Chapter 19) and reaction coordinate identification methods (Chapter 20) extend the reaction-coordinate-first strategy to systems with rugged potential energy surfaces where harmonic TST is not applicable.

[2] In fact, we diagonalized *to find* the dynamically separable motions.

Example: TST for $H_2 + H \rightarrow H + H_2$ at 420K

The symmetric H-exchange reaction between H_2 and H illustrates essentially all aspects of a harmonic TST calculation. Siegbahn and Liu [28] performed accurate *ab initio* FCI calculations for the $H_2 + H$ reaction with a basis set including s, p, and d atomic orbitals. Truhlar and Horowitz [29] used the Siegbahn and Liu results to parameterize a continuous model of the Born-Oppenheimer PES. Slices of the $H_2 + H$ potential at 180° (collinear) and 120° were shown in Figure 7.2.1. Key parameters for a TST calculation were extracted by Henriksen and Hansen [30]. The saddle point is collinear so the transition state has four internal frequencies.

	H	H_2	$[H \cdots H \cdots H]^\ddagger$
σ_{el}	2	1	2
σ_{sym}	1	2	2*
$V(kJ/mol)$		0.0	41.0
$\omega(cm^{-1})$	-	4395	2058, 909, 909, 1511i
d_{H-H}(Å)		0.74	0.93, 0.93

In this example, we compute the TST rate constant at $T = 1000K$. Note that only the masses, geometries, energies, and vibrational frequencies at the transition state and the reactant minimum are needed. For the $H \cdots H \cdots H$ transition state, the center of mass is at the central hydrogen nucleus, giving two degenerate inertial moments of $I_\ddagger = 2 \cdot (0.93\text{Å})^2 1.007amu = 1.742\text{Å}^2 amu$. For H_2, the two degenerate moments of inertia are $I_{HH} = 2 \cdot (0.74\text{Å}/2)^2 1.007amu = 0.276\text{Å}^2 amu$. The thermal de Broglie angle is $\varphi_T = h/\sqrt{2\pi I k_B T}$ which takes the value (1.004) for $I = 1.0\text{Å}^2 amu$ and $T = 300K$. The thermal angle is easily scaled to the appropriate values for $H \cdots H \cdots H$ and H_2 at $T = 420K$:

$$\varphi_\ddagger = 1.004 \frac{\sqrt{1.0 \cdot 300}}{\sqrt{1.742 \cdot 420}} = 0.643$$

$$\varphi_{HH} = 1.004 \frac{\sqrt{1.0 \cdot 300}}{\sqrt{0.276 \cdot 420}} = 1.615$$

We can similarly scale the thermal de Broglie wavelengths according to the masses and the temperature.

$$\lambda_\ddagger = 1.004\text{Å} \frac{\sqrt{1.0 \cdot 300}}{\sqrt{3 \cdot 1.007 \cdot 420}} = 0.488\text{Å}$$

$$\lambda_{HH} = 1.004\text{Å}\frac{\sqrt{1.0 \cdot 300}}{\sqrt{2 \cdot 1.007 \cdot 420}} = 0.598\text{Å}$$

$$\lambda_H = 1.004\text{Å}\frac{\sqrt{1.0 \cdot 300}}{\sqrt{1.007 \cdot 420}} = 0.846\text{Å}$$

The dimensionless frequencies and vibrational partition functions are also obtained by scaling:

ω	$\hbar\omega(kJ/mol)$	$\beta\hbar\omega$	$(1-e^{-\beta\hbar\omega})^{-1}$
$4395cm^{-1}$	52.5	15.0	1.00
$2058cm^{-1}$	24.6	7.04	1.00
$909cm^{-1}(2\times)$	10.9	3.11	1.05
$1511cm^{-1}$	-	5.17*i*	-

The zero point corrected energy difference is

$$\Delta V_{zp}^{\ddagger} = \Delta V + \frac{1}{2}\sum_{i}^{(\ddagger)} \hbar\omega_i - \frac{1}{2}\hbar\omega_{HH}$$

which gives $\Delta V_{zp}^{\ddagger} = 37.9kJ/mol$. The radicals are electronically degenerate (spin up or spin down). The electronic degeneracy has no consequence because it is present both in the transition state and in the reactants. On the other hand the symmetry factors must be handled carefully. Infinitessimal motion along the reaction coordinate breaks the symmetry of the transition state. Thus, the transition state symmetry number should be $\sigma_{sym}^{\ddagger} = 1$, not $\sigma_{sym}^{\ddagger} = 2$.

$$\begin{aligned} k_{TST} &= \frac{k_B T}{h}\frac{Q_{\ddagger}/\mathcal{V}_o}{(Q_H/\mathcal{V}_o)(Q_{HH}/\mathcal{V}_o)} \\ &= \frac{k_B T}{h}\sigma_{HH}\frac{\lambda_H^3\lambda_{HH}^3}{\lambda_{\ddagger}^3}\frac{\varphi_{HH}^2}{\varphi_{\ddagger}^2}\frac{\prod_i^{\ddagger}(1-e^{-\beta\hbar\omega_i})^{-1}}{(1-e^{-\beta\hbar\omega_{HH}})^{-1}}exp\left[-\beta\Delta V_{zp}^{\ddagger}\right] \\ &= 8.75\times 10^{12}\cdot 2\cdot\frac{(0.846)^3(0.598)^3}{(0.488)^3}\frac{(1.615)^2}{(0.643)^2}\frac{(1.0)(1.05)^2}{(1.0)}e^{-10.85}\frac{\text{Å}^3}{s} \end{aligned}$$

The resulting TST rate constant is $k_{TST} = 2.63 \cdot 10^9\text{Å}^3/reaction/s$. After some unit conversions, we obtain $k_{TST} = 1.6 \cdot 10^6 L/mol/s$. The experimental rate is about $k_{expt} = 2 \cdot 10^9 L/mol/s$. What went wrong? Recrossing effects would make our estimated rate even lower, but there are other culprits: (i) Tunneling, according to the Bell formula, gives a transmission coefficient of $\kappa_{QM} = |\beta\hbar\omega_{\ddagger}/2|/sin(|\beta\hbar\omega_{\ddagger}/2|) = 57$. But the real effects of tunneling, as we will see in Chapter 12, are probably much smaller. (ii) The classical partition function for rotation is not accurate for H_2. (iii) The most important source of error is probably the basis set used for the FCI calculation. Although d-functions are ample, the atomic orbitals were not constructed from triple or even double zeta primitives like modern basis functions.

10.4 Thermodynamic formulation

There are many ways to write the harmonic TST rate constant. As already mentioned in section 10.3, it is not necessary to use reference concentrations and volumes. For example, consider the generic reaction $\mathbf{A} + \mathbf{B} \rightleftharpoons \ddagger \rightarrow products$. One can write $k_{TST} = k_B T\, h^{-1} \cdot (\mathcal{Q}_{\ddagger}/\mathcal{V})/(\mathcal{Q}_{\mathbf{A}}/\mathcal{V})(\mathcal{Q}_{\mathbf{B}}/\mathcal{V})$ for any choice of the volume $\mathcal{V}$ as long as the same volume $\mathcal{V}$ is used to compute the partition functions. So why did we write equation (10.3.8) in terms of a specific reference concentration and volume? It allows us to separate the TST rate constant into two factors: (1) a true (dimensionless) equilibrium constant that can be converted to a standard activation free energy $\Delta G^{\ddagger}$, and (2) a prefactor that also depends on the reference concentration. Specifically, the equilibrium constant for $\mathbf{A} + \mathbf{B} \rightleftharpoons \ddagger$ is $K_{\ddagger} = \mathcal{Q}_{\ddagger}/\mathcal{Q}_{\mathbf{A}}\mathcal{Q}_{\mathbf{B}}$ and the activation free energy is

$$\Delta G^{\ddagger} = -k_B T \ln K_{\ddagger} \tag{10.4.1}$$

where the volume per molecule $\mathcal{V}_o$ must correspond to the reference concentration and to the volume in the translational partition functions. The TST rate constant from equation (10.3.8) can then be written as

$$k_{TST} = \frac{k_B T}{h} \mathcal{V}_o^{\nu-1} exp\left[-\frac{\Delta G^{\ddagger}}{k_B T}\right] \tag{10.4.2}$$

The standard Gibbs free energy can be separated into standard activation enthalpy and entropy contributions

$$\Delta G^{\ddagger} = \Delta H^{\ddagger} - T\Delta S^{\ddagger} \tag{10.4.3}$$

Using equation (10.4.3) in equation (10.4.2) yields the powerful result of Wynne-Jones and Eyring [12]

$$k_{TST} = \frac{k_B T}{h} \mathcal{V}_o^{\nu-1} e^{\Delta S^{\ddagger}/k_B} e^{-\beta\Delta H^{\ddagger}} \tag{10.4.4}$$

Equation (10.4.4) supplants the empirical Arrhenius law ($k = A\, e^{-E_a/k_B T}$ [2]) with an alternative in which the parameters have well-defined physical meanings. Specifically, Equation

(10.4.4) suggests a natural dimensionless rate constant $k\beta h V_0^{1-\nu}$ and the Eyring plot based on the relationship

$$ln\left[k_{TST}\beta h V_0^{1-\nu}\right] = \Delta S^{\ddagger}/k_B - \beta\Delta H^{\ddagger} \tag{10.4.5}$$

These plots present $ln[k\beta h V_0^{1-\nu}]$ vs. $1/T$. In the literature they are more commonly presented with units: $ln(k/T)$ vs. $1/T$. The slope in the Eyring plot gives the thermodynamic parameter $\Delta H^{\ddagger}$ instead of the empirical Arrhenius parameter E_a. The intercept in the Eyring plot identifies the thermodynamic parameter $\Delta S^{\ddagger}/k_B$ instead of the empirical Arrhenius parameter lnA. An example Eyring plot [31] is shown in Figure 10.4.1.

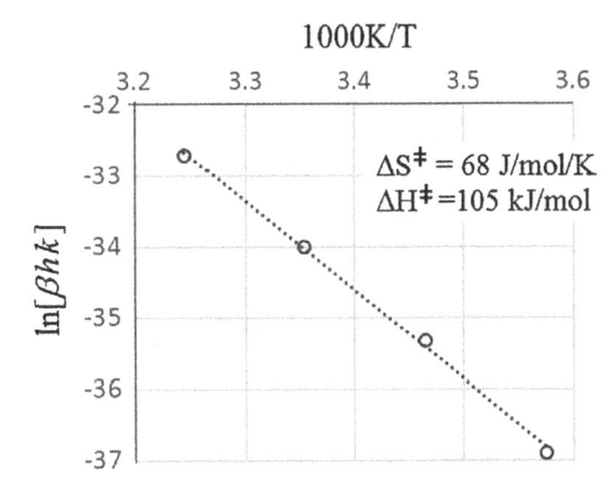

Figure 10.4.1: Sulfite (SO_3^{2-}) and mercuric ions (Hg^{2+}) in cloud droplets form ($HgSO_3$) which decomposes to $Hg^{(0)}$ and sulfate (S(VI)) species. Van Loon et al. *J. Phys. Chem.* A 104, 1621-26 (2000) used UV vis spectroscopy and TST to understand the temperature dependence of this reaction.

The thermodynamic formulation of TST is a powerful tool for extracting information about transition states and mechanisms from experiments. For example, the number of molecules that are assembled in the transition state can often be inferred from the size of $\Delta S^{\ddagger}$. For reactions in solution, the dependences of $\Delta H^{\ddagger}$ on the solvent dielectric constant and pH can reveal whether a transition state is protonated, deprotonated, or neutral.

Eyring plots are best used to interpret activation parameters for elementary steps. When applied to pseudo-rate constants that emerge from multistep mechanisms, the activation parameters from an Eyring plot may inadvertently inherit the characteristics of various adsorption constants and equilibrium constants in the overall mechanism. The following example shows how this situation can lead to (what appears to be) a negative activation enthalpy.

Example: Cracking kinetics in a zeolite

A simple model for alkane cracking in a zeolite involves two kinetically relevant steps. Cracking of an alkane (**A**) produces another alkane (**A′**) and an olefin (**O**). Cracking mechanisms have been examined in many theoretical studies [32,33]. At the simplest level, the cracking mechanism is

$$\mathbf{A} + * \overset{K}{\rightleftharpoons} \mathbf{A}*$$

$$\mathbf{A}* \overset{k}{\longrightarrow} \mathbf{A}' + \mathbf{O}$$

Both forward and backward directions in the first step are assumed to be very fast relative to the second cracking step. Other mechanistic details are not important for the current example. Using methods introduced in Chapter 4, the overall rate law is

$$\mathcal{R} = \frac{kK[*]_0[\mathbf{A}]}{1 + K[\mathbf{A}]}$$

where $[*]_0 = [*] + [\mathbf{A}*]$. Note: acid sites in a zeolite are dispersed and accessible throughout the volume, so they have a concentration rather than a surface density [32]. According to TST, $k = k_B T h^{-1} exp[\Delta S^{\ddagger}/k_B] exp[-\Delta H^{\ddagger}/k_B T]$. The equilibrium adsorption constant can also be written in terms of entropy and enthalpy contributions, $K = exp[\Delta S^o_{ads}/k_B] exp[-\Delta H^o_{ads}/k_B T]$. At high temperature conditions, most of the acid sites in the zeolite do not have an adsorbed alkane, i.e. $K[\mathbf{A}] \ll 1$. Therefore, the rate law simplifies to $\mathcal{R} \approx kK[*]_0[\mathbf{A}]$ where

$$kK = k_B T h^{-1} exp[(\Delta S^o_{ads} + \Delta S^{\ddagger})/k_B] exp[-(\Delta H^o_{ads} + \Delta H^{\ddagger})/k_B T]$$

If the overall rate is interpreted as a pseudo-first order reaction, the *apparent* enthalpy and entropy of activation would be $\Delta H^o_{ads} + \Delta H^{\ddagger}$ and $\Delta S^o_{ads} + \Delta S^{\ddagger}$. The chemical step activation parameters $\Delta H^{\ddagger}$ and $\Delta S^{\ddagger}$ are only weakly dependent on the chain length of the alkane. The adsorption parameters ΔH^o_{ads} and ΔS^o_{ads} are strongly dependent on chain length. The chemical step activation enthalpy is positive ($\Delta H^{\ddagger} > 0$) and the binding enthalpy is negative ($\Delta H^o_{ads} < 0$ to a degree that depends on chain length). For sufficiently large chains, the *apparent* activation enthalpy can be negative! (see Figure 10.4.2 [34]). Note that this is a special case, where the apparent activation parameters still have a clear interpretation. Usually, the rates from multistep mechanisms are simply not amenable to analysis with Eyring plots.

Some additional caveats should be remembered when interpreting Eyring plots of experimental data. First, remember that even elementary rate constants involve tunneling and recrossing effects that are not included in the TST expression. These non-TST effects can influence the apparent experimental values of $\Delta H^{\ddagger}$ and $\Delta S^{\ddagger}$. For example, tunneling can lower the apparent activation enthalpy [35], and recrossing effects can lower the apparent activation entropy.

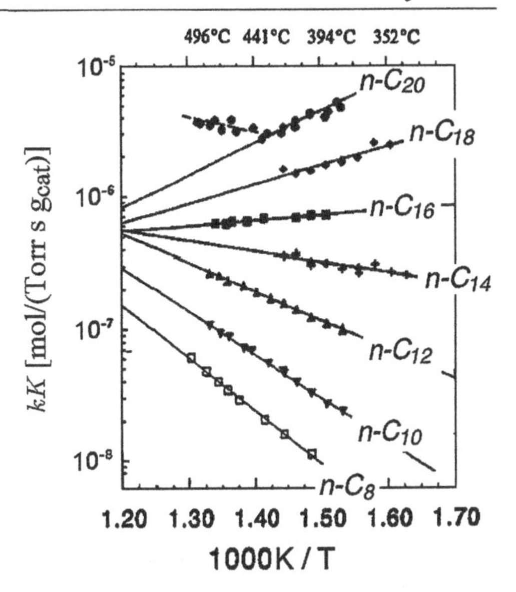

Figure 10.4.2: Kinetics of n-alkane cracking in zeolites with different chain lengths. The slopes suggest two contributions to the apparent activation enthalpy: an adsorption constant with an enthalpic contribution that becomes more negative with increasing chain length, and an activation enthalpy from the chemical step that is nearly independent of chain-length. [Adapted from Wei, *Chem. Eng. Sci.* 51, 2995-99 (1996).]

Later chapters discuss ways that non-TST factors can be included in rate calculations for proper comparisons to experiment.

Also note that $\Delta H^{\ddagger}$ and $\Delta S^{\ddagger}$ may appear to be temperature independent constants over small temperature intervals, but they both depend on temperature. Accordingly, the apparent value of $\Delta H^{\ddagger}$ and $\Delta S^{\ddagger}$ may depend strongly on the temperature interval used to construct the Eyring plot [36]. In a sense, activation parameters are analogous to Legendre transforms in thermodynamics [37]. They are a slope-intercept representation of the rate constant, and the values of the slope and intercept depend on where the tangent in the Eyring plot is drawn.

In principle, the activation enthalpy can be negative. However, this usually occurs when TST is used to analyze nucleation, protein folding, or other rate processes with rugged energy landscapes and numerous intermediates. See for example the protein folding results in Figure 10.4.3.

The dynamics of processes with rugged energy landscapes severely violate the no-recrossing assumption of TST, so much so that they more closely resemble diffusion over a barrier (see Chapters 13 and 18). Therefore the $k_B T/h$ prefactor (and therefore the $ln(k/T)$ vs. $1/T$ axes of the Eyring plot) are not well justified. Moreover, mobility along the nucleus size coordinate (in nucleation) and transitions between intermediates (in protein folding) may be thermally activated [38–40]. In the protein folding example, the negative activation enthalpy has been interpreted as the result of different heat capacities for the reactant states and transition states [41]. However, the temperature dependence of folding rates is rather weak, e.g. in Figure 10.4.3 the rate only changes by a factor of about three over the reported range. Such subtle temperature

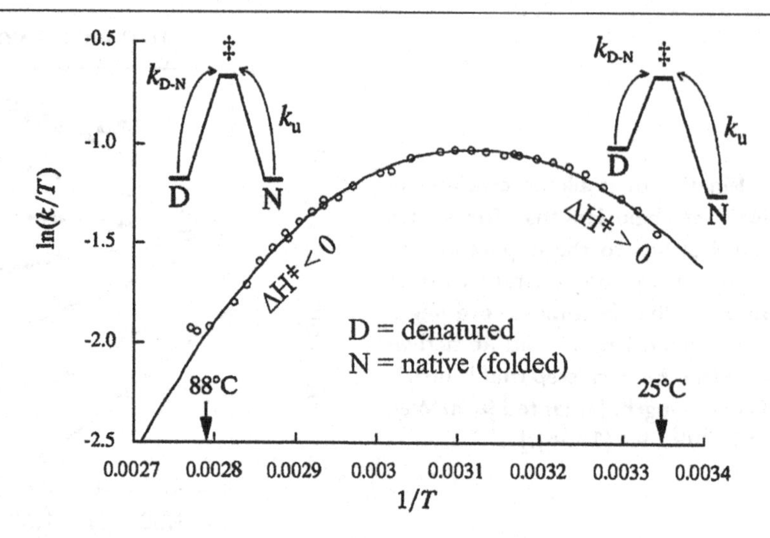

Figure 10.4.3: The apparent activation enthalpy for protein folding becomes negative at high temperatures. This has been attributed to different heat capacities for the unfolded states and transition states. Negative activation enthalpies may also emerge from temperature dependence in the stability of intermediates or from non-TST type temperature dependence in the mobility/prefactors. The insets depict free energy barriers that remain positive despite the negative activation enthalpy due to a loss of entropy upon folding. [From Oliveberg et al. *Proc. Nat. Acad. Sci. USA*, 92, 8926-29 (1995).]

effects, including the negative activation enthalpies, could result from any number of factors as outlined above.

10.5 Flux across a dividing surface

Wigner and Pelzer used the equilibrium distribution to compute the one-directional flux through a dividing surface that separates reactants from the products [6,18]. Their work predated the theory of Eyring, but the calculations were difficult to apply without modern computers and rare events methods. Now, the situation has changed, and Wigner's more general framework has become a practical computational tool. The theories are equivalent, but subtle differences between the prefactor and exponential parts of the two theories are sometimes overlooked.

Let $q(\mathbf{x})$ be the reaction coordinate. In general $q(\mathbf{x})$ may be a complicated function of the coordinates and the momenta. The previous section and also Chapter 20 discuss ways of identifying the appropriate coordinate $q(\mathbf{x})$. For practical reasons that are summarized elsewhere [20], we will assume that $q(\mathbf{x})$ only depends on the configuration space variables. Among the isosurfaces of $q(\mathbf{x})$, i.e. surfaces on which $q(\mathbf{x})$ is a constant, choose a special dividing surface $q(\mathbf{x}) = q_{\ddagger}$

which separates the reactant configurations from product configurations. Figure 10.5.1 illustrates the dividing surface between reactant states (A) and product states (B). Recall that A and B (not in boldface font) refer not to molecules, but more generally to regions of configuration space on opposite sides of the divide.

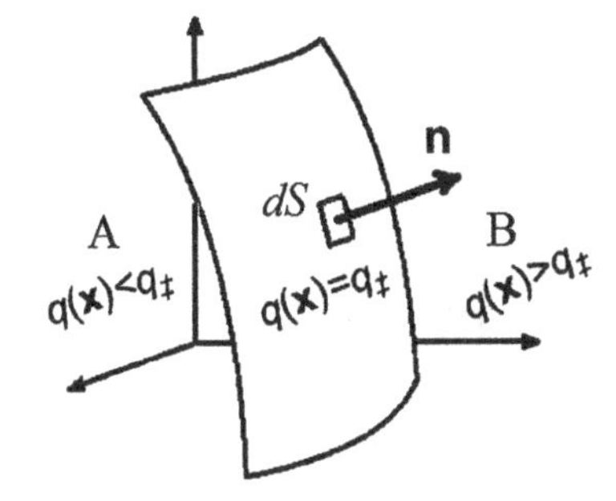

Figure 10.5.1: The dividing surface is a special isosurface of the reaction coordinate which separates reactant states $A = \{\mathbf{x}|q(\mathbf{x}) < q_{\ddagger}\}$ from product states $B = \{\mathbf{x}|q(\mathbf{x}) > q_{\ddagger}\}$. Transition state theory obtains the rate constant from the one-directional flux through each dividing surface element dS at equilibrium.

For convenience, define an indicator of configurations within B as

$$h_B(\mathbf{x}) = \begin{cases} 1 & q(\mathbf{x}) > q_{\ddagger} \\ 0 & otherwise \end{cases} \tag{10.5.1}$$

A similar indicator for states within A can be defined as $h_A(\mathbf{x}) = 1 - h_B(\mathbf{x})$. By definition the TST rate times the equilibrium population of reactants (state A) is the integrated one-directional flux from an equilibrium population of states on the surface $q(\mathbf{x}) = q_{\ddagger}$, i.e.

$$\kappa_{TST} \equiv \left\langle \overrightarrow{flux} \right\rangle_{\ddagger} / \langle h_A(\mathbf{x}) \rangle \tag{10.5.2}$$

Note that in equation (10.5.2), both the directed flux and the reactant population have been divided by the partition function to express the rate as a ratio of equilibrium averages. Also note that this flux actually has units of a frequency. The average directed flux can be computed from

$$\begin{aligned} \left\langle \overrightarrow{flux} \right\rangle_{\ddagger} &= \int_{q(\mathbf{x})=q_{\ddagger}} \rho_{eq}(\mathbf{x},\mathbf{p}) \frac{1}{2} |\dot{\mathbf{x}} \cdot n| \, dS \\ &= \frac{1}{2} \int_{q(\mathbf{x})=q_{\ddagger}} \rho_{eq}(\mathbf{x},\mathbf{p}) \, |\dot{\mathbf{x}} \cdot \nabla q| \, \|\nabla q\|^{-1} dS \\ &= \frac{1}{2} \int \rho_{eq}(\mathbf{x},\mathbf{p}) \, |\dot{q}| \, \delta\left[q(\mathbf{x}) - q_{\ddagger}\right] d\mathbf{x} d\mathbf{p} \\ &= \frac{1}{2} \left\langle |\dot{q}| \, \delta\left[q(\mathbf{x}) - q_{\ddagger}\right] \right\rangle \end{aligned} \tag{10.5.3}$$

In the first line, the absolute velocity has been multiplied by a factor 1/2 to discount those velocities initially directed from product to reactant.

Equations (10.5.2) and (10.5.3) result in $k_{TST} = \langle |\dot{q}|\, \delta\left[q(\mathbf{x}) - q_{\ddagger}\right]\rangle / 2\,\langle h_A(\mathbf{x})\rangle$. To obtain more familiar physical interpretations for the rate expression, multiply and divide by $\langle \delta\left[q(\mathbf{x}) - q_{\ddagger}\right]\rangle$. The result is

$$k_{TST} = \frac{1}{2}\,\langle |\dot{q}|\rangle_{\ddagger}\, \ell_q^{-1} exp\left[-\beta \Delta F_q^{\ddagger}\right] \tag{10.5.4}$$

Here

$$\ell_q^{-1} exp\left[-\beta \Delta F_q^{\ddagger}\right] = \frac{\langle \delta[q(\mathbf{x}) - q_{\ddagger}]\rangle}{\langle h_A(\mathbf{x})\rangle}, \tag{10.5.5}$$

ℓ_q is the unit of length along coordinate q, and

$$\langle |\dot{q}|\rangle_{\ddagger} = \frac{\langle |\dot{q}|\, \delta\left[q(\mathbf{x}) - q_{\ddagger}\right]\rangle}{\langle \delta\left[q(\mathbf{x}) - q_{\ddagger}\right]\rangle} \tag{10.5.6}$$

is a conditionally averaged absolute velocity along the reaction coordinate at the transition state.

The free energy difference in equation (10.5.5) has a different interpretation from those we have previously encountered. It effectively defines $\Delta F_q^{\ddagger} = F(q_{\ddagger}) - F_A$, where $F_A = -k_B T\, ln(\ell_q^{-1} \int_A dq\, exp[-\beta F(q)])$, and $F(q)$ is the Landau free energy as a function of q. Landau free energies are projections that provide the equilibrium probability density to find the system at position q via $\rho_{eq}(q) = \ell_q^{-1} exp[-\beta F(q)]$. See Chapter 11 for formal and computational aspects of Landau free energy calculations. Also see the exercises for alternative but equivalent formulations of the generalized TST rate expression.

Equation (10.5.3) of the generalized TST flux derivation integrates over $d\mathbf{x}d\mathbf{p}$ which implicitly suggests a canonical ensemble. However, equations (10.5.4)–(10.5.5) are just fluxes and probabilities phrased in terms of equilibrium averages, so with appropriate changes in sampling procedures they can be applied in other ensembles. Of course, it remains important to choose an appropriate ensemble for the rate process of interest. For example, a unimolecular gas phase reaction at low pressure would require NVE simulations, while bubble nucleation requires NPT simulations to accommodate the large activation volume.[3]

A few additional important points about the interpretation of equation (10.5.4) should be noted. The constant k_{TST} in equation (10.5.4) is an abstract first order TST rate constant for escape from state "A". With careful analysis, it can be converted into the familiar phenomenological rate constants and rate expressions for various types of processes.

1. First order rate constants: For reactions that consume a single reactant, conformer, or catalytic intermediate, state A corresponds to a single discrete entity. If the reaction occurs via

[3] But as explained in later chapters, TST is not appropriate for nucleation.

a single pathway, then equation (10.5.4) for k_{TST} immediately identifies the appropriate first order rate constant

$$k_{TST} = \frac{1}{2} \langle |\dot{q}| \rangle_{\ddagger} \ell_q^{-1} exp\left[-\beta \Delta F_q^{\ddagger}\right] \quad (unimolecular) \tag{10.5.7}$$

Matters are more complicated when a single discrete entity can react via multiple similar or symmetry equivalent pathways. For example in solid diffusion by vacancy hopping, any one of several neighboring atoms might hop into the vacancy via similar transition states. Likewise, any one of several equivalent ligands on an organometallic complex might undergo the same homolysis reaction.

(a) For symmetry equivalent pathways, it is usually easiest to compute the rate for one individual pathway and then apply symmetry factors to obtain the overall escape rate from A.

(b) When the pathways are non-equivalent but similar and well-separated in the sense that the ‡-ensemble from one pathway does not overlap with the ‡-ensemble from another, then one can separately compute the rate for each path. The parallel rates are then added to obtain the overall escape rate from A.

(c) When the pathways are similar with ‡-ensembles that do overlap, one can construct reaction coordinates from collective variables like coordination numbers or energy gaps. Collective variable reaction coordinates report progress along any one of the multiple reaction pathways. Sampling with collective variables can be difficult, but these calculations directly give an overall rate that accounts for the ensemble of paths and transition states.

2. Bimolecular reactions in solution may proceed via an associated complex. In this case the overall rate can be computed in stages. First compute an equilibrium constant for the association step. Then obtain k_{TST} for the chemical reaction step from equation (10.5.4) where state A is the associated complex. If the chemical reaction step is slow, then the association will be quasi-equilibrated. If the chemical step is extremely fast, then a PSSA rate law should be developed with the associated complex being formed at the rate of diffusion control.
3. Bimolecular rate constants for reactions that do not involve an associated complex of reactants require some conversions. Recall that k_{TST} in equation (10.5.4) is a first order decay rate from "state A". If the simulation has $n_{\mathbf{C}}$ and $n_{\mathbf{D}}$ bimolecular reactant molecules of type **C** and **D** then there are $n_{\mathbf{C}} n_{\mathbf{D}}$ equivalent ways to leave state A. Formally, $(k_{TST})_{eq.10.5.4} = k_{ext} n_{\mathbf{C}} n_{\mathbf{D}}$ where k_{ext} is the extensive number of events per time per **C** per **D**. Dividing $(k_{TST})_{eq.10.5.4} = k_{ext} n_{\mathbf{C}} n_{\mathbf{D}}$ by the simulation volume $\mathcal{V}$ gives a second order rate law:

$(k_{TST})_{eq.10.5.4}/\mathcal{V} = (k_{ext}\mathcal{V})(n_{\mathrm{C}}/\mathcal{V})(n_{\mathrm{D}}/\mathcal{V})$. Now we can solve for the quantity $k_{ext}\mathcal{V}$ to identify the bimolecular generalized TST rate constant as $(k_{TST})_{eq.10.5.4} \cdot \mathcal{V}/(n_{\mathrm{C}}n_{\mathrm{D}})$. Typically, there is just one reactant and one product in the simulation volume, i.e. $n_{\mathrm{C}} = n_{\mathrm{D}} = 1$. Thus

$$k_{TST} = \frac{1}{2}\mathcal{V}\langle|\dot{q}|\rangle_{\ddagger}\ell_q^{-1}exp\left[-\beta\Delta F_q^{\ddagger}\right] \quad (bimolecular) \tag{10.5.8}$$

These arguments can be generalized to higher reaction orders as necessary.

Why would anyone use rare events methods for a TST calculation when harmonic TST only requires DFT calculations and a pencil? The reason is that reactions in systems with rugged energy landscapes severely violate the assumptions of harmonic TST [42]. For example, consider chemical reactions in solution, in molten salts, and in enzymes. For each of these examples, the local quadratic approximation in harmonic TST cannot possibly describe the ensemble of saddle points and minima, but generalized TST might work well given a suitable reaction coordinate. Chapter 11 discusses some powerful rare events methods to compute $F(q)$ and $\langle|\dot{q}|\rangle_{\ddagger}$. Unfortunately, many rare events analyses do sophisticated calculations to get $F(q)$ and then incorrectly combine it with the k_BT/h prefactor of harmonic TST. Table 10.4 summarizes the proper pairings of prefactors and Boltzmann factors.

Table 10.4: Prefactor and exponential parts of TST rate constant for the Eyring and rare events formulations.

	Eyring	Rare events
kinetic prefactor	$k_BT\,\mathcal{V}_o^{\nu-1}/h$	$\frac{1}{2}\langle\lvert\dot{q}\rvert\rangle_{\ddagger}\ell_q^{-1}$
Boltzmann factor	$Q_{\ddagger}/\prod_{i=1}^{\nu}Q_{\mathbf{A}(i)}$	$exp[-\Delta F_q^{\ddagger}/k_BT]$
compute using:	saddle + *ab initio*	umbrella sampling

When using equations from Table 10.4, recall that ν is the overall reaction order, translational partition functions for harmonic TST should be computed using a volume $\mathcal{V}_o$ which matches that in the prefactor, and ℓ_q is the unit of length along the reaction coordinate. The example below illustrates a rare events based TST calculation for the diffusivity of methane within a natural gas hydrate crystal [43].

Example: Methane diffusion in gas hydrate

Natural gas hydrates are potential climate change amplifiers, natural gas resources, and CO_2 sequestration reservoirs [44,45]. Structurally, they are comprised of a crystalline matrix of water molecules hydrogen bonded into 12 and 14 sided polyhedra which tile space by sharing sides with each other [46]. The "faces" of the polyhedra are rings of 5 or 6 hydrogen bonded O-atoms. Within each polyhedral cage there is a void which can "host" a small hydrophobic "guest" molecule like CH_4 or CO_2. Diffusivities of guests in the host lattice are of interest because of efforts to displace CH_4 and sequester CO_2 in natural gas hydrates. Peters et al. computed the methane diffusivity in a natural gas hydrate using a molecular force field, free energy calculations, transition state theory, dynamical rate constant corrections, and kinetic Monte Carlo [43]. TST was used to estimate the rate at which methane hops from a donor cage (D) into neighboring vacant acceptor (A) cage.

The free energy landscape for donor-to-acceptor hopping was computed using an equilibrium path sampling method [43] (similar to Bolas [47] and hybrid MC/MD [48,49]). At each time step, the donor and acceptor cage centers were identified based on the ca. 24 water molecules surrounding each polyhedral cage. Spherical bipolar coordinates were used to track the donor-to-acceptor progress. The (approximate) reaction coordinate is therefore $q = ln[R_D/R_A]$ with R_D and R_A being the distance from the hopping methane (M) to the donor and acceptor cage centers, respectively [43]. The coordinate q is dimensionless, so $\ell_q \equiv 1$ with no units.

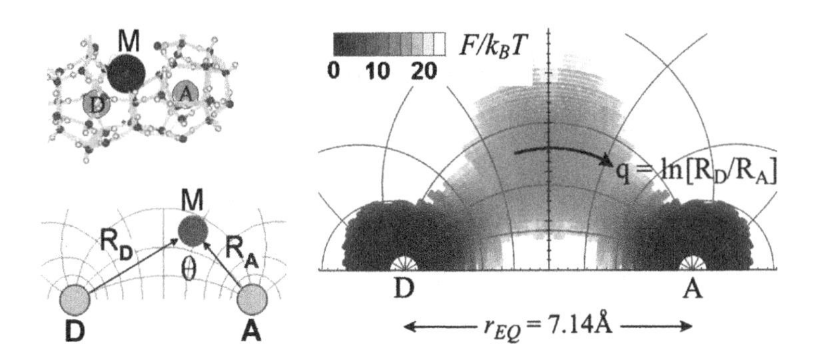

The figure shows: (lower left) the spherical bipolar coordinate system; (upper left) the water molecules that form the donor and acceptor cages, the methane molecule M, the donor cage center D, and the acceptor cage center A; (right) the free energy landscape along with curves of constant q and θ. The free energy can be further projected from $F(q, \theta)$ to $F(q)$ by numerical integration:

$$F(q) = -k_B T \, ln \int_0^{\pi} d\theta \, e^{-\beta F(q,\theta)}$$

The resulting one-dimensional free energy profile is shown below. The broad depression in the two-dimensional free energy landscape becomes a shallow intermediate at $q = 0$ on the one-dimensional free energy profile.

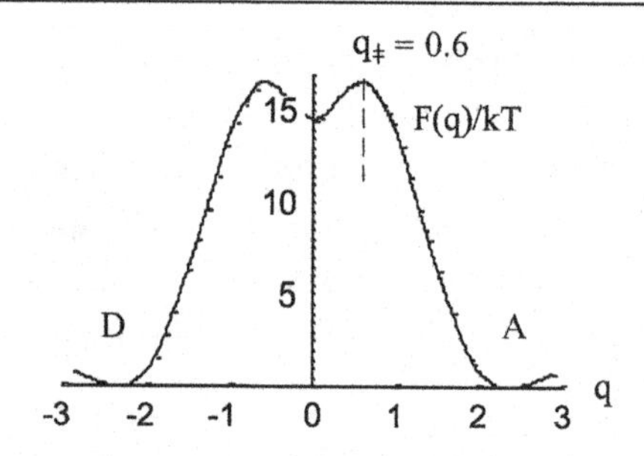

The coordinate system and also the location of the dividing surface along the reaction coordinate should be chosen carefully for a generalized TST calculation. If the dividing surface in this calculation is placed at $q = 0$, most trajectories will rattle around in the shallow intermediate and violate the no-recrossing assumption. The choice $q_\ddagger = 0.6$ mitigates the effects of recrossing. From this choice, Peters et al [43]. obtain $\Delta F_q^\ddagger = 16.4 k_B T$.

The prefactor can be approximated by an instructive analytic calculation. q depends on the position of the methane united atom and also the surrounding water molecules (via the ghost atoms). Denote the coordinates of methane as $\mathbf{x}_M$, and the positions of the donor and acceptor ghost atoms as $\mathbf{x}_D$ and $\mathbf{x}_A$. The ghost atom positions move slowly because they average the motions of 24 surrounding water molecules, so their contributions to $\langle|\dot{q}|\rangle_\ddagger$ can be neglected, i.e. [43]

$$\begin{aligned}\langle|\dot{q}|\rangle_\ddagger &= \left\langle\left|\frac{\partial q}{\partial(\mathbf{x}_M, \mathbf{x}_D, \mathbf{x}_A)} \cdot \frac{d(\mathbf{x}_M, \mathbf{x}_D, \mathbf{x}_A)}{dt}\right|\right\rangle_\ddagger \\ &\approx \left\langle\left|\frac{\partial q}{\partial \mathbf{x}_M} \cdot \frac{d\mathbf{x}_M}{dt}\right|\right\rangle_\ddagger = \sqrt{\frac{2k_BT}{\pi m_M}}\left\langle\left\|\frac{\partial q}{\partial \mathbf{x}_M}\right\|\right\rangle_\ddagger\end{aligned}$$

where m_M is the mass of a methane molecule (16amu).

The average of $\|\partial q/\partial \mathbf{x}_M\|$ can be computed using tabulated Lame coefficients, i.e. coordinate scalefactors for orthogonal coordinate systems [50]. In general, the Lame coefficient for a coordinate q is $h_q = \|\partial \mathbf{x}/\partial q\|$. Using the relationship between gradients and Lame coefficients [50], $\|\partial q/\partial \mathbf{x}_M\| = h_q^{-1}$. For the spherical bipolar coordinate q,

$$h_q = \frac{r_{eq}}{2(cosh(q) - cos\theta)}$$

where r_{eq} is the equilibrium distance between D and A. Finally, the prefactor is

$$\langle|\dot{q}|\rangle_\ddagger = \sqrt{\frac{8k_BT}{\pi m_M r_{eq}^2}}\,(cosh(q_\ddagger) - \langle cos\theta\rangle_\ddagger)$$

Numerical integration is required to evaluate $\langle cos\theta \rangle_{\ddagger}$:

$$\langle cos\theta \rangle_{\ddagger} = \frac{\int_0^{\pi} d\theta \, e^{-\beta F(q_{\ddagger},\theta)} cos\theta}{\int_0^{\pi} d\theta \, e^{-\beta F(q_{\ddagger},\theta)}}$$

In dimensionless form, the resulting prefactor is $\beta h \langle |\dot{q}| \rangle_{\ddagger} = 0.27$. Note that $k_B T/h$ is nearly a factor of four larger than the correct prefactor. Peters et al. used the computed hopping rates in a kinetic Monte Carlo simulation to estimate the diffusivity as approximately $D \approx 10^{-16} m^2/s$ for typical methane vacancy concentrations [43]. Similar diffusivities were later estimated from experiments by Kuhs and coworkers [51].

The generalized TST formula in equation (10.5.4) requires a calculation of the free energy as a function of reaction coordinate q. Instead of starting from the molecular PES, suppose that we begin with a model free energy profile along q. Direct models of the free energy are often useful for phenomena at the macromolecular and colloidal scale. For examples, see models for polymer translocation through a pore [52], dissociation barriers between colloids [53], protein unfolding in pulling experiments [54], and mechanochemical models for the action of cellulase enzymes [55]. The dynamics in these processes are not likely to obey the assumptions of TST. Nevertheless there is a simple "one-dimensional Vineyard" [56] version of TST that can provide crude rate estimates and trends. In addition, Vineyard's calculation is not ideal for chemical reactions because it treats all vibrations classically. Despite its limited accuracy, the 1D Vineyard TST is useful for back-of-the-envelope calculations and it provides some enlightening insight on TST prefactors.

Example: The one dimensional 'Vineyard' TST

Consider a potential of mean force $F(q)$ that is approximately harmonic in the reactant potential well as shown in the figure.

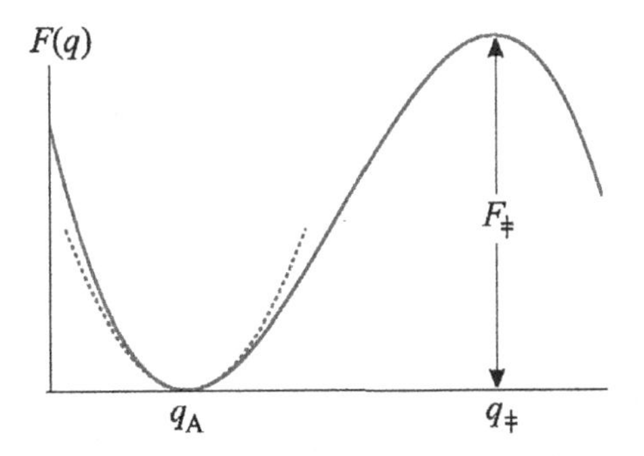

Let the free energy near the minimum in state A be approximated as $F(q) \approx \frac{1}{2}\mu\omega_A^2(q - q_A)^2$. Also let the free energy barrier measured *from the bottom* of state A be $F_\ddagger$. Throughout this book, free energy barriers measured from a minimum are denoted $F_\ddagger$, while those which are measured from the ensemble free energy of the reactants are denoted $\Delta F^\ddagger$. Within a normalization constant, the equilibrium probability of finding the system in a narrow interval of width δq around $q_\ddagger$ is

$$\delta q\, \ell_q^{-1} exp[-\beta F_\ddagger]$$

The equilibrium probability of finding the system in state A is

$$\int_{\cup} dq\, \ell_q^{-1} exp[-\beta F(q)] \approx \ell_q^{-1}\sqrt{\frac{2\pi k_B T}{\mu\omega_A^2}}$$

where the symbol $\cup$ indicates integration over the stable reactant well, and where the integral has been evaluated with a Gaussian approximation. Again we need the frequency at which transition states leave the narrow interval δq around $q_\ddagger$,

$$\nu_\ddagger = \frac{\langle|\dot{q}|\rangle_\ddagger}{2\delta q} = \frac{1}{2\delta q}\sqrt{\frac{2k_B T}{\pi\mu}}$$

Now the flux-over-population TST rate constant is

$$\begin{aligned} k_{TST} &= \frac{1}{2\delta q}\sqrt{\frac{2k_B T}{\pi\mu}}\left(\frac{\delta q\, \ell_q^{-1} exp[-\beta F_\ddagger]}{\ell_q^{-1}\sqrt{\frac{2\pi k_B T}{\mu\omega_A^2}}}\right) \\ &= \frac{\omega_A}{2\pi} exp[-\beta F_\ddagger] \end{aligned} \tag{10.5.9}$$

ℓ_q and the interval width δq vanish from the rate expression. Also note that Planck's h never even enters the derivation. The frequency ω_A in the prefactor is often interpreted as an "attempt frequency". However, the derivation shows that ω_A emerges from the partition function, i.e. ω_A emerges from the *thermodynamics* of the reactant state, not from the dynamics. See Vineyard [56] for a multidimensional version of this one dimensional result.

The 1D Vineyard TST derivation emphasizes that the units of the reaction coordinate have no bearing on the TST rate. Perhaps less obvious is that non-linear transformations of the reaction coordinate have no effect on the predicted TST rate. Let us examine the TST calculation for two variables $q(\mathbf{x})$ and $\varphi(\mathbf{x})$, where φ is some monotonic function of q. For example, we can often keep track of an angle $\varphi(\mathbf{x})$ or its cosine $q(\mathbf{x}) = cos\varphi(\mathbf{x})$. The probability of being in an interval $(q,\ q + dq)$ must be the same as the probability of being in the corresponding interval $(\varphi(q),\ \varphi(q) + \varphi'(q)dq)$. Thus, within normalization factors, the free energy profiles must be related by

$$\ell_q^{-1} e^{-\beta F_q(q)}\, |dq| = \ell_\varphi^{-1} e^{-\beta F_\varphi(\varphi)}\, |d\varphi| \tag{10.5.10}$$

Equation (10.5.10) immediately gives a relationship between the free energy profiles along q and φ. Within a constant, the relationship is

$$F_\varphi(\varphi) = F_q(q) + k_B T\, ln \left| \frac{d\varphi}{dq} \right| \tag{10.5.11}$$

Equation (10.5.11) shows that, when $d\varphi/dq$ is not a constant, the free energy maximum along q will not correspond to the free energy maximum along φ.

On the other hand, the TST rate constant computed using corresponding locations $q_\ddagger$ and $\varphi_\ddagger = \varphi(q_\ddagger)$ as dividing surfaces is exactly the same because the Jacobian term in equation (10.5.11) is exactly compensated by a change in the prefactor. Specifically,

$$\frac{\langle|\dot{\varphi}|\rangle_\ddagger}{\langle|\dot{q}|\rangle_\ddagger} = \left| \frac{d\varphi}{dq} \right|_\ddagger \quad and \quad \frac{exp[-\beta F_\varphi(\varphi_\ddagger)]}{exp[-\beta F_q(q_\ddagger)]} = \left| \frac{dq}{d\varphi} \right|_\ddagger \tag{10.5.12}$$

Thus the TST rate constant, which is a product of the absolute velocity prefactor and the Boltzmann factor, is influenced only by the dividing surface, and not by changes of scale nor by nonlinear transformations between mechanistically equivalent coordinates.

10.6 Variational transition state theory

Recall that harmonic TST (section 10.3) was based on a reaction coordinate with locally separable dynamics in the vicinity of the saddle point. The equation of motion for the unstable mode reaction coordinate was $d^2q/dt^2 = \left|\omega_\ddagger\right|^2 q$ so that trajectories with $q = 0$ at $t = 0$ will ever accelerate in the direction of their initial velocities. Thus, within the trust radius of the local quadratic approximation, the hyperplanar dividing surface $q = 0$ is perfect.

Of course, the dynamics of real reactions do not perfectly separate into uncoupled harmonic motions. Even reactions in the gas phase with a single well-defined reaction channel, have curved reaction pathways, anharmonic potential energy barriers, and transverse vibrational modes with frequencies that change along the reaction pathway. Pelzer and Wigner recognized that these features cause trajectories to violate the dynamical no-recrossing assumption of TST [6]. Chemical reactions in aqueous solution, in enzymes, and in molten salts show even more pernicious pathologies. These systems have rugged potential energy surfaces with countless inconsequential saddles and minima so that a *harmonic* TST approach is hopeless [57], but other TST frameworks may be reasonably accurate. Although the dynamics are not perfectly separable, chemical reactions in solution primarily involve only a few important solute and solvent coordinates. If the most important degrees of freedom can be identified and used to construct a dividing surface, then generalized TST might provide reasonably accurate rate constants.

What quantitative rule should guide the choice of coordinates and dividing surfaces? As Pelzer and Wigner noted, recrossing contributions always cause TST to overestimate the (classical) rate [58]. Figure 10.6.1 shows how recrossing influences the computed TST rate.

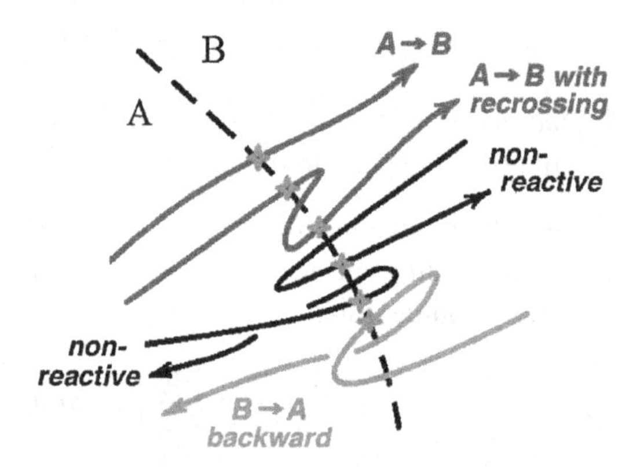

Figure 10.6.1: The dashed line depicts a dividing surface between states A and B. Small crosses mark each forward crossing event. The forward crossing events all contribute to the TST rate calculation, but only the direct $A \rightarrow B$ trajectory is consistent with the assumptions of TST. All other trajectories make "extra" contributions to the TST rate.

The fact that TST provides an upper bound on the rate immediately suggests a variational principle. Wigner initially suggested that the dividing surface should be positioned so that "...the total number of crossings is as small as possible." [58] Horiuti later suggested that the dividing surface should be varied *to minimize the rate* [59]. Note that minimizing the rate is potentially different from minimizing the *number* of recrossings, because the rate appropriately weights trajectories according to velocity. Keck performed the first VTST calculations [60,61], and the modern VTST methods are due to Garrett and Truhlar [62–64], Schenter et al. [65,66], and Pollak et al. [26,67,68].

Let trial dividing surfaces be parameterized by $q(\mathbf{x}) = q_\ddagger$, i.e. by a specific isosurface of some reaction coordinate, with both $q(\mathbf{x})$ and $q_\ddagger$ to be determined. The VTST rate constant is

$$k_{VTST} = \min_{q(\mathbf{x}),q_\ddagger} \frac{1}{2} \langle|\dot{q}|\rangle_\ddagger \ell_q^{-1} exp\left[-\beta\Delta F_q^\ddagger\right] \qquad (10.6.1)$$

where minimization is over all trial coordinates and all possible values of $q_\ddagger$ for each trial coordinate. Practical implementations of VTST often assume a reaction coordinate $q(\mathbf{x})$, and then only minimize over different $q_\ddagger$ values in $q(\mathbf{x}) = q_\ddagger$. A true VTST involves optimization of

$q_\ddagger$ and also a vastly more difficult functional optimization of $q(\mathbf{x})$ itself. This point is illustrated in Figure 10.6.2.

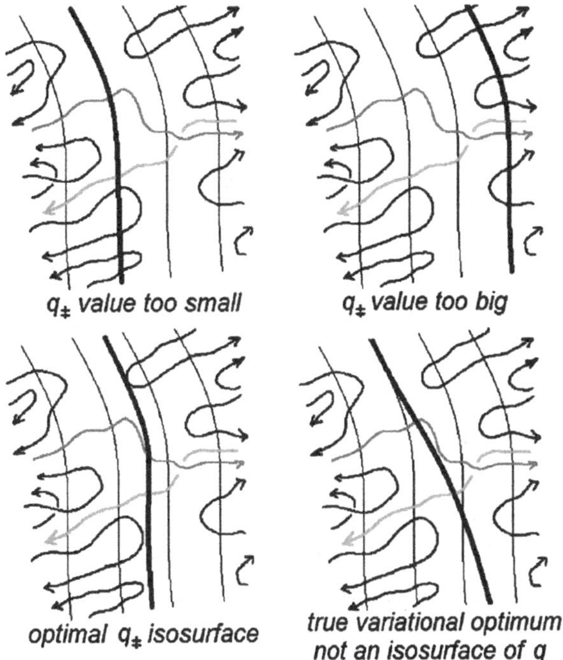

Figure 10.6.2: A schematic collection of trajectories near the dynamical bottleneck. A family of isosurfaces for some coordinate $q(\mathbf{x})$ is also shown. The value of $q_\ddagger$ in $q(\mathbf{x}) = q_\ddagger$ can be optimized, but the variationally optimal dividing surface is not an isosurface of $q(\mathbf{x})$.

One important and practical application for VTST is in computing rate constants for association/dissociation reactions that do not have a saddle point between reactants and products. For example, VTST has been used to find transition states and rates for barrierless $H + CH_3 \rightarrow CH_4$ and $O + OH \rightarrow OOH$ reactions [69,70]. The example below is an illustrative rate calculation for a dissociation $\mathbf{AB} \rightarrow \mathbf{A} + \mathbf{B}$ in which the potential of mean force between fragments $\mathbf{A}$ and $\mathbf{B}$ is a Morse potential.

Example: Dissociation with no saddle point

Consider a reaction $\mathbf{AB} \rightarrow \mathbf{A} + \mathbf{B}$ for which the separated fragments can recombine with no energetic barrier. Let r be the distance between $\mathbf{A}$ and $\mathbf{B}$ fragments, and let $U(r)$ be the potential of mean force, integrated over all orientations and internal degrees of freedom of $\mathbf{A}$ and $\mathbf{B}$. For purposes of illustration, suppose that $U(r)$ is a Morse potential,

$$U(r) = E_D \left\{ e^{-2a(r/r_0 - 1)} - 2e^{-a(r/r_0 - 1)} \right\}.$$

In the Morse potential, E_D is the dissociation energy, r_0 is the equilibrium separation distance between $\mathbf{A}$ and $\mathbf{B}$, $a = \sqrt{\mu\omega_0^2 r_0^2/(2E_D)}$, the reduced mass is $\mu^{-1} = m_{\mathbf{A}}^{-1} + m_{\mathbf{B}}^{-1}$, and ω_0 is the angular vibrational frequency at the energy minimum. The potential of mean force $U(r)$ is shown in the figure below.

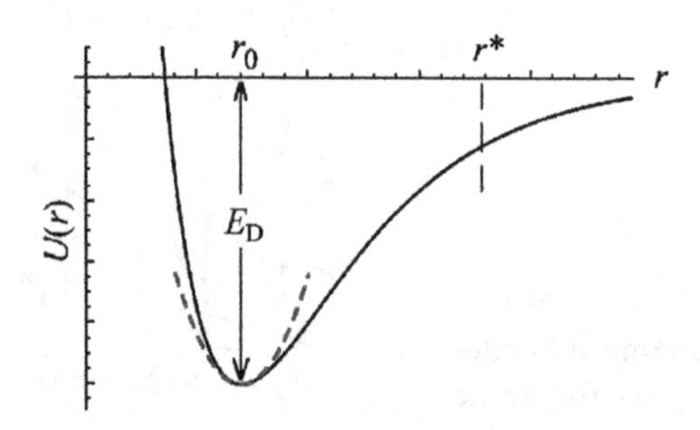

The last parameter that we need for a TST dissociation rate calculation is the dividing surface location along r. Because there is no saddle point, let r_* be a trial dividing surface location. Before proceeding with the calculation, note that RRKM should be used instead of TST if **A** and **B** are small fragments in a low pressure gas environment (see Chapter 9). Let us suppose the pressure is sufficiently high to ignore these concerns. Then for a dividing surface r_*, the TST dissociation rate constant is

$$k_{TST}(r_*) = \frac{1}{2}\langle|\dot{r}|\rangle_{r_*}\ell_r^{-1}exp\left[-\beta\Delta F_r(r_*)\right]$$

The absolute average velocity along r is

$$\langle|\dot{r}|\rangle_* = \sqrt{\frac{2k_BT}{\pi\mu}}$$

The Boltzmann factor is

$$\begin{aligned}\ell_r^{-1}exp\left[-\beta\Delta F_r(r_*)\right] &= \frac{\int d\mathbf{x_A}\int d\mathbf{x_B}\delta[r(\mathbf{x})-r_*]exp[-\beta U(r)]}{\int d\mathbf{x_A}\int d\mathbf{x_B}h_{\mathbf{AB}}[r(\mathbf{x})]exp[-\beta U(r)]}\\ &= \frac{V\int_0^{\infty}4\pi r^2\delta[r-r_*]exp[-\beta U(r)]dr}{V\int_0^{r_*}4\pi r^2 exp[-\beta U(r)]dr}\end{aligned}$$

Now use $U(r)\approx -E_D+\mu\omega_0^2(r-r_0)^2/2$ to approximate the integral in the denominator. The result is

$$\ell_r^{-1}exp\left[-\beta\Delta F_r(r_*)\right] = \frac{r_*^2 exp[-\beta(E_D+U(r_*))]}{\sqrt{2\pi}\left\{1+\beta\mu\omega_0^2r_0^2\right\}/(\beta\mu\omega_0^2)^{3/2}}$$

For fragments that bind strongly, $\beta\mu\omega_0^2 r_0^2 \gg 1$. The Boltzmann factor then simplifies to $\ell_r^{-1} exp[-\beta\Delta F_r(r_*)] \approx (r_*^2/r_0^2) exp[-\beta(E_D + U(r_*))]/\sqrt{2\pi/\beta\mu\omega_0^2}$. Combining the Boltzmann factor and the prefactor gives [71,72]

$$k_{TST}(r_*) = k_0 \cdot \frac{r_*^2}{r_0^2} e^{-\beta U(r_*)}$$

where

$$k_0 = \frac{\omega_0}{2\pi} e^{-\beta E_D}$$

The factor r_*^2 accounts for the surface area of the spherical dividing surface. The factor $exp[-\beta U(r_*)]$ accounts for the increasing energy required to separate $\mathbf{A}$ and $\mathbf{B}$ by a distance r_*. The opposing trends in these two factors give rise to an optimum dividing surface.

$$k_{VTST} \;=\; k_0 \min_{r_*} \frac{r_*^2}{r_0^2} exp[-\beta U(r_*)]$$

The figure below shows results of the variational minimization for the case $a = 1$ with $\bar{r} = r/r_0$. Note that the variational optimum r_*, is a decreasing function of temperature.

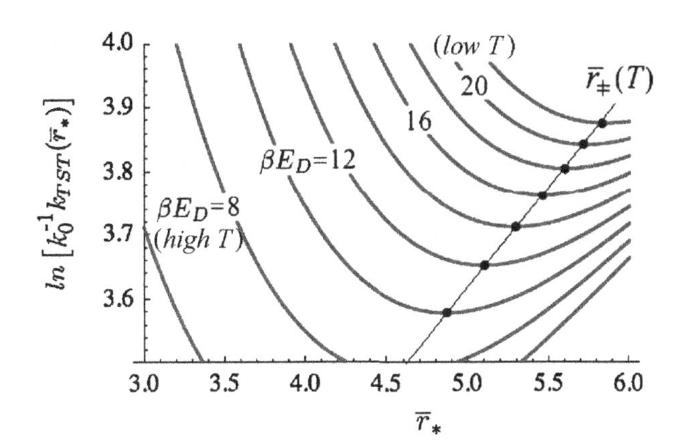

For the case $a = 1$, the optimal dividing surface location is several times longer than the equilibrium bond distance. The association rate constant can be determined from the equilibrium constant and the dissociation rate.

VTST for the intrinsic arclength reaction coordinate

In harmonic TST, we assumed that the optimal dividing surface is the hyperplane perpendicular to the unstable mode at the saddle point. This seemed like a natural choice, but the molecular geometry and transverse vibrational frequencies change with arclength s along the MEP. Accordingly, the changing contributions of rotational and vibrational free energy can shift the optimal dividing surface away from the hyperplane that passes through the saddle point. Isosurfaces of the arclength $s(\mathbf{x})$ form a series of hyperplanes which are perpendicular to the MEP (see Figure 10.6.3).

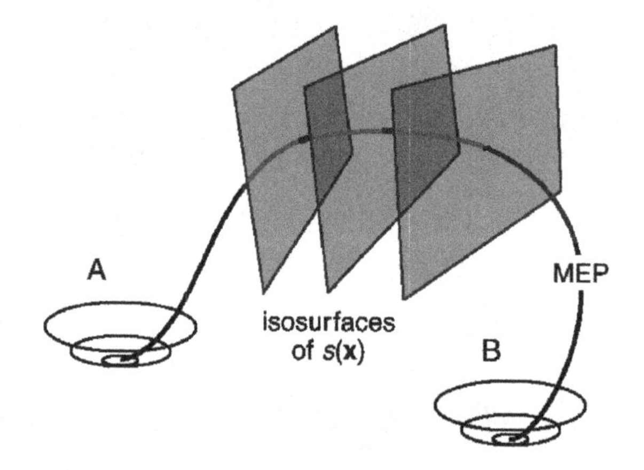

Figure 10.6.3: Isosurfaces of the intrinsic arclength reaction coordinate $s(\mathbf{x})$ form a series of hyperplanes perpendicular to the minimum energy path (MEP). These trial dividing surfaces are used in the VTST procedure of Truhlar and Garrett.

Garrett and Truhlar used VTST to identify the optimal hyperplane within this family of isosurfaces. For each isosurface of $s(\mathbf{x})$, they compute the transverse frequencies at s, the potential energy on the MEP at s, and the molecular geometry along the MEP at s. These quantities provide $\Delta G^o(s)$ in same way that $\Delta G^{\ddagger}$ is computed for harmonic TST (at $s = 0$), except that the $\Delta G^o(s)$ calculation omits the MEP tangent vector $\boldsymbol{a}'(s)$ instead of the unstable eigenvector at the saddle point. The result is a TST rate constant for each s-isosurface [62–64,73]

$$k_{TST}(s) = \frac{k_B T}{h} \nu_o^{\nu-1} exp\left[-\frac{\Delta G^o(s)}{k_B T}\right] \tag{10.6.2}$$

Finally, the arclength parameter s is varied to minimize the rate constant

$$k_{VTST} \approx \min_s k_{TST}(s) \tag{10.6.3}$$

One might expect curvature of the reaction pathway to influence this calculation, but it does not, like a Pappus' theorem for reaction pathways [74]. Note, however, that curvature does influence the extent of recrossing, and therefore it can influence the accuracy of the VTST rates.

As noted by Garrett and Truhlar [64], their procedure is still an approximate VTST calculation. A full VTST calculation would find optimal dividing surfaces that are not hyperplanes. For two dimensional potentials, Pechukas and Pollak constructed exact periodic orbit dividing surfaces (PODS) [26]. Others have used Lie canonical perturbation theory to obtain exact dividing surfaces in the full phase space [75–77]. Dividing surfaces that depend on momenta lead to prefactors that include averaged accelerations (forces). These surface eliminate recrossing, but the TST calculation for one of these surfaces is no longer amenable to simple pen-and-paper statistical mechanics.

Limitations

The functional nature of $k_{TST}[q(\mathbf{x}); q_{\ddagger}]$ makes VTST a non-trivial and effectively infinite-dimensional, optimization problem. To fully appreciate the difficulty, note that even a single evaluation of k_{TST} for a trial dividing surface $q(\mathbf{x}) = q_{\ddagger}$ requires free energy calculations and prefactor calculations. Thus, true variational dividing surface optimizations, considering *all* degrees of freedom, have only been performed for simple models.

Accurate approximations to VTST have been developed for gas phase reactions, but reactions in solution and enzymes pose more difficult challenges. There are too many degrees of freedom to treat explicitly, and it is often difficult to distinguish the important degrees of freedom from the bath. Thus reactions in solution or in enzymes require that VTST be performed with an incomplete space of *a priori* selected variables. Some promising methods based on collective variables have been developed. For example, Schenter et al. proposed VTST in the combined space of a solute arclength reaction coordinate and a collective solvent coordinate [66]. The String method in collective variables [78] and the Path-like collective variables version of Metadynamics [79] enable calculations that are analogous to the hyperplane optimization of Garrett and Truhlar. Pollak et al. proposed an approach to VTST [67,68,80] that iteratively improves an approximate dividing surface. All of these methods work within a small *a priori* selected subset of solute and solvent coordinates. See Zinovjev and Tunon [81] for a method that starts from a large list of variables and systematically wheedles it down to the best ones.

VTST is a powerful theoretical principle that profoundly shaped our understanding of reaction rate theory. In certain cases and within certain approximations, VTST can also be a practical computational framework. However, VTST remains limited in applications to reactions in solution and enzymes. The major difficulty is identifying an appropriate basis of reaction coordinates for the dividing surface optimization. As Klippenstein et al. noted, "The definition of the transition state dividing surface is equivalent to locally defining a reaction coordinate" [82]. A new group of methods, discussed in Chapter 20, has emerged for optimizing reaction coordinates. In some applications, these methods have surpassed VTST as a framework for dividing surface optimization. When accurate solute and solvent coordinates cannot be identified, we must be content to account for dividing surface errors (and dynamical recrossing) with transmission coefficients from reactive flux approaches (see Chapter 13).

10.7 Harmonic TST with internal coordinates

Harmonic TST is often used with *ab initio* continuum solvation models for reactions in solution. The procedure exactly parallels that of standard harmonic TST, except that the PES is altered by the presence of a continuum dielectric [83,84], i.e. an implicit solvent. Remarkable progress has been made with these relatively simple calculations [85,86], but they also have some serious limitations. Of course TST is not expected to capture dynamical solvent effects, but there are even more basic issues. For activated processes that involve desolvation, the energy landscape may not have a saddle point, c.f. ion-pair dissociation [87,88], ion-attachment at kink sites in crystal growth [89,90], or reactions involving solvent cavity formation [91,92]. Also reactions in which one or more water molecules shuttle protons pose major challenges [35, 93–95]: transition states are difficult to identify, and accurate activation parameters are often elusive. Even when a saddle point corresponding to the transition state can be identified, there are major challenges in computing accurate free energy barriers. For example, consider a reaction whose transition state includes an explicit water molecule to shuttle protons. To correctly model the activation entropy, the water molecule in the reactant state should be considered as a separately rotating and translating entity, but to correctly model the activation enthalpy the explicit water molecule should be allowed to interact with the reactants [96].

Are there more consistent strategies for modeling the reactants and transition states in solution? The ideal calculation would model the entire reaction process in an explicit solvent, e.g. using QM/MM methods with a selected reaction coordinate [97–99]. However, these calculations are also fraught with difficulties. Simulations with adaptive QM/MM regions fail to conserve momentum and/or energy as molecules cross the QM/MM boundary, and reaction coordinates are often difficult to identify for reactions in polar solvents. These problems have been circumvented with various approximate and application-specific fixes:

- The free energy surface or PES can be computed as a function of two (or possibly three) selected bond lengths. For example, Klahn et al. computed 2D reduced PESs to examine the creation of ATP from phosphate ions [100]. Rosta et al. took a step farther by computing true free energy surfaces as a function of two bond lengths in their study of RNase enzymes [101]. In these studies it is important to remember that differences in reduced masses will influence the dynamics and the reaction coordinate, e.g. the OH bonds and CC bonds selected by Rosta et al. [101] have a nearly three-fold difference in vibrational frequency. Additionally, solvent coordinates that are not obviously relevant can also make important contributions [102].
- Another approach reduces the problem to a single reaction coordinate at the outset [103–105]. For example, several studies have used the coordinate $q = b_1 - b_2$ where b_2 is a bond being formed and b_1 is a bond being broken. This approach ignores coordinates other than b_1 and b_2, and assumes equal contributions from coordinates b_1 and b_2 regardless of

their relative reduced masses and without knowing the shape of their 2D free energy landscape. Using one arbitrarily chosen coordinate is more risky than using two, but reduction to 1D is advantageous for sampling (especially with *ab initio* or QM/MM Hamiltonians).

In principle, the unstable eigenmode from harmonic TST with continuum solvation could incorporate changes in all (solute) bonds and angles, including the effects of different masses. As such it would be preferable to each of the procedures described above. Additionally, a biased QM/MM simulation using the unstable eigenmode coordinate would systematically identify how explicit solvation improves upon the continuum solvation model. Unfortunately, the unstable eigenmode for harmonic TST is obtained from the mass weighted Hessian in a fixed laboratory coordinate system. The lack of rotational invariance makes the unstable mode coordinate unsuitable (or at least terribly inconvenient) for biasing an all-atom simulation.

There is a way to combine the advantages of completely internal solute coordinates, mass-weighting, and local dynamical separability. The recipe borrows the blueprint of a harmonic TST calculation, but uses internal coordinates (**b**) and a PES with continuum solvation (V_{PCM}). With these ingredients, $\mathbf{f}=\mathbf{Ma}$ becomes [73]

$$-\mathbf{G}\left[\frac{\partial^2 V_{PCM}}{\partial\mathbf{b}\partial\mathbf{b}}\right]\Delta\mathbf{b}=\frac{d^2}{dt^2}\Delta\mathbf{b} \tag{10.7.1}$$

where $\mathbf{G}=\mathbf{BM}^{-1}\mathbf{B}^{\mathrm{T}}$, $\mathbf{B}$ is the Wilson-B matrix [106], and $\Delta\mathbf{b}$ is a vector displacement from the saddle in bondlength, angle, and dihedral coordinates. The matrix $\mathbf{G}\left[\partial^2 V_{PCM}/\partial\mathbf{b}\partial\mathbf{b}\right]$ is not symmetric. Therefore, the eigenvalues in $\Delta\mathbf{b}$-space are not necessarily orthogonal to each other (see Figure 10.7.1).

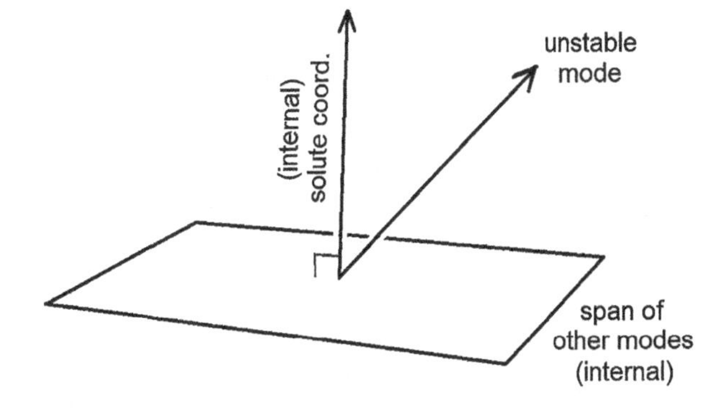

Figure 10.7.1: In internal coordinates, $\mathbf{f}=\mathbf{Ma}$ identifies an unstable mode that is not necessarily orthogonal to the stable modes. The correct dividing surface is an isosurface of the internal coordinate with separable dynamics.

The **G** matrix closely resembles the metric tensor from transition path theory (TPT) [78,107, 108]. The key difference is that **G** contains the different masses (i.e. dynamics) whereas TPT uses a purely geometric tensor $\mathbf{BB}^{\mathrm{T}}$ which has been thermally averaged over all degrees of freedom not included in **B** [78,107,108]. The non-symmetric structure of equation (10.7.1) also resembles the theory of Berezhkovskii and Szabo for diffusion over a saddle on the free energy landscape. [109].

It can be shown (see exercises) that a reaction coordinate with separable dynamics can be obtained from the unstable eigenvector of the commuted matrix product

$$\left[\frac{\partial^2 V_{PCM}}{\partial \mathbf{b} \partial \mathbf{b}}\right] \mathbf{G}\, \mathbf{u}_{\ddagger} = -\omega_{\ddagger}^2 \mathbf{u}_{\ddagger} \tag{10.7.2}$$

The vector $\mathbf{u}_{\ddagger}$ provides an internal rotationally invariant representation of the unstable mode reaction coordinate

$$q = \mathbf{u}_{\ddagger}^T \Delta \mathbf{b} \tag{10.7.3}$$

To appreciate what this coordinate is doing, consider the collinear $H_2 + D$ reaction in the gas phase. The "heavier" $H - D$ bond is more important in $q = \mathbf{u}_{\ddagger}^T \Delta \mathbf{b}$ than the lighter $H - H$ bond, as one expects from the more typical mass-weighted *Cartesian* analysis.[4] The coordinate $q = \mathbf{u}_{\ddagger}^T \Delta \mathbf{b}$ could dramatically improve results from QM/MM simulations of reactions in solution by reducing the sampling space to an accurate scalar internal coordinate. It would also provide a direct comparison of results from explicit and implicit solvent models. To my knowledge none of these calculations have been done.

10.8 Non-idealities

Non-idealities can significantly influence the rates of reactions in high pressure gases and reactions between charged species in solution. Should our theories and phenomenological expressions be phrased in terms of activities rather than concentrations? As written by Hinshelwood [110], "The question really is ... when $c_{\mathbf{A}} c_{\mathbf{B}}$ fails, whether $a_{\mathbf{A}} a_{\mathbf{B}}$ gives an empirically better approximation. ... Bronsted [111] turns the flank of the difficulty by making the rate proportional to the concentration of a critical complex **X** ... whence $[rate] = k c_{\mathbf{A}} c_{\mathbf{B}} f_{\mathbf{A}} f_{\mathbf{B}} / f_{\mathbf{X}}$." The $f_{\mathbf{A}}$, $f_{\mathbf{B}}$, and $f_{\mathbf{X}}$ in Bronsted's equation are activity coefficients, and his 1922 paper anticipated (before the development of TST) where they should appear.

To derive Bronsted's equation with the advantage of hindsight, recall how we derived the TST rate constant for a reaction $\mathbf{A} + \mathbf{B} \rightarrow products$ in an ideal solution. The derivation began with

[4] I thank Bryan Goldsmith for testing this claim.

$\mathcal{R} = \nu_\ddagger[\ddagger]$ and invoked partition functions to evaluate $K_\ddagger$ in $K_\ddagger \approx \mathcal{V}_o^{-1}[\ddagger]/[\mathbf{A}][\mathbf{B}]$. We then solved for $[\ddagger]$, and multipled by $\nu_\ddagger = \frac{1}{2}\langle|\dot{q}|\rangle_\ddagger/\lambda_q$ to obtain the TST rate constant.

Clearly, non-idealities can influence $[\ddagger]$. Again using $\mathcal{V}_o^{-1}$ as a reference concentration and now including activity coefficients,

$$K_\ddagger = \gamma_\ddagger[\ddagger]\mathcal{V}_o/(\gamma_\mathbf{A}[\mathbf{A}]\mathcal{V}_o\gamma_\mathbf{B}[\mathbf{B}]\mathcal{V}_o) \tag{10.8.1}$$

so that $[\ddagger] = \mathcal{V}_o^{\nu-1}(\gamma_\mathbf{A}\gamma_\mathbf{B}/\gamma_\ddagger)K_\ddagger[\mathbf{A}][\mathbf{B}]$ where $\nu = 2$ is the overall reaction order. Non-idealities should not influence the kinematic factors in $\nu_\ddagger$, so the rate with non-idealities included is

$$\mathcal{R}_{TST} = \left(\frac{k_B T}{h}\mathcal{V}_o^{\nu-1}K_\ddagger\frac{\gamma_\mathbf{A}\gamma_\mathbf{B}}{\gamma_\ddagger}\right)[\mathbf{A}][\mathbf{B}] \tag{10.8.2}$$

We leave it to the reader to construct analogous expressions for reactions of other molecularity, for reactions in the gas phase, and for reactions at surfaces. The rate expression (10.8.2) can be interpreted in two different ways [110].

1. Harned [112] suggested the use of elementary rate laws in which a rate constant multiplies the mass action product of activities [110].

$$\mathcal{R} = k\, a_\mathbf{A} a_\mathbf{B}$$

   The TST rate constant that would multiply activities is

$$k_{TST} = \left(k_B T\, h^{-1}\mathcal{V}_o^{-1}K_\ddagger/\gamma_\ddagger\right).$$

2. In the more common convention [113], an elementary rate constant multiples the reactant concentrations:

$$\mathcal{R} = k\,[\mathbf{A}][\mathbf{B}]$$

   The TST rate constant, according to this convention, becomes

$$k_{TST} = \left(k_B T\, h^{-1}\mathcal{V}_o^{\nu-1}K_\ddagger\gamma_\mathbf{A}\gamma_\mathbf{B}/\gamma_\ddagger\right).$$

At some level, what really matters is the rate itself, and we are free to choose for convenience whether it is a "constant" multiplied by activities or concentrations.[5] But both conventions yield

[5] Hougen and Watson advocated a third convention where rate constants are multiplied by mole fractions to obtain the rate. In the gas phase their convention leads to rate constants that depend on pressure, and in the liquid phase it leads to rate constants that depend on composition.

a rate constant that has absorbed the activity coefficient for the transition state – a quantity that is usually difficult to determine. Moreover, plotting rates as a function of activities would be cumbersome, except for those solvents and solutes where activity coefficients are well known. Ultimately, if the rate constant must absorb one activity coefficient, it might as well absorb them all. These are the practical arguments in favor of conventional rate constants that multiply concentrations.

The equations of Bronsted [111] and Bjerrum [114] show that a ratio of the rate constants at finite solute concentration to the rate constant at infinite dilution (where activities are unity) isolates the ratio of activity coefficients.

$$k_{TST}/k_{TST}^{(0)} = \gamma_{\mathbf{A}}\gamma_{\mathbf{B}}/\gamma_{\ddagger} \tag{10.8.3}$$

where k_{TST} and $k_{TST}^{(0)}$ are the rate constant at finite concentration and extrapolated to infinite dilution, respectively. Equation (10.8.3) can be used with experimental data to estimate $\gamma_{\ddagger}/(\gamma_{\mathbf{A}}\gamma_{\mathbf{B}})$. Consider the elementary reaction in aqueous solution

$$\mathbf{A} + \mathbf{B} \rightarrow products$$

in which the reactants and products have charge $z_{\mathbf{A}}$ and $z_{\mathbf{B}}$. The transition state must have charge $z_{\mathbf{A}} + z_{\mathbf{B}}$. The activity coefficient of an ion in solution is given (at sufficiently low concentrations) by the Debye-Huckel limiting law

$$ln\,\gamma_i = -\frac{1}{2} z_i^2 \sqrt{I} \tag{10.8.4}$$

where the ionic strength is $I = 0.5 \sum_i z_i^2 [i]$ for i = **A**, **B**, and products. The transition states have concentrations that are too small to influence I. Inserting the Debye-Huckel limiting law into equation (10.8.3) gives after some simplification

$$ln\left[k_{TST}/k_{TST}^{(0)}\right] \quad = \quad z_A z_B \sqrt{I} \tag{10.8.5}$$

Debye-Huckel theory only applies to dilute solutions, so the Bronsted-Bjerrum result should not be used for highly concentrated salt solutions. The predicted behavior has been verified for several reactions involving ionic species as shown in Figure 10.8.1 [115–117]. In particular, experiments confirm that increasing the ionic strength increases the rate constant relative to the rate constant at infinite dilution when ions have like charge. Increasing the ionic strength decreases the rate constant relative to the rate constant at infinite dilution when ions have opposite charges. The Bronsted-Bjerrum equation is wonderful, and in principle the same procedure could be used to understand non-idealities for all kinds of reactions.

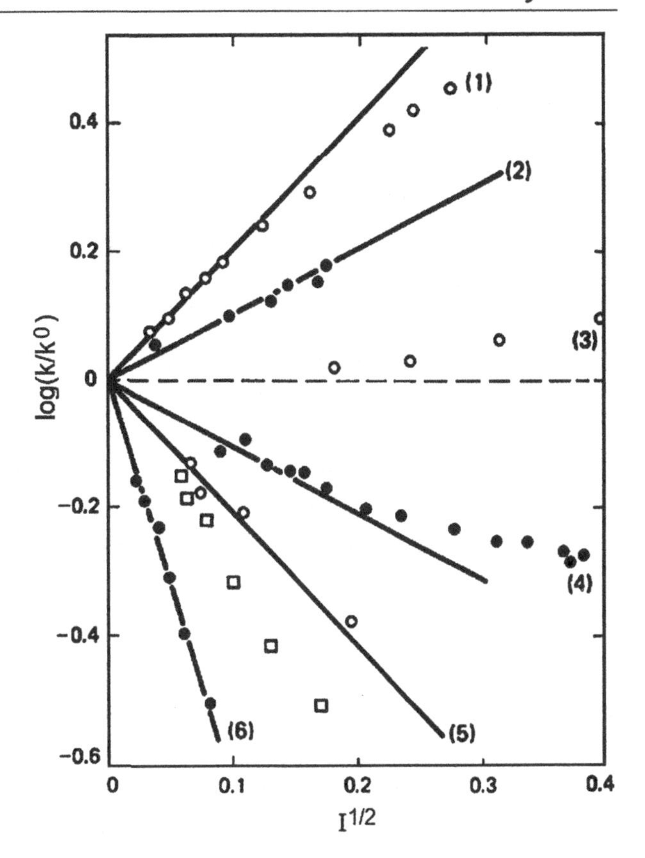

Figure 10.8.1: Rate constants relative to rate constants at infinite dilution vs. the square root of ionic strength. The reactants in the reactions corresponding to the numbered data series are (1) $BrCH_2COO^- + S_2O_3^{2-}$, (2) $e^-_{aq} + NO_2^-$, (3) $H_3O^+ + C_{12}H_{22}O_{11}$, i.e. acid catalyzed sucrose inversion, (4) $H^+ + Br^- + H_2O_2$, (5) $OH^- + Co(NH_3)_5Br^{2+}$ where circles are using NaBr to control ionic strength and squares are using Na_2SO_4, and (6) $Fe(H_2O)_6^{2+} + Co(C_2O_4)_3^{3-}$. The solid lines have slopes $z_A z_B$ as predicted by the Bronted-Bjerrum equation. [From Weston and Schwartz, "Chemical Kinetics" Prentice Hall (1972).]

Exercises

1. For a reaction $\mathbf{A} + \mathbf{B} \rightleftarrows \mathbf{C} + \mathbf{D}$ show that the forward and backward TST rates obey detailed balance. In later chapters we will develop corrections to TST to account for tunneling and recrossing. What must be true of the forward and backward corrections?
2. What symmetry numbers should be used for reactions of type $X^- + CH_3X \rightarrow XCH_3 + X^-$? Will your answer change if one hydrogen atom becomes a deuterium?
3. Estimate the activation entropy for a unimolecular reaction with frequency factor 10^{15}/s at 300K. Does your answer depend on the standard state?
4. Estimate the activation entropy for a bimolecular reaction with frequency factor 10^{10}L/mol/s at 300K. Does your answer depend on the standard state?
5. [Adapted from Moore and Pearson] Hydrogen gas at T > 25^oC is 25% para-hydrogen and 75% ortho-hydrogen. Farkas [*Z. Phys. Chem.* B10, 419 (1930)] monitored the conversion of pure para-hydrogen to an equilibrium mixture at 923K with no catalyst. The

pseudo-first order decay rate based on

$$ln\left(\frac{(\%para)_0 - 25}{\%para - 25}\right) = k_{pseudo}t$$

depends on H_2 pressure as

$P_{H2}/mmHg$	50	100	200	400
$k \times 10^3/s$	1.06	1.53	2.17	3.10

(a) What is the reaction order with respect to H_2, and what mechanism does the observed reaction order suggest?

(b) Use the data to calculate the elementary rate constant for the reaction $H + (para)H_2 \rightarrow (ortho)H_2 + H$. Hint: the equilibrium constant was given.

6. Redo the harmonic TST calculation for the $H_2 + H$ reaction, but this time use the quantum mechanical partition function for rotation of a linear molecule

$$Q_{rot} = \sigma_{sym}^{-1} \sum_{j=0}^{\infty} (2j+1)e^{-j(j+1)\varphi_T(I)^2/4\pi}.$$

Plot the ratio of TST rate constants with rotation treated quantum and classically up to 1000K.

7. Using information in the $H_2 + H$ example calculation, compute the rate constant for the $D_2 + D$ reaction. Hint: the frequencies and moments of inertia are rather easy to transform for this complete isotopic exchange.
8. Revisit the TST calculation for $H_2 + H$. Separately compute the translational, rotational, vibrational, electronic, and zero point energy contributions to the activation free energy. Apart from the difference in Born-Oppenheimer energies from reactant minimum to the saddle point, which degrees of freedom make the largest contributions to the activation free energy difference?
9. Use electronic structure calculations to compute rates for the reactions $H_2 + Cl\cdot \rightarrow HCl + H\cdot$ and $Cl_2 + H\cdot \rightarrow HCl + Cl\cdot$ in the reaction $H_2 + Cl_2 \rightarrow 2HCl$. Based on your calculations, what is the total overall reaction rate as a function of $[H_2]$, $[Cl_2]$, and the total radical population. What fraction of the total radical population at steady state is $Cl\cdot$?
10. The reaction $(C_2H_5)_3N + C_2H_5I \rightarrow (C_2H_5)_4N^+I^-$ in a benzene solvent has an unusually strong negative volume of activation ($\Delta V^{\ddagger}$= -50.2 cm^3/mol). Explain.
11. Zewail and coworkers use femtosecond chemistry to directly investigate the properties of transition states.[118,119] However, compare the definition of "transition state" in [Zewail, *J. Phys. Chem. B.* (1996)] [119] to more traditional definitions. Do they observe transition states according to the traditional definition?

12. TST is expected to give inaccurate nucleation rates because nucleation trajectories are not ballistic barrier crossings. Nevertheless, one could use equation (10.5.4) to estimate the homogeneous nucleation rate J [events per volume per time]. Provide the equation to obtain a TST estimate for J from k_{TST}. Hint: you will need a volume, but should it be the volume *before, during, or after* the nucleation event?
13. Some studies investigate biological proton transfer reactions by choosing a few select reaction coordinates and computing free energy surfaces, e.g. with QM/MM simulations. A schematic example of one of these free energy surfaces is shown in the figure below. Let us suppose that the correct subspace of coordinates has been chosen and critically examine other parts of calculation.

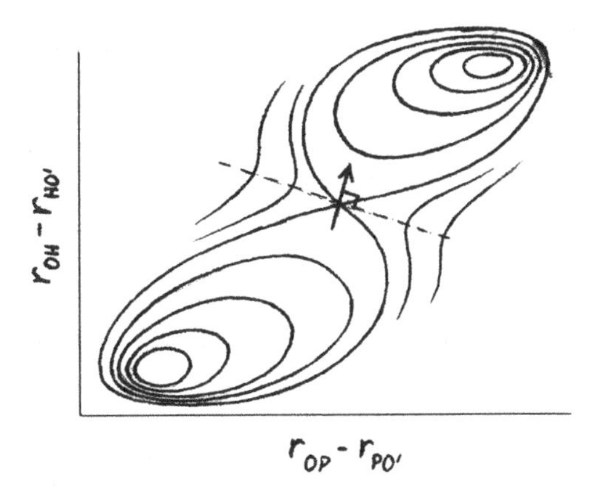

The figure depicts a typical procedure that is used to define the dividing surface. The optimized free energy of the dividing surface relative to the reactants' free energy is then used in a TST calculation with the formula $k_{TST} = k_B T/h\, exp[-\beta \Delta F^{\ddagger}]$. Find *three* errors in this all-too-common type of TST calculation.
14. Show that the internal coordinate q in equation (10.7.3) is dynamically separable. Hint: first show that if $\mathbf{v}_{\ddagger}$ is an eigenvector of $\mathbf{G}^{1/2}\left[\partial^2 V_{PCM}/\partial\mathbf{b}\partial\mathbf{b}\right]\mathbf{G}^{1/2}$ then $\mathbf{u}_{\ddagger} = \mathbf{G}^{-1/2}\mathbf{v}_{\ddagger}$ is an eigenvector of $\left[\partial^2 V_{PCM}/\partial\mathbf{b}\partial\mathbf{b}\right]\mathbf{G}$.
15. Consider a chain of n beads each of mass m in one dimension. Each bead is connected to its neighbors by a Morse potential with equilibrium bond length ℓ, force constant parameter a, and dissociation energy E_D. Now suppose the endpoints of the chain (beads 1 and n) are pulled in opposite directions with constant tension f.

$$V_f(x) = \sum_{i=1}^{N-1} E_D(1 - exp[-a(|x_{i+1} - x_i| - \ell)])^2 - (x_n - x_1) \cdot f$$

Are some bonds easier to break than others? Compute the overall rate at which the chain breaks using TST.

(a) Do the calculation for the largest specific chain length that you can manage. Ignore anharmonic effects for all bonds that are not near their breaking points. How does the answer scale with f for small forces.

(b) Read about methods to diagonalize symmetric tridiagonal Toeplitz matrices and matrices that are almost Toeplitz except for a single element along the diagonal. Devise an algorithm to solve this problem for any n.

16. Start from equation (10.5.2) and express k_{TST} as

$$k_{TST} = \frac{1}{2}\langle|\mathbf{v}\cdot\hat{\mathbf{u}}_q|\rangle_{||\nabla q||\delta[q(\mathbf{x})-q^\ddagger]}\, exp[-\beta\Delta F_{||\nabla q||\delta[q(\mathbf{x})-q^\ddagger]}] \qquad (10.8.6)$$

where $\mathbf{v}$ is the 3N-dimensional velocity vector, $\mathbf{u}_q = \nabla q/||\nabla q||$ is the unit normal vector to the dividing surface, $\langle\cdot\rangle_{||\nabla q||\delta[q(\mathbf{x})-q^\ddagger]}$ indicates the "hitting point" average as defined in Chapter 11, and

$$exp[-\beta\Delta F_{||\nabla q||\delta[q(\mathbf{x})-q^\ddagger]}] \equiv \frac{\int d\mathbf{x}e^{-\beta U(\mathbf{x})}||\nabla q||\delta[q(\mathbf{x})-q^\ddagger]}{\int d\mathbf{x}e^{-\beta U(\mathbf{x})}h_A(\mathbf{x})}$$

Comment on the similarities, differences, and units in equations (10.5.4) and equation (10.8.6). In particular, notice that both expressions are exact.

17. If q has units of length, then in the generalized TST of equation (10.5.4) we could *define* the free energy of activation using

$$\lambda_{T,q}^{-1} exp\left[-\beta\Delta F_q^\ddagger\right] = \frac{\langle\delta[q(\mathbf{x})-q_\ddagger]\rangle}{\langle h_A(\mathbf{x})\rangle},$$

where the de Broglie wavelength $\lambda_{T,q}$ has replaced ℓ_q. Express the resulting prefactor $\langle|\dot{q}|\rangle_\ddagger/2\lambda_{T,q}$ in terms of k_BT/h and ∇q. Discuss the procedure to compute $\lambda_{T,q}$, e.g. by using equipartition law to find the effective mass of q at the transition state. Explain why this formulation only makes sense when q has units of length.

18. The no-recrossing assumption of TST is not likely to be valid for several of the reactions in Figure 10.8.1. In fact, several of these reactions are probably diffusion controlled. Start from the Debye's expression for k_D [equation (5.3.3)] with a screened electrostatic potential

$$V(r) = \frac{z_A z_B e^2}{4\pi\varepsilon r} exp[-\kappa r]$$

where κ^{-1} is the Debye screening length. Invoke the approximation $exp[-\kappa r] \approx 1-\kappa r$ to obtain [117]

$$k_D = k_D^0 \cdot f(z_A z_B \kappa)$$

where k_D^0 is the diffusion controlled rate at zero ionic strength and where f includes all screening effects. Finally, use your expression to show that $ln(k_D/k_D^0) = z_A z_B\sqrt{I}$.

References

[1] E. Wigner, H. Eyring, Sci. Mon. 44 (1937) 564–567.
[2] S. Arrhenius, J. Phys. Chem. 4 (1889) 226.
[3] W.C.M. Lewis, J. Chem. Soc. 113 (1918) 471.
[4] W.H. Rodebush, J. Am. Chem. Soc. 45 (1923) 606–614.
[5] H. Eyring, M. Polanyi, Z. Phys. Chem. B 12 (1931) 279.
[6] H. Pelzer, E. Wigner, Z. Phys. Chem. B 15 (1932) 445.
[7] V.K. La Mer, J. Chem. Phys. 1 (1933) 289.
[8] M.G. Evans, M. Polanyi, Trans. Faraday Soc. 34 (1938) 11–24.
[9] H. Eyring, Chem. Rev. 17 (1935) 65–77.
[10] K.J. Laidler, M.C. King, J. Phys. Chem. 87 (1983) 2657–2664.
[11] H. Eyring, J. Chem. Phys. 3 (1935) 107.
[12] W.F.K. Wynne-Jones, H. Eyring, J. Chem. Phys. 3 (1935) 492.
[13] J.R.E.T. Pineda, S.D. Schwartz, Philos. Trans. R. Soc. Lond. B, Biol. Sci. 361 (2006) 1433–1438.
[14] F. Jensen, J. Am. Chem. Soc. 114 (1992) 1596–1603.
[15] S.R. Billeter, S.P. Webb, T. Iordanov, P.K. Agarwal, S. Hammes-Schiffer, J. Chem. Phys. 114 (2001) 6925.
[16] P.K. Agarwal, S.R. Billeter, S. Hammes-Schiffer, J. Phys. Chem. B 106 (2002) 3283–3293.
[17] D.R. Glowacki, J.N. Harvey, A.J. Mulholland, Nat. Chem. 4 (2012) 169–176.
[18] E. Wigner, Trans. Faraday Soc. 34 (1938) 29–41.
[19] W.H. Miller, Faraday Discuss. 110 (1998) 1–21.
[20] R.G. Mullen, J.-E. Shea, B. Peters, J. Chem. Phys. 140 (2014) 41104.
[21] D. Chandler, J. Chem. Phys. 68 (1978) 2959–2970.
[22] J. Pfaendtner, X. Yu, L.J. Broadbelt, Theor. Chem. Acc. 118 (2007) 881–898.
[23] T.L. Hill, An Introduction to Statistical Thermodynamics, Dover, New York, 1986.
[24] K.A. Dill, S. Bromberg, Molecular Driving Forces, Garland Science, New York, 2003.
[25] E. Pollak, P. Pechukas, J. Am. Chem. Soc. (23) (1978) 2984–2991.
[26] P. Pechukas, E. Pollak, J. Chem. Phys. 71 (1979) 2062.
[27] H. Hu, M.N. Kobrak, C. Xu, S. Hammes-Schiffer, J. Phys. Chem. A 104 (2000) 8058–8066.
[28] P. Siegbahn, B. Liu, J. Chem. Phys. 68 (1978) 2457.
[29] D.G. Truhlar, C.J. Horowitz, J. Chem. Phys. 68 (1978) 2466–2476.
[30] N.E. Henriksen, F.Y. Hansen, Theories of Molecular Reaction Dynamics, Oxford Press, 2008.
[31] L. Van Loon, E. Mader, S.L. Scott, J. Phys. Chem. A 104 (2000) 1621–1626.
[32] R.A. van Santen, G.J. Kramer, Chem. Rev. 95 (1995) 637–660.
[33] J.A. Swisher, N. Hansen, T. Maesen, F.J. Keil, B. Smit, A.T. Bell, J. Phys. Chem. C 114 (2010) 10229–10239.
[34] J. Wei, Chem. Eng. Sci. 51 (1996) 2995–2999.
[35] T. Hwang, B.R. Goldsmith, B. Peters, S.L. Scott, Inorg. Chem. 52 (2013) 13904–13917.
[36] D. Truhlar, A. Kohen, Proc. Natl. Acad. Sci. USA 98 (2001) 848–851.
[37] H.B. Callen, Thermodynamics and an Introduction to Thermostatistics, 2nd ed., Wiley, 1985.
[38] M. Karplus, J. Phys. Chem. B 104 (2000) 11–27.
[39] R. Phillips, Crystals, Defects, and Microstructures: Modeling Across Scales, Cambridge Press, Cambridge, UK, 2001.
[40] S.V. Krivov, M. Karplus, Proc. Natl. Acad. Sci. USA 101 (2004) 14766–14770.
[41] M. Oliveberg, Y.J. Tan, A.R. Fersht, Proc. Natl. Acad. Sci. USA 92 (1995) 8926–8929.
[42] D.G. Truhlar, B.C. Garrett, S.J. Klippenstein, J. Phys. Chem. 3654 (1996) 12771–12800.
[43] B. Peters, N.E.R. Zimmermann, G.T. Beckham, J.W. Tester, B.L. Trout, J. Am. Chem. Soc. 130 (2008) 17342–17350.
[44] J.P. Kennett, K.G. Cannariato, I.L. Hendy, R. Behl, Methane Hydrates in Quaternary Climate Change: The Clathrate Gun Hypothesis, American Geophysical Union, Washington DC, 2003.
[45] C.A. Koh, E.D. Sloan, A.K. Sum, D.T. Wu, Annu. Rev. Chem. Biomol. Eng. 2 (2011) 237–257.

[46] E. Dendy Sloan, Nature 426 (2003) 353.
[47] R. Radhakrishnan, T. Schlick, J. Chem. Phys. 121 (2004) 2436–2444.
[48] E. Leontidis, B.M. Forrest, A.H. Widmann, U.W. Suter, J. Chem. Soc. Faraday Trans. 91 (1995) 2355–2368.
[49] R. Faller, J.J. de Pablo, J. Chem. Phys. 116 (2002) 55–59.
[50] G.A. Korn, T.M. Korn, Mathematical Handbook for Scientists and Engineers, Dover, Mineola, NY, 2000.
[51] A. Falenty, A.N. Salamatin, W.F. Kuhs, J. Phys. Chem. C 117 (2013) 8443–8457.
[52] M. Muthukumar, J. Chem. Phys. 111 (1999) 10371–10374.
[53] J. Israelachvili, Intermolecular and Surface Forces, Academic Press, San Diego, 1992.
[54] K. Sneppen, G. Zocchi, Physics in Molecular Biology, Cambridge University Press, 2005.
[55] C.L. Ting, D.E. Makarov, Z.-G. Wang, J. Phys. Chem. B 113 (2009) 4970–4977.
[56] G.H. Vineyard, J. Phys. Chem. Solids 3 (1957) 121–127.
[57] G.A. Voth, R.M. Hochstrasser, J. Phys. Chem. 100 (1996) 13034–13049.
[58] E. Wigner, J. Chem. Phys. 5 (1937) 720–725.
[59] J. Horiuti, Bull. Chem. Soc. Jpn. 13 (1938) 210.
[60] J.C. Keck, J. Chem. Phys. 32 (1960) 1035–1050.
[61] J.C. Keck, Adv. Chem. Phys. 13 (1967) 85–121.
[62] B.C. Garrett, D.G. Truhlar, J. Am. Chem. Soc. 101 (1979) 4534–4548.
[63] D.G. Truhlar, B.C. Garrett, Acc. Chem. Res. 235 (1980) 440–448.
[64] B.C. Garrett, D.G. Truhlar, in: C. Dykstra (Ed.), Theory and Applications of Computational Chemistry: The First Forty Years, Elsevier, Amsterdam, 2005, pp. 84–87.
[65] D.G. Truhlar, G.K. Schenter, B.C. Garrett, J. Chem. Phys. 98 (1993) 5756–5770.
[66] G.K. Schenter, B.C. Garrett, D.G. Truhlar, J. Phys. Chem. B 105 (2001) 9672–9685.
[67] E. Pollak, J. Chem. Phys. 95 (1991) 533–539.
[68] G. Gershinsky, E. Pollak, J. Chem. Phys. 101 (1994) 7174–7176.
[69] S.N. Rai, D.G. Truhlar, J. Chem. Phys. 79 (1983) 6046–6059.
[70] W.L. Hase, R.J. Duchovic, J. Chem. Phys. 83 (1985) 3448–3453.
[71] P. Hanggi, P. Talkner, M. Borkovec, Rev. Mod. Phys. 62 (1990) 251–341.
[72] A. Fong, B. Peters, S.L. Scott, ACS Catal. 6 (2016) 6073–6085.
[73] A. Fernandez-Ramos, B.A. Ellingson, B.C. Garrett, D.G. Truhlar, Rev. Comput. Chem. 23 (2007) 125–232.
[74] S.C. Smith, J. Phys. Chem. A 104 (2000) 10489–10499.
[75] T. Komatsuzaki, R.S. Berry, J. Chem. Phys. 110 (1999) 9160–9173.
[76] T. Uzer, C. Jaffe, J. Palacian, P. Yanguas, S. Wiggins, Nonlinearity 15 (2002) 957–992.
[77] R. Hernandez, T. Uzer, T. Bartsch, Chem. Phys. 370 (2010) 270–276.
[78] L. Maragliano, A. Fischer, E. Vanden-Eijnden, G. Ciccotti, J. Chem. Phys. 125 (2006) 24106.
[79] A. Laio, F.L. Gervasio, Rep. Prog. Phys. 71 (2008) 126601.
[80] G. Gershinsky, E. Pollak, J. Chem. Phys. 103 (1995) 8501–8512.
[81] K. Zinovjev, I. Tunon, J. Chem. Phys. 143 (2015) 134111.
[82] S.J. Klippenstein, V.S. Pande, D.G. Truhlar, J. Am. Chem. Soc. 136 (2013) 528–546.
[83] S. Miertus, E. Scrocco, J. Tomasi, Chem. Phys. 55 (1981) 117–129.
[84] A. Klamt, Wiley Interdiscip. Rev. Comput. Mol. Sci. 1 (2011) 699–709.
[85] C.J. Cramer, D.G. Truhlar, Chem. Rev. 99 (1999) 2161–2200.
[86] C.J. Cramer, D.G. Truhlar, Acc. Chem. Res. 41 (2008) 760–768.
[87] P.L. Geissler, C. Dellago, D. Chandler, J. Phys. Chem. B 103 (1999) 3706–3710.
[88] R.G. Mullen, J.E. Shea, B. Peters, J. Chem. Theory Comput. 10 (2014) 659.
[89] D.N. Petsev, K. Chen, O. Gliko, P.G. Vekilov, Proc. Natl. Acad. Sci. USA 100 (2003) 792–796.
[90] S. Piana, F. Jones, J.D. Gale, J. Am. Chem. Soc. 128 (2006) 13568–13574.
[91] S. Sharma, P.G. Debenedetti, J. Phys. Chem. B 116 (2012) 13282–13289.
[92] S. Sharma, P.G. Debenedetti, Proc. Natl. Acad. Sci. USA 109 (2012) 4365–4370.
[93] P.L. Geissler, C. Dellago, D. Chandler, J. Hutter, M. Parrinello, Science 291 (2001) 2121–2124.
[94] J.-W. Chu, B.R. Brooks, B.L. Trout, J. Am. Chem. Soc. 126 (2004) 16601–16607.

[95] B. Peters, B.L. Trout, Biochemistry 45 (2006) 5384–5392.
[96] B.R. Goldsmith, T. Hwang, S. Seritan, B. Peters, S.L. Scott, J. Am. Chem. Soc. 137 (2015) 9604–9616.
[97] H. Lin, D.G. Truhlar, Theor. Chem. Acc. 117 (2006) 185–199.
[98] H. Hu, W. Yang, Annu. Rev. Phys. Chem. 59 (2008) 573–601.
[99] O. Acevedo, W.L. Jorgensen, Acc. Chem. Res. 43 (2010) 142–151.
[100] M. Klahn, E. Rosta, A. Warshel, J. Am. Chem. Soc. 128 (2006) 15310–15323.
[101] E. Rosta, H.L. Woodcock, B.R. Brooks, G. Hummer, J. Comput. Chem. 30 (2009) 1634–1641.
[102] E. Pollak, J. Chem. Phys. 85 (1986) 865–867.
[103] B. Ensing, E.J. Meijer, P.E. Blochl, E.J. Baerends, J. Phys. Chem. A 105 (2001) 3300.
[104] T.T. Trinh, A.P.J. Jansen, R.A. van Santen, E. Jan Meijer, J. Phys. Chem. C 113 (2009) 2647–2652.
[105] A. Pavlova, T.T. Trinh, R.A. van Santen, E. Jan Meijer, Phys. Chem. Chem. Phys. 15 (2013) 1123–1129.
[106] E.B. Wilson, J.C. Decius, P.C. Cross, Molecular Vibrations: The Theory of Infrared and Raman Vibrational Spectra, Dover, New York, 1980.
[107] E. Vanden-Eijnden, in: M. Ferrario, G. Ciccotti, K. Binder (Eds.), Computer Simulations in Condensed Matter: From Materials to Chemical Biology, in: Lecture Notes in Physics, vol. 703, Springer, Berlin, Heidelberg, 2006, pp. 439–478.
[108] W. E, E. Vanden-Eijnden, J. Stat. Phys. 123 (2006) 503–523.
[109] A. Berezhkovskii, A. Szabo, J. Chem. Phys. 122 (2005) 14503.
[110] C. Hinshelwood, Annu. Rep. Prog. Chem. 24 (1927) 314–317.
[111] J.N. Bronsted, Z. Phys. Chem. 102 (1922) 169.
[112] H.S. Harned, J. Am. Chem. Soc. 40 (1918) 1461–1481.
[113] J.N. Bronsted, K.J. Pederson, Z. Phys. 108 (1924) 185–235.
[114] N. Bjerrum, Z. Phys. Chem. 108 (1924) 82.
[115] C.W. Davies, in: G. Porter (Ed.), Progress in Reaction Kinetics, Pergamon Press, 1961, p. 161.
[116] K.J. Laidler, Reaction Kinetics: Reactions in Solution, vol. 2, Pergamon Press, Oxford, 1963.
[117] R.E. Weston, H.A. Schwarz, Chemical Kinetics, Prentice-Hall, Englewood Cliffs, NJ, 1972.
[118] J.C. Polanyi, A.H. Zewail, Acc. Chem. Res. 28 (1995) 119–132.
[119] A.H. Zewail, J. Phys. Chem. 100 (1996) 12701–12724.

CHAPTER 11

Landau free energies and restricted averages

> *"...although the choice of ΔU as the single argument of the weighting function w defined in (4) was a natural one for the free-energy difference problem, the general case in which w is any arbitrary function of the coordinates can extend the use of the biased sampling to a wider range of problems."*
>
> ***Torrie and Valleau, J. Comp. Chem. (1977)***

Free energy differences define many familiar equilibrium properties: binding coefficients, partition coefficients, equilibrium constants, chemical potentials, and more. Beyond static equilibria, free energy differences also play a central role in non-equilibrium processes like diffusion transport, membrane permeation, and chemical reactions. In reaction rate theory, free energies determine the overall fluxes through complex reaction networks and catalytic cycles. At a more detailed resolution, free energy differences along the reaction coordinate for each elementary reaction determine its forward and backward rate constants.

For most chemical reactions, stationary points on the *potential* energy surface (PES) provide the free energy of activation from standard translational, rotational, and vibrational partition functions. The situation is more complicated for systems with rugged energy landscapes and a multitude of saddle points, especially when none of the saddles are particularly important to the mechanism. Some examples of activated processes with rugged energy landscapes are

1. **Electron transfer**. Thermally activated electron transfer requires the preorganization of solvent molecules into a configuration where the reactant and product electronic states have the same energy. The small saddles associated with reorienting individual solvent molecules are largely irrelevant because so many solvent molecules are involved in preorganization.
2. **Reactions in solution or enzymes**. Aqueous solvents and enzymes can facilitate chemical reactions by electrostatically stabilizing the transition state or by conveying protons through 'wires' from donor to acceptor positions on the reactants. These reactions have large ensembles of transition states, each with too many low frequency modes to rely on harmonic TST. Instead, barriers and rates are computed from ensemble averages with collective variables as reaction coordinates [1].

3. **Nucleation**. Assembly of a nucleus typically involves repeated formation and dissolution of pre-critical nuclei by solute diffusion, desolvation, and incorporation at the nucleus surface. A nucleation trajectory may traverse many thousands of low-barrier saddle points on the PES, but the individual saddles are largely unimportant. In contrast, the free energy landscape is a smooth function of nucleus size.
4. **Protein folding**. The formation of specific native contacts between residues and the saddles traversed during their formation are largely irrelevant because there are hundreds of native contacts and thousands of possible non-native contacts. Folding progress and the relevant free energy barriers instead depend on the total number of native contacts that have been formed, perhaps with some contacts being more important than others.

Each of these processes has a highly collective reaction coordinate that is not related to the Cartesian coordinates by any simple coordinate transformation. Therefore the saddle search algorithms and minimum energy path optimizers that work seamlessly with harmonic TST are not applicable. Voth and Hochstrasser emphasized this point by writing [2] "A search for the minimum energy path in a liquid phase reaction would lead to the solid state!" We must somehow obtain free energy barriers for collective coordinates without relying on saddle search algorithms, minimum energy path optimizations, or standard translational, rotational, and vibrational partition functions.

Recall that the Helmholtz free energy is $F = -k_B T \, ln \Sigma_\nu exp[-\beta H(\nu)]$ where ν indicates a microstate in phase space. The Helmholtz free energy will not help us predict barriers or fluxes from one group of states to another, because it sums all states into a single number. Rate theories instead require a Landau type free energy, i.e. the free energy as a function of an order parameter. Landau free energies emerged in studies of phase transitions and critical phenomena where the order parameter is often a measure of fluid density, crystallinity, or magnetism [3, 4]. Landau free energies are now used in many applications of molecular simulation. Hansen and van Gunsteren [5] counted over 3500 papers on free energy computation between 2000 and 2010 with the publication rate increasing at 17% per year. Fortunately, many of these methods are variations on a few major themes: thermodynamic integration, umbrella sampling, thermodynamic perturbation theory, and non-equilibrium work theorems.

The various theoretical frameworks for computing free energies also require efficient algorithms for sampling phase space with constraints, restraints, or biased potentials. The machinery for sampling phase space includes Monte Carlo (MC), molecular dynamics (MD), and hybrid Monte Carlo/Molecular dynamics (hybrid MC/MD) methods. Our discussion focuses on methods for sampling the canonical $(N\mathcal{V}T)$ ensemble for two reasons. First, partition functions and detailed balance in the canonical ensemble are relatively simple. Second, many condensed phase systems can be treated as incompressible in which case the canonical $(N\mathcal{V}T)$ and isothermal-isobaric (NPT) ensembles are identical.

11.1 Monte Carlo, molecular dynamics, and hybrid sampling

Even for relatively small model systems, the configuration space is unimaginably large. For example, a two-dimensional 20 × 20 Ising model has $2^{400} \approx 2.6 \times 10^{120}$ states [6]. Today's fastest (petaflop) computers do approximately a mole of floating point operations per year. Even for this small 2D Ising model, petaflop power is still nearly 100 orders of magnitude too slow for an exhaustive sum over all states. Beyond the Ising model, this "curse of dimensionality" impacts molecular systems, coarse grained models, field theoretic models, etc. How can one hope to compute averages and free energies for any truly complex system?

Fortunately, an exhaustive sum over all microstates is unnecessary. Free energy differences, averages, and forces can instead be obtained from relatively small but representative samples of the equilibrium distribution. On the other hand, collecting a representative sample is nontrivial. The equilibrium distribution is usually dominated by a small fraction of the total microstates because of the exponential dependence on energy in the Boltzmann distribution. For example, in the canonical ensemble

$$p(\nu) = Q^{-1} exp[-\beta H(\nu)] \tag{11.1.1}$$

where H is the Hamiltonian and

$$Q = \Sigma_\nu exp[-\beta H(\nu)] \tag{11.1.2}$$

is the canonical partition function. If $\{\nu_1, \nu_2, \cdots, \nu_M\}$ is a representative sample from $p(\nu)$, then the formal average $\langle \mathcal{O} \rangle \equiv \sum_\nu \mathcal{O}(\nu) p(\nu)$ can be estimated as

$$\langle \mathcal{O} \rangle \approx \frac{1}{M} \sum_{i=1}^{M} \mathcal{O}(\nu_i) \tag{11.1.3}$$

By the central limit theorem [7], the sample estimate of $\langle \mathcal{O} \rangle$ *should be* accurate to about $\left\langle (\delta\mathcal{O})^2 \right\rangle^{1/2} / \sqrt{M}$. In practice, the accuracy also depends on the ability to generate a representative sample.

To generate a "small" but representative sample of microstates, Monte Carlo [8] methods employ a Markov chain [7] with state-to-state transition probabilities that obey detailed balance [9–12],

$$p_{eq}(\nu) p(\nu \to \nu') = p_{eq}(\nu') p(\nu' \to \nu). \tag{11.1.4}$$

Here $p(\nu' \to \nu)$ is the probability of a transition from state ν to ν' according to the sampling procedure. Detailed balance requires that all net state-to-state fluxes should vanish at equilibrium as shown in Figure 11.1.1. State-to-state transition probabilities that satisfy detailed balance (and ergodicity) are sufficient to ensure that a Markov chain generates the equilibrium distribution. This claim will be justified in Chapter 14, but for now we take the eventual approach to equilibrium as fact.

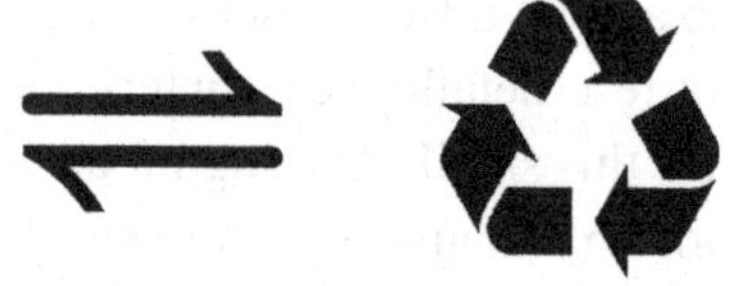

Figure 11.1.1: The symbol for equilibrium depicts detailed balance. The perpetual fluxes depicted by the recycling symbol violate detailed balance.

It should be noted that detailed balance is not absolutely necessary to ensure convergence to equilibrium. For example, schemes that generate the Boltzmann distribution may do so with perpetual currents or pre-ordained non-random move sequences [13]. However, steady currents and prescriptive event sequences do not occur at equilibrium in real systems [7]. Additionally, the easiest way to ensure/prove convergence to the correct equilibrium is to design schemes that do obey detailed balance [14].

Algorithm: Metropolis Monte Carlo

Let us begin with a microstate ν.

1. Use a random number generator to choose a trial microstate ν' symmetrically in the sense that $p_{gen}(\nu \rightarrow \nu') = p_{gen}(\nu' \rightarrow \nu)$.
2. Accept the transition to trial state ν' with probability $p_{acc}(\nu \rightarrow \nu') = min[1, p_{eq}(\nu')/p_{eq}(\nu)]$. Else reject the transition and keep ν as the current state.
3. Record the current state for computing averages.

Note that rejected trial states are not added to the sample. Instead rejections will result in duplicate sample entries. The Metropolis algorithm obeys detailed balance because

$$\begin{aligned} p_{eq}(\nu)p(\nu \rightarrow \nu') &= p_{gen}(\nu \rightarrow \nu')p_{eq}(\nu)min[1, p_{eq}(\nu')/p_{eq}(\nu)] \\ &= p_{gen}(\nu' \rightarrow \nu)p_{eq}(\nu')min[1, p_{eq}(\nu)/p_{eq}(\nu')] \\ &= p_{eq}(\nu')p(\nu' \rightarrow \nu) \end{aligned}$$

Where memory and/or writing to disk is a bottleneck, one can instead record the current state at regular intervals in the number of *attempted* moves.

Monte Carlo was originally developed for sampling the canonical ensemble, but the formulas above are readily adapted for other ensembles. Microstate probabilities and Monte Carlo acceptance rules for sampling other ensembles can be found in several other books [14–17].

Readers who wade into an application armed only with my absolute basic discussion of Monte Carlo will quickly encounter its limitations. Especially in simulations of condensed phase processes, the most easily constructed Monte Carlo moves will have negligible acceptance probabilities. Consider for example a polymer melt. Local displacements of individual segments must be extremely small to have any chance of acceptance, and navigating the configuration space of an entangled melt with small local displacements is excruciatingly slow. There are elegant theoretical frameworks and algorithms for efficient condensed phase Monte Carlo simulations. Among them, the Rosenbluth [18] and Configurational Bias [19] frameworks are quite general. However, these methods require libraries of rather complicated moves that must be specifically reformulated for each new application.

MD-based sampling methods

Is there a versatile and readily applicable strategy for sampling configuration space in condensed phase systems? In many cases, the answer is yes. A molecular dynamics (MD) trajectory naturally incorporates the collective motions by which atoms circumnavigate each other. Standard MD simulations sample the microcanonical ($N\mathcal{V}E$) ensemble, but for condensed phase systems we are usually more interested in canonical ($N\mathcal{V}T$) or isothermal-isobaric (NPT) ensembles. There are two prevalent strategies for using MD to sample the canonical ensemble.

The extended Lagrangian methods solve deterministic equations of motion much like a standard molecular dynamics simulation, but they include extra variables that mimic the effects of a temperature or pressure bath. Examples include the thermostats of Nose [20,21], Hoover [22] and Martyna et al. [23]. The second strategy uses stochastic equations of motion that include random forces and friction in addition to the usual deterministic forces. Examples include Langevin dynamics [24] and the Andersen thermostat [25]. The random forces and friction are balanced according to the fluctuation-dissipation theorem to ensure that a long time trajectory samples the equilibrium distribution. For details on numerical integrators for extended Lagrangian thermostats, Langevin dynamics, and other stochastic equations of motion see the texts by Leimkuhler [24] and Tuckerman [17].

Hybrid sampling methods

Sections 11.3–11.6 will describe several methods that use simulations with biased potentials and forces to compute Landau free energies: umbrella sampling, thermodynamic integration, metadynamics, adaptive biasing force, etc. Bias potentials are easily included in Monte Carlo simulations, and Monte Carlo is highly efficient for lattice models and certain off-lattice applications [14,16]. However, for off-lattice condensed phase systems, there is little room for any one atom to move. The only viable Monte Carlo moves must be highly coordinated collective motions of many atoms.

Off-lattice condensed phase systems are more easily sampled by the collective motions which are naturally generated in an MD simulation, but biased MD simulations require derivatives of the reaction coordinate to construct the biasing forces. The need for derivatives is a serious limitation when the optimal reaction coordinate is not differentiable, e.g. discrete cluster sizes for nucleation [26] and native contact counts for protein folding [27]. How does one impose a bias with non-differentiable coordinates?

Hybrid MC/MD [28–32] combines the best of Monte Carlo and MD. The central idea is to use short MD trajectories (or even individual timesteps) as trial Monte Carlo moves. The general strategy includes three basic steps. Starting from a configuration $\mathbf{x}_0$:

1. Generate random momenta: $p_{gen}(\mathbf{p}_0)$
2. Run a short dynamical trajectory from $\mathbf{x}_0, \mathbf{p}_0$ to $\mathbf{x}_t, \mathbf{p}_t$.
3. Accept or reject the new endpoint configuration: $p_{acc}(\mathbf{x}_0 \to \mathbf{x}_t)$

These three steps are repeatedly executed to generate a random walk in configuration space and averages are compiled as done in a standard Monte Carlo simulation.

Like other Monte Carlo procedures, the acceptance rule should be designed to make forward and backward moves/trajectories obey detailed balance. The acceptance rule depends on the bias potential, on the nature of the dynamical trajectories, and on the procedure for generating the initial momenta. The classic hybrid MC/MD algorithm is that of Mehlig et al. [30].

Algorithm: Hybrid MC/MD (Mehlig et al.)

Adopt a time step δt which is too large to conserve energy. Let each short trajectory contain n of these large time steps. Also adopt a deterministic, time-reversible, and measure-conserving integrator like the velocity Verlet algorithm [33].

1. From configuration $\mathbf{x}$, draw Boltzmann distributed momenta $\mathbf{p}$, i.e. $\rho_{gen}(\mathbf{p}) \propto exp[-\beta \mathbf{p}^{\dagger}\mathbf{M}^{-1}\mathbf{p}/2]$ where $\mathbf{M}$ is the diagonal matrix of atomic masses.
2. Run an MD trajectory from $(\mathbf{x}, \mathbf{p})$ to $(\mathbf{x}', \mathbf{p}') = (e^{\mathbf{L}(\delta t)})^n(\mathbf{x}, \mathbf{p})$ where $e^{\mathbf{L}(\delta t)}$ represents a velocity Verlet integrator with timestep δt.
3. Accept the move from $(\mathbf{x}, \mathbf{p})$ to $(\mathbf{x}', \mathbf{p}')$ with probability $min\{1, exp[-\beta \Delta H]\}$ where $\Delta H = H(\mathbf{x}', \mathbf{p}') - H(\mathbf{x}, \mathbf{p})$.

Repeated application of steps 1–3 generates the canonical ensemble in configuration space, i.e. $\rho_{eq}(\mathbf{x}) = Q^{-1}e^{-\beta V(\mathbf{x})}$ where Q is the configurational partition function.[a] The proof [30] hinges critically upon measure conservation: $d\mathbf{x}d\mathbf{p} = d\mathbf{x}'d\mathbf{p}'$ and time-reversibility: $\rho(\mathbf{x}_t, \mathbf{p}_t|\mathbf{x}_0, \mathbf{p}_0) = \rho(\mathbf{x}_0, -\mathbf{p}_0|\mathbf{x}_t, -\mathbf{p}_t)$. Therefore, hybrid MC/MD should only be performed with discretized integrators that obey these properties.

The size of the MD time step (δt) and the number of MD steps (n) per trajectory can be optimized to tune efficiency. As δt increases, the energy conservation errors will become large and the fraction of moves accepted will decrease. Large n also increases energy conservation error and lowers the rate of acceptance. However, a large $n\delta t$ also has the advantage of resulting in larger trial moves/trajectories. Usually, one chooses a small n to minimize force evaluations and δt so that the fraction of accepted trial moves is between 50–75%. Of course, modest energy conservation errors are essential for hybrid MC/MD and one should not seek to generate trajectories that conserve energy.

[a] Note the continuous configuration space probability density, $\rho_{eq}(\mathbf{x})$, rather than the discrete state probability p_{eq} of equation (11.1.1). Also note that the momentum generation task in step 1 is easy because $\mathbf{M}^{-1}$ is diagonal.

Our motivation for hybrid MC/MD was to sample a *biased* distribution *without* using derivatives of the order parameter. The algorithm of Mehlig et al. can do that *if the bias is a hard wall confining potential.* For example, to bias the sampling along a reaction coordinate q one could sample the distribution $exp[-\beta(V(\mathbf{x}) + V_{bias}(\mathbf{x}))]$ with

$$V_{bias}(\mathbf{x}) = \begin{cases} \infty & if \quad q(\mathbf{x}) \notin [q_{min}, q_{max}] \\ 0 & if \quad q(\mathbf{x}) \in [q_{min}, q_{max}] \end{cases}. \tag{11.1.5}$$

Then the requirement $q_{min} < q(\mathbf{x}') < q_{max}$ would be added to the acceptance rule of Mehlig et al.

Alternative hybrid MC/MD formulations can be concocted from different integrators and different initial momentum generators. One particularly versatile hybrid MC/MD algorithm can be applied to lattice or off-lattice simulations with or without a bias potential [34,35]. Suppose that we wish to sample the configurational distribution

$$\rho_{eq}(\mathbf{x})e^{-\beta V_{bias}(\mathbf{x})} \propto exp[-\beta(V(\mathbf{x}) + V_{bias}(\mathbf{x}))]$$

where $V_{bias}(\mathbf{x}) = \alpha(q(\mathbf{x}) - q*)^2/2$ is a harmonic restraining potential (rather than a hard wall confining potential). The detailed balance criterion accounting for the biased distribution is

$$\begin{aligned} &\rho_{eq}(\mathbf{x}_0)e^{-\beta V_{bias}(\mathbf{x}_0)} \int d\mathbf{p}_0 \int d\mathbf{p}_t \rho_{gen}(\mathbf{p}_0)\rho(\mathbf{x}_t, \mathbf{p}_t|\mathbf{x}_0, \mathbf{p}_0) p_{acc}(\mathbf{x}_0 \to \mathbf{x}_t) \\ &\quad = \rho_{eq}(\mathbf{x}_t)e^{-\beta V_{bias}(\mathbf{x}_t)} \int d\mathbf{p}_0 \int d\mathbf{p}_t \rho_{gen}(\mathbf{p}_t)\rho(\mathbf{x}_0, \mathbf{p}_0|\mathbf{x}_t, \mathbf{p}_t) p_{acc}(\mathbf{x}_t \to \mathbf{x}_0) \end{aligned} \tag{11.1.6}$$

Again, sample the initial momenta from the Boltzmann distribution so that

$$\rho_{gen}(\mathbf{p}) = \rho_{eq}(\mathbf{p}) \tag{11.1.7}$$

Instead of using coarsely discretized Hamiltonian dynamics like Mehlig et al., use a short unbiased Langevin dynamics trajectory to sample $\rho(\mathbf{x}_t, \mathbf{p}_t|\mathbf{x}_0, \mathbf{p}_0)$. The Langevin equation is stochastically time reversible in the sense that [7]

$$\rho_{eq}(\mathbf{x}_0, \mathbf{p}_0)\rho(\mathbf{x}_t, \mathbf{p}_t|\mathbf{x}_0, \mathbf{p}_0) = \rho_{eq}(\mathbf{x}_t, -\mathbf{p}_t)\rho(\mathbf{x}_0, -\mathbf{p}_0|\mathbf{x}_t, -\mathbf{p}_t) \qquad (11.1.8)$$

In equations (11.1.6)–(11.1.8), the bias potential is not included within $\rho_{eq}(\mathbf{x})$ nor in the dynamics $\rho(\mathbf{x}', \mathbf{p}'|\mathbf{x}, \mathbf{p})$. At this point the biasing factor is still waiting to become important. Using equations (11.1.7) and (11.1.8) in the detailed balance relation (11.1.6) gives

$$\begin{aligned} & e^{-\beta V_{bias}(\mathbf{x}_0)} p_{acc}(\mathbf{x}_0 \to \mathbf{x}_t) \int d\mathbf{p}_0 \int d\mathbf{p}_t \rho_{eq}(\mathbf{x}_0, \mathbf{p}_0)\rho(\mathbf{x}_t, \mathbf{p}_t|\mathbf{x}_0, \mathbf{p}_0) \\ &= e^{-\beta V_{bias}(\mathbf{x}_t)} p_{acc}(\mathbf{x}_t \to \mathbf{x}_0) \int d\mathbf{p}_0 \int d\mathbf{p}_t \rho_{eq}(\mathbf{x}_t, \mathbf{p}_t)\rho(\mathbf{x}_0, \mathbf{p}_0|\mathbf{x}_t, \mathbf{p}_t) \\ &= e^{-\beta V_{bias}(\mathbf{x}_t)} p_{acc}(\mathbf{x}_t \to \mathbf{x}_0) \int d\mathbf{p}_0 \int d(-\mathbf{p}_t) \rho_{eq}(\mathbf{x}_t, -\mathbf{p}_t)\rho(\mathbf{x}_0, -\mathbf{p}_0|\mathbf{x}_t, -\mathbf{p}_t) \end{aligned}$$

The integrals cancel leaving

$$e^{-\beta V_{bias}(\mathbf{x}_0)} p_{acc}(\mathbf{x}_0 \to \mathbf{x}_t) = e^{-\beta V_{bias}(\mathbf{x}_t)} p_{acc}(\mathbf{x}_t \to \mathbf{x}_0)$$

Finally, using the Metropolis function to construct the acceptance rule gives

$$p_{acc}(\mathbf{x}_0 \to \mathbf{x}_t) = min\{1, exp[-\beta(V_{bias}(\mathbf{x}_t) - V_{bias}(\mathbf{x}_0))\} \qquad (11.1.9)$$

Equations (11.1.6)–(11.1.9) justify the following hybrid MC/MD algorithm.

Algorithm: Hybrid MC/MD with Langevin trajectories

Use the velocity Verlet version of the Langevin integrator and adopt a fixed trajectory duration t.

1. From configuration $\mathbf{x}$, draw Boltzmann distributed momenta $\mathbf{p}$.
2. Run a Langevin dynamics trajectory of duration t from $(\mathbf{x}, \mathbf{p})$ to $(\mathbf{x}', \mathbf{p}') = e^{t\mathbf{L}}(\mathbf{x}, \mathbf{p})$. Here $e^{t\mathbf{L}}$ stands for a Langevin dynamics integrator.
3. Accept the move from $(\mathbf{x}, \mathbf{p})$ to $(\mathbf{x}', \mathbf{p}')$ with probability $min\{1, exp[-\beta\Delta V_{bias}]\}$ as defined in equation (11.1.9).

Repeated application of steps 1-3 generates $\rho(\mathbf{x}) \propto exp[-\beta(V(\mathbf{x}) + V_{bias}(\mathbf{x}))]$. See Joswiak et al. [35] for an extension of this to systems with rigid molecules (e.g. rigid water).

The Langevin based hybrid MC/MD with the acceptance rule (11.1.9) is important because of its tremendous versatility. It has been used in numerous umbrella sampling studies with coordinates that are non-differentiable and/or discrete. It is easily extended to multi-dimensional umbrella sampling calculations. It readily generalizes to a "hybrid MC/MC algorithm" [34,36] in which long MC trajectories are generated and then accepted or rejected based on the change in a bias potential. Hybrid MC/MC is useful in applications like nucleation where evaluating the order parameter of interest (nucleus size) is more costly than running the MC trajectories. Finally, note that many additional variations on hybrid MC/MD can be derived, e.g. see the exercises for a hybrid MC/MD scheme that samples the NVE ensemble.

Before closing this section, note Bolas [37] and other equilibrium path sampling [38–40] methods for computing free energies. These MC/MD schemes incorporate path sampling ideas to accept/reject entire paths (not just the endpoints) based on a detailed balance criterion. Their advantage is in partially avoiding the frequent starting/stopping of hybrid MC/MD. They are even more versatile than hybrid MC/MD because dynamical characteristics of the paths can be used in the acceptance rule. The most prominent example of this idea is the s-ensemble technique [41,42] which has been used to sample paths of unusually high mobility in jammed glassy systems. The bias employed in s-ensemble calculations is not a function of the endpoint, but rather a functional measure of mobility that is obtained by integrating along each trial trajectory.

11.2 Thermodynamic perturbation theory

Section 11.1 showed how importance sampling methods can generate a representative sample for computing ensemble averages. Can a representative sample help estimate free energies? Importance sampling methods cannot provide the absolute free energy, but they can help estimate free energy *differences*. Consider two systems "0" and "1" with different Hamiltonians H_0 and $H_1 = H_0 + \Delta H$. The partition function for system 1 is

$$\begin{aligned} Q_1 &\equiv \sum_\nu e^{-\beta H_1(\nu)} \\ &= \sum_\nu e^{-\beta H_0(\nu)} e^{-\beta \Delta H(\nu)} \\ &= Q_0 \left\langle e^{-\beta \Delta H} \right\rangle_0 \end{aligned}$$

where $Q_0 = \sum_\nu e^{-\beta H_0(\nu)}$ and $\langle \mathcal{O} \rangle_0 = Q_0^{-1} \sum_\nu \mathcal{O}(\nu) e^{-\beta H_0(\nu)}$ so that the exponential average $\left\langle e^{-\beta \Delta H} \right\rangle_0$ is computed by sampling microstates according to statistical weight $exp[-\beta H_0(\nu)]$. Taking logarithms on both sides gives the thermodynamic perturbation identity [43]:

$$\beta \Delta F = -ln \left\langle e^{-\beta \Delta H} \right\rangle_0$$

where $\Delta F = F_1 - F_0$ is the free energy difference between states 1 and 0.[1] The thermodynamic perturbation identity is formally exact. It is basically a reweighting formula [45,46]: states are sampled according to $exp[-\beta H_0(\nu)]$, but each one is reweighted by $exp[-\beta(H_1(\nu) - H_0(\nu))]$.

In practice, reweighting requires the phase space distributions of systems 0 and 1 to be similar, i.e. to have a strong overlap. When those states which dominate the Q_0 partition function are rarely visited in system 1, then the reweighting factor $exp[-\beta(H_1 - H_0)]$ will vary over many orders of magnitude making the exponential average difficult to converge. One remedy is to interpolate between H_0 and H_1 by introducing parameter λ in

$$H(\lambda) = H_0 + \lambda \Delta H$$

and a series of intermediate λ values:

$$0 = \lambda_0 < \lambda_1 < \lambda_2 \cdots < \lambda_n = 1$$

The one large perturbation is then replaced by a series of smaller perturbations

$$\beta \Delta F = -\sum_{i=0}^{n-1} ln \, \langle exp[-\beta(\lambda_{i+1} - \lambda_i)\Delta H] \rangle_i \tag{11.2.1}$$

where $\langle \cdot \rangle_i$ indicates an average using the distribution $exp[-\beta H(\lambda_i)]$. The stages $\lambda_0, \lambda_1, \lambda_2, \cdots, \lambda_n$ and corresponding ensembles must be designed carefully to ensure accuracy. In general, the accuracy of a free energy calculation depends on the phase space overlap between stages. Kofke and coworkers emphasized that having overlap is not sufficient and that the most reliable results are obtained from very strong overlap or by using overlapping regions as explicit stages [47,48]. Figure 11.2.1 depicts two viable strategies for a situation where the direct perturbation approach is not viable.

Analyses of accuracy and efficiency have resulted in many additional guidelines.

- Applying the Bennett acceptance ratio (BAR) formula [14,49] to estimate the free energy difference at each stage provides better accuracy than exponential averages [50]. BAR is a special two-state case of the WHAM equations which are presented in section 11.3.
- To ensure overlap between distributions at successive stages, the perturbation formula should be used to insert particles and not to delete them [51,52].

[1] For an account the history see Jorgensen and Thomas [44].

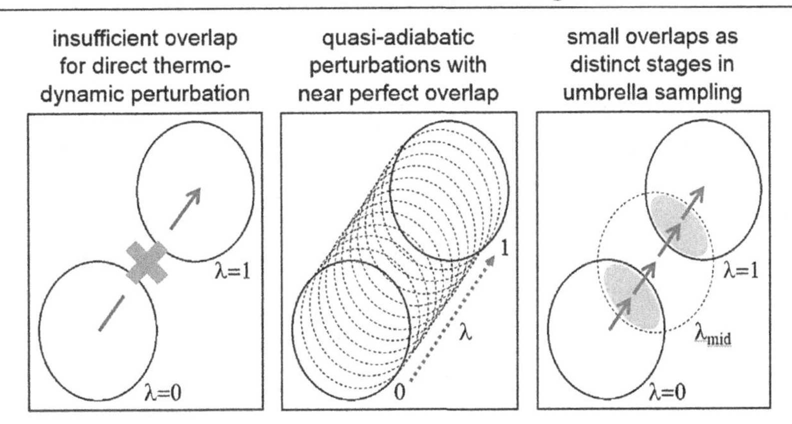

Figure 11.2.1: If systems 0 and 1 have insufficient phase space overlap, then the free energies from direct thermodynamic perturbation will be highly inaccurate. Instead, the free energy difference between $\lambda = 0$ and $\lambda = 1$ can be obtained from a family of intermediate perturbations so that each stage has near perfect overlap with its neighbors. Alternatively, the overlap between coarser stages can be treated as an explicit stage. This latter strategy will be used to match umbrella sampling windows in the next section.

- The first stage of insertion where the hard core and/or long range coulomb potentials are created requires special care [53–56].

Free energy perturbations have been used to compute the solvation free energy by turning on the interactions between a solute and the solvent in stages [57]. Another common application of thermodynamic perturbation theory is an alchemical transformation [55,56,58,59]. Alchemical transformation studies estimate the reversible work to transform one functional group or molecule to another in stages. Alchemical transformations have been incorporated into clever thermodynamic cycles to compute solubility differences [57], binding energy differences [59, 60], acidity differences [61], and interfacial free energy differences [62]. Alchemical transformations illustrate how simulations can make testable predictions via paths that have no natural analogue.

Thermodynamic perturbation theory is primarily used to compute free energy differences between systems with different Hamiltonians. However, it can also be an important tool for kinetics applications. In Chapter 22 we will learn that $\Delta\Delta G^{\ddagger}$ can often be predicted from $\Delta\Delta G^{o}$, i.e. that changes in the activation free energy can often be predicted from changes in the reaction free energy. Thermodynamic perturbation theory is a powerful tool for predicting $\Delta\Delta G^{o}$.

11.3 Projections

The previous sections discussed free energy changes that result from changes in the Hamiltonian, but in reaction rate theory we are concerned with the free energy differences between one part of phase space and another for a single Hamiltonian. Beyond free energy differences between endpoints, rates also depend on free energy barriers along the pathway from one state to another. We will need tools that can compute the free energy landscape between reactants and products, between donor and acceptor states, between folded and unfolded proteins, or between supersaturated solution and a postcritical nucleus. In each case, the free energy landscape that we require is a Landau free energy. To compute Landau free energies, we will employ restraints and/or constraints on collective variables, order parameters, and reaction coordinates.

In the simplest case, the Landau free energy projects $\rho_{eq}(\mathbf{x},\mathbf{p})$ onto a single order parameter. The Landau free energy for a reaction coordinate $q(\mathbf{x})$ in the canonical ensemble is[2]

$$\begin{aligned} F(q) &= -k_B T\, ln \int d\mathbf{x} d\mathbf{p}\, \rho_{eq}(\mathbf{x},\mathbf{p}) \ell_q \delta[q(\mathbf{x}) - q] \\ &= -k_B T\, ln \left\langle \ell_q \delta[q - q(\mathbf{x})] \right\rangle \end{aligned} \tag{11.3.1}$$

where $\rho_{eq}(\mathbf{x},\mathbf{p}) = Q^{-1} exp[-\beta H(\mathbf{x},\mathbf{p})]$, ℓ_q is the unit along q, and Q is the canonical partition function. Physically,

$$\rho_{eq}(q) = \ell_q^{-1} exp[-\beta F(q)]$$

and $\rho_{eq}(q)dq$ is the probability to find the system between q and $q + dq$ at equilibrium. Sometimes we will compute the Landau free energy as a function of two or three collective variables, order parameters, and/or reaction coordinates. For all of the following chapters, it will be useful to develop a strong formal and intuitive understanding of Landau free energies and projections. Let us start with the example below.

Example: Canonical projection

Consider a (classical) harmonic oscillator with mass m and vibrational frequency ω so that $H = m\omega^2 x^2/2 + p^2/2m$. The thermal equilibrium distribution for the oscillator is $\rho_{eq}(x, p) \propto exp[-\beta H]$. The factor

$$\int dx dp\, exp[-\beta H] = 2\pi k_B T/\omega$$

[2] We assume a purely configurational reaction coordinate for reasons given in Chapter 20.

normalizes the distribution, i.e.

$$\rho_{eq}(x,p) = \frac{\omega}{2\pi k_B T} exp[-\beta H(x,p)]$$

Now let us project $\rho(x,p)$ onto the energy E:

$$\begin{aligned}
\rho(E) &= \int_{-\infty}^{\infty} dx \int_{-\infty}^{\infty} dp\, \rho_{eq}(x,p)\delta[H(x,p)-E] \\
&= \int_{-\sqrt{2E/m\omega^2}}^{\sqrt{2E/m\omega^2}} dx \frac{\omega}{2\pi k_B T} e^{-\beta E} \frac{2}{\omega\sqrt{2E/m\omega^2 - x^2}} \\
&= \frac{2}{\pi k_B T} e^{-\beta E} \int_0^{\sqrt{2E/m\omega^2}} dx/\sqrt{2E/m\omega^2 - x^2} \\
&= \beta e^{-\beta E}
\end{aligned}$$

which, in hindsight, is a rather unsurprising answer. In the first step, we have used the identity $\int_\Omega dx f(x)\delta[g(x)-g_0] = \sum_k f(x_k)/|g'(x_k)|$ where the x_k are roots of $g(x) = g_0$ in the interval Ω.

Example: Microcanonical projection

Consider a harmonic oscillator with unit mass and unit vibrational frequency so that the energy is $E = (x^2 + p^2)/2$. The oscillator traces out a ring-shaped distribution $\rho(x,p) = (2\pi)^{-1}\delta[(x^2+p^2)/2 - E]$. The distribution is normalized:

$$\int dxdp(2\pi)^{-1}\delta[(x^2+p^2)/2 - E] = \int dr\, r\, \delta[r^2/2 - E] = \frac{\sqrt{2E}}{|\sqrt{2E}|} = 1$$

where we have again used the identity $\int_\Omega dx f(x)\delta[g(x)-g_0] = \sum_k f(x_k)/|g'(x_k)|$. To project $\rho(x,p)$ onto x:

$$\begin{aligned}
\rho(x) &= \int dx_* \int dp(2\pi)^{-1}\delta[\tfrac{1}{2}(x_*^2 + p^2) - E]\delta[x - x_*] \\
&= \int dx_*(2\pi)^{-1}\delta[x - x_*] \int dp\delta[\tfrac{1}{2}p^2 - (E - \tfrac{1}{2}x_*^2)] \\
&= \int dx_* \pi^{-1}\delta[x - x_*](2E - x_*^2)^{-1/2} \\
&= \pi^{-1}(2E - x^2)^{-1/2}
\end{aligned}$$

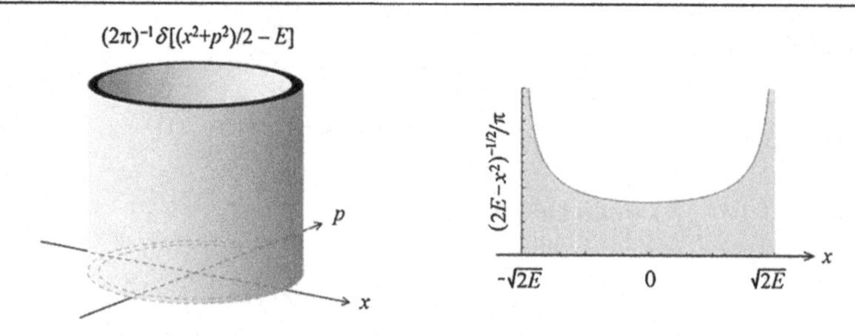

The projection onto x remains normalized, i.e. $\int \rho(x)dx = 1$. The general case where the mass and frequency differ from unity is left to the reader.

Obtaining a projection by formal integration is only possible for certain simple models and certain choices of the coordinate q. Thus, in practice, projections are often affected by binning the data from a simulation. The procedure is relatively straightforward – divide the q-axis into small intervals and record the number of times that each interval is visited. This results in a histogram that approximates $\rho_{eq}(q)$.

11.4 Non-Boltzmann sampling

In most areas of statistics, biased samples are things to avoid, but the methods in this section intentionally bias the sampling procedure and then *a posteriori* "fix" the bias to recover the unbiased statistics. Why are such indirect routes to the unbiased distribution useful? Unbiased simulations naturally sample states according to their equilibrium probability, and thus they naturally gravitate toward regions of *low* free energy. Thus we cannot efficiently sample high free energy barriers using straightforward Monte Carlo or MD-based importance sampling methods. The problem of sampling high barriers has inspired the development of several non-Boltzmann (or biased) sampling methods.

Suppose that the collective variable of interest is $q(\mathbf{x})$. Then if the potential energy is biased by any function of q the bias "slips through" the projection integral. To see this, let the real PES be $V(\mathbf{x})$ and let the biased PES be

$$V_b(\mathbf{x}) = V(\mathbf{x}) + \Delta V(q(\mathbf{x}))$$

The free energy of the biased system "b" (within constants) is

$$F_b(q) \quad = \quad -k_B T \, ln \int d\mathbf{x}\, e^{-\beta V(\mathbf{x})} e^{-\beta \Delta V(q(\mathbf{x}))} \ell_q \delta[q - q(\mathbf{x})]$$

$$
\begin{aligned}
&= \Delta V(q) - k_B T \ln \int d\mathbf{x} e^{-\beta V(\mathbf{x})} \ell_q \delta[q - q(\mathbf{x})] \qquad (11.4.1) \\
&= \Delta V(q) + F(q)
\end{aligned}
$$

Thus the basic strategy of non-Boltzmann sampling is:

1. choose a convenient bias $\Delta V(q)$,
2. perform simulations with the biased potential to get $F_b(q)$, and
3. recover the desired $F(q)$ by subtracting the bias.

States at the top of the free energy barrier are the most sparsely sampled in a standard calculation, so a natural choice for $\Delta V(q)$ is to exactly negate the barrier, i.e. $\Delta V(q) = -F(q)$. In practice, we do not *a priori* know $F(q)$, but we can still make use of property (11.4.1) by umbrella sampling [63,64].

Umbrella sampling with windows between hard walls

As for many rare events methods, umbrella sampling is a way to compute the free energy in stages. In umbrella sampling, each stage restrains the simulation to a small interval of interest (a "window") along the collective variable $q(\mathbf{x})$. The simplest bias potential for the ith window confines the system within hard walls

$$
\Delta V_i(\mathbf{x}) = \begin{cases} 0 & q_{i,min} \le q(\mathbf{x}) \le q_{i,max} \\ \infty & otherwise \end{cases} \qquad (11.4.2)
$$

Metropolis Monte Carlo with potential $V(\mathbf{x}) + \Delta V_i(\mathbf{x})$ will generate the Boltzmann distribution and the correct projection onto $F(q)$ within a constant in the window $q_{i,min} \le q(\mathbf{x}) \le q_{i,max}$. Running a series of these simulations with overlapping windows generates piecewise renormalized sections of $\rho(q)$. Let the piecewise fragments of the true distribution be denoted $\rho_i(q)$. Each piecewise renormalized $\rho_i(q)$ corresponds to a vertically shifted part of the free energy profile, $F_i(q)$.

In principle, each piecewise $\rho_i(q)$ section is related to the true distribution $\rho(q)$ by

$$
\rho_i(q) = e^{-\beta \Delta V_i(q)} \frac{Q}{Q_i} \rho(q) \qquad (11.4.3)
$$

where Q and Q_i are partition functions for the unbiased and biased systems respectively. In practice, the region where we can accurately sample $\rho(q)$ will have a vanishingly small overlap with some of the biased $\rho_i(q)$ distributions, so equation (11.4.3) is not directly useful. However, each biased distribution does overlap with neighboring biased distributions as depicted in Figure 11.4.1. The simple windows shown in the figure have been constructed so that they overlap by a single histogram bin.

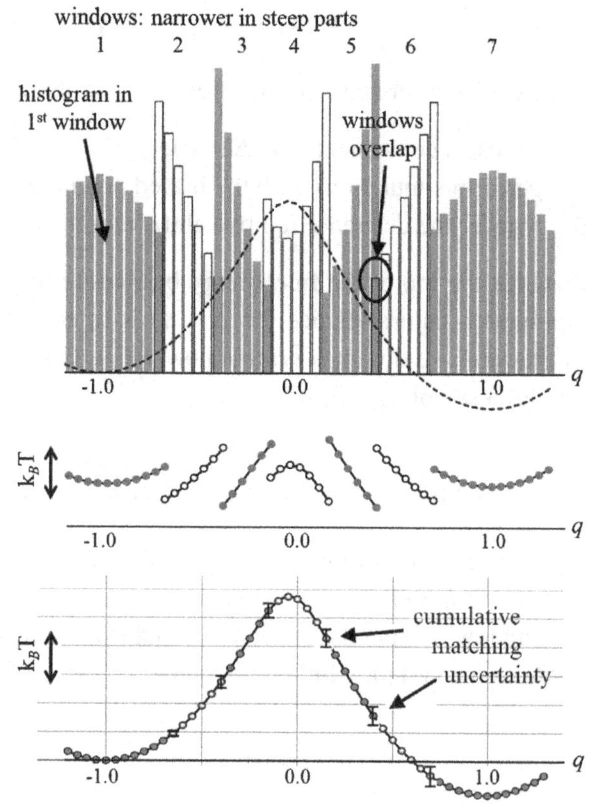

Figure 11.4.1: Schematic diagram showing the umbrella sampling procedure. Initially the free energy profile is unknown (dotted line), so the q-axis is divided into a series of overlapping windows. Monte Carlo simulations within each window generate piecewise probability distributions and piecewise free energy profiles. A continuous free energy profile is constructed by matching the overlapping window edges. Uncertainties in the piecewise distributions propagate through the matching procedure to give a cumulative uncertainty that grows with each additional matching window.

Both $\rho_i(q)$ and $\rho_{i+1}(q)$ are non-zero at $q = q_{i,max} = q_{i+1,min}$, i.e. where the rightmost bin of window i overlaps with the leftmost bin of window $i+1$. At this special q, the two distributions must be related by

$$Q_{i+1}\rho_{i+1}(q)\, e^{+\beta \Delta V_{i+1}(q)} = Q_i \rho_i(q)\, e^{+\beta \Delta V_i(q)} \qquad (11.4.4)$$

The factors of $e^{+\beta \Delta V_i(q)}$ and $e^{+\beta \Delta V_{i+1}(q)}$ are both unity in the hard wall confinement scheme. Recall that the goal was to find $\rho(q)$, so using equations (11.4.3) and ratios Q_{k+1}/Q_k from equation (11.4.4)

$$\rho(q) = \rho_i(q)\, e^{+\beta \Delta V_i(q)} \prod_{k=0}^{i-1} \frac{Q_{k+1}}{Q_k} \qquad (11.4.5)$$

Taking logarithms of equation (11.4.5)

$$F(q) = F_i(q) - \Delta V_i(q) - k_B T \sum_{k=0}^{i} ln\,(Q_{k+1}/Q_k) \qquad (11.4.6)$$

where the ith window has $\Delta V_i(q) = 0$. Equation (11.4.6) gives the unbiased free energy profile in terms of the local and accumulated biases. For square-well umbrella potentials, $Q_{k+1}/Q_k \approx \rho_k(q_{k,max})/\rho_{k+1}(q_{k+1,min})$, and the latter ratio is directly generated during the biased Monte Carlo simulations. The umbrella sampling procedure with square-well windows is depicted in Figure 11.4.1.

At each stage in the matching process, uncertainty in Q_{i+1}/Q_i (or equivalently in $\rho_i(q)/\rho_{i+1}(q)$) adds to uncertainty in the free energy profile. Suppose that the normalized window distributions ρ_i and ρ_{i+1} were estimated from M_i and M_{i+1} configurations, respectively. Then their rightmost and leftmost bins contain $M_i\rho_i(q)\Delta q$ and $M_{i+1}\rho_{i+1}(q)\Delta q$ configurations, respectively, where Δq is the bin width. The number of points in a particular bin given M_i trials is a binomial random variable with Bernoulli parameter $\rho_i(q)\Delta q$. Using the bin population to estimate the mean and using the binomial variance formula, gives the relative uncertainty in $\rho_i(q)$ as [65]

$$\frac{\delta\rho_i(q)}{\rho_i(q)} = \frac{\sqrt{1-\rho_i(q)\Delta q}}{\sqrt{M_i\rho_i(q)\Delta q}} \approx \frac{1}{\sqrt{M_i\rho_i(q)\Delta q}}$$

and similarly for the relative uncertainty in $\rho_{i+1}(q)$. The approximation is valid if the number of configurations in bin q is much smaller than the total number of configurations M_i. Using error propagation rules for independent variables gives the relative uncertainty for the individual ratios that appear in equation (11.4.6):

$$\frac{\delta(\rho_i(q)/\rho_{i+1}(q))}{\rho_i(q)/\rho_{i+1}(q)} = \left(\frac{1}{M_i\rho_i(q)} + \frac{1}{M_{i+1}\rho_{i+1}(q)}\right)^{1/2}$$

The free energy differences between points separated by several windows will accrue some uncertainty from matching of the windows between them. Specifically, for points q and q' the relative uncertainty in $\rho(q)/\rho(q')$ is

$$\frac{\delta(\rho(q)/\rho(q'))}{\rho(q)/\rho(q')} = \left(\sum_k\left[\frac{1}{M_k\rho_k(q_{k,k+1})} + \frac{1}{M_{k+1}\rho_{k+1}(q_{k,k+1})}\right]\right)^{1/2}$$

Here the sum runs from the window containing q to the window (before that) containing q' and where $q_{k,k+1}$ is the value of q in the shared bin where the kth and $(k+1)^{th}$ windows overlap. Umbrella sampling has been performed with hard wall confining bias potentials in many studies of nucleation [34,66,67].

Umbrella sampling with harmonic restraints

Umbrella sampling with hard wall confining potentials works well for Monte Carlo methods, but hard walls are not convenient for MD-based sampling strategies. For off-lattice condensed phases where MD-based sampling methods are preferred, it is more convenient to use quadratic/harmonic bias potentials. Consider a series of biased MD or hybrid MC/MD simulations with overlapping "umbrella" bias potentials of the form

$$\Delta V_i(\mathbf{x}) = \frac{1}{2}\alpha_i (q(\mathbf{x}) - q_i)^2$$

A series of these simulations with suitably chosen α_i and q_i generates a series of overlapping projected distributions $\rho_i(q)$ that are (again) related to $\rho(q)$ by

$$\rho_i(q) = (Q_i/Q)^{-1} exp[-\beta \Delta V_i(q)]\, \rho(q) \quad (11.4.7)$$

where $\rho(q) = Q^{-1}\ell_q^{-1} exp[-\beta F(q)]$ and $Q_i = \int d\mathbf{x}\, e^{-\beta \Delta V_i(\mathbf{x})} e^{-\beta V(\mathbf{x})}$. The umbrella potentials and the resulting distributions are shown in Figure 11.4.2.

Figure 11.4.2 and the summation in equation (11.4.8) include an unbiased "$\Delta V_0 = 0$" distribution in the reactant state and an unbiased "$\Delta V_{n+1} = 0$" distribution in the product state. Unbiased simulations in the reactant and product state are useful for selecting order parameters and for deciding where to place the first umbrella potentials. When the reactants and products are separated by a large barrier, the unbiased simulations will naturally remain in the reactant/product states. Once the umbrella simulations are complete, the unbiased data can be used along with the biased simulation data to construct the full distribution.[3] Unbiased simulation data should not be included if there are spontaneous barrier crossings. All of the results below remain valid if the unbiased reactant and product distributions are omitted.

Equation (11.4.7) would be exact except that $\rho_i(q)$ is estimated from a finite sample. Umbrella sampling with hard walls presented a similar difficulty that required an uncertain matching at the edges of neighboring windows. For umbrella sampling with harmonic umbrella potentials, the matching is more complicated because each distribution $\rho_i(q)$ at each q provides some information about $\rho(q)$. To construct the actual distribution, information from all of the windows must somehow be self-consistently blended. Ongoing efforts seek to improve estimates of $\rho(q)$ and $F(q)$ from umbrella sampling results [68,69], but the classic construction is the weighted histogram analysis method (WHAM) [65,70].

[3] When using WHAM with unbiased reactant and product windows, let $\Delta V_0(q) = 0$ for those reactant values of q which are spontaneously sampled and infinity otherwise. Similarly let $\Delta V_{n+1}(q) = 0$ for those product values of q which are spontaneously sampled and infinity otherwise.

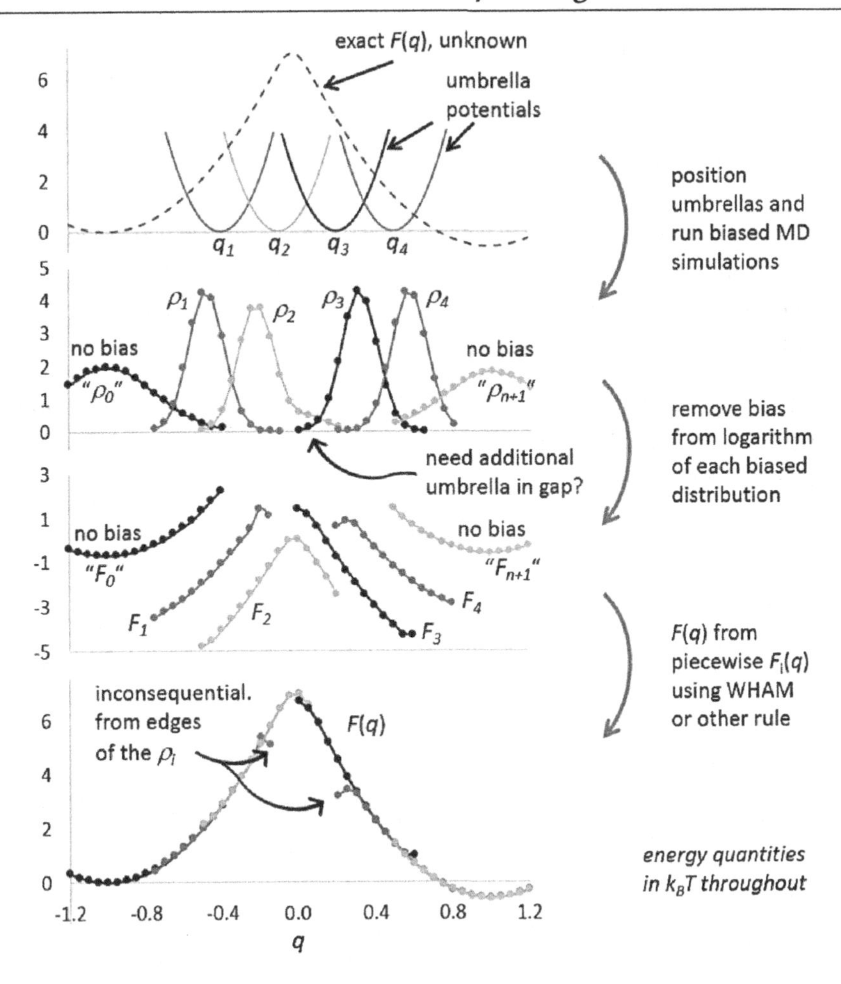

Figure 11.4.2: Schematic diagram showing umbrella sampling with harmonic restraints.

The estimate of $\rho(q)$ from an individual umbrella sampling window is

$$\rho(q) = \frac{Q_i}{Q} \rho_i(q) exp[+\beta \Delta V_i(q)]$$

where the ratio Q_i/Q is not known. WHAM takes the actual distribution to be a q-dependent weighted average of the individual estimates:

$$\rho(q) = \sum_{i=0}^{n+1} w_i(q) \frac{Q_i}{Q} \rho_i(q) exp[+\beta \Delta V_i(q)] \tag{11.4.8}$$

where the weights are normalized, i.e.

$$\sum_{i=0}^{n+1} w_i(q) = 1$$

Because $\rho(q)$ is constructed from the set of inexact $\rho_i(q)$ estimates, $\rho(q)$ will also be an estimate. But how can the estimation error in going from the set of $\rho_i(q)$ to the one $\rho(q)$ be minimized? Intuitively, the highest weights at each q should be ascribed to those simulations that generated the most data near q.

Without going through the full derivation, WHAM obtains these weights by minimizing the estimation error $\langle(\rho(q) - \hat{\rho}(q))^2\rangle$ over all possible choices of the $w_i(q)$. After solving for the optimal weights the WHAM result is

$$\rho(q) = \frac{\sum_{i=0}^{n+1} M_i \rho_i(q)}{\sum_{i=0}^{n+1} M_i e^{\beta \Delta F_i} e^{-\beta \Delta V_i(q)}} \tag{11.4.9}$$

where

$$e^{-\beta \Delta F_i} = \int dq\, \rho(q)\, e^{-\beta \Delta V_i(q)} \tag{11.4.10}$$

Equations (11.4.9) and (11.4.10) must be self consistently iterated to determine $\rho(q)$ and the ΔF_i. Typically, the iterations are started from $\Delta F_i = 0$.

Umbrella sampling is remarkably simple and effective. First, the only parameters to choose are the spring constants and placement of the umbrella potentials. Second, the individual $\rho_i(q)$ distributions converge quickly because diffusion across the window takes a small time. If variables other than q equilibrate quickly,[4] then the hard walls procedure will equilibrate within a window in $t \sim (q_{k,max} - q_{k,min})^2/D$, and the harmonic umbrella potential procedure will equilibrate in $t \sim 2k_BT/(D\alpha_i)$ where α_i was the spring constant in the umbrella potential. Third, each window can be sampled in parallel with no inter-process communication. Results from umbrella sampling and WHAM are fairly robust as long as the free energy within each window varies by no more than $2k_BT$. When the initial windows are placed too far apart, it is easy to add additional windows without redoing the entire calculation. Finally, WHAM can be extended to two dimensions by replacing the scalar q with a vector collective variable $\mathbf{q}$ in equations (11.4.9) and (11.4.10). In most applications, three dimensional umbrella sampling is a prohibitively expensive venture. Note, however, that free energy surfaces in two, three, or even higher dimensions can often be reconstructed by reweighting [35,71]. It is easy (in principle) to

[4] Sometimes they do not.

reconstruct the joint distribution in space (q_1, q_2) when an umbrella sampling calculation has been performed along the single variable q_1:

$$\rho(q_1, q_2) = \rho(q_1)\rho(q_2|q_1) \tag{11.4.11}$$

$\rho(q_2|q_1)$ can be constructed from the individual histograms $\rho_i(q_2|q_1)$ by using the same converged linear combination that was generated in equation (11.4.9). Reconstructed multivariable densities (and free energy landscapes) are extremely useful, but in practice the reweighting construction in equation (11.4.11) should be used carefully. The reconstruction only works when (i) q_2 has been recorded alongside each q_1 datum, and (ii) when holding q_1 constant does not cause non-ergodicity in the sampling of q_2. The second requirement effectively means that q_1 should be the correct reaction coordinate or else closely related to the correct reaction coordinate.

11.5 Thermodynamic integration

Let us briefly revisit the free energy difference ΔF between systems with different Hamiltonians. Again suppose that we can interpolate between systems with Hamiltonians H_0 and H_1 by introducing parameter λ:

$$H_\lambda = H_0 + \lambda \Delta H$$

where $\Delta H = H_1 - H_0$. Let the free energy of the system with fixed λ be

$$\beta F(\lambda) = -ln \sum_\nu exp[-\beta H_\lambda(\nu)]$$

Taking the derivative with respect to λ gives

$$\frac{\partial F}{\partial \lambda} = \left\langle \frac{\partial H}{\partial \lambda} \right\rangle_\lambda \tag{11.5.1}$$

where $\langle \cdot \rangle_\lambda$ indicates an average using the distribution $exp[-\beta H_\lambda]$. Now integrating from $\lambda = 0$ to $\lambda = 1$ gives the thermodynamic integration formula [17,72]

$$\Delta F = \int_0^1 \left\langle \frac{\partial H}{\partial \lambda} \right\rangle_\lambda d\lambda \tag{11.5.2}$$

Equation (11.5.2), like thermodynamic perturbation theory, relates a free energy difference to equilibrium averages at intermediate values of λ. In thermodynamic perturbation theory, the free energy was reduced to a discrete sum of exponential averages, while thermodynamic integration reduces the free energy difference to an integral over a generalized mean force.

Equation (11.5.2), has an important interpretation and several connections to other theoretical results. The reversible work required to pull the system at an infinitesimally slow rate from $\lambda = 0$ to $\lambda = 1$ equals the change in free energy. The free energy difference is a lower bound on the *average* work that must be performed along a real path from $\lambda = 0$ to $\lambda = 1$, i.e. $\Delta F \leq W$. One might assume from the inequality between average non-equilibrium work and free energy, that irreversible pulling experiments can at best provide an upper bound on ΔF. To the contrary, if we perform the irreversible pulling experiment many times then ΔF can be obtained from the Jarzynski *equality* [73]:

$$exp[-\beta \Delta F] = \langle exp[-\beta W_{0\to 1}]\rangle_0$$

Here the subscript "0" indicates an exponential average over work performed $W_{0\to 1}$ along trajectories starting from an equilibrium distribution of initial conditions with $\lambda = 0$. The Jarzynski equality and the closely related Crooks fluctuation theorem [74] have rapidly become cornerstone results in non-equilibrium statistical mechanics. The distribution of non-equilibrium work values can be directly simulated, e.g. using thousands of steered MD trajectories [75], and, remarkably, their exponential average provides the reversible work for the same (infinitely slow) transformation. From a computational standpoint, steered MD/non-equilibrium work analyses are always less efficient routes to free energy differences than thermodynamic integration and umbrella sampling methods [76]. However, non-equilibrium work theorems do provide a route to free energy differences directly from single-molecule pulling experiments [77].

Let us now re-focus on the Landau free energy $F(q)$ for a reaction coordinate $q(\mathbf{x})$. How can thermodynamic integration ideas help us compute $F(q)$? We could add a restraint to the Hamiltonian and then average $\partial H/\partial q$. However, $\partial H/\partial q$ is not easy to average because the kinetic energy will typically have a complicated dependence on curvilinear coordinates q. To avoid the kinetic energy operator, start by writing $F(q)$ in terms of purely configurational integrals and then consider its derivative with respect to q [5]:

$$\begin{aligned}
\frac{\partial F}{\partial q} &= -k_B T \frac{\partial}{\partial q} ln \frac{\int d\mathbf{x}\, e^{-\beta V(\mathbf{x})}\delta[q(\mathbf{x})-q]}{\int d\mathbf{x}\, e^{-\beta V(\mathbf{x})}} \\
&= -k_B T \frac{\int d\hat{q} d\mathbf{Q}\, |\mathbf{J}|\, e^{-\beta V} \frac{\partial}{\partial \hat{q}} \delta[q-\hat{q}]}{\int d\mathbf{x}\, e^{-\beta V(\mathbf{x})}\delta[q(\mathbf{x})-q]}
\end{aligned}$$

The coordinate transformation $\mathbf{x} \to (\hat{q}, \mathbf{Q})$ was invoked to enable an integration by parts in the steps below.

Presumably we know the $\hat{q}$-coordinate, but the other coordinates (those represented by $\mathbf{Q}$) are unspecified. Recall that the Jacobian for a point transformation in the full phase space

[5] Here, $q(\mathbf{x})$ becomes $\hat{q}$ when $\mathbf{x} \to (\hat{q}, \mathbf{Q})$, and $\partial\delta[\hat{q} - q]/\partial q = \partial\delta[q - \hat{q}]/\partial\hat{q}$.

is unity, but the Jacobian between configurational parts of a point transformation is not [78]. Let us proceed despite the looming obstacles associated with $|\mathbf{J}|$. Integration by parts for the $\hat{q}$-dependence gives

$$\frac{\partial F}{\partial q} = -k_B T \frac{\int d\hat{q} d\mathbf{Q}\, \delta[q - \hat{q}] \frac{\partial}{\partial \hat{q}} \left\{ |\mathbf{J}| e^{-\beta V} \right\}}{\int d\mathbf{x} e^{-\beta V(\mathbf{x})} \delta[q(\mathbf{x}) - q]}$$

Now to avoid specifying the unrestricted $\mathbf{Q}$ coordinates, change back to Cartesian variables, $(\hat{q}, \mathbf{Q}) \rightarrow \mathbf{x}$.

$$\frac{\partial F}{\partial q} = -k_B T \frac{\int d\mathbf{x}\, \delta[q(\mathbf{x}) - q]\, |\mathbf{J}|^{-1} \frac{\partial}{\partial q} \left\{ |\mathbf{J}| e^{-\beta V} \right\}}{\int d\mathbf{x} e^{-\beta V(\mathbf{x})} \delta[q(\mathbf{x}) - q]}$$

Completing the derivatives and writing all resulting terms as conditional averages gives [79,80]

$$\frac{\partial F}{\partial q} = \left\langle \frac{\partial V}{\partial q} \right\rangle_q - k_B T \left\langle \frac{\partial ln|\mathbf{J}|}{\partial q} \right\rangle_q \tag{11.5.3}$$

The $\partial ln|\mathbf{J}|/\partial q$ dependence accounts for changing entropic contributions to the projection along the q-axis. The Landau free energy is obtained by integrating the conditionally averaged mean forces (and Jacobians) along the q-axis

$$F(q) - F(q_0) = \int_{q_0}^{q} dq \left\langle \frac{\partial V}{\partial q} - k_B T \frac{\partial ln|\mathbf{J}|}{\partial q} \right\rangle_q \tag{11.5.4}$$

$\partial V/\partial q$ and $\partial ln|\mathbf{J}|/\partial q$ look simple enough, but they are deceptively complicated. Consider $\partial V/\partial q$:

$$\frac{\partial V}{\partial q} = \sum_j \left(\frac{\partial V}{\partial x_j} \right)_{\neq j} \left(\frac{\partial x_j}{\partial q} \right)_{\mathbf{Q}} \tag{11.5.5}$$

Because of the $(\partial x_j/\partial q)_{\mathbf{Q}}$ terms, the $\partial V/\partial q$ calculation would seem to require a complete set of $\mathbf{Q}$ coordinates. The situation is even worse for $|\mathbf{J}|$ which entirely depends on the complete set of $(q, \mathbf{Q})$ coordinates. On the other hand, we should not need to specify $\mathbf{Q}$ only to integrate them out again – the final answer must not depend on our choice of the $\mathbf{Q}$ coordinate system [79–81]. den Otter developed general expressions for the dilation/contraction of volume elements upon a small displacement in q. To keep track of the q-direction, he introduced a local coordinate system in the form of a vector field and obtained the general expression [82]

$$\frac{\partial F}{\partial q} = \left\langle \frac{\nabla q \cdot \nabla V}{|\nabla q|^2} - k_B T\, \nabla \cdot \left(\frac{\nabla q}{|\nabla q|^2} \right) \right\rangle_q \tag{11.5.6}$$

which relates the derivative of the free energy to a directional derivative of the potential and to the changing entropy from projection to varying volume elements. Ciccotti et al. [83] generalized the den Otter result to include multiple coordinates. Alternatively, van Gunsteren [84] suggested obtaining the mean force $\partial F/\partial q$ by an explicit average of the constraint forces in a simulation. Explicit force averages, which only require first derivatives of q, are important in the adaptive biasing force approaches [85]. Once $\partial F/\partial q$ is obtained at a series of locations along the reaction coordinate, it must then be inserted into equation (11.5.4) and numerically integrated.

Algorithm: Thermodynamic integration with harmonic restraints

1. Discretization: Choose integration slices at evenly spaced locations $q_0, q_1, q_2, \ldots, q_n$.
2. Simulations: Obtain samples for computing (approximate) conditional averages at each slice by running simulations in a harmonically *restrained* potential

$$V_i(\mathbf{x}) = V(\mathbf{x}) + \alpha(q(\mathbf{x}) - q_i)^2/2$$

   Here $q(\mathbf{x})$ is the reaction coordinate, α is a strong spring constant, and q_i is the target reaction coordinate value for the ith slice.
3. Conditional mean forces: During the restrained simulations for each slice i, compute $\partial F/\partial q$ according to equation (11.5.6).
4. Sampling error analysis: The restrained ensembles will not be exactly centered at the intended locations q_i. Specifically, the range of values that will be sampled is approximately

$$\left(q_i - \frac{F'(q_i)}{F''(q_i) + \alpha}\right) \pm \sqrt{\frac{2k_B T}{F''(q_i) + \alpha}}$$

   Note that the range of q-values sampled depends on α and on the local slope and curvature of the (*a priori* unknown) $F(q)$. To ensure that an appropriately centered and appropriately narrow range of values has been sampled, one should check *a posteriori* that the chosen value of η at each slice is large enough to satisfy $F''(q_i)/\alpha \ll 1$ and $F'(q_i)^2/(k_B T\alpha) \ll 1$.
5. Trapezoidal integration: Approximate the Landau free energy at each point q_i by computing

$$F(q_i) = F(q_0) + \frac{1}{2}\sum_{j=1}^{i-1}(q_{j+1} - q_j)\left(\left(\frac{\partial F}{\partial q}\right)_{j+1} + \left(\frac{\partial F}{\partial q}\right)_j\right)$$

The formula given here also works for non-uniformly spaced $q_0, q_1, q_2, ..., q_n$, but a uniform grid simplifies the integration error analysis. Specifically, the integration error from each interval scales as $\sim(\Delta q)^3$ where $\Delta q = q_{j+1} - q_j$. Additional errors will come from uncertainty in the averages to obtain $\partial F/\partial q$ and from the finite spring constant α.

In most cases, harmonic restraints can be made strong enough to give negligible displacements from the target coordinate location and non-zero widths of the sampling windows. However, the Blue Moon method offers an elegant way to perfectly eliminate these errors. The Blue Moon approach uses a sharp *constraint* to sample the $q(\mathbf{x}) = q_i$ surfaces. Moreover, the quantities to be averaged ($\partial V/\partial q$ and the Jacobian terms that depend on ∇q) are already being used to bias forces at each MD time step to maintain a constant value of q. Thus Blue Moon can be implemented with relatively little computational overhead. However, there is one important catch: constrained averages and conditional averages are different. In fact, reaction rate theory and rare events methods utilize *three* types of restricted averages: conditional, constrained, and hitting points averages.

Conditional average

The ensemble used for a conditional average is $\rho_{eq}(\mathbf{x})\delta[q(\mathbf{x}) - q]$. These averages have already been encountered in our previous discussions. For example, the TST formulation of equation (10.5.4) defines the prefactor using a conditional average. Conditional averages also define the Landau free energy. Conditional averages are defined by

$$\langle \mathcal{O} \rangle_q = \frac{\int d\mathbf{x} \rho_{eq}(\mathbf{x}) \mathcal{O}(\mathbf{x}) \delta[q(\mathbf{x}) - q]}{\int d\mathbf{x} \rho_{eq}(\mathbf{x}) \delta[q(\mathbf{x}) - q]}$$

and they are easily computed by histogram-based methods like umbrella sampling or by molecular dynamics with strong harmonic restraints. In practice (and in theory) the delta function is represented by an infinitesimally narrow interval along the q-axis, e.g. $q_i - \epsilon < q(\mathbf{x}) < q_i + \epsilon$. Although the precise δ-function is obtained from $\epsilon \to 0^+$, the surface in this limit still varies in *relative* thickness from one point to another. These changes in relative thickness effectively give different weights to different locations on the $q(\mathbf{x}) = q$ surface.

Hitting points average

The ensemble used for a hitting points average is $\rho_{eq}(\mathbf{x})|\nabla q|\delta[q(\mathbf{x})-q]$, and the hitting points average is

$$\langle \mathcal{O} \rangle_{q(\mathbf{x})=q} = \frac{\int d\mathbf{x} \rho_{eq}(\mathbf{x}) \mathcal{O}(\mathbf{x}) \, |\nabla q| \, \delta[q(\mathbf{x})-q]}{\int d\mathbf{x} \rho_{eq}(\mathbf{x}) \, |\nabla q| \, \delta[q(\mathbf{x})-q]}$$

Hitting point averages are true surface integrals – like the surface integrals that appear in textbook electromagnetism and fluid mechanics calculations, but the dimensionality in a molecular system is too high to explicitly parameterize the surface. Instead, the hitting points average contains factors of $|\nabla q|$ that give each point on the $q(\mathbf{x}) = q$ surface an equal weight in the volume integrals. Hitting point averages arise in transition path theory [86,87] and also in the prefactor of "Eyring's" TST, cf. equation (3.8) in Hanggi et al. [88]. The hitting points ensemble is so-named because it represents the ensemble of configurations which would be obtained from the points at which unconstrained canonically distributed trajectories pass through the surface $q(\mathbf{x}) = q$.

Constrained dynamics average

The Blue Moon approach [80,89] sharply enforces a constraint during each timestep by incorporating bias forces into the dynamics. A constrained dynamical simulation generates the ensemble

$$\rho_{eq}(\mathbf{x}) \left\{ \nabla q^T \mathbf{M}^{-1} \nabla q \right\}^{1/2} \delta[q(\mathbf{x}) - q]$$

so that the constrained dynamics average is

$$\langle \mathcal{O} \rangle_{q(\mathbf{x})=q,\, \dot{q}=0} = \frac{\int d\mathbf{x} \rho_{eq}(\mathbf{x}) \mathcal{O}(\mathbf{x}) \left\{ \nabla q^T \mathbf{M}^{-1} \nabla q \right\}^{1/2} \delta[q(\mathbf{x})-q]}{\int d\mathbf{x} \rho_{eq}(\mathbf{x}) \left\{ \nabla q^T \mathbf{M}^{-1} \nabla q \right\}^{1/2} \delta[q(\mathbf{x})-q]} \tag{11.5.7}$$

Here $\mathbf{M}$ is the diagonal matrix of atomic masses. Why should the average of a configurational quantity be influenced by the masses of atoms involved in coordinate $q(\mathbf{x})$? The mass dependence originates from the way the ensemble of configurations is collected. Formally, the constrained MD restricts both $q(\mathbf{x}) = q$ and $\dot{q} = 0$. The additional constraint on $\dot{q}$ requires no acceleration, and therefore it introduces a dependence on masses. A derivation of the constrained dynamics average can be found in Sprik and Ciccotti [80].

Figure 11.5.1 depicts the restricted ensembles which correspond to each of the three averages. The ensembles which are used in these three averages differ by reweighting factors: 1 (for conditional), $|\nabla q|$ (for hitting points), and $\left\{ \nabla q^T \mathbf{M}^{-1} \nabla q \right\}^{1/2}$ (for constrained MD). Fortunately,

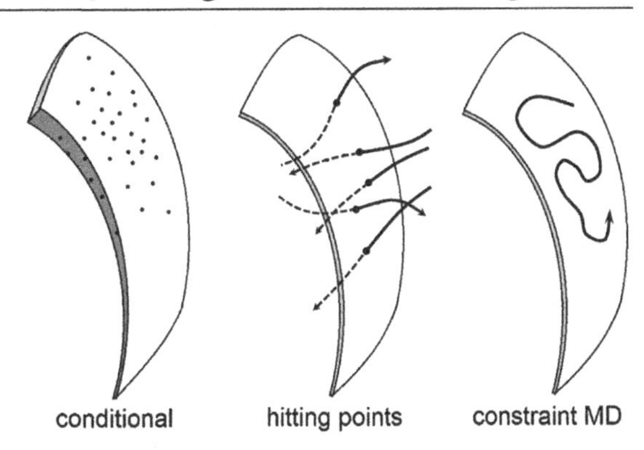

Figure 11.5.1: Three types of restricted averages. Masses are irrelevant for the conditional average and the hitting point average, but masses are important for the constrained average (except for cases where all masses are the same).

the reweighting factors tend to be relatively weak functions of the configuration $\mathbf{x}$, so that an average with respect to one distribution can be obtained by sampling another and then reweighting the results. In general, reweighted averages are related to unweighted averages by:

$$\begin{aligned}\langle \mathcal{O} \rangle_{V_0+\Delta V} &= \frac{\int d\mathbf{x}\, e^{-\beta(V_0(\mathbf{x})+\Delta V(\mathbf{x}))}\mathcal{O}(\mathbf{x})}{\int d\mathbf{x}\, e^{-\beta(V_0(\mathbf{x})+\Delta V(\mathbf{x}))}} \\ &= \frac{\left\langle e^{-\beta\Delta V(\mathbf{x})}\mathcal{O}(\mathbf{x})\right\rangle_{V_0}}{\left\langle e^{-\beta\Delta V(\mathbf{x})}\right\rangle_{V_0}} \end{aligned} \tag{11.5.8}$$

where $\langle\cdot\rangle_{V_0} = \int d\mathbf{x}\,(\cdot)\, e^{-\beta V_0(\mathbf{x})} / \int d\mathbf{x}\, e^{-\beta V_0(\mathbf{x})}$.

Equation (11.5.8) shows how a simulation guided by potential $V_0(\mathbf{x})$ can yield averages for a system with potential $V_0 + \Delta V$. The same strategy can be used to interconvert between constrained averages, conditional averages, and hitting point averages. For example, to obtain the conditional average from a constrained simulation [14],

$$\langle \mathcal{O}(\mathbf{x})\rangle_q^{cond} = \frac{\left\langle \left\{\nabla q^T \mathbf{M}^{-1}\nabla q\right\}^{-1/2} \mathcal{O}(\mathbf{x})\right\rangle^{constr}_{q(\mathbf{x})=q,\,\dot{q}=0}}{\left\langle \left\{\nabla q^T \mathbf{M}^{-1}\nabla q\right\}^{-1/2}\right\rangle^{constr}_{q(\mathbf{x})=q,\,\dot{q}=0}}$$

The factors of $\left\{\nabla q^T \mathbf{M}^{-1}\nabla q\right\}^{-1/2}$ serve to eliminate the factors of $\left\{\nabla q^T \mathbf{M}^{-1}\nabla q\right\}^{1/2}$ which emerge from sampling with constraints, and in the denominator the average of $\left\{\nabla q^T \mathbf{M}^{-1}\nabla q\right\}^{-1/2}$ serves to normalize the reweighted distribution.

Example: Reweighting factors for restricted averages

Consider the reaction $AB + C \rightarrow A + BC$. Let the reaction coordinate be $q = r_{AB} - r_{BC}$ and let the two other coordinates be θ and $\sigma = r_{AB} + r_{BC}$.

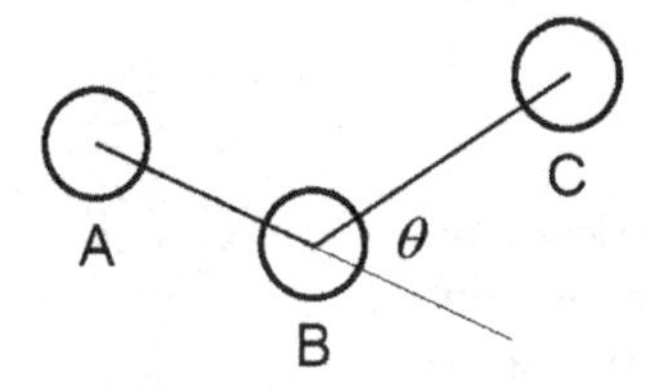

A hitting points average involves the factor

$$|\nabla q| = \sqrt{2(2 + cos\theta)}$$

A constrained MD average involves the closely related factor

$$\left\{\nabla q^T \mathbf{M}^{-1} \nabla q\right\}^{1/2} = \left(\frac{1}{m_A} + \frac{2}{m_B}(1 + cos\theta) + \frac{1}{m_C}\right)^{1/2}$$

When all atoms have the same mass, the constrained MD average becomes the same as the hitting points average.

11.6 Other methods for computing free energies

Adiabatic partitioning methods

These methods [90–92] impose a high temperature and strong friction (or a heavy mass) on the reaction coordinate(s) while sampling the bath coordinates with the actual temperature of interest and their unadulterated dynamics. The dynamics of a reaction coordinate q in this framework are somewhat strange. The force on q is that from the free energy landscape at the actual temperature, but barriers on the free energy landscape are easily overcome because the reaction coordinate(s) themselves are at an elevated temperature. The strong friction on the reaction coordinate ensures that its dynamics remain slow enough to adiabatically sample the bath configurations while the reaction coordinate remains effectively frozen. In this way, the reaction coordinate(s) can slowly evolve on a temperature-rescaled version of the true potential of mean force.

Let the reaction coordinate be q and the bath coordinates be $\mathbf{Q}$. If the coordinates are adiabatically separated with different temperatures and if the exploration of $\mathbf{Q}$ at fixed q is ergodic, then Rosso et al. [90] showed that the adiabatic free energy dynamics (AFED) samples the density $\rho_{\beta\beta_q}(q) \propto exp[-\beta_q F(q;\beta)]$ where $\beta = 1/k_B T$, $\beta_q = 1/k_B T_q$, and where $F(q;\beta) = -\beta^{-1} ln \int d\mathbf{x} exp[-\beta V(\mathbf{x})] \ell_q \delta[q(\mathbf{x}) - q]$ as per the usual definition. Note that $\beta_q/\beta < 1$ so that q evolves at the higher temperature. Taking the logarithm of $\rho_{\beta\beta_q}(q)$ and multiplying by $-\beta_q^{-1}$ gives [90]

$$F(q;\beta) = -\beta_q^{-1} ln \rho_{\beta\beta_q}(q)$$

where $F(q;\beta)$ is just the usual free energy at the actual temperature.

These approaches do not entirely flatten barriers, so some obstacles in the q-variables will persist unless an extremely high T_q/T is used. On the other hand, an extremely high T_q/T is ill-advised because the actual (low) temperature results must be recovered by rescaling the high temperature free energy surface by the factor T_q/T. Thus, any numerical errors in $\rho_{\beta\beta_q}(q)$ will be amplified by the scalefactor T_q/T. Most applications of AFED [90] and temperature accelerated molecular dynamics (TAMD) [92] have used $T_q/T \sim 2$–3.

Flooding, metadynamics, and related methods

These methods use the history of previously visited states to guide the sampling into regions that have not yet been sampled. The most widely used method in this category is Metadynamics [93–95] which builds upon several earlier works: taboo search [96], local elevation [97], bond boost methods [98], conformational flooding [99], and Calvo's generalization of the Wang-Landau method [100]. In the Metadynamics approach, the history of visited states during an MD simulation is used to construct a time-dependent bias potential $V_{bias}(\mathbf{q},t)$. The bias potential is constructed as a sum of Gaussians in the space of *a priori* selected collective variables (CVs). If the Gaussians are sufficiently small in amplitude and if they are deposited with a sufficiently high frequency [95], the bias potential effectively takes the form

$$V_{bias}(\mathbf{q},t) = \varpi \int_0^t dt' exp\left[-\sum_{i=1}^{n} \frac{(q_i - q_i(t'))^2}{2\sigma_i^2}\right] \tag{11.6.1}$$

where t is the time elapsed since the beginning of the simulation and n is the number of CVs included. ϖ is an energy per time deposition rate and σ_i is the width of the multidimensional Gaussian along the q_i-axis [95]. The bias potential accumulated at each location is approximately proportional to the extent of prior sampling at that location. As time progresses the overall landscape becomes approximately flat, basin-by-basin as shown in Figure 11.6.1. Once all basins are filled, the free energy is obtained by subtracting $V_{bias}(\mathbf{q},t)$ from $F(\mathbf{q}) + V_{bias}(\mathbf{q},t) \approx const$.

By design, the bias potential in equation (11.6.1) discourages the system from revisiting the most heavily sampled regions. Note that the bias is applied *in the CV space*, not in the space of atomistically detailed configurations. Thus configurations which are entirely new in the atomistically detailed $\mathbf{x}$-space are still discouraged if they share the same $\mathbf{q}$-coordinates with

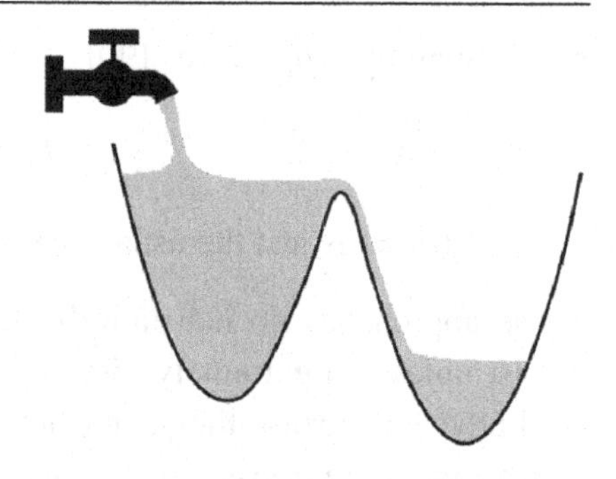

Figure 11.6.1: The history-dependent bias potential created by a Metadynamics simulation gradually fills basins and promotes transitions to neighboring basins. At long times, the bias potential approximates the negated free energy landscape.

configurations that have previously been encountered. This property is the brilliance of Metadynamics if good CVs are chosen, and the Achilles' heel of Metadynamics if poor CVs are chosen.

Nearly every review on Metadynamics properly emphasizes the critical importance of selecting good CVs [95,101,102], but these cautionary notes have been ignored in many applications of Metadynamics. For example, one authoritative review [102] includes illustrative examples, notes on pitfalls, and prescient recommendations for selecting CVs. Among these recommendations, the review says "...for the CVs to be dynamically meaningful [the CVs should] correspond to the reaction coordinate..." Here the authors cite Dellago et al. [103] who provided tests for reaction coordinate quality and demonstrated that inaccurate coordinates give free energy surfaces that misrepresent the dynamics, kinetics, and mechanisms. Then the "Applications" section of the same Metadynamics review [102] highlights dozens of studies (including many by the developers) that have used Metadynamics results to draw mechanistic conclusions without testing whether the chosen CVs are even crudely approximate reaction coordinates. Without tests, we cannot know which mechanistic conclusions are reliable and which mechanistic conclusions are not. Metadynamics itself is an extremely powerful sampling method, but (as for other biased sampling methods) mechanistic analyses with untested CVs are ill-advised.

There are some additional (and more easily avoided) pitfalls to beware of in Metadynamics simulations.

- Fast deposition rates fill basins and provide free energy estimates quickly, but they also increase errors in the free energy estimates. Slow moving collective variables in $\mathbf{q}$ may lead to Gaussians being deposited on top of each other [95]. In extreme cases, the simulation can "surf" on the mounting pile of Gaussians.
- The collective variables in $\mathbf{q}$ must be differentiable even if the optimal reaction coordinate(s) are not. Marinelli et al. [104] and Salvalaglio et al. [105] have rectified this problem by using reweighting methods to recover the free energy as a function of better reaction coordinates.

- Free energy estimates from the original version of Metadynamics do not actually converge. Even after all basins have been filled, the steadily increasing water level $F(\mathbf{q}) + V_{bias}(\mathbf{q}, t)$ will fluctuate in $\mathbf{q}$. The fluctuations can be mitigated by averaging profiles at several times [101], or by using the alternative Well-Tempered Metadynamics approach.

In the Well-Tempered Metadynamics (WTMetaD) approach [95,106], the bias potential grows according to the equation

$$\frac{\partial}{\partial t} V_{bias}(\mathbf{q}, t) = \varpi \, exp\left[-\frac{V_{bias}(\mathbf{q}(t), t)}{V_{fill}}\right] exp\left[-\sum_{i=1}^{n} \frac{(q_i - q_i(t'))^2}{2\sigma_i^2}\right] \tag{11.6.2}$$

where V_{fill} is an adjustable parameter. The effect of equation (11.6.2) is that Gaussians of smaller amplitude are deposited in regions which have accumulated a larger bias potential. Where and when the depth of the bias potential far exceeds V_{fill} the new Gaussians become vanishingly small and the bias potential stops growing. In the long time limit the bias potential created by WTMetaD *converges* to [106]

$$V_{bias}(\mathbf{q}) = -\frac{V_{fill}}{V_{fill} + k_B T} F(\mathbf{q}) + const.$$

Like the adiabatic partitioning methods [90], barriers in WTMetaD are only partially flattened. Approximately flattened barriers like those in the original Metadynamics method can be recovered by choosing V_{fill} comparable to or larger than the free energy barrier. Of course, free energy barriers are not *a priori* known and this has inspired adaptive Metadynamics/WTMetaD approaches [95].

Adaptive biasing force

Metadynamics adaptively constructs a bias potential from the frequency at which points are visited during a simulation. The adaptive biasing force (ABF) method [85,107] builds a history-dependent bias potential from a running average of the position dependent forces. At each point in time during a simulation the current estimate of the mean force is subtracted from the instantaneous forces, so that as time progresses the simulation feels no net force along the chosen collective variables.

11.7 Cautionary notes

This chapter focused on a few specific aspects of free energy calculations which are important for studying chemical reactions and other rare events. Of course the vast literature on free energy calculations [5,108–111] goes far beyond these topics. A ubiquitous challenge in free energy calculations is to efficiently and thoroughly sample phase space in systems with many barriers and stable intermediates. In fact, before embarking on a free energy calculation and rare events analysis, one should be aware of all potentially important products, intermediates, and off-pathway intermediates. Sometimes all of the stable and metastable states are *a priori* known with a fair degree of confidence. For example, in studies of nucleation there is typically one stable state and one metastable state with few if any stable intermediates between. On the other hand, for biomolecular processes like protein folding there may be hundreds of stable intermediates. Replica exchange methods [112–114] are a powerful and unbiased way to explore the energy landscape and discover stable intermediates.

Replica exchange methods emerged as a way to extract a well-defined (usually canonical) ensemble from a process that loosely resembles simulated annealing. In replica exchange methods, several simultaneous simulations ('replicas') are evolved at different temperatures. Joint statistics for the extended ensemble of replicas can be used to derive acceptance probabilities for the swapping of replicas between temperatures. These temperature swap moves have a significant acceptance probability if the temperatures are chosen so that the density of states generated by replicas at neighboring temperatures overlap. Replica exchange is not an efficient method for computing free energy barriers, but it is quite good at jumping over them as depicted in Figure 11.7.1. The advantages of replica exchange over other exploration strategies are (1) that the replicas individually explore a well-defined distribution and (2) the exploration is completely unbiased whereas states discovered by other approaches may be predetermined (or prevented) by the choice of collective variables.

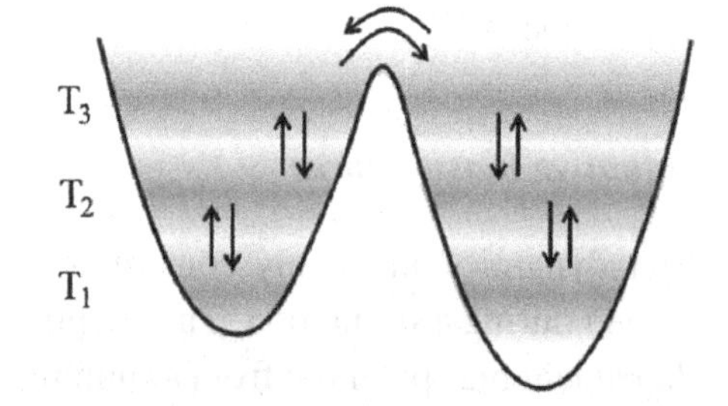

Figure 11.7.1: Without any imposed bias to influence the exploration, replica exchange methods can jump over barriers and accelerate the discovery of unforeseen products, intermediates, and off-pathway intermediates.

Computed free energies generally improve in accuracy with more extensive sampling. When judging convergence, recall that sample sizes should be adjusted for correlation times. If configurations generated by the sampling method decorrelate over millions of moves, then it will take billions of moves to collect a representative sample. See van Gunsteren et al. [108] and Klimovich et al. [111] for convergence tests, error estimates, and troubleshooting procedures.

In rare events applications, one of the first signs of trouble is a hysteresis in the free energy profile as depicted in Figure 11.7.2. For methods like umbrella sampling, the free energy profile should not depend on the direction of sampling. In some cases, a hysteresis is weak in the sense that it can be overcome by long sampling times. However, even a weak hysteresis suggests that the CVs are not sufficiently accurate for mechanistic analysis [115].

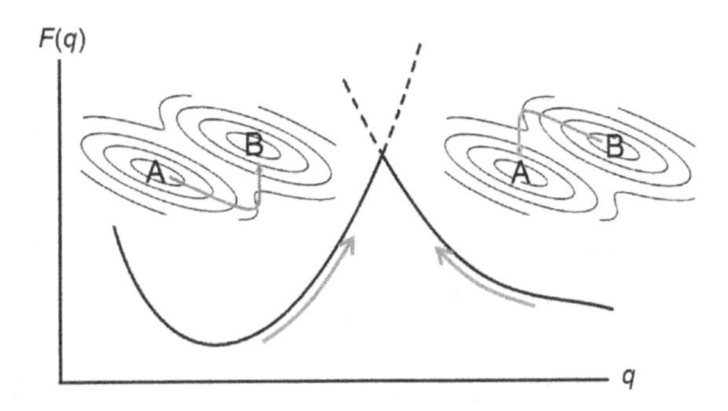

Figure 11.7.2: Schematic showing results of a free energy calculation along coordinate q. The two insets depict the two dimensional free energy surface with an important collective variable that is omitted in the $F(q)$ calculation. Sampling from A to B or from B to A along q can result in hysteresis. Even if a hysteresis can be fixed by additional sampling, it suggests an inadequate set of coordinates.

Conventional wisdom says that including extra CVs in a free energy calculation leads to better results as long as the additional dimension can still be sampled. However, extra variables can obscure a simple mechanistic picture [116], and sometimes extra variables can cause more serious problems. For example, see exercise 5 in Chapter 8 for a discussion of the artificial solid-solid transformation pathways that are generated by methods that use the solid lattice parameters as extra CVs without properly balancing the solid stresses.

Finally, remember that the free energy landscape $F(q)$ is not directly observable except in those special cases where the coordinate q can be "pulled" as in a single molecule pulling experiment [117] or extracted from a structure factor like the pair correlation function $g(r)$ [6]. Most of the coordinates we use in simulations cannot be directly pulled/driven. For example, we cannot experimentally pull along the arclength of a minimum energy path, along pathlike collective variables, along RMSD from a native structure, along dihedral angles, or from one Voronoi cell index to another. Thus the free energy surface as a function of these coordinates cannot be directly validated or falsified by any experiment. Accordingly, a free energy surface as a function of these variables should not be (but too often is) the central finding of a rare events analysis. Analyses that use these tools should further strive for testable predictions, e.g. by using the free energy surface to compute equilibrium constants, to compute absolute rates, or to predict an observable trend by performing calculations for a series of similar molecules or reaction conditions.

Exercises

1. Show that $\delta[q(\mathbf{x})-q] = |\nabla q|^{-1}\delta[(q(\mathbf{x})-q)/|\nabla q|]$ when $|\nabla q| \neq 0$. What does this result tell you about the ensemble of configurations that is used for computing a conditional average?

2. Compute the projections $\rho(x)$ and $\rho(p)$ from the ring distribution $\rho(x, p)$ in section 11.3. Use the projections to show that x and p are uncorrelated but not independent.
3. Compute the Landau free energy $F(q)$ for the Hamiltonian

$$H = \frac{p_q^2}{2m} + V(q) + \frac{1}{2}\sum_i \left\{ \frac{p_i^2}{m_i} + m_i\omega_i^2 \left(x_i - \frac{c_i}{m_i\omega_i^2} q \right)^2 \right\}$$

Assume that differential phase space volumes are given by $dqdp_q \prod_i dx_i dp_i$. The collection of x_i variables represent a bath of harmonic oscillators linearly coupled to q.
4. By definition, the probability to find an equilibrium system between q and $q + dq$ is $\ell_q^{-1} exp[-\beta F_q(q)]dq$ where $F_q(q)$ is the free energy from a projection onto q. Now suppose that one instead wants the free energy from a projection onto the related variable $\vartheta = \vartheta(q)$. Find $F_\vartheta(\vartheta)$ such that $\ell_\vartheta^{-1} exp[-\beta F_\vartheta(\vartheta)]d\vartheta$ gives the probability to find the same equilibrium system between ϑ and $\vartheta + d\vartheta$.
5. Consider pulling a molecule through solution from point $\mathbf{x}$ to point $\mathbf{x}'$ at constant velocity. Use the Jarzynski relation and the Stokes' drag formula to make a qualitative sketch of the distribution of non-equilibrium work values W and also the distribution of $exp[-\beta W]$. How do the distributions change with pulling velocity?
6. What determines the maximum permissible variance σ^2 for the Gaussians which are deposited during a Metadynamics calculation?
7. Read F. Pietrucci and W. Andreoni, *Phys. Rev. Lett.* 107, 085504 (2011). Explain how their Social Permutation Invariant (SPRINT) coordinates work. Show that, in combination with Metadynamics, the SPRINT coordinates can discover those chemical bonds in a complex system which are most easily broken and thereby discover new compounds and intermediates.
8. The usual projection onto q is $\ell_q^{-1} exp[-\beta F(q)] = \int d\mathbf{x} e^{-\beta V(\mathbf{x})}\delta[q(\mathbf{x}) - q]$, but a thermostatted MD simulation with a constraint $q(\mathbf{x}) = q$ samples the part of phase space with the metric-tensor-weighted distribution

$$\left(\nabla q^\dagger \mathbf{M}^{-1} \nabla q\right) e^{-\beta H(\mathbf{x},\mathbf{p})}\delta[q(\mathbf{x}) - q]\delta[dq/dt]$$

where $\mathbf{M}$ is the diagonal mass matrix and $\nabla \equiv \partial/\partial\mathbf{x}$. Use $dq/dt = \mathbf{p}\mathbf{M}^{-1}\nabla q$ to integrate away the momentum dependence and thereby derive equation (11.5.7).
9. Calculate the (scalar) metric tensor $\nabla q^\dagger \mathbf{M}^{-1} \nabla q$ and $\nabla^2 q$ for an end-to-end distance coordinate $q(\mathbf{x}) = |\mathbf{x}_2 - \mathbf{x}_1|$. This coordinate is common in studies of polymers and biomolecules. Check your work against Rathore et al. *J. Chem. Phys.* 120, 5781–5788 (2004).

10. Calculate the (scalar) metric tensor $\nabla q^{\dagger}\mathbf{M}^{-1}\nabla q$ for the cosine of an angle:

$$q(\mathbf{x}) = \frac{(\mathbf{x}_1 - \mathbf{x}_2) \cdot (\mathbf{x}_3 - \mathbf{x}_2)}{|\mathbf{x}_1 - \mathbf{x}_2||\mathbf{x}_3 - \mathbf{x}_2|}$$

11. The binding coefficient between two molecules is $K = \int e^{-\beta w(r)} 4\pi r^2 dr$ where $w(r)$ is the potential of mean force defined so that $lim_{r\to\infty} w(r) = 0$. Derive this formula and discuss the role of cutoffs in the calculation of K. Devise a way of using the formula without sampling all positions in a sphere around the binding site. For a discussion of two routes to these binding coefficients, see Deng and Roux, *J. Phys. Chem. B*, 113, 2234 (2009) and Shirts et al., *Ann. Rep. Comp. Chem.* 3, 41 (2007).
12. Develop a Hybrid MC/MD algorithm for sampling the microcanonical distribution. Swap the momenta for particles with equivalent masses to obtain equiprobable microstates without changing the energy [118]. Discuss the effects of swaps on the angular momentum for systems that are and are not in periodic boundary conditions.
13. A. Fong et al., *ACS Catal.* 5, 3360 (2015) studied ethylene polymerization. One of the mechanisms they studied, involves a site in which the alkyl chain bonded to a Cr center can loop around and coordinate via its vinyl end to the Cr center again.

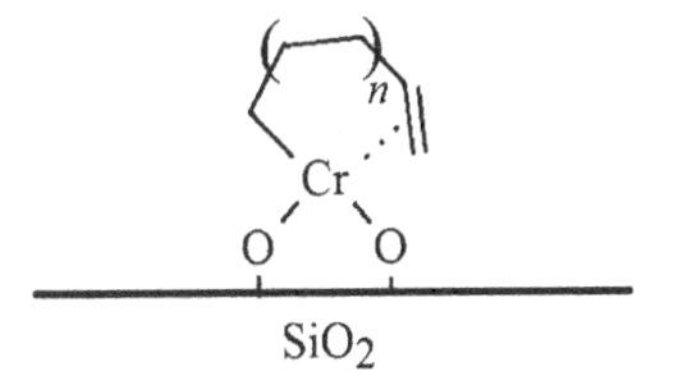

Suppose that the binding energy of the vinyl fragment at the active site is described by a Morse potential: $V(r) = E_D(1 - exp[-a(r - r_0)])^2$. If the Cr-site is located on a flat SiO_2 surface, find the probability and free energy to find the vinyl end at distance r from the site. Use ideal chain statistics and restrict the random walk of the polymer to $z > 0$.

14. Write a Monte Carlo code to sample the Boltzmann distribution for a 32×32 Ising model. Let the dimensionless Hamiltonian be $\beta H = -\beta J \sum_{\langle ij \rangle} s_i s_j$ where s_i and s_j are spins at lattice positions i and j and where each spin can take the value $+1$ or -1. Compute the average absolute value of the magnetization $M = (32)^{-2} \sum_i s_i$ as a function of βJ. Estimate the critical temperature T_c corresponding to the onset of spontaneous magnetization, and compare your estimate to the exact theoretical result.
15. Use your Ising model code and the umbrella sampling technique to compute $F(M)$ for temperatures $T = 0.800T_c$, $T = 0.900T_c$, $T = 0.950T_c$, and $T = 0.975T_c$. Plot $ln\Delta F$ vs. $ln(T_c - T)/T_c$ where $\Delta F = F(0) - F(\langle|M|\rangle)$. Explain why simulations with a different aspect ratio might reveal more about scaling laws.

16. Read A.D. Bruce et al., *Phys. Rev. E.* 61, 906 (2000). Explain how their Lattice Switch Monte Carlo method mixes concepts from thermodynamic perturbation theory (where perturbations change the Hamiltonian) and concepts from umbrella sampling (where collective variables are used to bias sampling).

References

[1] L. Masgrau, D.G. Truhlar, Acc. Chem. Res. 48 (2015) 431–438.
[2] G.A. Voth, R.M. Hochstrasser, J. Phys. Chem. 100 (1996) 13034–13049.
[3] P. Chaikin, T.C. Lubensky, Principles of Condensed Matter Physics, Cambridge University Press, Cambridge, 2000.
[4] H.E. Stanley, Introduction to Phase Transitions and Critical Phenomena, Oxford University Press, New York, 1971.
[5] N. Hansen, W.F.V. Gunsteren, J. Chem. Theory Comput. 10 (2014) 2632–2647.
[6] D. Chandler, Introduction to Modern Statistical Mechanics, Oxford Press, New York, 1997.
[7] N.G. van Kampen, Stochastic Processes in Physics and Chemistry, Elsevier, Amsterdam, 2007.
[8] N. Metropolis, A.W. Rosenbluth, M.N. Rosenbluth, A.H. Teller, E. Teller, J. Chem. Phys. 21 (1953) 1087–1092.
[9] J.C. Maxwell, Philos. Trans. R. Soc. Lond. 157 (1867) 49–88.
[10] L. Boltzmann, Sitzungsber. Kais. Akad. Wiss. 66 (1872) 275–370.
[11] R. Weigschieder, Monatsh. Chem. 32 (1901) 849–906.
[12] G.N. Lewis, Proc. Natl. Acad. Sci. USA 11 (1925) 179–183.
[13] V.I. Manousiouthakis, M.W. Deem, J. Chem. Phys. 110 (1999) 2753.
[14] D. Frenkel, B. Smit, Understanding Molecular Simulation: From Algorithms to Applications, Academic Press, San Diego, 2002.
[15] A. Leach, Molecular Modelling: Principles and Applications, Prentice Hall, Pearson Education, Englewood Cliffs, NJ, 2001.
[16] D.P. Landau, K. Binder, A Guide to Monte Carlo Simulations in Statistical Physics, Cambridge University Press, Cambridge, UK, 2009.
[17] M. Tuckerman, Statistical Mechanics: Theory and Molecular Simulation, Oxford University Press, Oxford, UK, 2010.
[18] M.N. Rosenbluth, A.W. Rosenbluth, J. Chem. Phys. 23 (1955) 356–359.
[19] J.I. Siepmann, D. Frenkel, Mol. Phys. 75 (1992) 59–70.
[20] S. Nose, M.L. Klein, Mol. Phys. 50 (1983) 1055.
[21] S. Nose, J. Chem. Phys. 81 (1984) 511.
[22] W.G. Hoover, Phys. Rev. A 31 (1985) 1695.
[23] G.J. Martyna, M.L. Klein, M. Tuckerman, J. Chem. Phys. 97 (1992) 2635–2643.
[24] B. Leimkuhler, Molecular Dynamics: With Deterministic and Stochastic Numerical Methods, Springer, Berlin, Heidelberg, 2015.
[25] H.C. Andersen, J. Chem. Phys. 72 (1980) 2384.
[26] G.T. Beckham, B. Peters, J. Phys. Chem. Lett. 2 (2011) 1133–1138.
[27] R.B. Best, G. Hummer, Proc. Natl. Acad. Sci. USA 102 (2005) 6732–6737.
[28] S. Duane, A.D. Kennedy, B.J. Pendleton, D. Roweth, Phys. Lett. B 195 (1987) 216–222.
[29] E. Leontidis, B.M. Forrest, A.H. Widmann, U.W. Suter, J. Chem. Soc. Faraday Trans. 91 (1995) 2355–2368.
[30] B. Mehlig, D.W. Heermann, B.M. Forrest, Phys. Rev. B 45 (1992) 679–685.
[31] R. Faller, J.J. de Pablo, J. Chem. Phys. 116 (2002) 55–59.
[32] E. Akhmatskaya, S. Reich, J. Comput. Phys. 227 (2008) 4934–4954.

[33] M.P. Allen, D.J. Tildesley, Computer Simulation of Liquids, Clarendon Press, Oxford, UK, 1987.
[34] S. Auer, D. Frenkel, Annu. Rev. Phys. Chem. 55 (2004) 333–361.
[35] M.N. Joswiak, N. Duff, M.F. Doherty, B. Peters, J. Phys. Chem. Lett. 4 (2013) 4267.
[36] B. Peters, J. Chem. Phys. 131 (2009) 244103.
[37] R. Radhakrishnan, T. Schlick, J. Chem. Phys. 121 (2004) 2436–2444.
[38] G. Adjanor, M. Athenes, J. Chem. Phys. 123 (2005) 234104.
[39] G. Adjanor, M. Athenes, F. Calvo, Eur. Phys. J. B 53 (2006) 47–60.
[40] B. Peters, N.E.R. Zimmermann, G.T. Beckham, J.W. Tester, B.L. Trout, J. Am. Chem. Soc. 130 (2008) 17342–17350.
[41] L.O. Hedges, R.L. Jack, J.P. Garrahan, D. Chandler, Science 323 (2009) 1309–1313.
[42] R.L. Jack, L.O. Hedges, J.P. Garrahan, D. Chandler, Phys. Rev. Lett. 107 (2011) 275702.
[43] R. Zwanzig, J. Chem. Phys. 22 (1954) 1420–1426.
[44] W.L. Jorgensen, L.L. Thomas, J. Chem. Theory Comput. 4 (2008) 869–876.
[45] W.W. Wood, J. Chem. Phys. 48 (1968) 415–434.
[46] A.Z. Panagiotopoulos, J. Phys. Condens. Matter 25 (2000) R25–R52.
[47] D. Wu, D.A. Kofke, J. Chem. Phys. 123 (2005) 54103.
[48] D. Wu, D.A. Kofke, J. Chem. Phys. 123 (2005) 085109.
[49] C.H. Bennett, J. Comput. Phys. 22 (1976) 245–268.
[50] M.R. Shirts, V.S. Pande, J. Chem. Phys. 122 (2005) 144107.
[51] N.D. Lu, D.A. Kofke, J. Chem. Phys. 114 (2001) 7303–7311.
[52] N.D. Lu, D.A. Kofke, J. Chem. Phys. 115 (2001) 6866–6875.
[53] M. Mezei, J. Comput. Chem. 13 (1992) 651–656.
[54] T.C. Beutler, A.E. Mark, R.C. van Shaik, P.R. Gerber, W.F. van Gunsteren, Chem. Phys. Lett. 222 (1994) 529–539.
[55] M.R. Shirts, D.L. Mobley, J.D. Chodera, Annu. Rep. Comput. Chem. 3 (2007) 41–58.
[56] J.D. Chodera, D.L. Mobley, M.R. Shirts, R.W. Dixon, K. Branson, V.S. Pande, Curr. Opin. Struct. Biol. 21 (2011) 150–160.
[57] W.L. Jorgensen, C. Ravimohan, J. Chem. Phys. 83 (1985) 3050–3054.
[58] L. Jorgensen, Acc. Chem. Res. 22 (5) (1989) 184–189.
[59] T. Straatsma, J.A. McCammon, Annu. Rev. Phys. Chem. 43 (1992) 407–435.
[60] D.L. Mobley, P.V. Klimovich, J. Chem. Phys. 137 (2012) 230901.
[61] W.L. Jorgensen, J.M. Briggs, J. Am. Chem. Soc. 111 (1989) 4190–4197.
[62] N. Duff, Y.R. Dahal, J.D. Schmit, B. Peters, J. Chem. Phys. 140 (2014) 014501.
[63] G.M. Torrie, J.P. Valleau, J. Comput. Phys. 23 (1977) 187–199.
[64] J. Kastner, Wiley Interdiscip. Rev. Comput. Mol. Sci. 1 (2011) 932–942.
[65] S. Kumar, D. Bouzida, R.H. Swendsen, J.M. Rosenbergl, P.A. Kollman, J. Comput. Chem. 13 (1992) 1011–1021.
[66] A.C. Pan, D. Chandler, J. Phys. Chem. B 108 (2004) 19681–19686.
[67] B.C. Knott, N. Duff, M.F. Doherty, B. Peters, J. Chem. Phys. 131 (2009) 224112.
[68] M.R. Shirts, J.D. Chodera, J. Chem. Phys. 129 (2008) 124105.
[69] F. Zhu, G. Hummer, J. Comput. Chem. 33 (2011) 453–465.
[70] A.M. Ferrenberg, R.H. Swendsen, Phys. Rev. Lett. 61 (1988) 2635–2638.
[71] R.G. Mullen, J.E. Shea, B. Peters, J. Chem. Theory Comput. 10 (2014) 659.
[72] J.G. Kirkwood, J. Chem. Phys. 3 (1935) 300.
[73] C. Jarzynski, Phys. Rev. Lett. 78 (1997) 2690.
[74] G. Crooks, Phys. Rev. E 61 (2000) 2361–2366.
[75] B. Isralewitz, M. Gao, K. Schulten, Curr. Opin. Struct. Biol. 11 (2001) 224–230.
[76] H. Oberhofer, C. Dellago, P.L. Geissler, J. Phys. Chem. B 109 (2005) 6902–6915.
[77] C. Bustamante, J. Liphardt, F. Ritort, Phys. Today (July 2005) 43–48.

[78] K. Wannier, Statistical Physics, Dover, New York, 1966.
[79] M.J. Ruiz-Montero, D. Frenkel, J.J. Brey, Mol. Phys. 90 (1997) 925–941.
[80] M. Sprik, G. Ciccotti, J. Chem. Phys. 109 (1998) 7737–7744.
[81] W.K. Den Otter, W.J. Briels, J. Chem. Phys. 109 (1998) 4139–4146.
[82] W.K. den Otter, J. Chem. Phys. (ISSN 0021-9606) 112 (2000) 7283.
[83] G. Ciccotti, R. Kapral, E. Vanden-Eijnden, ChemPhysChem 6 (2005) 1809–1814.
[84] W.F. van Gunsteren, in: W.F. van Gunsteren, P.K. Weiner (Eds.), Computer Simulations of Biomolecular Systems: Theoretical and Experimental Applications, vol. 1, Springer, Leiden, Netherlands, 1989, p. 27.
[85] J. Comer, J.C. Gumbart, J. Henin, T. Lelievre, A. Pohorille, C. Chipot, J. Phys. Chem. B 119 (2015) 1129–1151.
[86] E. Vanden-Eijnden, in: M. Ferrario, G. Ciccotti, K. Binder (Eds.), Computer Simulations in Condensed Matter: From Materials to Chemical Biology, in: Lecture Notes in Physics, vol. 703, Springer, Berlin, Heidelberg, 2006, pp. 439–478.
[87] W. E, E. Vanden-Eijnden, J. Stat. Phys. 123 (2006) 503–523.
[88] P. Hanggi, P. Talkner, M. Borkovec, Rev. Mod. Phys. 62 (1990) 251–341.
[89] E.A. Carter, G. Ciccotti, J.T. Hynes, R. Kapral, Chem. Phys. Lett. 156 (1989) 472–477.
[90] L. Rosso, P. Minary, Z. Zhu, M.E. Tuckerman, J. Chem. Phys. 116 (2002) 4389–4402.
[91] J. VandeVondele, U. Rothlisberger, J. Phys. Chem. B 106 (2002) 203–208.
[92] C.F. Abrams, E. Vanden-Eijnden, Proc. Natl. Acad. Sci. USA 107 (2010) 4961–4966.
[93] A. Laio, M. Parrinello, Proc. Natl. Acad. Sci. USA 99 (2002) 12562–12566.
[94] B. Ensing, M. De Vivo, Z. Liu, P. Moore, M.L. Klein, Acc. Chem. Res. 39 (2006) 73–81.
[95] G. Bussi, D. Branduardi, Rev. Comput. Chem. 28 (2015) 1–49.
[96] D. Cvijovic, J. Klinowski, Science 267 (1995) 664–666.
[97] T. Huber, A.E. Torda, W.F. van Gunsteren, J. Comput.-Aided Mol. Des. 8 (1994) 695–708.
[98] J.-C. Wang, S. Pal, K.A. Fichthorn, Phys. Rev. B 63 (2001) 085403.
[99] H. Grubmuller, Phys. Rev. E 52 (1995) 2893–2906.
[100] F. Calvo, Mol. Phys. 100 (2002) 3421–3427.
[101] A. Laio, F.L. Gervasio, Rep. Prog. Phys. 71 (2008) 126601.
[102] A. Barducci, M. Bonomi, M. Parrinello, Wiley Interdiscip. Rev. Comput. Mol. Sci. 1 (2011) 826–843.
[103] C. Dellago, P.G. Bolhuis, P.L. Geissler, Adv. Chem. Phys. 123 (2001) 1–84.
[104] F. Marinelli, F. Pietrucci, A. Laio, S. Piana, PLoS Comput. Biol. 5 (2009) e1000452.
[105] M. Salvalaglio, C. Perego, F. Giberti, M. Mazzotti, M. Parrinello, Proc. Natl. Acad. Sci. USA 112 (2015) E6–E14.
[106] A. Barducci, G. Bussi, M. Parrinello, Phys. Rev. Lett. 100 (2008) 020603.
[107] E. Darve, A. Pohorille, J. Chem. Phys. 115 (2001) 9169–9183.
[108] W.F. van Gunsteren, D. Bakowies, R. Baron, I. Chandrasekhar, M. Christen, X. Daura, P. Gee, D.P. Geerke, A. Glattli, P.H. Hunenberger, M. Kastenholz, C. Oostenbrink, M. Schenk, D. Trzesniak, N.F.A. van der Vegt, H.B. Yu, Angew. Chem., Int. Ed. Engl. 45 (2006) 4064–4092.
[109] C. Chipot, A. Pohorille, Free Energy Calculations, Springer-Verlag, Berlin, Heidelberg, 2007.
[110] C.D. Christ, A.E. Mark, W.F. van Gunsteren, J. Comput. Chem. 31 (2010) 1569–1582.
[111] P.V. Klimovich, M.R. Shirts, D.L. Mobley, J. Comput.-Aided Mol. Des. 29 (2015) 397–411.
[112] Y. Sugita, Y. Okamoto, Chem. Phys. Lett. 314 (1999) 141–151.
[113] A. Mitsutake, Y. Sugita, Y. Okamoto, Biopolymers Peptide Sci. 60 (2001) 96–123.
[114] Y.M. Rhee, V.S. Pande, Biophys. J. 84 (2003) 775–786.
[115] E. Rosta, H.L. Woodcock, B.R. Brooks, G. Hummer, J. Comput. Chem. 30 (2009) 1634–1641.
[116] B. Peters, J. Phys. Chem. B 119 (2015) 6349–6356.
[117] G. Hummer, A. Szabo, Proc. Natl. Acad. Sci. USA 107 (2010) 10–15.
[118] R.G. Mullen, J.-E. Shea, B. Peters, J. Chem. Theory Comput. 11 (2015) 2421–2428.

CHAPTER 12

Tunneling

"... if the de Broglie wave gets through, even though with some difficulty, it always smuggles a particle with it.... To evaluate this formula I had to calculate the integral of the expression $\sqrt{1 - a/r}dr$, and I did not know how to do it.... I expressed at the end of [the paper] my thanks to Kotshchin for his help with the mathematics. Later when the paper appeared, he wrote me that he had become a laughingstock among his colleagues, who had learned what kind of highbrow mathematical help he had given me."

Gamow, My World Line (1970)

Tunneling [1] refers to transmission through a barrier at energies below the classical reaction energy threshold. The hallmark of tunneling is a rate that becomes temperature independent at low temperatures. Tunneling can usually (but not always [2]) be ignored at high temperatures or when heavy atoms are important in the reaction coordinate. But for reactions involving proton or hydrogen transfer, tunneling can dramatically accelerate rates even at room temperature. Figure 12.0.1 from Benderskii et al. [3] compiles data for many reactions that exhibit strong tunneling contributions, some at temperatures as high as 200K. The figure also shows that the transition between thermally activated and tunneling dominated regimes can take place over a wide temperature interval.

The onset of tunneling can also be seen on an Arrhenius plot as the temperature interval where lnk vs. $1/k_BT$ changes from a slope of $-E_a$ to a slope of approximately zero. When only a small temperature interval is examined, the effects of tunneling are easily misinterpreted as artificially small activation energy or enthalpy parameters. At temperatures where tunneling is important, the experimental activation parameters are inextricably mixed with tunneling contributions. There is no way to change this situation, because the experiment cannot "turn off" tunneling to obtain a classical barrier estimate. Ignoring tunneling can lead to errors in certain applications, e.g. in testing new mechanisms by comparing predicted and observed activation enthalpies, or for identifying the rate determining step in a multistep reaction with proton or hydrogen transfer steps. For reactions that involve H-atom motion at moderate or low temperatures, the best practice is to check for the possible importance of tunneling. This chapter discusses simple tests and corrections to account for tunneling.

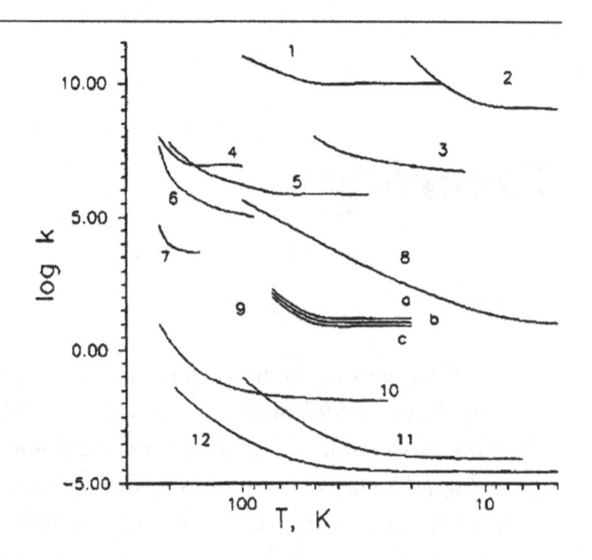

Figure 12.0.1: Tunneling leads to temperature independent rates below a tunneling crossover temperature. Signatures of tunneling have been observed for many reactions as shown in this compilation of results for several low temperature reactions. [From Benderskii et al. *Physics Reports*, 233, 195–339 (1993).]

12.1 One-dimensional tunneling models

Real molecules are never one-dimensional, but one-dimensional tunneling models are often surprisingly effective. One dimensional tunneling models assume the reaction coordinate is separable from all other degrees of freedom. In the vicinity of the saddle point, recall that the unstable mode coordinate of harmonic TST has locally separable dynamics. The one dimensional classical TST rate constant for passage over a barrier of height $V_{\ddagger}$ is $k_{TST} = k_B T\, h^{-1} exp[-V_{\ddagger}/k_B T]$. For one isolated degree of freedom, the TST rate constant can equally be written as [4]

$$k_{TST} = \frac{1}{h}\int_0^{\infty} P_{CL}(E)\, e^{-E/k_B T} dE \tag{12.1.1}$$

$$= \frac{k_B T}{h}\int_0^{\infty} P_{CL}(E)\, e^{-E/k_B T} d(E/k_B T) \tag{12.1.2}$$

where the lower integration bound starts at the zero point energy level in the reactant state and where the factor P_{CL} is the classical transmission probability

$$P_{CL}(E) = \begin{cases} 1 & for \quad E > V_{\ddagger} \\ 0 & for \quad E < V_{\ddagger} \end{cases} \tag{12.1.3}$$

The physics behind equation (12.1.3) is trivial. A classical point particle moving along the q-axis is reflected by the barrier if the energy is $E < V_{\ddagger}$ and transmitted if the energy is $E > V_{\ddagger}$.

Quantum mechanically, the same particle would be delocalized over some interval on the q-axis. Even in the classical world, a delocalized body can do something akin to tunneling over a barrier. For example, consider the pole-vaulter in Figure 12.1.1. Her center of mass never eclipses the barrier threshold, and yet she clears the bar anyway. Of course, the pole vaulter differs from quantum mechanical tunneling in many aspects, but some useful models of tunneling (especially ring-polymer molecular dynamics) exhibit barrier crossing trajectories much like those of the pole-vaulter.

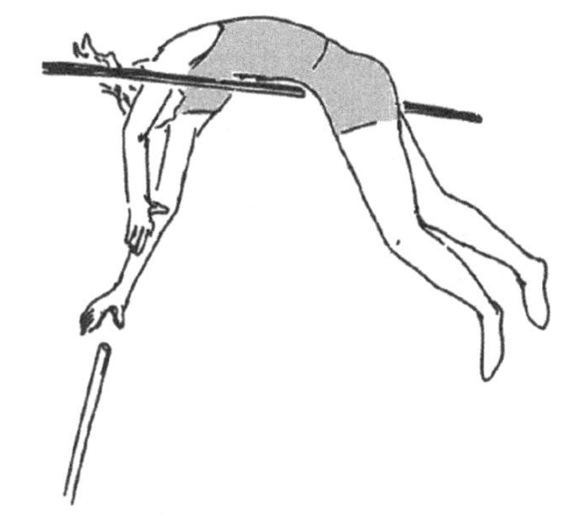

Figure 12.1.1: A good pole vaulter can outjump her own center-of-mass by contorting her body as it goes over the bar. A ring polymer tunnels over reaction barriers in the same manner.

Quantum mechanically, a trajectory can tunnel through the barrier at energies below $V_{\ddagger}$ and trajectories can also be reflected at energies above $V_{\ddagger}$. Given a model for the tunneling probability $P_{QM}(E)$, the rate constant is [4,5]

$$k = \frac{k_B T}{h} \int_0^\infty P_{QM}(E)\, e^{-E/k_B T} d(E/k_B T). \tag{12.1.4}$$

It is usually convenient to define a transmission coefficient κ_{QM} that isolates the effects of tunneling. The transmission coefficient is the ratio of the rates with (equation (12.1.4)) and without (equation (12.1.1)) the effects of tunneling.

$$\begin{aligned} \kappa_{QM} &= \frac{\int_0^\infty P_{QM}(E) e^{-E/k_B T} d(E/k_B T)}{\int_0^\infty P_{CL}(E) e^{-E/k_B T} d(E/k_B T)} \\ &= \int_0^\infty P_{QM}(E) e^{-(E-V_{\ddagger})/k_B T} d(E/k_B T) \end{aligned} \tag{12.1.5}$$

Thus far, this book has avoided serious quantum mechanics, but at least a minimal introduction is needed to explain the nature of the tunneling probability $P_{QM}(E)$ and to obtain quantitative expressions for κ_{QM}.

A standard instructive calculation [4] considers tunneling through a square-top barrier of height $V_{\ddagger}$. It is customary since the work of Gamow [6], but somewhat counterintuitive, to derive the transmission and reflectance probabilities (see Figure 12.1.2) using a time independent

Schrodinger equation.[1] How can a time-*independent* equation describe the incoming trajectories and the subsequent transmission and reflection amplitudes? Welcome to the odd world of quantum mechanics! The short answer is that a continuous beam of scatterers is no different from an individual (as long as they have the same energy and cannot interfere with each other). A better answer requires wavepacket scattering calculations [8], and those are beyond the scope of this book.

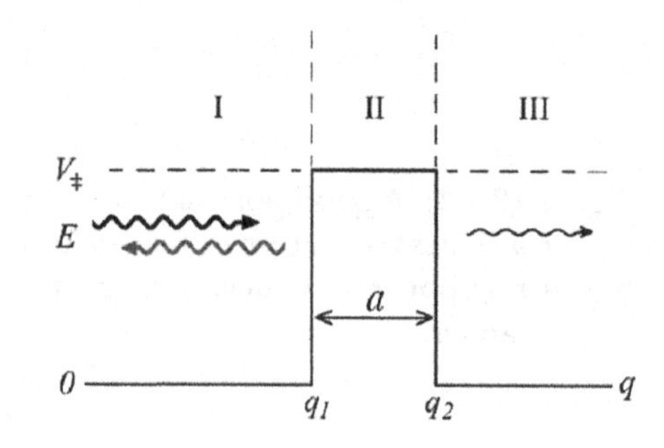

Figure 12.1.2: Tunneling through a square-top barrier of height $V_{\ddagger}$.

The quantum mechanical Hamiltonian is

$$\hat{H} = -\frac{\hbar}{2m}\frac{\partial^2}{\partial q^2} + V(q) \tag{12.1.6}$$

and the time-independent Schrodinger equation is

$$\hat{H}\Psi = E\Psi. \tag{12.1.7}$$

In regions I and III, the potential is $V(q) = 0$, and the momentum is a constant given by $E = p^2/2m$. The wavefunctions, with as yet undetermined coefficients, are a superposition of left ($-$) and right ($+$) moving waves:

$$\Psi_I(q) = A_{I+}e^{ipq/\hbar} + A_{I-}e^{-ipq/\hbar} \tag{12.1.8}$$

$$\Psi_{III}(q) = A_{III+}e^{ipq/\hbar} + A_{III-}e^{-ipq/\hbar} \tag{12.1.9}$$

In region II, the potential is $V(q) = V_{\ddagger}$, and for $E < V_{\ddagger}$ the wavefunction is

$$\Psi_{II}(q) = A_{II+}e^{|p_{CL}|q/\hbar} + A_{II-}e^{-|p_{CL}|q/\hbar} \tag{12.1.10}$$

where the classical momentum is imaginary

$$p_{CL} = \sqrt{2m(E - V_{\ddagger})} \tag{12.1.11}$$

[1] See Merzbacher's article on the history of tunneling [7].

Equations (12.1.8), (12.1.9), and (12.1.10) are easily verified solutions to equation (12.1.7) in the respective regions. Continuity and differentiability are requirements of an admissible wavefunction for a bounded potential. Thus we must choose the six coefficients so that Ψ is continuous and differentiable throughout. The matching conditions are thus $\Psi_I(q_1) = \Psi_{II}(q_1)$, $\Psi'_I(q_1) = \Psi'_{II}(q_1)$, $\Psi_{II}(q_2) = \Psi_{III}(q_2)$, and $\Psi'_{II}(q_2) = \Psi'_{III}(q_2)$. Four matching equations cannot identify six unknown coefficients, so two additional conditions are needed. One more constant can be specified based on the nature of the incident wave. If the incident trajectory is moving toward the right, it cannot result in a leftward traveling wave in region III, so $A_{III-} = 0$. Additionally, the amplitude of the incident wave is arbitrary, i.e. we are only concerned with the *fractional* transmittance and reflectance. For convenience, take the incident amplitude as $A_{I+} = 1$. Now there are six equations and six unknowns. After some algebra, the reflected and transmitted wave amplitudes can be obtained. The calculation can be repeated for cases where $E > V_{\ddagger}$ in which case the wave function is oscillatory in sections I, II, and III. After some more algebra to match solutions, the transmittance and reflectance amplitudes emerge also for classically allowed energies. In each case, $|A_{III+}|^2$ gives the transmittance probability as a function of E:

$$P_{QM}(E) = \begin{cases} \left[1 + \frac{V_{\ddagger}^2}{4E(E-V_{\ddagger})} sin^2(\frac{a}{\hbar}\sqrt{2m(E - V_{\ddagger})})\right]^{-1} & E > V_{\ddagger} \\ \left[1 + \frac{V_{\ddagger}^2}{4E(V_{\ddagger}-E)} sinh^2(\frac{a}{\hbar}\sqrt{2m(V_{\ddagger} - E)})\right]^{-1} & E < V_{\ddagger} \end{cases} \tag{12.1.12}$$

with the probability of reflection being $1 - P_{QM}(E)$. In the classically forbidden case ($E < V_{\ddagger}$), the quantity $2a\sqrt{2m(V_{\ddagger} - E)}$ that appears within $sinh$ is the classical action for one orbit in an inverted (upside-down) version of the barrier in Figure 12.1.2.

Contributions to the rate constant at each energy are proportional to $P_{QM}(E)$ and also to the Boltzmann factor $exp[-E/k_BT]$ as per equation (12.1.4). Figure 12.1.3 shows the contributions to k including tunneling and also the classical contributions to k_{TST}. It is convenient to write the tunneling probability in terms of dimensionless groups and familiar properties. The parameters that govern tunneling are the incoming energy relative to the barrier height and the width of the barrier relative to the thermal de Broglie wavelength. The barrier height in units of k_BT is also important for the Boltzmann factor. Therefore it is useful to write equation (12.1.12) in terms of $\lambda_T = \hbar/\sqrt{2\pi m k_B T}$, $u_0 = V_{\ddagger}/k_BT$, and $\varepsilon = (E - V_{\ddagger})/V_{\ddagger}$.

$$P_{QM}(\varepsilon) = \begin{cases} \left[1 + (4(\varepsilon + 1)\varepsilon)^{-1} sin^2(\frac{a}{\lambda_T}\sqrt{u_0\varepsilon/\pi})\right]^{-1} & \varepsilon > 0 \\ \left[1 - (4(\varepsilon + 1)\varepsilon)^{-1} sinh^2(\frac{a}{\lambda_T}\sqrt{u_0\varepsilon/\pi})\right]^{-1} & \varepsilon < 0 \end{cases} \tag{12.1.13}$$

The square top barrier is a good model for certain applications, e.g. electron tunneling through a molecular spacer or insulating layer. It is a rather poor model for proton transfer reactions, but we can use it to illustrate qualitative aspects of the tunneling probability. Recall from the discussion of TST that the thermal de Broglie wavelength at 300K and for an effective mass of $1amu$ is approximately 1Å. A representative distance a for proton transfer between donor and acceptor in a chemical reaction is around 3Å. Therefore, the values $u_0 = 12$ and $(a/\lambda_T)\sqrt{u_0/\pi} = 10$ are representative. Figure 12.1.3 below shows the classical and quantum transmission probability and also the integrands $exp[-E/k_BT]P_{CL}(E)$ and $exp[-E/k_BT]P_{QM}(E)$. There is some propensity for non-classical reflection. However, multiplication of $P_{QM}(E)$ by $exp[-E/k_BT]$ in equation (12.1.4) accentuates low energy phenomena making non-classical tunneling far more important than non-classical reflection.

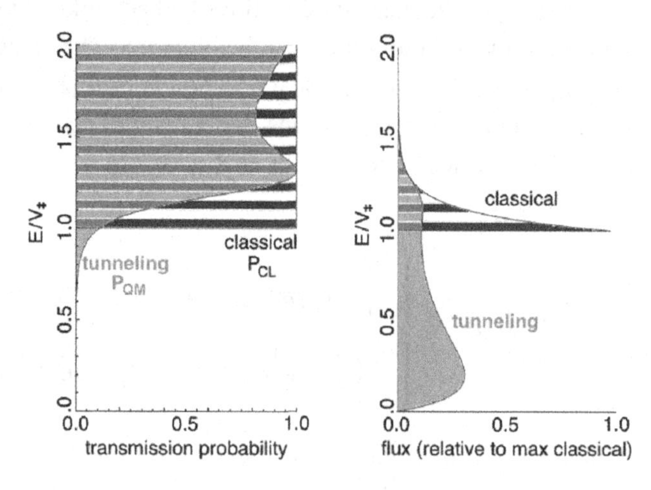

Figure 12.1.3: The transmission probability $P_{QM}(E)$ and the Boltzmann weighted contribution to the flux through the square-top barrier, $exp[-E/k_BT]P_{QM}(E)$. Representative dimensionless parameter values $u_0 = 12$ and $(a/\lambda_T)\sqrt{u_0/\pi} = 10$ were used to construct the plots.

The square-top barrier has uniform width from top to base, and the narrow base makes it especially permeable to deep tunneling at very low energies. The propensity for deep tunneling can be seen in Figure 12.1.3. Barriers for real chemical reactions tend to broaden at the base, so that the onset of tunneling is more gradual as temperature is lowered.

The calculation above is a quantum scattering calculation for a square topped one-dimensional barrier. The quantum mechanical scattering problem for real molecular potentials is far more difficult and worthy of an entirely separate book [9]. Fortunately, the Wentzel-Kramers-Brillouin (WKB) approximation [10] gives adequate estimates of the tunneling probability for hydrogen nuclei. The tunneling probability [6] according to a WKB approximation is

$$P_{SC} = \{1 + exp\,[2\theta(E)]\}^{-1} \tag{12.1.14}$$

where

$$\theta(E) = \frac{1}{\hbar}\int_{q_1(E)}^{q_2(E)} \sqrt{2m\,|V(q) - E|}dq \tag{12.1.15}$$

is a dimensionless barrier penetration integral. The subscript "SC" refers to semiclassical mechanics of which the WKB approximation is a special one-dimensional case. $q_1(E)$ and $q_2(E)$ are classical turning points which define the left and right edges of the classically forbidden region. The barrier penetration integral is the classical action for an orbit between $q_1(E)$ and $q_2(E)$ when the potential is turned upside down [11,12]. Recall that a similar classical-action-on-inverted-potential quantity emerged in the quantum transmission probabilities for the square top barrier [11,12]. To obtain the semi-classical transmission coefficient, use the WKB equation (12.1.5) with equation (12.1.14)

$$\kappa_{QM} = \int_0^\infty P_{SC}(E) e^{-(E-V_\ddagger)/k_B T} d(E/k_B T) \qquad (12.1.16)$$

where $E = 0$ corresponds to the reactants in their ground state energy. Equation (12.1.16) is a semiclassical approximation to the true quantum mechanical results. Some results for realistic barrier models are outlined below.

Parabolic barrier

In the same year (1932) of his seminal TST calculation [13], Wigner developed an approximation to the tunneling correction for a parabolic barrier [14]. The Wigner model approximates the barrier as an infinite downward parabola having an imaginary frequency with magnitude $\omega_\ddagger$. If the parabolic barrier is sufficiently broad, i.e. if $\hbar \left|\omega_\ddagger\right| / k_B T \ll 1$ is sufficiently small, then all tunneling will occur near the barrier top making details of the ground state irrelevant. To evaluate κ_{QM} we must first determine the barrier penetration integral $\theta(E)$. For the parabolic barrier, θ is proportional to the area under a half-circle as shown in Figure 12.1.4.

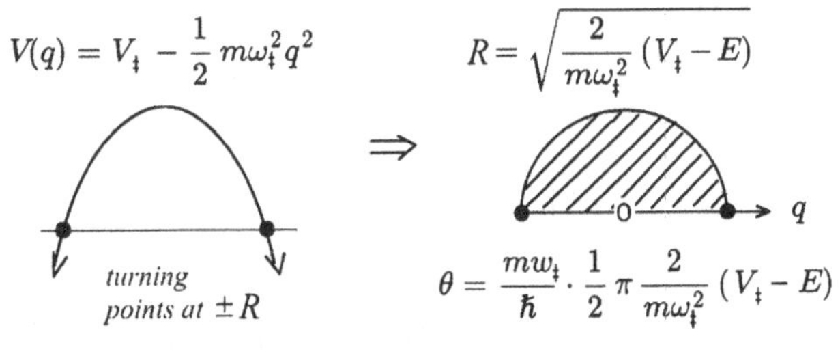

Figure 12.1.4: For a parabolic barrier, $\theta(E)$ corresponds, within constants, to half of the area under a circle with radius $R = (2(V_\ddagger - E)/(m\omega_\ddagger^2))^{(1/2)}$.

Following the calculation depicted in Figure 12.1.4, the barrier penetration integral as a function of energy for the parabolic barrier is

$$\theta(E) = \frac{\pi(V_\ddagger - E)}{\hbar\omega_\ddagger} \qquad (12.1.17)$$

Using equation (12.1.17) in equation (12.1.16), we can change the variable of integration to θ. The change of variables gives

$$\kappa_{QM} = \int_{-\infty}^{\infty} \frac{e^{\alpha_0 \theta/\pi}}{1+e^{2\theta}} \frac{\alpha_0}{\pi} d\theta \quad (12.1.18)$$

where $\alpha_0 = \beta\hbar\omega_\ddagger$. The mass appears to have vanished from the calculation, but it is still there within $\omega_\ddagger$ because different isotopes will give different imaginary frequencies. More will be said about this point in the discussion of isotope effects. At temperatures where $\alpha_0 < 2\pi$, the integral can be done exactly, giving

$$\kappa_{QM}(T) = \frac{\alpha_0/2}{sin(\alpha_0/2)} \quad (12.1.19)$$

Equation (12.1.19), the Bell tunneling correction [15], is the closed form result from an infinite parabolic barrier model.

For low temperatures where $\alpha_0 > 2\pi$, the integrand in equation (12.1.18) diverges as $\theta \to \infty$. This is an artifact from extending the parabolic potential to $E \to -\infty$. Wigner had previously suggested the approximation

$$\kappa_{QM}(T) \approx 1 + \alpha_0^2/24 \quad (12.1.20)$$

The Wigner correction is just the first two terms in a series expansion for the exact formula: $\kappa_{QM}(T) = 1 + \alpha_0^2/24 + 7\alpha_0^4/5760 + ...$ Wigner's correction can be computed for all temperatures, but the extended temperature range is achieved by truncating a series that diverges. The table below shows the Bell and Wigner formulas begin to diverge from each other for $\alpha_0 > \pi$, a sign that the higher order terms are already becoming important. The rightmost column compares the third and second terms of the expansion in $\beta\hbar\omega_\ddagger$.

α_0	$\frac{\alpha_0/2}{sin(\alpha_0/2)}$	$1+\frac{\alpha_0^2}{24}$	$\frac{7\alpha_0^4/5760}{\alpha_0^2/24}$	
0.1	1.00	1.00	0.0003	Bell more
1.0	1.04	1.04	0.03	justified than
3.2	1.58	1.42	0.3	Wigner
5.0	4.22	2.04	0.7	do not use
6.3	nonsense	2.66	1.2	Bell or Wigner

Neither the Wigner nor Bell versions of the parabolic barrier correction can qualitatively capture the low temperature rate behavior shown in Figure 12.0.1. Johnston and Rapp [16] recommend that the parabolic barrier tunneling corrections should only be used when $\beta\hbar\omega_\ddagger < 3.6$. Unfortunately, the temperatures at which the simple Bell and Wigner corrections are valid coincide with temperatures at which tunneling is largely unimportant. In practice, these are mostly useful as tests to gauge the importance of tunneling. If $\beta\hbar\omega_{\ddagger} > \pi$ then tunneling is important, and one of the more accurate tunneling models below should be used for accurate rates and activation parameters.

Shortcomings of the parabolic barrier model motivated several alternative one-dimensional tunneling models for low temperatures. The alternative calculations vary widely in the amount of information about the potential energy surface that is required. The most accurate corrections require the complete minimum energy pathway (MEP) and the vibrational frequencies of the transverse modes along the MEP [17]. Examples include direct applications of WKB theory using the potential as a function of arclength along the MEP [18] and the vibrationally adiabatic tunneling model [19]. The simplest low-temperature one-dimensional models of tunneling require no more information than what is already required for a harmonic TST calculation. These include the truncated parabolic barrier model [20,21] and corrections based on an Eckart barrier model [22–24].

The truncated parabolic barrier

The parabolic barrier tunneling correction diverged at low temperatures because the ground state energy was artificially extended to $E \to -\infty$. The truncated parabolic barrier model fixes the unphysical deep tunneling artifacts by replacing the lower integration bound in equation (12.1.18) with the ground state energy [20]. The truncated parabolic barrier is an extremely simple tunneling correction that remains well defined (but not necessarily accurate) at arbitrarily low temperatures. It can be derived using the same penetration integral (equation (12.1.17)) from the analysis of the infinite parabolic barrier.

Suppose that the force constant along the reaction coordinate at the barrier top is $\partial^2 V/\partial q^2|_{\ddagger} = -m\omega_{\ddagger}^2$. Also suppose that the forward and backward activation energy thresholds have been determined: $V_{\ddagger f}$ and $V_{\ddagger b}$. These should already account for zero point energies of the reactants, products, and transition states. The truncated parabolic barrier model based on parameters $V_{\ddagger f}$, $V_{\ddagger b}$, and $m\omega_{\ddagger}^2$ is shown in Figure 12.1.5. Note that tunneling changes rates, but it should not change equilibrium constants. The TST rate constant already satisfies detailed balance, so tunneling corrections for the forward and backward directions must be equivalent. A thermodynamically consistent tunneling correction can be obtained by computing one tunneling correction from the smaller of the forward and backward barriers. Here we assume that $V_{\ddagger f} < V_{\ddagger b}$.

The parabolic barrier is constructed to match the zero-point corrected activation energy in the forward and reverse directions as well as the second derivative of the potential along the reaction coordinate at the transition state. Where entrance or exit complexes have lower energy than

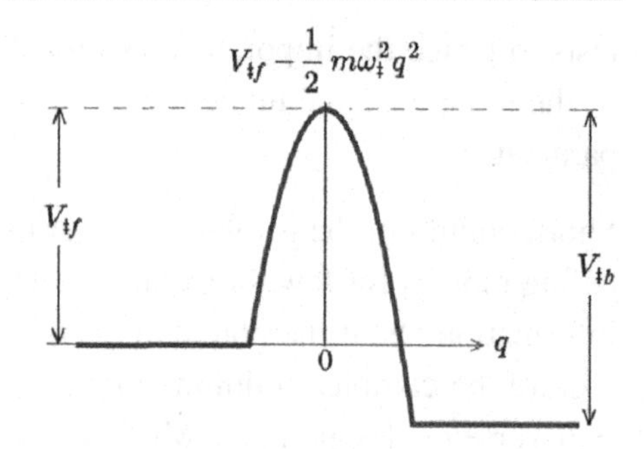

Figure 12.1.5: The truncated parabolic barrier model eliminates spurious tunneling contributions from energies below the reactant and product ground states. The tunneling correction is computed for the exothermic direction (the forward direction in the figure).

infinitely separated reactants or products, the energies of the entrance and exit complexes should be used to define $V_{\ddagger f}$ and $V_{\ddagger b}$ [25].

The truncated parabolic barrier tunneling correction is a function of two dimensionless parameters $\alpha_0 = \beta\hbar\omega_\ddagger$ and $\alpha_1 = 2\pi V_{\ddagger f}/\hbar\omega_\ddagger$:

$$\kappa_{QM} = \frac{\alpha_0/2}{sin[\alpha_0/2]} - \frac{\alpha_0/2\pi}{1-\alpha_0/2\pi} exp\left[\alpha_1\left(\frac{\alpha_0}{2\pi}-1\right)\right] \quad for \quad \alpha_0 \leq 2\pi \tag{12.1.21}$$

or

$$\kappa_{QM} = \frac{\alpha_0/2\pi}{\alpha_0/2\pi - 1}\left\{exp\left[\alpha_1\left(\frac{\alpha_0}{2\pi}-1\right)\right]-1\right\} \quad for \quad \alpha_0 \geq 2\pi \tag{12.1.22}$$

The piecewise transmission coefficient formula is a continuous smooth function of α_0. The truncated parabolic barrier correction tends to overestimate the importance of tunneling because the parabolic barrier is artificially narrow at low energies [26]. Nevertheless, it correctly describes the vanishing activation energy at low temperatures.

The asymmetric Eckart barrier

Eckart introduced a simple one dimensional barrier model for which the tunneling and reflectance probabilities could be determined exactly in the symmetric case [22]. Johnston and Heicklen later obtained tunneling corrections for the asymmetric Eckart model of exothermic and endothermic reactions [23]. Eckart models are usually the most accurate among those that do not require information about the potential energy surface beyond what is required by harmonic TST. The Eckart barrier has the form [22]

$$V(q) = \frac{Ay}{1+y} + \frac{By}{(1+y)^2} \tag{12.1.23}$$

where

$$y = exp[2\pi q/L] \tag{12.1.24}$$

A, B, and L are defined in terms of $V_f^\ddagger$, $V_b^\ddagger$, and the force constant along the reaction coordinate at the barrier top, $\partial^2 V/\partial q^2|_\ddagger = -m\omega_\ddagger^2$.

$$\begin{aligned} A &= V_{\ddagger f} - V_{\ddagger b} \\ B &= \left\{\sqrt{V_{\ddagger f}} + \sqrt{V_{\ddagger b}}\right\}^2 \\ \frac{L}{2\pi} &= \left\{\sqrt{m\omega_\ddagger^2/2V_{\ddagger f}} + \sqrt{m\omega_\ddagger^2/2V_{\ddagger b}}\right\}^{-1} \end{aligned} \tag{12.1.25}$$

Again, the barriers $V_{\ddagger f}$ and $V_{\ddagger b}$ should already account for zero point energies of the reactants, products, and transition states. And again, where entrance or exit complexes have lower energy than infinitely separated reactants or products, the energies of the entrance and exit complexes should be used to define $V_{\ddagger f}$ and $V_{\ddagger b}$ [25]. Figure 12.1.6 shows the Eckart barrier model along with a (truncated) parabolic barrier model for the same reaction.

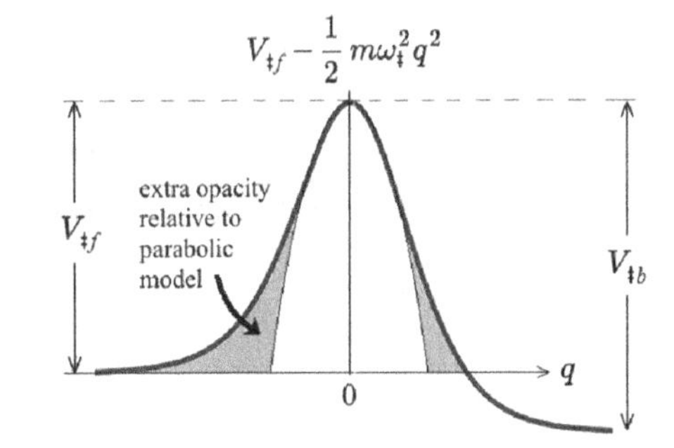

Figure 12.1.6: The asymmetric Eckart potential gives a realistic interpolated barrier that matches the zero-point corrected activation energy in the forward and reverse directions as well as the second derivative of the potential along the reaction coordinate at the transition state.

The relationships between $V_{\ddagger f}$, $V_{\ddagger b}$, and $m\omega_\ddagger^2$ and the Eckart potential parameters A, B, and L can also be inverted

$$\begin{aligned} m\omega_\ddagger^2 &= \pi^2(A^2 - B^2)/(2L^2B^2) \\ V_{\ddagger f} &= (A+B)^2/4B \\ V_{\ddagger b} &= (A-B)^2/4B \end{aligned} \tag{12.1.26}$$

Equations (12.1.25) are useful for converting *ab initio* results to an interpolating Eckart potential. Equations (12.1.26) are useful for comparing the Eckart barrier tunneling correction to results from other models.

As for the truncated parabolic barrier correction, the asymmetric Eckart barrier tunneling correction is based on the smaller of the forward and backward activation barriers. The same tunneling correction then applies to both the forward and backward TST rate constants. These conventions ensure detailed balance for the tunneling corrected rate constants. In the usual harmonic TST approximation starting from mass scaled coordinates, the appropriate mass is simply $m_0 = 1amu$. If other coordinates are used, procedures outlined by Garrett and Truhlar [27] or Johannesson and Jonsson [28] should be used to find the appropriate mass. As Garrett and Truhlar have emphasized, the results do not depend on conventions for scaling the reaction coordinate as long as the mass is correspondingly scaled [27].

The probability that a trajectory will tunnel through an Eckart barrier as a function of the incident energy E is

$$P_{QM}(E) = 1 - \frac{cosh[2\pi(a-b)] + cosh[2\pi d]}{cosh[2\pi(a+b)] + cosh[2\pi d]} \quad if \quad d \in \mathbb{R} \tag{12.1.27}$$

or

$$P_{QM}(E) = 1 - \frac{cosh[2\pi(a-b)] + cos[2\pi|d|]}{cosh[2\pi(a+b)] + cos[2\pi|d|]} \quad if \quad d \notin \mathbb{R} \tag{12.1.28}$$

where

$$2\pi a = 2(\alpha_1\xi)^{1/2}(\alpha_1^{-1/2} + \alpha_2^{-1/2})^{-1} \tag{12.1.29}$$

$$2\pi b = 2\{\alpha_1\xi + |\alpha_1 - \alpha_2|\}^{1/2}(\alpha_1^{-1/2} + \alpha_2^{-1/2})^{-1} \tag{12.1.30}$$

$$2\pi d = 2(\alpha_1\alpha_2 - \pi^2/4)^{1/2} \tag{12.1.31}$$

and where

$$\alpha_1 = 2\pi V_{\ddagger f}/\hbar\omega_\ddagger \tag{12.1.32}$$

$$\alpha_2 = 2\pi V_{\ddagger b}/\hbar\omega_\ddagger \tag{12.1.33}$$

$$\xi = E/V_{\ddagger f} \tag{12.1.34}$$

Don't be alarmed by all of those definitions. They are all emerging from the incident energy E and just three barrier characteristics $V_{\ddagger f}$, $V_{\ddagger b}$, and $m\omega_\ddagger^2$. It is convenient to think of $P_{QM}(E)$ as a function $P_{QM}(\xi|\alpha_1, \alpha_2)$ in which α_1, α_2 are parameters of the potential, and ξ will be integrated out in the calculation of κ_{QM}. Note that Johnston and Heicklen contains a critical typo in the equation for parameter b [23]. The formulas above work correctly provided the convention $V_{\ddagger f} < V_{\ddagger b}$ is maintained.

While the tunneling probability P_{QM} is exact (for the model potential), the tunneling transmission coefficient κ_{QM} must be computed numerically. This is not a major disadvantage. Numerical integration is required for all but the simplest parabolic barrier models. The tunneling transmission coefficient, which also depends on the usual parameter $\alpha_0 = \beta\hbar\omega_\ddagger$, is

$$\kappa_{QM} = \frac{\alpha_1\alpha_0}{2\pi}\int_0^\infty exp\left[-\frac{(\xi-1)\alpha_1\alpha_0}{2\pi}\right] P_{QM}(\xi|\alpha_1, \alpha_2)d\xi \tag{12.1.35}$$

The integration is fairly straightforward, but Figure 12.1.7 provides a table of computed values in terms of the three dimensionless parameters α_1, α_2, and $\alpha_0 = \beta\hbar\omega_\ddagger$. An early application of the Eckart tunneling correction can be seen in the work by Shultz and LeRoy [29] on $H + H_2 \rightleftarrows H_2 + H$, one of the first studies to clearly show a curved Arrhenius plot.

$\beta\hbar\omega_\ddagger = 2$

$\alpha_1 \backslash \alpha_2$	0.5	2.0	8.0	20.0
0.5	0.15	0.09	-0.01	-0.07
2.0		0.28	0.17	0.11
8.0			0.22	0.18
20.0				0.18

$\beta\hbar\omega_\ddagger = 4$

$\alpha_1 \backslash \alpha_2$	0.5	2.0	8.0	20.0
0.5	0.29	0.18	0.03	-0.05
2.0		0.65	0.48	0.39
8.0			0.71	0.69
20.0				0.74

$\beta\hbar\omega_\ddagger = 6$

$\alpha_1 \backslash \alpha_2$	0.5	2.0	8.0	20.0
0.5	0.44	0.29	0.10	0.00
2.0		1.06	0.86	0.74
8.0			1.51	1.53
20.0				1.81

$\beta\hbar\omega_\ddagger = 8$

$\alpha_1 \backslash \alpha_2$	0.5	2.0	8.0	20.0
0.5	0.59	0.41	0.19	0.08
2.0		1.52	1.29	1.16
8.0			2.62	2.74
20.0				3.83

$\beta\hbar\omega_\ddagger = 10$

$\alpha_1 \backslash \alpha_2$	0.5	2.0	8.0	20.0
0.5	0.74	0.54	0.29	0.17
2.0		1.99	1.77	1.62
8.0			4.04	4.31
20.0				7.05

$\beta\hbar\omega_\ddagger = 12$

$\alpha_1 \backslash \alpha_2$	0.5	2.0	8.0	20.0
0.5	0.88	0.66	0.40	0.27
2.0		2.49	2.27	2.11
8.0			5.73	6.16
20.0				11.2

$\beta\hbar\omega_\ddagger = 16$

$\alpha_1 \backslash \alpha_2$	0.5	2.0	8.0	20.0
0.5	1.18	0.93	0.63	0.49
2.0		3.53	3.33	3.15
8.0			9.65	10.4
20.0				21.2

$\beta\hbar\omega_\ddagger = 18$

$\alpha_1 \backslash \alpha_2$	0.5	2.0	8.0	20.0
0.5	1.32	1.06	0.75	0.61
2.0		4.06	3.88	3.69
8.0			11.8	12.6
20.0				26.6

Figure 12.1.7: Tabulated values of $ln(\kappa_{QM})$ from the asymmetric Eckart barrier as a function of α_1, α_2, and $\beta\hbar\omega_\ddagger$. In each graph, moving to the right corresponds to increasing $V_{\ddagger b}$, and moving downward corresponds to increasing $V_{\ddagger f}$. The diagonal corresponds to results for a symmetric Eckart barrier.

The results of Johnston and Heicklen inspired several other approaches that interpolate the energy along the reaction barrier based on properties of the reactants, products, and saddle point [24,30]. The table shows how extremely important tunneling corrections can become. Far beyond two- or ten-fold enhancements, tunneling at low temperatures can be responsible for thousand fold enhancements or more. Even at room temperature, large tunneling corrections are common. The importance of tunneling for room temperature enzyme reactions has been widely noted [31]. Garcia-Viloca et al. write: "In the transmission coefficient, recrossing events and nonequilibrium excitations appear to be unimportant, but tunneling can lead to a sizeable contribution to rate enhancements" [32,33]. For aromatic amine dehydrogenase, Masgrau et al. estimated that tunneling accounts for as much as 99.9% of room temperature reactivity [34].

12.2 Kinetic isotope effects

The potential energy surface in Cartesian coordinates is isotope independent, but isotopes do change the mass weighted landscape and the vibrational frequencies. A kinetic isotope effect (KIE) is a ratio between reaction rates with and without isotopic substitution. KIEs are particularly strong for protons (H) and deuterons (D) where the mass difference is a factor of two. For other elements, e.g. ^{12}C vs. ^{13}C, the isotopic mass differences are relatively small and accordingly their KIEs are more subtle. Figure 12.2.1 [35] illustrates how primary KIEs emerge from differences in ground state energies and tunneling probabilities [36].

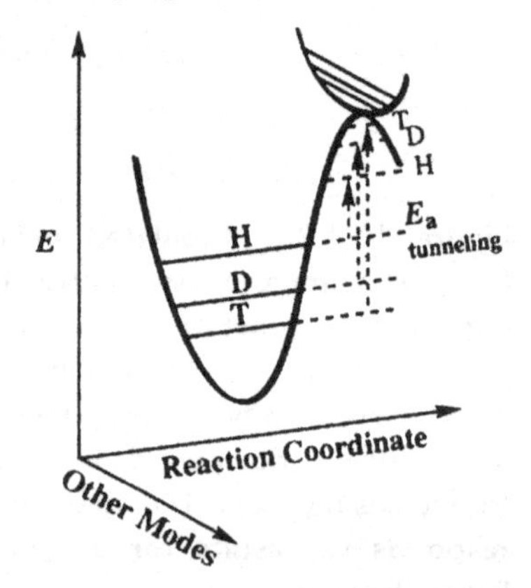

Figure 12.2.1: Different isotopes give different zero point energies, different transition state frequencies, and different propensities for tunneling. All of these factors can influence the activation energy on an Arrhenius plot. [From Kohen, Prog. Reaction Kinetics and Mechanism, 28, 119–156 (2003).]

KIEs can be directly and somewhat reliably predicted from *ab initio* calculations. Errors in the computed potential energy surface mostly cancel from the KIE, leaving only ratios of more reliably predicted tunneling and zero point energy differences. In Johnston's words "The kinetic isotope effect is independent of small errors in calculated activation energy, and it is a sharp test of the vibrational analysis and tunneling correction" [25]. Figure 12.2.2 compares KIE data to two theoretical models [16], one including zero point effects and one including both zero point effects and tunneling effects.

KIEs are easily computed from the ratio of tunneling corrected TST rate expressions. Many of the partition functions in the ratio nearly cancel. Historically, factors in the KIE with values near unity were neglected entirely. These approximate calculations are useful for developing a chemical intuition for interpreting the magnitude of a KIE. An example calculation, adapted and somewhat extended from the text by Espenson [37], is shown below.

Example: KIEs from selectivity

The H vs. D kinetic isotope effect has been studied for each of the two reactions shown below [37]. The H atom adjacent to the initial $C - Br$ bond in the first reactant has been exchanged with deuterium.

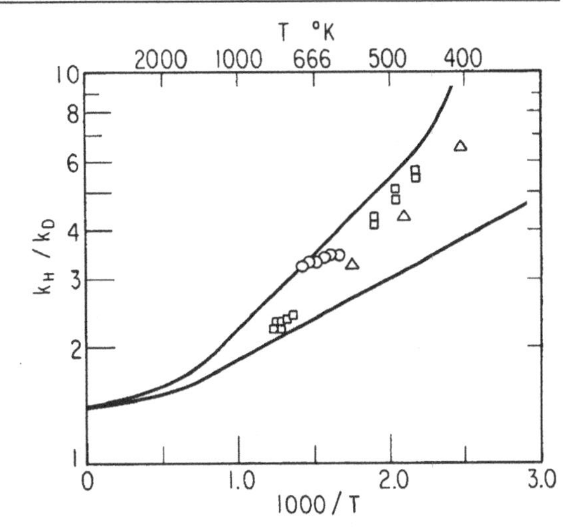

Figure 12.2.2: KIEs for H or D atom abstraction from the CH_3 groups of acetone and ethane by a methyl radical. The lower curve includes differences in zero point energies, but not tunneling. The upper curve includes an Eckart barrier tunneling correction. Johnston noted that these results must be the effect of high temperature tunneling because the $C-H$ and $C-D$ vibrational frequencies are precisely known. [From Johnston and Rapp, *J. Am. Chem. Soc.* 83, 1–9 (1961).]

Br, H + $^-$O ⟶ + HO + Br$^-$ KIE = 6.7

Br, H + $^-$O ⟶ O, H + Br$^-$ KIE ≈ 1.0

The absence of an appreciable KIE for the second reaction is not surprising, because the CH bond has not been broken. Because the second reaction has a negligible KIE, the KIE for the first reaction is evident from the product selectivity. No kinetics experiments are needed to estimate its KIE!

Espenson attributes the large KIE for the first reaction to a zero point difference between ground states because "the two transition states (where $R-H$ and $R-D$ are breaking) are at the same energy." The transition states are indeed at the same energy, but tunneling may alter the activation energies. Explicitly accounting for tunneling and zero point energy factors in the KIE gives

$$\frac{k_H}{k_D} = exp\left[\frac{\hbar(\omega_{CH} - \omega_{CD})}{2k_B T}\right] \cdot \frac{\kappa_{QM}(H)}{\kappa_{QM}(D)} \cdot (secondary\ KIE) \qquad (12.2.1)$$

where all other factors have been lumped into a secondary KIE. Secondary KIEs arise from subtle frequency shifts in the unaltered bonds and from changes in rotational and translational partition functions. The secondary KIE can be obtained from a complete analysis of translation, rotation, and vibration with and without isotopic substitution. Usually the secondary KIE ratio is near unity.

The exponential term (part of the primary KIE) involves the difference in zero point energies of the bonds that are being broken. A back-of-the-envelope estimate of the zero point differences can be obtained from the harmonic oscillator frequency formula:

$$\omega = \sqrt{k/\mu}$$

where k is a force constant and μ is a reduced mass. The force constant is just a second derivative of the potential energy, so $k_{CH} = k_{CD}$. If we *approximate* that simple rules for effective masses apply within a big molecule, $\mu_{CH} = m_H m_C/(m_H + m_C)$, and likewise for μ_{CD}. C is already much heavier than H or D, so $\mu_{CH} \approx m_H$ and $\mu_{CD} \approx m_D$. Therefore,

$$\frac{\omega_{CH}}{\omega_{CD}} \approx \sqrt{\frac{m_D}{m_H}} = \sqrt{2}$$

The aliphatic CH bond has a typical stretching frequency of $2900 cm^{-1}$, which gives $\hbar\omega_{CH} = 34.67 kJ/mol$. Dividing by $\sqrt{2}$ gives the corresponding CD stretch quanta $\hbar\omega_{CD} \approx 24.52 kJ/mol$. Plugging these values into equation (12.2.1) for a temperature of 298K gives

$$\frac{k_H}{k_D} \approx 7.8 \cdot \frac{\kappa_{QM}(H)}{\kappa_{QM}(D)} \cdot (secondary\, KIE)$$

7.8 is not far from the experimental value of 6.7, but our calculation was rather primitive. How should the result be interpreted? It is possible that the tunneling and secondary KIE factors are both nearly unity, that they partially cancel, or that our vibrational analysis in the primary KIE calculation is too simplistic. Without a full scale computational analysis of the rates with and without isotopic substitution, the KIE results should be interpreted with some caution. On a more optimistic note, the simple calculation has correctly identified a substantial KIE for one reaction and not for the other [37].

With modern *ab initio* calculations, it is easy to perform the full rate calculation including isotope effects on all frequencies, rotational partition functions, translational partition functions, and tunneling factors. By doing the calculation this way, one can be sure to account for both primary KIEs (those arising from isotope effects on the reaction coordinate degree of freedom), and secondary KIEs (those arising from effects on other degrees of freedom). Because of the relative ease with which KIEs can be computed, they have become an important link between experiment and theory. Our discussion has focused on KIEs for elementary reactions. However, they are perhaps most useful in understanding multistep mechanisms. When the reactants are labeled, the KIE can tell us whether the labeled atom is involved in the rate determining step. Moreover, experiments at varied reactant concentrations can have different rate determining steps. KIEs in this situation can reveal which elementary step is overtaking the role of kinetic bottleneck as concentrations change [38].

12.3 Tunneling or tunnel splitting

Each of the tunneling corrections above assume a continuum of energy levels. How should tunneling be interpreted when the potential resembles a double-well with two bound states? Before Gamow's model [6] of alpha decay, Hund developed a theory for tunneling between bound states of molecules [39,40]. To this day, many studies of tunneling examine molecules that have symmetric double-well type potentials: tropolone, malonaldehyde, ammonia inversion, etc. [41–43]. Figure 12.3.1 depicts the two symmetric bound states of malonaldehyde [42].

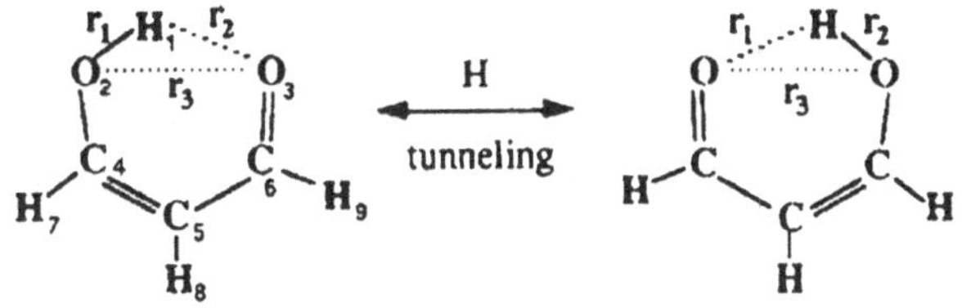

Figure 12.3.1: Intermolecular proton transfer in malonaldehyde involves tunneling through the barrier between quantized energy levels. [From Shida et al. *J. Chem. Phys.* 91, 4061 (1989).]

For a true one-dimensional symmetric double-well potential, the energy levels occur in narrowly separated pairs. Each pair corresponds to a state with amplitude on both the left and right hand side of the barrier as shown in Figure 12.3.2. The symmetric double-well requires that we abandon the model with a continuous stream of reflected and transmitted particles. Instead we must consider time-dependent quantum mechanics and state-to-state transitions. Suppose that at time $t = 0$, all amplitude is isolated on the left side of the double-well. All eigenstates have equal amplitude on the left and on the right, so the initial wavefunction is a mixture of energy eigenstates. If the initial condition is a mixture of just two narrowly separated states, then amplitude will shift from the left well to the right well and back with a period [4]

$$\tau = h/\Delta E \tag{12.3.1}$$

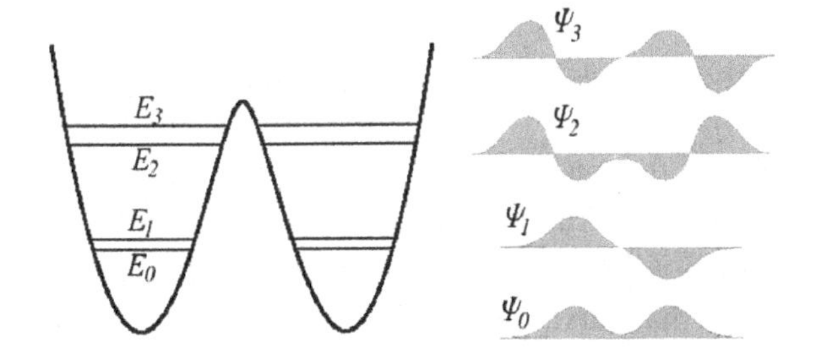

Figure 12.3.2: Symmetric double wells give discrete paired states with energies that differ by a narrow tunnel splitting energy ΔE. Symmetry allows the system to hop between the two wells faster than the rate predicted by tunneling estimates.

where ΔE is the tunnel splitting energy. The tunnel splitting energy, depicted in Figure 12.3.2, can be estimated using the WKB approximation as

$$\Delta E = \frac{\hbar\omega_0}{\pi} exp[-\theta(E)] \tag{12.3.2}$$

Here ω_0 is the frequency in the stable wells and $\theta(E)$ is the same barrier penetration integral defined in section 12.1. The value of E is intermediate between the two split energy levels. An energy that straddles the split pair of energy levels can also be identified by the WKB approximation. The (approximate) energy of the nth pair is given by the root E_n for the equation

$$\frac{1}{\hbar}\int_{q_1(E_n)}^{q_2(E_n)} dq\sqrt{2m(E_n - V(q))} = (n + 1/2)\pi \tag{12.3.3}$$

where the action integral spans the classically *allowed* region in one well, i.e. where $q_1(E_n)$ and $q_2(E_n)$ are the left and right classical turning points at energy E_n. For a one-dimensional model, the effects of other states can be ignored if their energy levels are far from those of the narrowly split pair [44]. Tunnel splittings can be probed experimentally, e.g. by microwave and NMR spectroscopy, so "double-well" tunneling processes provide unusually direct tests for computational predictions. For malonaldehyde, one dimensional *ab initio* models provide reasonable approximations to the observed tunnel splitting [41].

Tunnel splittings can also be computed from multidimensional models, although the calculations are significantly more difficult [43,45]. Fortunately, multidimensional molecular potential energy surfaces tend to have closely spaced and vibrationally coupled energy levels. These can often be treated as a continuum of energy levels, which allows the use of simple tunneling corrections like those in section 12.1. Figure 12.3.3 shows an example in which the Eckart tunneling correction accurately describes hydrogen transfer between two bound sites on a single molecule [46].

Successful results like that shown in Figure 12.3.3 must still be interpreted carefully. Computational KIE estimates are somewhat reliable, but absolute rate calculations are not. In cases of agreement, it is difficult to tell whether all parts of the absolute rate calculation are working correctly, or whether some errors are fortuitously compensating for others.

12.4 Multidimensional tunneling models

Recall that one dimensional tunneling models assumed dynamical separability of the reaction coordinate. Curved reaction pathways couple the reaction coordinate to transverse vibrational modes [47], causing assumptions of the one-dimensional tunneling models to break down. Note that curvature does not require atoms to follow a physically curved trajectory in space.

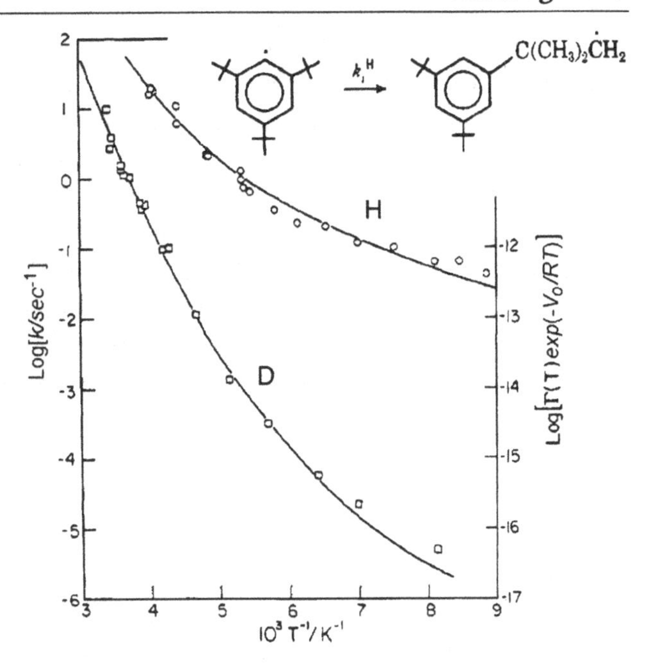

Figure 12.3.3: Intermolecular hydrogen transfer between aryl radical and alkyl radical states must involve a double well type potential, but is adequately described by an Eckart tunneling correction based on continuous energy levels. Note the extremely large kinetic isotope effect. [From Brunton et al. J. Am. Chem. Soc. 98, 6803-11 (1976).]

Even the collinear atom transfer reactions have curved reaction pathways because the changes in bondlengths are not perfectly concerted. Hydrogen transfer reactions with curved reaction pathways exhibit corner cutting tunnel paths and other intrinsically multidimensional tunneling mechanisms. This section provides a very brief (and vastly oversimplified) view of multidimensional tunneling theories.

Small curvature tunneling models

Small curvature tunneling models account for curvature by computing the barrier penetration integral along approximately optimal corner-cutting tunnel paths [19,48,49]. The tunneling probability and the barrier penetration integral are computed using the one dimensional formulas, $P_{SC} = (1 + exp[2\theta(s)])^{-1}$, and

$$\theta(s) = \frac{1}{\hbar}\int_{s_1}^{s_2} \sqrt{2\mu(s)\,|V(\mathbf{x}(s)) - E|}ds.$$

However, the action is computed for a multidimensional curve $\mathbf{x}(s)$ that does not follow the MEP. In these calculations, the path that minimizes the action maximizes the tunneling probability. Based on the structure of the action integral, the optimal path will be determined by a compromise between arclength and energy along the path. Paths that cut corners are short, but paths that follow the MEP traverse the lowest energy routes over the barrier.

Marcus and Coltrin [48] considered curvature effects in tunneling corrections for $\mathbf{A} + \mathbf{BC} \rightarrow \mathbf{AB} + \mathbf{C}$ reactions. They treated the transverse mode as an adiabatic ground state vibration at each point along the reaction pathway. They noted that the inner envelope of the vibrations naturally reduces the tunneling path length without increasing the energy level [19,48]. Truhlar, Garrett, and coworkers generalized the Marcus-Coltrin treatment to more complicated reactions in their small curvature tunneling theory [19,49]. In addition to properties required for a harmonic TST calculation, small curvature tunneling calculations require the vibrationally adiabatic ground state potential energy profile along the MEP and hence transverse vibrational frequencies along the MEP. Small curvature tunneling has provided accurate results for several cases [34,50], but there are also some indications that it overestimates the effects of curvature [26,51].

Large curvature tunneling models

Transfer of the hydrogen or proton between heavy donor and acceptor centers, e.g. in the $Cl + HCl \rightleftarrows Cl + HCl$ reaction, leads to a number of surprising deviations from classical TST. Recall that heavy-light-heavy atom transfer reactions have small skew angles and sharply curved transition pathways. Babamov and Marcus noted that the donor-acceptor approach motion is slow while the H-atom transfer between the donor and acceptor atoms is fast. They introduced coordinates to separate these motions and to build a non-adiabatic model. Their analysis showed that zero point energy in the H-transfer degree of freedom causes the entrance and exit channels to merge before the donor and acceptor distances reach the saddle point [52]. Their results revealed a strong "corner cutting" behavior in which the H-atom jumps from one channel to the other well before reaching the saddle point location.

Garrett et al. developed a large curvature tunneling model [53,54] which can be applied to more complex heavy-light-heavy reactions. Garrett and Truhlar assume that trajectories approach the saddle region adiabatically from $s = -\infty$ up to an arclength $s_1 < 0$ along the MEP. The arclength s_1 is that for which the vibrationally adiabatic ground state energy equals the incident energy, i.e. s_1 is determined by $E = V_0(s_1) + \sum_{transverse} \hbar\omega_i(s_1)/2$. The equation for s_1 has a second root that defines the other turning point $s_2 > 0$ [54]. A trajectory coming from $s = -\infty$ has opportunities to tunnel across the ridge to the other channel at each value of s leading up to s_1. If no tunneling occurs, it has a second opportunity to tunnel across the ridge as the donor and acceptor fragments recede back to $s = -\infty$. Trajectories with sufficiently high incident energy E can also jump from one channel to another without tunneling. Several variations of the large curvature tunneling method have been developed [30,55,56], with each version approximating the tunnel path from s_1 to s_2 in a different way. Figure 12.4.1 illustrates the original method of Garrett et al. in which a straight-line tunneling path was adopted. Figure 12.4.1 also shows (in the insets above) the non-classical corner cutting mechanism of Babamov and Marcus.

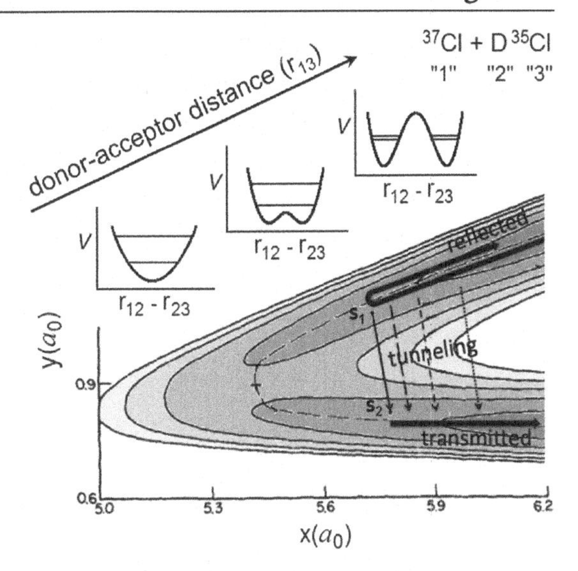

Figure 12.4.1: The large curvature tunneling correction approximates the tunneling probabilities as the donor and acceptor atoms approach (and retreat) at energy E in a heavy-light-heavy reaction. The insets above show how donor-acceptor distance modulates the potential for the H-transfer step, and in particular how zero point energy facilitates the H-transfer even for trajectories that do not reach the saddle point region. [Modified from Garrett et al., *J. Chem. Phys.* 78, 4400-13 (1983).]

The semiclassical transition state theory (SCTST) framework by Hernandez and Miller used spectroscopic constants to account for anharmonicity, reaction pathway curvature, and tunneling in the vicinity of the saddle point. Additionally, the SCTST formalism accounts for tunneling via all possible paths. Some applications suggest these methods can give accurate low temperature results [26,57], but the anharmonicity inputs are not easily obtained.

Exercises

1. Section 12.1 notes that the truncated parabolic barrier model correctly predicts the vanishing of activation energy at low temperature. Show how this happens by constructing the entire rate constant $k = k_{TST}\kappa_{QM}$ with κ_{QM} from Equation (12.1.22).
2. Reproduce the square top barrier results from Figure 12.1.3. Then create similar plots using the WKB approximation (Equations (12.1.14)–(12.1.16)). How do the exact results compare to those from the WKB approximation?
3. Equilibrium isotope effects are changes in the equilibrium constant due to isotopic substitution. Explain how KIEs for the forward and reverse reactions are related to the equilibrium isotope effect.
4. Develop a one-dimensional tunneling correction for the infinite triangular barrier: $V(x) = V_{\ddagger} - A|x|$. Where might this correction be applicable?
5. Perform an *ab initio* calculation to obtain the geometries, energies, and frequencies of the reactant and transition state configurations for the $H_2 + H \rightarrow H + H_2$ reaction. Compute the TST rate constant and the rate with an Eckart barrier tunneling correction. Construct

Eyring plots from the TST rate and from the tunneling corrected TST rate at three temperatures. How much does tunneling affect the apparent activation enthalpy and entropy?

6. RRKM theory can be extended to include the tunneling probability $P_{QM}(E)$ at each energy E. Let E_0 be the threshold energy, $\rho(E)$ the density of states at energy E, and $N^{\ddagger}(E - E_0)$ the number of ways to partition energy beyond the threshold among the non-reaction coordinate degrees of freedom at the transition state. The tunneling corrected RRKM formula is

$$k_{RRKM-QM}(E) = \frac{N^{\ddagger}_{QM}(E - E_0)}{h\rho(E)}$$

Here,

$$N^{\ddagger}_{QM}(E) = \int_0^E P_{QM}(E_t)\rho(E - E_t)dE_t$$

is a tunneling probability weighted cumulative density of states up to energy E.

(a) Replace $P_{QM}(E)$ with the Heaviside function $\theta(E - E_0)$, corresponding to perfect transmission (unity) at energies above the threshold and zero transmission at energies below. Show that the RRKM theory is recovered.

(b) Sketch $N^{\ddagger}_{QM}(E)$ and its 'classical' counterpart $N^{\ddagger}(E)$ vs. E/E_0. See Gray et al. for an application to the unimolecular reaction $HNC \rightarrow HCN$ [58].

References

[1] J. Frenkel, Wave Mechanics, Elementary Theory, Clarendon Press, Oxford, UK, 1932.
[2] P.S. Zuev, R.S. Sheridan, T.V. Albu, D.G. Truhlar, D.A. Hrovat, W. Borden, Science 299 (2003) 867–870.
[3] V.A. Benderskii, V. Goldanskii, D.E. Makarov, Phys. Rep. 233 (1993) 195–339.
[4] G.C. Schatz, M.A. Ratner, Quantum Mechanics in Chemistry, Dover, Mineola, NY, 2002.
[5] W.H. Miller, Faraday Discuss. 110 (1998) 1–21.
[6] G. Gamow, Z. Phys. 51 (1928) 204–212.
[7] E. Merzbacher, Phys. Today (2002) 44–49.
[8] D.J. Tannor, Annu. Rev. Phys. Chem. 51 (2000) 553–600.
[9] R.D. Levine, Molecular Reaction Dynamics, Cambridge University Press, Cambridge, UK, 2005.
[10] J.L. Dunham, Phys. Rev. 41 (1932) 713–720.
[11] W.H. Miller, J. Chem. Phys. 83 (1979) 960–963.
[12] J. Ankerhold, Quantum Tunneling in Complex Systems: The Semiclassical Approach, Springer, Berlin, Heidelberg, 2007.
[13] H. Pelzer, E. Wigner, Z. Phys. Chem. B 15 (1932) 445.
[14] E. Wigner, Z. Phys. Chem. B 19 (1932) 203.
[15] R.P. Bell, Trans. Faraday Soc. 55 (1) (1959).
[16] H.S. Johnston, D. Rapp, J. Am. Chem. Soc. 83 (1961) 1–9.
[17] A. Fernandez-Ramos, D.G. Truhlar, C. Corchado, J. Espinosa-Garcia, J. Phys. Chem. A 106 (2002) 4957–4960.
[18] B.C. Garrett, D.G. Truhlar, J. Phys. Chem. 83 (1979) 2921–2926.
[19] R.T. Skodje, D.G. Truhlar, B.C. Garrett, J. Chem. Phys. (ISSN 0021-9606) 77 (1982) 5955.

[20] R.T. Skodje, D.G. Truhlar, J. Phys. Chem. 85 (1981) 624–628.
[21] J.T. Fermann, S. Auerbach, J. Chem. Phys. (ISSN 0021-9606) 112 (2000) 6787.
[22] C. Eckart, Phys. Rev. 127 (1930) 1303–1309.
[23] H.S. Johnston, J. Heicklen, J. Chem. Phys. 66 (1962) 532–533.
[24] I. Shavitt, J. Chem. Phys. (ISSN 0021-9606) 49 (1968) 4048.
[25] H.S. Johnston, Adv. Chem. Phys. 83 (1961) 1–9.
[26] B. Peters, A.T. Bell, A. Chakraborty, J. Chem. Phys. 121 (2004) 4461–4466.
[27] D.G. Truhlar, B.C. Garrett, J. Phys. Chem. A 107 (2003) 4006.
[28] G.H. Johannesson, H. Jonsson, J. Chem. Phys. 115 (2001) 9644.
[29] W.R. Schulz, D.J. Le Roy, J. Chem. Phys. (ISSN 0021-9606) 42 (1965) 3869.
[30] A. Gonzalez-Lafont, T.N. Truong, D.G. Truhlar, J. Chem. Phys. 95 (1991) 8875–8894.
[31] Q. Cui, M. Karplus, J. Am. Chem. Soc. 124 (2002) 3093–3124.
[32] M. Garcia-Viloca, J. Gao, M. Karplus, D.G. Truhlar, Science 303 (2004) 186–195.
[33] J. Pu, J. Gao, D.G. Truhlar, Chem. Rev. 106 (2006) 3140–3169.
[34] L. Masgrau, A. Roujeinikova, L.O. Johannissen, P. Hothi, J. Basran, K.E. Ranaghan, A.J. Mulholland, M.J. Sutcliffe, N.S. Scrutton, D. Leys, Science 312 (2006) 237–241.
[35] A. Kohen, Prog. React. Kinet. Mech. 28 (2003) 119–156.
[36] K.B. Wiberg, Chem. Rev. (22) (1955) 713–743.
[37] J.H. Espenson, Chemical Kinetics and Reaction Mechanisms, McGraw-Hill, New York, 1995.
[38] E.M. Simmons, J.F. Hartwig, Angew. Chem., Int. Ed. Engl. (ISSN 1521-3773) 51 (Mar. 2012) 3066–3072.
[39] F. Hund, Z. Phys. 40 (1927) 742.
[40] F. Hund, Z. Phys. 43 (1927) 805.
[41] J. Bicerano, H.F. Schaefer, W.H. Miller, J. Chem. Phys. 105 (1983) 2550–2553.
[42] N. Shida, P.F. Barbara, J.E. Almlof, J. Chem. Phys. 91 (1989) 4061.
[43] T.D. Sewell, Y. Guo, D.L. Thompson, J. Chem. Phys. (ISSN 0021-9606) 103 (1995) 8557.
[44] M.D. Harmony, Chem. Soc. Rev. 1 (1972) 211–228.
[45] N. Makri, W.H. Miller, J. Chem. Phys. 91 (1989) 4026.
[46] G. Brunton, D. Griller, L.R.C. Barclay, K.U. Ingold, J. Am. Chem. Soc. 98 (1976) 6803–6811.
[47] W.H. Miller, N.C. Handy, J.E. Adams, J. Chem. Phys. 72 (1980) 99–112.
[48] R.A. Marcus, M.E. Coltrin, J. Chem. Phys. 67 (1977) 2609–2613.
[49] W.T. Duncan, R.L. Bell, T.N. Truong, J. Comput. Chem. 19 (1998) 1039–1052.
[50] D.G. Truhlar, J. Gao, C. Alhambra, M. Garcia-Viloca, J. Corchado, M.L. Sánchez, J. Villà, Acc. Chem. Res. 35 (2002) 341–349.
[51] A.D. Isaacson, J. Chem. Phys. 107 (1997) 3832–3839.
[52] V.K. Babamov, R.A. Marcus, J. Chem. Phys. 74 (1981) 1790.
[53] B.C. Garrett, D.G. Truhlar, A.F. Wagner, T.H. Dunning, J. Chem. Phys. 78 (1983) 4400.
[54] B.C. Garrett, T. Joseph, T.N. Truong, D.G. Truhlar, Chem. Phys. 136 (1989) 271–283.
[55] Y.-P. Liu, D.-H. Lu, A. Gonzalez-Lafont, D.G. Truhlar, B.C. Garrett, J. Am. Chem. Soc. 115 (1993) 7806–7817.
[56] A. Fernandez-Ramos, D.G. Truhlar, J. Chem. Theory Comput. 1 (2005) 1063–1078.
[57] R. Hernandez, W.H. Miller, Chem. Phys. Lett. (ISSN 0009-2614) 214 (Oct 1993) 129–136.
[58] S.K. Gray, W.H. Miller, Y. Yamaguchi, H.F. Schaefer, J. Chem. Phys. (ISSN 0021-9606) 73 (1980) 2733.

CHAPTER 13

Reactive flux

Counting the number of persons that cross the border from A to B within a certain time interval defines the migration rate, but this overcounts for tourists who keep their nationality A and only visit country B for a short time, and likewise, the B nationals on their way back from a tourist visit to A. Moreover, the true emigrants could cross the border more than once (for instance to move their furniture).

Bolhuis and Dellago, Rev. Comp. Chem. (2011)

A reactive trajectory requires both activation and successful passage through the transition state and on to products. A solvent, an enzyme, or even intramolecular vibrational coupling can alter the dynamics and disrupt trajectories *en route* from the reactants to the products. Technically, before we can discuss 'altered dynamics' we must define the reference dynamics. In reaction rate theory, the reference dynamical model is that of transition state theory (TST). TST assumes a thermal equilibrium ensemble at the dividing surface and that all trajectories which cross the dividing surface complete the journey to products without ever recrossing. The actual dynamics can severely violate these assumptions depending on the strength of coupling between the reaction coordinate and the other "bath" degrees of freedom.

Chapters 16 and 17 focus on simple dynamical models for the effects of bath coupling on the reaction coordinate motion. This chapter focuses on numerically exact reactive flux methods that quantify recrossing effects by simulating the full multidimensional dynamics. The reactive flux approach takes advantage of the time scale separation [1]: the long time for activation is circumvented by launching trajectories directly from the transition state and simulating the dynamics only during the short relaxation time. When combined with efficient methods for computing free energy barriers and TST rate constants, the reactive flux approach exactly accounts for all classical non-TST effects: dynamical recrossing, non-equilibrium effects, and dividing surface inaccuracies.

Chapter 12 focused on tunneling corrections that do not require any simulations of the barrier crossing dynamics. However, tunneling and dynamical recrossing are, in principle, entangled with each other. We will briefly discuss path integral methods for (approximate) quantum mechanical reactive flux calculations at the end of this chapter. Let us first account for the effects of recrossing within the framework of classical mechanics and phenomenological rate laws.

13.1 Phenomenological rate laws and time correlations

In non-equilibrium statistical mechanics, Green-Kubo relations give transport coefficients in terms of equilibrium time correlation functions. For example, diffusivities can be obtained from equilibrium velocity-velocity time correlation functions [2] and conductivities can be obtained from equilibrium heat-flow time correlation functions [3]. This section shows how rate constants emerge from the equilibrium time correlations of the reactant and product populations. Consider an isomerization reaction

$$\mathbf{A} \rightleftarrows \mathbf{B} \tag{13.1.1}$$

The phenomenological rate laws, assuming an elementary isomerization mechanism, are

$$d[\mathbf{A}]/dt = k_{\mathbf{B}\to\mathbf{A}}[\mathbf{B}] - k_{\mathbf{A}\to\mathbf{B}}[\mathbf{A}] \tag{13.1.2}$$

$$d[\mathbf{B}]/dt = k_{\mathbf{A}\to\mathbf{B}}[\mathbf{A}] - k_{\mathbf{B}\to\mathbf{A}}[\mathbf{B}] \tag{13.1.3}$$

where $k_{\mathbf{A}\to\mathbf{B}}$ and $k_{\mathbf{B}\to\mathbf{A}}$ are the (true) rate constants for interconversion. Recall from Chapter 3 that the solutions to these equations can be written in terms of deviations from equilibrium

$$\Delta c_{\mathbf{A}}(t) = \Delta c_{\mathbf{A}}(0) exp[-t/\tau_{rxn}] \tag{13.1.4}$$

where $\Delta c_{\mathbf{A}}(t) = [\mathbf{A}](t) - [\mathbf{A}]_{eq}$ and where the reaction time is

$$\tau_{rxn} = (k_{\mathbf{A}\to\mathbf{B}} + k_{\mathbf{B}\to\mathbf{A}})^{-1} \tag{13.1.5}$$

The two rate constants can be determined experimentally from τ_{rxn} along with the equilibrium constant $K_{eq} = k_{\mathbf{A}\to\mathbf{B}}/k_{\mathbf{B}\to\mathbf{A}} = [\mathbf{B}]_{eq}/[\mathbf{A}]_{eq}$.

In a simulation of the same process, let the numbers of **A** and **B** molecules as a function of time be given by $n_{\mathbf{A}}(t)$ and $n_{\mathbf{B}}(t)$. If **A** and **B** can only interconvert with each other, then $n_{\mathbf{A}}(t) + n_{\mathbf{B}}(t) = n$ will be constant while $n_{\mathbf{A}}(t)$ and $n_{\mathbf{B}}(t)$ will fluctuate around their averages. The time evolution of $n_{\mathbf{A}}(t)$ and $n_{\mathbf{B}}(t)$ in a long *equilibrium* simulation can provide the rate constant. To see how, let $h_{\mathbf{A}}(\mathbf{x})$ be the indicator for a molecule in state **A**

$$h_{\mathbf{A}}(\mathbf{x}) = \begin{cases} 1 & q(\mathbf{x}) < q_{\ddagger} \\ 0 & otherwise \end{cases} \tag{13.1.6}$$

with $\mathbf{x}$ now being the coordinates for a particular molecule and $q_{\ddagger}$ being the dividing surface location along reaction coordinate $q(\mathbf{x})$. For a simulation with n interconverting molecules

$$n_{\mathbf{A}} = \sum_{i=1}^{n} h_{\mathbf{A}}(\mathbf{x}^{(i)}) \tag{13.1.7}$$

where $\mathbf{x}^{(i)}$ are the full coordinates of the ith molecule.

Also consider a perturbed Hamiltonian which can be used to prepare a system that is initially enriched in component **A** where the degree of enrichment depends on a parameter ε [4].

$$H \equiv H_0 - \varepsilon \sum_{i=1}^{n} h_{\mathbf{A}}(\mathbf{x}^{(i)}) \tag{13.1.8}$$

Here H_0 is the equilibrium unperturbed Hamiltonian and ε is the perturbation strength. The initial distribution prepared with $\varepsilon > 0$ is

$$\rho_\varepsilon = Q(\varepsilon)^{-1} exp[-\beta H_0] exp[\beta \varepsilon n_{\mathbf{A}}] \tag{13.1.9}$$

where

$$Q(\varepsilon) = \int d\mathbf{x}^{(1)} d\mathbf{x}^{(2)} \cdots d\mathbf{x}^{(n)} exp[-\beta H_0(\{\mathbf{x}^{(i)}\})] exp[\beta \varepsilon n_{\mathbf{A}}(\{\mathbf{x}^{(i)}\})]. \tag{13.1.10}$$

Experimentally, an **A**-enriched system could be prepared, e.g. by a sudden temperature jump or sudden addition of a co-solvent that favors **B**. Immediately following the co-solvent addition, the system appears to have a non-equilibrium enrichment of component **A**.

The non-equilibrium relaxation as a function of time after the turning off the perturbation is [4]

$$\overline{n}_{\mathbf{A}}(t) = Q(\varepsilon)^{-1} \int d\mathbf{x}_0^{(1)} d\mathbf{x}_0^{(2)} \cdots d\mathbf{x}_0^{(n)} e^{-\beta H_0(\{\mathbf{x}_0^{(i)}\})]} exp[\beta \varepsilon n_{\mathbf{A}}(\{\mathbf{x}_0^{(i)}\})] n_{\mathbf{A}}(\{\mathbf{x}_t^{(i)}\}) \tag{13.1.11}$$

where starting from $t = 0^+$, each of the variables $x_t^{(i)}$ evolves according to the unperturbed ($\varepsilon = 0$) Hamiltonian. Multiplying and dividing by $Q(0)$ gives

$$\overline{n}_{\mathbf{A}}(t) = \frac{\left\langle exp[\beta \varepsilon n_{\mathbf{A}}(\{\mathbf{x}_0^{(i)}\})] n_{\mathbf{A}}(\{\mathbf{x}_t^{(i)}\}) \right\rangle_0}{\left\langle exp[\beta \varepsilon n_{\mathbf{A}}(\{\mathbf{x}_0^{(i)}\})] \right\rangle_0} \tag{13.1.12}$$

where $\overline{n}_{\mathbf{A}}(t)$ is the average non-equilibrium relaxation behavior, $\langle\,\rangle_0$ indicates an equilibrium average in the absence of the perturbation. Expanding in powers of $\beta\varepsilon$ gives $\overline{n}_{\mathbf{A}}(t) = \langle n_{\mathbf{A}} \rangle_0 + \beta\varepsilon \langle \delta n_{\mathbf{A}}(0) \delta n_{\mathbf{A}}(t) \rangle_0 + \cdots$ where $\delta n_{\mathbf{A}}(t) = n_{\mathbf{A}}(t) - \langle n_{\mathbf{A}} \rangle_0$. To linear order this is the expected relaxation according to the Onsager regression hypothesis [1,4]

$$\frac{\overline{n}_{\mathbf{A}}(t) - \langle n_{\mathbf{A}} \rangle_0}{\overline{n}_{\mathbf{A}}(0) - \langle n_{\mathbf{A}} \rangle_0} \approx \frac{\langle \delta n_{\mathbf{A}}(0) \delta n_{\mathbf{A}}(t) \rangle_0}{\left\langle (\delta n_{\mathbf{A}})^2 \right\rangle_0} \tag{13.1.13}$$

Equation (13.1.13) is a general result for weak perturbations, but for chemical reactions, Dellago et al. [4] derived the response to arbitrarily strong perturbations. Specifically, they showed that the relaxation from a strong perturbation is also proportional to the relaxation of equilibrium fluctuations,

$$\overline{n}_{\mathbf{A}}(t) - \langle n_{\mathbf{A}} \rangle_0 = \frac{n\left(e^{\beta\varepsilon} - 1\right)}{\left(e^{\beta\varepsilon} - 1\right)\langle h_{\mathbf{A}} \rangle_0 + 1} \langle \delta h_{\mathbf{A}}(\mathbf{x}_0) \delta h_{\mathbf{A}}(\mathbf{x}_t) \rangle_0 \tag{13.1.14}$$

where n is the total number of **A** or **B** molecules. Note some important aspects of this result.

1. Equation (13.1.14), when normalized by the zero time limit in $\overline{n}_{\mathbf{A}}(0) - \langle n_{\mathbf{A}} \rangle_0$ still satisfies the Onsager Regression Hypothesis.
2. Although the response to ε is nonlinear, the time dependent part is exactly that of the equilibrium correlation function, $\langle \delta h_{\mathbf{A}}(\mathbf{x}_0) \delta h_{\mathbf{A}}(\mathbf{x}_t) \rangle_0$. By definition, $\overline{n}_{\mathbf{A}}(t) - \langle n_{\mathbf{A}} \rangle_0$ is just $\Delta c_{\mathbf{A}}(t) \cdot \mathcal{V}$ where $\mathcal{V}$ is the simulation volume. Therefore, both $\Delta c_{\mathbf{A}}(t)$ and $\langle \delta h_{\mathbf{A}}(\mathbf{x}_0) \delta h_{\mathbf{A}}(\mathbf{x}_t) \rangle_0$ must exponentially decay at rate $1/\tau_{rxn}$. Equation (13.1.14) demonstrates that equilibrium correlation functions can fully describe relaxation even for highly non-equilibrium initial compositions.

While the links between equilibrium time correlation functions and the phenomenological rate laws are satisfying, they do not directly help compute rate constants. The problem is that an equilibrium distribution will place nearly all initial conditions in the stable reactant or product states. For any appreciable fraction of the trajectories to escape, the simulations would have to run for a long time τ_{rxn} which could be seconds, hours, or years.

13.2 Reactive flux formalism

Section 13.1 considered correlation functions for the probability to remain in state A some time after being in state A. It is customary to instead consider the appearance of state B starting from an equilibrium ensemble within A. Because equation (13.1.14) is valid for arbitrarily large deviations from equilibrium, let us start from a system of pure A. Letting $\varepsilon \to \infty$ and using $h_B(\mathbf{x}) = 1 - h_A(\mathbf{x})$ to simplify equation (13.1.14) gives

$$\overline{h}_B(t) = \frac{\langle h_A(\mathbf{x}_0) h_B(\mathbf{x}_t) \rangle_0}{\langle h_A \rangle} \tag{13.2.1}$$

where the new object $\overline{h}_B(t)$ emerges from $(n - \overline{n}_A(t))/n$. The left side of equation (13.2.1) is the time-dependent non-equilibrium conversion to state B starting from pure A at time $t = 0$. The right side of equation (13.2.1) is the equilibrium conditional probability to reach state B at time t given initial conditions within state A. The subscript zero in $\langle h_A(\mathbf{x}_0) h_B(\mathbf{x}_t) \rangle_0$ indicates an equilibrium distribution of initial conditions within state A. If we used the notation of conditional probability, the right side of equation (13.2.1) could be written $p_{eq}(B, t|A, 0) =$

$p_{eq}(B, t \cap A, 0)/p_{eq}(A, 0)$. The phenomenological arguments in section 13.1 would suggest that

$$\overline{h}_B(t) = \langle h_B \rangle (1 - exp[-t/\tau_{rxn}]) \tag{13.2.2}$$

However, a close examination of the equilibrium correlation function reveals important deviations from the phenomenological description. (See Figure 13.2.1.)

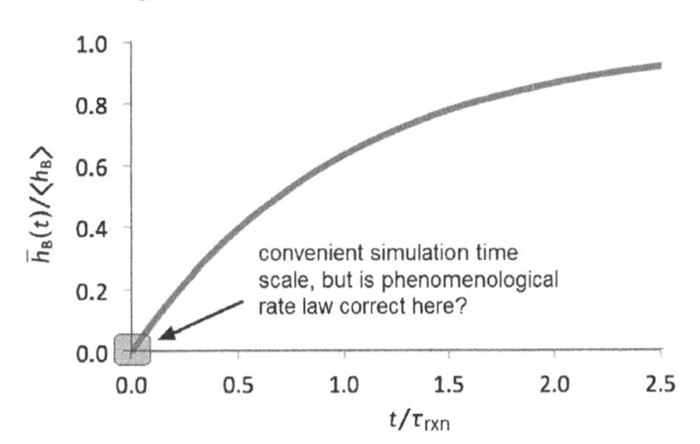

Figure 13.2.1: The phenomenological rate law predicts exponential relaxation at long times, but there may be deviations for $t \ll \tau_{rxn}$.

$\langle h_A(\mathbf{x}_0) h_B(\mathbf{x}_t) \rangle_0$ is a side-side time correlation function [5,6] because h_A and h_B indicate two sides of the dividing surface. To understand the behavior of $\overline{h}_B(t)$ at very short times, we differentiate to obtain a flux-side time correlation function [1,6]. A few identities are needed to obtain the desired result. First, the time derivative of the discontinuous function $h_A(\mathbf{x}_t)$ is

$$\frac{d}{dt} h_A(\mathbf{x}_t) = \frac{dh_A}{dq}\frac{dq}{dt} = -\delta[q(t) - q_\ddagger]\dot{q}(t) \tag{13.2.3}$$

Second, time translational symmetry allows a shift in the origin of time within an equilibrium correlation function, e.g. $\langle X(0)Y(t) \rangle_0 = \langle X(-t)Y(0) \rangle_{-t}$ where the subscript $-t$ indicates an equilibrium average at time $-t$. In the calculation below time translational symmetry is used to shift the time origin before and then after differentiation [1].

$$\begin{aligned}\frac{d}{dt}\overline{h}_B(t) &= \langle h_A \rangle^{-1} \frac{d}{dt} \langle h_A(\mathbf{x}_{-t}) h_B(\mathbf{x}_0) \rangle_{-t} \\ &= \langle h_A \rangle^{-1} \left\langle \delta[q(\mathbf{x}_{-t}) - q_\ddagger]\dot{q}(\mathbf{x}_{-t}) h_B(\mathbf{x}_0) \right\rangle_{-t} \\ &= \langle h_A \rangle^{-1} \left\langle \delta[q(\mathbf{x}_0) - q_\ddagger]\dot{q}(\mathbf{x}_0) h_B(\mathbf{x}_t) \right\rangle_0 \end{aligned} \tag{13.2.4}$$

The flux-side correlation function $d\overline{h}_B/dt$ relates the initial flux through the dividing surface $q(\mathbf{x}) = q_\ddagger$ to the later presence of the trajectory on the product side. Equation (13.2.4) shows how some of the initial flux through $q(\mathbf{x}) = q_\ddagger$ can be negated by transient dynamical recrossing events. Within some short molecular relaxation time (τ_{mol}), recrossing ceases and all trajectories will settle into one of the two stable states. That portion of the initial flux which remains committed to the product state beyond the time τ_{mol} is the reactive flux [1].

The behavior of the flux-side correlation function at short times provides insight on TST, recrossing effects, and their relation to the rate constants of the phenomenological rate law. For trajectories initiated at the transition state, the "side" in the limit as $t \to 0^+$ is perfectly determined by the initial velocity along q, i.e. if $\dot{q}(0) > 0$ then $h_B(\mathbf{x}_{0^+}) = 1$. If $\dot{q}(0) < 0$ then $h_B(\mathbf{x}_{0^+}) = 0$. This allows a remarkable simplification of $d\bar{h}_B/dt$ [1]

$$
\begin{aligned}
\left.\frac{d}{dt}\bar{h}_B(t)\right|_{0^+} &= \langle h_A\rangle_0^{-1}\left\langle \delta[q(0)-q_\ddagger]\dot{q}(0)h_B(\mathbf{x}_{0^+})\right\rangle_0 \\
&= \langle h_A\rangle_0^{-1}\left\langle \delta[q(0)-q_\ddagger]\cdot\frac{1}{2}\,|\dot{q}(0)|\right\rangle_0 \\
&= \frac{1}{2}\langle|\dot{q}(0)|\rangle_\ddagger\,\ell_q^{-1} exp\left[-\Delta F_\ddagger/k_BT\right] \\
&= k_{TST}
\end{aligned}
\qquad (13.2.5)
$$

Thus at very short times the slope of $\bar{h}_B(t)$ is the TST rate constant that we derived in section 10.5. Equations (13.2.4) and (13.2.5) can now be combined to give a concrete definition for the (classical) transmission coefficient [1].

$$
\kappa_{CL} = \frac{k_{A\to B}}{k_{TST}} = \frac{\left\langle \delta[q(0)-q_\ddagger]\dot{q}(0)h_B(\tau_{mol})\right\rangle_0}{\frac{1}{2}\left\langle \delta[q(0)-q_\ddagger]\,|\dot{q}(0)|\right\rangle_0} \qquad (13.2.6)
$$

Apart from ignoring quantum effects, the analysis has made no approximations. The short time behavior of the side-side and flux-side correlation functions is shown in Figure 13.2.2.

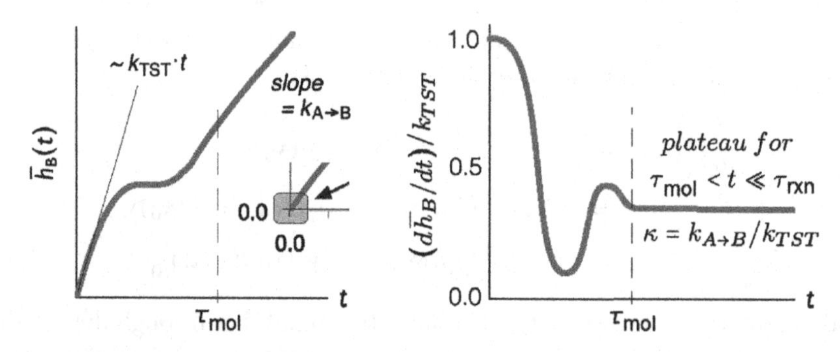

Figure 13.2.2: Initially, the slope of $\bar{h}_B(t)$ is the TST rate constant. However, some of the initial flux is from non-reactive recrossing trajectories. After a short molecular relaxation time (τ_{mol}) the slope of $\bar{h}_B(t)$ becomes the true rate constant. Accordingly the correlation function $(d\bar{h}_B/dt)/k_{TST}$ plateaus at the value of the transmission coefficient κ for $\tau_{mol} < t \ll \tau_{rxn}$. At longer times, the correlation function decays at rate $(\tau_{rxn})^{-1}$.

Timescale separation

Chandler's reactive flux formalism emphasized the importance of a separation between the activation (τ_{rxn}) and relaxation (τ_{mol}) time scales [1]. As discussed in section 10.1, the rate constant only exists if there is a clear separation of time scales between τ_{mol} and τ_{rxn}. All memory of the specific initial conditions within A should decay on a time scale much shorter than the reaction time. Given an adequate time scale separation, all states within A will have the same (exponential) distribution of lifetimes. Moreover, all trajectories initiated within A will have time to explore a representative sample of equilibrium states within A before they escape.

Non-equilibrium effects

Equation (13.1.13) relates a non-equilibrium relaxation process to equilibrium fluctuations. What then, can reactive flux correlation functions say about non-equilibrium effects? To answer this question we must separately discuss deviations from equilibrium along the reaction coordinate (in direction $\nabla q(\mathbf{x})$) and in directions perpendicular to the reaction coordinate (directions that span $q(\mathbf{x}) = q_{\ddagger}$).

Deviations from equilibrium *in directions that span the* $q(\mathbf{x}) = q_{\ddagger}$ *surface* have been proposed in numerous contexts, but they are minor effects and close scrutiny in some cases suggests that they do not exist. For example, many authors have suspected a 'bobsled effect' in which transition states on the outside of a curved pathway are more populous than transition states at the inside of the bend. Indeed, classical trajectory analysis finds many trajectories that approach the bend with no transverse vibrational excitation and are then thrown to the outside of the bend like a bobsled [7]. However, the same (classical) simulations showed just as many incoming trajectories with transverse vibrations that cut the corner.[1] For classical mechanics, we cannot conclusively dismiss non-equilibrium distributions of transition states, but strong evidence for their existence is, as yet, lacking.

In contrast, deviations from equilibrium *along the reaction coordinate* must exist for a net reaction to occur [8,9]. For this type of non-equilibrium effect, the transmission coefficient can be viewed as either a correction for non-equilibrium effects or as a correction for dynamical recrossing. In fact, these two effects are essentially the same. The starting distribution in the side-side correlation function is $h_A[\mathbf{x}]exp[-\beta H]$ which contains a strong non-equilibrium discontinuity at the boundary of state A. $\overline{h}_B(t)$ was intended to represent relaxation from a system of pure reactants, but obviously states which are initially just to the left of $q_{\ddagger}$ and which have positive velocities $\dot{q}(0)$ will become products within a very short time τ_{mol}. Equivalently there

[1] Note that quantum tunneling and zero point energies in vibrationally adiabatic models can qualitatively influence the typical reaction pathways as described in section 12.4.

are states within A having positions just to the left of $q_‡$ and negative velocities $\dot{q}(0)$ which have emerged from the product state in the recent past – but the existence of product states contradicts our assumptions. Because there are no products, no combination of position and momenta which originates from the product state can be part of the initial distribution. The distribution $h_A[\mathbf{x}]exp[-\beta H]$, which switches discontinuously from equilibrium to zero at $q_‡$, is therefore not "the correct one." If the initial distribution instead included only states which were reactants a time τ_{mol} in the past, then the correlation function would immediately decay as $exp[-t/\tau_{rxn}]$, i.e. there would be no transient effects of recrossing. The distribution in state A would then leak to create the distribution of state B, but as the populations of B and A grow and shrink the distributions would change only their size, not their shape, even from the earliest times [10]. This behavior is precisely the exponential decay of the phenomenological law.

Reactive flux calculations

Keck [11], Anderson [7], and Bennett [12] developed the earliest methods that initiated trajectories at the dividing surface $q(\mathbf{x}) = q_‡$ to investigate recrossing.[2] The later methods developed by Chandler and others [14–17] generate transmission coefficients according to the reactive flux correlation function (equation (13.2.6)). Reactive flux calculations begin with the same preliminary tasks that are required in a TST calculation:

1. Select an appropriate reaction coordinate $q(\mathbf{x})$,
2. Compute the Landau free energy $F(q)$,
3. Choose the dividing surface location $q_‡$,
4. Compute $\Delta F^‡ = F(q_‡) + k_B T\, ln \int_A \ell_q^{-1} exp[-\beta F(q)]dq$, and
5. Compute the TST prefactor $\frac{1}{2}\langle|\dot{q}(0)|\rangle_‡$.

The remaining steps in computing a reactive flux correlation function require some clever modifications of equation (13.2.6). As written, equation (13.2.6) is a ratio of two extremely small quantities, each proportional to $exp[-\beta F(q_‡)]$. Dividing the numerator and denominator by $\langle\delta[q(\mathbf{x}) - q_‡]\rangle$ transforms the ratio of rare events to a ratio of easily computed conditional averages [1,15]:

$$\kappa_{CL} = \frac{\langle\dot{q}(0)h_B(\tau_{mol})\rangle_‡}{\frac{1}{2}\langle|\dot{q}(0)|\rangle_‡} \tag{13.2.7}$$

Equation (13.2.7) shows that the transmission coefficient is a dynamical prefactor correction that can be computed separately from the activation free energy and TST rate constant.

$$k_{A\to B} = \kappa_{CL} k_{TST}^{A\to B} \tag{13.2.8}$$

[2] Anderson's work is now appreciated as the origin of the powerful effective positive flux approach [13].

Thus the reactive flux approach provides dynamically corrected rate constants from a combination of short trajectories (τ_{mol}) and efficient importance sampling methods for computing $\Delta F^{\ddagger}$ and k_{TST}. There is no need to simulate the dynamics during the long waiting time to obtain exact rates. This section focuses on two key steps in the calculation of a transmission coefficient: (1) sampling initial conditions at the dividing surface, and (2) running short MD trajectories from the sampled initial conditions.

Frenkel and Smit discuss a version of the reactive flux method that uses hard constraints to sample initial conditions. One should beware that hard constraints introduce a sampling bias when applied to nonlinear coordinates (see chapter 11). The bias is appropriately corrected in the algorithm of Frenkel and Smit, but the correction adds unnecessary complexity to the calculation. Additionally, sampling with hard constraints requires derivatives of the coordinate $q(\mathbf{x})$. In some cases, the ideal reaction coordinate may not be easily differentiated, e.g. when it involves on-the-fly numerical optimization of ghost-atom positions [18,19].

As discussed in Chapter 11, the distribution $\rho_{eq}(\mathbf{x})\delta[q(\mathbf{x}) - q]$ is more easily generated by sampling with a harmonic *restraint*, i.e. by sampling with a bias $a(q(\mathbf{x}) - q_{\ddagger})^2/2$ added to the Hamiltonian. The restrained sample can be collected with Monte Carlo [20,21], molecular dynamics [22], or hybrid MC/MD [23–25]. For sufficiently large a the sampling interval $q_{\ddagger} \pm \sqrt{2k_B T/a}$ will be so narrow that all trajectories pass through it ballistically with no appreciable change in velocity along q. It is often useful to sample a wider interval $q_- < q(\mathbf{x}) < q_+$ and collect the sub-portion of the sample within the narrower region $q_{\ddagger} \pm \delta q$. Sometimes, it is possible to create an equilibrium sample within $q_{\ddagger} \pm \delta q$ by reweighting [26] data from importance sampling along an entirely different coordinate [27].

Algorithm: Computing the reactive flux correlation function

Choose a reaction coordinate $q(\mathbf{x})$ and transition state location $q_{\ddagger}$. Estimate a molecular relaxation time τ_{mol}. The correct relaxation time will be identified in the final stages of the calculation, but initially overestimating τ_{mol} will avoid the need to run trajectories over again. A few picoseconds is usually sufficient.

1. Collect a sample of n configurations from the equilibrium distribution in a narrow interval $q_{\ddagger} \pm \delta q$ around the transition state. This can be done using the procedures for computing a conditional average as described in Chapter 11.
2. For each of the n saved configurations, generate random initial velocities $\dot{\mathbf{x}}_0$ from the Boltzmann distribution for all atoms. Compute the initial velocity along q from $\dot{q}_0 = \dot{\mathbf{x}}_0 \cdot \nabla q$. Save the value of $|\dot{q}_0|$. Run a short constant energy MD trajectory from $t = 0$ to about $t = 3\tau_{mol}$. At each timestep, save the product $\dot{q}(0)h_B(\mathbf{x}_t)$.

3. The average absolute velocity in equation (13.2.7) is

$$\langle |\dot{q}(0)| \rangle_{\ddagger} = n^{-1} \sum_{k=1}^{n} \{|\dot{q}(0)|\}_k$$

where k indicates data from the kth trajectory. Now compute for each time $0 < t < 3\tau_{mol}$ the functions

$$\langle \dot{q}(0) h_B(\mathbf{x}_t) \rangle_{\ddagger} = n^{-1} \sum_{k=1}^{n} \{\dot{q}(0) h_B(\mathbf{x}_t)\}_k$$

and the reactive flux correlation function $\kappa(t)$:

$$\kappa(t) = \frac{\langle \dot{q}(0) h_B(t) \rangle_{\ddagger}}{\frac{1}{2} \langle |\dot{q}(0)| \rangle_{\ddagger}}$$

Several notations and terminologies have been used for $\kappa(t)$. In later sections, the corresponding quantum correlation function (without normalization) will be called a flux-side correlation function.

4. Plot $\kappa(t)$ i.e. the time-dependent quantity $k_{TST}^{-1} d\overline{h}_B/dt$ in Figure 13.2.2. Identify the molecular relaxation time τ_{mol} and the classical transmission coefficient $\kappa_{CL} = \kappa(\tau_{mol})$ from the plateau in $\kappa(t)$. If no plateau is visible in $\kappa(t)$, perform the following time scale separation test. Start a small collection of trajectories in state A (B) and attempt to compute the mean first passage time to state B (A).
   (a) If both mean first passage times ($A \to B$ and $B \to A$) exceed the previous estimate of τ_{mol} by more than an order of magnitude then abort the trajectories. Increase the initial estimate of τ_{mol} and repeat steps (2), (3), and (4).
   (b) If either of the mean first passage times ($A \to B$ or $B \to A$) is comparable to the previous estimate of τ_{mol} then A and B are not kinetically separated from each other. The state with the shorter lifetime is effectively part of the more stable state.

The derivation of equation (13.2.8) shows that the true rate constant is given by $k_{A \to B} = \kappa_{CL} k_{TST}^{A \to B}$, regardless of which dividing surface is used in the calculations of κ_{CL} and $k_{TST}^{A \to B}$ [28]. However, recall from variational TST that a poorly chosen dividing surface results in an artificially large TST rate constant and an artificially small transmission coefficient. A transmission coefficient near unity means that the dynamics nearly conform to the no-recrossing assumption of TST *and* that $q(\mathbf{x}) = q_{\ddagger}$ is a nearly optimal dividing surface. A small transmission coefficient is not so easily interpreted. It may indicate a rugged energy landscape and/or strong coupling between the reaction coordinate and the bath. Alternatively, it may just be the result of a poorly chosen dividing surface. Optimized dividing surfaces allow an unambiguous interpretation of a small transmission coefficient as a characteristic of the dynamics [29]. Later sections of this chapter will show that the number of trajectories required to accurately estimate κ_{CL} increases sharply as κ_{CL} becomes small. Thus, while $\kappa_{CL} k_{TST}^{A \to B}$ is formally dividing surface independent, variationally optimized dividing surfaces improve computational efficiency and enable unambiguous conclusions about the dynamics. The following example shows how an accurate reaction coordinate and dividing surface yield a free energy profile and dynamical trajectories that are mutually consistent sources of mechanistic insight.

Example: Methane diffusion in gas hydrate revisited

Section 10.5 showed an example TST calculation for the methane diffusivity by cage-to-cage hopping in a natural gas hydrate [18]. The simulations employed a donor-acceptor coordinate system and found a symmetric free energy barrier with a shallow intermediate located midway along the reaction pathway. The dividing surface was placed at $q_{\ddagger} = 0.6$. The free energy profile and the flux-side correlation function for the dividing surface at $q_{\ddagger} = 0.6$ are shown in the figure below. The resulting transmission coefficient is approximately $\kappa = 0.29$ [18].

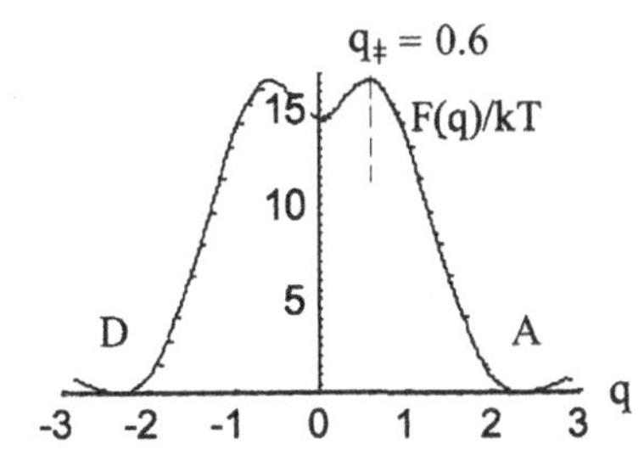

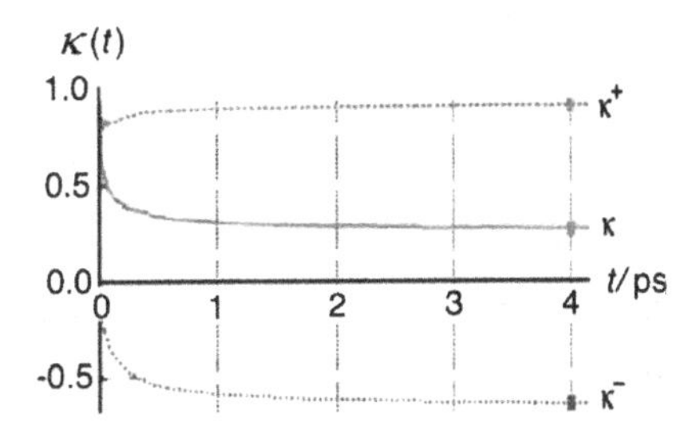

The plot showing $\kappa(t)$ also shows the separate contributions [14] from positive (κ^+) and negative (κ^-) initial velocities along $q(\mathbf{x})$. Note that the two opposing contributions largely cancel to give a rather small transmission coefficient. The κ^+ and negative κ^- contributions are particularly interesting in this case because of the shape of the free energy barrier. Trajectories launched from $q_{\ddagger}$ toward the basin on the right directly relax to the product (acceptor) state. Hence $\kappa^+ \approx 1$. Trajectories launched from $q_{\ddagger}$ toward the left will be temporarily trapped in the shallow intermediate state. The trapped trajectories can escape in either of the two directions. Thus approximately half of the initially leftward flux at $q_{\ddagger}$ ultimately contributes to the forward rate constant giving $\kappa^- \approx -0.5$. Both κ^+

and κ^- are slightly smaller than these idealized explanations would suggest, but nonetheless they are useful ways to understand that the transmission coefficient is small mostly because of the κ^- contribution.

The multidimensional dynamics reflect features of the free energy profile because the reaction coordinate and dividing surface were well-chosen. To further demonstrate this point, the figure below shows the time-evolving swarm of trajectories as a time-dependent probability density. The nonlinear q-isosurfaces and isosurfaces of an orthogonal θ-coordinate are also shown with foci at the centers of the donor and acceptor cages.

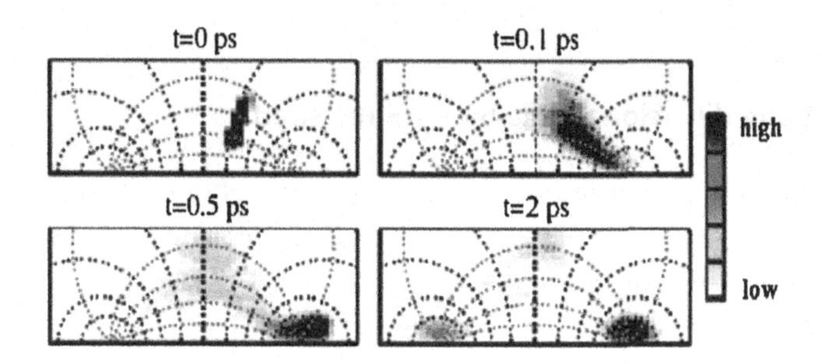

The swarm of trajectories was initiated from the $q(\mathbf{x}) = q_{\ddagger}$ surface at $t = 0ps$. Trajectories initially going to the right stabilize in the product (acceptor) state after just $0.1ps$. Trajectories initially going to the left get trapped in the shallow intermediate state. Nearly all trajectories have stabilized in one of the two basins at $\tau_{mol} \approx 2ps$. [Figures adapted from Peters et al. *J. Am. Chem. Soc.* 130, 17342-50 (2008).]

The computational advantage of the reactive flux approach over brute force simulations is clear from a scaling comparison. The computational effort to observe spontaneous activation during standard brute force molecular dynamics scales like $\sim exp[\Delta F^{\ddagger}/k_B T]$. Computing the free energy barrier by importance sampling requires an effort that scales like $\sim \Delta F^{\ddagger}/k_B T$. The additional transmission coefficient calculation requires a small additional computational effort that is approximately independent of barrier height.

Let us now determine the number n of configurations (and short trajectories) required for an accurate estimate of the transmission coefficient. For the reactive flux algorithm, the transmission coefficient estimate is

$$\hat{\kappa}_n = \frac{2}{n\langle|\dot{q}|\rangle_{\ddagger}} \sum_{i=1}^{n} \{\dot{q}(0) h_B(\tau_{mol})\}_i \tag{13.2.9}$$

where τ_{mol} is long enough for the reactive flux correlation function to reach a non-zero plateau. The variance in the estimate of κ can be computed by starting from equation (13.2.9):

$$\begin{aligned}\sigma_{\kappa,n}^2 &= \left\langle \hat{\kappa}_n^2 \right\rangle - \kappa^2 \\ &= \frac{4\left\langle \dot{q}(0)^2 h_B(t)^2 \right\rangle_{\ddagger}}{n \left\langle |\dot{q}| \right\rangle_{\ddagger}^2} + \frac{n-1}{n}\left\langle \{\hat{\kappa}_1\}_i \{\hat{\kappa}_1\}_j \right\rangle_{\ddagger} - \kappa^2\end{aligned} \tag{13.2.10}$$

The second equality uses the single datum estimator $\hat{\kappa}_1$ to express averages over terms like $\{\dot{q}(0)h_B(t)\}_i \{\dot{q}(0)h_B(t)\}_j$ which involve different trajectories from different configurations. The middle term can be simplified by introducing an estimator for the deviation from κ based on a single datum:

$$\hat{\kappa}_1 = \kappa + (\widehat{\delta\kappa_1}) \tag{13.2.11}$$

Because different configurations and trajectories are independent of each other, terms of type $\left\langle (\widehat{\delta\kappa_1})_i (\widehat{\delta\kappa_1})_j \right\rangle_{\ddagger}$ are zero. Therefore, equation (13.2.10) simplifies to

$$\sigma_{\kappa,n}^2 = \frac{4}{n \left\langle |\dot{q}| \right\rangle_{\ddagger}^2} \left\langle \dot{q}(0)^2 h_B(t) \right\rangle_{\ddagger} - \frac{1}{n}\kappa^2 \tag{13.2.12}$$

where the identity $h_B(t)^2 = h_B(t)$ has been used. This error estimate can easily be computed along with the transmission coefficient to obtain confidence intervals. The only approximation made thus far is that in practice one would use $\hat{\kappa}_n$ instead of κ in equation (13.2.12).

If the transmission coefficient is near 1.0, those trajectories that become products must be mostly those which were initiated with positive values of $\dot{q}(0)$. The first term on the right side of equation (13.2.12) then becomes $2\left\langle \dot{q}^2 \right\rangle_{\ddagger} / n \left\langle |\dot{q}| \right\rangle_{\ddagger}^2$ so that the maximum standard error in the transmission coefficient is

$$\sigma_{\kappa,n}^2 \approx \frac{1}{n}\left(\pi - \kappa^2\right) \tag{13.2.13}$$

The coefficient π results from a simplistic model in which positive initial velocities lead to products and negative velocities lead back to reactants. Of course, this model would implicitly suggest that $\kappa = 1$, and therefore might seem inappropriate in an error analysis for small transmission coefficients. However, Chapter 17 shows that the probability of committing to the product state given an initial velocity $\dot{q}(0)$ has a sigmoid dependence on $\dot{q}(0)$. The symmetry of that sigmoid function (surprisingly), also gives the coefficient as π when the analysis is done more carefully. Thus, the relative standard error in the transmission coefficient is

$$\frac{\sigma_{\kappa,n}}{\hat{\kappa}_n} \approx \frac{1}{\sqrt{n}} \frac{\sqrt{\pi - \kappa^2}}{\kappa} \tag{13.2.14}$$

Equation (13.2.14) reveals a serious efficiency problem in the reactive flux approach for small transmission coefficients. Figure 13.2.3 shows the number of trajectories required to achieve 10% relative accuracy for different values of κ.

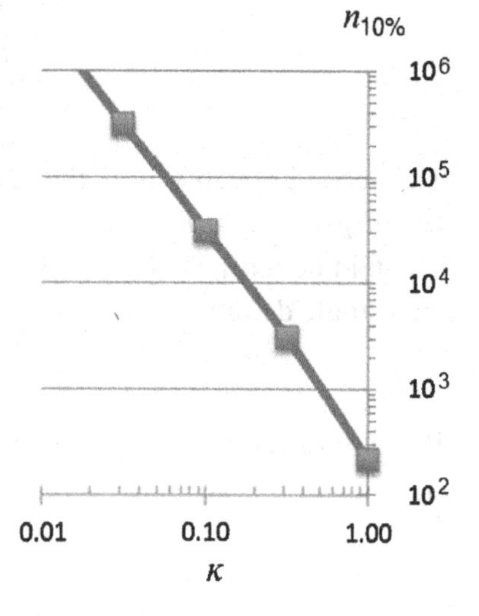

Figure 13.2.3: The cost of accurately computing the transmission coefficient from a reactive flux correlation function increases quickly as the transmission coefficient becomes small. The figure shows the number of trajectories required to achieve 10% relative uncertainty in κ as a function of κ.

In the language of statistics, the flux-side correlation estimator for the transmission coefficient is unbiased but inefficient. How can a more efficient estimator be constructed? It is useful to examine the source of inefficiency in the flux-side estimator when κ is small. The estimate can be broken into two contributions:

$$\hat{\kappa}_{n+}^{(+)} = \frac{2}{n_+ \langle |\dot{q}| \rangle_\ddagger} \sum_{i \ni \dot{q}(0)_i > 0}^{n_+} \{\dot{q}(0) h_B(t)\}_i \tag{13.2.15}$$

$$\hat{\kappa}_{n-}^{(-)} = \frac{2}{n_- \langle |\dot{q}| \rangle_\ddagger} \sum_{i \ni \dot{q}(0)_i < 0}^{n_-} \{\dot{q}(0) h_B(t)\}_i \tag{13.2.16}$$

with $\hat{\kappa}_n = \hat{\kappa}_{n+}^{(+)} + \hat{\kappa}_{n-}^{(-)}$ and where n_+ and n_- are the numbers of trajectories with initially positive and negative velocities along q. Cases where the transmission coefficient is large, i.e. where $\kappa \approx 1$, must emerge from $\hat{\kappa}_{n+}^{(+)} \approx 1$ and $\hat{\kappa}_{n-}^{(-)} \approx 0$. If $q_\ddagger$ has been chosen optimally, then cases where the transmission coefficient is extremely small must emerge from $\hat{\kappa}_{n+}^{(+)} \approx 0.5$ and $\hat{\kappa}_{n-}^{(-)} \approx -0.5$. Thus the estimator becomes inefficient for $\kappa \ll 1$ because it is a small difference of two large numbers. The following section shows how a more efficient estimator can be constructed by following trajectories in forward *and* backward time and by simultaneously using more stringent definitions of states A and B.

13.3 Effective positive flux

Reactive flux methods compute transition rates between apposite states A and B with only the surface $q(\mathbf{x}) = q_\ddagger$ between them: $A = \{\mathbf{x}|q(\mathbf{x}) < q_\ddagger\}$ and $B = \{\mathbf{x}|q(\mathbf{x}) > q_\ddagger\}$. The effective positive flux (EPF) approach [13,30,31] adopts more restrictive A and B definitions that include only the typical fluctuations within the stable reactant and product basins. The smaller A and B states no longer share a boundary as shown in Figure 13.3.1. By leaving a large "no man's land" around the transition state region, EPF dramatically improves upon the efficiency of reactive flux for small transmission coefficients. EPF classifies the points at which dynamical trajectories cross $q(\mathbf{x}) = q_\ddagger$ according to "effective" (E), "positive" (P), and "first" (F) criteria. Positive (P), first (F), and effective (E) crossing points are defined as follows:

- A crossing point is positive (P) if $\dot{q}_0 = \dot{\mathbf{x}}_0 \cdot \nabla q > 0$.
- A crossing point $(\mathbf{x}_0, \dot{\mathbf{x}}_0)$ is first (F) if the time-reversed trajectory from $(\mathbf{x}_0, \dot{\mathbf{x}}_0)$ goes directly back to state A without crossing back to the B-side of $q(\mathbf{x}) = q_\ddagger$. Note that $(\mathbf{x}_0, \dot{\mathbf{x}}_0)$ can only be first if its crossing point was positive.
- A crossing point $(\mathbf{x}_0, \dot{\mathbf{x}}_0)$ is effective (E) if its time reversed trajectory goes back to A and its forward time trajectory goes to state B.

These definitions are illustrated in Figure 13.3.1. The effective positive flux contribution from a crossing point $(\mathbf{x}_0, \dot{\mathbf{x}}_0)$ contributes to the forward rate if

$$\begin{aligned} \chi_{AB}^{EPF}(\mathbf{x}_0, \dot{\mathbf{x}}_0) &= 1 \quad \text{if trajectory from } (\mathbf{x}_0, \dot{\mathbf{x}}_0) \text{ is } E,\ P, \text{ and } F \\ &= 0 \quad \text{otherwise} \end{aligned} \tag{13.3.1}$$

The transmission coefficient can be equivalently formulated as [13,30,31]

$$\kappa_{CL} = \frac{\left\langle \dot{q}_0 \chi_{AB}^{EPF}(\mathbf{x}_0, \dot{\mathbf{x}}_0) \right\rangle_\ddagger}{\frac{1}{2}\langle |\dot{q}| \rangle_\ddagger} \tag{13.3.2}$$

In contrast to the flux-side correlation function that emerged from the reactive flux formalism, there is no time dependent correlation function. Instead, the trajectories used to evaluate $\chi_{AB}^{EPF}(\mathbf{x}_0, \dot{\mathbf{x}}_0)$ are continued until they fail one of the criteria P, F, and E, or until A and B have been reached in backward and forward time. Note that by testing properties in the order P, F, then E much of the work in computing dynamical trajectories can be avoided [13].

Why does EPF work? A conceptual sketch can be given in lieu of the proof by van Erp [13]. Using the conventional h_B and $h_A = 1 - h_B$ definitions from reactive flux, Chandler proved that $\langle \dot{q}_0 h_A(-t) h_B(t) \rangle_\ddagger$ = $\langle \dot{q}_0 h_B(t) \rangle_\ddagger$ - $\langle \dot{q}_0 h_B(-t) h_B(t) \rangle_\ddagger$ = $\langle \dot{q}_0 h_B(t) \rangle_\ddagger$. The result by Chandler shows that requiring "effective" trajectories is redundant with simply requiring forward trajectories go to state B. A crossing point $(\mathbf{x}_0, \dot{\mathbf{x}}_0)$ cannot be "first" without also being "positive". Thus we need only explain what is special about first crossing points. Actually, nothing

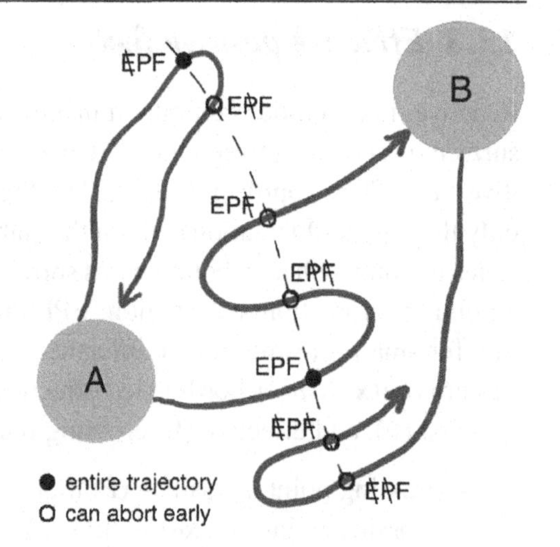

Figure 13.3.1: The effective positive flux approach employs a dividing surface between restrictive and non-adjoining states A and B. Each point sampled on the dividing surface is classified as (P) positive if $\dot{q} > 0$ so that the velocity is directed toward B, (F) first if the time reversed trajectory visits A without recrossing $q(\mathbf{x}) = q_\ddagger$, and (E) effective if the trajectory connects A to B in the forward direction. The classification is done in the order PFE because trajectories can be terminated as soon as one requirement is not met.

makes them special. An effective trajectory always crosses the dividing surface an odd number of times with the first, third, fifth, etc. being crossings with positive effective flux. EPF could just as easily sum over the last or middle crossing point, but counting the first crossing point makes the algorithm particularly simple.

It may seem from the preceding arguments like EPF has just introduced a redundant requirement for trajectories to satisfy. In a sense that is true, but the extra requirement saves a tremendous amount of computation for systems with small transmission coefficients. This can be seen from Figure 13.3.2. From each crossing point, the reactive flux approach would generate a trajectory long enough to reach A or B. By contrast, many of the crossing points fail either or both of the P and F tests and their trajectories can be terminated early. Figure 13.3.1 is just a schematic, but crossing points that give early termination amount to five of the seven depicted crossing points. Additionally, in the EPF approach there are no negative contributions to κ_{CL}. All contributions are zero or positive. Thus EPF does not rely on a cancellation of large positive and negative contributions to obtain a small κ, and hence it requires far fewer trajectories to converge.

The convergence behavior of EPF can be estimated for small κ. If n crossing points are examined and m of them have $\chi_{AB}^{EPF}(\mathbf{x}_0, \dot{\mathbf{x}}_0) = 0$, then after normalization by $\frac{1}{2}\langle|\dot{q}|\rangle_\ddagger$ the remaining $(n - m)$ are positive contributions of size ~ 1. Using a binomial estimate, the variance in the EPF estimate of κ is $\sigma_{\kappa,n}^2 \sim \kappa(1-\kappa)/n$. The tremendous advantage of EPF emerges in the relative error expression, which *for small transmission coefficients* is

$$\frac{\sigma_{\kappa,n}}{\kappa} \sim 1/\sqrt{\kappa n} \tag{13.3.3}$$

Figure 13.3.2 compares the efficiency of EPF and reactive flux methods for estimating κ. To understand whether κ_{CL} is small due to dividing surface error or due to dynamical friction, one can additionally perform a committor test on the dividing surface accuracy (see Chapter 20). Stirling et al. [32] developed an EPF-like method that simultaneously generates the transmission coefficient and performs a committor test with no additional effort.

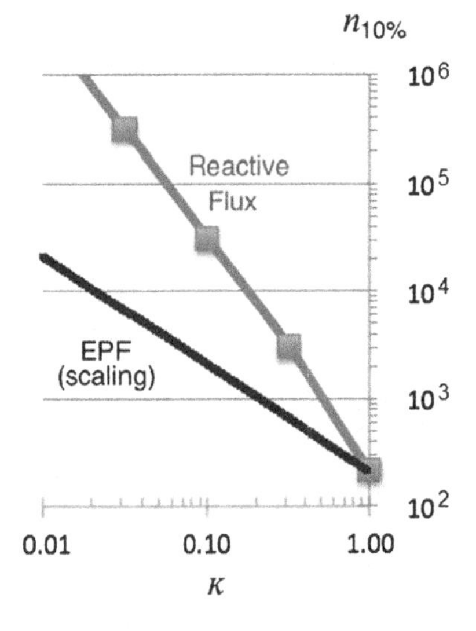

Figure 13.3.2: Achieving 10% relative uncertainty in κ with effective positive flux (EPF) requires far fewer trajectories than with the reactive flux approach. In addition, many of the trajectories in EPF can terminated long before they reach A or B.

13.4 Quantum dynamical correlation functions

The previous sections considered classical flux correlation functions of the type $\langle \dot{q}(0) h_B(t) \rangle_\ddagger$. Fluxes were computed for an ensemble of initial conditions on the dividing surface, and correlated to the destination (A or B) after classical time evolution according to Hamilton's equations. Time correlation functions for quantum dynamics require some important changes. Because of the uncertainty principle, we cannot simultaneously specify the initial position $q_\ddagger$ and momentum $\dot{q}(0)$, i.e. the central ingredients of TST are unknowable. Several quantum transition state theories have been developed [33–38], but they are fundamentally different from classical TST and beyond the scope of this book.

Miller and others developed quantum time correlation functions to obtain exact rates that account for the uncertainty in initial conditions and for quantum dynamics. [5,6,39]. Quantum versions of the reactive flux approach project a Boltzmann average of the quantum mechanical flux operator onto the time-evolved state of the system in the far-distant future. Most recent work builds on the exact quantum mechanical flux-flux, flux-side, and side-side correlation functions from the 1983 work of Miller et al. [5]. The quantum mechanical flux operator is

$$\hat{f} = \frac{i}{\hbar}[\hat{H}, \hat{h}_B] = \delta[q - q_\ddagger]\frac{\hat{p}}{m} \tag{13.4.1}$$

where $\hat{H}$ is the Hamiltonian, and $i[\hat{H}, \hat{X}]/\hbar$ is the standard commutator formula for the time derivative $d\hat{X}/dt$ of an observable X. In the Heisenberg representation [40],

$$\hat{h}_B(t) = e^{+i\hat{H}t/\hbar}\hat{h}_B e^{-i\hat{H}t/\hbar}$$

is a time-dependent "side" operator. Using these definitions, the flux-side correlation function is [5]

$$C_{FS}(t) = Q_A^{-1} tr\left[e^{-\beta\hat{H}/2}\hat{f}e^{-\beta\hat{H}/2}e^{+i\hat{H}t/\hbar}\hat{h}_B e^{-i\hat{H}t/\hbar}\right] \tag{13.4.2}$$

and the flux-flux correlation function is [5]

$$C_{FF}(t) = Q_A^{-1} tr\left[e^{-\beta\hat{H}/2}\hat{f}e^{-\beta\hat{H}/2}e^{+i\hat{H}t/\hbar}\hat{f}e^{-i\hat{H}t/\hbar}\right]. \tag{13.4.3}$$

Here $Q_A = tr\left[exp(-\beta H)\right]$ is the partition function for the reactants, and $tr[\hat{X}]$ is the quantum mechanical trace of an operator $\hat{X}$, i.e. a sum over diagonal matrix elements $\left\langle \psi_i \middle| \hat{X} \middle| \psi_i \right\rangle$ in (any) basis $\psi_0, \psi_1, \psi_2, \ldots$ The flux-side correlation function is analogous to the classical reactive flux correlation function $\kappa(t)$ except that $C_{FS}(t)$ has not been normalized by the TST rate.

The rate constant is obtained from the flux-side correlation function as

$$k = lim_{t\to\infty} C_{FS}(t) \tag{13.4.4}$$

and from the flux-flux correlation function as

$$k = \int_0^\infty C_{FF}(t)dt \tag{13.4.5}$$

The quantum correlation functions converge after a short time [6,39,41,42], like their classical counterparts. However, even for perfectly separable dynamics, the quantum correlation functions take approximately $\hbar\beta$ to converge, whereas their classical counterparts would converge instantly. The following example and Figure 13.4.1 show flux-flux correlation functions and rate calculations for a model potential and for a simple reaction. Note that in 1960, preceding even classical reactive flux methods, Yamamoto [43] proposed a Kubo transformed version of equation (13.4.5) that integrates to the same rate as the correlations of Miller et al. [5].

Example: Flux-flux correlation function and rate constant

Consider quantum mechanical motion across the top of a very broad barrier. If displacement from the barrier top by one de Broglie wavelength leads to a very small energy change, i.e. if $-m\omega_\ddagger^2\lambda_T^2 \ll k_B T$, then we can approximate the barrier as a constant potential $V_{\ddagger}$. The Hamiltonian in the vicinity of the barrier top is then

$$\hat{H} = \frac{\hat{p}^2}{2m} + V_{\ddagger}$$

The flux-flux correlation function can be written as

$$C_{FF}(t) = Q_A^{-1} tr\left[\hat{f} e^{+i\hat{H}t_c^*/\hbar} \hat{f} e^{-i\hat{H}t_c/\hbar}\right]$$

which follows from equation (13.4.3) after a cyclic permutation within the trace. Combining the root-Boltzmann factor $e^{\beta H/2}$ and the time propagator $e^{-i\hat{H}t/\hbar}$ into one operator $e^{-i\hat{H}t_c/\hbar}$ effectively defines the complex time

$$t_c = t - i\beta\hbar/2. \tag{13.4.6}$$

Miller et al. [5] showed that the flux-flux correlation function can be written in terms of position-space matrix elements as [44]

$$C_{FF}(t) = Q_A^{-1}\left(\frac{\hbar}{2m}\right)^2 \left\{\frac{\partial^2}{\partial q' \partial q}\left|\left\langle q' \left| e^{-iHt_c/\hbar} \right| q\right\rangle\right|^2 - 4\left|\frac{\partial}{\partial q'}\left\langle q' \left| e^{-iHt_c/\hbar} \right| q\right\rangle\right|^2\right\}_{q',q=0}$$

The one required matrix element is

$$\begin{aligned}
\left\langle q' \left| e^{-iHt_c/\hbar} \right| q\right\rangle &= \int dp \left\langle q' \right| e^{-iHt_c/\hbar} \left| p\right\rangle \left\langle p \left| q\right.\right\rangle \\
&= e^{-iV_{\ddagger}t_c/\hbar} \int dp \left\langle q' | p\right\rangle \left\langle p | q\right\rangle e^{-ip^2 t_c/(2m\hbar)} \\
&= \frac{e^{-iV_{\ddagger}t_c/\hbar}}{2\pi\hbar} \int dp\, e^{ip(q'-q)/\hbar} e^{-ip^2 t_c/(2m\hbar)} \\
&= e^{-iV_{\ddagger}t_c/\hbar} \left(\frac{m}{2\pi i\hbar t_c}\right)^{1/2} e^{i(q-q')^2 m/(2\hbar t_c)}
\end{aligned}$$

The matrix element is an even function of both q and q', so the term involving $\partial\left\langle q' \left| e^{-iHt_c/\hbar} \right| q\right\rangle / \partial q'$ is zero. The second derivative term survives evaluation at $q = q' = 0$, resulting in [45]

$$C_{FF}(t) = \frac{Q_A^{-1} exp[-\beta V_{\ddagger}]/4\pi}{\left[(2t/\beta\hbar)^2 + 1\right]^{3/2} (\beta\hbar/2)^2}$$

As shown in the figure below, the correlation function decays to zero after a time on the order $\beta\hbar$. Thus numerical schemes for computing $C_{FF}(t)$ in systems with separable dynamics need only a few multiples of $\beta\hbar$ to obtain the rate. At 300K, $\beta\hbar = 25 fs$. If coupling between the reaction coordinate and the bath causes recrossing, the required time may be much longer.

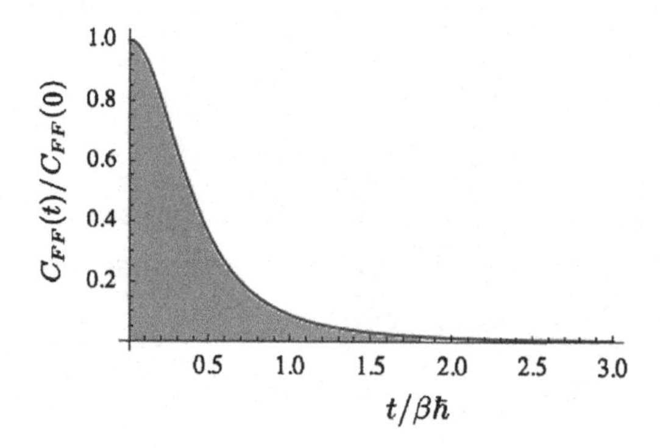

The final step is to use equation (13.4.5) to compute the rate constant. Note how closely the calculation resembles the Green-Kubo route to a diffusivity. The result for this featureless potential is

$$k = \frac{k_B T}{h} Q_A^{-1} exp[-\beta V_{\ddagger}]$$

This is exactly the classical TST result because there can be no tunneling through the infinitely broad and flat barrier top, and because there are no features on the potential to scatter an incoming wavepacket.

For reactions with sharp barriers, large curvature coupling, or other anharmonic features the quantum flux-flux correlation function becomes more interesting [6,39]. Figure 13.4.1 shows the flux-flux correlation function for a heavy-light-heavy hydrogen transfer reaction in the gas phase [46]. The initial positive lobe in the correlation function is from a delocalized wavepacket that is initially crossing in the positive direction. The second lobe labeled "recrossing" may include contributions from classically allowed and classically forbidden paths.

A common assumption in the literature [47] is that dynamical recrossing gives a transmission coefficient $\kappa_{CL} < 1$ which can be multiplied by a separate correction $\kappa_{QM} > 1$ to account for tunneling. The usual procedure is to then multiply the TST rate by both corrections: $k = \kappa_{CL}\kappa_{QM}k_{TST}$. Tunneling and classical recrossing are not actually independent of each other [48]. Where quantum effects are important, both tunneling and recrossing are actually rooted in quantum dynamics. It is not yet clear how much the widely used $k = \kappa_{CL}\kappa_{QM}k_{TST}$ approximation deviates from the full quantum dynamical transmission coefficient.

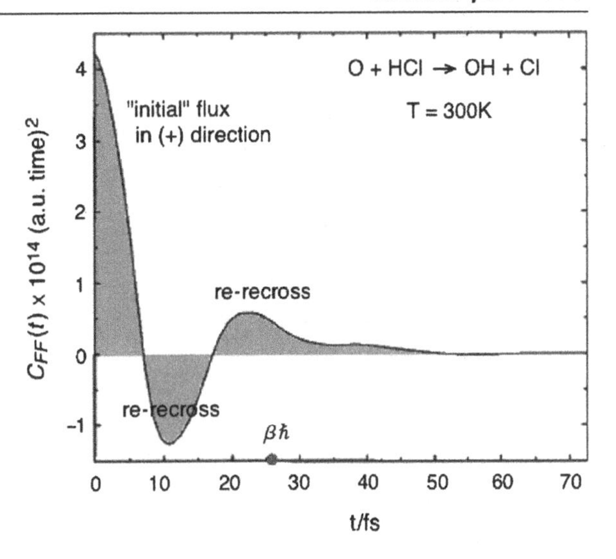

Figure 13.4.1: Numerically computed flux-flux correlation function for a heavy-light-heavy hydrogen transfer reaction. There are significant recrossing contributions even within the time $\beta\hbar$ which is the expected time for an unimpeded quantum mechanical crossing. [From Thompson and Miller, *J. Chem. Phys.* 106, 142-50 (1997).]

Quantum dynamics from classical simulations

Several methods can approximate quantum dynamics with classical molecular simulations: surface hopping [49–51], mixed quantum/classical Liouville (MQCL) methods [52,53], semi-classical initial value representation (SC-IVR) approaches [6,54,55], ring polymer molecular dynamics (RPMD) [56,57], centroid dynamics [35,36], etc. The methods differ widely in their applicability to large complex systems, in their ability to correctly predict equilibrium properties, and in the degree to which they can describe quantum coherence. Figure 13.4.2 depicts the classical, mixed quantum/classical, and the RPMD approaches.

This section focuses on the RPMD approach [56–58], which was inspired by Feynman's path integral formulation of quantum mechanics. RPMD is a versatile extended-ensemble framework for approximately including quantum effects in purely classical simulations. *For a single quantum degree of freedom* with classical Hamiltonian

$$H = \frac{p^2}{2m} + V(x), \tag{13.4.7}$$

the corresponding 1D ring-polymer Hamiltonian with n-beads is

$$H_n(\mathbf{p},\mathbf{x}) = \sum_{i=1}^{n} \left\{ \frac{p_i^2}{2m} + V(x_i) + \frac{1}{2} m \omega_{RP}^2 (x_i - x_{i-1})^2 \right\} \tag{13.4.8}$$

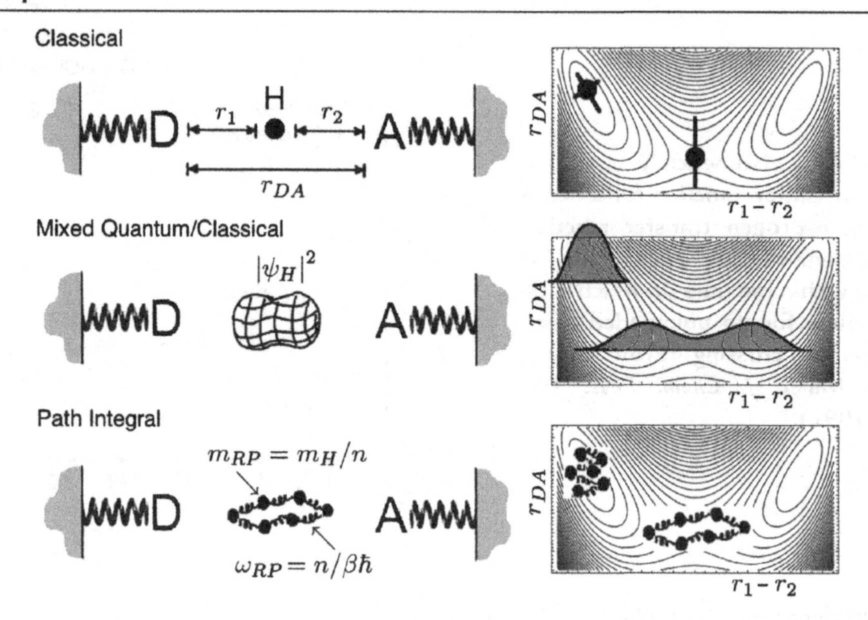

Figure 13.4.2: Schematic depicting the classical, mixed quantum/classical, and path integral approaches for a proton transfer between a donor and acceptor. Also shown is a schematic potential energy surface and the nature of the reactant and transition states for each of the frameworks.

where $x_0 = x_n$ and $\omega_{RP} = n/\beta\hbar$. Averages in the extended RPMD ensemble are defined as [57]

$$\langle \mathcal{O} \rangle = \frac{1}{(2\pi\hbar)^n Q_n} \int d\mathbf{x} d\mathbf{p}\, e^{-(\beta/n) H_n(\mathbf{x},\mathbf{p})} \frac{1}{n} \sum_{i=1}^{n} \mathcal{O}(x_i, p_i) \tag{13.4.9}$$

and the partition function Q_n is that for the extended ensemble with the elevated temperature nT,

$$Q_n = \frac{1}{(2\pi\hbar)^n} \int d\mathbf{x} d\mathbf{p}\, e^{-\beta_n H_n(\mathbf{x},\mathbf{p})} \tag{13.4.10}$$

Generalizing RPMD to systems with many atoms is straightforward. Each atom becomes a 3D ring polymer with n beads. Each bead interacts only with the corresponding beads on the ring polymers for other atoms as shown in Figure 13.4.3.

Note that equation (13.4.9) uses the average of $\mathcal{O}$-values along the ring polymer, not the value of $\mathcal{O}$ at the ring polymer centroid. These differences will, in general, give different averages for operators that depend nonlinearly on the positions of many atoms. However, Manolopoulos and coworkers note that the flux-side correlation function in the long term limit is unaffected by these subtle changes [57–59]. The most convenient definitions of the reaction coordinate for

Figure 13.4.3: The diagram at left shows two classical atoms i and j interacting via a pair potential $u(r)$. The diagram at right shows how ring polymer versions of the atoms interact with each other only through corresponding beads.

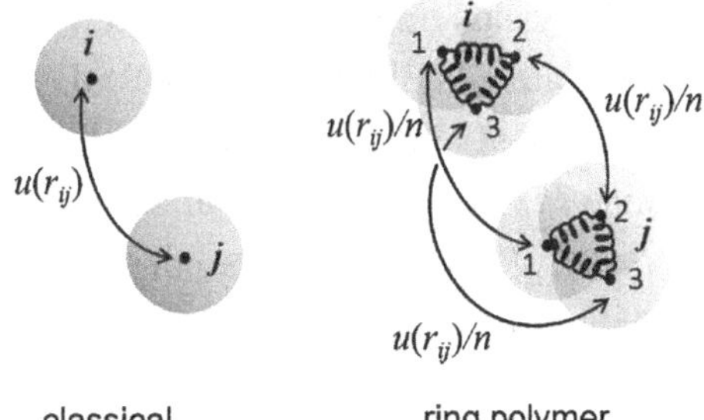

the ring polymer are the averages

$$\overline{q} = \frac{1}{n}\sum_{i=1}^{n} q_i \tag{13.4.11}$$

and velocities

$$\dot{\overline{q}} = \frac{1}{n}\sum_{i=1}^{n} \dot{q}_i \tag{13.4.12}$$

The averaged values are then used to construct $\dot{\overline{q}}_0\delta[\overline{q}_0 - q_\ddagger]h_B(\overline{q}_t)$ (rather than averaging the flux-side operator along the ring polymers). The resulting flux-side correlation function is

$$\widetilde{C}_{FS}(t) = \frac{1}{(2\pi\hbar)^n Q_{n,A}} \int d\mathbf{q}_0 d\mathbf{p}_0 e^{-\beta_n H_n(\mathbf{q}_0,\mathbf{p}_0)} \dot{\overline{q}}_0 \delta[\overline{q}_0 - q_\ddagger] h_B(\overline{q}_t) \tag{13.4.13}$$

The twiddle overbar in $\widetilde{C}_{FS}(t)$ is there because time correlation functions based on RPMD simulations are Kubo transforms of their regular time correlation counterparts. The proper flux-side correlation function $C_{FS}(t)$ can be obtained from the relationship between the Fourier transforms of a function and its Kubo transform [57].

$$\hat{C}_{FS}(\omega) = \frac{\beta\hbar\omega}{1 - e^{-\beta\hbar\omega}} \hat{\widetilde{C}}_{FS}(\omega) \tag{13.4.14}$$

The Fourier transform and inversion to obtain $C_{FS}(t)$ would provide direct insight on the relaxation dynamics, but these steps are not necessary to compute the rate because $\widetilde{C}_{FS}(t)$ and $C_{FS}(t)$ have the same long time limit. Thus the rate is

$$k = \widetilde{C}_{FS}(\tau_{mol}) \tag{13.4.15}$$

which can be normalized by the classical TST rate to obtain an RPMD transmission coefficient.

Equations (13.2.4)–(13.2.6) showed how the classical rate calculation can be performed in three efficient stages: a free energy calculation, a TST calculation, and a transmission coefficient.

RPMD yields transmission coefficients via a similar strategy. The TST and reactive flux calculations are initially done with RPMD, i.e. compute $F(\overline{q})$, $k_{RPMD-TST}[\overline{q}]$, and then $\kappa[\overline{q}]$. Here $F(\overline{q})$ is the Landau free energy obtained by umbrella sampling along $\overline{q}$ with the RPMD Hamiltonian. $k_{RPMD-TST}$ is obtained from $\frac{1}{2}\langle|\dot{\overline{q}}|\rangle_{\ddagger}$ and $F(\overline{q})$ (see section 10.5). Now computing $\widetilde{C}_{FS}(\tau_{mol})$ allows us to compute a transmission coefficient relative to the $k_{RPMD-TST}$. This is not the real transmission coefficient. The last step is to compute $k_{TST}[q]$ without RPMD. The final transmission coefficient is

$$\kappa_{RPMD} = \frac{\widetilde{C}_{FS}(\tau_{mol})}{k_{RPMD-TST}} \cdot \frac{k_{RPMD-TST}}{k_{TST}}$$

Within the ratio $k_{RPMD-TST}/k_{TST}$ there is a difference between the $F(\overline{q})$ and $F(q)$ free energy barriers. The barrier in $F(\overline{q})$ is lower than the barrier in $F(q)$ because the ring polymer can drape over the barrier top at the transition state. Recall the pole vaulter in Figure 12.1.1 who can jump "over" a bar even though her center of mass does not. In addition to the difference in free energy barriers, the RPMD centroid has "noisier" dynamics than motion of the single classical particle because of the fast vibrational modes in the ring polymer. The internal ring polymer dynamics influence both the ratio of kinematic factors in $k_{RPMD-TST}/k_{TST}$ as well as the correlation function $\widetilde{C}_{FS}(\tau_{mol})/k_{RPMD-TST}$. Figure 13.4.4 from Collepardo-Guevara et al. [60], shows $F(\overline{q})$ and the time correlation $\widetilde{C}_{FS}(t)/k_{RPMD-TST}$ from RPMD simulations of a proton and a deuteron transfer reaction in a polar solvent.

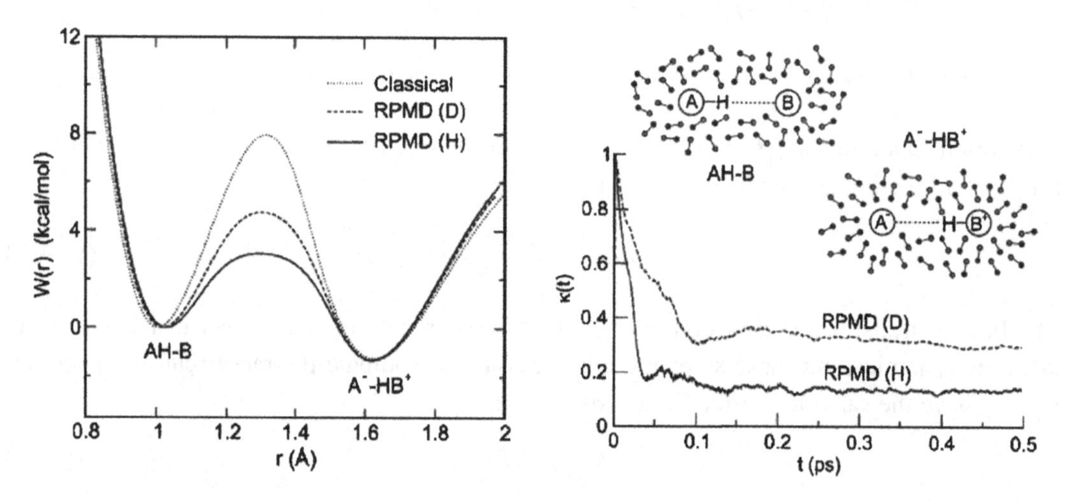

Figure 13.4.4: The RPMD free energy profile and transmission coefficients for a simple model of proton transfer between phenol (AH) and trimethylamine (B) in the aprotic methyl chloride solvent. Results are shown for proton and for deuteron transfer. [From Collepardo-Guevara et al. J. Chem. Phys. 128, 144502 (2008).]

RPMD has a number of satisfying properties for computing rate constants. (i) It exactly preserves the quantum Boltzmann distribution [57,61] so the correct equilibria and detailed balance are assured. (ii) Rates from RPMD become exact in the high temperature limit. (iii) The short time limit coincides with the quantum TST approximation of centroid dynamics [35, 36]. (iv) RPMD correctly predicts the quantum transmission coefficient ("Bell's formula") for a parabolic barrier down to the temperature at which the latter diverges. (v) RPMD rates from equation (13.4.15) are independent of variations in the dividing surface. In contrast, WKB-type tunneling corrections from Chapter 12 depend on the chosen reaction coordinate through its effect on the barrier height and the imaginary frequency.

RPMD also has some known disadvantages. For example, spurious high frequencies appear in the spectrum from internal modes of the ring polymers [61]. The accuracy of RPMD correlation functions is uncertain for complex systems [57] because the true quantum mechanical time correlation functions cannot be evaluated. RPMD does not include quantum phase information, and therefore cannot capture coherences beyond a time $\beta\hbar$ [57]. Reactive flux correlations can persist for longer ps-time scales, but these are in part from "classical coherences" [62] which RPMD should describe well. Another issue is that the most accurate classical force fields are parameterized to match experimental properties. These force fields already (attempt to) include quantum effects, so using them in RPMD simulations would double count the quantum effects. Finally, RPMD (despite being one of the easiest ways to approximate quantum dynamics) is still far more demanding than a semiclassical WKB-type tunneling correction. As we have already noted, it would be useful to know when the simple WKB-type corrections are sufficient.

Exercises

1. Section 13.2 argues that recrossing and deviations from equilibrium along the q-coordinate are essentially two ways of looking at the same phenomena. What happens when there truly is no recrossing? Consider the $(q, \dot{q})$ phase space for a one-dimensional parabolic barrier and Hamiltonian $H = \frac{1}{2}m\dot{q}^2 - \frac{1}{2}m\omega_{\ddagger}^2 q^2$. Sketch the distribution $h_A[q]exp[-\beta H]$ and also sketch the quasi-steady state distribution that would emerge after a short relaxation time τ_{mol}. Also in the $(q, \dot{q})$ space, sketch the trajectories which are being counted by TST. Are there deviations from equilibrium along the q-coordinate when there is no recrossing? Why does TST work perfectly for this case?
2. Write a code to sample initial velocities and run short inertial Langevin dynamics trajectories from the top of a one-dimensional parabolic barrier. Use the trajectories to compute the reactive flux correlation function. Choose a realistic reduced mass, imaginary frequency, and coupling strength (friction) for your calculation. Obtain the TST rate constant from the limit as $t \to 0^+$ and the transmission coefficient from longer times.

3. Repeat exercise (2) for a wide range of friction constants. Read ahead to see that the Kramers theory transmission coefficient passes through a maximum at intermediate friction. Why does the transmission coefficient for this system not pass through a maximum?
4. Some studies have computed transmission coefficients for processes with diffusive dynamics by discretizing the trajectories in time so that they appear to have well-defined velocities. Explain why the real transmission coefficient for overdamped Brownian dynamics is zero.
5. Derive equation (13.2.14) including the factor involving π.
6. Based on equation (13.2.12), it may seem like more accurate estimates of κ could be obtained by placing the dividing surface far to the left of the barrier top. Indeed, this simple fix would reduce the value of $\langle \dot{q}(0)^2 h_B(t) \rangle$ and thereby reduce the factor of π in equation (13.2.14). Explain why this would ultimately make the accuracy worse. Hint: what happens to κ?
7. Develop a code to simulate a polymer chain with n-segments connected by harmonic springs and with each segment evolving according to an inertial Langevin dynamics. To simplify the model, ignore excluded volume interactions between segments which are not bonded to each other. Additionally, let the end segments interact with each other via a Morse potential.
   (a) Compute $F(q)$ where q is the end-to-end distance. Choose the Morse potential parameters, polymer segment spring constants, and number of segments to obtain two stable states in the end-to-end distance variable. Hint: this calculation can be done analytically.
   (b) Compute the forward and backward TST rate constants using a dividing surface of constant end-to-end distance.
   (c) Compute the transmission coefficient for several values of the friction.
   (d) Explain how your results in part (c) are related to solvent viscosity.
   (e) As viscosity approaches zero, does the transmission coefficient approach one? Explain.
8. Consider a system with TST rate constant k_{TST}. Suppose that each time the dividing surface is crossed, the probability of becoming trapped in the new state without recrossing is P, regardless of the position or velocity on the dividing surface. Compute the transmission coefficient. Hint: the solution involves two geometric series.

References

[1] D. Chandler, J. Chem. Phys. 68 (1978) 2959–2970.
[2] R. Zwanzig, Nonequilibrium Statistical Mechanics, Oxford University Press, Oxford, 2001.
[3] P. Keblinski, S.R. Phillpot, S.U.S. Choi, J.A. Eastman, Int. J. Heat Mass Transf. 45 (2002) 855–863.
[4] C. Dellago, P.G. Bolhuis, P.L. Geissler, in: Erice Proceedings, vol. 1, 2005, pp. 1–38, chap. 1.
[5] W.H. Miller, S.D. Schwartz, J.W. Tromp, J. Chem. Phys. 79 (1983) 4889–4898.
[6] W.H. Miller, Faraday Discuss. 110 (1998) 1–21.
[7] J.B. Anderson, J. Chem. Phys. 58 (1973) 4684.

[8] H.A. Kramers, Physica A 7 (1940) 284–304.
[9] J.B. Anderson, Adv. Chem. Phys. 91 (1995) 381–431.
[10] M.J. Ruiz-Montero, D. Frenkel, J.J. Brey, Mol. Phys. 90 (1997) 925–941.
[11] J.C. Keck, Discuss. Faraday Soc. 33 (1962) 173–182.
[12] C.H. Bennett, in: A.S. Nowick, J.J. Burton (Eds.), Diffusion in Solids: Recent Developments, Academic Press, New York, 1975, pp. 73–113.
[13] T.S. van Erp, Adv. Chem. Phys. 151 (2012) 27–60.
[14] R.A. Kuharski, D. Chandler, J.A. Montgomery, F. Rabii, S.J. Singer, J. Phys. Chem. 92 (1988) 3261–3267.
[15] J.E. Straub, M. Borkovec, B.J. Berne, J. Phys. Chem. 92 (1988) 3711–3725.
[16] M.A. Wilson, D. Chandler, Chem. Phys. 149 (1990) 11–20.
[17] Z. Zhang, K. Haug, H. Metiu, J. Chem. Phys. 93 (1990) 31–34.
[18] B. Peters, N.E.R. Zimmermann, G.T. Beckham, J.W. Tester, B.L. Trout, J. Am. Chem. Soc. 130 (2008) 17342–17350.
[19] R.G. Mullen, J.-E. Shea, B. Peters, J. Chem. Theory Comput. 11 (2015) 2421–2428.
[20] N. Metropolis, A.W. Rosenbluth, M.N. Rosenbluth, A.H. Teller, E. Teller, J. Chem. Phys. 21 (1953) 1087–1092.
[21] D. Frenkel, B. Smit, Understanding Molecular Simulation: From Algorithms to Applications, Academic Press, San Diego, 2002.
[22] M. Tuckerman, Statistical Mechanics: Theory and Molecular Simulation, Oxford University Press, Oxford, UK, 2010.
[23] B. Mehlig, D.W. Heermann, B.M. Forrest, Phys. Rev. B 45 (1992) 679–685.
[24] E. Leontidis, B.M. Forrest, A.H. Widmann, U.W. Suter, J. Chem. Soc. Faraday Trans. 91 (1995) 2355–2368.
[25] R. Faller, J.J. de Pablo, J. Chem. Phys. 116 (2002) 55–59.
[26] A.M. Ferrenberg, R.H. Swendsen, Phys. Rev. Lett. 61 (1988) 2635–2638.
[27] G. Bussi, D. Branduardi, Rev. Comput. Chem. 28 (2015) 1–49.
[28] W.H. Miller, Acc. Chem. Res. 9 (1976) 306–312.
[29] R.G. Mullen, J.-E. Shea, B. Peters, J. Chem. Phys. 140 (2014) 41104.
[30] T.S. van Erp, P.G. Bolhuis, J. Comput. Phys. 205 (2005) 157–181.
[31] T.S. van Erp, J. Chem. Phys. 125 (2006) 174106.
[32] J. Daru, A. Stirling, J. Chem. Theory Comput. 10 (2014) 1121–1127.
[33] E. Wigner, Z. Phys. Chem. B 19 (1932) 203.
[34] F.J. McLafferty, P. Pechukas, Chem. Phys. Lett. 27 (1974) 511.
[35] G.A. Voth, D. Chandler, W.H. Miller, J. Chem. Phys. 91 (1989) 7749.
[36] J. Cao, G.A. Voth, J. Chem. Phys. 105 (1996) 6856.
[37] E. Pollak, J. Chem. Phys. 108 (1998) 8711.
[38] T.J.H. Hele, S.C. Althorpe, J. Chem. Phys. 138 (2013) 084108.
[39] W.H. Miller, J. Phys. Chem. A 102 (1998) 793–806.
[40] G.C. Schatz, M.A. Ratner, Quantum Mechanics in Chemistry, Dover, Mineola, NY, 2002.
[41] H. Wang, W.L. Hase, Chem. Phys. 212 (1996) 247–258.
[42] K. Yamashita, W.H. Miller, J. Chem. Phys. 82 (1985) 5475.
[43] T. Yamamoto, J. Chem. Phys. 33 (1960) 281–289.
[44] N.E. Henriksen, F.Y. Hansen, Theories of Molecular Reaction Dynamics, Oxford Press, 2008.
[45] N.E. Henriksen, F.Y. Hansen, Phys. Chem. Chem. Phys. 4 (2002) 5995–6000.
[46] W.H. Thompson, W.H. Miller, J. Chem. Phys. 106 (1997) 142–150.
[47] D.G. Truhlar, J. Gao, C. Alhambra, M. Garcia-Viloca, J. Corchado, M.L. Sánchez, J. Villà, Acc. Chem. Res. 35 (2002) 341–349.
[48] S.R. Billeter, S.P. Webb, T. Iordanov, P.K. Agarwal, S. Hammes-Schiffer, J. Chem. Phys. 114 (2001) 6925.
[49] J.C. Tully, J. Chem. Phys. 93 (1990) 1061–1071.
[50] E. Tapavicza, I. Tavernelli, U. Rothlisberger, Phys. Rev. Lett. 98 (2007) 023001.
[51] H.M. Jaeger, S. Fischer, O.V. Prezhdo, J. Chem. Phys. 137 (2012) 22A545.

[52] R. Kapral, Annu. Rev. Phys. Chem. 57 (2006) 129.
[53] S. Bonella, D.F. Coker, D. Mac Kernan, R. Kapral, G. Ciccotti, in: I. Burghardt, V. May, D.A. Micha, E.R. Bittner (Eds.), Energy Transfer Dynamics in Biomaterial Systems, vol. 93, Springer, Berlin, 2009, pp. 415–436.
[54] N. Makri, W.H. Miller, J. Chem. Phys. 91 (1989) 4026.
[55] W.H. Miller, J. Phys. Chem. A 105 (2001) 2942–2955.
[56] D. Chandler, P.G. Wolynes, J. Chem. Phys. 74 (1981) 4078.
[57] S. Habershon, D.E. Manolopoulos, T.E. Markland, T.F. Miller, Annu. Rev. Phys. Chem. 64 (2013) 387–413.
[58] I.R. Craig, D.E. Manolopoulos, J. Chem. Phys. 121 (2004) 3368–3373.
[59] I.R. Craig, D.E. Manolopoulos, J. Chem. Phys. 122 (2005) 84106.
[60] R. Collepardo-Guevara, I.R. Craig, D.E. Manolopoulos, J. Chem. Phys. 128 (2008) 144502.
[61] J.O. Richardson, S.C. Althorpe, J. Chem. Phys. 131 (2009) 214106.
[62] W.H. Miller, J. Chem. Phys. 136 (2012) 210901.

CHAPTER 14

Discrete stochastic variables

"...the second law of thermodynamics is continually being violated, and that to a considerable extent in any sufficiently small group of molecules."

Maxwell, Nature, (1879)

A stochastic process is the time evolution of a random variable or a collection of random variables. The range of all possible values is called the state space. Depending on the nature of a random variable, its state space may be continuous or discrete. Examples of discrete random variables in physical chemistry include the configurations of a lattice model, quantized vibrational energy levels, nodes of a network representing configurations of a protein, or integer numbers of species in a biochemical reaction network. Examples of continuous random variables include position along a dihedral angle coordinate or kinetic energies of gas molecules.

Much has been made about the arrow of time and the inexorable increase of entropy. Some irreversible macroscopic phenomena defy an easy explanation, but we can understand most non-equilibrium phenomena at the molecular scale by using the theory of stochastic processes. We will use stochastic modeling techniques for two purposes: (i) to look at the present and make probabilistic statements about the future, and (ii) to look at the present and make statements about the likely past. Whether we look forward or backward in time, we will see that the certainties about the present decay to an increasingly diverse ensemble of pasts and futures. Both types of analyses are important for reaction rate theory and rare events research.

14.1 Basic definitions

This chapter briefly introduces discrete stochastic processes, and the next chapter focuses on continuous stochastic processes. Here, discrete and continuous pertain to the nature of the state space, but time can also be modeled as a discrete or continuous variable. These cases lead to four prototypical types of stochastic processes:

- Continuous state space with continuous time, e.g. a Fokker-Planck equation [1].
- Continuous state space with discrete time, e.g. Greens function reaction dynamics [2].

- Discrete state space with continuous time, e.g. a discrete master equation [3].
- Discrete state space with discrete time, e.g. a Markov state model [4].

This chapter only considers the four basic types of models above, but there are processes that involve a mixture of discrete and continuous state variables. For example, in surface hopping simulations the system jumps between two potential energy surfaces each with a continuous state space.

Markov property

A sequence is Markovian if the future depends only on the present, independent of any additional sequence history.[1] For example, trajectories in the full phase space $x = (\mathbf{x}, \dot{\mathbf{x}})$ are clearly Markovian, i.e. the state at the next time step depends only on the present position and momenta. Mathematically, a process is Markovian if and only if the conditional probabilities have the following forgetfulness property:

$$\rho(x_j,\ t_j | x_{j-1}, t_{j-1};\ x_{j-2}, t_{j-2},\ \cdots,\ x_0, t_0) = \rho(x_j,\ t_j | x_{j-1}, t_{j-1}). \tag{14.1.1}$$

For a Markov process, we can systematically factor the probability of a discretized "trajectory" originating from (x_0, t_0) as follows:

$$\rho(x_n, t_n;\ x_{n-1}, t_{n-1};\ \cdots\ x_1, t_1 | x_0, t_0) = \prod_{j=1}^{n} \rho(x_j, t_j | x_{j-1}, t_{j-1}) \tag{14.1.2}$$

The above definitions of a Markov process are sufficient, but let us also state an equivalent alternative definition. If and only if the process is Markovian, then the Chapman-Kolmogorov equation applies to the transition probabilities, i.e.

$$\rho(x_3, t_3 | x_1, t_1) = \int \rho(x_3, t_3 | x_2, t_2) \rho(x_2, t_2 | x_1, t_1) dx_2 \tag{14.1.3}$$

for any $t_1 < t_2 < t_3$ (see Figure 14.1.1). Equation (14.1.3) is a special statement about Markovian transition probabilities, as opposed to $\rho(x_2, t_2) = \int \rho(x_2, t_2 | x_1, t_1) \rho(x_1, t_1) dx_1$ which is true for any stochastic process. Equations (14.1.1)–(14.1.3) are written for continuous probability densities and transition probabilities. They are straightforwardly generalized to discrete state spaces: simply replace integration by summation and replace the density $\rho(x)$ by a vector of probabilities $\mathbf{p}$.

In the following chapters, we will encounter some dynamical models which are naturally Markovian and some on which the Markovian property is imposed.

[1] Andrey Markov developed the theory of Markov chains to analyze character sequences in "Onegin" [5]. Markov chains did not become important in poetry, but they are useful in many other contexts.

Figure 14.1.1: The Chapman-Kolmogorov equation is exactly satisfied for any Markovian process. The figures at left and right illustrate the Chapman-Kolmogorov equation for continuous and discrete state spaces, respectively.

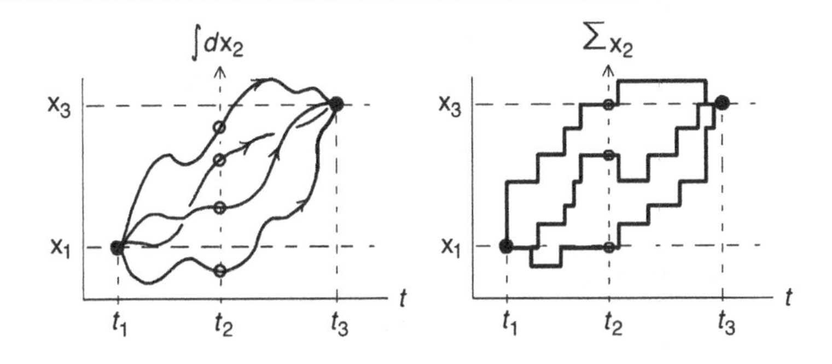

Temporal homogeneity

A stochastic process is temporally homogeneous if the transition probabilities depend only on relative time elapsed and not on specific absolute times, i.e., if $\rho(x_2, t_2|x_1, t_1) = \rho(x_2, t_2 - t_1|x_1, 0)$ for any t_2, t_1. Temporal homogeneity, i.e. time-translational symmetry, is important in the theory of time-correlation functions, as we have already seen in the chapter on reactive flux methods. Molecular systems are temporally homogeneous unless they are externally perturbed. Of course, many experiments employ external perturbations: laser pulses, temperature jumps, changes in a pulling force, etc. Both Markov and non-Markovian processes can be temporally homogeneous or inhomogeneous.

Ergodicity

A system is ergodic if no subspaces are dynamically isolated from others in state space. It is important to recognize that ergodicity can effectively depend on the time available for observation. For example, consider the pitch drop experiment at University of Queensland. Pitch, i.e. bitumen coal tar, seems like a brittle solid, but it drips through a funnel in the Queensland experiment at a painfully slow rate of approximately once per nine years. Based on the drip frequency, the viscosity of pitch is about 10^{11} times that of water. Any molecular process requiring motion of the surrounding pitch would appear non-ergodically trapped. However, given enough time, pitch is an ergodic fluid system.

Stationary and equilibrium distributions

A system that is Markovian, ergodic, and temporally homogeneous generates a unique stationary distribution in the long time limit. For a Markovian dynamics $x(t)$ that ergodically samples the stationary distribution $\rho_{SS}(x)$, the stationary average for any function $f(x)$ can be obtained from an infinitely long trajectory $x(t)$ or from an ensemble average:

$$\langle f \rangle = \lim_{\tau \to \infty} \frac{1}{\tau} \int_0^{\tau} f(x(t))dt = \int f(x)\rho_{SS}(x)dx \tag{14.1.4}$$

This property is commonly exploited in molecular simulations, e.g. by using Hamilton's equations to sample the microcanonical ensemble, or by using Metropolis Monte Carlo to sample the canonical ensemble. In contrast, we will also encounter some Markovian but non-ergodic models with absorbing states that irreversibly absorb all of the initial probability over time. In these systems, time averages are only meaningful when given as functions of time starting from some initial probability distribution.

Time reversal properties

A process $x(t)$ can be *stochastically time reversible* in one of two senses depending on the nature of the dynamics. For dynamics without momenta, e.g. overdamped diffusive dynamics, stochastic time reversibility is simply a temporal detailed balance criterion [3]

$$\rho_{eq}(\mathbf{x})\rho(\mathbf{y}, \tau|\mathbf{x}, 0) = \rho_{eq}(\mathbf{y})\rho(\mathbf{x}, \tau|\mathbf{y}, 0) \tag{14.1.5}$$

For dynamics with momenta, e.g. inertial Langevin dynamics, stochastic time reversibility involves a momentum reversal [3]

$$\rho_{eq}(\mathbf{x}, \dot{\mathbf{x}})\rho(\mathbf{y}, \dot{\mathbf{y}}, \tau|\mathbf{x}, \dot{\mathbf{x}}, 0) = \rho_{eq}(\mathbf{y}, -\dot{\mathbf{y}})\rho(\mathbf{x}, -\dot{\mathbf{x}}, \tau|\mathbf{y}, -\dot{\mathbf{y}}, 0) \tag{14.1.6}$$

and $\rho_{eq}(\mathbf{x}, -\dot{\mathbf{x}}) = \rho_{eq}(\mathbf{x}, \dot{\mathbf{x}})$. These dynamical statements of detailed balance must apply to equilibrium molecular systems. Otherwise, the equilibrium dynamics would include non-zero "loop currents" that could be continuously exploited to do work in violation of the second law.

Note that stochastic time reversibility is different from, but related to, time-reversibility for the deterministic Hamilton's equations, in which a state $(\mathbf{x}, \mathbf{p})$ evolves in forward time τ to $(\mathbf{x}', \mathbf{p}')$ if and only if $(\mathbf{x}', -\mathbf{p}')$ evolves in forward time τ to $(\mathbf{x}, \mathbf{p})$. Intuitively, deterministic time reversibility is a special limiting case of stochastic time reversibility for an inertial process, but the limit is non-trivial [6]. While the transition probabilities $\rho(\mathbf{y}, \dot{\mathbf{y}}, \tau|\mathbf{x}, \dot{\mathbf{x}}, 0)$ change gradually *en route* to the deterministic limit, the equilibrium distribution $\rho_{eq}(\mathbf{x}, \dot{\mathbf{x}})$ must change precipitously from $Q^{-1}exp[-\beta H(x)]$ to $\Omega(E)^{-1}$ at the point where friction finally vanishes.

Metastability

Continuous state spaces are often approximated by discrete state models. The fidelity of a coarsely discretized model to the true continuous dynamics depends on the degree to which the discrete states are metastable. Metastable states can be prepared and equilibrated for long times even though they are not globally stable. Recall from previous chapters that rate constants are well-defined only when the reaction time scale is adequately separated from faster relaxation time scales. Metastability and time scale separation are illustrated in Schematic 14.1.2 below, taken from a study of H-atom migration on Ni(100). In contrast to the naturally discretizable state space for H-atoms on the solid surface, diffusion in a fluid phase cannot be meaningfully discretized. Of course, one can introduce a grid with cells of size L and set the cell-to-neighboring-cell hopping rate to $\nu = D/L^2$. The resulting discrete Markov model would reproduce the desired diffusivity, but the grid itself would be meaningless because transitions from one cell to another in a fluid are not actually Markovian.

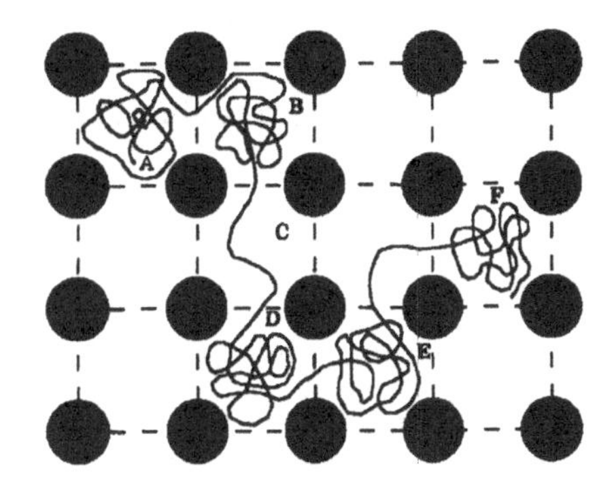

Figure 14.1.2: Schematic illustration of H migration on a Ni(100) surface. The H-atom spends long times in hollow sites with occasional hops to a new hollow site. The figure depicts transitions like $A \to B$ and $C \to D$ between neighboring sites and also transitions like $B \to D$ and $E \to F$ that effectively hop over nearest neighbors into more distant sites. [From Metiu et al. *J. Chem. Phys.* 93, 31–34 (1990).]

14.2 The master equation

All properties of a Markovian stochastic model can be derived from the master equation and its transition probabilities.[2] Consider a system with n discrete states and denote the probability of being in state j at time t by $p_j(t)$. Also denote the transition frequency ***from*** state i ***to*** state j by $w_{j\leftarrow i}$. The units of $w_{j\leftarrow i}$ are events per unit time, but they are not always elemen-

[2] Master equations were introduced by Pauli (Sommerfeld at 60 Festschrift, Leipzig, 1928) in the context of quantum correlation functions.

tary first order rate constants. The $w_{j\leftarrow i}$ may instead be a rate law that depends on properties of state i. For example, the rate at which bacteria are born in a colony [7] might be modeled as $w_{i+1\leftarrow i} = k \cdot i$ where k is the frequency at which a single bacterium divides and i is the number of bacteria in the colony. The time derivative of $p_j(t)$ is given by the master equation

$$\frac{dp_j}{dt} = \sum_{i \neq j} w_{j\leftarrow i} p_i - p_j \sum_{i \neq j} w_{i\leftarrow j} \tag{14.2.1}$$

where i in both sums goes from 1 to n. In matrix notation

$$\frac{d}{dt}\mathbf{p} = -\mathbf{W}\mathbf{p} \tag{14.2.2}$$

where

$$\mathbf{W} = \begin{bmatrix} \sum_{i=1}^{n} w_{i\leftarrow 1} & -w_{1\leftarrow 2} & \cdots & -w_{1\leftarrow n} \\ -w_{2\leftarrow 1} & \sum_{i=1}^{n} w_{i\leftarrow 2} & \cdots & -w_{2\leftarrow n} \\ \vdots & \vdots & & \vdots \\ -w_{n\leftarrow 1} & -w_{n\leftarrow 2} & \cdots & \sum_{i=1}^{n} w_{i\leftarrow n} \end{bmatrix} \tag{14.2.3}$$

with $w_{j\leftarrow j} \equiv 0$ for notational convenience. The solution to the master equation can formally be written in terms of a matrix exponential

$$\mathbf{p}(t) = e^{-\mathbf{W}t}\mathbf{p}(0) \tag{14.2.4}$$

The eigenvalues and eigenvectors of $\mathbf{W}$ identify the relaxation rates and the corresponding modes, much like the analysis of coupled rate equations in Chapter 3. The example below models proton wire transport through a carbon nanotube between two reservoirs.

Example: Proton wire transport through nanotube

Carbon nanotube diameters can be tuned to accommodate a single one-dimensional chain of water molecules [8,9]. Suppose the nanotube bridges an impermeable membrane with a water wire connecting two water reservoirs as shown in the figure below. An excess proton is depicted between two water molecules near the right end of the proton wire. [Figure modified from Peng et al., *J. Phys. Chem. B*. 119, 9212-18 (2014).]

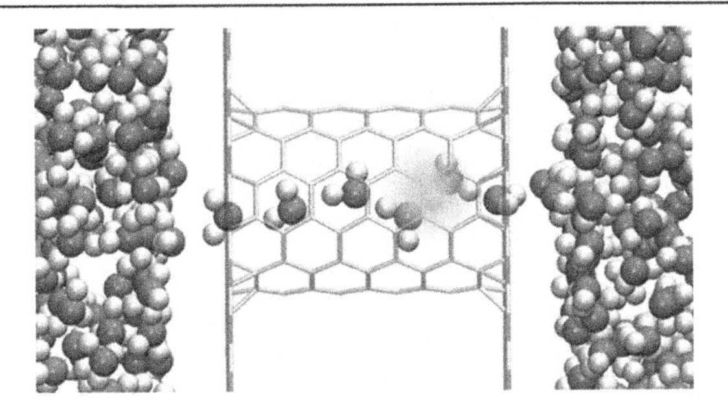

The proton is more soluble in bulk water than in the nanotube interior, because the water wire can only solvate the proton in one of the three directions. Once inside the nanotube, the excess proton moves left or right in a random walk by a Grotthus-type wire transfer mechanism. The proton hops left with rate k_- or right with rate k_+ until it reaches the water reservoirs which can be treated as absorbing states. The left and right hopping rates may be different if there is a potential across the membrane. Let positions 0 and n correspond to the bulk water reservoirs and let states 1 through $(n-1)$ correspond to numbered *gaps between* oxygen atoms as shown in the figure above. The Markov chain model for this system is

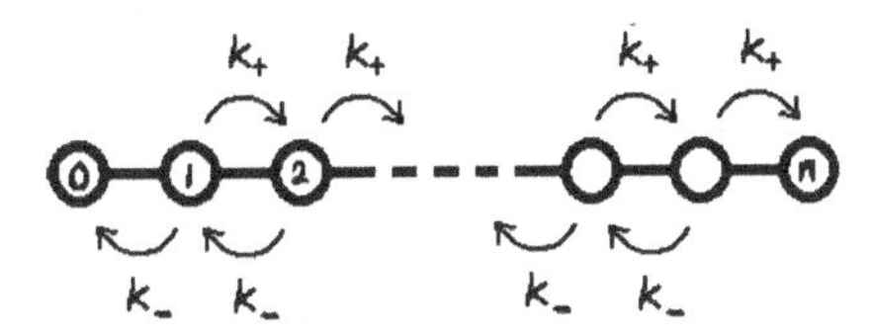

Note that the proton can never escape from states 0 and n, and therefore the model is not ergodic. The transition rate matrix $\mathbf{W}$ for the discretized model is

$$\mathbf{W} = \begin{bmatrix} 0 & -k_- & 0 & \cdots & & 0 \\ 0 & k_+ + k_- & -k_- & 0 & \cdots & 0 \\ 0 & -k_+ & k_+ + k_- & \ddots & & \vdots \\ \vdots & 0 & -k_+ & \ddots & -k_- & 0 \\ & \vdots & & \ddots & k_+ + k_- & 0 \\ 0 & 0 & \cdots & 0 & -k_+ & 0 \end{bmatrix}$$

In terms of $\theta = k_-/(k_+ + k_-)$, the dimensionless transition rate matrix with $n = 6$ is

$$\mathbf{W}/(k_+ + k_-) = \begin{bmatrix} 0 & -\theta & 0 & 0 & 0 & 0 & 0 \\ 0 & 1 & -\theta & 0 & 0 & 0 & 0 \\ 0 & \theta - 1 & 1 & -\theta & 0 & 0 & 0 \\ 0 & 0 & \theta - 1 & 1 & -\theta & 0 & 0 \\ 0 & 0 & 0 & \theta - 1 & 1 & -\theta & 0 \\ 0 & 0 & 0 & 0 & \theta - 1 & 1 & 0 \\ 0 & 0 & 0 & 0 & 0 & \theta - 1 & 0 \end{bmatrix}$$

The case $n = 6$ can be solved exactly. The eigenvalues of the transition rate matrix are natural relaxation rates, τ_m^{-1}. The nondimensionalized eigenvalues are shown in the figure below as a function of the dimensionless parameter θ.

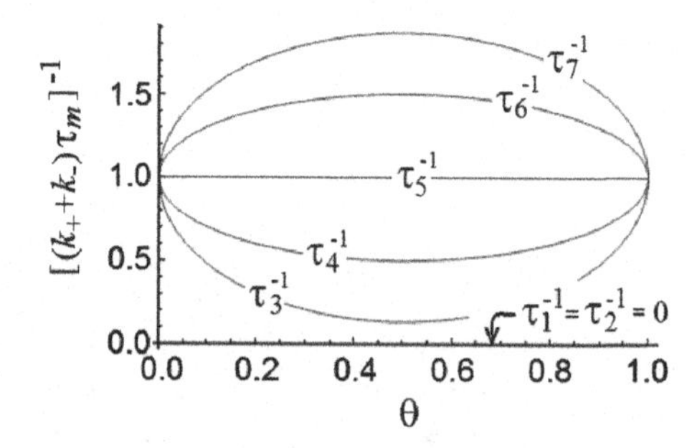

Note that we numbered states from zero, so $\mathbf{W}$ has seven eigenvalues. Two of the eigenvalues are zero, corresponding to all of the initial probability density being absorbed into the two endpoints. The remaining five eigenvalues are all positive, confirming that all trajectories are eventually consumed by the absorbing states. The solution $\mathbf{p}(t) = e^{-\mathbf{W}t}\mathbf{p}(0)$ is shown below for the unbiased case $\theta = 0.5$ at $t = 0$, $t = 2/(k_+ + k_-)$, and $t = 8/(k_+ + k_-)$.

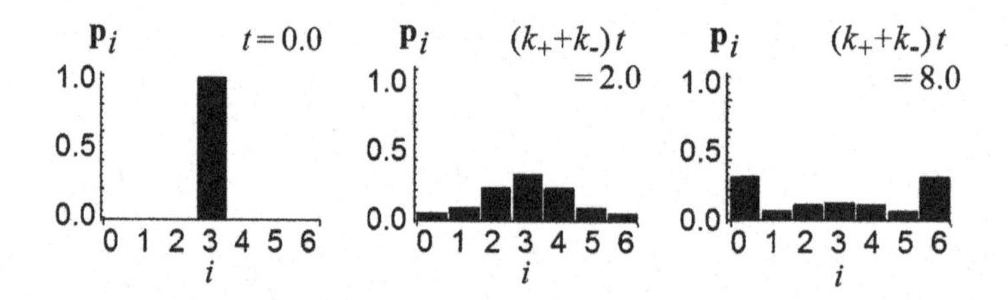

The example above considered transitions between a finite number of states. However, many stochastic processes have an infinite number of states. For example, sizes of nuclei, length of a growing polymer chain, the number of signaling proteins in a cell, etc. All of these are, in principle, unbounded state spaces. Even when the states cannot be enumerated, the master equation can still be formulated by specifying a rule for the state-to-state transition frequencies. Moreover, many properties can be obtained from an infinite master equation without actually solving or simulating its stochastic dynamics. For example, moments of the stochastic process as a function of time can often be computed exactly. The following examples illustrate the direct calculation of moments and steady-state distributions from the master equation.

Example: Moments of a birth-death process

Suppose that protein A is created within a cell of constant volume at a constant rate $\mathcal{r}_0$ and consumed by excess proteases with a pseudo-first order rate constant k. For a low-concentration protein in the small volume of a cell, the typical population may be fewer than 10 molecules. At such low populations, fluctuations are too large to rely on deterministic rate equations like $dc_{\mathbf{A}}/dt = \mathcal{r}_0 - kc_{\mathbf{A}}$. To capture the effects of fluctuations, let us treat the number of protein molecules n as a discrete stochastic variable with an infinite number of states. Multiplying $dc_{\mathbf{A}}/dt = \mathcal{r}_0 - kc_{\mathbf{A}}$ by the cell volume $\mathcal{V}$ gives the deterministic law

$$dn/dt = r - k\,n$$

where $r = \mathcal{r}_0\mathcal{V}$ and $n = c_A\mathcal{V}$. The new equation for n is still deterministic, but now the coefficients identify event-per-time rates: $\omega_{n+1\leftarrow n} = r$ and $\omega_{n-1\leftarrow n} = k\,n$. The transition frequencies define the infinite dimensional transition matrix for the stochastic evolution of n

$$\mathbf{W} = \begin{bmatrix} r & -k & 0 & \cdots & & \\ -r & r+k & -2k & 0 & \cdots \\ 0 & -r & r+2k & & \\ \vdots & 0 & -r & \ddots & \\ & \vdots & & \ddots & \end{bmatrix}$$

where the state space is $\{0, 1, 2, \ldots\}$. The time dependence of the mean for this stochastic process is

$$\begin{aligned}\frac{d}{dt}\langle n\rangle &= \frac{d}{dt}\sum_n n\mathbf{p}_n \\ &= \sum_n n\frac{d\mathbf{p}_n}{dt} = -\sum_n n(\mathbf{W}\mathbf{p})_n \\ &= \sum_n r\mathbf{p}_n - k\sum_n n\mathbf{p}_n \\ &= r - k\langle n\rangle\end{aligned}$$

After a long time, the stochastic average is $\langle n\rangle = r/k$. Interestingly, the mean has the same time dependence and long-time average as predicted by the deterministic law. This does not always happen. Averages preserve the deterministic behavior only when the transition rates are linear functions of the state index n [10].

The steady-state distribution satisfies

$$\frac{d}{dt}\mathbf{p}_{SS} = -\mathbf{W}\mathbf{p}_{SS} = \mathbf{0} \tag{14.2.5}$$

In contrast to a true equilibrium, the steady-state fluxes do not necessarily have to be zero. Later chapters on mean first passage times, classical nucleation theory, Kramers theory, and Grote-Hynes theory will derive rates from non-equilibrium steady-state fluxes. We will see that equation (14.2.5) also defines the splitting probability (i.e. the committor) and thereby identifies rates, currents, and dynamical bottlenecks for essentially any rare event. We save that discussion for Chapters 18 and 20.

The master equation is particularly easy to analyze for "one-step" processes where transitions only occur between neighboring states in a sequentially ordered state space. For these processes, it is useful to write the steady-state condition as

$$(\text{flux between } n,\ n+1) = (\text{flux between } n-1,\ n)$$

for all n. This flux balance starting point often yields a closed form expression for the steady-state distribution.

Example: Steady state from flux balance

Again let protein **A** be created at a constant rate r and consumed by proteases with an effective first order rate constant k. Let the random variable n be the number of proteins in the system. Then at steady state we can write $(flux\ between\ n,\ n+1) = (flux\ between\ n-1,\ n)$ as

$$r\mathbf{p}_{SS,n} - k(n+1)\mathbf{p}_{SS,n+1} = r\mathbf{p}_{SS,n-1} - kn\mathbf{p}_{SS,n}$$

We have already shown that $d\langle n\rangle/dt = r - k\langle n\rangle$, so $\langle n\rangle_{SS} = r/k$. For notational convenience, let $\lambda \equiv \langle n\rangle_{SS}$. Dividing the steady-state balance equation by k and replacing all factors of r/k by λ gives

$$(n+1)\mathbf{p}_{SS,n+1} - \lambda\mathbf{p}_{SS,n} = n\mathbf{p}_{SS,n} - \lambda\mathbf{p}_{SS,n-1}$$

In this case, the steady-state fluxes must be zero to obtain a distribution that can be normalized. Setting the flux between n and $n-1$ to zero gives recurrence relations:

$$\mathbf{p}_{SS,n} = \frac{\lambda}{n}\mathbf{p}_{SS,n-1} = \frac{\lambda^2}{n(n-1)}\mathbf{p}_{SS,n-2} = \cdots = \frac{\lambda^n}{n!}\mathbf{p}_{SS,0}$$

The normalization condition is

$$1 = \sum_n \mathbf{p}_{SS,n} = \mathbf{p}_{SS,0}\sum_n \frac{\lambda^n}{n!} = \mathbf{p}_{SS,0}exp[\lambda]$$

which sets the last unknown parameter: $\mathbf{p}_{SS,0} = exp[-\lambda]$. The steady-state distribution of a birth-death process is therefore a Poisson distribution

$$\mathbf{p}_{SS,n} = \lambda^n exp[-\lambda]/n!$$

with $\lambda \equiv \langle n\rangle_{SS} = r/k$. For a Poisson distribution, the mean and variance are both equal to λ. Thus the size of fluctuations in n relative to the mean is

$$\frac{\left\langle(\delta n)^2\right\rangle_{SS}^{1/2}}{\langle n\rangle_{SS}} = \frac{1}{\sqrt{\langle n\rangle_{SS}}}$$

We have seen that the deterministic model correctly predicts $\langle n\rangle_{SS}$ and the averaged dynamics, but fluctuations around the mean are relatively large when $\langle n\rangle_{SS}$ is small as shown in the figure below.

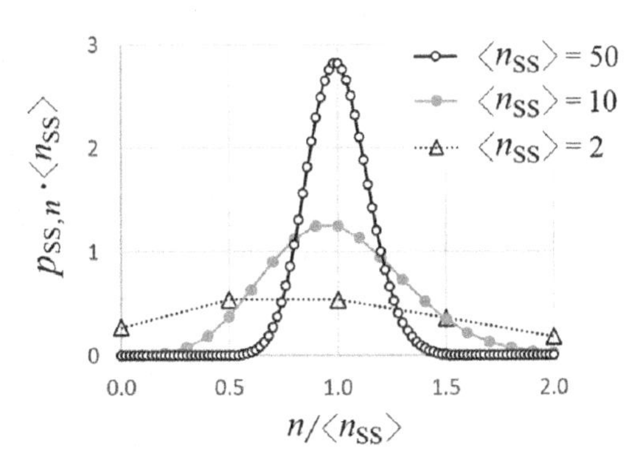

Readers may recall from equilibrium statistical mechanics that fluctuations are important in small systems. The situation is generally similar for small stochastically evolving populations.

14.3 Classical nucleation theory

Nucleation is a stochastic process that initiates the formation of a stable phase from a metastable phase [11,12]. Figure 14.3.1 illustrates the metastable zone on a typical two component phase diagram. Solute precipitate nucleation can occur when a solution of A and B is cooled into the metastable zone. The same diagram shows freezing points for pure A and pure B liquids. The pure substances undergo a single component liquid-to-solid nucleation when the liquid is supercooled below its freezing point.

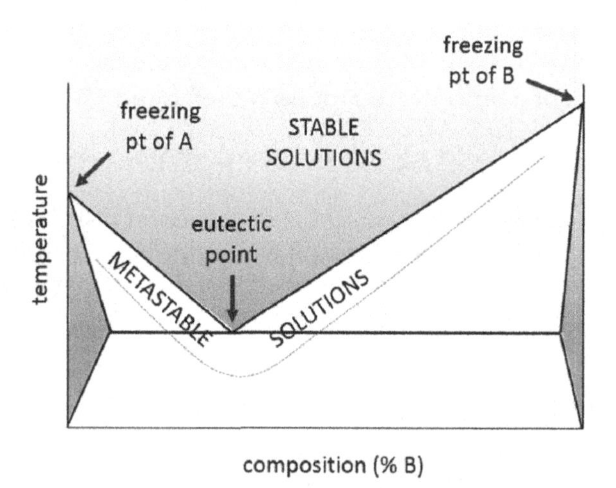

Figure 14.3.1: Schematic phase diagram for an A–B mixture. The boundary between metastable and unstable solutions is only known approximately.

In the absence of a seed or template for the new phase, spontaneous nucleation can take a very long time. The typical waiting period is called an induction time, and, like reactions, nucleation events are rare. However, the kinetics of nucleation are more strongly dependent on concentrations and temperatures than the kinetics of chemical reactions. Some 200 years after Fahrenheit's observations of supercooled water, the classical nucleation theory (CNT) [13,14] provided an explanation for long induction times in nucleation processes. CNT gives the free energy to form a nucleus of the new phase within a metastable phase as the sum of two terms.

$$F_{CNT}(n) = -n\Delta\mu + a\gamma n^{2/3} \tag{14.3.1}$$

where $\Delta\mu$ is the chemical potential difference between the nucleating phase and the metastable parent phase, n is the number of atoms in the nucleus, γ is the interfacial free energy per unit area, and the shape factor $a = 4\pi(3v_0/4\pi)^{2/3}$ makes $an^{2/3}$ equal to the surface area of a spherical nucleus of n atoms. Solute precipitate nucleation [15,16] is driven by a supersaturation, i.e. a non-equilibrium solute concentration beyond that of a saturated solution. The driving force is $\Delta\mu$ which depends on the solute activity in the supersaturated solution relative to that in a saturated solution. For solutions that are nearly ideal, the driving force is

$$\Delta\mu \approx k_B T \, ln S \tag{14.3.2}$$

Here $S = c_0/c_{sat}$ is the supersaturation, a ratio of the supersaturated (c_0) and saturated (c_{sat}) solute concentrations. Nucleation requires $c_0 > c_{sat}$, $S > 1$, and by convention $\Delta\mu > 0$.

For small nuclei, the (positive) interfacial free-energy term drives clusters to redissolve. For sufficiently large nuclei, the bulk free energy term (negative) dominates so that nuclei tend to grow. The surface and bulk driving forces balance at the critical size $n^{\ddagger}$ where $\partial F_{CNT}/\partial n = 0$. The maximum in free energy occurs at

$$n^{\ddagger} = (2\gamma a/3\Delta\mu)^3 \tag{14.3.3}$$

giving a nucleation barrier

$$F_{CNT}(n^{\ddagger}) = \frac{4(\gamma a)^3}{27\Delta\mu^2} \tag{14.3.4}$$

These results assume that nuclei are spheres with the fixed macroscopic values of γ and $\Delta\mu$ even for small sizes down to a single atom. These unrealistic assumptions limit the quantitative accuracy of the theory, but despite its simplicity CNT still makes some reliable predictions about kinetic trends. Figure 14.3.2 illustrates the free energy barrier according to CNT.

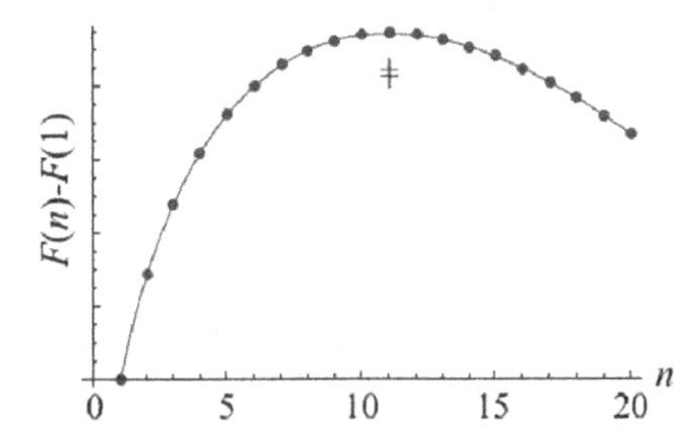

Figure 14.3.2: Schematic free energy barrier according to CNT. The smallest possible nucleus size (for solute precipitation) is $n = 1$.

Obtaining a rate from the free energy barrier requires some discussion of its interpretation. The free energy of equation (14.3.1) represents the reversible work to transform a nucleus from a single solute ($n = 1$) to some larger size n. The populations of nuclei with size n relative to those of size $n = 1$ are given by [14,17]

$$F_{CNT}(n) - F_{CNT}(1) = -k_B T \, ln\left[c_{eq}(n)/c_{eq}(1)\right] \tag{14.3.5}$$

Thus equation (14.3.3) for $n^{\ddagger}$ identifies the critical nucleus size, and equations (14.3.4) and (14.3.5) give the (equilibrium) concentration of critical nuclei relative to the concentration of monomers. The rate calculation additionally requires some information about the dynamics at the top of the barrier, i.e. the rate at which solutes attach and detach from nuclei of near critical sizes. CNT assumes that solutes attach and detach one solute atom (or molecule) at a time [18], as depicted in Figure 14.3.3.

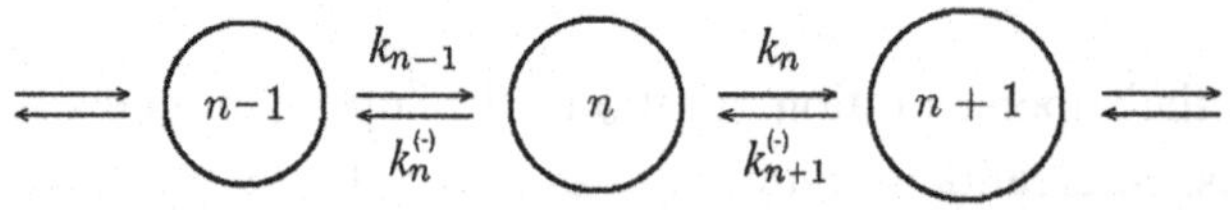

$$\rightleftharpoons \; (n-1) \; \underset{k_n^{(-)}}{\overset{k_{n-1}}{\rightleftharpoons}} \; (n) \; \underset{k_{n+1}^{(-)}}{\overset{k_n}{\rightleftharpoons}} \; (n+1) \; \rightleftharpoons$$

Figure 14.3.3: CNT assumes that solutes attach and detach from the nucleus one at a time. The rate of attachment to a nucleus of size n is k_n. The rates of detachment, $k_n^{(-)}$, are determined by detailed balance.

For a given model of the attachment rates, the free energies specify the corresponding detachment rates by detailed balance.

$$k_{n+1}^{(-)} c_{eq}(n+1) = k_n c_{eq}(n) \tag{14.3.6}$$

Equations (14.3.1)–(14.3.6) define properties of a hypothetical ensemble of nuclei with populations that are in equilibrium with each other at the supersaturated solute conditions.[3] Now we use the equilibrium properties to write a discrete master equation for the steady-state non-equilibrium populations at each size n

$$\frac{dc_n}{dt} = k_{n-1}c_{n-1} - k_n^{(-)}c_n - k_n c_n + k_{n+1}^{(-)}c_{n+1} \tag{14.3.7}$$

This is a rather simple "one-step" type master equation, but it has an infinite number of states *and* the equilibrium value of c_n diverges as $n \to \infty$. Still, we can compute the nucleation rate under steady-state conditions where large post-critical nuclei are removed from solution and redissolved to replenish the metastable solute concentration. These conditions may seem like an artificial construct, but they are not so unlike nucleation in a continuous flow crystallizer.

At steady state, the net flux from size n to size $n+1$ is the same for all n. Therefore the rate is

$$J = c_{SS}(n)k_n - c_{SS}(n+1)k_{n+1}^{(-)} \tag{14.3.8}$$

where $c_{SS}(n)$ is the steady-state concentration of nuclei with size n. Rewrite equation (14.3.8) using detailed balance and the crossover function

$$\xi_{SS}(n) \equiv c_{SS}(n)/c_{eq}(n). \tag{14.3.9}$$

[3] This ensemble cannot actually exist under supersaturated conditions.

to obtain the rate expression

$$J = c_{eq}(n)k_n \{\xi_{SS}(n) - \xi_{SS}(n+1)\} \tag{14.3.10}$$

Note that the expression for J is essentially that of Fick's law in the theory of diffusion: $J = -k_n \partial c_{SS}/\partial n$ where k_n and $\partial c_{SS}/\partial n$ are the diffusivity and concentration gradients along a generalized coordinate n. For the steady-state nucleation process, the appropriate boundary conditions are

$$\xi_{SS} = 1 \quad at \quad n = 1 \tag{14.3.11}$$

and

$$\xi_{SS} \to 0 \quad as \quad n \to \infty \tag{14.3.12}$$

These state that large nuclei are vanishingly rare (because they are being removed and redissolved), and that nuclei of size $n = 1$ are at the supersaturated reference concentration for the hypothetical equilibrium.

At this point, a marvelously simple trick exploits the boundary conditions to obtain an equation for the rate

$$\sum_{n=1}^{\infty} \frac{J}{c_{eq}(n)k_n} = \sum_{n=1}^{\infty} \{\xi_{SS}(n) - \xi_{SS}(n+1)\} = 1 \tag{14.3.13}$$

Equation (14.3.13) is then easily solved to obtain the Becker-Doering equation [18]

$$J = \left[\sum_{n=1}^{\infty} \frac{1}{c_{eq}(n)k_n} \right]^{-1} \tag{14.3.14}$$

Thus far there was no mention of a critical nucleus size in the derivation. However, the sum is dominated by terms at the top of the free energy barrier where the equilibrium concentration of nuclei is smallest. The Becker-Doering equation also places no specific requirements on the barrier shape or the form of the attachment rates k_n. Thus, given equilibrium free energies and attachment rates, the Becker-Doering equation can equally be used for other one-step processes. Examples include 1D nucleation in crystal growth, amyloid fibril growth, supramolecular polymerization, etc.

14.4 Kinetic Monte Carlo

The previous sections focused on simple stochastic processes that are amenable to exact theoretical analysis, but many important processes are too complicated for hand calculations. Some stochastic processes are too complicated to even write down the transition matrix. Instead, we may only have rules for generating potential transitions and their rates "on the fly" from the present state of the system. Kinetic Monte Carlo (kMC) algorithms use these rules to generate stochastic trajectories.

kMC is a powerful way of accelerating simulations of phenomena that involve rare events. For a standard MD simulation, the vast majority of effort is spent on dynamics within metastable states with only rare and brief barrier crossings. kMC uses the state-to-state transition frequencies to bypass the long waiting time between interesting transitions. The acceleration over standard MD can be enormous. Accordingly, kMC has become a standard simulation tool for reaction-diffusion processes [19–22], solid-state diffusion [23–25], nucleation and growth models [26], and biochemical networks [7]. Moreover, kMC provides an essentially exact dynamics if all states are metastable, all states and transitions are included, and correlated transitions are negligible. Where kMC is applicable, its fidelity to the true dynamics is an important advantage over atom-to-united-atom or atom-to-field coarse graining schemes which generally do not preserve the dynamics.

This section presents the original and simplest kMC algorithm [27,28]. Apart from a complete set of state-to-state transition rates,[4] we only need to know that transitions between metastable states have well known forgetfulness properties and exponentially distributed waiting times. Then the entire molecular dance between reaction events can be replaced with a single and remarkably simple kMC "time step". Suppose that a particular transition occurs with frequency w. For very short time intervals dt, the probability that the reaction occurs is wdt. The probability that the reaction will *not* occur during a waiting time t satisfies

$$p(t) = exp[-wt]$$

which decays to zero as the waiting time increases to infinity. The probability density for observing the first event between time t and $t + dt$ satisfies $\rho(t)dt = p(t) - p(t + dt)$. Thus, the (negative) derivative of the waiting time distribution,

$$\rho(t) = w\, exp[-wt],$$

is the waiting time probability density.

[4] The most difficult part of kMC is obtaining a complete and accurate set of state-to-state transition rates.

Now suppose there are many independent competing transitions from state i to other states with frequencies $w_{1\leftarrow i}, w_{2\leftarrow i}, w_{3\leftarrow i}, \ldots$, the net rate of transition is

$$w_{i,tot} = \sum_{j \neq i} w_{j\leftarrow i} \tag{14.4.1}$$

The probability that no transition occurs during a period t is $p(t) = exp[-w_{i,tot}t]$ and the probability density for the first occurrence as a function of time t is

$$\rho(t) = w_{i,tot}exp[-w_{i,tot}t]. \tag{14.4.2}$$

The joint probability that the next transition occurs at time t and that the next transition is $i \rightarrow j$ is given by $\rho_{joint}(t, j) = \rho(t) \cdot p(j\,|t)$ with

$$p(j\,|t) = w_{j\leftarrow i}/w_{i,tot}. \tag{14.4.3}$$

Equations (14.4.2) and (14.4.3) form the basis of the kMC algorithm. To generate the stochastic trajectory, kMC (also known as the Gillespie algorithm) uses two random numbers per event:

1. When does the next reaction occur? (sample t from density (14.4.2))
2. Which reaction occurs next? (sample j from distribution (14.4.3))

Algorithm: Kinetic Monte Carlo [27,28]

Repeatedly execute the following steps, always starting from the current state of the system. From state i:

1) compute rates $w_{1\leftarrow i}, w_{2\leftarrow i}, \ldots$ for all possible transitions out of state i

2) compute $w_{i,tot} = w_{1\leftarrow i} + w_{2\leftarrow i} + \ldots$

3) sample x_1 and x_2 from a uniform distribution on $[0, 1]$

4) compute $t = -ln[x_1]/w_{i,tot}$

5) choose j such that $\sum_{k=1}^{j-1} w_{k\leftarrow i}/w_{i,tot} < x_2 \leq \sum_{k=1}^{j} w_{k\leftarrow i}/w_{i,tot}$

6) change the state of the system from i to j and increment clock by time t

The left and right figures below show how random numbers x_1 and x_2 are used to generate t and j, respectively.

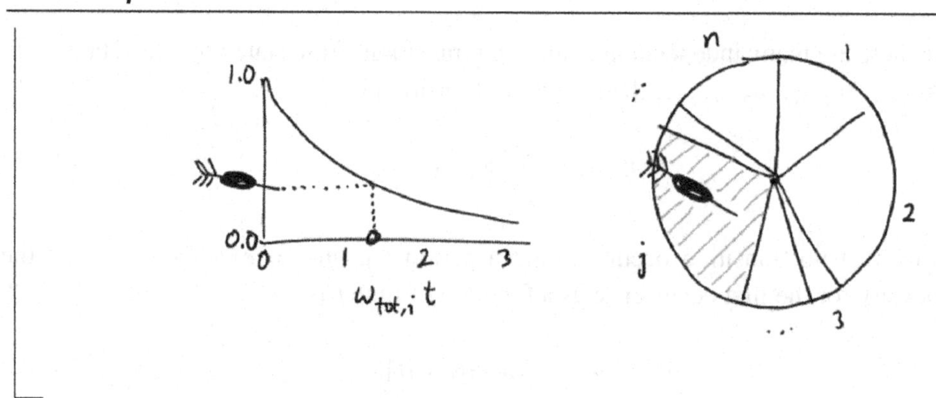

This basic kMC algorithm can exactly solve the master equation in the sense that it generates representative stochastic trajectories with essentially no approximation to the dynamics. The accrued time generated during a kMC simulation can be interpreted as a real time axis, unlike the Monte Carlo step count generated in Metropolis Monte Carlo [29]. The following example illustrates kMC simulations of a simple bistable chemical reaction network.

Example: kMC applied to the Schlogl [30] reaction network

Consider a chemical system in which **A** is created at rate r_1. **A** can degrade with first order rate $r_2 = k_2 n$ where n is the number of **A** molecules. As the concentration of **A** grows it begins to react with itself to generate a new species **B**. We will assume that species **B** is abundant and does not change appreciably during these reactions. The rates at which these reactions occur are

$$
\begin{array}{lclcl}
\varnothing & \rightarrow & \mathbf{A} & r_1 & \\
\mathbf{A} & \rightarrow & \varnothing & r_{-1} & = k_{-1}n \\
3\mathbf{A} & \rightarrow & 2\mathbf{A}+\mathbf{B} & r_2 & = k_2 n(n-1)(n-2) \\
2\mathbf{A}+\mathbf{B} & \rightarrow & 3\mathbf{A} & r_{-2} & = k_{-2}n(n-1)
\end{array}
$$

Note that we write the rates using combinatorial expressions. If $n = 2$, then the third reaction cannot happen. If there are $n > 3$ molecules of type **A**, then reaction three can happen with three collision partners which can be chosen in $n(n-1)(n-2)/3!$ ways. When $n = 6 \cdot 10^{23}$ we can replace $n(n-1)(n-2)$ with n^3, but n^3 and $n(n-1)(n-2)$ are quite different when n is a small number. Erban et al. [31] simulated this system using values $r_1 = 2250$, $k_{-1} = 37.5$, $k_2 = 0.18$, and $k_{-2} = 2.5 \cdot 10^{-4}$. These parameters give rise to two steady-state populations in the deterministic limit: $n_{SS} = 100$ or $n_{SS} = 400$. The deterministic equations just settle to one of the two steady states (depending on the initial conditions) and remain there permanently. The stochastic behavior is far more interesting. The system fluctuates around the two deterministic steady states and occasionally switches between them. The figure below [from Erban et al. SIAM (2009)] shows the kMC trajectories.

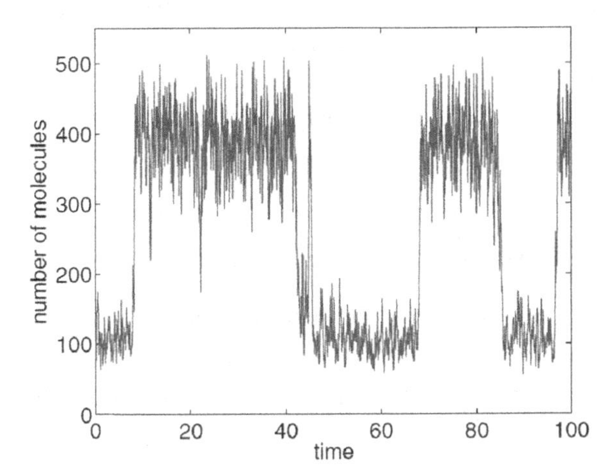

Clearly, there is a dynamical bottleneck between two stable basins, but it is quite different from the transition states that we encountered in earlier chapters. For example, there is no Hamiltonian and no guarantee that transition paths in one direction are time reversed versions of transition paths in the opposite direction.

kMC simulations have been used to understand many processes where deviations from the deterministic and/or mean field behavior are important. In general, the findings reveal that kMC simulations for small system sizes exhibit qualitative deviations from deterministic kinetic models, especially in the appearance of stochastic transitions between stable states [31]. Stochastically stable attractor states can also emerge where deterministic analyses do not predict the presence of a stable steady state [32,33]. These observations are both related to a hierarchical time scale separation. Specifically, the time scales are

$$\tau(\text{molecular relaxation}) \ll \tau(\text{reaction events}) \ll \tau(\text{switching events})$$

kMC bypasses the need to simulate the detailed molecular relaxation processes by using waiting times for individual reactions as a sort of stochastic time step. In cases where rare switching events require billions of kMC steps, one still needs methods beyond basic kMC. For example, the exact "next event algorithm" by Gibson and Bruck [34] accelerates large scale spatiotemporal kMC simulations in which reactions that occur only influence rates of reactions between nearby species. There are also approximate methods to accelerate kMC by merging states that rapidly interconvert into one superbasin state, by advancing multiple reactions at a time, or by running separate regions of a spatiotemporal kMC simulation in parallel [35–39].

Hybrid algorithms that combine kMC with transition path sampling bypass the waiting time between switching events while exactly preserving the exact kMC description of their dynamics [40,41]. Graph transformation methods directly generate the switching rates (without actually simulating switching trajectories) [42]. Finally, note that kMC can also be used to generate stochastic events during otherwise deterministic simulations. For example, Simmons et al. generated stochastic nucleation events during phase field simulations of eutectic solidification and microstructure evolution [26].

14.5 Markov state models

A Markov state model (MSM) is a discrete time-integrated version of a continuous time master equation. MSMs have become the leading simulation approach for modeling biomolecular conformational dynamics [4]. This section presents some fundamental properties of MSMs and a basic introduction to methods for constructing them. The central idea behind the MSM is to partition configuration space so that a continuous trajectory can be assigned to one of the partitions at each point in time. Figure 14.5.1 illustrates the partitioning and the discretized representation of the trajectory [43].

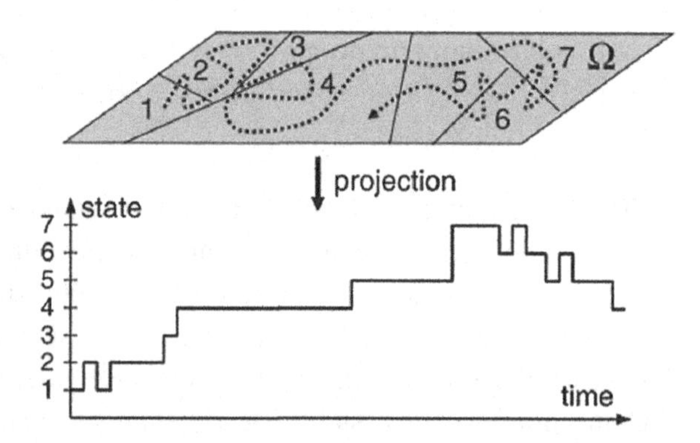

Figure 14.5.1: The upper plane depicts a configuration space Ω that has been partitioned into regions, i.e. into states of an MSM. The dynamics of a continuous trajectory projected into the discrete space appears to be a series of instantaneous state-to-state jumps. [Prinz et al. *J. Chem. Phys.* 134, 174105 (2011)]

Formal properties

Time evolution in an MSM is affected by a difference equation involving a transition matrix:

$$\mathbf{p}(t+\tau) = \mathbf{T}(\tau)\mathbf{p}(t) \tag{14.5.1}$$

Compare the discrete τ-step in equation (14.5.1) to the equivalent integration from 0 to τ in a master equation involving a rate matrix $\mathbf{W}$.[5] According to section 14.2, integration of the

[5] Many authors use $\mathbf{K}$ for the rate matrix instead of $\mathbf{W}$ because (for typical biomolecular applications) the rate matrix elements are elementary first order rate constants.

master equation gives $\mathbf{p}(t+\tau) = e^{-\mathbf{W}\tau}\mathbf{p}(t)$ [44]. Therefore, the transition matrix **T** and the rate matrix **W** must be related by

$$\mathbf{T}(\tau) = e^{-\mathbf{W}\tau} \tag{14.5.2}$$

Typically, a constant "lag time" τ is selected and time is then discretely advanced in multiples of τ by repeated applications of matrix **T**. Thus time evolution directly uses an integrated statement of the Chapman-Kolmogorov equation:

$$\mathbf{T}(n\tau) = \mathbf{T}(\tau)^n \tag{14.5.3}$$

Once the matrix $\mathbf{T}(\tau)$ has been diagonalized, the time evolution is quickly and analytically performed by raising the diagonalized version of $\mathbf{T}(\tau)$ to high powers. The eigenvectors and eigenvalues are defined by

$$\mathbf{T}(\tau)\boldsymbol{\psi}_m^R = \lambda_m \boldsymbol{\psi}_m^R \tag{14.5.4}$$

The eigenvectors of **T** are the same as those of **W**. The eigenvalues of **T** (the λ_m's) are related to those of **W** (the $1/\tau_m$'s) by the equation

$$ln\lambda_m(\tau) = -\tau/\tau_m \tag{14.5.5}$$

Note that the λ_m's depend on τ, while the τ_m's (if the lag time is sufficiently long) do not.

Any appropriately structured transition matrix satisfies the Chapman-Kolmogorov equation (14.5.3). However, if $\mathbf{T}(\tau)$ is the transition matrix for simulations of length τ, then the transition probabilities in actual simulations of length $n\tau$ may not match those from $\mathbf{T}(\tau)^n$ [43]. Inevitably, projection from the full phase space to a discretely partitioned configuration space leads to some non-Markovian memory in the transition probabilities. How long ago was the current state entered? What had been the previous state before that? A stochastic model that accounts for the recent history could be constructed, but it would be very difficult to parameterize. Alternatively, one could avoid non-Markovianity by choosing τ to be very long. In the extreme case where τ exceeds all relaxation times, the dynamics would be perfectly Markovian (but useless) because everything would equilibrate during a single τ-step. A useful model should retain enough space and time resolution to enable further analyses, e.g. simulations of time resolved NMR or FRET data. The key challenges are to judiciously choose τ and to carefully partition the configuration space so that the actual dynamics (especially those of the slowest motions) are accurately described by the MSM [4].

Constructing an MSM

In practice MSMs are constructed using a two stage discretization process [45,46]. First a sample of configurations is collected from a long MD trajectory or from a collection of short MD trajectories. The sample of configurations is then clustered into small groups containing similar structures. Then non-overlapping regions are defined around each cluster, e.g. Voronoi cells [47,48]. Finally, transition rates between the regions are estimated to complete the MSM.

Clustering sampled configurations

The similarity metric used for the initial clustering of configurations is critical to the success of the MSM approach. The Markovian approximation requires that trajectories which visit a cluster should quickly "forget" how and when they arrived. Arrival by two different routes should not lead to different dynamics or lifetimes within the cluster. Problems can occur when a similarity metric groups two configurations into one cluster when they are actually separated by a large barrier. For example, protein configurations which are similar according to an RMSD metric may have different proline backbone angles, and proline rotations are slower than some protein folding processes [49].

Apart from proline rotations (which might be explicitly included as different states), crude structure-similarity metrics like RMSD are often sufficient for an initial clustering of configurations. As long as the initial metric adequately distinguishes the dominant metastable states, then subsequent analyses can group the initial collection of fine clusters into coarser metastable states with Markovian interconversions. Note that more systematic ways of building the initial basis for distinguishing and clustering configurations are being developed [50–52]. Improved metrics for the initial clustering of states will help build models which also remain Markovian even for the faster relaxation processes. Regardless of the distance metric, the first step is an initial clustering of configurations into a collection of discrete states. A commonly used algorithm for this purpose is k-centers [4], described below.

Algorithm: k-centers clustering

The goal of k-centers clustering is to generate clusters that are uniformly dispersed throughout configuration space.

1. Select a random configuration to be the center of the initial cluster, i.e. cluster 1. Assign all other states to be members of cluster 1.
2. For each configuration x_i that is not a cluster center, compute the distance to the closest cluster center. Call this distance d_i.
3. Create a new cluster having as its center the configuration with the largest d_i from step 2.

4. Compute distances between all configurations and the new cluster center. Reassign configurations to the new cluster if they are closer to its center than to the center of their previous cluster.
5. Repeat steps 2–5 until the desired number of clusters have been created.

The algorithm is depicted for generation of four clusters in the figure below.

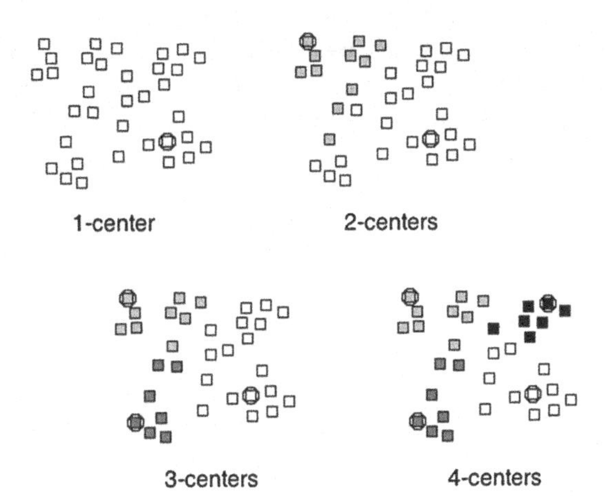

Note that the k-centers algorithm does not require that clusters have similar numbers of member configurations.

In practice, the initially sampled configurations should be grouped into *hundreds or thousands of clusters* to ensure a fine discretization of space. A fine discretization is required to provide accurate dynamical descriptions at later stages, i.e. to avoid clustering dynamically dissimilar configurations. Ultimately, the number of clusters (k) will be limited by the available data for estimating the state-to-state transition matrix.

After the initial clustering, a Voronoi construction[6] defines the regions corresponding to each cluster. Specifically, Voronoi cells are created around each configuration and "colored" according to the previous clustering analysis. Each cluster of configurations generated using the k-centers algorithm now corresponds to a union of its Voronoi cells. The Voronoi construction partitions the entire space into non-overlapping regions. A configuration belongs to region i if it falls within the Voronoi cell surrounding any member of cluster i. Figure 14.5.2 shows the Voronoi construction and the grouping of Voronoi cells according to the cluster analysis.

[6] Starting from a collection of points in configuration space and a distance metric, a Voronoi construction partitions the space such that the region within each Voronoi cell is closer to the configuration at its "center" than to any of the other points.[47,48].

The regions created by the Voronoi construction and grouping are the initial "states" of the MSM.

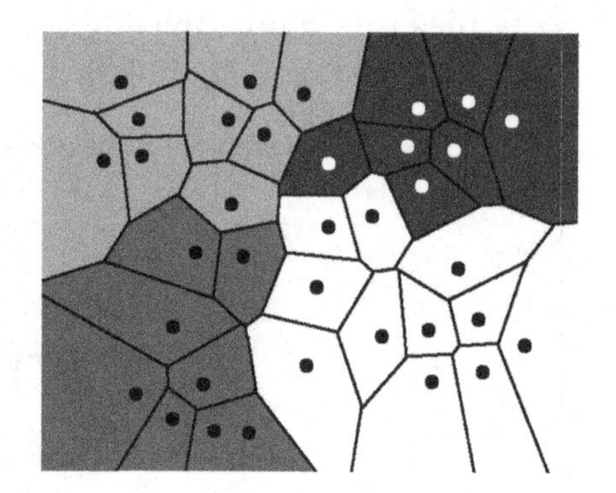

Figure 14.5.2: Illustrating a Voronoi construction for the same set of points that was used to illustrate the k-centers algorithm. The Voronoi cells have been colored to create four regions corresponding to the four-centers. These regions (usually far more numerous than four) will become "states" in the MSM.

Constructing the transition matrix

This step is more complicated than one might expect, because of detailed balance constraints, boundary recrossing effects, and considerations related to the finite trajectory data that is available [4].

First, the MSM transition rates must obey detailed balance. Otherwise, the MSM will not have a nontrivial equilibrium state for which all fluxes vanish. To enforce detailed balance, the number of observed state-to-state transitions should be symmetrized such that i-to-j transitions are equal in number to j-to-i transitions. Let $\hat{C}_{i \leftarrow j}$ be the number of j-to-i transitions observed in the raw data series and then create a symmetrized count of the transitions using the formula

$$C_{i \leftarrow j} = \left(\hat{C}_{i \leftarrow j} + \hat{C}_{j \leftarrow i}\right)/2 \tag{14.5.6}$$

If $C_{i \leftarrow j}$ and $\hat{C}_{i \leftarrow j}$ are significantly different, i.e. more different than Poisson statistics would suggest, then the trajectory data is insufficient.

Second, note that counting transitions, i.e. computing the $\hat{C}_{i \leftarrow j}$, is not as simple as it might seem. The clustering rules and distance metrics (at least using current algorithms) are based on geometrical, not dynamical, considerations. Therefore, the frequency of boundary crossings between i and j over-estimates the true rate constant for conversion from i to j. As shown in the chapter on reactive flux methods, the true transition rate is the fraction of the crossing flux that remains in the product state after a short relaxation time τ_{mol}. The MSM literature accounts for transient effects similarly by using a lag time τ.

MSMs are constructed from configurations sampled during a long MD trajectory, not from trajectories initiated at the interfaces between states as in the reactive flux approach. How then does one determine the reactive flux after a lag time τ? MSM constructions use a sliding lag-time window to count transitions [4]. The sliding lag-time window approach counts $i(t)$ to $j(t+\tau)$ transitions and then divides the total number by $\tau/\Delta t$ where Δt is the time interval between saved configurations from a long trajectory. Figure 14.5.3 illustrates the sliding lag-time window approach for counting transitions.

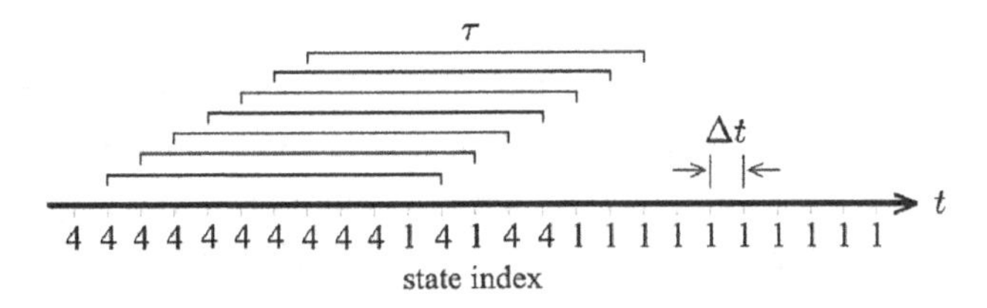

Figure 14.5.3: Counting each transition from one Δt timestep to the next would suggest there are three transitions from state 4 to state 1, and two transitions from 1 to 4. The sliding interval approach counts $\tau/\Delta t$ net crossings from states 4 to 1 as it passes the time interval with the flurry of correlated recrossings. After division by $\tau/\Delta t$ the sliding interval approach counts just one crossing event from state 4 to state 1.

The lag time should not be too short, else it will count correlated jump events and thereby overestimate the transition probabilities. Because a single lag time is adopted for all state-to-state transitions, the selected lag time is typically much longer than a molecular relaxation time. Long lag times are perfectly acceptable when the slowest relaxations are of interest, e.g. millisecond or longer processes like folding. Of course, all information about the dynamics of processes that relax faster than τ is lost, i.e. one cannot recover the short time dynamics by using $\mathbf{T}(\tau/n) = \mathbf{T}(\tau)^{1/n}$. An alternative to the sliding lag-time approach is the cores approach. Core sets approaches [53–55] count only those transitions which reach the inner core region of a state after leaving the core of another state. The counting of transitions between cores is much like the effective positive flux (EPF) approach [56] in Chapter 13.

After the state-to-state transitions have been counted, the transition matrix elements are estimated as[7]

$$T_{ji}(\tau) = \frac{C_{j\leftarrow i}}{\sum_k C_{k\leftarrow i}} \tag{14.5.7}$$

Pande et al. showed that this is the maximum likelihood estimate for T_{ji} and developed its associated error formulas [4]. The rate matrix of the corresponding master equation can be estimated from the same data as

$$w_{j\leftarrow i} = \frac{C_{j\leftarrow i}}{p_{eq}(i)t_{tot}} \tag{14.5.8}$$

[7] Mathematicians use a transposed transition matrix that operates from the right: $\mathbf{p}(t+\tau) = \mathbf{p}(t)\mathbf{T}(\tau)$.

where $p_{eq}(i)$ is the fraction of time spent in state i and t_{tot} is the total duration of the time series. For a sample of configurations obtained from an equilibrium trajectory, the free energy of each cluster can be approximated as $-kT\, ln\, p_{eq}(i)$. Notice that the $w_{j \leftarrow i}$ obey detailed balance because of the requirement that $C_{j \leftarrow i} = C_{i \leftarrow j}$.

Validation procedures

If the lag time is sufficiently long, then Chapman-Kolmogorov says $\mathbf{T}(\tau)^n = \mathbf{T}(n\tau)$ where $\mathbf{T}(n\tau)$ is an independently constructed transition matrix using an n-fold longer lag time $n\tau$. Testing for parity of each individual matrix element is tedious and is expected to reveal a few discrepancies even for the most carefully constructed MSM. More important characteristics to preserve are the long time behaviors.

In section 14.6 we will see that the eigenvector $\boldsymbol{\psi}_2^R$ corresponds to the slowest transition and that reactants and products are naturally distinguished by the sign of $\psi_2^R(i)$, i.e. by the value of $\boldsymbol{\psi}_2^R$ at the ith state in the MSM model. Define state A as a union of the reactant states, i.e.

$$A = \left\{ i \,\middle|\, \psi_2^R(i) < 0 \right\}$$

Prinz et al. suggest the following "relaxation test" [43]

$$\varphi(i) = \begin{cases} p_{eq}(i) / \sum_{j \in A} p_{eq}(j) & if \quad i \in A \\ 0 & if \quad i \notin A \end{cases} \tag{14.5.9}$$

An MSM constructed using an appropriate τ should be able, for any integer n, to predict $p(A, n\tau | A, 0)$ using $\mathbf{T}(\tau)^n$ applied to the vector $\boldsymbol{\varphi}$. The MSM prediction is compared to the direct trajectory correlation function $p_{MD}(A, n\tau | A, 0)$, i.e. the test is

$$\mathbf{h}_A^{\dagger} \mathbf{T}(\tau)^n \boldsymbol{\varphi} \stackrel{?}{=} p_{MD}(A, n\tau | A, 0) \tag{14.5.10}$$

where $h_A(i) = 1$ if state $i \in A$ and 0 otherwise. The relaxation test directly checks the Chapman-Kolmogorov equation, but only for the slowest relaxation.

Alternatively, one can compute $\mathbf{T}(\tau)$ for several different τ and then plot the predicted time scales τ_i obtained from diagonalizing $\mathbf{T}(\tau)$. The slowest time scales will depend on τ if the lag time is too small, but they should become constant for sufficiently large τ. The time scale test simultaneously checks for accuracy of the model at multiple time scales and also systematically reveals the smallest appropriate lag time for modeling the process of interest [43]. Note that the slowest transitions may be observed too infrequently to estimate rates with data from a single long MD trajectory. The adaptive sampling and replica exchange sampling procedures of Pande and coworkers [57], or discrete transition path sampling approaches [58–60] can help access rare transitions.

Coarse grained dynamics

The initial MSM created after clustering and parameterizing the transition matrix will contain hundreds or thousands of states. For mechanistic interpretation, the dynamics and state space can be further simplified. The goal in simplifying the dynamics is to (1) preserve the dynamics of the critical slow variables and experimental observables, and (2) discard non-essential details. Spectral gaps in the sequence of eigenvalues (time scales) naturally identify the slow modes to include when coarse graining. When an n-state discrete stochastic model is coarse grained to an m-state discrete stochastic model, the coarse graining errors are governed by the spectral gap, τ_{m+1}/τ_m. See Figure 14.6.1 for an illustration of spectral gaps in discrete master equations and in Markov State Models. Some of the main approaches for coarse graining MSMs are outlined below.

- Perron-Cluster-Cluster-Analysis (PCCA) [61,62]: PCCA successively divides space according to the nodes of eigenvectors, i.e. into regions where the eigenvectors are positive vs. negative. PCCA may mistakenly fragment certain long lived states because boundaries created at an early stage cannot be adjusted in later stages. For example, the optimal two-state model in a system with reactant (A), product (B), and intermediate (I) may cut the intermediate state. The cut through state I becomes permanent even though A|I and I|B cuts would be better. Readers are directed to the PCCA+ method [63] which resolves some problems with PCCA.
- Metastability maximization: The algorithm by Chodera et al. [64] iteratively groups states according to an overall metastability criterion. Overall metastability is defined as the sum of metastabilities for individual states: $M(\tau) = \sum_i T_{ii}(\tau)$ where $\mathbf{T}$ is the transition matrix from an MSM. The algorithm involves a reclustering of the original configurations within each microstate making it possible to achieve very fine resolution of the optimal lumped state boundaries.
- Bayesian Agglomerative Clustering Engine (BACE) [4,65]: BACE uses a Kullback-Leibler divergence criterion to identify clusters which are part of rapidly mixing groups of states. Iterative application of the algorithm constructs hierarchically lumped states that preserve the relative time scales of different transitions. Additionally, the Bayesian framework of BACE allows the lumping process to account for uncertainties in the state-to-state rate parameters.

Applications of MSMs

Once an MSM that faithfully reproduces the dynamics has been parameterized, it can be used for many further analyses: computing the overall rate, interpreting experimental time correlation functions, identifying transitions that are dynamical bottlenecks, and computing fluxes through specific pathways in the network of intermediates. Figure 14.5.4 shows an MSM for a relatively small protein [66]. The MSM in Figure 14.5.4 has compressed thousands of degrees of freedom, local minima, and pathways into an interconnected network of just 14 states. The MSM approach provides a compact model that correctly describes both the thermodynamics and the dynamics.

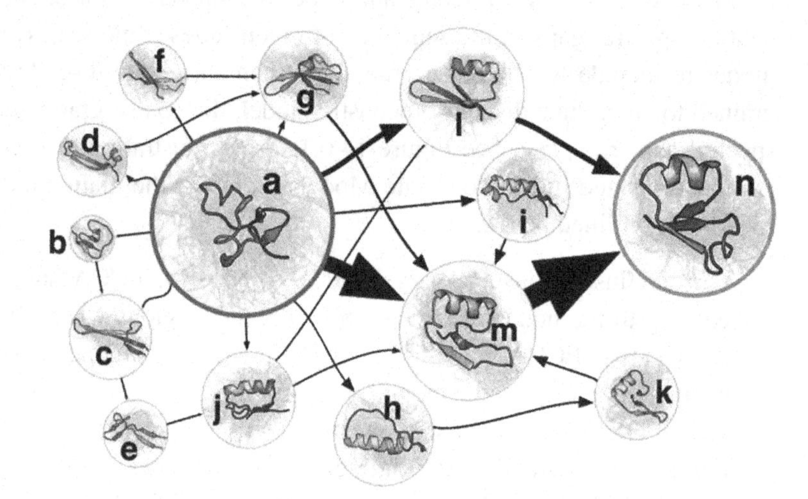

Figure 14.5.4: Showing a "small" 14 state MSM from millisecond time scale folding simulations of NTL9. Arrows convey the contributions of each path to the overall folding rate [Voelz et al. *J. Am. Chem. Soc.* 132, 1526 (2010)].

Limitations of MSMs

Current MSM constructions for biomolecular conformational transitions begin with fully atomistic MD simulations at a specific temperature, ionic strength, pH, etc. Whether the simulation data comes from one long MD run or from adaptively generated MD runs, the state-to-state transition rates are basically a summary of the spontaneous events observed in MD simulations. The MSM does not reveal the nature of the transition states along pathways between states in the MSM. MSMs therefore cannot (yet) predict how the transition rates will change with temperature, ionic strength, pH, etc. [67], except via different MSMs from simulation data at different conditions.

In simulations where the solvent and protein temperature are separately controlled, the protein structure fluctuations were more sensitive to solvent temperature than to protein temperature [68]. The interpretation ascribed to these results is that solvent fluctuations induce conformational motions of proteins [69].[8] This interpretation is potentially troubling for MSMs that omit solvent degrees of freedom at the very first stage of clustering. Despite the solvent motions, the

[8] An alternative explanation is that the two ways of creating a strange T-imbalance may have two different effects on the protein-water interfacial free energy.

slow large scale motions are probably well-described given an adequate lag time. Chapman-Kolmogorov tests do suggest this is the case [43]. On the other hand, metastable states that involve solvent drying transitions [70] might become lumped or hidden (in the Markov model sense) by starting from a state space defined only by protein backbone atoms. There are now methods to explicitly include solvent degrees of freedom in the state definitions [50]. Computational methods for building, validating, and using MSMs and discrete master equations are changing rapidly. Superior methods will surely supplant those this introductory chapter before this book is published.

14.6 Spectral theory

Spectral theories concern the eigenvalues and eigenstates of matrices and linear operators. The most important spectral theorem for our purposes is that operators which are Hermitian (or Hermitian-similar) [71] yield real eigenvalues and a complete basis of orthogonal eigenstates. As shown below, the spectral theorem is applicable to all finite-dimensional master equations with state-to-state transitions that obey detailed balance. Spectral analyses of master equations have a long history in chemical physics, starting with works by Shuler [3,72–74]. Spectral theories are now used in kinetic lumping strategies [75,76], in analysis of time correlation functions [44,77], and in analysis of Markov State Models [4,43].

Consider a system for which the equilibrium distribution is known and normalized, i.e. for which $\sum_i p_{eq}(i) = 1$. Further suppose that all states have a non-zero probability at equilibrium, $p_{eq}(i) > 0$ for all i. Start with the detailed balance relation, $w_{i \leftarrow j} p_{eq}(j) = w_{j \leftarrow i} p_{eq}(i)$, and divide both sides by $\left[p_{eq}(i) p_{eq}(j)\right]^{1/2}$. The result is an equation

$$p_{eq}^{-1/2}(i)\, w_{i \leftarrow j}\, p_{eq}^{+1/2}(j) = p_{eq}^{-1/2}(j)\, w_{j \leftarrow i}\, p_{eq}^{+1/2}(i) \tag{14.6.1}$$

which suggests the definition [78]

$$w_{ij}^S \equiv p_{eq}^{-1/2}(i)\, w_{i \leftarrow j}\, p_{eq}^{+1/2}(j) \tag{14.6.2}$$

where the matrix elements are symmetric: $w_{ij}^S = w_{ji}^S$. Define the corresponding symmetrized matrix $\mathbf{W}^S \equiv \mathbf{P}_{eq}^{-1/2} \mathbf{W} \mathbf{P}_{eq}^{+1/2}$ with $\mathbf{P}_{eq} = diag[p_{eq}(1), p_{eq}(2), \cdots, p_{eq}(n)]$. After the transformation, the master equation becomes

$$\frac{d}{dt}\mathbf{p}^S = -\mathbf{W}^S \mathbf{p}^S \tag{14.6.3}$$

where $\mathbf{p}^S \equiv \mathbf{P}_{eq}^{-1/2}\mathbf{p}$. The symmetrized formulation reveals, via the spectral theorem, that $\mathbf{W}^S$ has real eigenvalues and orthogonal eigenvectors:

$$\mathbf{W}^S \boldsymbol{\psi}_m^S = \tau_m^{-1} \boldsymbol{\psi}_m^S \tag{14.6.4}$$

with the first eigenvector (by convention) being $\psi_1^S(i) = p_{eq}^{1/2}(i)$. The first eigenvalue is zero and the rest are positive with increasing magnitudes

$$0 = \tau_1^{-1} < \tau_2^{-1} \le \tau_3^{-1} \le \tau_4^{-1} \cdots \le \tau_n^{-1} \tag{14.6.5}$$

so that the nth eigenvector corresponds to the fastest relaxation and the first (equilibrium) eigenvector never relaxes. The $\mathbf{W}^S$ eigenvectors and eigenvalues can be used to compute averages, time-correlation functions, and other properties. The symmetric formulation is computationally convenient because of special algorithms for working with symmetric matrices. However, all vectors in the symmetric formulation are scaled by factors of $\mathbf{P}_{eq}^{1/2}$ which are rather hard to interpret.

Alternatively, we can define a new inner product, right eigenvectors, and left eigenvectors so that $\mathbf{W}$ itself behaves like a self-adjoint operator. First, let us resolve a slight notational problem. We have been using $p(i)$ to indicate the ith component of vector $\mathbf{p}$. To avoid yet additional subscripts, remember that

- "0" and "t" are times,
- "i", "j", and "n" refer to components of a vector, and
- "m", "k", and "ℓ" refer to different eigenvectors.

The eigen*values* of $\mathbf{W}^S$ are identical to those of $\mathbf{W}$ because equation (14.6.2) is a similarity transformation. The left- and right- eigen*vectors* of the original matrix $\mathbf{W}$ can be identified by inserting $\mathbf{W}^S = \mathbf{P}_{eq}^{-1/2}\mathbf{W}\mathbf{P}_{eq}^{+1/2}$ into equation (14.6.4). Rearranging the factors of $\mathbf{P}_{eq}^{1/2}$ gives

$$\mathbf{W}\boldsymbol{\psi}_m^R = \tau_m^{-1}\boldsymbol{\psi}_m^R \quad with \quad \psi_m^R(i) = p_{eq}^{1/2}(i)\,\psi_m^S(i) \tag{14.6.6}$$

and

$$\boldsymbol{\psi}_m^L\mathbf{W} = \boldsymbol{\psi}_m^L\tau_m^{-1} \quad with \quad \psi_m^L(i) = p_{eq}^{-1/2}(i)\,\psi_m^S(i) \tag{14.6.7}$$

Note the three representations of the equilibrium eigenstate: $\psi_1^R(i) = p_{eq}(i)$, $\psi_1^S(i) = p_{eq}(i)^{1/2}$, and $\psi_1^L(i) = 1$. The $\boldsymbol{\psi}_m^R$ are wonderful for visualization of dynamical modes and time scales, as illustrated in Figure 14.6.1 from Prinz et al. [43] Note that the eigenfunctions change sign at the location corresponding to the bottleneck in the corresponding relaxation process. Later chapters will show that the left eigenfunctions (which also change sign at bottlenecks) provide natural reaction coordinates.

Let us define inner products for any two vectors $\boldsymbol{\varphi}$ and $\boldsymbol{\vartheta}$ by [3,73,79]

$$\langle\varphi|\,\vartheta\rangle \equiv \sum_i \varphi(i)\vartheta(i)/p_{eq}(i) \tag{14.6.8}$$

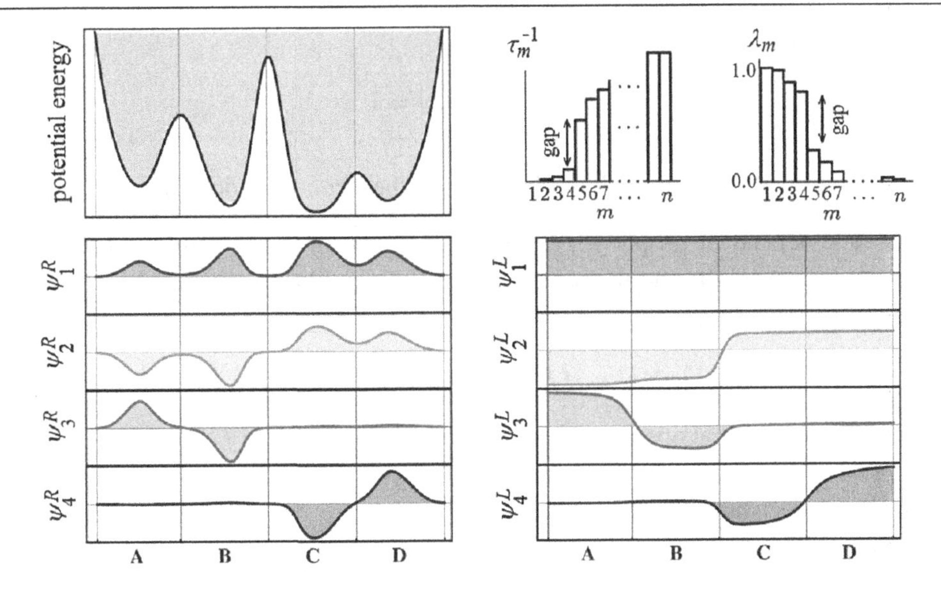

Figure 14.6.1: Eigenvectors and eigenvalues from diffusion on a one dimensional potential with four metastable states, A, B, C, and D. The state space has been discretized using 100 evenly spaced points along the x-axis. The 96 fastest modes (not shown) are separated by a wide spectral gap from the first four modes. ψ_1^R is the equilibrium distribution. ψ_2^R corresponds to relaxation from the lumped (A, B) basin to the lumped (C, D) basin via the transition state between B and C. ψ_3^R corresponds to relaxation between A and B. Finally, ψ_4^R corresponds to relaxation between C and D. The left eigenvectors and (schematic) eigenvalues of W and T are also shown. [Adapted from Prinz et al. *J. Chem. Phys.* (2011).]

Consistent with the left and right eigenvector definitions above, the "ket" vector has unscaled elements,

$$|\boldsymbol{\vartheta}\rangle = (\vartheta(1), \cdots, \vartheta(n)),$$

and the "bra" includes the factors of $p_{eq}(i)^{-1}$,

$$\langle\boldsymbol{\varphi}| = (\varphi(1)/p_{eq}(1), \cdots, \varphi(n)/p_{eq}(n)).$$

For example, the left and right eigenvectors of $\mathbf{W}$ are $\langle\boldsymbol{\psi}_m|$ and $|\boldsymbol{\psi}_m\rangle$. The formal master equation solution $\mathbf{p}(t) = e^{-\mathbf{W}t}\mathbf{p}(0)$ can also be written in terms of the left and right eigenvectors:

$$|\mathbf{p}(t)\rangle = \sum_{m=1}^{n} e^{-t/\tau_m} |\boldsymbol{\psi}_m\rangle\langle\boldsymbol{\psi}_m|\mathbf{p}(0)\rangle \tag{14.6.9}$$

From this equation, one can identify the discrete state-to-state Green's function as the ji-element of $\sum_{m=1}^{n} |\boldsymbol{\psi}_m\rangle\langle\boldsymbol{\psi}_m| e^{-t/\tau_m}$ [44,72,74]. The special case $t = 0$ corresponds to the useful completeness relation

$$\sum_m |\psi_m\rangle\langle\psi_m| = \mathbf{1}$$

i.e. the i, j element of the operator $\sum_m |\psi_m\rangle\langle\psi_m|$ is the Kronecker delta δ_{ij}.

Consider a state-dependent property $f(i)$ where the state of the system $i = 1, 2, 3, ..., $ or n evolves in time according to the master equation. The equilibrium average of f can be computed as

$$\langle f\rangle = \langle\psi_1|\,\hat{f}\,|\vartheta\rangle \tag{14.6.10}$$

where $\hat{f}$ acts like a diagonal operator on any vector ϑ, e.g. $\hat{f}\,|\vartheta\rangle = (f(1)\vartheta(1), \cdots, f(n)\vartheta(n))$ and/or $\langle\vartheta|\,\hat{f} = (f(1)\vartheta(1)/p_{eq}(1), \cdots, f(n)\vartheta(n)/p_{eq}(n))$. With the bra-ket notation, it is important to distinguish between thermal averages which are denoted $\langle\cdot\rangle$, and inner products which are denoted $\langle\cdot|\,\cdot\rangle$

Time correlation functions can also be computed using the eigenvectors and eigenvalues [44].

$$\begin{aligned}\langle g(t)f(0)\rangle &= \sum_{i,j=1}^{n} g(j)p(j,t|i,0)f(i)p_{eq}(i) \\ &= \sum_m \langle\psi_1|\,\hat{g}\,|\psi_m\rangle\langle\psi_m|\,\hat{f}\,|\psi_1\rangle e^{-t/\tau_m}\end{aligned} \tag{14.6.11}$$

The Markov State Model (MSM) literature uses equivalent expressions, but with different notations [4,43]. The main required change is that the eigenvalues of the transition matrix $\mathbf{T}(\tau) = exp[-\mathbf{W}\tau]$ are the lag-time dependent quantities $\lambda_m(\tau) = exp[-\tau/\tau_m]$.

Example: Single exponential decay of $\langle\psi_k^L(t)\psi_k^L(0)\rangle$

Let us more generally examine time correlations of the left eigenfunctions, e.g. $\langle\psi_k^L(t)\psi_\ell^L(0)\rangle$. Here $\psi_k^L(t)$ refers to the value of $\psi_k^L(i(t))$, so that $\psi_k^L(t)\psi_\ell^L(0)$ is a product of two scalars, and not a vector dot product. However, equation (14.6.11) for the correlation function does require the inner product

$$\langle\psi_1|\,\hat{\psi}_k^L\,|\psi_m\rangle = \sum_i \psi_1^L(i)\psi_k^L(i)\psi_m^R(i) = \delta_{km}.$$

Note that $\psi_1^L(i) = 1$ has helped to simplify $\langle\psi_1|\,\hat{\psi}_k^L\,|\psi_m\rangle$. For similar reasons,

$$\langle\psi_m|\,\hat{\psi}_\ell^L\,|\psi_1\rangle = \delta_{m\ell}.$$

Now inserting these inner products into equation (14.6.11) gives

$$\left\langle \psi_k^L(t)\psi_\ell^L(0)\right\rangle = \sum_m \delta_{km}\delta_{m\ell}e^{-t/\tau_m} = \delta_{k\ell}e^{-t/\tau_k}.$$

The left eigenfunctions therefore have autocorrelation functions that decay as pure single exponentials. Moreover cross correlations of the left eigenfunctions identically vanish, indicating that motion along each left eigenfunction is dynamically separable from motions along the other left eigenfunctions. Note that the longest time scales are associated with motions that change the value of ψ_2^L. Thus $\psi_2^L(i)$ is the ideal reaction coordinate for a discrete master equation or MSM.

Variational two-state theory (V2ST)

Recall that variational transition state theory (VTST) optimized the dividing surface so as to maximize the TST rate constant. We also saw in Chapter 13 that an optimized dividing surface maximizes the transmission coefficient, i.e. the plateau in a reactive flux correlation function. The reactive flux formalism quantifies *inertial* reactive flux along a continuous reaction coordinate. The reactive flux formalism cannot be directly used to analyze a discrete master equation. However, a similar correlation function analysis does provide a systematic way to partition a discrete MSM into an optimal two-state model [44,80,81].

The goal of variational two-state theory (V2ST) is to optimally lump the n discrete states of an MSM into two mutually exclusive groups, set "A" and set "B". To find the optimal grouping, consider a trial grouping such that all states i in the model belong to one and only one of the lumped sets A and B. Define h_B as

$$h_B(i) = \begin{cases} 0 & if \quad i \in A \\ 1 & if \quad i \in B \end{cases} \tag{14.6.12}$$

where i is one of the discrete states in the master equation or MSM. In the discussion that follows, we ask the reader to infer the nature (time, state index, or eigenvector) of arguments and subscripts of h_B and ψ based on context and on the conventions listed in section 14.6. Equation (14.6.11) can be used to compute a correlation function which identifies the lifetime of state B:

$$\langle h_B(t)h_B(0)\rangle = \sum_{m=1}^{n} \left\langle \psi_1 \middle| \hat{h}_B \middle| \psi_m \right\rangle^2 exp[-t/\tau_m]$$

Here $\hat{h}_B$ acts like the diagonal operator $\hat{h}_B |\vartheta\rangle = (h_B(1)\vartheta(1), \cdots, h_B(n)\vartheta(n))$. The limiting values of $\langle h_B(t)h_B(0)\rangle$ can be used to construct a normalized autocorrelation function [44]:

$$\begin{aligned} C_{BB}(t) &= \frac{\langle h_B(t)h_B(0)\rangle - \langle h_B(\infty)h_B(0)\rangle}{\langle h_B(0)h_B(0)\rangle - \langle h_B(\infty)h_B(0)\rangle} \\ &= \frac{\sum_{m=2}^{n} \left\langle \psi_1 \middle| \hat{h}_B \middle| \psi_m \right\rangle^2 e^{-t/\tau_m}}{\left\langle h_B^2 \right\rangle - \langle h_B \rangle^2} \end{aligned} \tag{14.6.13}$$

An initial fast decay in $C_{BB}(t)$ is followed by a long decay over time scale τ_2, reminiscent of the time correlation functions for inertial barrier crossing (see Figure 14.6.2). The optimal division into two sets A and B is that which maximizes the amplitude of the most slowly relaxing $m = 2$ mode, i.e.

$$\max_{trial\ h_B(i)} \frac{\left\langle \psi_1 \middle| \hat{h}_B \middle| \psi_2 \right\rangle^2}{\langle h_B \rangle (1 - \langle h_B \rangle)} \tag{14.6.14}$$

Buchete and Hummer note that approximately optimal sets can be constructed based on the sign of $\psi_2^R(i)$:

$$h_{B,V2ST}(i) \approx \begin{cases} 0 & if \quad \psi_2^R(i) < 0 \\ 1 & if \quad \psi_2^R(i) \geq 0 \end{cases} \tag{14.6.15}$$

Even for the exactly optimal division into A and B sets, the correlation function $\langle h_B(t)h_B(0)\rangle$ contains several relaxation times [44]. Thus no sharp boundary between lumped states A and B in a discrete master equation can perfectly replicate the slowest relaxation of the ψ_2 eigenfunction.

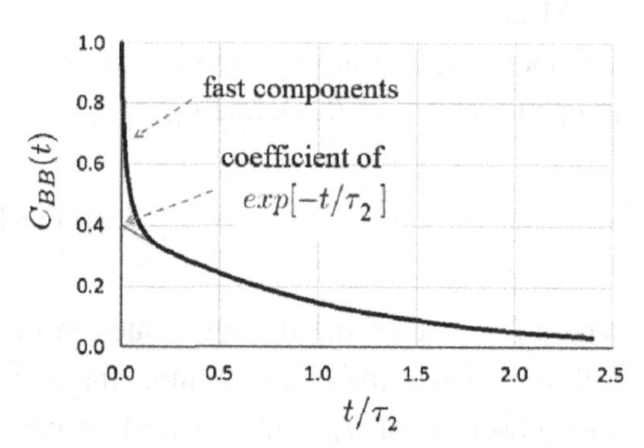

Figure 14.6.2: Components of ψ_3, ψ_4, ... in the $C_{BB}(t)$ correlation function decay rapidly leaving the slowly decaying ψ_2 component. Choosing the A, B boundary to coincide with sign changes of ψ_2 approximately maximizes the coefficient of $exp[-t/\tau_2]$.

Exercises

1. Consider a set of first order (reversible) reactions in a constant volume batch reactor. Show that the mole fractions $\mathbf{x}(t)$ evolve according to an equation $\mathbf{x}'(t) = -\mathbf{K}\mathbf{x}(t)$ which

is exactly analogous to the discrete master equation $\mathbf{p}'(t) = -\mathbf{W}\mathbf{p}(t)$. Show that evolution of $\mathbf{p}(t)$ [or $\mathbf{x}(t)$] from any initial condition $\mathbf{p}(0)$ [or $\mathbf{x}(0)$] proceeds entirely within the hyperplane passing through each of the pure states. See the figure below for an illustration. Hint: consider inner products with $\langle \boldsymbol{\psi}_1 |$.

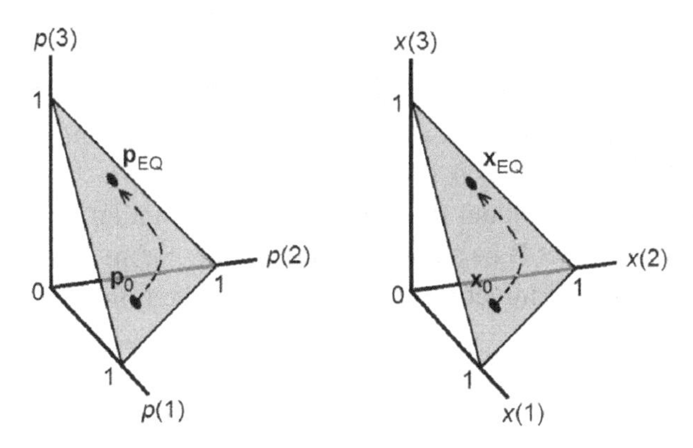

2. Consider a continuous flow stirred tank reactor in which a set of first order reactions occur between reactants in a liquid phase.
   (a) Show that the concentrations in the reactor satisfy

$$\frac{d\mathbf{c}}{dt} = -\mathbf{K}\mathbf{c} - \tau^{-1}\mathbf{c} + \tau^{-1}\mathbf{c}_{in} \tag{14.6.16}$$

   where $\mathbf{K}$ is the same rate matrix as that for the batch reactor, $\mathbf{c}$ is the vector of species concentrations in the reactor, $\mathbf{c}_{in}$ is a constant vector of concentrations in the inlet stream, and τ is the residence time, i.e. the ratio of reactor volume to volumetric flowrate through the reactor.
   (b) Show that the eigenvectors of $\mathbf{K}$, i.e. the $\boldsymbol{\psi}_m^R$, are also eigenvectors of $\mathbf{K} + \tau^{-1}\mathbf{1}$, and give the new spectrum of eigenvalues in terms of the eigenvalues of $\mathbf{K}$.
   (c) Solve equation (14.6.16) for $\mathbf{c}(t)$. How would your solution change if $\mathbf{c}_{in}$ was a time-varying input?
3. Develop a master equation for the number $n_{\mathbf{AB}}$ of $\mathbf{AB}$ molecules in a small system where $\mathbf{A}$ and $\mathbf{B}$ molecules associate via the reversible reaction $\mathbf{A} + \mathbf{B} \rightleftarrows \mathbf{AB}$. Let the forward and backward rate constants be k_f and k_b, respectively. Let the initial numbers of $\mathbf{A}$, $\mathbf{B}$ and $\mathbf{AB}$ molecules be $n_{\mathbf{A}}$, $n_{\mathbf{B}}$, and zero, respectively. Find the equilibrium distribution, the slowest relaxation time, and the most slowly relaxing vector.
4. Revisit exercise 3 in Chapter 4 which explored the model of Monod et al. for allostery and cooperativity in tetrameric enzymes. That question (and Monod et al.) suggest the quasi-equilibrium rate law as $\mathfrak{r} = k_{max}Y$ where k_{max} is the rate of reaction per occupied R-type site. Suppose that adsorption of substrate $\mathbf{A}$ at each site is diffusion controlled

with rate constant k_D. Suppose that the forward rate of conversion from R to T is k_f with $k_f \ll k_D$.

(a) Write down the rate equations (rows of a master equation) that you would use to model the stochastically evolving state of a single enzyme. Hint: count but don't overcount the number of ways to choose which site becomes occupied.

(b) Let the states ($\mathbf{A}_4R$), ($\mathbf{A}_3R$), ($\mathbf{A}_2R$), ($\mathbf{A}_1R$), (R), (T), ($\mathbf{A}_1T$), ($\mathbf{A}_2T$), ($\mathbf{A}_3T$), and ($\mathbf{A}_4T$) correspond to positions -5, -4, -3, -2, -1, +1, +2, +3, +4, and +5 along an x-axis. Sketch the stochastic trajectory x vs. t for an individual enzyme in the case where $K = 1$, $c = 1$, $y \gg 1$, and $k_f < k_{max} < k_D[\mathbf{A}]$.

5. Consider the following model for infection in a population of size n. Let **A** be an uninfected individual and let **B** be an infected individual. Suppose that the processes by which individuals contract the infection and recover have the following rates.

$$\begin{array}{llllll} \mathbf{B} & \rightarrow \mathbf{A} & r_0 & = & k_R n_\mathbf{B} & \text{(recovery)} \\ \mathbf{A}+\mathbf{B} & \rightarrow 2\mathbf{B} & r_2 & = & k_I n_\mathbf{A} n_\mathbf{B} & \text{(infection)} \end{array}$$

where $n_\mathbf{A}$ is the number of healthy individuals, and $n_\mathbf{B}$ is the number of infected individuals. Use the rate equations to write down a deterministic ODE for the fraction $x = n_\mathbf{A}/n$. According to the deterministic model, what values of $k_I n/k_R$ give rise to two steady states? What does a stability analysis say about each of the two steady states? How/why does the discrete stochastic model contradict the deterministic stability analysis. Choose several values of $k_I n/k_R$ and run kMC simulations starting from a 100% infected population. For each value of $k_I n/k_R$, estimate the mean time for the disease to spontaneously eradicate itself.

6. [M. Joswiak] For crystal growth at low supersaturations, solute molecules attach and detach with rates j^+ and j^- at "kink sites" along terrace edges (see figure (b)). Attachment at kink sites is favorable under growth conditions, but kink sites must first be created by the unfavorable attachment of a first solute to the straight portion of a terrace edge (see figure (a)). The attachment and detachment rates for the kink creation step are j_a^+ and j_a^- respectively.

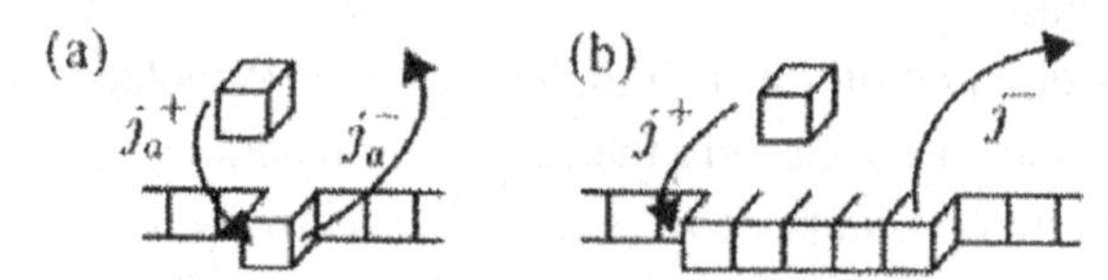

Contiguous solutes on a terrace edge form a 1-dimensional (1D) nucleus. While the nucleus is short, there is a chance that all progress will unravel back to the original straight edge. Let n be the number of solutes in a 1D nucleus. Let the free energy as a function of nucleus size be $F(n=0) = 0$ and $F(n) = 2\phi - n\Delta\mu$ for $n \geq 1$ where $2\phi > 0$ is the energy to create the two kink sites and $\Delta\mu > 0$ is the chemical potential driving force for transfer of solutes from solution to the crystal. Assume that $j_a^+ = j^+$. Construct a master equation and use it to compute the rate of 1D nucleation. Hint: use the Becker-Doering equation.

7. Chapter 13 discussed methods for computing transmission coefficients from $\overline{h}_B(t)$ and the plateau in its derivative $d\overline{h}_B(t)/dt$ after a short molecular relaxation time. Some investigators have instead estimated the transmission coefficient from the plateau in $\overline{h}_B(t)/t$. Compare this method for computing the transmission coefficient to the sliding window approach for counting transitions in the construction of MSMs.
8. Consider time resolved measurements for two state-dependent properties $f(i)$ and $g(i)$ where the state of the system $i = 1, 2, 3, \ldots$, or n evolves in time according to the master equation. If property g is measured at varying times after f, then define the correlation function

$$C_{fg}(t) \equiv \frac{\langle f(0)g(t)\rangle - \langle f\rangle \langle g\rangle}{\langle fg\rangle - \langle f\rangle \langle g\rangle}$$

Express $C_{fg}(t)$ in terms of eigenvalues, eigenfunctions, and inner products. Check your answer against the result in Buchete and Hummer [44].

9. Show that a transition rate matrix $\mathbf{W}$ obeys detailed balance if and only if $\langle \mathbf{f} | \mathbf{W}\mathbf{g}\rangle = \langle \mathbf{W}\mathbf{f} | \mathbf{g}\rangle$ for all $\mathbf{f}$ and $\mathbf{g}$ where the inner product is that of equation (14.6.8).
10. [G. Poon] Survival probability models are often used to predict the distribution of waiting times before a reaction/event happens.
    (a) Show that the probability that an event with first order rate constant k will not occur during a waiting time t satisfies a Poisson distribution, $p_0(t) = \exp[-kt]$. Give the corresponding $\mathbf{W}$-matrix for a discrete master equation with two states (event happened or event has not yet happened).
    (b) Bell's formula (see Chapter 22.3) predicts that the rate constant k can be accelerated by a force f directed along the reaction coordinate, $k(f) = k(0)\exp[\beta f q_0^{\ddagger}]$ where $q_0^{\ddagger}$ is the location of the transition along the reaction coordinate with zero force. What is the distribution of waiting times under conditions where the force gradually varies with time, i.e. where $f = f(t)$?

References

[1] R. Zwanzig, Nonequilibrium Statistical Mechanics, Oxford University Press, Oxford, 2001.
[2] J. van Zon, P.R. ten Wolde, Phys. Rev. Lett. 94 (2005) 128103.
[3] N.G. van Kampen, Stochastic Processes in Physics and Chemistry, Elsevier, Amsterdam, 2007.
[4] G.R. Bowman, V.S. Pande, F. Noe, in: G.R. Bowman, V.S. Pande, F. Noe (Eds.), An Introduction to Markov State Models and Their Application to Long Timescale Molecular Simulation, Springer, Berlin, Heidelberg, 2013.
[5] B. Hayes, Am. Sci. 101 (2013) 92–97.

[6] L.P. Kadanoff, Statistical Physics: Statics, Dynamics, and Renormalization, World Scientific, Singapore, 2000.
[7] S.S. Andrews, T. Dinh, A.P. Arkin, in: Encyclopedia of Complexity and Systems Science, Springer, New York, 2009, pp. 8730–8749.
[8] C. Dellago, M. Naor, G. Hummer, Phys. Rev. Lett. (ISSN 0031-9007) 90 (Mar. 2003) 105902.
[9] Y. Peng, J.M.J. Swanson, S.-G. Kang, R. Zhou, G.A. Voth, J. Phys. Chem. B 119 (2014) 9212–9218.
[10] D.T. Gillespie, Markov Processes: An Introduction for Physical Scientists, Academic Press, Boston, 1992.
[11] D. Kashchiev, Nucleation: Basic Theory with Applications, Butterworth-Heinemann, Oxford, 2000.
[12] K.F. Kelton, A.L. Greer, Nucleation in Condensed Matter: Applications in Materials and Biology, Elsevier, Amsterdam, 2010.
[13] J.W. Gibbs, Trans. Conn. Acad. Arts Sci. 3 (1876) 108–248.
[14] J.W. Gibbs, Trans. Conn. Acad. Arts Sci. 16 (1878) 343–524.
[15] D. Kashchiev, G.M. van Rosmalen, Cryst. Res. Technol. 38 (2003) 555–574.
[16] V. Agarwal, B. Peters, Adv. Chem. Phys. 155 (2014) 97–160.
[17] M. Volmer, A. Weber, Z. Phys. Chem. 119 (1926) 277–301.
[18] R. Becker, W. Doring, Ann. Phys. 416 (1935) 719–752.
[19] S. Raimondeau, D.G. Vlachos, Comput. Chem. Eng. 26 (2002) 965–980.
[20] K. Reuter, M. Scheffler, Phys. Rev. B 73 (2006) 45433.
[21] J. Rogal, K. Reuter, M. Scheffler, Phys. Rev. B 77 (2008) 155410.
[22] L. Xu, D. Mei, G. Henkelman, J. Chem. Phys. 131 (2009) 244520.
[23] G. Henkelman, H. Jonsson, J. Chem. Phys. 115 (2001) 9657.
[24] B. Peters, N.E.R. Zimmermann, G.T. Beckham, J.W. Tester, B.L. Trout, J. Am. Chem. Soc. 130 (2008) 17342–17350.
[25] A. Pedersen, H. Jonsson, Acta Mater. 57 (2009) 4036–4045.
[26] J.P. Simmons, Y. Wen, C. Shen, Y.Z. Wang, Mater. Sci. Eng. A 365 (2004) 136–143.
[27] A.B. Bortz, M.H. Kalos, J.L. Lebowitz, J. Comput. Phys. 18 (1975) 10–18.
[28] D.T. Gillespie, J. Comput. Phys. 22 (1976) 403–434.
[29] M.P. Allen, D.J. Tildesley, Computer Simulation of Liquids, Clarendon Press, Oxford, UK, 1987.
[30] F. Schlogl, Z. Phys. 253 (1972) 147–161.
[31] R. Erban, S.J. Chapman, I.G. Kevrekidis, T. Vejchodsky, SIAM J. Appl. Math. 70 (2009) 984–1016.
[32] K. Reuter, D. Frenkel, M. Scheffler, Phys. Rev. Lett. 93 (2004) 116105.
[33] D. Schultz, A.M. Walczak, J.N. Onuchic, P.G. Wolynes, Proc. Natl. Acad. Sci. USA 105 (49) (2008) 19165–19170.
[34] M.A. Gibson, J. Bruck, J. Phys. Chem. A 104 (2000) 1876–1889.
[35] A. Chatterjee, D.G. Vlachos, J. Comput.-Aided Mater. Des. 14 (2007) 253–308.
[36] E. Martinez, J. Marian, M.H. Kalos, J.M. Perlado, J. Comput. Phys. 227 (2008) 3804–3823.
[37] J.P. Nilmeier, J. Marian, Comput. Phys. Commun. 185 (2014) 2479–2486.
[38] M.A. Novotny, Phys. Rev. Lett. 74 (1995) 1–5.
[39] K.A. Fichthorn, Y. Lin, J. Chem. Phys. 138 (2013) 164104.
[40] B. Harland, S.X. Sun, J. Chem. Phys. 127 (2007) 104103.
[41] N. Eidelson, B. Peters, J. Chem. Phys. 137 (2012) 94106.
[42] S.A. Trygubenko, D.J. Wales, J. Chem. Phys. 124 (2006) 234110.
[43] J.-H. Prinz, H. Wu, M. Sarich, B. Keller, M. Senne, M. Held, J.D. Chodera, C. Schutte, F. Noe, J. Chem. Phys. 134 (2011) 174105.
[44] N.-V. Buchete, G. Hummer, J. Phys. Chem. B 112 (2008) 6057.
[45] K.A. Beauchamp, G.R. Bowman, T.J. Lane, L. Maibaum, I.S. Haque, V.S. Pande, J. Chem. Theory Comput. 7 (2011) 3412–3419.
[46] M.K. Scherer, B. Trendelkamp-Schroer, F. Paul, G. Perez-Hernandez, M. Hoffmann, N. Plattner, C. Wehmeyer, J.-H. Prinz, F. Noe, J. Chem. Theory Comput. 11 (2015) 5525–5542.
[47] G.L. Dirichlet, J. Reine Angew. Math. 40 (1850) 209–227.

[48] G. Voronoi, J. Reine Angew. Math. 133 (1907) 97–178.
[49] M. Levitt, J. Mol. Biol. 145 (1981) 251–263.
[50] C. Gu, H.-W. Chang, L. Maibaum, V.S. Pande, G.E. Carlsson, L.J. Guibas, BMC Bioinform. 14 (2013) S8.
[51] C.R. Schwantes, V.S. Pande, J. Chem. Theory Comput. 11 (2015) 600–608.
[52] F. Vitalini, F. Noe, B.G. Keller, J. Chem. Theory Comput. 11 (2015) 3992–4004.
[53] P. Majek, R. Elber, J. Chem. Theory Comput. 6 (2010) 1805–1817.
[54] S. Kirmizialtin, R. Elber, J. Phys. Chem. A 115 (2011) 6137–6148.
[55] C. Schutte, F. Noe, J. Lu, M. Sarich, E. Vanden-Eijnden, J. Chem. Phys. 134 (2011) 204105.
[56] T.S. van Erp, Adv. Chem. Phys. 151 (2012) 27–60.
[57] X. Huang, G.R. Bowman, S. Bacallado, V.S. Pande, Proc. Natl. Acad. Sci. USA 106 (2009) 19765–19769.
[58] D.J. Wales, Mol. Phys. 100 (2002) 3285.
[59] D.J. Wales, Mol. Phys. 102 (2004) 891–908.
[60] J.M. Carr, S.A. Trygubenko, D.J. Wales, J. Chem. Phys. 122 (2005) 234903.
[61] P. Deuflhard, W. Huisinga, A. Fischer, C. Schutte, Linear Algebra Appl. 315 (2000) 39–59.
[62] F. Noe, I. Horenko, C. Schutte, J.C. Smith, J. Chem. Phys. 126 (2007) 155102.
[63] P. Deuflhard, M. Weber, Linear Algebra Appl. 398 (2005) 161.
[64] J.D. Chodera, N. Singhal, V.S. Pande, K.A. Dill, W.C. Swope, J. Chem. Phys. 126 (2007) 155101.
[65] G. Bowman, J. Chem. Phys. 137 (2012) 134111.
[66] V.A. Voelz, G.R. Bowman, K.A. Beauchamp, V.S. Pande, J. Am. Chem. Soc. 132 (2010) 1526–1528.
[67] B. Peters, J. Phys. Chem. B 119 (2015) 6349–6356.
[68] D. Vitkup, D. Ringe, G.A. Petsko, M. Karplus, Nature 7 (2000) 34–38.
[69] M. Karplus, J.A. McCammon, Nat. Struct. Biol. 9 (2002) 646–652.
[70] B.J. Berne, J.D. Weeks, R.H. Zhou, Annu. Rev. Phys. Chem. 60 (2009) 85–103.
[71] G.A. Korn, T.M. Korn, Mathematical Handbook for Scientists and Engineers, Dover, Mineola, NY, 2000.
[72] E.W. Montroll, K.E. Shuler, Adv. Chem. Phys. 1 (1958) 361–399.
[73] K.E. Shuler, Phys. Fluids 2 (1959) 442.
[74] N.G. van Kampen, J. Stat. Phys. 17 (1977) 71–88.
[75] J. Wei, J.C.W. Kuo, Ind. Eng. Chem. Fundam. 8 (1969) 114–123.
[76] N. Lempesis, D.G. Tsalikis, G.C. Boulougouris, D.N. Theodorou, J. Chem. Phys. 135 (2011) 204507.
[77] G.C. Boulougouris, D.N. Theodorou, J. Chem. Phys. 130 (2009) 044905.
[78] E.P. Wigner, J. Chem. Phys. 22 (1954) 1912.
[79] A. Debnath, R. Chakrabarti, K.L. Sebastian, J. Chem. Phys. 124 (2006) 204111.
[80] D.J. Bicout, A. Szabo, Protein Sci. 9 (2000) 452–465.
[81] A. Berezhkovskii, G. Hummer, A. Szabo, J. Chem. Phys. 130 (2009) 205102.

CHAPTER 15

Continuous stochastic variables

> *"...now suppose that the masses are initially at rest, and examine the manner in which they acquire velocity under the impact of the projectiles. ...the general equation, applicable not merely to the initial and final, but to all stages of the acquirement of motion. ...$df/dt = d^2 f/du^2 + 2hd(uf)/du$."*
>
> ***Rayleigh, Philos. Mag. (1891)***

Periodic boundary conditions, however elegant, are an artificial construct. Every real system is part of a larger extended system, and at the boundary of every subsystem there are inevitable interactions with the surroundings. It can be shown [1] that even the slightest random interactions with the bath suffice to create a Boltzmann distribution from a subsystem with otherwise Hamiltonian dynamics. Therefore, real systems are intrinsically stochastic even if we ignore quantum mechanics.

This chapter begins with the historically and practically important Langevin equation, the first and simplest stochastic equation of molecular motion [2]. We will examine its properties and show how it can be equivalently studied using the parallel language of Fokker-Planck equations [2,3]. We will further show how the Langevin equation behaves in the "overdamped" limit of large friction [3]. We outline how discrete stochastic models from Chapter 14 can be converted into Fokker-Planck equations when all transitions are between closely neighboring states [4]. Finally, we revisit the discussion of spectral theory, now starting from a Fokker-Planck equation.

15.1 Inertial Langevin dynamics

Langevin's stochastic equation of motion began as an effort to model Brownian diffusion [5]. His original equation looked somewhat different from the one that bears his name today. There was no potential of mean force (PMF), but the central new elements were already present: friction and random forces. In modern applications, the Langevin equation is used to model a diverse range of phenomena in physics, biology, chemistry, and even economics. To model the dynamics of a small system coupled to its environment the Langevin equation needs a PMF, friction, and random forces. If the small system is the scalar reaction coordinate q, then [2]

$$m\ddot{q} = -\frac{\partial F}{\partial q} - m\gamma\dot{q} + R(t) \tag{15.1.1}$$

where $F(q)$ is the PMF, γ is a drag (friction) coefficient, and $R(t)$ is a randomly fluctuating force due to interactions of q with its environment. For now, assume that the random force $R(t)$ is perfectly Markovian, i.e. that the random force has no memory of its own history nor of the history of $q(t)$. The random forces on average are zero,

$$\langle R(t)\rangle = 0. \tag{15.1.2}$$

The fluctuation-dissipation theorem requires an additional relationship between the random kicks, the friction, and the temperature. Intuitively there are several reasons to suspect such a relationship. First, the random kicks and dissipative forces are both associated with coupling to the bath, and the bath properties depend on the temperature. Second, friction in the absence of random kicks would eventually drain the system of all energy until it reached an energy minimum corresponding to $T = 0$ K. To reach a thermal equilibrium, the random kicks must somehow compensate for the tendency of friction to dissipate thermal energy. To balance the friction, the strength of the kicks must increase in proportion to the friction and the temperature. The random forces must also depend on the mass of the particle because equipartition requires $\langle m(dq/dt)^2\rangle = k_B T$. The specific relationship required to balance fluctuations and dissipation is

$$\langle R(0)R(t)\rangle = 2m\gamma k_B T\delta[t] \tag{15.1.3}$$

Note that equation (15.1.1) uses $-\partial F/\partial q$ instead of the more typical notation $-\partial V/\partial q$. Assuming that the bath response is fast and independent of q is tantamount to assuming that the force is averaged over the rapid fluctuations in the bath. Therefore the PMF, especially for condensed phase processes, is more appropriate for the internal forces term. Accordingly, a Langevin model based on simulation data should be constructed from the free energy profile or PMF along the coordinate q and not from raw instantaneous forces.[1]

Equation (15.1.3) describes an infinitely fast noise, but in a simulation, the random force cannot decorrelate any faster than the integration time step. The usual practice is to generate the random force as a Gaussian random variable with mean zero and variance $2m\gamma k_B T/\Delta t$ [6]. Upon integration over one time step, the effect is the same as an instantaneous impulse of size $2m\gamma k_B T$. The Langevin equation is then numerically solved as a pair of coupled first order equations. A crude algorithm is

[1] Differences between the PMF and the Landau free energy can be important, especially when the variable q is not slow compared to other variables. For example, the PMF as a function of distance r between two molecules approaches zero as $r \to \infty$, while the free energy diverges to $-\infty$ as $r \to \infty$. Because diffusion along r is no slower than diffusion on the sphere of constant r, a thermodynamically and dynamically correct model should use the PMF and also include the "angle" variables in the spherical coordinate system.

$$v(t+\Delta t) = v(t) - \gamma \Delta t\, D\frac{\partial \beta F}{\partial q} - \gamma \Delta t\, v(t) + \sqrt{2m^{-1}k_B T \gamma \Delta t} \cdot \xi$$

$$q(t+\Delta t) = q(t) + v(t)\Delta t$$

where $D = k_B T/m\gamma$ and ξ is a Gaussian random number with zero mean and unit variance. In general, Langevin dynamics algorithms, like deterministic molecular dynamics algorithms, use forces at the current position to extrapolate the new positions and new velocities after a finite time step. The dimensionless time step $\gamma \Delta t$ should be small enough that the deterministic part of the force changes only slightly between time steps [7]. Langevin dynamics may be applied for a single degree of freedom (as described above), or for an assembly of particles (as described below). The Langevin dynamics algorithm below is from Allen and Tildesley [8].

Algorithm: Langevin dynamics

First define parameters that depend on the timestep Δt, friction γ, mass m, and temperature T:

$D = k_B T/m\gamma$

$c_0 = exp[-\gamma \Delta t]$

$c_1 = (1.0 - c_0)/(\gamma \Delta t)$

$c_2 = (1.0 - c_1)/(\gamma \Delta t)$

$\sigma_r^2 = \Delta t\, D[2 - (3 - 4c_0 + c_0^2)/(\gamma \Delta t)]$

$\sigma_v^2 = \gamma D(1 - c_0^2)$

$c_{rv} = D(1 - c_0)^2/(\sigma_r \sigma_v)$

Then, for each timestep,

and for each coordinate,

$r_1 = ran(0, 1)$
$r_2 = ran(0, 1)$
$g_1 = (-2lnr_1)^{1/2} cos(2\pi r_2)$
$g_2 = (-2lnr_1)^{1/2} sin(2\pi r_2)$
$g_r = \sigma_r g_1$
$g_v = \sigma_v (c_{rv} g_1 + (1 - c_{rv}^2)^{1/2} g_2)$
$f_j = -\partial V/\partial q_j$
$\Delta q_j = c_1 \Delta t\, v_j + c_2 \Delta t^2 f_j/m + g_r$
$\Delta v_j = (c_0 - 1.0)v_j + c_1 \Delta t\, f_j/m + g_v$
update positions and velocities

$$q_j = q_j + \Delta q_j$$
$$v_j = v_j + \Delta v_j$$

Langevin dynamics can be used in atomistic simulations to generate trajectories that sample the canonical distribution. Shirts [10] and Leimkuhler [7] examined the extent to which various Langevin dynamics algorithms and other thermostats sample the correct states when used with finite timesteps. Leimkuhler and coworkers have also developed metrics to gauge the degree to which thermostatted dynamics deviate from undisturbed energy conserving microcanonical trajectories [7]. Their analyses have identified more accurate (and still relatively simple) Langevin dynamics algorithms. Note that hydrodynamic coupling between particles is not included in simple Langevin dynamics algorithms, so the Langevin friction cannot accurately replace that from a real solvent. See Ermak and McCammon for an algorithm that includes hydrodynamic interactions [9].

We have seen that the fluctuation-dissipation theorem requires a special balance between friction and random forces. However, it does not prescribe an appropriate value of the friction constant. As the example below shows, the value chosen for the friction has a strong influence on the nature of dynamical trajectories.

Example: Langevin dynamics on a model surface

The four panels show Langevin dynamics trajectories initiated from the saddle point of the Muller-Brown model potential energy surface at four values of the friction coefficient γ. Paths were propagated forward and backward in time until they reached a critical distance from the minimum energy locations. All other parameters are unchanged between the four cases.

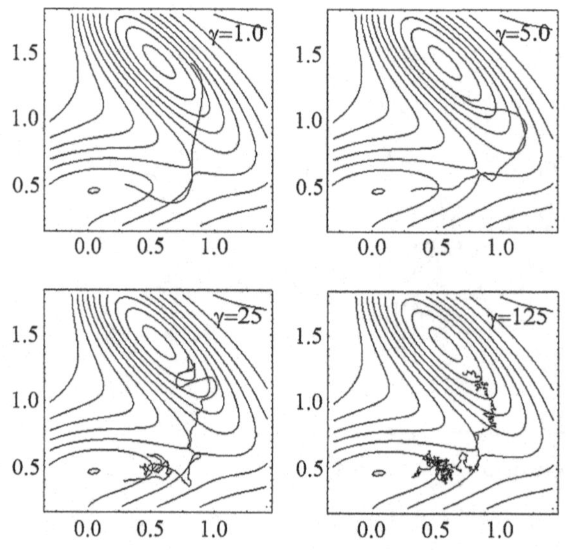

As γ increases the dynamics gradually lose their inertial characteristics and begin to resemble a random walk (diffusion) on the free energy landscape.

So what is the most appropriate Langevin friction? The Langevin thermostat provides a mechanism for heat transfer to/from the simulation box. However, dissipation by the Langevin thermostat does not occur via heat conduction or convection through the boundaries as in a natural system. Instead, dissipation occurs via a drag on every atom throughout the bulk simulation volume. Therefore the Langevin thermostat cannot precisely mimic real constant temperature processes with any choice of the friction constant. To model the real dynamics as closely as possible, some advocate using no friction at all. Deterministic dynamics are indeed necessary for testing certain dynamical properties [11,12], but recall from the opening remarks of this chapter that perfectly deterministic trajectories are also unrealistic. Most commercial and open source molecular dynamics codes recommend a Langevin friction that is independent of the simulation box size. However, the most natural choice for the dissipation rate depends on the thermal conductivity, heat capacity, and dimensions of the simulation box. For example, the friction should ideally be tuned so that the kinetic temperature fluctuations decay on time scale $\Delta t \sim \ell^2 C\rho/k$ where k is the thermal conductivity, ρ is the density, C is the heat capacity, and ℓ is the box side length. This prescription ensures that dissipation of spontaneous fluctuations occurs at a rate that matches the natural rate of heat dissipation by conduction. According to this prescription, the ideal friction constant should scale as $\gamma \sim \ell^{-2}$, consistent with intuition that thermostats are unnecessary for extremely large systems.

The Langevin equation (15.1.1) assumes that the bath decorrelation is instantaneous, an assumption that limits the types of processes that it can model. In the gas phase, the frequency of kicks from the bath is the collision frequency. One might model gas phase dynamics by setting the Langevin friction so that velocity correlations decay on the collision frequency time scale. However, the relaxation towards thermal equilibrium would then occur slowly and gradually, whereas the real system would evolve microcanonically between occasional abrupt collisions [13]. Clearly, Langevin dynamics is not realistic as a kinetic theory of gases even though it would generate the proper canonical ensemble [14].

More typically, the Langevin equation is used to model condensed phase dynamics. The fastest bath time scales are set by the intermolecular distance [~Å] and the thermal root mean square velocity [$(2k_BT/m)^{1/2} \sim 100$ m/s for small molecules at room temperature]. The typical collision time in the condensed phase is therefore $\sim 10^{-12}$ s. Whether the *ps* time scale can be considered fast depends on the process being modeled. 1.0 ps is extremely fast relative to the conformational transition time of a large biomolecule, so a Langevin model would be justified. However, 1.0 ps exceeds molecular vibrational periods and relaxation times. Therefore a simple Langevin model will not accurately describe the dynamics of bond breaking and bond formation in solution. On the other hand, chemical reaction dynamics can be modeled using a *generalized* Langevin equation in which the bath forces persist over a non-zero memory time. Generalized Langevin equations and the Grote-Hynes theory of reactions in solution are discussed in Chapter 17.

The solution to equation (15.1.1) for the special case of a constant force $\partial F/\partial q$ is particularly illuminating. Let us anticipate the emergence of a drift velocity[2]

$$v_D = -D\frac{\partial \beta F}{\partial q} \tag{15.1.4}$$

where the diffusion constant is that given by Einstein [15]

$$D = \frac{k_B T}{m\gamma} \tag{15.1.5}$$

Note that for constant $\partial F/\partial q$, there is no q-dependence in the Langevin equation. Therefore, in this special case, it can be written as a first order equation for the velocity

$$\dot{v} = -\gamma(v - v_D) + R(t)/m$$

where $v = \dot{q}$. The equation can be solved using the usual techniques for first order linear differential equations. The solution is

$$v(t) = v_0 e^{-\gamma t} + v_D(1 - e^{-\gamma t}) + \int_0^t dt' e^{-\gamma(t-t')} R(t')/m \tag{15.1.6}$$

Averaging over realizations of the random noise gives

$$\langle v(t)\rangle_R = v_D + (v_0 - v_D)e^{-\gamma t}$$

where the subscript R indicates an average over realizations of the random force, but not over the initial velocity. The velocity never actually settles to a constant average, but in the long time limit it fluctuates around v_D.

What is the mean squared displacement as a function of time? Recall that we must account for the non-zero limiting drift velocity. Starting from $\delta q(t) = \int_0^t v(t')dt'$, using equation (15.1.6), gives

$$\delta q(t) = v_D t + (v_0 - v_D)\gamma^{-1}(1 - e^{-\gamma t}) + \int_0^t dt' \int_0^{t'} dt'' e^{-\gamma(t'-t'')} R(t'')/m$$

[2] When D depends on q, the drift velocity will become $v_D = -D_q \partial\beta F/\partial q + \partial D_q/\partial q$.

Now to obtain the mean squared displacement as a function of time, $\delta q(t)$ must be squared and averaged over both the noise $R(t)$ and over the initial conditions v_0. The $\langle\cdot\rangle_R$ average can be evaluated using the random force properties: $\langle R(t)\rangle = 0$ and $\langle R(0)R(t)\rangle = 2m\gamma k_B T\delta[t]$. Several terms vanish because they linear in $R(t)$ and/or in v_0. However, a quadruple integral involving $R(t'')R(\tau'')$ must be completed. Fortunately, the same integral was done by Uhlenbeck and Ornstein in their analysis of the special $v_D = 0$ case [16]. Their noise averaged result, generalized to the case of a constant but non-zero force, is

$$\left\langle (\delta q(t))^2 \right\rangle_R = \left\{ v_D t + (v_0 - v_D)\gamma^{-1}(1 - e^{-\gamma t}) \right\}^2 + \frac{k_B T}{m\gamma^2}\left\{ 2\gamma t - 3 + 4e^{-\gamma t} - 2e^{-2\gamma t} \right\}$$

which has not yet been averaged over initial velocities. The first term arises from initial velocities and deterministic forces, while the second term arises from random forces. Regardless of the initial velocity, the drift velocity will dominate the mean squared displacement at long times. For the special case where the drift velocity is zero, the average over v_0 gives [16]

$$\left\langle (\delta q)^2 \right\rangle = \frac{2k_B T}{m\gamma^2}\left(\gamma t - 1 + e^{-\gamma t}\right) \tag{15.1.7}$$

Thus in the absence of a potential energy gradient, the Langevin equation predicts ballistic (inertial) motion for times $t < \gamma^{-1}$ and diffusion with $D = k_B T / m\gamma$ for longer times. (See Figure 15.1.1.)

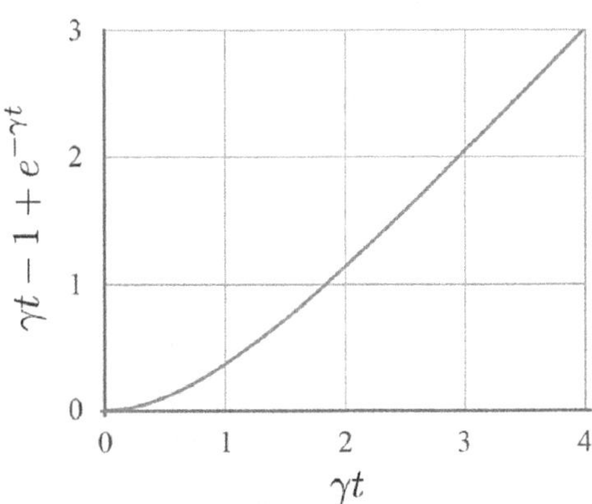

Figure 15.1.1: The Langevin equation gives ballistic motion for $\gamma t < 1$ and the linear relationship between mean squared displacement and time is recovered for $\gamma t > 1$.

15.2 Overdamped Langevin dynamics

For a flat potential energy surface, the Langevin equation *with any non-zero friction* gives trajectories that resemble Brownian motion when observed over times $t \gg \gamma^{-1}$. The magnitude of the friction only changes the effective size of the steps in the random walk, i.e. friction controls the time scale and the ballistic distance traveled during velocity-velocity decorrelation.[3] The typical distance that is traveled in a ballistic fashion before the system 'forgets' its initial velocity is

$$(\delta q)_{bal.} = \gamma^{-1}\sqrt{k_B T/m}. \tag{15.2.1}$$

As friction increases, the ballistic distance shrinks.

For a PMF with hills and valleys, what happens when the ballistic distance becomes too short to appreciably change the force? This question is particularly important in reaction rate theory. According to transition state theory, ballistic motion along the reaction coordinate should persist from the moment of barrier crossing to a point far enough down from the barrier top to make the recrossing probability negligible. At the other extreme are systems in which typical reactive trajectories cross the dividing surface many times as they diffuse over the barrier top. These limiting dynamical regimes are both addressed by the Kramers theory (see Chapter 16). The important point for our present discussion is that the relative sizes of $(\delta q)_{bal.}$ and length scales like the width of the barrier top determine whether we should think of the dynamics as inertial or diffusive.

Now let us develop quantitative guidelines for omitting the inertial term in the Langevin dynamics. Displacement by a distance $(\delta q)_{bal.}$ changes the (deterministic) forces by an amount $|(\delta q)_{bal.}\partial^2 F/\partial q^2|$. Meanwhile the same displacement sees the friction forces change by an approximate amount $m\gamma\sqrt{k_B T/m}$. The deterministic acceleration term in the Langevin equation can be omitted when, for all locations q, the friction forces change much more than the systematic PMF forces do, i.e. when $m\gamma\sqrt{k_B T/m} \gg |(\delta q)_{bal.}\partial^2 F/\partial q^2|$. After simplification, the criterion becomes

$$\text{if} \quad \frac{m\gamma^2}{|\partial^2 F/\partial q^2|} \gg 1, \quad \text{the dynamics are overdamped.}$$

For example, in a harmonic vibrational well with free energy $F(q) = \frac{1}{2}m\omega^2 q^2$, the inertial Langevin equation can be replaced with an overdamped Langevin equation if $\gamma^2 \gg \omega^2$. The overdamped oscillator will not exhibit any oscillatory behavior, whereas trajectories in the inertial regime will exhibit oscillations that gradually dephase.

As friction increases, the typical magnitude of the acceleration term $m\ddot{q}$ becomes quite large (not small as sometimes argued). However, the sign of the acceleration is rapidly changing because of the strong random forces and friction so that all inertial character is damped out. To model this overdamped motion, we simply omit the acceleration term from the Langevin

[3] From equation (15.1.6), a flat PMF gives $\langle v(t)v(0)\rangle = \left\langle v(0)^2\right\rangle e^{-\gamma t}$.

equation. We *cannot* eliminate the random forces because all trajectories from the equation $dq/dt = -D\partial\beta F/\partial q$ would deterministically evolve toward minima. In the high friction limit, the overdamped Langevin equation takes the form

$$\frac{dq}{dt} = -D\frac{\partial\beta F}{\partial q} + \widetilde{R}(t) \tag{15.2.2}$$

where $\langle\widetilde{R}\rangle = 0$ and

$$\langle\widetilde{R}(0)\widetilde{R}(t)\rangle = 2D\delta[t] \tag{15.2.3}$$

Note that the random forces have been renormalized to $\widetilde{R}(t) = R(t)/\gamma m$ and that the separate m, γ, and k_BT parameters are now collapsed into the one parameter $D = k_BT/m\gamma$. The mass/inertia now appears only in the diffusivity.

Up to this point, our discussion has focused entirely on cases with coordinate independent friction and diffusion. These textbook situations are rarely encountered in practice. What is a coordinate dependent diffusivity? Consider for example, models of nucleation where the nucleus size n is the stochastic variable of interest. The diffusivity along coordinate n, i.e. D_n, is the frequency of monomer attachment and detachment events. It often scales with nucleus surface area, so that $D_n \sim n^{2/3}$. Coordinate dependent diffusion is also common in biomolecular conformational transitions [17,18]. Coordinate dependent diffusivities and random forces lead to versions of equations (15.2.2) and (15.2.3) where D and $\widetilde{R}(t)$ are replaced by D_q and $\widetilde{R}_q(t)$. The Euler-Maruyama algorithm (below) for simulating overdamped dynamics is applicable to coordinate dependent diffusion.

Algorithm: Euler-Maruyama integration

Overdamped trajectories are generated by the extremely simple algorithm [19]

$$q(t+\Delta t) = q(t) - D_q\frac{\partial\beta F}{\partial q}\Delta t + \sqrt{2D_q\Delta t}\cdot\xi \tag{15.2.4}$$

where at each time step ξ is a Gaussian random number with zero mean and unit variance. The timesteps must be small, otherwise the force will change between steps and then the assumptions of our overdamped equation will be violated. In some cases, the forces and friction explicitly depend on time as well as position. These can also be integrated with an Euler-Maruyama algorithm: $q(t+\Delta t) = q(t) + v_D(q(t), t)\Delta t + \sqrt{2D(q(t), t)\Delta t}\cdot\xi$ where $D(q, t)$ is the effective diffusivity and $v_D(q(\tau), \tau)$ is the systematic drift velocity. An appropriate time step [20] should satisfy $\Delta t \cdot |\partial v_D/\partial q| < \varepsilon$ and $\Delta t^{1/2}\cdot|\partial\sqrt{2D(q,t)}/\partial q| < \varepsilon$ where ε is an error tolerance. For accurate results the trajectories need not appear continuous when plotted, but visually continuous trajectories also require that $\Delta t \leq$

$2\varepsilon^2 D(q,t)/v_D^2$ [20]. See Leimkuhler and Matthews for a similarly simple algorithm that has superior accuracy vs. time step properties [21].

15.3 Fokker-Planck equations

The Langevin equation describes a single trajectory $q(t)$ under the influence of random forces and friction. Each Langevin dynamics trajectory is different, so time averages and correlation functions must be computed from a long Langevin dynamics trajectory or from an ensemble of Langevin dynamics trajectories. Today, stochastic trajectories can be generated and numerically analyzed with little difficulty, but early pioneers of stochastic processes did not have this luxury. Instead of simulating trajectories, they [22–25] developed Fokker-Planck equations to directly model the time evolution of a probability density.

The Fokker-Planck equation is an approximation, but it is highly accurate for Markov processes that have small individual jumps [20,26–28]. In the overdamped Langevin equation jumps are infinitesimally small $(-v_D(q)dt + \sqrt{2D_q dt} \cdot \xi)$, and accordingly the Fokker-Planck equation for overdamped Langevin dynamics is exact [31]. Master equations for nucleation assume that nucleus size evolves by attachment and detachment of single molecules (jumps of size one), and accordingly these master equations can be *approximately* reformulated as Fokker-Planck equations [29,30].

The Fokker-Planck equation for a single variable q is

$$\frac{\partial \rho(q,t)}{\partial t} = -\frac{\partial}{\partial q}\left[v_D(q)\rho(q,t) - \frac{\partial}{\partial q}\left(D_q \rho(q,t)\right)\right] \tag{15.3.1}$$

It describes the evolution of the probability density $\rho(q,t)$ from an ensemble of stochastic trajectories initiated with density $\rho(q,0)$. The Fokker-Planck equation does not explicitly invoke random forces, but their effects are evident in the diffusion term which drives sharp initial conditions to decay toward a stationary distribution at long times. In writing a Fokker-Planck equation (or a Langevin equation) for the variable q we have implicitly assumed that

(i) $q(t)$ evolves continuously or by small jumps in the state space for q,
(ii) $q(t)$ is dynamically separable from all other variables in the system, and
(iii) the dynamics of $q(t)$ are not only separable, but also Markovian.

Most arbitrarily selected variables will not satisfy (i), (ii), and (iii), so the Fokker-Planck equation will not describe their dynamics correctly. Variables that do fulfill requirements (i), (ii), and (iii) are very special. If an appropriate variable q can be identified (see Chapter 20), then its dynamics within a complex multibody dynamical system are conveniently reduced to equation (15.3.1). Coarse graining the dynamics down to a single variable provides a concise description of the stationary distribution and the relaxation dynamics from any initial distribution.

As we already saw for the (overdamped) Langevin equation, the coefficients in the Fokker-Planck equation can be interpreted as a drift velocity

$$\lim_{\Delta t \to 0} \frac{\langle \Delta q \rangle_q}{\Delta t} = v_D(q) \tag{15.3.2}$$

and a diffusivity

$$\lim_{\Delta t \to 0} \frac{\left\langle (\Delta q)^2 \right\rangle_q}{2\Delta t} = D_q \tag{15.3.3}$$

Because it accounts for coordinate dependent diffusion, definition 15.3.2 is more general than (15.1.4). Also note that equations (15.3.2) and (15.3.3) are quite useful simplifications. The complete transition matrix may be terribly complicated, but the drift velocity and diffusivity are all that remains of the dynamics after conversion from a master equation to a Fokker-Planck equation.

Fokker-Planck equation from the master equation

The Fokker-Planck equation can be derived via the master equation and via the Langevin equation. Both are useful transformations in practice, so we briefly outline them here. Start from the master equation (in integral form) [25]

$$\frac{\partial \rho}{\partial t} = \int w_{q \leftarrow q+\Delta q} \rho(q+\Delta q, t) d\Delta q - \int w_{q+\Delta q \leftarrow q} \rho(q, t) d\Delta q$$

where $w_{q \leftarrow q+\Delta q}$ and $w_{q+\Delta q \leftarrow q}$ are continuous counterparts of the $\mathbf{W}_{i \leftarrow j}$ rate matrix elements. The two integrals in this master equation represent flow into and flow out of state q. The transition rates are written in terms of q and Δq (as opposed to $w_{q' \leftarrow q}$) to exploit property (i): if only small jumps in q are allowed, then $w_{q \leftarrow q+\Delta q}$ and $w_{q+\Delta q \leftarrow q}$ should have compact support near $\Delta q = 0$. By comparison, $w_{q \leftarrow q+\Delta q}$, $w_{q+\Delta q \leftarrow q}$, and $\rho(q, t)$ must vary more gradually with changes in q. Let us Taylor expand the q-dependence in the “flow into q” integral of the master equation:

$$\begin{aligned}
&\int w_{q \leftarrow q+\Delta q} \rho(q+\Delta q, t) d\Delta q \\
&= \int w_{q-\Delta q \leftarrow q} \rho(q, t) d\Delta q - \int \Delta q \frac{\partial}{\partial q} [w_{q-\Delta q \leftarrow q} \rho(q, t)] d\Delta q \\
&\quad + \frac{1}{2} \int (\Delta q)^2 \frac{\partial^2}{\partial q^2} [w_{q-\Delta q \leftarrow q} \rho(q, t)] d\Delta q + \cdots
\end{aligned}$$

The zeroth order integral in the Taylor expansion is exactly the same as the "flow out of q" integral. Therefore the master equation simplifies to

$$\begin{aligned} \frac{\partial \rho}{\partial t} &= -\frac{\partial}{\partial q}\left(\rho(q,t)\int \Delta q\, w_{q-\Delta q \leftarrow q} d\Delta q\right) \\ &+ \frac{1}{2}\frac{\partial^2}{\partial q^2}\left(\rho(q,t)\int (\Delta q)^2\, w_{q-\Delta q \leftarrow q} d\Delta q\right) \end{aligned} \tag{15.3.4}$$

where $\rho(q,t)$ and $\partial/\partial q$ have been pulled outside of the integrals over Δq. Terms of order $(\Delta q)^3$ and higher have been neglected.

The integrals that remain in equation (15.3.4) are surely complicated, but let us consider what they mean. One is the average displacement per time from location q, i.e. the drift velocity

$$v_D(q) \equiv \int \Delta q\, w_{q+\Delta q \leftarrow q} d\Delta q$$

The other is the average squared displacement per time from location q, i.e. twice the diffusivity

$$2D_q \equiv \int (\Delta q)^2\, w_{q+\Delta q \leftarrow q} d\Delta q$$

Inserting these identities[4] for the drift velocity and diffusivity into equation (15.3.4) gives the nonlinear Fokker-Planck equation (15.3.1). If we had retained the higher order terms we would have obtained the (exact) Kramers-Moyal expansion instead of the (approximate) Fokker-Planck equation.

Detailed balance and the Smoluchowski equation

In many applications to chemistry and physics, the dynamics obey a detailed balance relation. Detailed balance relates transition rates to *equilibrium* probability distributions, so a pedagogical discussion on detailed balance in *non-equilibrium* rate processes is warranted. Detailed balance does not imply equilibrium, nor stationarity, nor an absence of sinks, sources, and probability current. It is a relationship between transition rates and *hypothetical* equilibria. In some cases, the hypothetical equilibria are impossible to attain, e.g. the supersaturated conditions that drive nucleation are incompatible with a normalized equilibrium distribution of nuclei. Additionally, many rate calculations involve probability sources and sinks at boundaries that create non-equilibrium currents. Nevertheless, we can invoke detailed balance between hypothetical equilibrated microstates for all states except those whose microscopic transition rates are directly perturbed at the boundaries.

[4] Both identities require additional changes of sign before we can use them.

We have already seen that the Fokker-Planck equation no longer includes the detailed state-to-state interconversion rates. They are instead coarse grained to a simpler description of drift velocities and diffusivities at each point on the q-axis. So how can we still incorporate detailed balance? The Fokker-Planck equation is a continuity equation for probability,

$$\partial\rho/\partial t = -\partial j/\partial q,$$

where $j \equiv v_D(q)\rho(q,t) - \partial[D_q\rho(q,t)]/\partial q$. At equilibrium, all probability currents should vanish. Therefore detailed balance is imposed by requiring that $v_D(q)$, D_q, and the equilibrium distribution satisfy [28]

$$v_D(q)\rho_{eq}(q) - \frac{d}{dq}\left(D_q\rho_{eq}(q)\right) = 0 \tag{15.3.5}$$

Inserting the detailed balance relationship into the nonlinear Fokker-Planck equation (15.3.1) yields the Smoluchowski equation [3,32],

$$\frac{\partial\rho(q,t)}{\partial t} = \frac{\partial}{\partial q}\left[\rho_{eq}(q)D_q\frac{\partial}{\partial q}\left(\frac{\rho(q,t)}{\rho_{eq}(q)}\right)\right].$$

As noted in the discussion above, the hypothetical equilibrium distribution invoked in ρ_{eq} need not be attainable or normalizable. Thus in some cases it is more than a convenience to write the Smoluchowski equation as

$$\frac{\partial\rho(q,t)}{\partial t} = \frac{\partial}{\partial q}\left[e^{-\beta F(q)}D_q\frac{\partial}{\partial q}\left(e^{\beta F(q)}\rho(q,t)\right)\right] \tag{15.3.6}$$

where we have used $\rho_{eq}(q) \propto exp[-\beta F(q)]$. In contrast to equation (15.3.1), note that equation (15.3.6) has D *between* the derivatives as a consequence of detailed balance. The Smoluchowski equation is a natural starting point for analyses of overdamped barrier crossings (see Chapter 18).

A useful generalization of equation (15.3.6) begins with a *system* of overdamped Langevin equations and results in a multidimensional Smoluchowski equation with a vector force and a diffusion tensor:

$$\frac{\partial\rho}{\partial t} = \frac{\partial}{\partial\mathbf{q}}\cdot\left[e^{-\beta F(\mathbf{q})}\mathbf{D}(\mathbf{q})\frac{\partial}{\partial\mathbf{q}}\left(e^{\beta F(\mathbf{q})}\rho(\mathbf{q},t)\right)\right] \tag{15.3.7}$$

Many interesting phenomena can arise depending on the nature of the diffusion tensor **D**. For example, a highly anisotropic **D** indicates that some coordinates have very high mobilities relative to others. In other cases, diffusion tensors with non-zero off-diagonal coupling elements indicate motions of coordinates that are correlated to each other. Chapter 18 will explore the structure of **D** and its effects on barrier crossings.

Fokker-Planck equations from simulation data

Analyses in the physics and chemistry literature often begin with a Fokker-Planck equation based on a simple phenomenological model for the free energy surface and the diffusivity. These studies are powerful sources of insight and they often yield testable predictions. On the other hand, Fokker-Planck equations that are constructed in this way ultimately contain the physics that is built into them.

An alternative strategy is to start from a complex many-body dynamical system (an MD simulation), identify an appropriately Markovian and separable variable q, and then use simulation data to construct its Fokker-Planck equation. Assuming that an appropriate variable q can be identified, the basic idea is to extract the local drift velocity and diffusivity from swarms of short trajectories initiated at each position q_0 along the q-axis. Several computational frameworks have implemented this strategy [33–35]. Estimators for the drift velocity and diffusivity are loosely based on equations (15.3.2) and (15.3.3), but some modifications are necessary. Simulation trajectories have a finite duration, so we cannot simply square the displacement and take the $\Delta t \to 0$ limit to estimate D_{q_0} (as equation (15.3.3) might seem to suggest). We must account for the drifting mean position in the swarm of trajectories at each q_0. Define the drift corrected displacement:

$$\delta q(t) \equiv q(t) - \langle q(t) \rangle_{q_0}$$

where the average is over a swarm of trajectories initiated at location $q(0)$. The drift velocity is that of the moving average [33,36]

$$v_D(q_0) = \frac{d}{dt} \langle q(t) \rangle_{q_0} \tag{15.3.8}$$

and the diffusivity is computed from the same swarm of trajectories as

$$2D_{q_0} = \frac{d}{dt} \left\langle (\delta q(t))^2 \right\rangle_{q_0} \tag{15.3.9}$$

The quantities on the left in equations (15.3.8) and (15.3.9) are the numerically averaged rates at which a swarm of trajectories drifts and spreads when started at q_0. To construct the Fokker-Planck equation, initiate swarms of trajectories at a series of q_0-values. For each q_0, compute $d \langle q(t) \rangle_{q_0} / dt$ and $d \left\langle (\delta q(t))^2 \right\rangle_{q_0} / dt$ from the trajectory swarm data. Then solve (15.3.8) and (15.3.9) for $v_D(q)$ and D_q at each of the q_0-values. In some cases, the data can be used to identify unknown parameters in simple models for $F(q)$ and D_q. For example, the appropriate variable for studies of nucleation is nucleus size n and physically motivated models of the free energy and the diffusivity (attachment frequency) have the form $F(n) = -n\Delta\mu + \phi\gamma n^{2/3}$ and $D_n = D_1 n^{2/3}$. The seeding approach to nucleation uses short trajectory swarms to identify the unknown parameters $\phi\gamma$ and D_1, and thereby parameterizes the Zeldovich-Frenkel (Fokker-Planck) equation [34,35,37,38].

There are some important technical issues to consider in choosing an appropriate swarm duration. First, equations for the overdamped regime are being used, so the duration should be significantly longer than the velocity decorrelation time. Second, the duration should be short so that the force $\partial\beta F/\partial q$ does not change during the swarm evolution. In practice, it can be difficult to straddle these two opposing requirements, and one must often adopt a swarm duration that is too long to ignore changes in $\partial\beta F/\partial q$. Therefore, Hummer and Kevrekidis [33] recommend alternative versions of equations (15.3.8) and (15.3.9). Specifically, they estimate $D(\overline{q})$ and $\partial\beta F/\partial q|_{\overline{q}}$ instead of $D(q_0)$ and $\partial\beta F/\partial q|_{q0}$ where $\overline{q}$ is the time-averaged position of q over the duration of the swarm.

15.4 From discrete models to Fokker-Planck equations

We have been considering stochastic processes with continuous state spaces, but some processes with intrinsically discrete states are more easily modeled using continuous variables. Here we show, by way of example, how models with discrete states can be transformed to models with continuous variables while retaining nearly all behaviors of the discrete model. The tools which are illustrated in this section are applicable when transitions occur only between closely neighboring states with similar properties. The following shows a familiar biased random walk example.

Example: A biased and unbounded random walk

A random walker takes steps to the right on a one-dimensional lattice with rate constant k_+ and steps to the left with rate constant k_-. The master equation for this process is

$$\frac{dp_n}{dt} = k_+ p_{n-1} + k_- p_{n+1} - (k_+ + k_-)p_n$$

The matrix $\mathbf{W}$ in this case is of infinite dimension. $\mathbf{W}$ would be tridiagonal, and thus easy enough to write down, but still difficult to analyze. Quite often we can instead regard n as a continuous variable and introduce raising and lowering operators for functions of n as

$$e^{m\partial/\partial n} f(n) = f(n) + m\frac{\partial f}{\partial n} + \frac{1}{2}m^2\frac{\partial^2 f}{\partial n^2} + \cdots = f(n+m)$$

Now writing p_{n-1} and p_{n+1} as expansions around p_n gives

$$\begin{aligned}\frac{\partial p_n}{\partial t} &= -k_+(p_n - p_{n-1}) + k_-(p_{n+1} - p_n) \\ &= -k_+(1 - e^{-\partial/\partial n})p_n + k_-(e^{\partial/\partial n} - 1)p_n\end{aligned}$$

If changes in p_n are sufficiently gradual, then the expansion of $e^{\partial/\partial n}$ can be truncated at the second order. Discussions on the limitations and validity of these truncated expansions can be found elsewhere [2,28,39]. In this case, m is one and δp between neighboring states is clearly much smaller than one. Thus using

$$1 - e^{\pm\partial/\partial n} \approx \mp\frac{\partial}{\partial n} - \frac{1}{2}\frac{\partial^2}{\partial n^2}$$

gives

$$\frac{\partial p_n}{\partial t} = (k_- - k_+)\frac{\partial p_n}{\partial n} + \frac{1}{2}(k_- + k_+)\frac{\partial^2 p_n}{\partial n^2}$$

The right side of this partial differential equation describes drift and diffusion, respectively, on the discrete lattice. Note that the differential equation which has emerged has the form of a Fokker-Planck equation.

To eliminate the lattice entirely, we can reintroduce the physical lattice spacing Δq by the relation $q = n\Delta q$. Then $dq = \Delta q\, dn$ and $\rho(q)dq = p_n$ is the continuous probability density as a function of q. By correspondence with the more familiar drift-diffusion equation,

$$\frac{\partial \rho}{\partial t} = -v_D\frac{\partial \rho}{\partial q} + D\frac{\partial^2 \rho}{\partial q^2}, \tag{15.4.1}$$

we see that

$$v_D = -\Delta q \cdot (k_- - k_+) \quad and \quad D = \frac{(\Delta q)^2}{2}(k_- + k_+) \tag{15.4.2}$$

which gives a clear interpretation for drift and diffusion in terms of discrete step sizes, directions, and frequencies.

The above example shows how transition rates in a lattice simulation can be chosen to match diffusion coefficients and drift velocities for an off-lattice model. The discrete random walk model also illustrates that Fokker-Planck equations are special types of continuum master equations that result from processes that only involve local transitions. In other words Fokker-Planck equations correspond to master equations with nearly diagonal transition matrices.

Let us briefly take stock of the different methods we have already seen for studying reaction networks. In Chapter 3 we learned to formulate, simplify, and solve the deterministic rate equations for a complex reaction network. In Chapter 14 we learned about chemical master equations and kinetic Monte Carlo simulations that capture fluctuations around the deterministic solutions and spontaneous switches between stable states. We saw that fluctuations and spontaneous transitions between stable states are often important for reactions in small systems. Because each reaction event changes the species populations by only one or two molecules, all transitions in a chemical master equation are between closely neighboring states. Thus it should (usually) be possible to convert a chemical master equation to a Fokker-Planck description. The next two examples return to the chemical master equation of Schlogl to illustrate how Fokker-Planck equations can be constructed and used to understand fluctuations and multiple steady states in chemical reaction networks.

Example: Chemical Fokker-Planck equation for the Schlogl system

The Schlogl system is a biochemical network model with kinetics

$$
\begin{array}{llll}
\varnothing & \rightarrow \mathbf{A} & r_1 & \\
\mathbf{A} & \rightarrow \varnothing & r_{-1} & = k_{-1}n \\
3\mathbf{A} & \rightarrow 2\mathbf{A}+\mathbf{B} & r_2 & = k_2 n(n-1)(n-2) \\
2\mathbf{A}+\mathbf{B} & \rightarrow 3\mathbf{A} & r_{-2} & = k_{-2}n(n-1)
\end{array}
$$

where n is the number of $\mathbf{A}$ molecules in the system. Details about the birth, death, and $\mathbf{A}$, $\mathbf{B}$ interconversion kinetics in this model are given in section 14.4. The master equation governing n is

$$
\begin{aligned}
\frac{dp_n}{dt} &= r_1 p_{n-1} - r_1 p_n \\
&\quad + k_{-1}(n+1)p_{n+1} - k_{-1}np_n \\
&\quad + k_2(n+1)n(n-1)p_{n+1} - k_2 n(n-1)(n-2)p_n \\
&\quad + k_{-2}(n-1)(n-2)p_{n-1} - k_{-2}n(n-1)p_n
\end{aligned}
$$

It is convenient to abbreviate the master equation using the definitions

$$
\begin{aligned}
h_1(n) &= (r_1 + k_{-2}n(n-1))p_n(t) \\
h_2(n) &= (k_{-1}n + k_2 n(n-1)(n-2))p_n(t)
\end{aligned}
$$

Then the master equation takes the compact form

$$
\frac{dp_n}{dt} = h_2(n+1) - h_2(n) + h_1(n-1) - h_1(n)
$$

Next, expand the off-diagonal terms in the master equation to second order in the variable n.

$$
h_1(n-1) = h_1(n) - \frac{\partial h_1}{\partial n} + \frac{1}{2}\frac{\partial^2 h_1}{\partial n^2}
$$

$$h_2(n+1) = h_2(n) + \frac{\partial h_2}{\partial n} + \frac{1}{2}\frac{\partial^2 h_2}{\partial n^2}$$

Insert these expressions into the master equation to obtain

$$\frac{\partial p_n}{\partial t} = -\frac{\partial}{\partial n}[h_1(n) - h_2(n)] + \frac{\partial^2}{\partial n^2}\left[\frac{h_1(n) + h_2(n)}{2}\right]$$

Technically, the transformation to a continuum stochastic partial differential equation is done. However, it is useful to write the Fokker-Planck equation in terms of the drift and diffusion rates. The drift rate is

$$\begin{aligned} v_D(n) &= [h_1(n) - h_2(n)]/p_n(t) \\ &= -n(n-1)(n-2)k_2 + n(n-1)k_{-2} - nk_{-1} + r_1 \end{aligned}$$

and the coordinate-dependent diffusion rate is

$$\begin{aligned} D(n) &= [h_1(n) + h_2(n)]/(2p_n(t)) \\ &= \frac{1}{2}\{n(n-1)(n-2)k_2 + n(n-1)k_{-2} + nk_{-1} + r_1\} \end{aligned}$$

Now the Fokker-Planck equation takes the usual form

$$\frac{\partial p_n}{\partial t} = -\frac{\partial}{\partial n}\left[v_D(n) p_n\right] + \frac{\partial^2}{\partial n^2}\left[D(n) p_n\right] \tag{15.4.3}$$

The operator approach in this section can be used to create Fokker-Plank equations for many discrete systems where the transitions are local and where the stationary probabilities vary slowly between states. Stochastic integral equations can be obtained from discrete systems with highly non-local transitions, but these require techniques beyond the scope of this book [28]. Armed with some ways of deriving Fokker-Planck equations (where they are applicable), we are ready to examine the properties of their solutions.

15.5 Stationary solutions of Fokker-Planck equations

Chapter 14 included a few examples of discrete master equations with steady-state distributions. Fokker-Planck equations with time-independent drift and diffusion coefficients also yield stationary distributions in many familiar contexts. Chapters 16–18 will make extensive use of non-equilibrium steady-state solutions to Fokker-Planck equations. Diffusion on a bounded free energy landscape (equation (15.3.6)) gives the equilibrium distribution at long times. Here we show more generally that when the drift and diffusion rates are time independent, and when there are no sources or sinks, then the stationary distribution resembles an equilibrium distribution for some effective potential and density of states. We begin with the Fokker-Planck equation

$$\frac{\partial \rho}{\partial t} = -\frac{\partial}{\partial q}\left\{ v_D(q)\rho - \frac{\partial}{\partial q}\left(D(q)\rho\right)\right\} \tag{15.5.1}$$

where $D(q)$ is a coordinate-dependent diffusivity and $v_D(q)$ is a force-induced drift rate. Recall that the Fokker-Planck equation is a probability continuity equation, i.e. $\partial\rho/\partial\tau = -\partial j/\partial q$ where the flux j is the part of equation (15.5.1) within curly brackets. Steady-state distributions $\rho_{SS}(q)$ must satisfy

$$j_{SS} = v_D(q)\rho_{SS}(q) - \frac{d}{dq}\left(D(q)\rho_{SS}(q)\right) \tag{15.5.2}$$

where j_{SS} is a constant. In later chapters, several rate theories will be developed by making j_{SS} a nonzero constant.

If there are no sources or sinks (even at $\pm\infty$) then j_{SS} must be zero. Let $y(q) = D(q)\rho_{SS}(q)$ to obtain the separable equation $y\, v_D/D = y'$. Then integrate to obtain $ln\, y = \int (v_D/D)dq + C$. Finally, solve for ρ_{SS} and choose the integration constant so that the distribution is normalized. The result is

$$\rho_{SS}(q) = Q^{-1}\frac{1}{D(q)} exp \int_{-\infty}^{q} \left[v_D(q')/D(q')\right] dq' \tag{15.5.3}$$

with

$$Q = \int_{-\infty}^{\infty} dq \frac{1}{D(q)} exp \int_{-\infty}^{q} dq' \left[v_D(q')/D(q')\right] \tag{15.5.4}$$

Note how the distribution $\rho_{SS}(q)$ resembles an equilibrium distribution even if the underlying dynamics is a driven non-equilibrium system. Specifically, equation (15.5.3) can be interpreted as an effective equilibrium distribution for a system with potential $\varphi(q) = -\int (v_D/D)dq$ and density of states $\Omega(q) = 1/D(q)$. If $v_D(q) = -D\partial\beta F/\partial q$ with constant D, then equation (15.5.3) yields a proper equilibrium distribution, $\rho_{SS}(q) = Q^{-1}exp[-\beta F(q)]$.

Example: Chemical Fokker-Planck equation for Schlogl system

Again consider the Schlogl system as an example with the random variable of interest being n, the number of **A** molecules. Section 14.4 presented stochastic simulation (kinetic Monte Carlo) results from Erban et al. [40] with values $r_1 = 2250$, $k_{-1} = 37.5$, $k_2 = 0.18$, and $k_{-2} = 2.5\ 10^{-4}$. These parameters gave rise to two steady deterministic states: $n = 100$ or $n = 400$. From a long simulation run, Erban et al. computed the stationary probability of finding n molecules of type **A** in the system.

It should also be possible to predict the probability to find n molecules of type **A** from a Fokker-Planck equation without performing the kinetic Monte Carlo simulation. The Fokker-Planck equation for $p(n, t)$ was given in equation (15.4.3). The figure below, adapted from Erban et al., compares a portion of the stochastic simulation results (right) to the stationary distribution obtained from the Fokker-Planck equation.

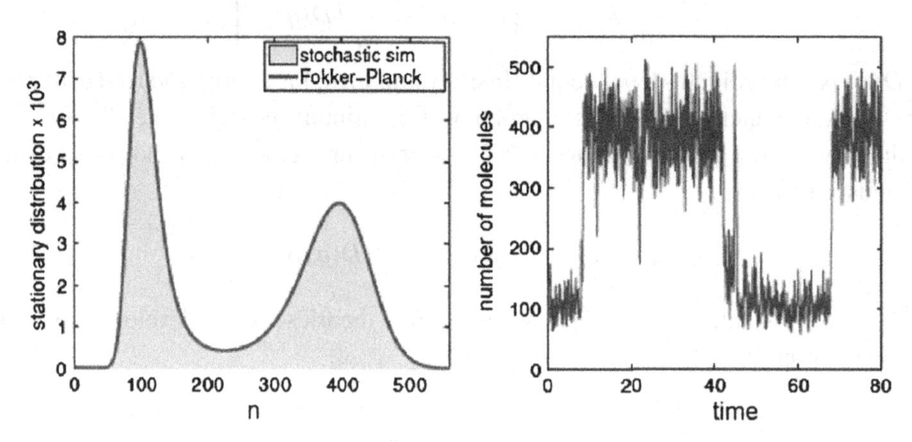

The stationary distribution predicted from the Fokker-Planck equation perfectly matches the histogram of kinetic Monte Carlo results (gray) from a long stochastic simulation. Figures from Erban et al. *SIAM J. Appl. Math.* 70, 984–1016 (2009).

The Schlogl system has bimodal stationary distribution with infrequent transitions between two metastable states. We have seen how spectral analyses of the master equation can predict the time scale associated with these transitions. Similar spectral analyses can be performed by starting from the Fokker-Planck equation.

15.6 Spectral theory revisited

For ergodic systems with a bounded partition function, the Fokker-Planck equation gives rise to a discrete spectrum of eigenvalues and eigenfunctions.[5] The spectral representation is useful for developing several formal expressions. This section outlines the spectral representation for

[5] The results of this section are not limited to Fokker-Planck equations. They more generally apply to any continuous master equation

$$\frac{\partial p(q,t)}{\partial t} = \int \left\{\omega(q|q')p(q',t) - \omega(q'|q)p(q,t)\right\} dq'$$

with transition probabilities that obey detailed balance. The Fokker-Planck is a special limiting case that arises when the $\omega(q|q')$ only have support over $|q - q'|$ distances that are small compared to the widths of peaks and overdamped dynamics in a continuous configuration space. The discussion is terse because the derivations closely parallel the spectral rate theory for discrete master equations (Chapter 14).

Let L be the operator in the Smoluchowski (forward overdamped Fokker-Planck) equation so that equation (15.3.7) becomes

$$\partial\rho/\partial t = L\rho$$

Additionally, let the operator L have "right" eigenfunctions (ψ_m^R) and eigenvalues ($-1/\tau_m$): $L\psi_m^R(\mathbf{q}) = -\tau_m^{-1}\psi_m^R(\mathbf{q})$. The first eigenfunction is the equilibrium distribution $\psi_1^R(\mathbf{q}) \sim exp[-\beta F(\mathbf{q})]$ with eigenvalue $1/\tau_1 = 0$. The right eigenfunctions are orthogonal with weight function $1/\psi_1^R(\mathbf{q})$, i.e.

$$\int d\mathbf{q}\psi_m^R(\mathbf{q})\psi_\ell^R(\mathbf{q})/\psi_1^R(\mathbf{q}) = \delta_{m\ell}$$

which assumes an appropriate normalization of each $\psi_m^R(\mathbf{q})$. All higher eigenvalues are negative, so their contributions to the time-dependent solution will gradually decay. As we did for the discrete master equation, we can adopt bra-ket notation for the right and left eigenfunctions. Let the right eigenfunction be the ket

$$|\psi_m\rangle = \psi_m^R(\mathbf{q})$$

and let the left eigenfunction of L be the bra

$$\langle\psi_m| = \psi_m^R(\mathbf{q})/\psi_1^R(\mathbf{q})$$

In bra-ket notation, nearly all formulas from section 14.6 directly extend to the case of a continuous master equation. For example, the orthogonality relation is $\langle\psi_m|\psi_\ell\rangle = \delta_{m\ell}$. Time-dependent solutions can again be expanded as $|\rho(\mathbf{q},t)\rangle = \sum_m e^{-t/\tau_m}\,|\psi_m(\mathbf{q})\rangle\,\langle\psi_m(\mathbf{q}_0) \times |\rho(\mathbf{q}_0,0)\rangle$ where the sum now runs over an infinite series of eigenfunctions. For the sharp initial condition $\rho(\mathbf{q},0) = \delta[\mathbf{q}-\mathbf{q}_0]$, the time dependent solution becomes the Greens function

$$\rho(\mathbf{q},t\,|\mathbf{q}_0,0) = \sum_{m=1}^{\infty} a_m e^{-t/\tau_m}\,|\psi_m(\mathbf{q})\rangle \tag{15.6.1}$$

with expansion coefficients

$$a_m = \langle\psi_m(\mathbf{q})|\delta[\mathbf{q}-\mathbf{q}_0]\rangle = \psi_m^R(\mathbf{q}_0)/\psi_1^R(\mathbf{q}_0)$$

minima in the equilibrium distribution [41]. We have focused on Fokker-Planck equations and Langevin equations because they are more important for the other topics in this book.

The completeness relation for the eigenfunctions of a continuous Fokker-Planck equation is

$$\sum_m |\psi_m(\mathbf{q})\rangle \langle\psi_m(\mathbf{q}')| = \delta[\mathbf{q} - \mathbf{q}']$$

Exact solutions of the Fokker-Planck equation for a high dimensional system are impossibly difficult, but new computational methods can approximate the slowest eigenfunctions directly from simulation data. The first (right) eigenfunction is the stationary equilibrium distribution. The rest of the eigenfunctions correspond to non-equilibrium relaxation processes with ever faster relaxation times. In many applications, there are one or more spectral gaps corresponding to a clear separation of time scales:

$$\cdots \tau_{k-1} > \tau_k \gg \tau_{k+1} > \tau_{k+2} \cdots$$

Spectral gaps allow an accurate low dimensional model to be constructed by simply discarding everything beyond the first k eigenfunctions. The result is a simple model that accurately describes the k slowest modes and their dynamics. The slowest (left) eigenfunction is particularly important because it provides a numerically exact reaction coordinate that increases monotonically as one moves through the high dimensional space from reactants to products [42–46]. The algorithm below shows how the eigenfunctions can be constructed using molecular simulation data.

Algorithm: Diffusion map

Let $S = \{\mathbf{x}_i\}_{i=1,\ldots,n}$ be a large collection of configurations sampled from a long molecular dynamics trajectory. The trajectory should be sufficiently long to see several examples of the slowest transition. Diffusion map results depend critically on a distance threshold parameter ε. It should be just small enough that no configurations within a distance ε are separated by a slow transition. See Singer for additional recommendations on the choice of ε [43,47].

1. For each $\mathbf{x}_i \in S$ compute

$$\rho_\varepsilon(\mathbf{x}_i) = \sum_j K(\mathbf{x}_i, \mathbf{x}_j)$$

where [42]

$$K(\mathbf{x}, \mathbf{y}) = exp[-||\mathbf{x} - \mathbf{y}||^2/2\varepsilon^2]$$

$\rho_\varepsilon(\mathbf{x}_i)$ quantifies the local density of the sample S at location $\mathbf{x}_i$.

2. Construct the symmetric matrix [42]

$$\widetilde{\mathbf{K}}_{ij} = \frac{K(\mathbf{x}_i, \mathbf{x}_j)}{\sqrt{\rho_\varepsilon(\mathbf{x}_i)\rho_\varepsilon(\mathbf{x}_j)}}$$

   To the extent that nearby configurations have similarly fast interconversion rates, the matrix $\widetilde{\mathbf{K}}_{ij}$ is analogous to the symmetrically renormalized transition matrix from the discrete spectral theory.
3. For each $\mathbf{x}_i \in S$, compute the diagonal matrix with elements $\mathbf{D}_{ii} = \sum_j \widetilde{\mathbf{K}}_{ij}$. $\mathbf{D}_{ii}$ indicates the total (renormalized) mobility in and out of state i. Use the $\mathbf{D}_{ii}$ to compute the symmetrized Markov matrix [42]

$$\mathbf{M} = \mathbf{D}^{-1/2}\widetilde{\mathbf{K}}\mathbf{D}^{-1/2}$$

4. Compute the first few eigenvalues and eigenvectors of $\mathbf{M}$, ideally up to a natural spectral gap. The actual "diffusion maps" are eigenvectors of $\mathbf{D}^{-1/2}\mathbf{M}\mathbf{D}^{1/2}$, but the eigenvectors of the symmetric matrix $\mathbf{M}$ are easier to obtain with numerical methods. The mth diffusion map eigenvector is obtained from $\psi_m^L \approx \mathbf{D}^{-1/2}\psi_m^S$ where ψ_m^S is the mth eigenvector of the symmetric matrix $\mathbf{M}$ [42].

The right eigenvectors are obtained by multiplying the left eigenvectors with the equilibrium distribution. The accuracy of the eigenvectors depends on the initial amount of data in S and an appropriate choice of parameter ε.

If the diffusion map procedure has worked correctly, then the following interpretations can be ascribed to the diffusion map eigenvectors.

- The first eigenvector corresponds to equilibrium: $1/\tau_1 = 0$ and $\psi_1^L(\mathbf{x}) \approx const.$
- The reaction coordinate is ψ_2^L. Reactant configurations are at $\psi_2^L(\mathbf{x}) < 0$ and products are at $\psi_2^L(\mathbf{x}) > 0$. Transition states are configurations with $\psi_2^L(\mathbf{x}) = 0$ and the rate of reactant-product interconversion is $1/\tau_2$. For barriers that are much larger than $k_B T$, the value of ψ_2^L changes rapidly as one moves across the transition state. Plateau regions where $\psi_2^L(\mathbf{x})$ is approximately constant correspond to reactants, products, or perhaps to metastable intermediate basins.
- $\psi_3^L(\mathbf{x})$ is the reaction coordinate for the next process in the time scale hierarchy. Often the $\psi_3^L(\mathbf{x})$ process is a relaxation within the reactant basin or within the product basin.

The free energy as a function of ψ_2^L or as a function of ψ_2^L and ψ_3^L provides a visual map of the free energy landscape as a function of the slowest dynamical modes. Figure 15.6.1 shows the diffusion map coordinates from analysis of conformational transitions of a Beta3s peptide [48].

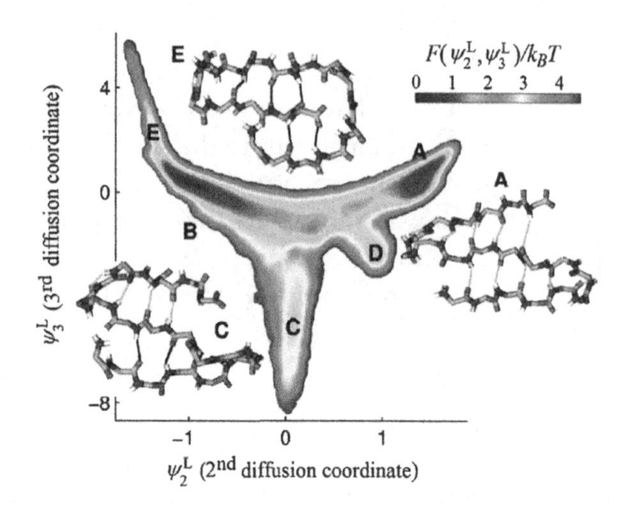

Figure 15.6.1: The (locally scaled) diffusion map approach was applied to conformational transitions in the Beta3s peptide. Key structures are shown along with the free energy ($-ln\psi_1^R$) as a function of the first and second diffusion map coordinates. [Modified with permission from Zheng et al. *J. Phys. Chem. B*. (2011).]

Diffusion maps and MSM eigenvectors are closely related, but there are differences in their computational details and conceptual starting points. Diffusion maps emerged from Fokker-Planck equations for diffusion on a high dimensional continuous manifold. The ψ_m^L coordinates that emerge from MSMs began from a discrete picture. Operationally, both approaches ultimately use discretized states. The MSM approach is probably more robust because it directly counts transition frequencies rather relying on a geometric proximity criterion. If ε is not chosen correctly, then diffusion map may assume facile interconversion between structures which are dynamically far apart. MSM will correctly describe all interconversions on time scales longer than the time-lag parameter. Of course most MSM implementations also employ a preliminary clustering procedure based on geometric distance, and therefore a poorly chosen distance metric can also corrupt MSM results.

Eigenfunction-based reaction coordinates have many of the properties of an ideal reaction coordinate (see Chapter 20). Furthermore, the eigenfunctions can be systematically generated without *a priori* defined reactant and product basins in configuration space. An eigenfunction-based reaction coordinate accurately summarizes the time scales and transitions observed during long molecular dynamics trajectories. However, eigenfunctions also have some practical disadvantages as reaction coordinates.

- In most implementations, eigenfunctions are constructed from long unbiased trajectories that must spontaneously cross over all significant barriers. Thus, applications to processes with barriers much larger than k_BT are still a major challenge. Methods are being developed that use importance sampling methods in the construction of diffusion maps and MSMs [49–51].

- Eigenfunction-based reaction coordinates only exist for systems that have a well-defined reactant-product equilibrium. They do not exist for processes like nucleation in which the product free energy diverges [52].
- There is (as yet) no reliable way to relate the eigenfunctions to physical variables that have a clear mechanistic interpretation [53]. However, there are ongoing efforts to build the eigenfunctions from bases of physically meaningful collective variables [54,55].

Because of these limitations, spectral analyses based on MD trajectory data at one condition (temperature, ionic strength, etc.) cannot yet predict results across different conditions or across a family of related reactions. Reaction coordinates with a clear physical interpretation are essential ingredients of models that predict such trends [53]. It seems likely that some combination of dimensionality reduction techniques [42,43,45,46], high throughput hypothesis testing techniques [53,56,57], and spectral analyses [58–60] will yield practical reaction coordinates for the most complex transitions in the near future.

Exercises

1. Compute the correlation function $\langle q(t)q(0)\rangle_R$ as a function of the initial velocity $v(0)$ and position $q(0)$ for a damped harmonic oscillator with (inertial) Langevin equation $m\ddot{q} = -m\omega^2 q - m\gamma\dot{q} + R(t)$.
2. Compute $\langle q(t)q(0)\rangle_R$ as a function of the initial position $q(0)$ for a harmonic oscillator with overdamped Langevin equation $\dot{q} = -D\beta m\omega^2 q + \widetilde{R}(t)$.
3. A particle diffuses freely from the origin in one dimension. The probability density to find the particle at location x at time t, satisfies

   $$\frac{\partial P}{\partial t} = D\frac{\partial^2 P}{\partial x^2}$$

   with $P(x,0) = \delta[x]$. For any $x \neq 0$ and $t > 0$, $P(x,t)$ also satisfies the integral equation

   $$P(x,t) = \int_0^t F(x,t')P(0,t-t')dt'$$

   where $F(x,t)$ is the probability that the particle *first* visits x at time t' with $t' < t$ [61].

   (0, 0) (x, t') (0, t-t')

   What is $\hat{F}(x,s)$, i.e. the Laplace transform of $F(x,t)$. Find, (contour) plot, and interpret $F(x,t)$.

4. Show that the Fokker-Planck equation for free inertial Brownian motion is

$$\frac{\partial \rho}{\partial t} = \gamma \frac{\partial}{\partial v} \left\{ v\rho + \frac{k_B T}{m} \frac{\partial \rho}{\partial v} \right\}$$

and that

$$\rho(v,t) = \left(2\pi k_B T\,(1-e^{-2\gamma t})/m\right)^{-1/2} exp[-\frac{mv^2(1-(v/v_0)e^{-\gamma t})^2}{2k_B T\,(1-e^{-2\gamma t})}]$$

is a valid solution.

5. Use the method of characteristics to obtain the solution in problem (4). Hint: first Fourier transform the equation to obtain a first order PDE.
6. This exercise is a guided derivation of the Smoluchowski equation from the overdamped Langevin equation.
   (a) Starting from $\partial\rho/\partial t + \partial j/\partial q = 0$, show that $\partial\rho/\partial t + \hat{A}_q\rho = \hat{B}_q(t)\rho$ where $\hat{A}_q\rho = \frac{\partial}{\partial q}(v_D(q)\rho)$ and $\hat{B}_q(t)\rho = -\frac{\partial}{\partial q}\widetilde{R}_q(t)\rho$.
   (b) Obtain an implicit solution to $\partial\rho/\partial t + \hat{A}_q\rho = \hat{B}_q(t)\rho$ in the form of an integral equation:

$$\rho(q,t) = e^{-\hat{A}_q t}\rho(q,0) + \int_0^t dt' e^{-\hat{A}_q(t-t')}\hat{B}_q(t')\rho(q,t')$$

   Hint: $y'(t) + ay(t) = b(t)$ has solution $y = e^{-at}y_0 + \int_0^t dt' e^{-a(t-t')}b(t')$.
   (c) Use the implicit solution to show that

$$\frac{\partial\rho(q,t)}{\partial t} + \hat{A}_q\rho(q,t) = \hat{B}_q(t)e^{-\hat{A}_q t}\rho(q,0) + \hat{B}_q(t)\int_0^t dt' e^{-\hat{A}_q(t-t')}\hat{B}_q(t')\rho(q,t')$$

   (d) Now average over the random forces to obtain

$$\frac{\partial\rho(q,t)}{\partial t} = -\frac{\partial}{\partial q}\left[v_D(q)\rho\right] + \frac{\partial}{\partial q}\left[D_q\frac{\partial\rho}{\partial q}\right]$$

   Comment on some of the subtle aspects of the noise averaging steps. What happens to $\hat{B}_q(t)e^{-\hat{A}_q t}\rho(q,0)$? Why does $\int_0^t dt'\widetilde{R}_q(t)\widetilde{R}_q(t') = \mathbf{1}\times D_q$? Why does noise that influenced $\rho(q,t')$ at times before t' not overlap with the noise at $\widetilde{R}_q(t)$? Finally, why doesn't the derivative in $\int_0^t dt'\widetilde{R}(t)\partial_q[\widetilde{R}_q(t')\rho(q,t')]$ lead to an additional term?
7. Solve the Smoluchowski equation

$$\frac{\partial\rho}{\partial t} = D\frac{\partial}{\partial q}\left\{\frac{\partial\rho}{\partial q} + \rho\frac{\partial\beta F}{\partial q}\right\}$$

for a harmonic oscillator potential of mean force $F(q) = m\omega^2 q^2/2$. First, solve for the initial condition $\rho(q,0) = \delta[0]$, then try the initial condition $\rho(q,0) = \delta[q-q_0]$. Hint: use the equilibrium solution to transform this into an eigenvalue problem.

8. Solve the Smoluchowski equation of problem (7) for a parabolic barrier potential of mean force $\beta F = -m\omega^2 q^2/2$ where $\omega^2 > 0$. First, solve for the initial condition $\rho(q,0) = \delta[0]$, then try the initial condition $\rho(q,0) = \delta[q - q_0]$.
9. Multiply the Smoluchowski equations of problems (7) and (8) by q^n and integrate to obtain a series of ordinary differential equations for the moments, i.e. construct a series of ODEs to obtain $\langle q \rangle_t = \int \rho(q,t) q dq$, $\left\langle (q - \langle q \rangle_t)^2 \right\rangle_t = \int \rho(q,t)(q - \langle q \rangle_t)^2 dq$, etc.
10. A Fokker-Planck equation is always linear in the mathematical sense. However, we refer to them as linear when $v_D(q)$ is linear in q and $D(q)$ is a constant in q, and nonlinear otherwise. Extend exercise 9 to show that in the general nonlinear case $d\langle q \rangle_t /dt = \langle v_D(q) \rangle_t$. Then show that for a linear Fokker-Planck equation the first moment evolves according to $d\langle q \rangle_t /dt = v_D(\langle q \rangle_t)$. Note that the same result emerged for the discrete birth-death process in section 14.2.
11. An example in Chapter 14 analyzed a master equation for the discrete random variable n in a birth-death process. n became a Poisson distributed random variable in the stationary limit. Convert the discrete master equation to a Fokker-Planck equation, solve for the stationary distribution, and compare to the exact Poisson result. What parameter controls the accuracy of the results from the transformation to a Fokker-Planck equation?
12. Polymer translocation through a narrow pore involves intermediate low-entropy polymer configurations as shown in the figure below [adapted from Muthukumar, J. Chem. Phys. 111, 10371 (1999)].

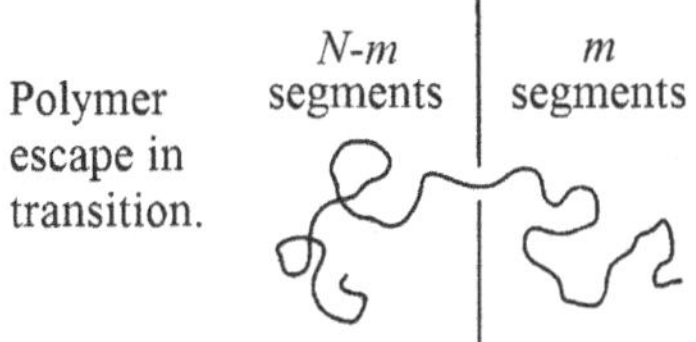

The free energy is $\beta F(m) = 2^{-1} ln[m(N - m)]$ where N is the number of segments on the polymer chain and m is the number of segments on the right side of the orifice. Read and reproduce the analysis by Muthukumar [62] to obtain the Fokker-Planck equation for $\rho(m,t)$:

$$\frac{\partial \rho}{\partial t} = k_0 \frac{\partial}{\partial m} \left\{ \frac{\partial \beta F}{\partial m} \rho + \frac{\partial}{\partial m} \rho \right\}$$

where k_0 is the m-independent frequency at which segments jump rightward across the orifice.

(a) Consider the alternative "linear free energy relationship" model for the transition frequencies from state m to $m+1$,

$$\omega_{m \to m+1} = k_0 exp[-\beta \Delta F_{m \to m+1}/2]$$

where $\Delta F_{m\to m+1} = F(m+1) - F(m)$. Compute the transition frequency from $m+1$ to m to show that detailed balance is "built in" to this construction.

(b) Show that

$$k_0^{-1}\frac{\partial p_m}{\partial t} = (e^{-\partial/\partial m} - 1)p_m e^{-\beta\Delta F_{m\to m+1}/2} + (e^{\partial/\partial m} - 1)p_m e^{-\beta\Delta F_{m\to m-1}/2}$$

Group the first derivative terms and second derivative terms to identify the drift velocity and diffusivity as functions of m.

(c) Show that

$$\lim_{\delta\to 0}(e^{-\beta\Delta F_{m\to m-\delta/2}} - e^{-\beta\Delta F_{m\to m+\delta/2}})/\delta = \frac{d\beta F}{dm}$$

and that for small δ the diffusivity term can be simplified using $(e^{-\beta\Delta F_{m\to m-\delta/2}} + e^{-\beta\Delta F_{m\to m+\delta/2}})/2 \approx 1$.

(d) Use the results from parts (a)–(c) to obtain the Fokker-Plank equation for a continuous m and large N. Is your Fokker-Planck equation the same or different from that of Muthukumar? Explain.

(e) Repeat the analysis in steps (a)–(d) for the general case: $\omega_{m\to m+1} = k_0 exp[-\alpha\beta\Delta F_{m\to m+1}]$ where $0 \le \alpha \le 1$. Do the results depend on α?

13. Analyze the Schlogl reaction system as follows.

(a) For the parameters given in the example of section 15.5, write down the macroscopic deterministic rate equations for a constant volume batch reactor. Find the two nullclines, i.e. along one curve in [**A**], [**B**] space $d[\mathbf{A}]/dt = 0$, and along the other $d[\mathbf{B}]/dt = 0$. Characterize the stationary points, i.e. the points where nullcline curves intersect. Use linear stability analysis (or sketch flowlines if preferred) to characterize the stability of each stationary point.

(b) Use the change of variables $y(n,t) = D(n)p(n,t)$ to convert the Fokker-Planck equation (15.4.3) into a Smoluchowski equation. Then make a second change of variables to $\varphi(n,t) = y(n,t)/\sqrt{y_{SS}(n)}$. Show that the operator which results is a self-adjoint Schrodinger-type equation (with imaginary time). You may find some hints in exercise 15.

(c) Explain how these results could be used to estimate the time to relax from any initial condition to the steady state $p_{SS}(n)$. Also explain how you might determine whether it was okay to develop and use a Fokker-Planck equation that ignores the number of [**B**] molecules.

14. Consider a property $A(\mathbf{q})$ and denote its time-evolving average started from initial condition $\mathbf{q}_0$ as $\bar{A}(\mathbf{q}_0, t)$, i.e.

$$\bar{A}(\mathbf{q}_0, t) = \langle A(\mathbf{q}(t))\rangle_{\mathbf{q}(0)=\mathbf{q}_0}$$

where the average is over all realizations of the stochastic noise. Let the probability distribution $\rho(\mathbf{q},t)$ evolve according to a Fokker-Planck equation $\partial\rho/\partial t = L\rho$. Show that $\bar{A}(\mathbf{q}_0,t)$ can be obtained (i) from $\langle A(\mathbf{q})\psi_1(\mathbf{q})|\rho(\mathbf{q},t)\rangle$ where $|\rho(\mathbf{q},t)\rangle$ is given by equation (15.6.1) or (ii) from $\left\langle\psi_1(\mathbf{q})|\hat{A}(t)|\delta[\mathbf{q}-\mathbf{q}_0]\right\rangle$ where $\hat{A}(t) = \hat{A}\sum_m |\psi_m\rangle\langle\psi_m| e^{-t/\tau_m}$. Compare routes (i) and (ii) to the Schrodinger and Heisenberg approaches in quantum mechanics, respectively.

15. Consider the Smoluchowski equation $\partial\rho/\partial t = L\rho(q,t)$ where

$$\hat{L}\rho \equiv D\frac{\partial}{\partial q}\left\{e^{-\beta F(q)}\frac{\partial}{\partial q}\left(e^{\beta F(q)}\rho\right)\right\}$$

(a) Let $\rho(q,t) = e^{-\beta F(q)/2}\phi(q,t)$. Show that $\hat{L}\rho = -De^{-\beta F(q)/2}\hat{L}_S\phi$ where $\hat{L}_S$ is the self-adjoint Schrodinger-like operator [63]

$$\hat{L}_S\phi \equiv -\frac{\partial^2\phi}{\partial q^2} + V(q)\phi$$

where the effective potential $V(q)$ is related to the original $F(q)$ by

$$V(q) = -\frac{1}{2}\frac{\partial^2\beta F}{\partial q^2} + \frac{1}{4}\left(\frac{\partial\beta F}{\partial q}\right)^2. \tag{15.6.2}$$

(b) Show that eigenfunctions of $\hat{L}$ are the functions $\psi_i(q) = e^{-\beta F(q)/2}\phi_i(q)$ where the $\phi_i(q)$ are eigenfunctions of the self-adjoint operator $\hat{L}_S$. Verify that the first eigenfunction of $\hat{L}_S$ is $\phi_1(q) = e^{-\beta F(q)/2}$. Is this consistent with the result in part (a)? [74]

(c) The two free energy surfaces below are both symmetric with cusp-like barriers of the same height and minima at the same location. Also assume the same diffusivity D.

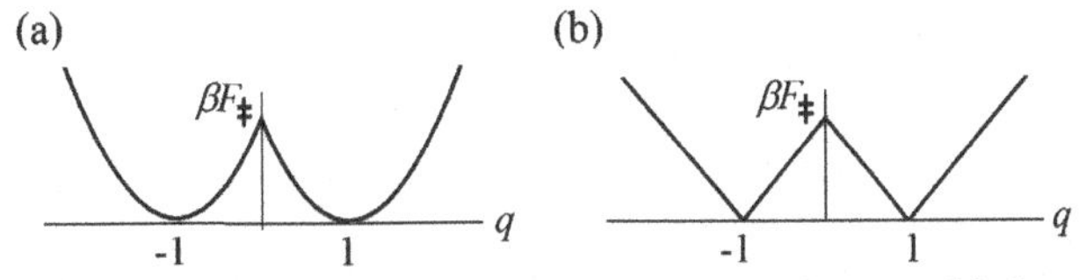

Before working out the two solutions, which one do you think will have the faster relaxation rate? Why? Find the equilibrium eigenfunction and the most slowly relaxing eigenfunction for each of these two state systems.

References

[1] L.P. Kadanoff, Statistical Physics: Statics, Dynamics, and Renormalization, World Scientific, Singapore, 2000.
[2] W.A. Coffey, Y.P. Kalmykov, J.T. Waldron, The Langevin Equation, World Scientific, Singapore, 2004.

[3] A. Nitzan, Dynamics in Condensed Phases: Relaxation, Transfer, and Reactions in Condensed Molecular Systems, Oxford University Press, Oxford, 2006.
[4] R. Erban, S.J. Chapman, Phys. Biol. 6 (2009) 46001.
[5] P. Langevin, C. R. Acad. Sci. Paris 146 (1908) 530–533.
[6] M. Tuckerman, Statistical Mechanics: Theory and Molecular Simulation, Oxford University Press, Oxford, UK, 2010.
[7] B. Leimkuhler, Molecular Dynamics: With Deterministic and Stochastic Numerical Methods, Springer, Berlin, Heidelberg, 2015.
[8] M.P. Allen, D.J. Tildesley, Computer Simulation of Liquids, Clarendon Press, Oxford, UK, 1987.
[9] D.L. Ermak, J.A. McCammon, J. Chem. Phys. 69 (1978) 1352.
[10] M.R. Shirts, J. Chem. Theory Comput. 9 (2013) 909–926.
[11] M. Grunwald, C. Dellago, P.L. Geissler, J. Chem. Phys. 129 (2008) 194101.
[12] R.G. Mullen, J.-E. Shea, B. Peters, J. Chem. Phys. 140 (2014) 41104.
[13] H.C. Andersen, J. Chem. Phys. 72 (1980) 2384.
[14] P. Hanggi, P. Jung, Adv. Chem. Phys. 89 (1995) 239–326.
[15] A. Einstein, Investigations on the Theory of the Brownian Movement, Courier Corporation, 1956.
[16] G.E. Uhlenbeck, L.S. Ornstein, Phys. Rev. 36 (1930) 823.
[17] G. Hummer, New J. Phys. 7 (2005) 34.
[18] R.B. Best, G. Hummer, Proc. Natl. Acad. Sci. USA 107 (2010) 1088–1093.
[19] D.J. Higham, SIAM Rev. 43 (2001) 525–546.
[20] D.T. Gillespie, Markov Processes: An Introduction for Physical Scientists, Academic Press, Boston, 1992.
[21] B. Leimkuhler, C. Matthews, Appl. Math. Res. Express 2013 (2013) 34–56.
[22] L. Rayleigh, Philos. Mag. 32 (1891) 424–444.
[23] A.D. Fokker, Ann. Phys. 348 (1914) 810–820.
[24] M. von Smoluchowski, Z. Phys. 17 (1916) 557–585.
[25] M. Planck, Sitz.ber. Preuss. Akad. Wiss. 24 (1917).
[26] C.W. Gardiner, Handbook of Stochastic Methods, 2nd ed., Springer-Verlag, Berlin, Heidelberg, 1985.
[27] R. Zwanzig, Nonequilibrium Statistical Mechanics, Oxford University Press, Oxford, 2001.
[28] N.G. van Kampen, Stochastic Processes in Physics and Chemistry, Elsevier, Amsterdam, 2007.
[29] Y.B. Zeldovich, Acta Physicochim. URSS 18 (1943) 1–22.
[30] J. Frenkel, Kinetic Theory of Liquids, Oxford University Press, London, 1946.
[31] A. Kolmogorov, Math. Ann. 104 (1931) 415–458.
[32] R. Zwanzig, Phys. Rev. 124 (1961) 983.
[33] G. Hummer, I.G. Kevrekidis, J. Chem. Phys. 118 (2003) 10762.
[34] B.C. Knott, V. Molinero, M.F. Doherty, B. Peters, J. Am. Chem. Soc. 134 (2012) 1–6.
[35] J.R. Espinosa, C. Vega, C. Valeriani, E. Sanz, J. Chem. Phys. 144 (3) (2016) 034501.
[36] W. Im, B. Roux, J. Mol. Biol. 319 (2002) 1177–1197.
[37] E. Sanz, C. Vega, J.R. Espinosa, R. Caballero-Bernal, J.L.F. Abascal, C. Valeriani, J. Am. Chem. Soc. 135 (2013) 15008–15017.
[38] N.E.R. Zimmermann, B. Vorselaars, D. Quigley, B. Peters, J. Am. Chem. Soc. 137 (2015) 13352–13361.
[39] I. Oppenheim, Adv. Chem. Phys. 15 (1969) 1–11.
[40] R. Erban, S.J. Chapman, I.G. Kevrekidis, T. Vejchodsky, SIAM J. Appl. Math. 70 (2009) 984–1016.
[41] N.G. Van Kampen, Adv. Chem. Phys. 15 (1969) 65–77.
[42] R.R. Coifman, I.G. Kevrekidis, S. Lafon, M. Maggioni, B. Nadler, SIAM J. Multiscale Model. Simul. 7 (2008) 842–864.
[43] A. Singer, R. Erban, I.G. Kevrekidis, R.R. Coifman, Proc. Natl. Acad. Sci. USA 106 (2009) 16090–16095.
[44] P.J. Ledbetter, C. Clementi, J. Chem. Phys. 135 (2011) 44116.
[45] A.L. Ferguson, A.Z. Panagiotopoulos, I.G. Kevrekidis, P.G. Debenedetti, Chem. Phys. Lett. 509 (2011) 1–11.
[46] M.A. Rohrdanz, W. Zheng, C. Clementi, Annu. Rev. Phys. Chem. 64 (2013) 295–316.

[47] A. Singer, Appl. Comput. Harmon. Anal. 21 (2006) 128–134.
[48] W. Zheng, B. Qi, M.A. Rohrdanz, A. Caflisch, A.R. Dinner, C. Clementi, J. Phys. Chem. B 115 (2011) 13065–13074.
[49] X. Huang, G.R. Bowman, S. Bacallado, V.S. Pande, Proc. Natl. Acad. Sci. USA 106 (2009) 19765–19769.
[50] K.A. Beauchamp, G.R. Bowman, T.J. Lane, L. Maibaum, I.S. Haque, V.S. Pande, J. Chem. Theory Comput. 7 (2011) 3412–3419.
[51] W. Zheng, M.A. Rohrdanz, C. Clementi, J. Phys. Chem. B 117 (2013) 12769–12776.
[52] J.S. Langer, Systems Far from Equilibrium, Springer, Berlin, Heidelberg, 1980, pp. 12–47.
[53] B. Peters, J. Phys. Chem. B 119 (2015) 6349–6356.
[54] C.R. Schwantes, V.S. Pande, J. Chem. Theory Comput. 11 (2015) 600–608.
[55] F. Vitalini, F. Noe, B.G. Keller, J. Chem. Theory Comput. 11 (2015) 3992–4004.
[56] B. Peters, B.L. Trout, J. Chem. Phys. 125 (2006) 54108.
[57] B. Peters, Mol. Simul. 36 (2010) 1265–1281.
[58] K.E. Shuler, Phys. Fluids 2 (1959) 442.
[59] N.-V. Buchete, G. Hummer, J. Phys. Chem. B 112 (2008) 6057.
[60] J.-H. Prinz, H. Wu, M. Sarich, B. Keller, M. Senne, M. Held, J.D. Chodera, C. Schutte, F. Noe, J. Chem. Phys. 134 (2011) 174105.
[61] S. Redner, A Guide to First-Passage Processes, Cambridge University Press, Boston, 2001.
[62] M. Muthukumar, J. Chem. Phys. 111 (1999) 10371–10374.
[63] N.G. van Kampen, J. Stat. Phys. 17 (1977) 71–88.

CHAPTER 16

Kramers theory

The very fact that a reaction is going on means that [transition states] will never be in exact temperature equilibrium with state A.

Kramers, Physica VII (1940)

Kramers developed an important rate theory that spans a number of dynamical regimes [1]. At one extreme, Kramers theory describes a quasi-microcanonical dynamics where the rate is limited by slow energy transfer processes that cannot activate reactants nor quench activated products. At the other extreme, Kramers theory describes processes like nucleation and biomolecular conformational transitions where the dynamics resemble overdamped diffusion over a barrier. An intermediate regime of Kramers theory describes trajectories that cross a critical dividing surface and continue along the reaction coordinate to the product state with few barrier recrossings, almost like the dynamics assumed in transition state theory. These *qualitative* results have been validated in numerous simulations and even in some experiments. In the high friction limit, the Kramers theory is quantitatively correct. In particular, the high friction Kramers theory corroborates Pontryagin's theory of diffusion over barriers [2], classic theories of nucleation [3], theories of polymer relaxation dynamics [4], and some predictions about protein folding rates [5].

However, the Kramers theory has also been criticized because it does not accurately predict transmission coefficients (nor absolute rates) for most chemical reactions [6,7]. The shortcomings of Kramers theory are most severe in the weak-coupling limit, where it ostensibly models unimolecular dissociation reactions in low pressure gases. In this limit, Kramers invokes a steady weak friction that is markedly different from the strong but infrequent collisions that activate real unimolecular decay processes [4,7,8]. Additionally, the inertial Langevin equation at the heart of Kramers theory cannot accurately describe coupling between a slow solvent and rapid motion along a reaction coordinate [9,10].

Nevertheless, Kramers theory does *qualitatively* explain how reaction dynamics and rates depend on coupling between the reaction coordinate and other degrees of freedom. Moreover, Kramers theory inspired several theories that do accurately model reaction dynamics in condensed phases [4,11]. For example, the inertial Langevin equation in the Kramers theory can be generalized to obtain the Grote-Hynes theory which more accurately predicts transmission coefficients for reactions in solution [9]. Kramers theory has also been generalized to multiple dimensions which extends its domain of validity by explicitly including the dynamics of several slow variables [12,13].

Kramers theory continues to evolve, but experiments that test its predictions about solvent viscosity and chemical reaction rates in solution remain difficult [7,14,15]. Efforts to modify the solvent viscosity invariably also make small changes in solvation free energies and therefore in the free energy barrier. Simulations and particularly rare events methods, where dynamics and thermodynamics can be separately manipulated, provide one of the best arenas to test theories of reaction dynamics in the condensed phase. Additionally, there are continuing efforts to forge a better understanding of the abstract parameters in the Kramers theory and thereby to enable more easily tested predictions [16–20].

16.1 Intermediate and high friction

The starting point in Kramers theory is the inertial Langevin equation [1,21]. The inertial Langevin equation and its properties were discussed in Chapter 15, but let us briefly recap the main points. The inertial Langevin equation is

$$m\ddot{q} = -\frac{\partial V}{\partial q} - m\gamma\dot{q} + R(t) \tag{16.1.1}$$

Here q represents a reaction coordinate, m is the reduced mass/inertia for the reaction coordinate, γ is the friction, and $V(q)$ is a potential of mean force (PMF).[1] $R(t)$ is a random force that models the effects of bath degrees of freedom on the reaction coordinate. The random force along coordinate q has zero mean, $\langle R(t)\rangle = 0$, and a delta correlated variance as required by the fluctuation dissipation theorem [21]: $\langle R(t)R(0)\rangle = 2m\gamma k_B T\delta[t]$.

In principle, the Langevin equation can be constructed from molecular simulations. For example, the effective mass associated with a coordinate q can be obtained from equipartition: $m\langle\dot{q}^2\rangle = k_B T$. The effective friction can also be computed using clamped simulations as described in Chapter 17. The PMF can be computed using umbrella sampling or other methods for computing free energies. In real systems, these calculations are complicated because the effective mass m, the friction γ, and of course $V(q)$ are functions of the location along q. Let us postpone computational procedures until the next chapter. Here we follow Kramers in considering a constant mass m, a constant friction γ, and an idealized parabolic barrier model for the PMF.

[1] Systems described by Kramers theory span the condensed phase (where we might obtain $-\partial V/\partial q$ by differentiating a free energy profile $F(q)$) to a nearly isolated degree of freedom (where we might obtain $-\partial V/\partial q$ by differentiating the Born-Oppenheimer potential).

First, note that the inertial Langevin equation (16.1.1) can be rewritten as a system of first order equations for the position q and velocity v along the reaction coordinate. The pair of equations is $dq/dt = v$ and $m\, dv/dt = -\partial V/\partial q - m\gamma v + R(t)$. These equations can be converted to a special Klein-Kramers type Fokker-Planck equation for the probability density $\rho(q, v, t)$ [21, 22]:

$$\frac{\partial \rho}{\partial t} = \left\{ -v\frac{\partial}{\partial q} + \gamma \frac{\partial}{\partial v}\left(v + \frac{k_B T}{m}\frac{\partial}{\partial v} \right) + \frac{1}{m}\frac{\partial V}{\partial q}\frac{\partial}{\partial v} \right\} \rho \tag{16.1.2}$$

Kramers' theory models the barrier as a parabolic potential

$$V(q) = V_{\ddagger} - \frac{1}{2} m\omega_{\ddagger}^2 (q - q_{\ddagger})^2 \tag{16.1.3}$$

Let us assume for convenience that the value of $q_{\ddagger} = 0$, i.e. that transition states are located at the origin on the q-axis. We can always shift q by $q_{\ddagger}$ to make the origin coincide with the transition state. Later in our discussion we will make another harmonic approximation to the PMF in the reactant minimum. Kramers theory assumes nothing about the product state except that it should irreversibly absorb reactive trajectories. The PMF and the two harmonic approximations are shown in Figure 16.1.1.

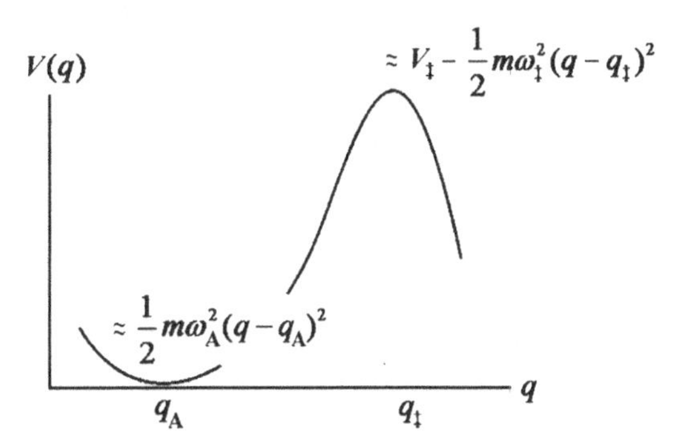

Figure 16.1.1: Harmonic approximations at the reactant minimum (q_A) and barrier top ($q_{\ddagger}$) along a potential energy profile $V(q)$. The bottom of the reactant well is the zero of energy.

Using the parabolic model for $V(q)$ near $q_{\ddagger}$ in the Klein-Kramers equation gives

$$\frac{\partial \rho}{\partial t} = -v\frac{\partial \rho}{\partial q} - \omega_{\ddagger}^2 q \frac{\partial \rho}{\partial v} + \gamma \frac{\partial}{\partial v}(v\rho) + \frac{\gamma k_B T}{m}\frac{\partial^2 \rho}{\partial v^2}$$

TST computes the rate as though the transition states are at equilibrium with the reactants. Instead, Kramers envisioned a steady-state non-equilibrium flux of trajectories over the barrier [1]. If the barrier is large compared to $k_B T$, then an ensemble prepared on the reactant side of the barrier will quickly reach a local equilibrium within the reactant state, perturbed only slightly by a slow leak over the barrier. If we "rescue" each trajectory that escapes to the product side of the barrier and "replace" it on the reactant side, the non-equilibrium system will develop a steady-state distribution, $\rho_{SS}(q, v)$. Deep within the reactant well, the steady-state distribution will resemble the equilibrium distribution of reactants. Trajectories are always removed from the product side, so ρ_{SS} must approach zero deep within the product basin. For all points between, the non-equilibrium steady-state density must satisfy the equation

$$0 = -v\frac{\partial \rho_{SS}}{\partial q} - \omega_{\ddagger}^2 q \frac{\partial \rho_{SS}}{\partial v} + \gamma \frac{\partial}{\partial v}(v\rho_{SS}) + \frac{\gamma k_B T}{m}\frac{\partial^2 \rho_{SS}}{\partial v^2} \tag{16.1.4}$$

with boundary conditions

$$\rho_{SS}(q, v) \to 0 \quad as \quad q \to \infty \tag{16.1.5}$$

and

$$\rho_{SS}(q, v) \to \rho_{eq}(q, v) \quad as \quad q \to -\infty \tag{16.1.6}$$

We cannot properly normalize $\rho_{eq}(q, v)$ because $exp[-\beta V(q)]$ diverges as $q \to \infty$. Kramers fixed this problem with a crossover function $\xi_{SS}(q, v)$, defined as the ratio of steady-state and equilibrium distributions [1]

$$\xi_{SS}(q, v) \equiv \rho_{SS}(q, v)/\rho_{eq}(q, v) \tag{16.1.7}$$

Substituting the Kramers' crossover definition into equation (16.1.4) for $\rho_{SS}(q, v)$ gives an equation for $\xi_{SS}(q, v)$

$$v\frac{\partial \xi_{SS}}{\partial q} + \omega_{\ddagger}^2 q \frac{\partial \xi_{SS}}{\partial v} + \gamma v \frac{\partial \xi_{SS}}{\partial v} = \frac{\gamma k_B T}{m}\frac{\partial^2 \xi_{SS}}{\partial v^2} \tag{16.1.8}$$

with new asymptotic boundary conditions

$$\xi_{SS}(q, v) \to 0 \quad as \quad q \to \infty$$

and

$$\xi_{SS}(q, v) \to 1 \quad as \quad q \to -\infty$$

The method of characteristics (or an inspired guess [1]) reveals that the dependence on q and v can be collapsed to dependence on a single variable $u = q - av$. To identify the appropriate constant a, use the chain rule to obtain the derivatives $\partial \xi_{SS}/\partial q = \xi'_{SS}(u)$, $\partial \xi_{SS}/\partial v = -a\xi'_{SS}(u)$, and $\partial^2 \xi_{SS}/\partial v^2 = a^2 \xi''_{SS}(u)$. Substitute these into equation (16.1.8) and simplify to obtain

$$-a\omega_{\ddagger}^2 \left(q - v\frac{1 - \gamma a}{a\omega_{\ddagger}^2} \right) \xi'_{SS}(u) = \frac{\gamma k_B T}{m} a^2 \xi''_{SS}(u) \tag{16.1.9}$$

The dependence on q and v will collapse to a dependence on the combined variable $u \equiv q - av$ if we can choose a such that equation (16.1.9) contains no independent factors of q and v. Specifically, if we choose a such that the term in parentheses becomes u, we will have eliminated all separate factors of q and v, while still having a linear equation. Therefore we must solve

$$q - v\frac{1-\gamma a}{a\omega_{\ddagger}^2} = q - av$$

to find the appropriate value of a. This equation has two solutions

$$a = -\frac{\gamma}{2\omega_{\ddagger}^2}\left(1 \pm \sqrt{1 + 4\omega_{\ddagger}^2/\gamma^2}\right)$$

For these two choices of a, equation (16.1.9) becomes

$$-\frac{m\omega_{\ddagger}^2}{\gamma k_B T a} u \xi_{SS}'(u) = \xi_{SS}''(u)$$

The general solution is

$$\xi_{SS}(u) = c_2 + c_1 \text{erf}\left[u\sqrt{\frac{m\omega_{\ddagger}^2}{2\gamma k_B T a}}\right]$$

where c_1 and c_2 are integration constants to be determined by the boundary conditions on $\xi_{SS}(u)$. As $q \to \infty$, $u \to \infty$ for any v, and as $q \to -\infty$, $u \to -\infty$ for all v. Therefore, the boundary conditions have become

$$\xi_{SS}(u) \to 0 \quad as \quad u \to \infty$$

and

$$\xi_{SS}(u) \to 1 \quad as \quad u \to -\infty$$

Remembering our convention that $\omega_{\ddagger}^2 > 0$, we must also require $a > 0$ for solutions that do not diverge.[2] Therefore,

$$a = \frac{\gamma}{2\omega_{\ddagger}^2}\left(\sqrt{1 + 4\omega_{\ddagger}^2/\gamma^2} - 1\right)$$

[2] $i\,\text{erf}(i\,u)$ is a real valued function for real u, but it diverges as $u \to \pm\infty$.

From properties of the erf function and the boundary conditions, the limit as $u \to \infty$ gives $\xi_{SS}(u) \to c_2 + c_1$. Therefore, $c_1 = -c_2$. Then as $u \to -\infty$, we find that $\xi_{SS}(u) \to 2c_2$, and therefore to match the left boundary condition $c_2 = 1/2$. Finally, we obtain [1]

$$\xi_{SS}(u) = \frac{1}{2}\text{erfc}\left[\sqrt{\frac{m\omega_{\ddagger}^2}{2\gamma k_B T a}} \cdot u\right] \tag{16.1.10}$$

where $\text{erfc}(x) \equiv 1 - \text{erf}(x)$. The solution for the crossover function is shown in Figure 16.1.2.

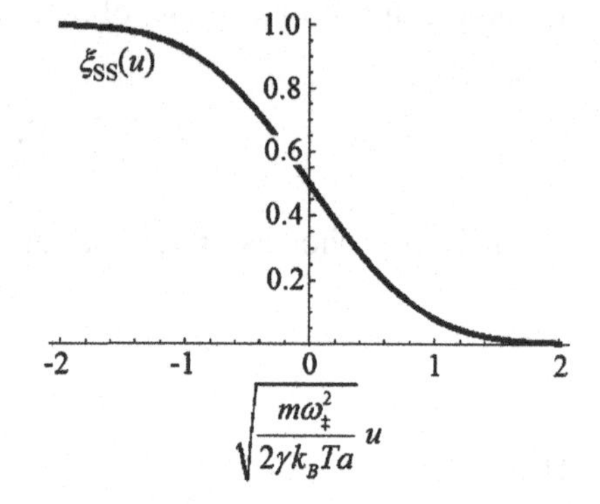

Figure 16.1.2: The Kramers crossover function in terms of the combined variable u.

Now we can use $\xi_{SS}(u)$ to reconstruct the (unnormalized) steady-state distribution

$$\rho_{SS} = \frac{1}{2}exp\left[-\frac{mv^2}{2k_B T} - \frac{1}{k_B T}\left(V_{\ddagger} - \frac{1}{2}m\omega_{\ddagger}^2 q^2\right)\right] \cdot \text{erfc}\left[\sqrt{\frac{m\omega_{\ddagger}^2}{2\gamma k_B T a}}(q - av)\right]$$

With some algebra, $\rho_{SS}(q, v)$, $\rho_{eq}(q, v)$, and $\xi_{SS}(q, v)$ can each be written in terms of dimensionless variables $\underline{q} = m^{1/2}\omega_{\ddagger} q/(k_B T)^{1/2}$ and $\underline{v} = m^{1/2} v/(k_B T)^{1/2}$. This leaves only one parameter, $\gamma/\omega_{\ddagger}$, in the steady-state distribution. In Figures 16.1.3 and 16.1.4 we have made this change of variables and plotted $\xi_{SS}(\underline{q}, \underline{v})$ and $\rho_{SS}(\underline{q}, \underline{v})$.

Several features are noteworthy in these plots. $\rho_{SS}(q, v)$ vanishes as we move toward positive values of q and toward negative values of v. The region where ρ_{SS} vanishes penetrates into the reactant state ($q < 0$) because there are no trajectories returning from the product state. $\rho_{SS}(q, v)$ cannot have support in regions that correspond to trajectories returning from the product state due to the rescue and replace boundary conditions. Additionally, note that isosurfaces of $\rho_{SS}(q, v)$ bend upward as they cross the barrier top because trajectories in the nonequilibrium ensemble tend to accelerate as the barrier is crossed. As friction increases, the velocity at the barrier top becomes increasingly irrelevant. At $\gamma/\omega_{\ddagger} = 4.0$ the steady-state distribution reveals that much of the forward flux is canceled by recrossing, i.e. by trajectories whose velocities are damped out in the barrier region before they fall down into the product state.

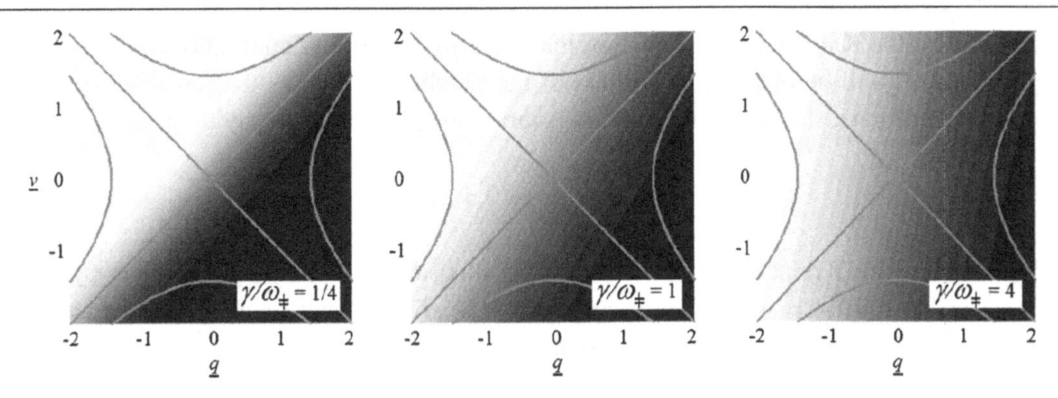

Figure 16.1.3: The contour lines show the total energy as a function of dimensionless position and velocity in the vicinity of the barrier top. The shading shows the Kramers crossover function $\xi_{SS}(\underline{q},\underline{v})$. The energy and crossover function are shown for three different values of the dimensionless parameter $\gamma/\omega_{\ddagger}$.

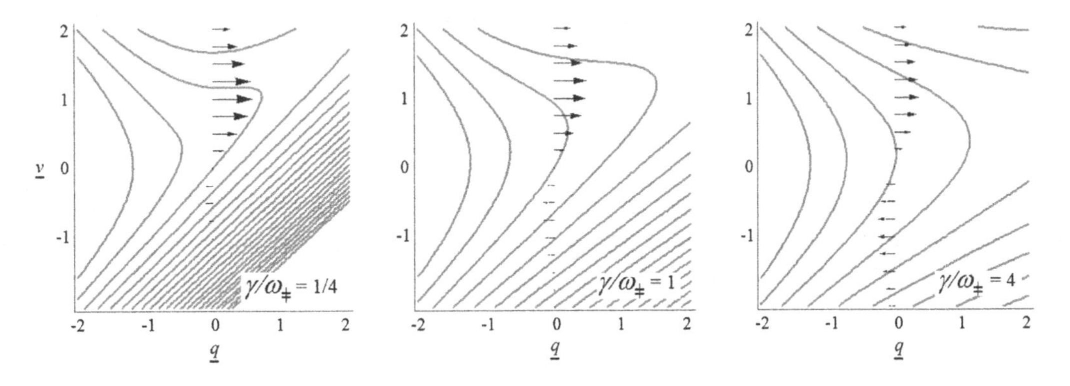

Figure 16.1.4: The contour lines show the steady-state distribution $\rho_{SS}(\underline{q},\underline{v})$ in the vicinity of the barrier top. The distribution diverges to infinity to the left of the barrier top, and becomes vanishingly small on the product side. The distribution is shown for three different values of the dimensionless parameter $\gamma/\omega_{\ddagger}$. Contours are separated by factors of two in ρ_{SS}. Arrows show the local flux crossing each point on the dividing surface $q = 0$.

To obtain the Kramers rate constant we must normalize $\rho_{SS}(q,v)$ and then compute the flux through a dividing surface [1]. $\rho_{SS}(q,v)$ has negligible support near the barrier top and in the product region. Essentially all support comes from the reactant state where $\rho_{SS}(q,v) \propto \rho_{eq}(q,v)$. Therefore the normalized density is

$$\rho_{SS}(q,v) = \frac{e^{-\beta m v^2/2 - \beta V(q)}\xi_{SS}(q,v)}{\int e^{-\beta m v^2/2 - \beta m\omega_A^2(q-q_A)^2}dq\,dv} = \frac{m\omega_A}{2\pi k_B T}e^{-\beta m v^2/2 - \beta V(q)}\xi_{SS}(q,v) \quad (16.1.11)$$

Additionally, the Klein-Kramers equation that gave $\rho_{SS}(q, v)$ is a continuity equation, so an equivalent rate will be obtained regardless of the dividing surface as long as it separates the reactant and product states. The calculation is most easily done for the dividing surface $q = 0$.

$$k_K = \int_{-\infty}^{\infty} v\rho_{SS}(0, v)dv \tag{16.1.12}$$

where k_K is the Kramers rate constant. The flux integral can be simplified by using $vexp[-mv^2/(2k_BT)] = -(k_BT/m)\partial exp[-mv^2/(2k_BT)]/\partial v$ and then integrating by parts. The result of the integration is the Kramers rate constant

$$k_K = \frac{\gamma}{\omega_\ddagger}\left(\sqrt{\frac{1}{4}+\frac{\omega_\ddagger^2}{\gamma^2}}-\frac{1}{2}\right)\frac{\omega_A}{2\pi}exp\left[-\beta V_\ddagger\right]$$

Recall that classical TST in one dimension gave $k_{TST} = (\omega_A/2\pi)exp(-V_\ddagger/k_BT)$. Thus the Kramers theory rate is $k_K = \kappa_K k_{TST}$ where κ_K is the Kramers transmission coefficient [1]

$$\kappa_K = \frac{\gamma}{\omega_\ddagger}\left(\sqrt{\frac{1}{4}+\frac{\omega_\ddagger^2}{\gamma^2}}-\frac{1}{2}\right) \tag{16.1.13}$$

Figure 16.1.5 shows κ_K as a function of $\gamma/\omega_\ddagger$. If we studied a bimolecular reaction, then the reactant partition function would change so as to equally modify $\rho_{SS}(x, v)$ and the TST rate constant. Therefore the transmission coefficient would still be that given in equation (16.1.13) [7].

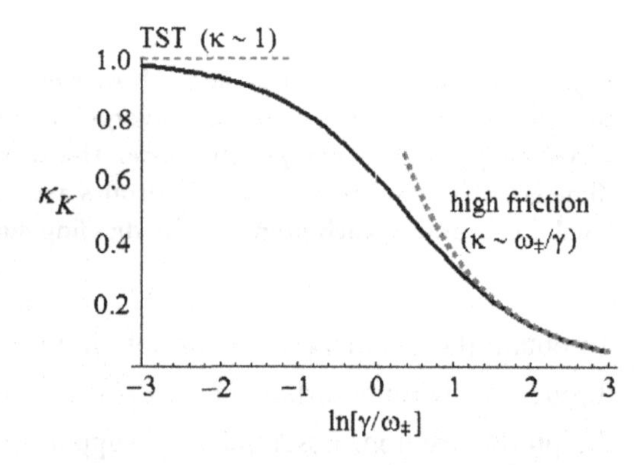

Figure 16.1.5: The transmission coefficient from Kramers theory for intermediate friction and assuming a parabolic barrier. At low friction the transmission coefficient of equation (16.1.13) becomes unity and at high friction it becomes inversely proportional to the friction.

At high friction, the dynamics along the reaction coordinate begin to resemble diffusion along the reaction coordinate as shown in Figure 16.1.6 and k_K becomes the mean first passage rate:

$$k_{MFP} = \frac{\omega_\ddagger}{\gamma}k_{TST} \tag{16.1.14}$$

The Kramers transmission coefficient does not explicitly depend on temperature, but strong temperature dependence may enter through the friction. γ is related to diffusivity along the reaction coordinate via the Einstein relation: $D = k_B T / m\gamma$. Based on theories of diffusion in liquids and glasses, γ should increase approximately in proportion to the viscosity. For solute precipitate nucleation in solids, γ is inversely proportional to the rate of solute diffusion - a process that typically occurs by vacancy hopping and/or other activated processes. Thus both viscosity and solid-state diffusivities can lead to a strong, even Arrhenius-like, temperature dependence *within the prefactor* of the high friction Kramers rate constant. See Chapter 18 for more details on these effects.

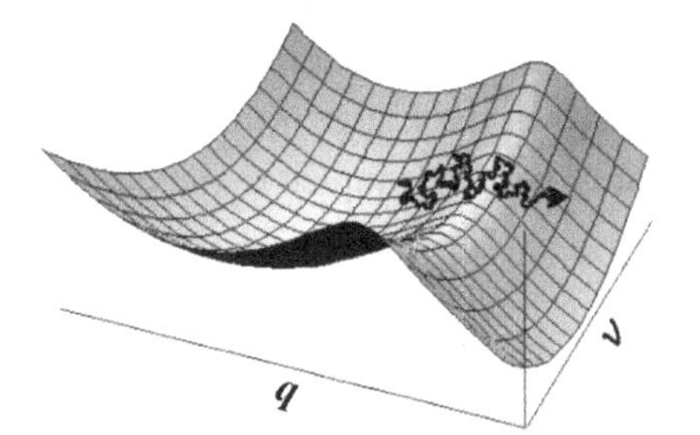

Figure 16.1.6: When $\gamma \gg \omega_\ddagger$ the dynamics resemble diffusion over the barrier top, and the Kramers transmission coefficient becomes $\omega_\ddagger/\gamma$.

This section has shown that Kramers theory encompasses regimes from the direct dynamics of TST to the overdamped dynamics of processes like protein folding and nucleation. As the next section shows, Kramers theory can also qualitatively describe dynamical effects of slow energy relaxation at low friction.

16.2 Low friction: the energy diffusion limit

At low friction, the transmission coefficient in equation (16.1.13) approaches unity, suggesting that the Kramers rate constant should become that of TST. However, the validity of the TST limit at low friction depends on the nature of the energy landscape. If motion along the reaction coordinate can continue indefinitely in the forward and backward directions, as in a bimolecular reaction, then the low friction limit of Kramers theory does reduce to TST [7]. Reactive trajectory studies for some bimolecular reactions in the gas phase confirm that TST is essentially correct [23], but significant crossing is observed for reactions with highly curved paths [24,25]. Several other scenarios can also cause recrossing, even in the limit of low friction.

Chapter 9 discussed non-TST rate expressions for unimolecular reactions in low pressure gases where activation (and also deactivation) occurs by strong but infrequent collisions with other gas molecules. Kramers [1] instead models the effects of a small steady friction that limits the rate at which trajectories can acquire (and dissipate) activation energy. Trajectories in this weak friction limit oscillate, or "orbit", the reactant well many times before they can appreciably change their energy level. Although Kramers' weak steady Langevin friction differs from the strong, infrequent, and prolonged collisions that occur in real gases, the model of Kramers qualitatively preserves the expected time scale hierarchy for weak coupling:

$$\omega_A^{-1} \ll \gamma^{-1} \ll k^{-1}$$

These time scales are the vibrational period in the reactant well, the time for vibrational energy dissipation (in lieu of rare collisions), and the escape time. Reactive trajectories for such a system will qualitatively resemble those shown in Figure 16.2.1 Note that the previous section envisioned $V(q)$ as a PMF or a free energy profile because the intermediate to high friction limit is most appropriate for processes in condensed phases. In the low friction limit, we envision $V(q)$ in the Langevin equation as the potential energy profile for a degree of freedom that is only weakly coupled to the other modes.

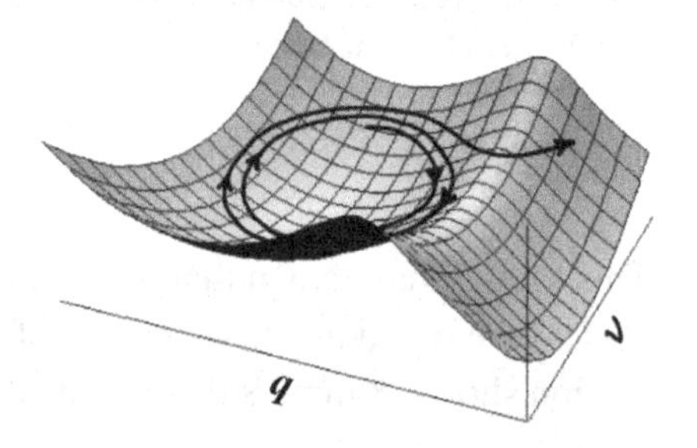

Figure 16.2.1: When $\gamma \ll \omega_A$ trajectories can orbit the reactant basin many times before dissipating energy on the order k_BT. This energy diffusion limited regime also lowers the transmission coefficient.

For low friction dynamics, the (q, v) coordinates are not ideal. Both q and v will oscillate wildly each time the system nearly acquires the transition state energy. Energy or action variables are more convenient because the rapid phase oscillations can be averaged out leaving the slowly varying energy or action variables [1,22]. Again we begin with an inertial Langevin equation, but now we multiply all terms by the velocity.

$$mv\dot{v} + v\frac{\partial V}{\partial q} = -m\gamma v^2 + vR(t) \tag{16.2.1}$$

If the total energy is $E = mv^2/2 + V(q)$, then the left hand side of equation (16.2.1) is dE/dt. Averaging over an oscillation period for a perfectly harmonic reactant well gives

$$\frac{dE}{dt} = -\gamma E + \overline{vR(t)}$$

The period-averaged random noise term continues to provide random kicks, but with an altered variance. Beyond this point the derivation is not easy. The Langevin equation is ultimately converted into a Smoluchowski equation for $\rho(E,t)$. The derivation requires a relationship between the classical action and the energy: $dE/dS = \omega(E)$ [26]. We skip all of the details [27], and go straight to the energy diffusion Smoluchowski equation

$$\frac{\partial \rho}{\partial t} = \frac{\partial}{\partial E}\left\{\gamma S(E)\left(1 + k_B T \frac{\partial}{\partial E}\right)\Omega(E)^{-1}\rho\right\} \tag{16.2.2}$$

where $\Omega(E)$ is the density of states in the reactant well at energy E. The quantity $\gamma S(E)$ is the rate of energy diffusion into and out of the reaction coordinate.

We again seek a non-equilibrium steady-state solution to equation (16.2.2). In the energy diffusion limit, trajectories escape to the product state as soon as they acquire the necessary activation energy. Note the difference between Kramers and RRKM theory here. RRKM theory explicitly considers all modes of the reactant and their total energy. Upon sudden activation to a total energy that exceeds the threshold, the RRKM theory predicts a random waiting time for sufficient energy to become focused into the reaction coordinate. The low friction Kramers model explicitly considers only one degree of freedom. The waiting time in the Kramers model is for threshold activation of the one explicitly modeled reaction coordinate, and the reaction then promptly occurs. The low friction Kramers model therefore invokes the boundary condition

$$\rho_{SS}(V_{\ddagger}) = 0 \tag{16.2.3}$$

Trajectories that escape are replaced at the bottom of the reactant well to maintain a steady-state distribution. A calculation similar to the mean first passage time calculation for the high friction limit gives the energy dissipation limited (EDL) rate

$$k_{EDL}^{-1} = \beta \int_0^{V_{\ddagger}} dE\,\Omega(E) exp\,(-\beta E) \int_E^{V_{\ddagger}} \frac{dE'}{\gamma S(E')} exp[+\beta E']$$

The integral over E' primarily contributes when E' reaches values near $V_{\ddagger}$. The integral over E primarily contributes near the reactant minimum. An approximate evaluation of the integrals gives

$$k_{EDL} \approx \beta\gamma S(V_{\ddagger}) \frac{\omega_A}{2\pi} exp[-\beta V_{\ddagger}] \tag{16.2.4}$$

Normalizing by the TST rate, the energy diffusion limited transmission coefficient is

$$\kappa_{EDL} \approx \beta\gamma S(V_{\ddagger}) \tag{16.2.5}$$

Equation (16.2.5) says that the transmission coefficient is proportional to the action integral over the reactant well at the transition state energy $S(V_{\ddagger}) \approx V^{\ddagger}/\omega_A$. Qualitatively, the number of recrossings is the ratio of the time to dissipate $k_B T$ of energy and the vibrational period of the stable well. As friction γ gets weaker, the time for dissipation grows, the number of recrossings grows, and the transmission coefficient goes down. Thus for unimolecular reactions, Kramers theory predicts a transmission coefficient that increases for low friction and then decreases for high friction. This result, shown in Figure 16.2.2, is the Kramers turnover.

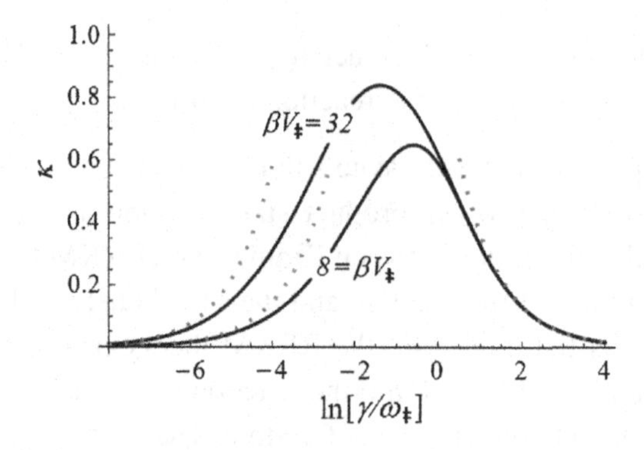

Figure 16.2.2: For purposes of illustration, suppose that the reactant well is harmonic and that $\omega_A = \omega_\ddagger$. The solid curves show the Melnikov-Meshkov transmission coefficient (κ_{MM}) for different values of $\gamma/\omega_\ddagger$ and $\beta V\ddagger$. The rightmost dotted curve shows the overdamped asymptotic transmission coefficient (κ_{MFP}). The dotted curves on the left show Kramers asymptotic low friction transmission coefficient (κ_{EDL}). Note that the low friction transmission coefficient depends on the barrier height but the high friction transmission coefficient does not.

Note that Kramers did not actually solve for the escape rate in the turnover friction regime. Mel'nikov and Meshkov [27,28] and Pollak et al. [29] developed the connection formulas that bridge the low and high friction regimes. Their routes to the connection formulas differ, but both consider the distribution of energy losses for trajectories that orbit the reactant basin at the barrier energy. Their common result is a Gaussian transition probability for the energy E_2 after one orbit beginning at energy E_1.

$$p(E_2|E_1) = (4\pi k_B T \gamma S_\ddagger)^{-1/2} exp\left[-\frac{(E_2 - E_1 + \gamma S_\ddagger)^2}{4k_B T \gamma S_\ddagger}\right]$$

In accordance with intuition, orbits started at $V_\ddagger$ tend to dissipate energy because $\gamma S_\ddagger > 0$. Mel'nikov and Meshkov used the energy transition probabilities to derive a transmission coefficient that bridges the high and low friction limits [28]

$$\kappa_{MM} = \frac{\gamma}{\omega_\ddagger}\left(\sqrt{\frac{1}{4}+\frac{\omega_\ddagger^2}{\gamma^2}} - \frac{1}{2}\right) exp\left\{\frac{1}{\pi}\int_0^\infty d\varsigma \frac{ln\left[1 - exp\left(\beta\gamma S(V_\ddagger)(\varsigma^2 + 1/4)\right)\right]}{\varsigma^2 + 1/4}\right\} \tag{16.2.6}$$

Mel'nikov and Meshkov also found that trajectories crossing the transition state have non-equilibrium velocity distributions. The average kinetic energy along q at the transition state increases from zero in the low friction limit to the equipartition value (i.e. to equilibrium) in the high friction limit [27]. The non-equilibrium distribution of kinetic energy at $q_{\ddagger}$ is also evident in Figure 16.1.4.

Because of the boundary condition in equation (16.2.3), the average kinetic energy (along the reaction coordinate) of trajectories as they cross the saddle region must approach zero as γ approaches zero. This finding stands in stark contrast to the perfectly equilibrium distribution of the velocities for the overdamped regime. Recall that reactive flux calculations use as initial conditions on the dividing surface an equilibrium velocity distribution, regardless of the friction and/or coupling strength [30,31]. The energy diffusion limit of Kramers theory pertains to a rather unrealistic model, but nevertheless this inconsistency with the reactive flux formalism might seem troubling. The apparent inconsistency can be resolved by considering the separate contributions [32] of the κ^+ (initially positive flux) and κ^- (initially negative flux) to κ as friction approaches zero for escape from a single well like that shown in Figure 16.2.1. Also revisit section 13.2 which explains that the transmission coefficient can be interpreted as the effects of recrossing for an ensemble of equilibrium transition states, or equivalently as a correction for non-equilibrium effects.

16.3 Insights and limitations

As a computational tool, Kramers theory is only occasionally useful – usually in analyses of simple models for which its assumptions are true by construction. However, Kramers theory was a conceptual breakthrough because it provides one unified framework for understanding how dynamics influence reaction rates. The key lessons from Kramers theory include:

A spatial diffusion limit: At high friction the rate decreases as γ^{-1} and approaches Pontryagin's [2] rate expression for diffusion over a barrier in the high friction limit (see Chapter 18).

An energy diffusion limit: At extremely low friction, escape from a potential well becomes limited by slow energy influx and dissipation. The transmission coefficient for these processes increases in proportion to the friction and the action per orbit, $\gamma S(E_{\ddagger})$ [1]. See Hanggi [26] and Mel'nikov [27] for an analysis of related effects on the kinetics of interconversion between two potential wells (e.g. in isomerization).

A turnover region: The transmission coefficient reaches a maximum for moderate friction. Transmission coefficients near the turnover are less than but comparable to unity [26,27].

There are many pitfalls to avoid in using Kramers theory and in interpreting its results. First, we should not expect an energy diffusion limit for atom exchange reactions or S_N2 reactions in the gas phase. Although such bimolecular reaction trajectories may proceed ballistically with little friction, the reactants and products can travel indefinitely backward and forward along the reaction coordinate. Simulation evidence [23–25,33,34] suggests that when barrier recrossing occurs in these reactions it arises for other reasons, e.g. from pathways that curve sharply near the saddle point.

There are quantitative problems with the energy diffusion limit even for reactions that do involve bound wells where slow energy transfer can be limiting. Kramers' weak, steady Langevin friction model is not accurate for energy transfer by strong and sudden, but infrequent collisions with other molecules. In fact, Kramers' energy diffusion limit is valid only when the energy transfer per orbit is less than k_BT [26,27]. Several investigators have developed alternative theories for the weak coupling limit [4,35,36].

Kramers theory describes a one-dimensional reaction coordinate coupled to a bath. Especially for intermediate friction, the bath is frequently envisioned to be the solvent. Technically, the bath must also include intramolecular modes other than the reaction coordinate, and therefore the friction includes "internal friction" from coupling to these modes [14,16,37]. Especially for reactions involving large molecules like proteins, intramolecular friction can be more important than friction from the surrounding solvent. Thus models and predictions where friction exclusively stems from solvent viscosity should be used with caution.

In the Kramers model, friction is a constant at every point along the reaction coordinate, but intramolecular contributions to the friction are often coordinate dependent [38–41]. These studies suggest that dynamics along different modes at a saddle point may be separable, but farther along the pathway the coupling (friction) may become large and help to quench reactive trajectories.

Kramers theory assumes a delta-correlated friction and random forces, i.e. a bath that responds infinitely fast. For typical bond-breaking/bond-making reaction in solution, the barrier crossing time is $\omega_\ddagger^{-1} \sim 100$ fs. A bath of internal modes and solvent molecules typically includes a multitude of slower vibrational frequencies and librational motions that cannot adiabatically equilibrate to such rapid changes in the reaction coordinate. Therefore the Kramers model, in which the bath responds the same way regardless of the dynamical history, cannot be correct for reactions that break/make bonds. Grote-Hynes theory (see Chapter 17) uses a non-Markovian friction and random forces to provide a more accurate description of chemical reaction dynamics in condensed phases [7].

In the high friction (overdamped) limit, the barrier crossing time becomes much longer than the bath relaxation time, $\gamma/\omega_\ddagger \gg 1$. As examples, protein folding [19,42–44] electron transfer [45], and nucleation [46,47] are often modeled as overdamped processes. Each of these processes involve concerted motions of hundreds or thousands of atoms, and in some cases the barrier crossing times can be nanoseconds. With so much time atop the barrier, faster motions (both solvent and internal non-reaction coordinates) will remain quasi-equilibrated throughout the barrier crossing event. The entire rate calculation reduces to a mean first passage time involving only a diffusion along the reaction coordinate and a free energy barrier. This high friction limit is where quantitative results from Kramers theory are most reliable. However, one must be wary of reaction coordinate error. Even in the limit of perfectly diffusive dynamics, choosing an incorrect reaction coordinate renders the one-dimensional description of Kramers theory inaccurate [17,48]. More will be said about this problem in Chapters 18 and 20.

Exercises

1. Nondimensionalize the inertial Langevin and Klein-Kramers equations using the rescaled variables: $\tau = \gamma t$, $\underline{v} = v\sqrt{m/k_B T}$, and $\underline{q} = \gamma q\sqrt{m/k_B T}$. After nondimensionalization, what happens to the variance of random noise $\langle \mathfrak{R}(0)\mathfrak{R}(t)\rangle = (?)\delta[t]$ in the inertial Langevin equation?
2. Write a code to compute the reactive flux correlation function from inertial Langevin trajectories on the potential $\beta V(q) = \beta V_{\ddagger}(1 - 3q^2 - 2q^3)$.
   (a) Separately compute the κ^+ and κ^- contributions to κ at γ = 0.01, 0.1, 1.0, 10.0, and 100.0.
   (b) Explain the observed trends in the κ^+ and κ^- contributions to κ.
   (c) Compare your numerical results to the Kramers turnover expression from Mel'nikov and Mesh'kov.
3. Begin with an inertial Langevin equation and compute the rate of crossing over a square topped barrier of size ΔF and width L. Assume a perfectly absorbing boundary condition at the right edge of the barrier and equilibrium for states beyond the left edge of the barrier. Compare your result to the transition state theory rate.
4. Compute the transmission coefficient for crossing over a sharp cusp shaped barrier. Hint: this problem is outlined in Kramers' 1940 paper. Complete the details.
5. Read Shoup and Szabo, Biophys. J. 40, 33 (1982). Describe in detail the relationship between the Debye model for diffusion control and the overdamped limit of Kramers theory.
6. Read Vekilov, Cryst. Growth & Design 7, 2796–2810 (2007) on crystal growth kinetics. Derive the mass dependences for an Eyring-like (inertial) attachment dynamics and a Kramers/Debye-like diffusional attachment. Check your predictions against those of Vekilov and the evidence cited therein.

References

[1] H.A. Kramers, Physica A 7 (1940) 284–304.
[2] L.S. Pontryagin, A.A. Andronov, A.A. Vitt, Zh. Eksp. Teor. Fiz. 3 (1933) 165–180.

[3] V. Agarwal, B. Peters, Adv. Chem. Phys. 155 (2014) 97–160.
[4] J.L. Skinner, P.G. Wolynes, J. Chem. Phys. 69 (1978) 2143.
[5] M. Jacob, M. Geeves, G. Holtermann, F.X. Schmid, Nat. Struct. Biol. 6 (1999) 923–926.
[6] R.I. Masel, Chemical Kinetics and Catalysis, Wiley, New York, 2001.
[7] J.T. Hynes, Annu. Rev. Phys. Chem. 36 (1985) 573–597.
[8] P.G. Wolynes, in: D. Stein (Ed.), Complex Systems, in: SFI Studies in the Sciences of Complexity, Addison-Wesley, Longman, 1989, pp. 355–387.
[9] R.F. Grote, J.T. Hynes, J. Chem. Phys. 73 (1980) 2715–2732.
[10] J.T. Hynes, in: O. Tapia, J. Bertran (Eds.), Solvent Effects and Chemical Reactivity, Kluwer, Amsterdam, 1996, pp. 231–258.
[11] E. Pollak, J. Chem. Phys. 85 (1986) 865–867.
[12] J.S. Langer, Ann. Phys. 54 (1969) 258–275.
[13] A.M. Berezhkovskii, E. Pollak, V. Yu, J. Chem. Phys. 97 (1992) 2422–2437.
[14] S.J. Hagen, Curr. Protein Pept. Sci. 999 (2010) 1–11.
[15] J.M. Anna, K.J. Kubarych, J. Chem. Phys. 133 (2010) 174506.
[16] J.J. Portman, S. Takada, P.G. Wolynes, J. Chem. Phys. 114 (2001) 5082.
[17] A. Berezhkovskii, A. Szabo, J. Chem. Phys. 122 (2005) 14503.
[18] C.D. Snow, Y.M. Rhee, V.S. Pande, Biophys. J. 91 (2006) 14–24.
[19] S.V. Krivov, M. Karplus, Proc. Natl. Acad. Sci. USA 105 (2008) 13841–13846.
[20] B. Peters, Chem. Phys. Lett. 554 (2012) 248–253.
[21] W.A. Coffey, Y.P. Kalmykov, J.T. Waldron, The Langevin Equation, World Scientific, Singapore, 2004.
[22] A. Nitzan, Dynamics in Condensed Phases: Relaxation, Transfer, and Reactions in Condensed Molecular Systems, Oxford University Press, Oxford, 2006.
[23] H. Hu, M.N. Kobrak, C. Xu, S. Hammes-Schiffer, J. Phys. Chem. A 104 (2000) 8058–8066.
[24] Y.J. Cho, S.R. Vande Linde, L. Zhu, W.L. Hase, J. Chem. Phys. 96 (1992) 8275.
[25] W.L. Hase, Science 266 (1994) 998–1002.
[26] P. Hanggi, J. Stat. Phys. 42 (1986) 105–148.
[27] V.I. Mel'nikov, Phys. Rep. 209 (1991) 1–71.
[28] V.I. Mel'nikov, S.V. Meshkov, J. Chem. Phys. 85 (1986) 1018.
[29] E. Pollak, H. Grabert, P. Hanggi, J. Chem. Phys. 91 (1989) 4073–4087.
[30] D. Chandler, J. Chem. Phys. 68 (1978) 2959–2970.
[31] J.E. Straub, M. Borkovec, B.J. Berne, J. Chem. Phys. 84 (1986) 1788–1794.
[32] R.A. Kuharski, D. Chandler, J.A. Montgomery, F. Rabii, S.J. Singer, J. Phys. Chem. 92 (1988) 3261–3267.
[33] T. Su, H. Wang, W.L. Hase, J. Phys. Chem. A 102 (1998) 9819–9828.
[34] L. Sun, W.L. Hase, K. Song, J. Am. Chem. Soc. 123 (2001) 5753–5756.
[35] E.W. Montroll, K.E. Shuler, Adv. Chem. Phys. 1 (1958) 361–399.
[36] J. Troe, Annu. Rev. Phys. Chem. 29 (1978) 223–250.
[37] A. Soranno, B. Buchli, D. Nettels, R.R. Cheng, S. Muller-Spath, S.H. Pfeil, A. Hoffmann, E.A. Lipman, D.E. Makarov, B. Shuler, Proc. Natl. Acad. Sci. USA 109 (2012) 17800–17806.
[38] B. Carmeli, A. Nitzan, Chem. Phys. Lett. 102 (1983) 517–522.
[39] G.R. Haynes, G.A. Voth, J. Chem. Phys. 103 (1995) 10176.
[40] H. Wang, W.L. Hase, Chem. Phys. 212 (1996) 247–258.
[41] B. Peters, A.T. Bell, A. Chakraborty, J. Chem. Phys. 121 (2004) 4453–4460.
[42] J.N. Onuchic, Z. Luthey-Schulten, P.G. Wolynes, Annu. Rev. Phys. Chem. 48 (1997) 545–600.
[43] D.J. Bicout, A. Szabo, Protein Sci. 9 (2000) 452–465.
[44] S. Beccara, T. Skrbic, R. Covino, P. Faccioli, Proc. Natl. Acad. Sci. USA 109 (2012) 2330–2335.
[45] D.F. Calef, P.G. Wolynes, J. Phys. Chem. 87 (1983) 3387–3400.
[46] D. Kashchiev, Nucleation: Basic Theory with Applications, Butterworth-Heinemann, Oxford, 2000.
[47] K.F. Kelton, A.L. Greer, Nucleation in Condensed Matter: Applications in Materials and Biology, Elsevier, Amsterdam, 2010.
[48] Y.M. Rhee, V.S. Pande, J. Phys. Chem. B 109 (2005) 6780–6786.

Grote-Hynes theory

Kramers theory predicted that the reaction rate would continuously decrease, and ultimately vanish, as the environment's friction or viscosity increased. But certainly there are many reactions not stopped in solids!

Hynes, Ann. Rev. Phys. Chem. (2015)

Kramers' theory of barrier crossing under the influence of friction is based on a simple Langevin equation with Markovian friction and random forces coupled to reaction coordinate motion [1]. The random forces and friction represent the 'bath', i.e. every degree of freedom except for the explicitly modeled 'reaction coordinate.' In the Kramers model, the response by the bath to a change in the reaction coordinate is both immediate and immediately forgetful. Real solvents require time to reshape the solute cavity, to rearrange the hydrogen bond network, and to reorient polar solvent molecules. These solvent relaxation processes can take many picoseconds to complete. By comparison, chemical bonds can be broken and made in just fractions of a picosecond because typical reactions only require a few atoms to move by a few Angstroms. Clearly, Markovian friction models cannot accurately describe solvent dynamics during a chemical reaction.

Problems with Markovian friction models are evident even in situations that do not involve a barrier crossing. For example, consider the velocity-velocity autocorrelation function for molecules diffusing in a liquid. The simple Langevin equation (15.1.1) with Markovian friction and a constant potential of mean force (PMF) predicts $\langle v(0)v(t)\rangle = \langle v(0)^2\rangle exp[-\gamma t]$. In contrast to an exponential decay, the velocity-velocity correlations from MD simulations in a liquid show inertial behavior, caging phenomena, and long-lived hydrodynamic effects. Generalized Langevin equations (GLEs) with non-Markovian friction provide a better description of dynamics in real solvents. The GLE for motion along a reaction coordinate is [2,3]

$$m\ddot{q} = -\frac{\partial V}{\partial q} - m\int_0^t \eta(t-\tau)\dot{q}(\tau)d\tau + R(t) \qquad (17.0.1)$$

where $\eta(t)$ is a time-dependent friction. Specifically, $dt\, m\, \eta(t)\, \dot{q}(0)$ is the t-delayed differential force exerted by the solvent on q. The integral sums up these differential force contributions from the velocity history of q at all previous times. The random force $R(t)$ has zero mean

$$\langle R(t)\rangle = 0 \tag{17.0.2}$$

and variance that is related to the time dependent friction $\eta(t)$ by [2]

$$\langle R(0)R(t)\rangle = mk_B T\eta(t) \tag{17.0.3}$$

with $\eta(t)$ decaying over some non-zero time. The GLE reduces to a standard Markovian Langevin equation if the velocity remains constant during the time it takes the random force correlations to decay. In such cases, we say that the friction forces are simultaneous [4] with $v(t)$, i.e. the friction is Markovian if there is a γ such that $\gamma v(t) \approx \int_{-\infty}^{t} \eta(t')v(t-t')dt'$ for any $v(t)$. For a Markovian friction, we can insert $\eta(t) = \gamma\delta[t-0]$ into the GLE and recover a standard Langevin equation.

The causal [4] (delayed) friction forces from a non-Markovian friction depend on the different frequency components in $v(t)$. The time dependence of $\eta(t)$ can be summarized by an integrated friction strength γ [3,5]:

$$\gamma \equiv \int_0^\infty \eta(t)dt \tag{17.0.4}$$

and a bath memory time τ_M:

$$\tau_M = \eta(0)^{-1} \int_0^\infty \eta(t)dt \tag{17.0.5}$$

These two characteristics of the friction will help us understand many of the predictions from Grote-Hynes theory. In some cases, it will be necessary to represent $\eta(t)$ as a weighted superposition of responses at different frequencies.

17.1 The Grote-Hynes equations

The aim of Grote-Hynes (GH) theory is to understand how non-Markovian dynamics of the bath response can alter the fast barrier crossing dynamics of chemical reactions in solution. GH theory does not require a pre-optimized reaction coordinate or a variationally optimized dividing surface. In most implementations the coordinate $q(\mathbf{x})$ is some convenient bond length or difference between bond lengths. To avoid confusion with more precise reaction coordinate definitions, this chapter will call q the "solute coordinate". We assume that the PMF $V(q)$ reaches a maximum at location zero. Because q is not necessarily an accurate reaction coordinate, it should be remembered that $q(\mathbf{x}) = 0$ does not necessarily coincide with the true transition state ensemble. Accordingly, we also amend our previous notation ($\omega_{\ddagger}$) to instead use ω_b for the adiabatic barrier frequency. More will be said on these important caveats in section 17.2.

To focus on frictional effects during motion over the barrier, Grote and Hynes assumed an infinite parabolic barrier model [6]:

$$V(q) = V_b - \frac{1}{2} m \omega_b^2 q^2$$

The one-dimensional potential energy profile $V(q)$ represents the PMF after integration over all bath degrees of freedom.

Define the reaction probability as the probability that a trajectory initiated at q_0 with initial velocity v_0 will commit to the product side of the barrier at later times [7]:

$$p_{RX}(q_0, v_0) = \lim_{t \to \infty} \int_{-\infty}^{\infty} dq \int_{-\infty}^{\infty} dv \, \rho(q, v, t \,|\, q_0, v_0) h_B(q) \tag{17.1.1}$$

Here $h_B(q) = 1$ for $q > 0$ and 0 otherwise, and $\rho(q, v, t \,|\, q_0, v_0)$ is a transition probability density. Definition 17.1.1 also assumes that the reactant and product states behave as absorbing sinks. The absorbing sink approximation is consistent with the infinite parabolic barrier model, but it is a source of error in some applications to systems with finite barriers [8,9].

The transition state theory rate and the transmission coefficient are easily obtained by integration of $p_{RX}(q_0, v_0)$. But to obtain $p_{RX}(q_0, v_0)$ we must first obtain $\rho(q, v, t \,|\, q_0, v_0)$. If the noise $R(t)$ is Gaussian, then the transition probability density $\rho(q, v, t \,|\, q_0, v_0)$ is also Gaussian [7,10].

$$\rho(q, v, t \,|\, q_0, v_0) = (2\pi)^{-1} \left| det\mathbf{A} \right|^{-1/2} exp\left[-\frac{1}{2} \mathbf{y}^{\mathbf{T}} \mathbf{A}^{-1} \mathbf{y} \right] \tag{17.1.2}$$

where

$$\mathbf{y}(t) = \begin{bmatrix} q(t) - \langle q(t) \rangle_{q_0, v_0} \\ v(t) - \langle v(t) \rangle_{q_0, v_0} \end{bmatrix} \tag{17.1.3}$$

Matrix **A** is the time dependent covariance

$$A_{ij}(t) = 2 \left\langle \mathbf{y}_i(t) \mathbf{y}_j(t) \right\rangle_{q_0, v_0} \tag{17.1.4}$$

Integrating equation (17.1.1) with $\rho(q, v, t \,|\, q_0, v_0)$ as given in equation (17.1.2) yields [7]

$$p_{RX}(q_0 = 0, \, v_0) = \lim_{t \to \infty} \frac{1}{2} \text{erfc} \left[-\frac{\langle q(t) \rangle_{q_0, v_0}}{2 \left\langle \delta q(t)^2 \right\rangle_{q_0, v_0}^{1/2}} \right] \tag{17.1.5}$$

where $\delta q(t) = q(t) - \langle q(t) \rangle$. The averages in equation (17.1.5) have brackets with subscripts that indicate the initial conditions of an evolving swarm of trajectories.

The importance of equation (17.1.5) is that the entire problem has been reduced to the first two moments of the distribution along coordinate q. We no longer need the properties of individual trajectories. Figure 17.1.1, adapted from Kohen and Tannor [7], depicts the evolving packet of trajectories, i.e. $\rho(q, v, t|q_0, v_0)$, initially moving forward along the $+q$-axis with part of the packet recrossing at later times.

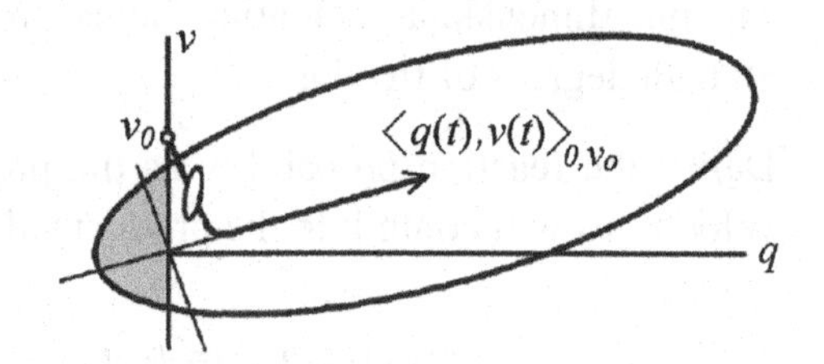

Figure 17.1.1: A packet of GLE trajectories initially moving forward along the q-axis. Part of the packet, shown in gray, recrosses at later times causing a reduction in the rate constant. [From Kohen and Tannor, *J. Chem. Phys.* 103, 6013-20 (1995).]

To make use of equation (17.1.5), the long time behavior of the mean $\langle q(t) \rangle_{q_0, v_0}$ and standard deviation $\left\langle \delta q(t)^2 \right\rangle_{q_0, v_0}^{1/2}$ are needed. For initial conditions at the top of the parabolic potential, Grote and Hynes [6] showed that the long time behavior must be exponentially diverging in time with

$$\langle q(t) \rangle_{q_0=0, v_0} = C(v_0) exp[\lambda t] \tag{17.1.6}$$

and

$$\left\langle \delta q(t)^2 \right\rangle_{q_0=0, v_0} = C(v_0)^2 \frac{k_B T}{m} \left(\frac{\omega^2}{\lambda^2} - 1 \right) exp[2\lambda t] \tag{17.1.7}$$

In these formulas, the coefficient $C(v_0)$ carries the effects of the complicated v_0-relaxation that happens at short times. At long times $\langle q(t) \rangle_{q_0, v_0}$ and $\left\langle \delta q(t)^2 \right\rangle_{q_0, v_0}^{1/2}$ diverge in the same way so that $p_{RX}(q_0 = 0,\ v_0)$ in equation (17.1.5) has a well-defined value. Inserting equation (17.1.6) into the GLE with a parabolic potential,

$$\ddot{q} = \omega_b^2 q - \int_0^t dt' \eta(t - t') \dot{q}(t') + R(t)/m \tag{17.1.8}$$

and eliminating common factors of $C(v_0) exp[\lambda t]$ gives [6]

$$\lambda^2 = \omega_b^2 - \lambda \int_0^t dt' \eta(t') exp[-\lambda t'] \tag{17.1.9}$$

Note that the random force term is omitted because realizations of the noise have already been averaged in obtaining $\langle q(t) \rangle_{q_0, v_0}$ and $\left\langle \delta q(t)^2 \right\rangle_{q_0, v_0}^{1/2}$. As $t \to \infty$ the term on the right hand side of equation (17.1.9) becomes $\hat{\eta}(\lambda)$, the Laplace transform of $\eta(t)$. The result is the Grote-Hynes equation for the effective barrier frequency λ in the presence of non-Markovian friction.

$$\lambda^2 + \lambda\hat{\eta}(\lambda) - \omega_b^2 = 0 \tag{17.1.10}$$

The GH equation has one root in the interval $0 < \lambda \leq \omega_b$. This positive root is the reactive frequency. What is the meaning of the reactive frequency?

- For the case of zero friction: $\eta(t) = 0$, the reactive frequency is $\lambda = \omega_b$, and the transmission coefficient is $\kappa = 1.0$.
- For Markovian friction: $\eta(t) = \gamma\delta[t]$, the reactive frequency is $\lambda = (\gamma/2)\{(1 + 4\omega_b^2/\gamma^2)^{1/2} - 1\}$, and the transmission coefficient is $\kappa = \lambda/\omega_b$ as obtained by Kramers.
- For non-Markovian friction, the transmission coefficient can be derived from the reaction probability $p_{RX}(q_0, v_0)$ at the transition state $q_0 = 0$. The calculations, left as an exercise, require expressions (17.1.6) and (17.1.7) for $\langle q(t)\rangle_{q_0,v_0}$ and $\langle \delta q(t)^2\rangle_{q_0,v_0}$ in terms of λ. The transmission coefficient is obtained by integrating the reaction probability

$$\kappa_{GH} = \frac{\int_{-\infty}^{\infty} v_0 p_{RX}(q_0 = 0, v_0)\rho_{eq}(v_0)dv_0}{\int_{-\infty}^{\infty} v_0\theta(v_0)\rho_{eq}(v_0)dv_0} \tag{17.1.11}$$

where $\theta(v)$ is the Heaviside function and where $\rho_{eq}(v_0) \propto exp[-mv_0^2/2k_BT]$. The result is the Grote-Hynes transmission coefficient:

$$\kappa_{GH} = \lambda/\omega_b \tag{17.1.12}$$

It can be shown that $\kappa_{GH} < 1$. The relationship between Kramers theory and GH theory becomes clear when equation (17.1.12) is rewritten as

$$\kappa_{GH} = \frac{\hat{\eta}(\lambda)}{2\omega_b}\left\{\left(1 + 4\omega_b^2/\hat{\eta}(\lambda)^2\right)^{1/2} - 1\right\} \tag{17.1.13}$$

Equation (17.1.13) *is* the Kramers intermediate-to-high friction result, except that the Markovian friction has been replaced by the Laplace transform of the non-Markovian friction at the reactive frequency λ. Equations (17.1.13) and (17.1.10) show precisely how the non-Markovian friction influences the transmission coefficient. For strong friction at the reactive frequency, $\hat{\eta}(\lambda) \gg \omega_b$ and $\kappa_{GH} \ll 1$. For weak friction at the reactive frequency, $\hat{\eta}(\lambda) \ll \omega_b$ and $\kappa_{GH} \approx 1$.

Can we anticipate how friction will impact the transmission coefficient based on its integrated strength (γ) and memory time (τ_M)? The time-dependent friction $\eta(t)$ generally decays with time, perhaps in a non-monotonic fashion as depicted in Figure 17.1.2. Accordingly, the frequency-dependent friction also tends to decrease with increasing frequency. Because the

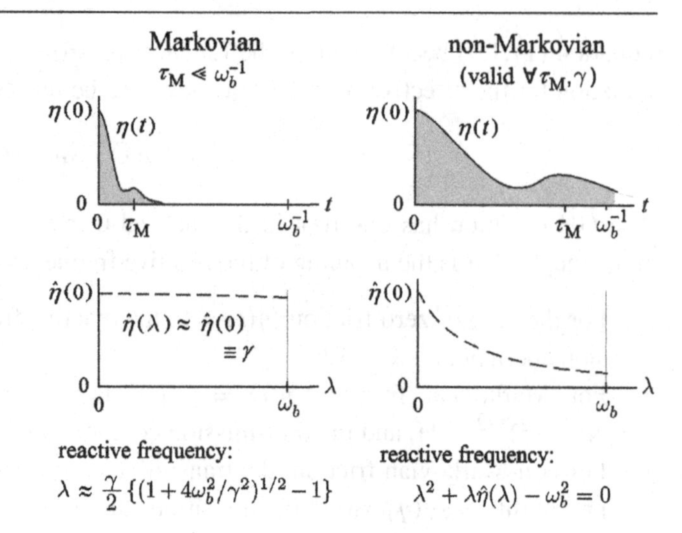

Figure 17.1.2: (Left) The Markovian friction assumption in Kramers theory is valid when $\tau_M \ll \omega_b^{-1}$. Because e^{0+} and $e^{-\omega_b t}$ decay slowly relative to $\eta(t)$, the frequency dependent friction is approximately $\hat{\eta}(0)$ for $0 < \lambda \leq \omega_b$. (Right) For non-Markovian friction, τ_M is comparable to or greater than ω_b^{-1}. The frequency-dependent friction then varies within the interval $0 < \lambda \leq \omega_b$, and the GH equation must be used to find λ.

reactive frequency must be in the interval $0 < \lambda \leq \omega_b$, the values of $\hat{\eta}(0)$ and $\hat{\eta}(\omega_b)$ approximately bracket the frequency dependent friction at the reactive frequency. Figure 17.1.2 shows (schematic) time- and frequency-dependent frictions for Markovian and non-Markovian cases.

Many results of GH theory can be understood in terms of the instantaneous friction $\eta(0)$ and the zero-frequency friction $\hat{\eta}(0)$. As anticipated, $\eta(0)$ and $\hat{\eta}(0)$ can be related to the friction strength and memory time:

- The zero frequency friction is $\hat{\eta}(0) = \int_0^\infty \eta(t)e^0 dt = \gamma$.
- The memory time is $\tau_M = \hat{\eta}(0)/\eta(0)$.
- The instantaneous friction is $\eta(0) = \gamma/\tau_M$.

For $\tau_M \ll \omega_b^{-1}$ the GH theory essentially recovers Kramers' results for Markovian friction.[1] At the other extreme when $\tau_M \gg \omega_b^{-1}$ the bath is too slow to respond to the moving solute coordinate. The solvent appears "frozen" [11,12] compared to the rapid motion of q, and $\eta(t)$ can be approximately replaced with its initial value:

$$\begin{aligned} \ddot{q} &= \omega_b^2 q - \int_0^t dt' \eta(t-t')\dot{q}(t') + R(t)/m \\ &\approx \omega_b^2 q - \eta(0)\int_0^t dt' \dot{q}(t') + R(t)/m \\ &\approx \omega_b^2 q - \gamma \tau_M^{-1} q + R(t)/m \end{aligned} \tag{17.1.14}$$

[1] Grote-Hynes theory is analogous to the intermediate-to-high friction results of Kramers theory. Even for very small γ and/or $\hat{\eta}(\lambda)$, GH theory has no energy diffusion limit because its parabolic barrier approximation provides no mechanism for trajectories to orbit the reactant basin and re-emerge [8,9].

The random forces vanish in the average behavior. Thus while the bath appears to be frozen, a swarm of trajectories launched from the transition state will evolve, on average, according to

$$\ddot{q} \approx (\omega_b^2 - \gamma \tau_M^{-1}) q$$

The averaged dynamics of q thus resemble motion on a parabolic potential with an effective non-adiabatic frequency [13]

$$\omega_{NA} = (\omega_b^2 - \gamma/\tau_M)^{1/2} \tag{17.1.15}$$

Depending on the strength of the zero-time (instantaneous) friction: $\eta(0) = \gamma/\tau_M$, the frozen solvent gives rise to two different limiting behaviors.

- **Non-adiabatic** (frozen solvent, weak friction): When the instantaneous friction is weak, $\gamma/\tau_M \ll \omega_b^2$, the GHT becomes a model of non-equilibrium solvation. The solvent degrees of freedom are effectively frozen, but ω_{NA} remains real so that the solvent friction does not severely impede the transitions. In this case, the transmission coefficient may remain close to unity despite the frozen solvent. van der Zwan and Hynes [11] showed that the frozen solvent approximation (17.1.14) with weak friction results in the non-adiabatic transmission coefficient

$$\kappa_{NA} = \omega_{NA}/\omega_b \tag{17.1.16}$$

  The non-adiabatic transmission coefficient is a lower bound on the GH transmission coefficient, $\kappa_{NA} \leq \kappa_{GH} \leq 1$, because of high frequency parts of the friction spectrum that can relax during the barrier crossing event.
- **Dynamical caging** (frozen solvent, strong friction): When the instantaneous friction is strong, $\gamma/\tau_M \gg \omega_b^2$, the non-adiabatic frequency becomes imaginary and the solute coordinate becomes trapped in a narrow solvent cage at the barrier top. The cage is not a thermodynamic feature of the PMF along q, but rather a transient dynamical feature. At long times ($t > \tau_M$) the bath relaxes allowing the solute coordinate to move away from the barrier top, but at shorter times ($t < \tau_M$) the solute coordinate oscillates within the solvent cage [11]. The oscillations cross the $q = 0$ surface many times resulting in a small transmission coefficient. There is no simple estimate for the size of the transmission coefficient in the dynamical caging regime.

Figure 17.1.3 illustrates the non-adiabatic and dynamical caging regimes for the case of a frozen solvent. More detailed analyses of the non-adiabatic and dynamical caging regimes follow below. Figure 17.1.3 also shows how the reactive flux correlation function $\kappa(t)$ relaxes toward the transmission coefficient in each of the regimes we have discussed. In the Markovian limit, the friction and random forces are instantaneous. Accordingly, $\kappa(t)$ immediately begins to drop from unity. In the nonadiabatic case, $\kappa(t)$ may look similar except for a short ballistic time during which the $\kappa(t)$ remains near unity. The dynamical caging regime results in a $\kappa(t)$ that oscillates, perhaps even becoming negative, before reaching a plateau value.

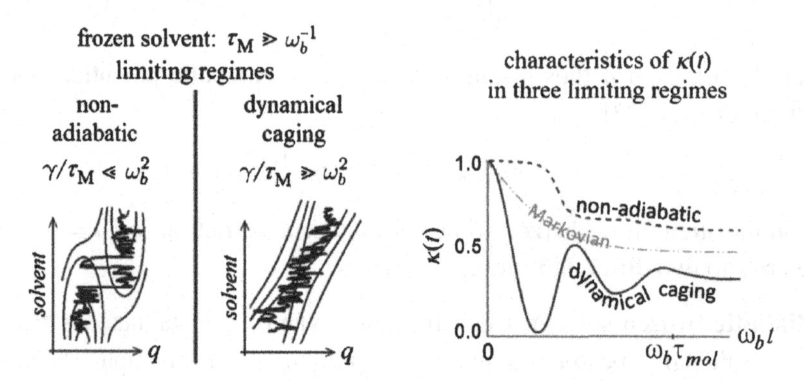

Figure 17.1.3: (Left) Schematic energy landscape, dynamics, and non-Markovian friction for a frozen solvent ($\tau_M \gg \omega_b^{-1}$). In the non-adiabatic regime, the frozen solvent is weakly coupled to the solute coordinate q. In the dynamical caging regime, the frozen solvent is strongly coupled to q. (Right) Typical characteristics of the reactive flux correlation functions in three different dynamical regimes. The figure is not intended to suggest that non-adiabatic transmission coefficients are large or that dynamical caging transmission coefficients are small.

In most applications, the friction kernel $\eta(t)$ must be determined numerically to obtain γ, τ_M, $\hat{\eta}(\lambda)$, and λ. However, Kohen and Tannor derived a number of approximate analytical results for an exponentially decaying model of $\eta(t)$.

Example: Exponentially decaying friction [7]

Consider an exponentially decaying friction of the form

$$\eta(t) = \gamma\tau_M^{-1} exp[-t/\tau_M]$$

It is easily verified that $\gamma = \int_0^\infty \eta(t)dt$ and $\tau_M = \eta(0)^{-1}\int_0^\infty \eta(t)dt$, consistent with our prior parameter definitions. The Laplace transform of $\eta(t)$ is

$$\hat{\eta}(\lambda) = \gamma\tau_M^{-1}/(\lambda + \tau_M^{-1}).$$

Inserting $\hat{\eta}(\lambda)$ into the GH equation gives a cubic equation for the reactive frequency λ. Dividing each term by ω_b^3 converts the cubic GH equation into a cubic equation for the transmission coefficient:

$$\kappa^3 + (\kappa^2 - 1)(\tau_M\omega_b)^{-1} + \kappa\{\gamma/\tau_M\omega_b^2 - 1\} = 0 \tag{17.1.17}$$

This equation always has one real and positive root.

Markovian limit: When $\tau_M\omega_b \ll 1$ the second term in equation (17.1.17) is much larger than the first. If we additionally assume a large instantaneous friction such that $\gamma/\tau_M \gg \omega_b^2$, the GH equation approximately simplifies to

$$(\kappa^2 - 1) + \kappa\gamma/\omega_b \approx 0.$$

The resulting transmission coefficient is that of Kramers theory for intermediate friction: $\kappa = (\gamma/2\omega_b)\{(1 + 4\omega_b^2/\gamma^2)^{1/2} - 1\}$.

Frozen solvent: When $\tau_M\omega_b \gg 1$ the cubic equation for κ approximately simplifies to

$$\kappa^2 + (\gamma\tau_M^{-1}/\omega_b^2 - 1) \approx 0 \qquad (17.1.18)$$

The simplified equation for κ now splits into two further cases depending on $\gamma\tau_M^{-1}/\omega_b^2$.

- When the friction (coupling) is weak ($\gamma\tau_M^{-1} < \omega_b^2$) both roots of equation (17.1.18) are real. The positive root is the non-adiabatic transmission coefficient: $\kappa_{NA} \approx \sqrt{1 - \gamma/(\tau_M\omega_b^2)}$.
- When the friction (coupling) is strong ($\gamma\tau_M^{-1} > \omega_b^2$) both roots of equation (17.1.18) are imaginary: $\pm i[\gamma/(\tau_M\omega_b^2) - 1]^{1/2}$. These are the trapped oscillations with frequency $[\gamma/(\tau_M\omega_b^2) - 1]^{1/2}$ in the dynamical caging regime. To find the transmission coefficient, we must go back to equation (17.1.17). The third root is $\kappa_{DC} \approx (\omega_b\tau_M)^{-1}/[\gamma/(\tau_M\omega_b^2) - 1]$ [7].

Roots for each case are shown below in the complex plane for numerical values in different regimes using the Kohen and Tannor example.

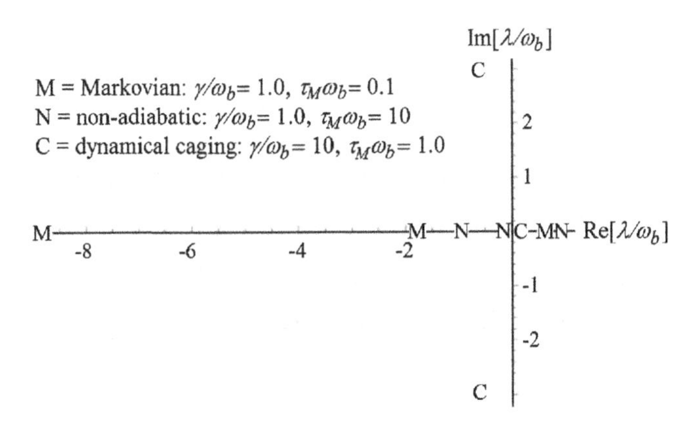

The roots are labeled as letters on the complex plane: C = roots for the caging regime, N = roots for the non-adiabatic regime, and M = roots for the Markovian regime. Notice that each case has only a single real positive root, and this corresponds to κ_{GH}.

Applications of GH theory with atomistic simulations usually employ the "clamping approximation" [14]. Specifically, the time-dependent friction is computed for a trajectory which is constrained to the $q = q*$ dividing surface. The friction obtained at $q*$ is assumed to be representative of the friction at other points along the coordinate q. GH theory within the clamping approximation is implemented as follows:

1. Choose a solute coordinate q. As described in section 17.2 and Chapter 20, q should be chosen carefully because the friction strength and memory will depend on which coordinate is chosen.
2. Compute the free energy barrier $F(q)$. The calculation should be extremely accurate near the maximum at $q*$ so that an adiabatic barrier frequency ω_b can be computed.
3. Let $\hat{u}_q = (\partial q/\partial \mathbf{x})/||\partial q/\partial \mathbf{x}||$ and record $R(t) = -(\partial V/\partial \mathbf{x}) \cdot \hat{u}_q\big|_t$ for each point in time during a constrained MD simulation. Compute the TST prefactor $\langle|\dot{q}|\rangle_{q*}$ and the time correlation function $\langle R(0)R(t)\rangle_{q*}$.

The required average $\langle|\dot{q}|\rangle_{q*}$ is most naturally done using harmonic restraints, e.g. running dynamics with a modified Hamiltonian $H' = H + k(q(\mathbf{x}) - q*)^2/2$. However, the average $\langle R(0)R(t)\rangle_{q*}$ should be computed using rigid constraints [15,16]. Otherwise, oscillations associated with the spring constant k will lead to artificial high-frequency components of $\eta(t)$. Because both $\langle|\dot{q}|\rangle_{q*}$ and $\langle R(0)R(t)\rangle_{q*}$ are needed, it is sensible to compute both in one calculation with constrained MD. Remember that averages with rigid constraints require a subtle correction related to the Jacobian of the constraint (see Chapter 11) [17–19].

GH theory has achieved impressive agreement with brute force simulation results and reactive flux simulations for various models [9,20]. It has also been successful in estimating transmission coefficients for several reactions in solution including an S_N2 reaction $[Cl - +CH_3Cl]$ [12,13,21], an S_N1 reaction $[t - BuCl \rightarrow t - Bu^+ + Cl^-]$ [22], ion-pair dissociation $[NaCl_{aq} \rightarrow Na^+_{aq} + Cl^-_{aq}]$ [18,23], ligand exchange rates [24,25], proton transfer reactions in solution [26], electron transfer in solution [27], and several enzyme reactions [28–30]. The example below highlights a study of transmission coefficients for enzyme catalysis.

Example: Reductive methylation in thymidylate synthase

Kanaan et al. [30] used GH theory to examine reductive methylation of 2'-deoxyuridine 5'-monophosphate (dUMP) by an ethylfolate donor in thymidylate synthase. The substrate-enzyme complex and the products are shown below.

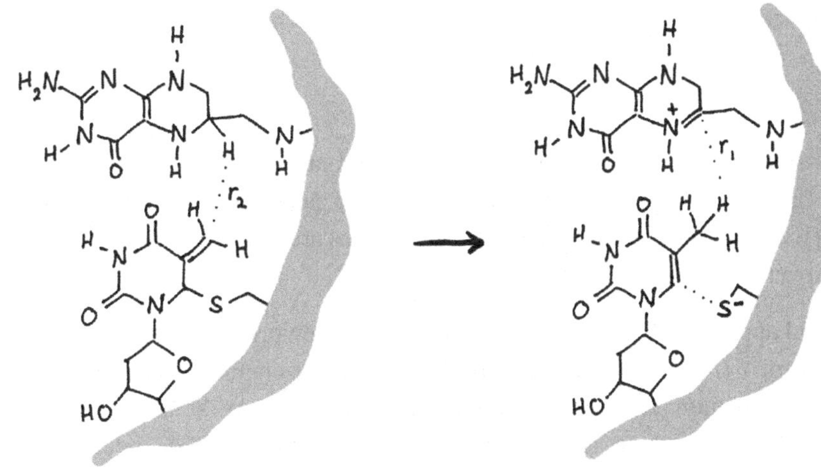

QM/MM simulations were used with the shaded region depicting the approximate boundary between the MM and QM regions in their model. Motion of the hydrogen nucleus was described classically. The solute coordinate was a difference of bond lengths $q = r_1 - r_2$ where r_1 and r_2 are the labeled bond lengths in the figure. Kanaan et al. [30] computed $F(q)$ and $\eta(t)$ at 293K. Kanaan et al. also report the numerically correct transmission coefficient κ_{RF} from reactive flux simulations. Their results are summarized in the following table. Note that I have extracted γ and κ_{DC} from the reported data for purposes of illustration.

ω_b/c	**3700cm^{-1}**	κ_{RF}	**0.30**
ω_b^{-1}	9.0fs	κ_{GH}	0.44
τ_M	96fs	κ_{NA}	0.28i
γ	(0.127fs)$^{-1}$	κ_{DC}	0.016
λ/c	1640cm^{-1}	κ_{KR}	0.014

The friction is strong ($\gamma/\tau_M\omega_b^2 \approx 6.64$) but slow ($\tau_M\omega_b \approx 10.6$). Because of the non-Markovian nature of the friction, the Kramers transmission coefficient errs by more than an order of magnitude. Because the friction is strong, the non-adiabatic estimate for the transmission coefficient is imaginary. One might expect the dynamical caging formula to apply, but it also provides a poor estimate of κ, perhaps because the time-dependent friction is non-exponential [30]. While the limiting approximations to κ_{GH} do not provide an accurate description, the κ_{GH} based on equation (17.1.12) is similar to the correct numerical transmission coefficient. These results illustrate how the fine details of the non-Markovian friction can influence rate constants.

GH transmission coefficients κ_{GH} are often a close match to exact results from reactive flux simulations. Failures of GH theory primarily arise in applications with anharmonic barriers and/or energy diffusion limitations [8,9]. These failures are to be expected given that an infinite parabolic barrier model is used in the derivation.

Computational applications of the Kramers and GH frameworks require similar input quantities from simulation: $F(q)$, k_{TST}, and $\eta(t)$ (to define γ). For chemical reactions in solution, GH theory provides far more accurate transmission coefficients, so the advantages of GH theory over Kramers theory are clear. The computational requirements of GH theory and reactive flux are also comparable, but their relative merits are a matter of priorities. GH theory provides more information about the dynamics of the solute coordinate and its coupling to the bath, but the reactive flux results are exact.

The following practical and computational issues should be considered when choosing between a basic TST calculation, a GH calculation, or a reactive flux approach:

1. Calculations using GH theory (or reactive flux) are significantly more costly than a TST calculation. Chemical reactions with a well-chosen coordinate q often have transmission coefficients in the range $0.1 < \kappa < 1.0$. In cases where TST alone provides order of magnitude accuracy, additional effort might best be invested to improve the TST calculation, e.g. with better electronic structure calculations or force fields. Hynes himself wrote "In certain perspectives (e.g. age, salary, number of noses), a factor of two has considerable significance. But ... TST is in fact an excellent rate constant theory for reactions..." [31].
2. Classical simulations have sometimes been used in GH calculations for proton, hydrogen, and hydride transfers with tunneling included via a separately computed WKB-type correction. Tunneling can be influenced by the action of the bath [32,33]. Moreover the WKB-type tunneling corrections are highly sensitive to ω_b [34], and ω_b depends on the chosen q-coordinate. Thus, situations that involve tunneling require especially careful analysis.
3. In interpreting the sizes of the quantities $\eta(t)$, $\hat{\eta}(\lambda)$, and κ_{GH}, remember that they are not fundamental properties of the reaction. These properties each depend on the choice of the solute coordinate q. (See section 17.2.)
4. Solvent modes with frequencies near the reactive frequency λ are likely culprits for dynamical recrossing effects [28–30]. However, in most cases there are many modes with frequencies near λ, and it is difficult to ascribe importance to any one mode.

5. Many reactions and activated processes exhibit coordinate-dependent friction [14,35,36]. These will require coordinate-dependent friction models and computational analyses beyond the clamping approximation [14,37,38].

GH theory is a powerful computational method, but its most important contributions are advances in conceptual understanding. GH theory explains why Kramers theory works poorly for chemical reactions in liquids: much of the friction spectrum is too slow to influence the fast barrier crossing event. GH theory also provides a physical interpretation for features in reactive flux correlation functions, e.g. the oscillations that come from dynamical caging. Finally, several alternative derivations of GH theory clarify the parallels between different terminologies and perspectives on reaction dynamics. The GH theory has been derived from each of the following starting points:

1. By Hanggi and Mojtabai via a flux-over-population analysis of the non-equilibrium current [39,40].
2. By Grote and Hynes via analysis of the long time dynamics in a generalized Langevin equation [6].
3. By Kohen and Tannor using the (equilibrium) reactive flux formalism [7].
4. By van der Zwan, Hynes, and Pollak [11,41] using TST with a multidimensional harmonic oscillator bath (section 17.2).

Calculation (1) is a non-equilibrium analysis. Calculations (2) and (3) assume equilibrium at the transition state with dynamics that exhibit complex recrossing behavior. Calculation (4) invokes extra degrees of freedom, equilibrium transition states, and TST-like dynamics. The equivalence of these very different routes to the same rate expression is profound and illuminating. In particular it confirms (as noted in Chapter 13) that equilibrium flux · transmission coefficient = non-equilibrium steady-state current.[2]

17.2 Multidimensional models and interpretations

Grote and Hynes used a GLE to account for solvent friction in the dynamics of a solute coordinate. Alternatively, the friction and random forces in the GLE can be recast as a multidimensional bath coupled to the solute coordinate motion [11,41–43]. Consider a multidimensional system with the Hamiltonian

$$H = \frac{p_q^2}{2m} + V(q) + \frac{1}{2}\sum_i \left\{ \frac{p_i^2}{m_i} + m_i\omega_i^2 \left(x_i - \frac{c_i}{m_i\omega_i^2} q \right)^2 \right\} \tag{17.2.1}$$

[2] These quasi-equilibrium expressions for k do break down under extreme non-stationary conditions that violate the inequality $k^{-1}\tau_{mol} dk/dt \ll 1$. Here dk/dt is a rate of change in k due to changing conditions in time, and τ_{mol} is the molecular relaxation time.

The first two terms are the kinetic energy and the PMF for a degree of freedom q. The summation includes kinetic and potential energy contributions for a bath of bilinearly coupled harmonic oscillators (BCHO). Each bath degree of freedom x_i is coupled to q such that, as the reaction coordinate changes position, the harmonic well of the bath mode adopts a new equilibrium position. The coupling coefficient c_i determines the amount that x_i must move per displacement along q to maintain equilibrium. It is easily shown from equation (17.2.1) that the free energy along q after integration over all x_i is simply $V(q)$. A two dimensional mechanical model for the BCHO Hamiltonian is considered in the following example.

Example: Solute coordinate coupled to one oscillator

Consider a swing mounted on a wheeled cart (see figure) as an example of the Hamiltonian in equation (17.2.1). The position of the cart corresponds to the solute coordinate q, and the position of the swinger corresponds to a single bath coordinate x. The position and velocity of the swinger, as well as those of the cart, are required to determine whether the barrier will be surmounted.

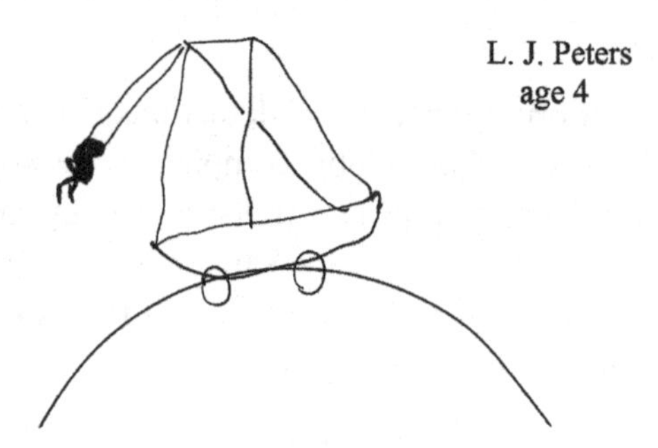

In the vicinity of the saddle point, the PMF for a solute (cart) coordinate q coupled to just one bath (swinger) mode x_1 is approximately

$$V = V_b - \frac{1}{2}m\omega_b^2 q^2 + \frac{1}{2}m_1\omega_1^2\left(x_1 - \frac{c_1}{m_1\omega_1^2}q\right)^2$$

V can be nondimensionalized as

$$\beta V = \beta V_b - \frac{1}{2}\bar{q}^2 + \frac{1}{2}(\bar{x} - \bar{c}\cdot\bar{q})^2$$

where

$$\begin{aligned}\bar{q} &= \sqrt{\beta m \omega_b^2}\, q \\ \bar{x} &= \sqrt{\beta m_1 \omega_1^2}\, x_1 \\ \bar{c} &= c_1 / \left(\omega_1 \omega_b \sqrt{m_1 m}\right)\end{aligned}$$

The potential energy surface changes depending on the coupling parameter $\bar{c}$ as shown in the contour plots below.

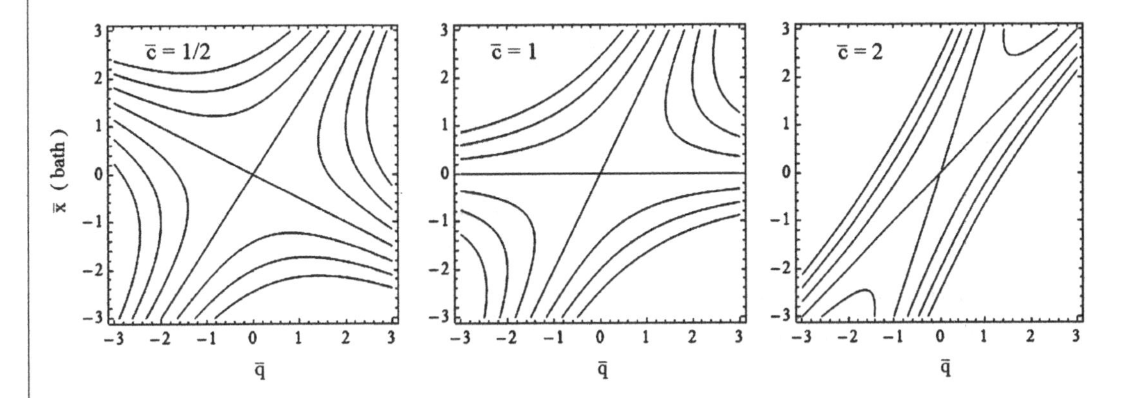

Some analogies with the GH theory are evident. Consider the case where $\bar{c} = 1/2$ with a frozen bath coordinate. The nonadiabatic regime prevails because at the saddle point $\partial^2 V/\partial q^2 < 0$ and therefore a non-adiabatic frequency can be computed from equation (17.1.15). In contrast for the case $\bar{c} = 2$, a frozen bath coordinate would confine q to a narrow region around the transition state. Therefore the case $\bar{c} = 2$ will more closely follow the dynamical caging model.

For the multidimensional model, Hamilton's equations can be used to derive the second order equations of motion:

$$m\ddot{q} = -\frac{\partial V}{\partial q} + \sum_i c_i \left(x_i - \frac{c_i}{m_i \omega_i^2} q \right)$$

and

$$m_i \ddot{x}_i = -m_i \omega_i^2 x_i + c_i q.$$

The dynamics of the bath degrees of freedom can be solved (by variation of parameters) in terms of the solute coordinate motion:

$$\begin{aligned} x_i(t) &= x_i(0)cos(\omega_i t) + \frac{p_i(0)}{m_i\omega_i}sin(\omega_i t) \\ &+\frac{c_i}{m_i\omega_i^2}q(t) - \frac{c_i}{m_i\omega_i^2}\int_0^t cos[\omega_i(t-t')]\dot{q}(t')dt' \end{aligned} \tag{17.2.2}$$

Substituting equation (17.2.2) into the equation of motion for $q(t)$ gives a generalized Langevin equation [42,44]

$$m\ddot{q} = -\frac{\partial V}{\partial q} - m\int_0^t \eta(t-t')\dot{q}(t')dt' + R(t) \tag{17.2.3}$$

where the friction is

$$\eta(t) = \frac{1}{m}\sum_i \frac{c_i^2}{m_i\omega_i^2}cos(\omega_i t) \tag{17.2.4}$$

and the random force is the sum of q-independent terms

$$R(t) = \sum_i c_i\left\{x_i(0)cos(\omega_i t) + \frac{p_i(0)}{m_i\omega_i}sin(\omega_i t)\right\}. \tag{17.2.5}$$

The random force defined in this manner is formally deterministic. However, we generally have no initial information $x_i(0)$ and $p_i(0)$ about the bath modes and there are moles of them. Thus the force at any time is essentially a random variable.

In a simulation, the initial conditions for the bath are sampled from a Boltzmann distribution. Specifically, the means are $\langle x_i(0)\rangle = c_i q/m_i\omega_i^2$ and $\langle p_i(0)\rangle = 0$. The covariances are $\langle x_i(0)p_j(0)\rangle = 0$, $\langle p_i(0)p_j(0)\rangle = m_i k_B T\delta_{ij}$, and $m_i\omega_i^2\langle x_i(0)x_j(0)\rangle = k_B T\delta_{ij}$. These choices preserve the fluctuation dissipation theorem for the friction and 'random' forces:

$$\begin{aligned} \langle R(0)R(t)\rangle &= \sum_{i,j} c_i c_j \langle x_i(0)x_j(0)\rangle cos(\omega_j t) \\ &= \sum_i c_i^2\frac{k_B T}{m_i\omega_i^2}cos(\omega_i t) \\ &= mk_B T\eta(t) \end{aligned} \tag{17.2.6}$$

Thus a generalized Langevin equation can always be represented by an equivalent multidimensional BCHO model [42,44]. To match the two models, one only needs to choose the coupling coefficients, masses, and frequency distributions appropriately.

GH theory also assumes a parabolic PMF for the solute coordinate: $V(q) = V_b - m\omega_b^2 q^2/2$. Thus, the equivalent multidimensional model is entirely comprised of coupled harmonic oscillators with one unstable mode and a single saddle point. Now consider the results of a harmonic TST calculation for the multidimensional BCHO model. The diagonalization in mass-weighted coordinates mixes the q and x_i coordinates to obtain an optimal coordinate with both solute and solvent components. The dynamics of the mixed coordinate is perfectly separable from the other coordinates so that recrossing is entirely eliminated. The PMF along the optimal mixed coordinate is parabolic with a renormalized imaginary frequency λ. The calculations are left as an exercise, but the final result is that harmonic TST applied to the multidimensional model gives exactly the GH theory rate [11,41]:

$$k_{hTST}[q, x_i] = \kappa_{GH}[q] k_{TST}[q] \tag{17.2.7}$$

In hindsight, this result is not surprising. By construction, the BCHO bath *must* have the same effect on the dynamics of q as did the non-Markovian friction. And because the bath and $V(q)$ are both harmonic, the harmonic TST calculation *must* work exactly.

So what does the (un)surprising equivalence (17.2.7) mean for real multidimensional systems with their non-linear and anharmonic complications? The question was debated for several decades. Many took the view that GH theory must essentially be a version of multidimensional VTST [41,45–47]. These authors asserted that successful rate predictions by GH theory imply a bath with properties like the BCHO. They further asserted that, when GH theory gives accurate transmission coefficients, a VTST calculation (if it could be performed) would give the correct rate with no further dynamical corrections. This viewpoint was supported by successful GH theory calculations [12,21] and VTST calculations [48,49] for the S_N2 reaction $Cl^- + CH_3Cl$ in water and its BCHO model, respectively.

Hynes argued [45] that GH theory is more general than a multidimensional harmonic TST, and that the exactly solvable BCHO model is just an illustrative example. Clearly, Hynes is correct that a real solvent bath is not created from harmonic oscillators. Moreover, real bath modes are coupled to each other [50] while the BCHO modes exchange energy only via their shared coupling to q. Slow solvent dynamics almost certainly results from the labyrinth of passages through configuration space, rather than from slow harmonic oscillations.

With no way to implement an exact VTST calculation, assertions of a VTST/GH theory equivalence could not be definitively tested. And so the debate stood unresolved for some time. Mullen et al. [51] devised a test of the assertion that VTST and GH theory are equivalent when GH theory is accurate. Without doing the VTST calculation, we know that it can only be exact if there is a perfect dividing surface that completely eliminates recrossing. If a perfect dividing surface exists in configuration space, then all trajectories that start in A and reach the dividing surface must continue directly to state B. Accordingly, trajectories that go from $A \to \mathbf{x} \to A$

cannot coexist with $B \to \mathbf{x} \to B$ trajectories (where both visit the same point $\mathbf{x}$). Mullen et al. called these "recrossing pairs". If recrossing pairs do exist, then a perfect dividing surface cannot (see Figure 17.2.1).

Figure 17.2.1: If there exists a perfect dividing surface for which the no-recrossing assumption is true, then there can never be excursions from $A \to \mathbf{x} \to A$ and also excursions $B \to \mathbf{x} \to B$ via the same intermediate state x. Transition path sampling provides data which can be inspected for the existence of these 'recrossing pairs'.

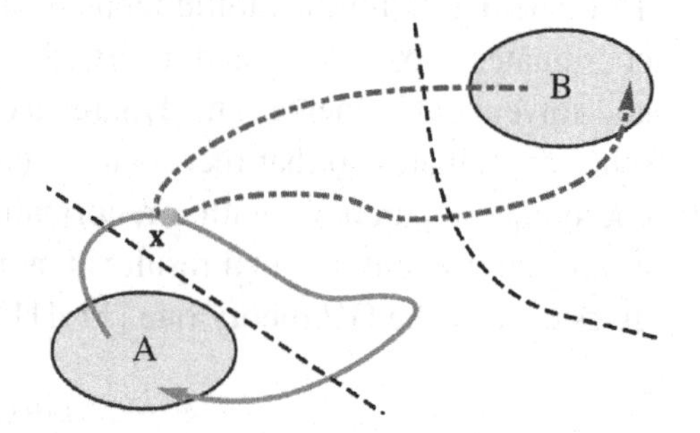

Mullen et al. [51] looked for recrossing pairs in the transition state region for ion-pair dissociation. For this system it had been specifically suggested in earlier work that VTST would eliminate recrossing because reactive flux and GH theory had both obtained transmission coefficients near $\kappa = 0.4$. Mullen et al. found that nearly 10% of the trajectories from transition path sampling were part of a recrossing pair. Therefore, no dividing surface from VTST can fully eliminate recrossing. In a related study, Mullen et al. [52] established an upper bound of $\kappa \lesssim 0.47$ for variationally optimized dividing surfaces.

The results of Mullen et al. illustrate that BCHOs are models of the solvent friction and not models of the solvent itself. It now appears clear that GH theory *can* provide accurate rates for a broader class of problems than can VTST. Note, however, that ion-pair association may be an unusual case because the barrier is entirely due to the solvent [18,23]. For reactions where the solvent merely perturbs an existing gas-phase barrier, VTST may indeed yield dividing surfaces that almost entirely eliminate recrossing.

Although GH theory has broader scope than VTST, some critiques of GH theory have merit. In particular, the GLE at the foundation of GH theory has only a heuristic justification. There is no proper derivation to justify the linear GLE for realistic solvents with complex anharmonic potentials [53]. Moreover, the solute coordinate is often chosen on an *ad hoc* basis. The resulting friction $\eta(t)$ and barrier frequency ω_b have little mechanistic significance because they are not related to the dynamics or free energy along the reaction coordinate. To draw mechanistic conclusions from GH theory, one should begin with an accurate reaction coordinate.

Exercises

1. Express the reaction probability $p_{RX}(q_0, v_0)$ in terms of the reactive frequency λ. Use the expressions of Kohen and Tannor [7] for $\langle \delta q(t) \rangle$ and $\langle \delta q(t)^2 \rangle$.

2. Describe the differences between the oscillatory dynamics of the dynamical caging regime and those of the energy diffusion limit in Kramers theory.
3. Given a spectral density

$$J(\omega) = \frac{\pi}{2}\sum_i \frac{c_i^2 \delta[\omega - \omega_i]}{m_i \omega_i^2}$$

the time-dependent friction is

$$\eta(t) = \frac{2}{\pi m}\int_0^\infty d\omega\, J(\omega)\omega^{-1} cos[\omega t].$$

A common model for liquids is the Ohmic bath with spectral density

$$J(\omega) = a\omega e^{-\omega/\omega_c}$$

where ω_c models an upper limit of the frequency spectrum and where a is the limit of $J(\omega)/\omega$ as $\omega \to 0$. For an Ohmic bath, compute the time dependent friction $\eta(t)$, the memory time τ_M, the time-integrated friction strength γ, and the GH theory transmission coefficient.
4. Show that

$$\hat{\eta}(s) = \frac{2}{\pi}\int_0^\infty \frac{J(\omega)}{\omega}\frac{s}{\omega^2 + s^2} d\omega$$

5. Convert Hamiltonian (17.2.1) to mass weighted coordinates $\overline{q} = m^{1/2}q$ and $\overline{x}_i = m_i^{1/2} x_i$ and diagonalize the mass-weighted Hessian at the saddle point to obtain the decoupled harmonic Hamiltonian [41]

$$H = \frac{1}{2}\dot{\overline{q}}^2 + V_\ddagger - \frac{1}{2}\lambda_\ddagger^2 \overline{q}^2 + \frac{1}{2}\sum_i (\dot{\overline{x}}_i^2 + \lambda_i^2 \overline{x}_i^2)$$

where

$$\lambda_\ddagger^2 = \omega_b^2 \left\{1 + \frac{1}{m}\sum_i c_i^2 \left[m_i \omega_i^2 (\omega_i^2 + \lambda_\ddagger^2)\right]^{-1}\right\}^{-1}$$

and

$$\lambda_j^2 = \omega_b^2 \left\{-1 + \frac{1}{m}\sum_i c_i^2 \left[m_i \omega_i^2 (\lambda_j^2 - \omega_i^2)\right]^{-1}\right\}^{-1}.$$

6. In the expression for $\lambda_{\ddagger}^2$ in problem (5), rewrite the summation in terms of an integral over the spectral density. Use the Laplace transform identity from problem (4) to obtain the GH equation for the reactive frequency of the harmonic bath model. Finally, use harmonic TST (classical, without quantization) to show that

$$k_{hTST} = \frac{\lambda_{\ddagger}}{\omega_b} k_{TST}[q]$$

where k_{hTST} is the multidimensional harmonic TST calculation and where $k_{TST}[q]$ is the one dimensional harmonic TST calculation using only the PMF $V(q)$ and mass m.

7. Simulations of generalized Langevin dynamics are more complicated than Langevin dynamics simulations with Markovian friction. The random forces must be correlated to match the time-dependent friction. A series of appropriately correlated forces can be generated by

$$R(k\Delta t) = \sum_{i=0}^{M-1} \{a_i sin[2\pi ik/M] + b_i cos[2\pi ik/M]\}$$

where each pair a_i and b_i are random Gaussian numbers from the distribution

$$\rho(a_i, b_i) = (2\pi\sigma_i^2)^{-1} exp[-(a_i^2 + b_i^2)/2\sigma_i^2]$$

with

$$\sigma_i^2 = \frac{mk_BT}{M} \sum_{k=0}^{M-1} \eta(k\Delta t) cos[2\pi i\, k/M].$$

M should be chosen large enough that $M\Delta t$ exceeds the simulation time. Develop a simulation code for generalized Langevin dynamics on an infinite parabolic barrier with an exponential time-dependent friction. Compute the reactive flux correlation function for the cases examined in Kohen and Tannor [7] and compare to their theoretical results.

References

[1] H.A. Kramers, Physica A 7 (1940) 284–304.
[2] R. Kubo, M. Toda, N. Hashitsume, Statistical Physics II, Springer-Verlag, Berlin, Heidelberg, 1985.
[3] A. Nitzan, Dynamics in Condensed Phases: Relaxation, Transfer, and Reactions in Condensed Molecular Systems, Oxford University Press, Oxford, 2006.
[4] R. Zwanzig, Phys. Rev. 124 (1961) 983.
[5] M. Tuckerman, Statistical Mechanics: Theory and Molecular Simulation, Oxford University Press, Oxford, UK, 2010.
[6] R.F. Grote, J.T. Hynes, J. Chem. Phys. 73 (1980) 2715–2732.
[7] D. Kohen, D.J. Tannor, J. Chem. Phys. 103 (1995) 6013–6020.

[8] J.E. Straub, M. Borkovec, B.J. Berne, J. Chem. Phys. 83 (1985) 3172–3174.
[9] J.E. Straub, M. Borkovec, B.J. Berne, J. Chem. Phys. 84 (1986) 1788–1794.
[10] S.A. Adelman, J. Chem. Phys. 64 (1976) 124–130.
[11] G. van der Zwan, J.T. Hynes, J. Chem. Phys. 78 (1983) 4174–4185.
[12] J.P. Bergsma, B.J. Gertner, K.R. Wilson, J.T. Hynes, J. Chem. Phys. 86 (1987) 1356–1376.
[13] B.J. Gertner, J.P. Bergsma, K.R. Wilson, S. Lee, J.T. Hynes, J. Chem. Phys. 86 (1987) 1377–1386.
[14] G.R. Haynes, G.A. Voth, J. Chem. Phys. 103 (1995) 10176.
[15] J.-P. Ryckaert, G. Ciccotti, J.C. Berendsen, J. Comput. Phys. 341 (1977) 327–341.
[16] H.C. Andersen, J. Comput. Phys. 52 (1983) 24–34.
[17] E.A. Carter, G. Ciccotti, J.T. Hynes, R. Kapral, Chem. Phys. Lett. 156 (1989) 472–477.
[18] G. Ciccotti, M. Ferrario, J.T. Hynes, R. Kapral, J. Chem. Phys. 93 (1990) 7137–7147.
[19] D. Frenkel, B. Smit, Understanding Molecular Simulation: From Algorithms to Applications, Academic Press, San Diego, 2002.
[20] J.E. Straub, M. Borkovec, B.J. Berne, J. Chem. Phys. 89 (1988) 4833–4847.
[21] B.J. Gertner, K.R. Wilson, J.T. Hynes, J. Chem. Phys. 90 (1989) 3537–3558.
[22] W.P. Kierstead, K.R. Wilson, J.T. Hynes, J. Chem. Phys. 95 (1991) 5256–5267.
[23] R. Rey, E. Guardia, J. Phys. Chem. 96 (1992) 4712–4718.
[24] L.X. Dang, H.V.R. Annapureddy, J. Chem. Phys. 139 (2013) 2011.
[25] H.R.V. Annapureddy, L.X. Dang, J. Phys. Chem. B 118 (2014) 7886–7891.
[26] D. Borgis, J.T. Hynes, J. Chem. Phys. 94 (1991) 3619–3628.
[27] D.A. Zichi, G. Ciccotti, J.T. Hynes, M. Ferrario, J. Phys. Chem. 93 (1989) 6261–6265.
[28] M. Roca, V. Moliner, I. Tunon, J.T. Hynes, J. Am. Chem. Soc. 128 (2006) 6186–6193.
[29] J.J. Ruiz-Pernia, I. Tunon, V. Moliner, J.T. Hynes, M. Roca, J. Am. Chem. Soc. 130 (2008) 7477–7488.
[30] N. Kanaan, M. Roca, I. Tunon, S. Marti, V. Moliner, J. Phys. Chem. B 114 (2010) 13593–13600.
[31] J.T. Hynes, Annu. Rev. Phys. Chem. 66 (2015) 1–20.
[32] A.O. Caldeira, A.J. Leggett, Ann. Phys. 149 (1983) 374–456.
[33] L.H. Yu, Phys. Rev. A 54 (1996) 3779–3782.
[34] R.T. Skodje, D.G. Truhlar, J. Phys. Chem. 85 (1981) 624–628.
[35] J.E. Straub, B.J. Berne, B. Roux, J. Chem. Phys. 93 (1990) 6804–6812.
[36] H. Wang, W.L. Hase, Chem. Phys. 212 (1996) 247–258.
[37] B. Carmeli, A. Nitzan, Chem. Phys. Lett. 102 (1983) 517–522.
[38] G.A. Voth, J. Chem. Phys. 97 (1992) 5908–5910.
[39] P. Hanggi, F. Mojtabai, Phys. Rev. A 26 (1982) 1168–1170.
[40] P. Hanggi, J. Stat. Phys. 42 (1986) 105–148.
[41] E. Pollak, J. Chem. Phys. 85 (1986) 865–867.
[42] R. Zwanzig, J. Stat. Phys. 9 (1973) 215–220.
[43] A.M. Levine, M. Shapiro, E. Pollak, J. Chem. Phys. 88 (1988) 1959–1966.
[44] J.M. Deutch, R. Silbey, Phys. Rev. A 3 (1971) 2049–2052.
[45] P. Barbara, J.T. Hynes, D. Chandler, M.C.R. Symons, M.H. Abraham, C.F. Wells, M. Henchman, M.J. Blandamer, D. Bush, K.H. Halawani, J. Schroeder, J. Troe, P.G. Wolynes, P. Suppan, H.L. Friedman, Faraday Discuss. Chem. Soc. 85 (1988) 341–364.
[46] E. Pollak, J. Chem. Phys. 95 (1991) 533–539.
[47] D.G. Truhlar, B.C. Garrett, J. Phys. Chem. B 104 (2000) 1069–1072.
[48] G. Gershinsky, E. Pollak, J. Chem. Phys. 101 (1994) 7174–7176.
[49] G. Gershinsky, E. Pollak, J. Chem. Phys. 103 (1995) 8501–8512.
[50] Y.I. Dakhnovskii, A.A. Ovchinnikov, Phys. Lett. A 113 (1985) 147–150.
[51] R.G. Mullen, J.-E. Shea, B. Peters, J. Chem. Phys. 140 (2014) 41104.
[52] R.G. Mullen, J.E. Shea, B. Peters, J. Chem. Theory Comput. 10 (2014) 659.
[53] J.E. Straub, M. Borkovec, B.J. Berne, J. Phys. Chem. 91 (1987) 4995.

CHAPTER 18

Diffusion over barriers

"The solution to this problem – the Greens function of eq. (2) – is rather complicated. Fortunately, as an equation for $[e^{\beta F(q)}\rho_{SS}(q)]$*, eq. (2) is self-adjoint ..."*

Onsager, J. Chem. Phys. (1938)

At this point in the book, we have learned several strategies for computing rate constants. Whether we look back to collision theory, transition state theory, RRKM theory, reactive flux, Kramers theory, Grote-Hynes theory, etc., the velocity at the barrier top, in some guise, was always a part of the final rate expression. The theories in this chapter are completely different. Trajectories from an overdamped (diffusion) process are continuous, but not differentiable, so there are no well-defined velocities. For irreversible phenomena like nucleation, we cannot even use spectral theories. Therefore we must start from entirely different assumptions in deriving the rate. This chapter outlines two general approaches: mean first passage times (MFPTs) and expressions based on committors (splitting probabilities). These closely related approaches yield a flux-over-population rate [1] from the steady-state population density with "rescue and replace" boundary conditions. These boundary conditions create a non-equilibrium steady-state current leading from the source (the reactant basin) to the sink (the product state). A similar construct was used in the discussions of classical nucleation theory (Chapter 14) and Kramers theory (Chapter 16). The steady-state rescue and replace construct at first seems to be rather artificial, but when the boundary conditions are imposed appropriately it has a strong theoretical foundation (*vide infra*).

In the modern literature, the overdamped limit where barrier crossing occurs by diffusion is often called the Kramers regime [2]. Theories for overdamped barrier crossings actually originated earlier, e.g. the mean first passage time (MFPT) expression of Pontryagin et al. [3] and Onsager's theory for the splitting probability [4]. The earlier frameworks are, in fact, more general than the rather prescriptive parabolic barrier and constant friction assumed in the Kramers theory. Theories based on MFPTs and committors are readily generalized to higher dimensions, anharmonic barriers, and coordinate dependent diffusivities.

This chapter begins with some mathematical results on forward and backward Fokker-Planck (Kolmogorov) equations that will be indispensable in what follows. Later sections present the

MFPT expression for a barrier in one dimension, Langer's multidimensional MFPT calculation, results on committors (splitting probabilities), and the close formal relationship between committors and MFPTs. This chapter concludes with a discussion of committors and related results for discrete master equations.

18.1 The forward and backward equations

The forward and backward Fokker-Planck equations, also known as forward and backward Kolmogorov equations, are integral to the calculations of committors and rates in the following sections. The forward and backward equations are differential statements of the Chapman-Kolmogorov equation. The forward equation describes the Green's function for forward time evolution. The backward equation describes the dependence on initial conditions in the Green's function. The forward equation is obtained from the Chapman-Kolmogorov equation with the intermediate timeslice positioned infinitesimally after the final time. The backward equation is obtained from the Chapman-Kolmogorov equation with the intermediate timeslice positioned infinitesimally after the initial time [5]. The starting point for the forward equation is

$$\rho(q, t+\Delta t \mid q_0, t_0) = \int d\Delta q \, \rho(q, t+\Delta t \mid q+\Delta q, t)\rho(q+\Delta q, t \mid q_0, t_0)$$

and for the backward equation,

$$\rho(q, t \mid q_0, t_0) = \int d\Delta q \, \rho(q, t \mid q_0+\Delta q, t_0+\Delta t)\rho(q_0+\Delta q, t_0+\Delta t \mid q_0, t_0)$$

The endpoints and intermediate timeslices for each case are shown in Figure 18.1.1. As Δt becomes small, the probability density on the intermediate timeslice becomes focused near the endpoint, i.e. near q for the forward equation and near q_0 for the backward equation.

In the limit $\Delta t \to 0$, the construction depicted in Figure 18.1.1(a) gives the "forward" Kolmogorov equation, i.e. an equation for $\rho(q, t \mid q_0, t_0)$ with fixed q_0 and t_0. The construction depicted in Figure 18.1.1(b) gives the "backward" equation [6], i.e. a Fokker-Planck equation for $\rho(q, t \mid q_0, t_0)$ with fixed q and t [5,7,8].

Chapter 15 showed that, for temporally homogeneous Markov processes with continuous variables, the transition probabilities are defined by a drift velocity $v_D(q)$ and a diffusivity $D(q)$. If the system at time t is known to be at location q, then an infinitesimal time later its position will be a Gaussian with center $q + v_D(q)dt$ and standard deviation $\sqrt{2D(q)dt}$. The corresponding forward equation is just the regular nonlinear Fokker-Planck equation

$$\frac{\partial \rho}{\partial t} = -\frac{\partial}{\partial q}\{v_D(q)\rho\} + \frac{\partial^2}{\partial q^2}\{D(q)\rho\} \tag{18.1.1}$$

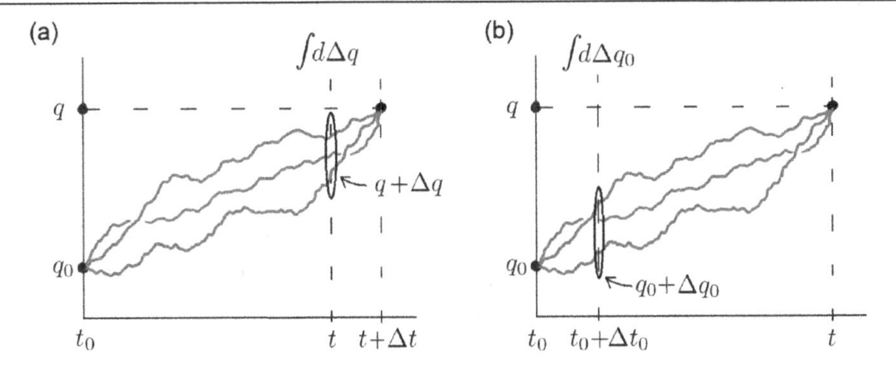

Figure 18.1.1: The forward and backward Kolmogorov equations are derived by considering the Chapman-Kolmogorov equation for timeslices that are infinitesimally close to the (a) final and (b) initial times, respectively.

but in this context ρ is a Green's function $\rho(q,t|q_0,t_0)$, i.e. $\rho(q,t)$ with initial conditions given by $\rho(q,0)=\delta[q-q_0]$. The backward equation is

$$-\frac{\partial \rho}{\partial t_0} = v_D(q_0)\frac{\partial \rho}{\partial q_0} + D(q_0)\frac{\partial^2 \rho}{\partial q_0^2} \tag{18.1.2}$$

where ρ is again the Green's function $\rho(q,t|q_0,t_0)$, but now the equation describes its dependence on the initial conditions with fixed q and t.

Equations (18.1.1) and (18.1.2) are quite general. However, the remainder of our discussion focuses on dynamics that obey detailed balance, like those of the overdamped Langevin equation. As shown in Chapter 15, imposing detailed balance on the forward equation yields the Smoluchowski equation.

$$\frac{\partial \rho(q,t|q_0,t_0)}{\partial t} = \hat{L}\rho(q,t|q_0,t_0) \tag{18.1.3}$$

Here the operator $\hat{L}$ is defined by

$$\hat{L}(\cdot) = \frac{\partial}{\partial q}e^{-\beta F(q)}D(q)\frac{\partial}{\partial q}e^{\beta F(q)}(\cdot) \tag{18.1.4}$$

and it is understood that operators act on everything to their right. Solving the forward equation gives the transition probabilities, i.e. the Green's functions for the dynamics in forward time. The interpretation is straightforward and now familiar from many earlier applications. The backward equation for overdamped Langevin dynamics is [9,10]

$$-\frac{\partial \rho(q,t|q_0,t_0)}{\partial t_0} = \hat{L}^\dagger \rho(q,t|q_0,t_0) \tag{18.1.5}$$

where

$$\hat{L}^{\dagger}\rho(q,t|q_0,t_0)=e^{\beta F(q_0)}\frac{\partial}{\partial q_0}D(q_0)e^{-\beta F(q_0)}\frac{\partial}{\partial q_0}\rho(q,t|q_0,t_0) \tag{18.1.6}$$

Note that $\hat{L}^{\dagger}$ is the adjoint [11] operator to $\hat{L}$.

The backward equation is *almost* the answer to the question "where did you come from given that you're here now?" The answer to that question now just requires another appeal to detailed balance.

$$\rho(q,t\,|\,q_0,t_0)\rho_{eq}(q_0)=\{\rho(q_0,t_0\,|\,q,t)\}\,\rho_{eq}(q) \tag{18.1.7}$$

The object in curly brackets indeed describes the distribution of original locations given the current location.

$$\{\rho(q_0,t_0\,|\,q,t)\}=\rho(q,t\,|\,q_0,t_0)\rho_{eq}(q_0)/\rho_{eq}(q) \tag{18.1.8}$$

So we cannot travel back in time, but we can solve the backward equation to see what the past might have been like. In retrospect (pardon the pun), this is not so surprising. We have essentially confirmed the stochastic time reversibility property for overdamped Langevin dynamics.

Now comes a fact that will be extremely important. For temporally homogeneous processes, i.e. for any physical process not driven by an external time-dependent stimuli, only the time difference $t-t_0$ matters in the Green's function. Therefore [5],

$$\rho(q,t|q_0,t_0)=\rho(q,t-t_0|q_0,0) \tag{18.1.9}$$

which implies a relationship between derivatives in final and initial time: $\partial\rho/\partial t=-\partial\rho/\partial t_0$. Thus we can equally write the backward equation for temporally homogeneous and overdamped Langevin dynamics as

$$\frac{\partial\rho(q,t|q_0,t_0)}{\partial t}=\hat{L}^{\dagger}\rho(q,t|q_0,t_0) \tag{18.1.10}$$

This equation is an interesting hybrid between the backward and forward equations: it relates the backward operation on q_0-dependence in $\rho(q,t|q_0,t_0)$ to the forward time derivative of $\rho(q,t|q_0,t_0)$. As the following sections will show, the hybrid equation is extremely useful.

18.2 Mean first passage times

Consider a swarm of stochastic trajectories initiated at a common point $\mathbf{x}_0$ with random initial momenta and independent random force realizations. Suppose that trajectories are terminated at the moment when they first reach some region B. Along each trajectory the route and time to reach B is different. The mean first passage time (MFPT) is the average time required to reach B. It depends, sometimes quite strongly, on the initial location $\mathbf{x}_0$.

MFPTs can be obtained numerically by following thousands of trajectories and averaging the times to reach B [12]. This "brute force" approach to obtain the MFPT, illustrated in Figure 18.2.1, has been widely used in simulations. For the brute force MFPT calculation, the process must be fast so that thousands of spontaneous escape trajectories can be simulated. Thus MFPTs from direct simulations are limited to simple models, to processes with small barriers, and (all too often) to simulations at unrealistically high temperatures, supersaturations, or pulling forces where the process is extremely fast relative to the real process of interest. Beyond efficiency considerations, Jungblut and Dellago noted that non-Markovian dynamics and imperfect progress coordinates can cause accuracy issues in brute force MFPT calculations [13].

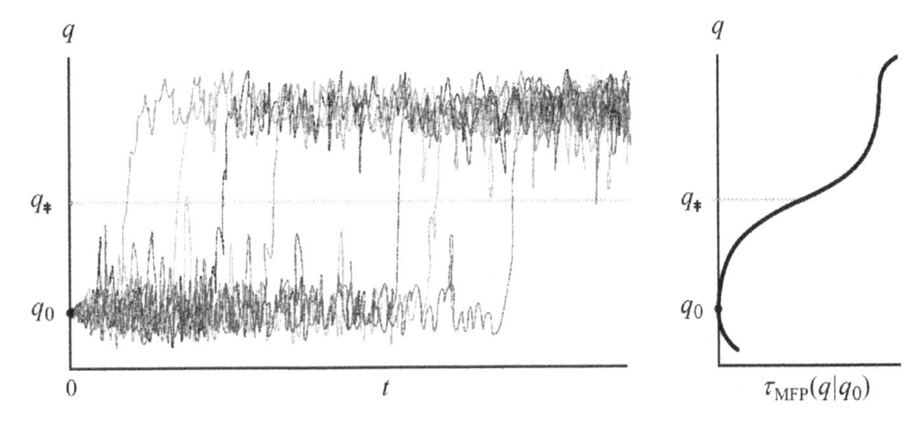

Figure 18.2.1: (Schematic) The left panel depicts a swarm of trajectories initiated from q_0 crossing through the transition state at $q_{\ddagger}$ and going on to a product state at random times. The right panel depicts the mean first passage time $\tau_{MFP}(q|q_0)$ as a function of q that would result from a numerical average over many trajectories.

This section shows that MFPTs can instead be computed from equilibrium properties and relatively simple results from non-equilibrium statistical mechanics. The MFPT derivation is conceptually challenging [3], but important for two reasons. First, the MFPT calculation for diffusion over a barrier in one-dimension is a blueprint for building theories of diffusion over barriers in higher dimensional landscapes [14,15]. Second, the free energies and coordinate diffusivities that enter an MFPT calculation are relatively easy to compute with standard rare events methods.[1]

The Green's function $\rho(q, t|q_0, 0)$ for overdamped dynamics (diffusion) on a potential of mean force is given by a Smoluchowski equation: $\partial\rho/\partial t = \hat{L}\rho$ where $\hat{L}$ acts on the

[1] Perhaps the main challenge is identifying a suitable reaction coordinate q.

q-dependence of ρ. Section 18.1 showed that $\rho(q,t|q_0,0)$ also satisfies the hybrid equation $\partial\rho(q,t|q_0,0)/\partial t = \hat{L}^{\dagger}\rho(q,t|q_0,0)$ where $\hat{L}^{\dagger}$ acts on the q_0-dependence of ρ. Here we use the hybrid equation to obtain the MFPT between two points along a reaction coordinate.

The MFPT depends on the shape of the free energy surface $F(q)$, the mobility/diffusivity on the free energy surface, the starting point (q_0), and the final location. Usually the final location of interest is the barrier top or a point on the opposite side of the barrier from q_0. The analysis below is valid for any two configurations. After obtaining a general MFPT formula, we will adapt it to the problem of barrier crossing rates. For now, let the final location be $q = q_F$. Without loss of generality, let q_0 be located to the left of q_F and let A be the set of all states to the left of q_F, i.e. $A = \{q\,|\,q < q_F\}$. To focus on *first* passage times, we impose an absorbing boundary condition at q_F,

$$\rho(q_F,t|q_0) = 0. \tag{18.2.1}$$

The absorbing boundary condition removes from consideration those trajectories which have already reached q_F in the past.

If we could solve for $\rho(q,t|q_0)$, we would integrate over all $q \in A$ to obtain the fraction of trajectories remaining in A after a time t:

$$\phi(t|q_0) \equiv \int_{-\infty}^{q_F} dq\, \rho(q,t|q_0)$$

For $q_0 \in A$ the initial condition on ϕ is

$$\phi(0|q_0) = 1$$

and because all trajectories eventually reach q_F, ϕ decays to zero for $t \to \infty$ as depicted in Figure 18.2.2.

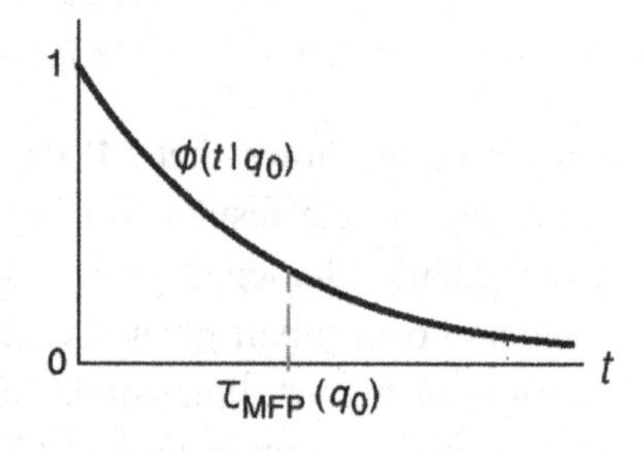

Figure 18.2.2: Schematic behavior of $\phi(t|q_0)$, i.e. the probability to remain in a metastable state A for a time t after initiation at position q_0. $\phi(t|q_0)$ may exhibit more complex transient decay behavior if q_0 is near the boundary of state A or if state A is not metastable.

The differential change in ϕ from time t to $t+dt$ is the probability that the first passage from A occurs between t and $t+dt$, i.e. the probability density of first passage times is $-\partial\phi(t|q_0)/\partial t$. Therefore, the MFPT from q_0 to q_F is

$$\tau_{MFP}(q_F|q_0) \equiv \int_0^{\infty} t \cdot \left[\frac{-\partial\phi(t|q_0)}{\partial t}\right] dt$$

Integrate by parts to obtain

$$\tau_{MFP}(q_F|q_0) = \int_0^\infty \phi(t|q_0)dt \tag{18.2.2}$$

from which the boundary terms $-(t \cdot \phi(t|q_0))_0^\infty$ have vanished. This equation cannot yet be evaluated because solving for $\rho(q,t|q_0)$ is usually a difficult task, and therefore we also do not know $\phi(t|q_0)$.

To avoid solving for $\rho(q,t|q_0)$, integrate $\partial\rho(q,t|q_0)/\partial t = \hat{L}^\dagger \rho(q,t|q_0)$ over all positions in A at time t,

$$\frac{\partial}{\partial t}\int_{-\infty}^{q_F} dq\, \rho(q,t|q_0) = \hat{L}^\dagger \int_{-\infty}^{q_F} dq\, \rho(q,t|q_0)$$

Recall that $\hat{L}^\dagger$ only acts upon the q_0-dependence in $\rho(q,t|q_0)$, so the integration over q has been moved inside the $\hat{L}^\dagger$ operator. Each of the integrals can now be replaced by $\phi(t|q_0)$ to give

$$\frac{\partial\phi(t|q_0)}{\partial t} = \hat{L}^\dagger\phi(t|q_0). \tag{18.2.3}$$

The q-dependence has been eliminated, and the equation that remains (18.2.3) describes how the probability to remain in A depends on t and q_0. It can be used to find the MFPT and also higher moments of the first passage time distribution. Let's proceed with the MFPT calculation by integrating over time

$$\phi(\infty|q_0) - \phi(0|q_0) = \hat{L}^\dagger \int_0^\infty \phi(t|q_0)dt$$

Using the limits $\phi(\infty|q_0) = 0$ and $\phi(0|q_0) = 1$, as well as $\int_0^\infty \phi(t|q_0)dt = \tau_{MFP}(q_F|q_0)$ gives

$$-1 = \hat{L}^\dagger \tau_{MFP}(q_F|q_0)$$

Again recall that $\hat{L}^\dagger$ acts via derivatives of the q_0-dependence in $\tau_{MFP}(q_F|q_0)$, and that q_F was just the final destination at which we imposed the absorbing boundary condition $\rho(q,t|q_0)|_{q=q_F} = 0$. The corresponding boundary condition on $\tau_{MFP}(q_F|q_0)$ is

$$\tau_{MFP}(q_F|q_F) = 0 \tag{18.2.4}$$

i.e. the MFPT from $q_0 = q_F$ to q_F must be zero. Thus the MFPT calculation has been reduced to an ordinary differential equation [3,16,17],

$$-1 = e^{\beta F(q_0)}\frac{d}{dq_0}D(q_0)e^{-\beta F(q_0)}\frac{d}{dq_0}\tau_{MFP}(q_F|q_0) \tag{18.2.5}$$

Formally, equation (18.2.5) is second order, so it would seem to require a second boundary condition. Typically, the nature of the free energy profile naturally prevents trajectories from wandering to $-\infty$ along the q-axis, and therefore no left boundary condition is needed. Sometimes, an additional reflecting boundary condition is needed at some leftmost value of q. For example, in nucleation we are often interested in the first passage time to reach the critical nucleus size $n_F = n_\ddagger$, but also no nucleus can become smaller than zero. Therefore an MFPT for nucleation would require an additional reflecting boundary condition at $n = 0$.

Equation (18.2.5) for the MFPT is easily solved in two stages of integration for any initial position $q_0 < q_F$. First,

$$D(q_0)e^{-\beta F(q_0)}\frac{d}{dq_0}\tau_{MFPT}(q_F|q_0) = -\int_{-\infty}^{q_0} dq' e^{-\beta F(q')}$$

then

$$\tau_{MFP}(q_F|q_0) = \int_{q_0}^{q_F} dq'' \frac{e^{\beta F(q'')}}{D(q'')}\int_{-\infty}^{q''} dq' e^{-\beta F(q')} \qquad (18.2.6)$$

For overdamped Langevin dynamics, equation (18.2.6) provides the MFPT from any point q_0 to any other point q_F as long as $q_F > q_0$. Figure 18.2.3 shows how the MFPT changes as a function of the reaction coordinate for a simple double well potential.

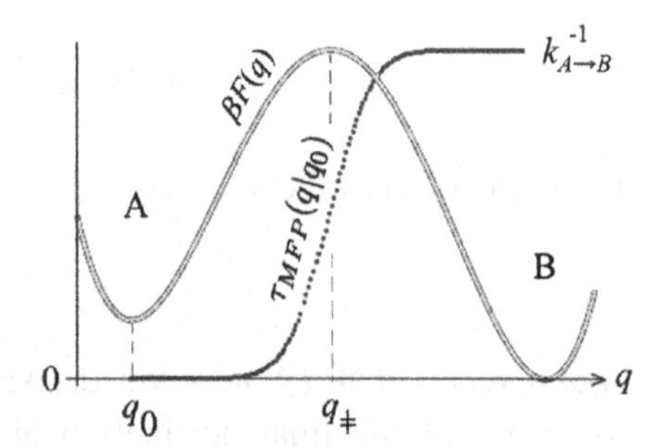

Figure 18.2.3: Showing $F(q)$ and the MFPT with constant D from q_0 to various final locations q. The MFPT is very small for all q deep within state A. The MFPT increases rapidly with q near the transition state, and then becomes approximately constant with value $k^{-1}_{A\to B}$ throughout state B.

However, when the barrier is much higher than $k_B T$ some simplifying approximations can be made because the predominant contributions to the integral containing $exp[-\beta F(q)]$ are near the minimum, and the predominant contributions to the integral containing $exp[+\beta F(q)]$ are near the barrier top. Specifically, note that

1. The integral over $e^{+\beta F(q)}$ contributes negligibly until q'' is near $q_\ddagger$.
2. For q'' near $q_\ddagger$ the integral over $e^{-\beta F(q)}$ spans the entire minimum near q_0.

3. The MFPT to any point q near the minimum in state B is approximately constant for all q_0 near the minimum in state A.

Based on these arguments, the MFPT *to reach the product basin from the reactant basin* is approximately

$$\tau_{MFP} \approx \int_{\cup} dq\, e^{-\beta F(q)} \int_{\cap} dq\, D(q)^{-1} e^{+\beta F(q)} \tag{18.2.7}$$

where subscript "∪" indicates integration over the reactant well and subscript "∩" indicates integration over the barrier top. Approximation (18.2.7) becomes increasingly accurate for high barriers, and it is equally valid for high barriers of any shape: parabolic barrier, asymmetric cusp, square top, etc. Note that the quantity $dq\, D(q)^{-1} exp[-\beta F(q)]$ is invariant to transformations that stretch or compress the coordinate q [18]. This invariance property can be used to obtain coordinates for which the diffusivity D is a constant at each point along the reaction pathway.

For examples of the MFPT calculation in high dimensions, see work by Adam and Delbruck [22] as well as models of protein folding by Bicout and Szabo [23]. Here we consider a simple example starting with the free energy profile and diffusivity along a reaction coordinate q.

Example: MFPT for parabolic approximation to barrier top

Consider a case where diffusivity is approximately constant near the top of a high barrier. Near the minimum the free energy resembles $F(q) = m\omega_A^2 q^2/2$ and in the transition state region, it resembles $F(q) = \Delta F_{\ddagger} - m\omega_{\ddagger}^2 (q - q_{\ddagger})^2/2$.

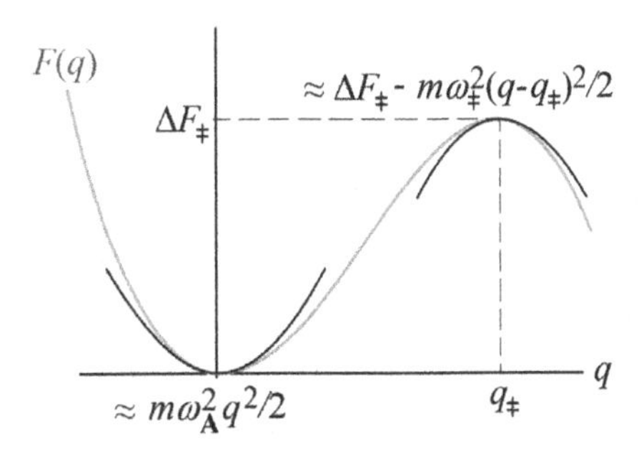

With these approximations, the MFPT formula becomes a product of two Gaussian integrals:

$$\tau_{MFP} = D \int_{\cup} dq e^{-\beta m\omega_A^2 q^2/2} \int_{\cap} dq e^{\beta \Delta F_{\ddagger} - \beta m\omega_{\ddagger}^2 (q-q)_{\ddagger}^2/2}$$

Doing the integrals and inverting the MFPT to obtain the rate constant gives

$$k_{MFPT} = \frac{D\sqrt{m\omega_{\ddagger}^{2} m\omega_{A}^{2}}}{2\pi k_{B}T} exp\left[-\beta\Delta F_{\ddagger}\right]$$

Now use $D = k_B T/m\gamma$ to recover the Kramers high friction result [19]

$$k_{MFPT} = \frac{\omega_A \omega_{\ddagger}}{2\pi\gamma} exp\left[-\beta\Delta F_{\ddagger}\right] \tag{18.2.8}$$

γ and $\omega_{\ddagger}$ are in the prefactor because they determine the mobility across the transition region and the width of the transition region, respectively. ω_A appears in the prefactor to account for entropy of the reactant well. For an application of this formula, see the example on pulling experiments [20,21] in Chapter 22.

When using these formulas, remember that a one dimensional Smoluchowski equation is a *model* for the *multidimensional* dynamics along a reaction coordinate in a high dimensional space. The results of this section are not valid for multidimensional dynamics unless one has chosen an accurate reaction coordinate $q(\mathbf{x})$ [24]. To ensure accurate results, one should carefully identify the reaction coordinate q or use multiple coordinates. Section 18.3 outlines the multidimensional MFPT calculation.

18.3 Langer's multidimensional theory

Most one-dimensional Langevin equations and Fokker-Planck equations are models of processes that involve many degrees of freedom. These models provide an accurate description of the dynamics only when constructed around an accurate reaction coordinate, and in many cases the correct reaction coordinate is not known. Sometimes, even when the correct reaction coordinate is known, it remains interesting to investigate multiple coordinates. Finally, there are problems like polymorph selection during nucleation which *require* analysis of at least two or more important coordinates [25,26]. Each of these situations requires analysis of diffusion over barriers in multiple dimensions.

The one-dimensional MFPT calculation required both a free energy profile and a mobility (or diffusivity) along the reaction coordinate. The multidimensional rate calculation requires a free energy surface and a mobility (or diffusion) tensor. Suppose that the dynamics of an activated process are thoroughly described by the free energy and mobility as functions of a multidimensional coordinate vector $\mathbf{q}$. Let the saddle point on the free energy surface be $\mathbf{q}_{\ddagger}$ and let the reactant minimum on the free energy landscape be at location $\mathbf{q}_0$. Langer expanded the free energy to second order near the saddle point $\mathbf{q}_\ddagger$ as [15]

$$F(\mathbf{q}_\ddagger + \Delta\mathbf{q}) \approx F(\mathbf{q}_\ddagger) + \frac{1}{2}\Delta\mathbf{q}^\dagger \mathbf{A} \Delta\mathbf{q} \tag{18.3.1}$$

The matrix $\mathbf{A}$ should have $n-1$ positive eigenvalues and one negative eigenvalue, where n is the number of coordinates within vector $\mathbf{q}$.

The mobility may be fast in some directions and slow in others, an aspect of the problem that was noted but not included in an early multidimensional MFPT calculation by Landauer and Swanson [14]. Langer extended their calculation to include the multidimensional diffusion tensor $\mathbf{D}$. Even more generally, the diffusion tensor may be a function of $\mathbf{q}$, but here we only consider a coordinate independent diffusion tensor.

As depicted in Figure 18.3.1, the diffusion tensor is easily obtained from an average of the covariance matrix for swarms of short molecular dynamics trajectories

$$2\mathbf{D}_{ij}t = \left\langle \delta q_i(t) \delta q_j(t) \right\rangle_\ddagger \tag{18.3.2}$$

where $\delta q_i(t) = q_i(t) - \langle q_i(t) \rangle_\ddagger$. The subscript $\ddagger$ indicates that the short trajectories are initiated from an equilibrium ensemble at the saddle point, i.e. $\mathbf{q}(0) = \mathbf{q}_\ddagger$. Note some subtleties in equation (18.3.2). First, the terms $\langle q_i(t) \rangle$ are subtracted in the $\delta q_i(t)$ to remove systematic drift. Ideally, there should be no systematic forces at the saddle point, but $\langle q_i(t) \rangle$ should be removed anyway to prevent small drift velocities from contributing to the mean squared displacements. Second, the trajectory duration should be short enough that the mean force does not change over the course of the short trajectory. For atomistic simulations the trajectory duration should also be long enough to transition from the inertial to the diffusive regime. Notes on the selection of an appropriate swarm trajectory duration can be found in the literature [27–29].

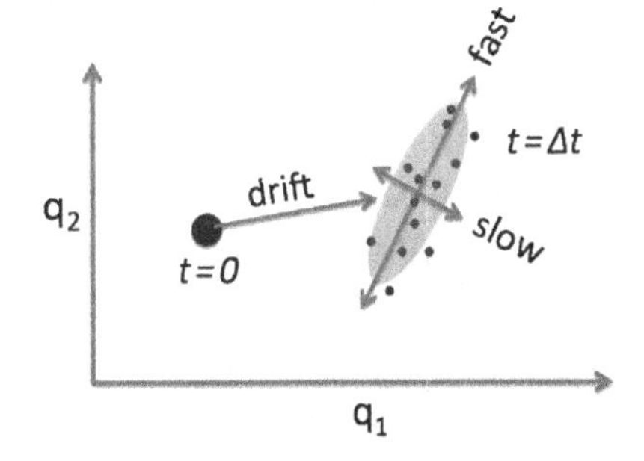

Figure 18.3.1: Computing the diffusion tensor via short trajectory swarms. The swarm drifts to a displacement $-\Delta t \mathbf{D} \partial F/\partial \mathbf{q}$ from the starting location. Additionally, the swarm spreads around its drifting mean as depicted by the scattered dots. The arrows labeled fast and slow depict eigenvectors of the diffusion tensor with large and small eigenvalues, respectively.

As in the one-dimensional MFPT calculation, the free energy in the reactant basin is also expanded around the stable minimum [15]

$$F(\mathbf{q}_0 + \Delta\mathbf{q}) \approx \frac{1}{2}\Delta\mathbf{q}^\dagger \mathbf{A}_0 \Delta\mathbf{q} \tag{18.3.3}$$

Here $\mathbf{A}_0$ is the matrix of second derivatives of the free energy at the minimum and accordingly it should have all positive eigenvalues. Note that equation (18.3.3) implicitly defines the zero of free energy as $F(\mathbf{q}_0) = 0$.

$F(\mathbf{q}_\ddagger)$, $\mathbf{A}_0$, $\mathbf{A}$, and $\mathbf{D}$ constitute the ingredients of the Langer theory, and the derivation can be found in his seminal paper [15]. Here we just report the main results and comment on its requisite assumptions. The average *current* of trajectories escaping over the saddle is directed along the lone unstable eigenvector of matrix **DA**,[2]

$$\mathbf{DAu} = -\lambda_+ \mathbf{u} \tag{18.3.4}$$

where $\lambda_+ > 0$. The rate constant for escape over the barrier from the reactant state is

$$k = \frac{1}{2\pi}\left(\frac{det\mathbf{A}_0}{|det\mathbf{A}|}\right)^{1/2} \lambda_+ exp\left[-\beta F(\mathbf{q}_\ddagger)\right] \tag{18.3.5}$$

It is a rate constant because it has been normalized by the population of reactants. Accordingly, the rate is the population of reactants (equilibrium or otherwise) multiplied by k. Langer's calculation assumes:

1. A steady-state distribution under rescue and replace boundary conditions.
2. A sufficiently large barrier (more properly a sufficient time scale separation) to establish quasi-equilibrium conditions within the reactant basin.
3. Validity of the harmonic approximation to the free energy landscape near $\mathbf{q}_\ddagger$.
4. A diffusion tensor that is independent of $\mathbf{q}$ on the free energy landscape.

Where the diffusion tensor is coordinate dependent and/or where the harmonic approximation is not valid, the full Smoluchowski equation can be constructed using atomistic simulations and numerically solved [25,30].

When one eigenvalue of the diffusion tensor is much smaller than others, the corresponding direction is dynamically frozen relative to diffusion along other directions. Berezhkovskii and Zitserman showed that severe diffusion anisotropy leads to trajectories that escape along the fast directions while remaining frozen in the slow directions. In extreme cases, diffusion anisotropy causes a breakdown of quasi-equilibrium in the reactant basin and to "saddle-point avoidance" where trajectories no longer escape via the saddle region [31,32]. Instead, the escape route and the MFPT become strongly dependent on the initial position of the frozen slow variable. Figure 18.3.2 shows how different degrees of dynamical anisotropy influence the path of escape from the metastable state.

[2] **D** is positive definite, so **DA** has just one negative eigenvalue emanating from **A**.

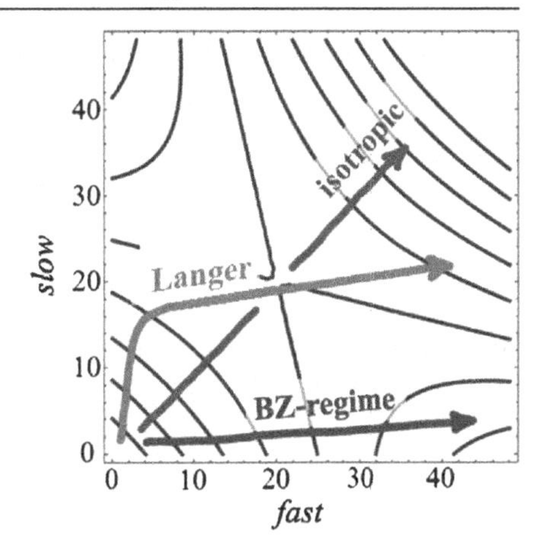

Figure 18.3.2: For isotropic diffusion, transition paths approximately follow the minimum free energy path. For anisotropic diffusion, transition paths tend to escape along the faster degree of freedom after a fluctuation in the slow variable lowers the barrier height. For extreme diffusion anisotropy (the Berezhkovskii-Zitserman regime), trajectories cross the barrier along the fast variable before the slow variable can move. [From Peters et al., *J. Chem. Phys.* 138, 054106 (2013).]

18.4 Committors (splitting probabilities)

The committor (also known as the splitting probability [33,34] and p-fold [35,36]) is the fraction of trajectories launched from an all-atom configuration $\mathbf{x}_0$ that reach the product state B before visiting the reactant state A when initiated with Boltzmann distributed initial velocities [37,38]. The committor was first introduced in a 1938 study of ion-pair dissociation by Onsager [4]. Chapter 20 discusses the committor in the context of fully atomistic simulations. Here, we consider its formal properties in systems whose dynamics have already been coarse grained to the level of a Fokker-Planck equation.

Suppose that states A and B are metastable states separated by a free energy barrier as shown in Figure 18.4.1. Let $A = \{q \,|\, q < q_{\partial A}\}$ and $B = \{q \,|\, q > q_{\partial B}\}$ as shown in Figure 18.4.1. Let I be that portion of space which is neither in A nor B, i.e. $I = \{q \,|\, q \notin A \text{ and } q \notin B\}$.

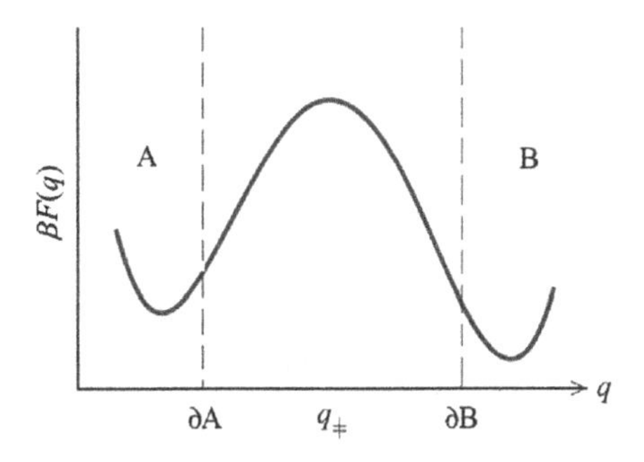

Figure 18.4.1: One-dimensional free energy profile $F(q)$ separating two states A and B. The boundaries of states A and B are at ∂A and ∂B, respectively.

Consider a swarm of stochastic trajectories initiated at an intermediate point $q_0 \in I$ and evolved with independent realizations of the random forces. A fraction $p_B(q_0)$ of those trajectories will first reach B, and $1 - p_B(q_0)$ of the trajectories will first reach A. Along each trajectory the route and time to first reach A or B is different. The probability density $\rho(q,t|q_0)$ for the not-yet-absorbed trajectories obeys a Smoluchowski equation $\partial\rho(q,t|q_0)/\partial t = \hat{L}\rho(q,t|q_0)$. The initial condition $\rho(q,0|q_0) = \delta[q-q_0]$ is implicit in the notation, and the absorbing boundary conditions at $q = \partial A$ and $q = \partial B$ are

$$\rho(q_{\partial A},t|q_0) = 0 \quad and \quad \rho(q_{\partial B},t|q_0) = 0. \tag{18.4.1}$$

If we could obtain $\rho(q,t|q_0)$, then we could compute the time-dependent flux into state B as

$$f_B(t|q_0) = -D(q)\left.\frac{\partial\rho(q,t|q_0)}{\partial q}\right|_{\partial B}.$$

Then we would obtain $p_B(q_0)$ by integrating $f_B(t|q_0)$ from $0 \le t < \infty$.

The complete solution of $\partial\rho/\partial t = \hat{L}\rho$ is difficult, but it can be circumvented by using the hybrid equation (18.1.10):

$$\partial\rho(q,t|q_0)/\partial t = \hat{L}^\dagger\rho(q,t|q_0)$$

with absorbing boundary conditions (equation (18.4.1)) at $q = \partial A$ and $q = \partial B$. $\hat{L}^\dagger$ only acts on the q_0-dependence of $\rho(q,t|q_0)$, so the operation $-D(q)\partial/\partial q$ which defines the flux into B now commutes with $\hat{L}^\dagger$. This allows us to write

$$\partial f_B(t|q_0)/\partial t = \hat{L}^\dagger f_B(t|q_0)$$

The q-dependence has now been eliminated, leaving an equation that describes how the flux into state B depends on t and q_0. Now we can integrate the equation for $f_B(t|q_0)$ over all time,

$$f_B(\infty|q_0) - f_B(0|q_0) = \hat{L}^\dagger\int_0^\infty dt\ f_B(t|q_0)$$

By definition

$$p_B(q_0) \equiv \int_0^\infty dt\ f_B(t|q_0)$$

Moreover, for any $q_0 \in I$ the initial flux is zero, and as $t \to \infty$, we have $f_B(t|q_0) \to 0$. Thus we arrive at the Onsager equation [4,17,39]

$$0 = \hat{L}^\dagger p_B(q_0) \tag{18.4.2}$$

with boundary conditions

$$p_B(\partial B) = 1 \tag{18.4.3}$$

and

$$p_B(\partial A) = 0 \tag{18.4.4}$$

For this one-dimensional model, the derivatives in operator $L^\dagger$ can be sequentially integrated to give an explicit solution for p_B in terms of the free energy profile and the diffusivity. The resulting expression for the committor is [17,39]

$$p_B(q) = \frac{\int_{\partial A}^{q} D(q)^{-1} exp[+\beta F(q)]dq}{\int_{\partial A}^{\partial B} D(q)^{-1} exp[+\beta F(q)]dq} \tag{18.4.5}$$

Figure 18.4.2 shows the committor as a function of the reaction coordinate for a simple double well potential. For a parabolic barrier with constant diffusivity, the committor is

$$p_B(q) = \frac{1}{2}\text{erfc}\left[-(q - q_\ddagger)\sqrt{\left|\beta F''(q_\ddagger)\right|/2}\right]$$

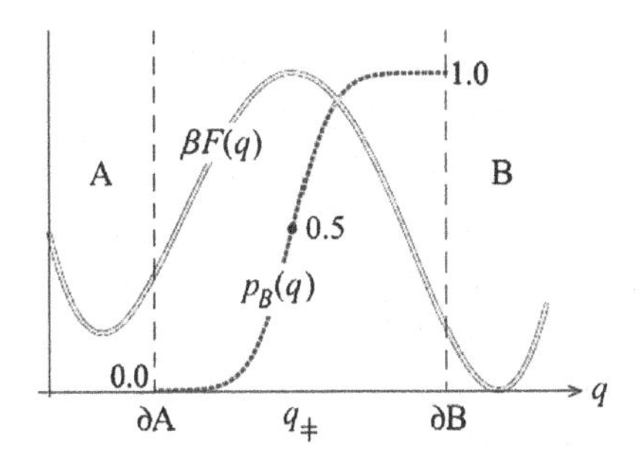

Figure 18.4.2: Showing the free energy profile $F(q)$ along with the resulting committor $p_B(q)$ for a coordinate independent diffusivity. $p_B(q)$ is not sensitive to the precise boundary locations ∂A or ∂B as long as these are far from the barrier top.

When basin definitions include only the typical fluctuations within the most stable states A and B, the committor throughout region I is a shifted and scaled version of the second Fokker-Planck eigenfunction [40,41]. See Figure 18.4.3 for an illustration of the committor-eigenfunction correspondence. This is not only true for simple one dimensional models. The isosurfaces of the committor and those of the second eigenfunction coincide with each other even in complex multidimensional systems (when basins A and B are properly defined). Recall from the spectral theory (Chapters 14 and 15) that the second eigenvector of the master equation is a natural reaction coordinate. Accordingly, the committor (when basins A and B are properly defined) is also a natural reaction coordinate. Relative to spectral analyses of the natural unbiased dynamics, the committor has the disadvantage of requiring *a priori* definitions of the reactant A and product B states. However, the A, B definitions also enable the use of many rare events methods that can efficiently overcome large barriers unlike the unbiased simulations that are usually used to parameterize Markov state models, diffusion maps, and other starting points for spectral analyses.

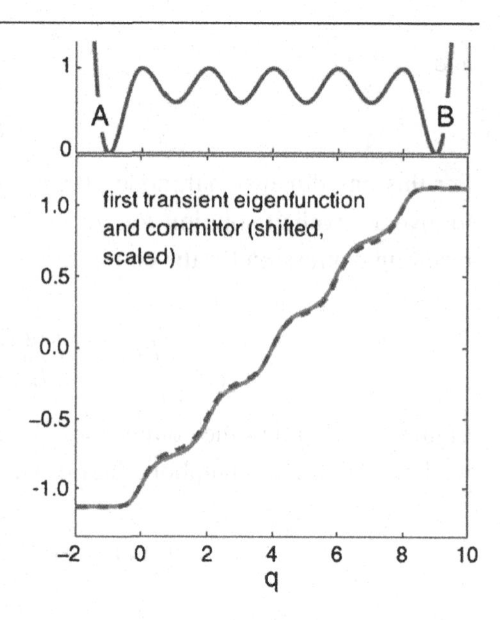

Figure 18.4.3: A model potential with two deep wells separated by five barriers and four shallow intermediates. The committor (when scaled and shifted) closely overlays the first transient eigenfunction of the Smoluchowski equation for diffusion on this potential. Note that the committor changes steeply near the barriers and becomes approximately flat near their minima. [Prinz et al. *J. Chem. Phys.* 134, 174105 (2011)]

Again, remember that the starting point was a one-dimensional Smoluchowski equation, a *model* for the *multidimensional* dynamics. The Onsager equation correctly predicts the committor only when constructed from the free energy and diffusivity along an accurate one dimensional reaction coordinate. The connection between a one-dimensional model and the high dimensional problem, in practice, must be achieved through an appropriate reaction coordinate $q(\mathbf{x})$ and definitions of states A, B, and I, e.g. $A = \{\mathbf{x}|\, q(\mathbf{x}) < q_{\partial A}\}$, $B = \{\mathbf{x}|\, q(\mathbf{x}) > q_{\partial B}\}$, and $I = \{\mathbf{x}|\, \mathbf{x} \notin A \, and \, \mathbf{x} \notin B\}$. Methods that identify accurate reaction coordinates from simulation data [42] are discussed in Chapter 20. The following section outlines a variational theory that identifies the optimal reaction coordinate from a multidimensional free energy landscape and diffusion tensor.

18.5 Berezhkovskii and Szabo: back to one dimension

For many decades, it was common and accepted practice to compute free energy barriers and/or rate constants with reaction coordinates of unknown accuracy. Do highly accurate free energy calculations and coordinate diffusivities compensate for using an inaccurate reaction coordinate? Berezhkovskii and Szabo showed that the answer is no [43]. They began with the multidimensional theory of Langer, so we briefly reiterate its assumptions here.

Assume that $\mathbf{q}$ is a vector of collective variables that includes all mechanistically important degrees of freedom. Also assume that the diffusion tensor in the $\mathbf{q}$-space is only mildly anisotropic to avoid saddle point avoidance behavior. Let $F(\mathbf{q})$ be the multidimensional free energy surface with a single saddle point at $\mathbf{q}_\ddagger$ and a single minimum at $\mathbf{q}_0$ in the reactant well. Let the free energy near $\mathbf{q}_\ddagger$ be $F(\mathbf{q}_\ddagger + \Delta\mathbf{q}) \approx F(\mathbf{q}_\ddagger) + \Delta\mathbf{q}^\dagger \mathbf{A} \Delta\mathbf{q}/2$ where matrix $\mathbf{A}$ has a single negative eigenvalue. Let the free energy near $\mathbf{q}_0$ be $F(\mathbf{q}_0 + \Delta\mathbf{q}) \approx \Delta\mathbf{q}^\dagger \mathbf{A}_0 \Delta\mathbf{q}/2$ where matrix $\mathbf{A}_0$ is positive definite. Finally, let $\mathbf{D}$ be the diffusion tensor in the vicinity of the saddle point. Based on these assumptions and definitions, the rate constant for escape from the reactant state is [15]

$$k = \frac{1}{2\pi}\left(\frac{det\mathbf{A}_0}{|det\mathbf{A}|}\right)^{1/2} \lambda_+ exp\left[-\beta F(\mathbf{q}_\ddagger)\right]$$

where $-\lambda_+ < 0$ is the one negative eigenvalue in $\mathbf{DAu} = -\lambda_+\mathbf{u}$. Also recall that $\mathbf{u}$ was the direction taken by typical trajectories as they cross over the saddle point.

Berezhkovskii and Szabo considered the rate that would result from an MFPT calculation with the free energy $F(\mathbf{q})$ and the diffusivity $\mathbf{D}$ projected onto a naively chosen scalar coordinate [43]. Specifically, they constructed a trial (scalar) reaction coordinate $q = (\mathbf{q} - \mathbf{q}_\ddagger) \cdot \mathbf{e}$ that quantifies displacement from the saddle point $\mathbf{q}_\ddagger$ along an arbitrarily chosen unit vector $\mathbf{e}$ in the $\mathbf{q}$ space. A minimal requirement on the choice of $\mathbf{e}$ is that the projected free energy $F(q)$, within the harmonic approximation, should have a maximum at $q = 0$. If the coordinate direction $\mathbf{e}$ is chosen so badly that $F(q)$ has a minimum at $q = 0$ then the rate calculation cannot be performed at all. For admissible choices of $\mathbf{e}$, Berezhkovskii and Szabo obtained

$$k(\mathbf{e}) = \frac{1}{2\pi}\left(\frac{det\mathbf{A}_0}{|det\mathbf{A}|}\right)^{1/2}\left(\frac{\mathbf{e}^\dagger\mathbf{De}}{|\mathbf{e}^\dagger\mathbf{A}^{-1}\mathbf{e}|}\right) exp\left[-\beta F(\mathbf{q}_\ddagger)\right] \tag{18.5.1}$$

For most choices of $\mathbf{e}$ the rate $k(\mathbf{e})$ differs from the correct multidimensional result, and importantly

$$k(\mathbf{e}) \geq k \tag{18.5.2}$$

for all admissible $\mathbf{e}$. Furthermore, $k(\mathbf{e})$ becomes the correct (Langer) multidimensional rate constant k if $\mathbf{e}$ is the unstable eigenvector of $\mathbf{AD}$ [43]:

$$\mathbf{ADe} = -\lambda_+\mathbf{e} \tag{18.5.3}$$

Henceforth, the symbol $\mathbf{e}$ refers to the optimal direction obtained via equation (18.5.3). While the eigenvalues of $\mathbf{AD}$ and $\mathbf{DA}$ are the same, their eigenvectors are not. The direction $\mathbf{e}$ and the direction of the current $\mathbf{u}$ are different except for two special cases: (i) where the diffusion tensor is isotropic so that $\mathbf{AD} = \mathbf{DA}$, and (ii) when an eigenvector of $\mathbf{D}$ coincides with the unstable eigenvector of $\mathbf{A}$.

The analysis of Berezhkovskii and Szabo identifies **e** as the reaction coordinate, i.e. the one direction that must be retained for a correct description of the activation barrier and the rate. They further showed that [43]

$$p_B(\mathbf{q}) = \frac{1}{2}\text{erfc}[-\mathbf{e}^\dagger(\mathbf{q}-\mathbf{q}_\ddagger)/\{2\left|\mathbf{e}^\dagger\mathbf{A}^{-1}\mathbf{e}\right|\}^{1/2}] \tag{18.5.4}$$

where $p_B(\mathbf{q})$ is the committor in the multidimensional space. Therefore, the reaction coordinate direction **e** is the gradient of the committor ∇p_B in the multidimensional **q**-space.

At first, the non-equivalence of **e** and **u** is strikingly counterintuitive. One might suspect that the direction taken by typical trajectories as they cross through the saddle region (**u**) is the reaction coordinate. Intuition says that we can watch movies of reactive trajectories to visually discern the reaction coordinate, but escaping trajectories follow direction **u** and not **e**. Thus, the characteristics of escaping trajectories do not reliably indicate the reaction coordinate. Figure 18.5.1 shows how mobility and characteristics of the free energy surface influence the two directions **e** and **u**.

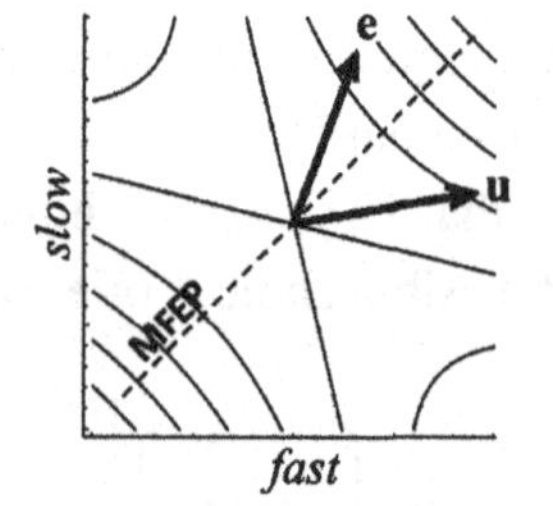

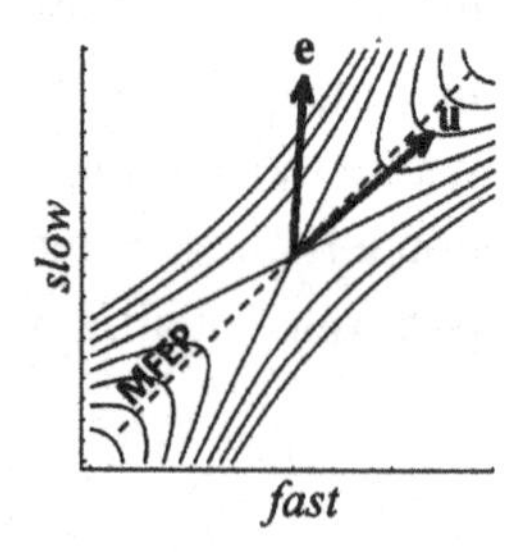

Figure 18.5.1: Qualitative depictions of e and u for anisotropic diffusion on two free energy surfaces. The left figure shows a free energy surface with a broad ridge between states *A* and *B*. The right figure shows a narrow tube type free energy surface.

The minimum free energy path (MFEP) is also shown in Figure 18.5.1. The MFEP is a steepest descent path from $\mathbf{q}_\ddagger$ on the free energy surface. Several methods have been developed to compute the MFEP including the string method in collective variables [48,49] and metadynamics with path-like variables [50,51]. MFEP optimizations are widely used because they remain efficient even when used with many collective variables. However, the free energy profile along the MFEP in several dimensions is not the free energy that enters the rate calculation. To obtain proper free energy barriers and rates one must further integrate away all off-path points with the same value of the committor [53,54]. Only a few studies [53–55] have undertaken the additional and highly non-trivial calculations to obtain activation barriers and rate constants from an optimized MFEP.

The rates and barriers are difficult to compute correctly, because the committor direction ($\mathbf{e} = \nabla p_B$) depends on both the free energy landscape *and the dynamics*, i.e. on which coordinates are fast and which are slow [52]. In general, the direction of the MFEP is different because it only depends on the free energy landscape. As seen in Figure 18.5.1, the reaction coordinate direction **e** deviates from the MFEP toward the direction of the slower coordinate(s). The flux direction **u** deviates from the MFEP toward the direction of the faster coordinate(s). This is true whether the bottleneck is a broad ridge or a saddle point along a narrow tube on the free energy surface.

For the free energy surface with the broad ridge between A and B, the directions **e** and **u** may be almost orthogonal to each other. For the narrow tube surface, **e** can even point uphill from the saddle point! Nevertheless, for both free energy surfaces the projection onto the **e**-direction yields a one-dimensional free energy surface with a maximum corresponding to the saddle point.

When the most important collective variables can be identified *a priori*, the Berezhkovskii-Szabo result provides a practical way to identify the scalar reaction coordinate, the activation barrier, and the escape rate. The Berezhkovskii-Szabo theory has been used in applications to peptide conformation transitions [28] in models of pulling experiments [56], and in nucleation [25,57]. A key limitation in applications is that even the few most important coordinates are rarely known. In principle, one could just include many coordinates to ensure a complete description, but starting with many coordinates requires an unfeasible high dimensional free energy calculation.

The Berezhkovskii-Szabo result is the overdamped counterpart to harmonic transition state theory (hTST). Both frameworks provide variationally optimized reaction coordinates within the harmonic approximation at a saddle point. Additionally, both frameworks account for dynamical differences between the various degrees of freedom. Dynamics influence the Berezhkovskii-Szabo results through the diffusion tensor. Dynamics influence the hTST results through the mass weighting of the Hessian. Finally, both frameworks identify an unstable mode with (locally) separable dynamics by diagonalization of a dynamics-modified second derivative matrix at the saddle point.

18.6 Classical nucleation theory revisited

Nucleation is the birth of a new stable phase from a metastable mother phase. The free energy to create a nucleus of size n depends on the chemical potential driving force for the phase transition and the interfacial free energy between the stable and metastable phases. As described in Chapter 14, the free energy as a function of nucleus size in classical nucleation theory (CNT) is $F_{CNT}(n) = -n\Delta\mu + a\gamma n^{2/3}$ [58,59]. The dynamics were modeled as a discrete stochastic process with an infinite series of one-molecule-at-a-time attachment and detachment steps. Chapter 14 also derived the exact rate expression of Becker and Doering [60].

The Becker-Doering expression is general in the sense that it makes no explicit reference to the shape of the barrier, to the shape of the nucleus, or to specific models for the solute attachment and detachment rates. With such a powerful result in hand, one might think that no further analysis is needed. However, the Becker-Doering rate expression is an infinite sum of terms which makes it difficult to recognize trends like the temperature dependence, solute concentration dependence, etc.

Instead of solving the discrete rate master equation, note that the one-molecule-at-a-time attachment and detachment events are small changes in nucleus size. Accordingly, the master equation can be converted into a Fokker-Planck equation in which n is a continuous variable. To avoid redundancy, this section appeals to several results and procedures which are explained in Chapter 15. The discussion of CNT was placed here to emphasize the close parallels between the CNT rate calculation, the MFPT formula, and Onsager's equation for the committor.

Starting from the master equation of Becker and Doering, equation (14.3.7), and applying the raising and lowering operators gives the Zeldovich-Frenkel (Smoluchowski) equation [61,62]:

$$\frac{\partial c(n,t)}{\partial t} = \frac{\partial}{\partial n}\left\{k_n c_{eq}(n)\frac{\partial}{\partial n}\left(\frac{c(n,t)}{c_{eq}(n)}\right)\right\} \tag{18.6.1}$$

Recall that k_n is the rate of solute attachment to nuclei of size n, $c_{eq}(n)$ is the equilibrium concentration of clusters of size n, and the equilibrium concentrations are arbitrarily normalized so that $c_{eq}(1)$ coincides with the isolated solute (monomer) concentration.

Following the same route as the discrete Becker-Doering analysis, define a crossover function $\xi(n,t) = c(n,t)/c_{eq}(n)$. Also write the concentrations of nuclei with different sizes in terms of the CNT free energy: $F_{CNT}(n) - F_{CNT}(1) = -k_B T \, ln\left[c_{eq}(n)/c_{eq}(1)\right]$. The result is the backward equation,

$$\frac{\partial \xi}{\partial t} = e^{\beta F_{CNT}(n)}\frac{\partial}{\partial n}\left(e^{-\beta F_{CNT}(n)}k_n\frac{\partial \xi}{\partial n}\right), \tag{18.6.2}$$

in which the attachment rate k_n occupies the usual place for a diffusivity.

The nature of the attachment kinetics depends on the specific application. Turnbull and Fischer developed expressions for k_n in solids using transition state theory [63]. Kashchiev gave expressions for k_n based on diffusion controlled attachment [62]. Zimmermann et al. [64] demonstrated that desolvation barriers determine k_n for ionic crystal precipitation. In many cases, the attachment rate can be obtained from the Einstein diffusion relation with simulation data [65].

$$\left\langle \delta n^2(t)\right\rangle \approx 2k_n t \tag{18.6.3}$$

where $\delta n(t) = n(t) - \langle n(t)\rangle$. Figure 18.6.1 shows how short molecular dynamics trajectories at the barrier top can be used to estimate k_n for critical nuclei.

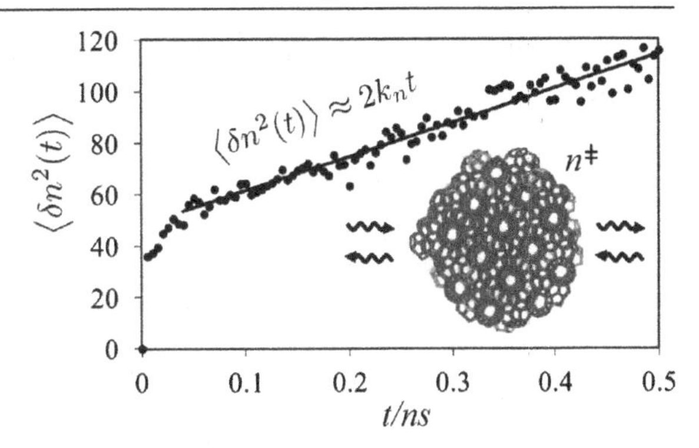

Figure 18.6.1: The attachment rate k_n can be estimated using short trajectories launched from the barrier top. The Einstein relation suggests that, after a transient time, $\langle \delta n^2(t) \rangle \approx 2k_n t$ where $\delta n(t) = n(t) - \langle n(t) \rangle$ and where $n(0) = n^\ddagger$. [Adapted from Knott et al. *J. Am. Chem. Soc.* 134, 19544-47 (2012).]

Now the rate must be computed for a steady-state situation in which large nuclei are removed, redissolved, and injected back into the metastable solution. Accordingly, we must solve [66]

$$0 = \frac{\partial}{\partial n}\left(e^{-\beta F_{CNT}(n)} k_n \frac{\partial \xi_{SS}}{\partial n}\right), \tag{18.6.4}$$

with boundary conditions

$$\xi_{SS} \to 1 \quad for \quad n \ll n^\ddagger \tag{18.6.5}$$

and

$$\xi_{SS} \to 0 \quad for \quad n \gg n^\ddagger \tag{18.6.6}$$

An exact solution is possible, but more easily used and interpreted results emerge from an expansion of $F_{CNT}(n)$ around the critical nucleus size.

$$F_{CNT}(n^\ddagger + \Delta n) \approx F_{CNT}(n^\ddagger) - \pi k_B T Z^2 \Delta n^2 \tag{18.6.7}$$

where $\Delta n = n - n^\ddagger$ and where Z is the Zeldovich factor [61,62]

$$Z = \sqrt{\frac{1}{2\pi}\left|\frac{d^2 \beta F_{CNT}}{dn^2}\right|_\ddagger} \tag{18.6.8}$$

The Zeldovich factor conveniently combines several factors which will appear in the prefactor. The approximate expansion will introduce only small errors as long as the nucleation barrier is large compared to $k_B T$. Incorporating Δn and Z into equation (18.6.4), integrating, and then matching boundary conditions gives

$$\xi_{SS}(n) = \frac{1}{2} erfc\left[Z\sqrt{\pi}(n - n^\ddagger)\right] \tag{18.6.9}$$

Following the same arguments used in the discrete Becker-Doering calculation, the steady state rate is

$$\begin{aligned} J &= -k_n \left.\frac{\partial c_{SS}}{\partial n}\right|_{\ddagger} \\ &= -k_n \left.\frac{\partial \xi_{SS}}{\partial n}\right|_{\ddagger} c_{eq}(n^{\ddagger}) \\ &= c_{SS}(1) k_n Z\, exp[-\beta(F_{CNT}(n^{\ddagger}) - F_{CNT}(1))] \end{aligned} \tag{18.6.10}$$

The prefactor depends weakly on supersaturation because of cancellations and logarithmic dependences in Z. Details are found in the review by Agarwal and Peters [66]. Often the supersaturation dependence in the prefactor is ignored resulting in a nucleation rate of the form

$$J = A\, exp[-\beta \Delta F_e^{\ddagger}/ln^2 S] \tag{18.6.11}$$

where the prefactor $A = c_{SS}(1) k_n Z\, exp[\beta F_{CNT}(1)]$ and where $\beta \Delta F_e^{\ddagger}$ is the dimensionless free energy barrier when $S = e$. In the nucleation literature $\beta \Delta F_e^{\ddagger}$ is often denoted B. After some simplification

$$\beta \Delta F_e^{\ddagger} = 4\,(\beta \gamma a/3)^3 \tag{18.6.12}$$

These equations show that classical nucleation theory predicts absolute nucleation rates as well as specific coefficients A and $\beta \Delta F_e^{\ddagger}$ that govern the supersaturation dependence of the nucleation kinetics. In practice, the absolute rate predictions are far from experimental rate measurements. However, an approximately linear relationship between $ln J$ and $1/ln^2 S$ is often observed [67–69]. See for example the data shown in Figure 18.6.2. Beyond correctly predicting an exponential dependence on $1/ln^2 S$, the expression for $\beta \Delta F_e^{\ddagger}$ has been used to predict and interpret the effects of solvents, salts, and other additives on the interfacial free energies of nuclei [57,68,70].

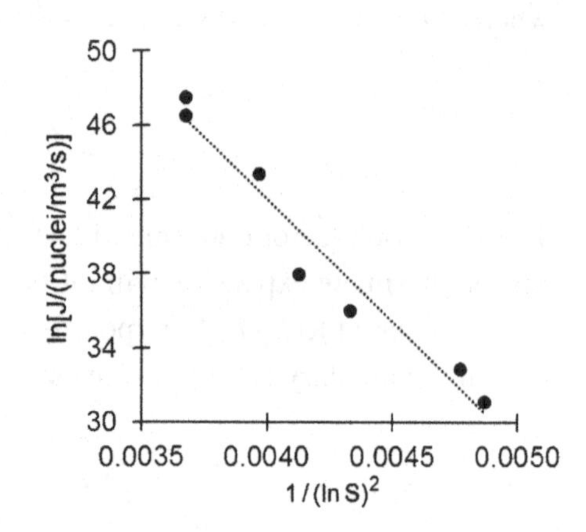

Figure 18.6.2: Data from Nielsen et al. for nucleation of $BaSO_4$ from aqueous solution at 298K. The data confirms the linear relationship between $ln J$ and $1/ln^2 S$ where S is the supersaturation ratio. See Kashchiev and van Rosmalen, *Cryst. Res. Technol.* 38, 555 (2003) for the same data plotted as $ln(J/S)$ vs. $1/ln^2 S$, which accounts for weak S-dependence within the prefactor.

Why has this discussion of CNT been placed between a section about committors and a section about extracting rates from the committor? Look back at equations (18.6.4) and (18.4.2). The equations are identical, with similar boundary conditions and similar solutions. Clearly $p_B(q)$ and $\rho_{SS}(q)/\rho_{eq}(q)$ are closely related to each other. The connections between rates, committors, and the crossover function $[\rho_{SS}(q)/\rho_{eq}(q)]$ are discussed in the next section.

18.7 Rates from the committor

The committor has been identified as the reaction coordinate in transition path sampling [35,44], in the Berezhkovskii-Szabo theory [43], and in transition path theory [45,46,71]. It can be shown that the committor is closely related to alternative definitions based on MFPTs [72], minimum cuts [73,74], and spectral theory definitions [41,75–78]. This section examines the *dynamics* of the committor in detail for those systems with overdamped (diffusive) dynamics [79–81]. Following Berezhkovskii and Szabo [81] we begin with a general Smoluchowski equation for diffusion on a high dimensional free energy landscape.

$$\frac{\partial \rho}{\partial t} = \frac{\partial}{\partial \mathbf{q}} \cdot \mathbf{D}(\mathbf{q}) e^{-\beta F(\mathbf{q})} \frac{\partial}{\partial \mathbf{q}} e^{\beta F(\mathbf{q})} \rho \tag{18.7.1}$$

where the diffusion tensor $\mathbf{D}(\mathbf{q})$ may be coordinate-dependent.

Consider two metastable regions A and B which are separated by an intermediate region I. Within region A the system is essentially at equilibrium, and within region B the population is zero. At steady state these boundary conditions lead to a flux from A to B through the intermediate region. By now, these steady-state analyses are familiar enough to proceed quickly. The steady-state probability density satisfies

$$0 = \frac{\partial}{\partial \mathbf{q}} \cdot \mathbf{D}(\mathbf{q}) e^{-\beta F(\mathbf{q})} \frac{\partial}{\partial \mathbf{q}} e^{\beta F(\mathbf{q})} \rho_{SS} \tag{18.7.2}$$

The steady-state rate can be computed by integrating the flux through any dividing surface S in region I

$$J_{SS} = -\int_S \mathbf{n}(\mathbf{q}) \cdot \mathbf{D}(\mathbf{q}) e^{-\beta F(\mathbf{q})} \frac{\partial}{\partial \mathbf{q}} e^{\beta F(\mathbf{q})} \rho_{SS} dS \tag{18.7.3}$$

The rate that results from this calculation does not depend on the choice of dividing surface. That statement must be interpreted very carefully! The surface invariant calculation integrates the flux through a surface *in the full-space* $\mathbf{q}$. The invariance is lost if the free energy and diffusivity are projected onto a scalar coordinate $q(\mathbf{q})$ *before* computing the flux through $q = q_{\ddagger}$.

Consider the multidimensional "lift-up" of the one-dimensional $q = q_{\ddagger}$ dividing surface.[3] The multidimensional lift-up surface gives a perfectly accurate rate in the multidimensional calculation, but the one-dimensional calculation projects the diffusion and free energy information into one dimension before computing the flux through $q = q_{\ddagger}$. Information about differences in the free energy, diffusivity, and flux at different locations on the multidimensional surface is lost in the projection onto q, and (except for the committor isosurfaces) [81] the result from the one-dimensional calculation is an incorrect rate.

Equation (18.7.2) shows that $e^{\beta F(\mathbf{q})}\rho_{SS}(\mathbf{q})$ (or equivalently $\rho_{SS}(\mathbf{q})/\rho_{eq}(\mathbf{q})$) satisfies the same Onsager equation that gives $p_B(\mathbf{q})$. Moreover, these two quantities have very similar boundary conditions. The table below outlines the analogies between the equations and boundary conditions for $p_B(\mathbf{q})$ and $\rho_{SS}(\mathbf{q})/\rho_{eq}(\mathbf{q})$.

Location	Committor	Steady state
$\mathbf{q} \notin A, B$	$0 = L^{\dagger} p_B(\mathbf{q})$	$0 = L^{\dagger} \rho_{SS}(\mathbf{q})/\rho_{eq}(\mathbf{q})$
$\mathbf{q} \in \partial A$	$p_B(\mathbf{q}) \to 0$	$\rho_{SS}(\mathbf{q})/\rho_{eq}(\mathbf{q}) \to 1$
$\mathbf{q} \in \partial B$	$p_B(\mathbf{q}) \to 1$	$\rho_{SS}(\mathbf{q})/\rho_{eq}(\mathbf{q}) \to 0$

The relationships between these equations and boundary conditions implies a simple relationship between $p_B(\mathbf{q})$ and $\rho_{SS}(\mathbf{q})$ [81]:

$$\rho_{SS}(\mathbf{q}) = (1 - p_B(\mathbf{q}))\rho_{eq}(\mathbf{q}) \tag{18.7.4}$$

As indicated in the epigraph, Onsager seems to have recognized these connections long before the rest of us [4]. Solving the Onsager equation for the committor also provides the steady-state distribution, the steady-state current, and the rate.

Vanden-Eijnden and coworkers illustrated solutions of the Onsager equation for several model free energy surfaces and diffusion tensors [47]. A two-dimensional example from their work is shown in Figure 18.7.1. Notice some characteristics of the committor isosurfaces:

1. The isosurfaces of p_B naturally form a foliation, i.e. a series of surfaces in which the early isosurfaces open around state A and the late isosurfaces close around state B.
2. The isosurface $p_B = 1/2$ passes through a saddle along the lower transition pathway, but passes through an intermediate minimum along the upper transition pathway.
3. p_B changes quickly as one moves along the unstable mode at a saddle point. p_B changes slowly within the metastable intermediate state, because trajectories launched from the intermediate state "forget" their initial conditions before escaping.

The committor can be defined for arbitrarily high dimensions. It can also be generalized to inertial reactions [38], giving what the older literature called a reaction probability [82].

[3] A one dimensional model of a tennis court is a line with a single point corresponding to the location of the net. In the more familiar three dimensions, the net is "lifted up" to become the two dimensional net at center-court.

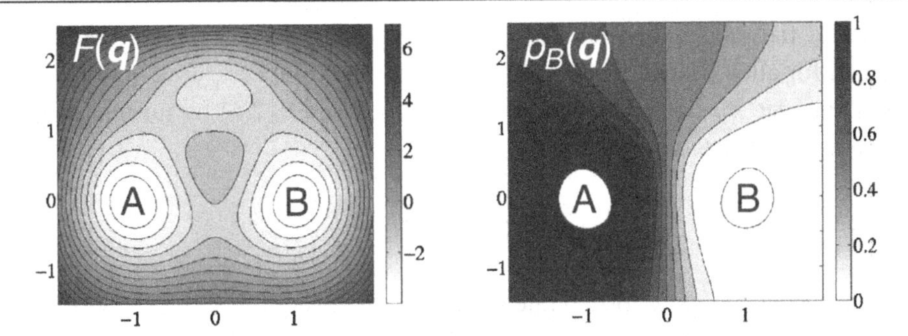

Figure 18.7.1: (Left) A contour map of a two-dimensional model free energy surface in $k_B T$ units. (Right) The corresponding contour map of the committor. There are two equally stable minima with two paths between them. One pathway involves a single saddle point. The other pathway traverses a stable intermediate via two slightly lower barriers. [From Metzner et al. J. Chem. Phys. 125, 84110 (2006).]

Let us now redo the flux calculation of equation (18.7.3) using an isosurface of the committor, i.e. for a surface

$$S = \{\mathbf{q} \mid p_B(\mathbf{q}) = p_B\} \tag{18.7.5}$$

First, use equation (18.7.4) to write the surface integral for the flux as a volume integral

$$\begin{aligned} J_{SS} &= -\int_S \mathbf{n}(\mathbf{q}) \cdot \mathbf{D}(\mathbf{q}) e^{-\beta F(\mathbf{q})} \frac{\partial}{\partial \mathbf{q}} e^{\beta F(\mathbf{q})} \rho_{SS} dS \\ &= \int_S \mathbf{n}(\mathbf{q}) \cdot \mathbf{D}(\mathbf{q}) \rho_{eq}(\mathbf{q}) \frac{\partial p_B}{\partial \mathbf{q}} dS \\ &= \int \delta[p_B(\mathbf{q}) - p_B] \frac{\partial p_B}{\partial \mathbf{q}} \cdot \mathbf{D}(\mathbf{q}) \rho_{eq}(\mathbf{q}) \frac{\partial p_B}{\partial \mathbf{q}} d\mathbf{q} \end{aligned} \tag{18.7.6}$$

Equation (18.7.6) says the rate is [81]

$$J_{SS} = \rho_{eq}(p_B) D_{p_B p_B}$$

where $\rho_{eq}(p_B) = \int \delta[p_B(\mathbf{q}) - p_B] \rho_{eq}(\mathbf{q}) d\mathbf{q}$, and $D_{p_B p_B}$ is a conditional average of the diffusivity along $p_B(\mathbf{q})$ at location p_B

$$D_{p_B p_B} = \left\langle \frac{\partial p_B}{\partial \mathbf{q}} \cdot \mathbf{D}(\mathbf{q}) \frac{\partial p_B}{\partial \mathbf{q}} \right\rangle_{p_B}$$

Thus projection of the Boltzmann distribution and multidimensional mobility onto the committor does preserve the rate information.

Now recall that the rate J_{SS} is the same for any p_B-isosurface. Therefore, an average over rates obtained for all p_B-isosurfaces has no effect on the rate, i.e. $J_{SS} = \int_0^1 J_{SS} dp_B / \int_0^1 dp_B$. The integration over p_B does, however, simplify equation (18.7.6) by turning the conditional average into an unrestricted average [81]

$$J_{SS} = \left\langle \frac{\partial p_B}{\partial \mathbf{q}} \cdot \mathbf{D}(\mathbf{q}) \frac{\partial p_B}{\partial \mathbf{q}} \right\rangle \tag{18.7.7}$$

Thus the steady-state non-equilibrium rate is an unrestricted equilibrium average of the diffusivity along the committor. The contributions to this average are uniformly dispersed across p_B, but where do the greatest contributions come from in the **q**-space? It might appear that they come from states A and B where the free energy is lowest. However, the committor within states A and B is identically zero or one, so that its gradient identically vanishes in the reactant and product states. All contributions come from intermediate regions, and the greatest contributions are from the vicinity of the bottleneck. Illustrations of this property can be seen in the works by Vanden-Eijnden and coworkers [38,47,83]. Figure 18.7.2 below illustrates the committor for a schematic multidimensional landscape alongside the integrand for the rate calculation in equation (18.7.7).

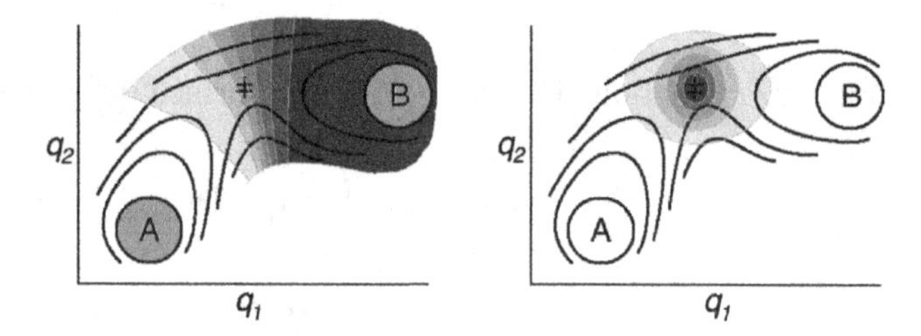

Figure 18.7.2: (Left) Schematic free energy landscape shaded according to contours of the committor $p_B(\mathbf{q})$. (Right) Schematic showing the integrand $e^{-\beta F(\mathbf{q})}||\nabla_{\mathbf{q}} p_B||^2$, i.e. the contributions to the rate in equation (18.7.7).

Several methods that use swarms of short trajectories to infer the dynamics at longer times [84–88] implicitly assume dynamical separability of the reaction coordinate(s), i.e. that the reaction coordinate predicts its own dynamical evolution regardless of other variables. Formally, the dynamical separability property [24] requires that $\rho(p_B, t|\mathbf{q}_0)$ is entirely determined by $p_B(\mathbf{q}_0)$ so that the committor itself, regardless of other details in $\mathbf{q}_0$, is sufficient to predict its own dynamical evolution along the p_B-coordinate. Dynamical separability determines whether a Fokker-Planck equation for $\rho(p_B, t)$ accurately predicts the multidimensional dynamics. However, even when p_B is not dynamically separable, equations (18.7.6)–(18.7.7) show that the free energy and mobility along p_B are still sufficient to obtain an accurate rate [38,80,81].

This section presented several new and potentially useful theories based on properties of committor. How does one use these results in practice? The committor itself is too expensive to compute directly, so we must identify more easily computed collective variables as reaction coordinates. In principle, we can compute the free energy surface and diffusion tensor in a space of collective variable reaction coordinates $\mathbf{q}$ and then solve the Onsager equation to obtain $p_B(\mathbf{q})$. However, rare events methods for computing free energies and coordinate diffusivities become prohibitively expensive in high dimensions. Thus we are forced to choose a low dimensional set of coordinates $\mathbf{q}$, but choosing the few most important reaction coordinates requires a strong *a priori* understanding of the mechanism. Ideally, mechanistic understanding should emerge *from* a simulation. It should not be a required input. Chapter 20 describes methods that can identify the few most important reaction coordinates for constructing models of the committor.

18.8 Discrete committors and rates

Each of the results in the previous section has a discrete master equation analogue. The discrete versions, summarized below, are known as "first step analyses" in the stochastic processes literature [89]. First step analyses are powerful tools in connection with Markov State Models and discrete master equations [41,71,90–92]. Consider a master equation with a finite number of states and rate matrix $\mathbf{W}$. Recall from Chapter 14 that in matrix notation the master equation is

$$d\mathbf{p}/dt = -\mathbf{W}\mathbf{p}$$

The off-diagonal elements are $\mathbf{W}_{ij} = -\omega_{i\leftarrow j}$ and the diagonal elements are $\mathbf{W}_{jj} = \sum_i \omega_{i\leftarrow j}$. Suppose that the discrete states can be divided into three groups: A and B form highly metastable superbasins and a third intermediate group I divides them. Further suppose that no transition goes directly from A to B without visiting at least one intermediate state in I. Groups A, B, and I are shown in Figure 18.8.1.

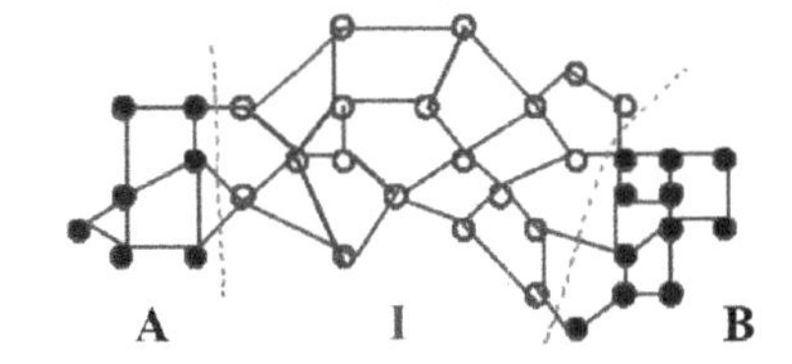

Figure 18.8.1: A graph representing transitions between states with metastable sets A and B as well as an intermediate set I.

Given the state-to-state transition rates, one can exactly compute the mean time to reach set B from an equilibrium distribution within set A. The calculation begins with a more specific calculation of the mean time to reach set B from any state i. Thus far we do not require $i \in I$. The mean time to reach B from i is the mean residence time in state i plus a sum accounting for the probability to visit states j "next" after state i, and the mean times to reach B from each of the possible "first steps" to states j.

$$\tau_{MFP}(i) = \tau_{RES}(i) + \sum_j^{\neq i} \frac{\omega_{j\leftarrow i}}{\omega_{tot}(i)} \tau_{MFP}(j) \tag{18.8.1}$$

where

$$\omega_{tot}(i) = \sum_{j \neq i} \omega_{j \leftarrow i} \tag{18.8.2}$$

and the mean residence time in state i is

$$\tau_{RES}(i) = \omega_{tot}(i)^{-1}. \tag{18.8.3}$$

$\tau_{MFP}(j) = 0$ for any state $j \in B$ because it cannot take any time to reach B from B. Multiplying the entire equation by $k_{tot}(i)$ gives

$$-1 = -\tau_{MFP}(i) \sum_{j}^{\neq i} \omega_{j \leftarrow i} + \sum_{j}^{\neq i} \omega_{j \leftarrow i} \tau_{MFP}(j) \tag{18.8.4}$$

which in matrix form is $1 = \tau^{\dagger}_{MFP}\mathbf{W}$, or

$$-1 = (-\mathbf{W})^{\dagger} \tau_{MFP}. \tag{18.8.5}$$

Note that the vector of MFPTs is obtained from the adjoint of the rate matrix, a result that closely resembles equation (18.2.5) for first passage times with continuous overdamped dynamics.

First step analysis can also be used to compute committors. For any state i, the committor is the probability of reaching the product state B before state A. Following similar logic as that for MFPTs, the committor for state i is a sum involving the probability of first steps to states j after i and the committors at the states j:

$$p_B(i) = \sum_{j}^{\neq i} \frac{\omega_{j \leftarrow i}}{\omega_{tot}(i)} p_B(j) \tag{18.8.6}$$

which in matrix notation becomes

$$\mathbf{0} = p^{\dagger}_{B}\mathbf{W} \tag{18.8.7}$$

This is a discrete version of Onsager's equation for the committor [81,90,91]

Equations (18.8.5) and (18.8.7) can be exactly solved for discrete networks with a finite number of states. However, many applications involve discrete networks with an uncountable number of states. For example, in models of nucleation and complex biochemical networks there is no limit to the size of a nucleus or to the numbers of biomolecules. Certain special cases with infinite transition matrices can still be solved exactly. However, most applications with infinite transition matrices require kinetic Monte Carlo (kMC) [93] simulations. When large barriers separate metastable sets A and B, e.g. in nucleation, then one must additionally combine kMC with path sampling [94,95] or graph reduction [96–98] techniques. These kMC based simulation methods provide committors which can then provide overall fluxes from A to B using the methods below.

Rates from discrete committor

Metzner et al. [91] and Berezhkovskii et al. [90] derived the overall rate from A to B in terms of committors for each of the states in I:

$$J = \sum_{m}^{p_B(m)\leq p_B^*} \sum_{n}^{p_B(n)\geq p_B^*} \omega_{n\leftarrow m}\, p_{eq}(m)\, \{p_B(n) - p_B(m)\} \tag{18.8.8}$$

where p_B* is any intermediate value of the committor between 0 and 1. This equation, where $p_B = p_B*$ has been used as a dividing surface, is the discrete analog of equation (18.7.3). The fraction of the total flux carried by each edge in the network is $J_{n\leftarrow m}/J$ where

$$J_{n\leftarrow m} = \omega_{n\leftarrow m}\, p_{eq}(m)\{p_B(n) - p_B(m)\} \tag{18.8.9}$$

for each m and n such that $p_B(n) > p_B(m)$. Echoing their results for the continuous case, Berezhkovskii and Szabo showed that lumping the discrete states according to their committor probabilities yields some simplifications [81].

Suppose that the discrete intermediate states are lumped into groups of states with similar committors, e.g. group $\mathbf{M}$ includes all states m with $p_B(m) \approx p_B(\mathbf{M})$, group $\mathbf{N}$ includes all states n with $p_B(n) \approx p_B(\mathbf{N})$, etc. Further adopt the definition

$$\bar{\omega}_{\mathbf{N}\leftarrow\mathbf{M}}\bar{p}_{eq}(\mathbf{M}) \equiv \sum_{m\in\mathbf{M}}\sum_{n\in\mathbf{N}} \omega_{n\leftarrow m}\, p_{eq}(m) \tag{18.8.10}$$

where $\bar{p}_{eq}(\mathbf{M}) = \sum_{m\in\mathbf{M}} p_{eq}(m)$. The definition of $\bar{\omega}_{\mathbf{N}\leftarrow\mathbf{M}}$ in equation (18.8.10) ensures that the lumped states obey detailed balance. With this definition the rate expression simplifies to

$$J = \sum_{\mathbf{M}}^{p_B(\mathbf{M})\leq p_B^*} \sum_{\mathbf{N}}^{p_B(\mathbf{N})\geq p_B^*} \bar{\omega}_{\mathbf{N}\leftarrow\mathbf{M}}\bar{p}_{eq}(\mathbf{M})\, \{\bar{p}_B(\mathbf{N}) - \bar{p}_B(\mathbf{M})\} \tag{18.8.11}$$

which remains exact despite the coarse grained state space.

Let us now further suppose that the discrete evolution along the p_B coordinate has a discrete dynamical separability property, i.e. that

$$p(\mathbf{N}, t\,|\,m, 0) = p(\mathbf{N}, t\,|\,\mathbf{M}, 0) \quad \forall m \in \mathbf{M} \tag{18.8.12}$$

Discrete dynamical separability requires that transitions from one value of p_B to another are slower than transitions within groups of states that have the same committor. When the dynamical separability holds, the discrete master equation becomes the coarsened (but still dynamically accurate) master equation

$$d\bar{\mathbf{p}}/dt = -\bar{\mathbf{W}}\bar{\mathbf{p}}$$

where the off-diagonal elements of $\bar{\mathbf{W}}$ are $\bar{\mathbf{W}}_{\mathbf{NM}} = -\bar{\omega}_{\mathbf{N}\leftarrow\mathbf{M}}$ and the diagonal elements are $\bar{\mathbf{W}}_{\mathbf{MM}} = \sum_{\mathbf{N}} \bar{\omega}_{\mathbf{N}\leftarrow\mathbf{M}}$ [81]. Even without dynamical separability, the committor-coarsened master equation still preserves the overall rate between the groups A and B that define the committor. See Banushkina and Krivov for a variational principle and algorithm that affects the committor-based coarsening transformation by starting with data from a long simulation trajectory [99].

Exercises

1. Consider a system with reaction coordinate q and free energy $\beta F(q) = \beta F_{\ddagger}\{1 - 3q^2 - 2q^3\}$. Suppose the diffusivity D along the q-axis is constant. Compute the MFPT for escape from the metastable state.
2. Show how the free energy barrier and diffusivity from problem (1) are altered upon coordinate transformation to $z = q^3$. Show that the MFPT is not affected by the change of variables.
3. Use equation (18.2.3) to develop an expression for the variance in the first passage time distribution. What variance can always be anticipated for barriers that are much higher than $k_B T$?
4. Compute the rate of diffusion over a square topped barrier of size ΔF and width L. Assume a perfectly absorbing boundary condition at the right edge of the barrier and equilibrium with states to the left of the barrier at its left edge. Compare each term in your result to terms in the expression for flux of solutes across a permeable membrane: $j = DHC_L/L$ where D is the diffusivity, H is a partition coefficient, and C_L is the concentration of solutes to the left of the membrane [100].
5. Compare in detail the arguments of erfc in equations (18.5.4) and (18.6.9).
6. Exercise (6) of Chapter 14 considered 1D nucleation events in crystal growth. Return to the model presented there, and derive an expression for the committor as a function of the nucleus length n. Let the boundary conditions be $p_B(n) = 0$ for $n = 0$ and let $p_B(n) \to 1$ when $n \to \infty$. Use the committor and the transition frequencies to give an expression for the 1D nucleation rate.
7. Revisit exercise 15.12 on the dynamics of polymer translocation through a pore [101]. Complete the details of the MFPT calculation as described by Muthukumar. Identify and

estimate the additional quantities that are needed to compute the overall translocation rate given the bulk polymer concentrations on the left and right sides of the membrane.

8. Construct a Master equation and Fokker-Planck equation for the following birth-death model of a population

$$\omega_{n \to n+1} = k_O n(1 - n/n_O)$$
$$\omega_{n \to n-1} = k_D n$$

Use the procedure described in Chapter 15 to find an effective free energy $F(n)$ and an effective diffusivity $D(n)$. Let state B correspond to extinction: $B = \{n = 0\}$. Let state A be the set of thriving populations: $A = \{n | n > n_O(1 - k_O/k_D)\}$. Compute $p_{SS}(n)$, $p_B(n)$, and the rate of extinction. Use the values $n_O = 100$, $k_O = 3/yr$, and $k_D = 4/yr$ to obtain numerical values.

9. [Adapted from D. Makarov, *J. Chem. Phys.* 141, 241103 (2014).] Single protein molecule pulling experiments attach colloidal beads to a protein via molecular tethers as shown in the figure below.

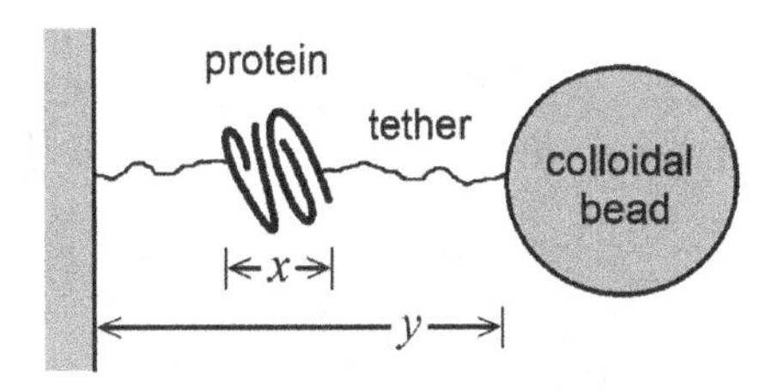

Using the beads as optical tweezers, the experimentalist can exert a specific force f on the bead. Let this force pull against a coordinate y, and let y be coupled to the extension of the protein x as shown in the figure. Let the unperturbed potential for the protein (P) in the vicinity of the transition state be

$$V_P(x) = V_{\ddagger} - m|\omega|^2(x - x_0)^2/2$$

where x is the ideal reaction coordinate in the absence of a pulling force. Let the potential for the tether (T), and bead (B) be

$$V_T(x, y) = m_T \omega_T^2 (y - x)^2/2$$

and $V_B(y) = -f \cdot y$, respectively. Finally, let the total potential be a sum of these contributions:

$$V(x, y) = V_P(x) + V_T(x, y) + V_B(y)$$

(a) Write down coupled overdamped Langevin equations for variables x and y. Convert your Langevin equations to the form

$$\begin{pmatrix} \dot{x} \\ \dot{y} \end{pmatrix} = -\mathbf{DA} \begin{pmatrix} x - x_{\ddagger} \\ y - y_{\ddagger} \end{pmatrix} + \begin{pmatrix} R_x(t) \\ R_y(t) \end{pmatrix}$$

Be sure to identify the elements of matrix **D**, the elements of matrix **A**, the location of the saddle point $(x_{\ddagger}, y_{\ddagger})$, and the properties of the two random forces R_x and R_y.

(b) Use Langer's theory to compute the rate of unfolding as a function of the pulling force f. Compare your result to a calculation in which all properties are projected onto the (measured) coordinate y.

(c) Do the tether properties and the pulling force influence the optimal reaction coordinate in the (x, y) space? If so, should the ideal tether (for probing the protein) be stiff or yielding?

10. Consider the free energy landscape below with two well-separated channels leading from state A to state B along the q-direction (horizontal). Suppose that the diffusivities along q in the upper and lower channels are D and D', as shown in the figure.

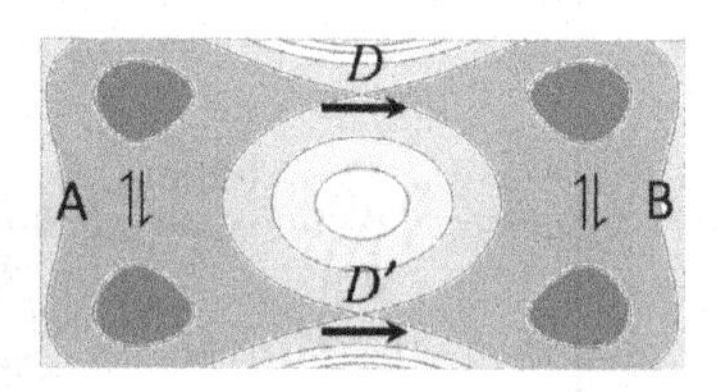

Will the projections onto $F(q)$ and $D(q)$ lead to a Fokker-Planck equation that correctly describes the dynamics of q? Can $F(q)$ and $D(q)$ be used to obtain the rate? Explain.

References

[1] L. Farkas, Z. Phys. Chem. 125 (1927) 236–242.
[2] P. Hanggi, P. Talkner, M. Borkovec, Rev. Mod. Phys. 62 (1990) 251–341.
[3] L.S. Pontryagin, A.A. Andronov, A.A. Vitt, Zh. Eksp. Teor. Fiz. 3 (1933) 165–180.
[4] L. Onsager, Phys. Rev. 54 (1938) 554–557.
[5] D.T. Gillespie, Markov Processes: An Introduction for Physical Scientists, Academic Press, Boston, 1992.
[6] A. Kolmogorov, Math. Ann. 104 (1931) 415–458.
[7] W.A. Coffey, Y.P. Kalmykov, J.T. Waldron, The Langevin Equation, World Scientific, Singapore, 2004.
[8] N.G. van Kampen, Stochastic Processes in Physics and Chemistry, Elsevier, Amsterdam, 2007.
[9] A. Nitzan, Dynamics in Condensed Phases: Relaxation, Transfer, and Reactions in Condensed Molecular Systems, Oxford University Press, Oxford, 2006.
[10] R. Zwanzig, Nonequilibrium Statistical Mechanics, Oxford University Press, Oxford, 2001.
[11] J.C. Burkill, The Theory of Ordinary Differential Equations, Oliver and Boyd, Edinburgh and London, 1962.
[12] J. Wedekind, R. Strey, D. Reguera, J. Chem. Phys. 126 (2007) 134103.
[13] S. Jungblut, C. Dellago, J. Chem. Phys. 142 (2015) 064103.
[14] R. Landauer, J.A. Swanson, Phys. Rev. 121 (1961) 1668–1674.

[15] J.S. Langer, Ann. Phys. 54 (1969) 258–275.
[16] A. Szabo, K. Schulten, Z. Schulten, J. Chem. Phys. 72 (1980) 4350–4357.
[17] K. Schulten, Z. Schulten, A. Szabo, J. Chem. Phys. 74 (1981) 4426–4432.
[18] S.V. Krivov, M. Karplus, Proc. Natl. Acad. Sci. USA 105 (2008) 13841–13846.
[19] H.A. Kramers, Physica A 7 (1940) 284–304.
[20] H. Karcher, S.E. Lee, M.R. Kaazempur-Mofrad, R.D. Kamm, Biophys. J. 90 (2006) 2686–2697.
[21] O.K. Dudko, G. Hummer, A. Szabo, Proc. Natl. Acad. Sci. USA 105 (2008) 15755–15760.
[22] G. Adam, M. Delbruck, in: A. Rich, N. Davidson (Eds.), Structural Chemistry and Molecular Biology, Freeman, San Francisco, 1968, p. 198.
[23] D.J. Bicout, A. Szabo, Protein Sci. 9 (2000) 452–465.
[24] B. Peters, P.G. Bolhuis, R.G. Mullen, J.-E. Shea, J. Chem. Phys. 138 (2013) 54106.
[25] B. Peters, J. Chem. Phys. 131 (2009) 244103.
[26] M. Salvalaglio, C. Perego, F. Giberti, M. Mazzotti, M. Parrinello, Proc. Natl. Acad. Sci. USA 112 (2015) E6–E14.
[27] W. Im, B. Roux, J. Mol. Biol. 319 (2002) 1177–1197.
[28] A. Ma, A. Nag, A.R. Dinner, J. Chem. Phys. 124 (2006) 144911.
[29] B.C. Knott, N. Duff, M.F. Doherty, B. Peters, J. Chem. Phys. 131 (2009) 224112.
[30] R.B. Best, G. Hummer, Proc. Natl. Acad. Sci. USA 107 (2010) 1088–1093.
[31] A.M. Berezhkovskii, V.Y. Zitserman, Chem. Phys. Lett. 158 (1989) 369.
[32] A.M. Berezhkovskii, V.Y. Zitserman, Physica A 166 (1990) 585–621.
[33] N.G. van Kampen, Prog. Theor. Phys. 64 (1978) 389–401.
[34] J.D. Chodera, V.S. Pande, Phys. Rev. Lett. 107 (2011) 98102.
[35] R. Du, V.S. Pande, A.Y. Grosberg, T. Tanaka, E.S. Shakhnovich, J. Chem. Phys. 108 (1998) 334.
[36] Y.M. Rhee, V.S. Pande, J. Phys. Chem. B 109 (2005) 6780–6786.
[37] P.G. Bolhuis, D. Chandler, C. Dellago, P.L. Geissler, Annu. Rev. Phys. Chem. 53 (2002) 291–318.
[38] E. Vanden-Eijnden, in: M. Ferrario, G. Ciccotti, K. Binder (Eds.), Computer Simulations in Condensed Matter: From Materials to Chemical Biology, in: Lecture Notes in Physics, vol. 703, Springer, Berlin, Heidelberg, 2006, pp. 439–478.
[39] M. Tachiya, J. Chem. Phys. 69 (1978) 2375.
[40] C. Schutte, F. Noe, J. Lu, M. Sarich, E. Vanden-Eijnden, J. Chem. Phys. 134 (2011) 204105.
[41] J.-H. Prinz, H. Wu, M. Sarich, B. Keller, M. Senne, M. Held, J.D. Chodera, C. Schutte, F. Noe, J. Chem. Phys. 134 (2011) 174105.
[42] B. Peters, Annu. Rev. Phys. Chem. 67 (2016) 669–690.
[43] A. Berezhkovskii, A. Szabo, J. Chem. Phys. 122 (2005) 14503.
[44] P.L. Geissler, C. Dellago, D. Chandler, J. Phys. Chem. B 103 (1999) 3706–3710.
[45] W. E, E. Vanden-Eijnden, in: S. Attinger, P. Koumoutsakos (Eds.), Multiscale Modeling & Simulation, vol. 5, Springer, 2004, pp. 35–68.
[46] W. E, W. Ren, E. Vanden-Eijnden, Chem. Phys. Lett. 413 (2005) 242–247.
[47] P. Metzner, C. Schutte, E. Vanden-Eijnden, J. Chem. Phys. 125 (2006) 84110.
[48] L. Maragliano, A. Fischer, E. Vanden-Eijnden, G. Ciccotti, J. Chem. Phys. 125 (2006) 24106.
[49] L. Maragliano, E. Vanden-Eijnden, Chem. Phys. Lett. 446 (2007) 182–190.
[50] D. Branduardi, F.L. Gervasio, M. Parrinello, J. Chem. Phys. 126 (2007) 54103.
[51] M. Bonomi, D. Branduardi, F.L. Gervasio, M. Parrinello, J. Am. Chem. Soc. 130 (2008) 13938–13944.
[52] M.E. Johnson, G. Hummer, J. Phys. Chem. B 116 (2012) 8573–8583.
[53] V. Ovchinnikov, M. Karplus, E. Vanden-Eijnden, J. Chem. Phys. 134 (2011) 85103.
[54] G. Diaz Leines, B. Ensing, Phys. Rev. Lett. 109 (2012) 20601.
[55] A. Bucci, T.-Q. Yu, E. Vanden-Eijnden, C.F. Abrams, J. Chem. Theory Comput. 12 (2016) 2964–2972.
[56] D.E. Makarov, in: A.F. Oberhauser (Ed.), Single Molecule Studies of Proteins, Springer, New York, 2013, pp. 235–268.

[57] G.G. Poon, B. Peters, J. Phys. Chem. B 120 (2016) 1679–1684.
[58] J.W. Gibbs, Trans. Conn. Acad. Arts Sci. 3 (1876) 108–248.
[59] J.W. Gibbs, Trans. Conn. Acad. Arts Sci. 16 (1878) 343–524.
[60] R. Becker, W. Doring, Ann. Phys. 416 (1935) 719–752.
[61] Y.B. Zeldovich, Acta Physicochim. URSS 18 (1943) 1–22.
[62] D. Kashchiev, Nucleation: Basic Theory with Applications, Butterworth-Heinemann, Oxford, 2000.
[63] D. Turnbull, J.C. Fisher, J. Chem. Phys. 17 (1949) 71–73.
[64] N.E.R. Zimmermann, B. Vorselaars, D. Quigley, B. Peters, J. Am. Chem. Soc. 137 (2015) 13352–13361.
[65] S. Auer, D. Frenkel, Annu. Rev. Phys. Chem. 55 (2004) 333–361.
[66] V. Agarwal, B. Peters, Adv. Chem. Phys. 155 (2014) 97–160.
[67] D. Kashchiev, G.M. van Rosmalen, Cryst. Res. Technol. 38 (2003) 555–574.
[68] C.P.M. Roelands, J.H. Ter Horst, H.J.M. Kramer, P.J. Jansens, Cryst. Growth Des. 6 (2006) 1380–1392.
[69] R.J. Davey, S.L.M. Schroeder, J.H. ter Horst, Angew. Chem., Int. Ed. Engl. (ISSN 1521-3773) 52 (Feb. 2013) 2166–2179.
[70] N. Duff, Y.R. Dahal, J.D. Schmit, B. Peters, J. Chem. Phys. 140 (2014) 014501.
[71] F. Noe, C. Schutte, E. Vanden-Eijnden, L. Reich, T.R. Weikl, Proc. Natl. Acad. Sci. USA 106 (2009) 19011–19016.
[72] S. Muff, A. Caflisch, J. Chem. Phys. 130 (2009) 125104.
[73] R. Scalco, A. Caflisch, J. Chem. Theory Comput. 8 (2012) 1580–1588.
[74] S.V. Krivov, J. Phys. Chem. B 115 (2011) 11382–11388.
[75] N.-V. Buchete, G. Hummer, J. Phys. Chem. B 112 (2008) 6057.
[76] A. Singer, R. Erban, I.G. Kevrekidis, R.R. Coifman, Proc. Natl. Acad. Sci. USA 106 (2009) 16090–16095.
[77] W. Zheng, B. Qi, M.A. Rohrdanz, A. Caflisch, A.R. Dinner, C. Clementi, J. Phys. Chem. B 115 (2011) 13065–13074.
[78] P.J. Ledbetter, C. Clementi, J. Chem. Phys. 135 (2011) 44116.
[79] W. E, E. Vanden-Eijnden, J. Stat. Phys. 123 (2006) 503–523.
[80] S.V. Krivov, J. Chem. Theory Comput. 9 (2013) 135–146.
[81] A.M. Berezhkovskii, A. Szabo, J. Phys. Chem. B 117 (2013) 13115–13119.
[82] R.F. Grote, J.T. Hynes, J. Chem. Phys. 73 (1980) 2715–2732.
[83] W. E, E. Vanden-Eijnden, Annu. Rev. Phys. Chem. 61 (2010) 391–420.
[84] G. Hummer, I.G. Kevrekidis, J. Chem. Phys. 118 (2003) 10762.
[85] S. Sriraman, I.G. Kevrekidis, G. Hummer, J. Phys. Chem. B 109 (2005) 6479–6484.
[86] S. Yang, J.N. Onuchic, A.E. Garcia, H. Levine, J. Mol. Biol. 372 (2007) 756–763.
[87] A.C. Pan, D. Sezer, B. Roux, J. Phys. Chem. B 112 (2008) 3432–3440.
[88] B.C. Knott, V. Molinero, M.F. Doherty, B. Peters, J. Am. Chem. Soc. 134 (2012) 1–6.
[89] N. Privault, Understanding Markov Chains, Springer, Berlin, Heidelberg, 2013.
[90] A. Berezhkovskii, G. Hummer, A. Szabo, J. Chem. Phys. 130 (2009) 205102.
[91] P. Metzner, C. Schutte, E. Vanden-Eijnden, SIAM J. Multiscale Model. Simul. 7 (2009) 1192–1219.
[92] P.G. Bolhuis, W. Lechner, J. Stat. Phys. 145 (2011) 841–859.
[93] A.B. Bortz, M.H. Kalos, J.L. Lebowitz, J. Comput. Phys. 18 (1975) 10–18.
[94] B. Harland, S.X. Sun, J. Chem. Phys. 127 (2007) 104103.
[95] N. Eidelson, B. Peters, J. Chem. Phys. 137 (2012) 94106.
[96] M.A. Novotny, Phys. Rev. Lett. 74 (1995) 1–5.
[97] S.A. Trygubenko, D.J. Wales, J. Chem. Phys. 124 (2006) 234110.
[98] S.A. Trygubenko, D.J. Wales, Mol. Phys. 104 (2006) 1497–1507.
[99] P.V. Banushkina, S.V. Krivov, J. Chem. Phys. 143 (2015) 184108.
[100] E.L. Cussler, Diffusion: Mass Transfer in Fluid Systems, 3rd ed., Cambridge Press, Cambridge, UK, 2009.
[101] M. Muthukumar, J. Chem. Phys. 111 (1999) 10371–10374.

CHAPTER 19

Transition path sampling

Transition path sampling is Monte Carlo in path space. Transition interface sampling is umbrella sampling in path space.

van Erp, Leiden Workshop (2012)

In contrast to chemical reactions that break and make a few strong bonds, many activated processes that occur in condensed phases disrupt and reorganize hundreds of weak molecular interactions [1–4]. The motion of individual molecules may only present barriers of order $k_B T$, but transitions that require many degrees of freedom to simultaneously adopt unlikely and/or unfavorable configurations can still present large overall barriers. These processes are often described as having rugged potential energy surfaces (PESs) because the transition paths take many routes over many small barriers, rather than routes that all traverse the same col. (See Figure 19.0.1.)

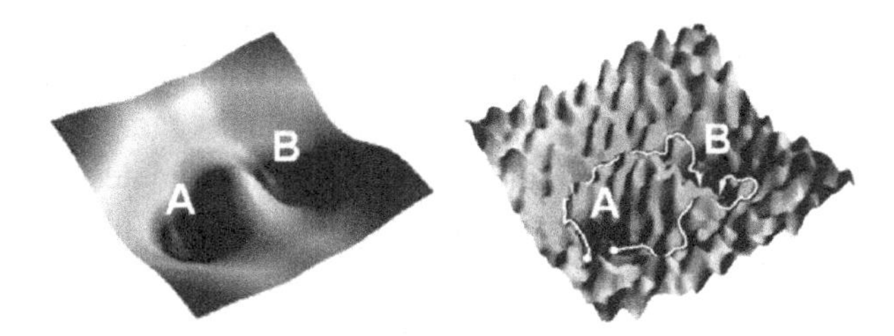

Figure 19.0.1: (Left) On a smooth PES, transition states can be defined based on properties and locations of a few saddle points. Saddles are many $k_B T$ over adjacent minima and only a few Angstroms separate reactants and products in configuration space. (Right) On a rugged PES, there are numerous saddles and pathways between the reactant A and product B states. Saddles are only a few $k_B T$ over the adjacent minima and long paths through configuration space separate reactants and products. Transition path sampling, while applicable to both situations, is not necessary for smooth PESs. [From Dellago et al. Adv. Chem. Phys. 123, 1–84 (2003).]

On a smooth potential with a few cols, the rate through each pathway can be reasonably approximated by transition state theory. On a rugged PES, the momenta are forgotten too quickly to carry trajectories over the barrier region. Prior to the development of transition path sampling, rare events approaches dealt with rugged PESs by projecting them onto a few collective variables, order parameters, or reaction coordinates.[1] Projection yields a low dimensional free energy landscape that can – if the coordinates are chosen correctly – yield a simple and intuitive understanding of complex processes with rugged PESs. Indeed, many of the rate theories from earlier chapters relied on *a priori* specified reaction coordinates.

For transition state theory (TST), reactive flux methods, and Grote-Hynes theory, we saw that a poorly chosen reaction coordinate can impair the accuracy and/or efficiency of a rate calculation. In principle, reactive flux methods can recover an accurate rate even after projection onto a poorly chosen coordinate, but certain mechanistic insights are lost. For example, saddles on the resulting free energy landscape will not correspond to transition states, and the dynamics of the inaccurate coordinate will be unnecessarily complex with long non-Markovian memory artifacts [5].

For overdamped barrier crossings, a good reaction coordinate is crucial for obtaining an accurate rate. Herein lies a serious catch-22. Rate theories that assume overdamped dynamics model barrier crossings on rugged PESs, so they require accurate reaction coordinates – but accurate reaction coordinates are notoriously elusive for systems with rugged PESs. Reaction coordinates remain unknown for many activated processes in condensed phases. Even for relatively simple processes in polar solvents like ion-pair dissociation [6] and conformational transitions of an alanine dipeptide [7], the seemingly obvious coordinates have turned out to be incorrect. Clearly, chemical intuition is not always reliable.

Transition path sampling (TPS) and several related methods [8–10] were designed to compute rate constants and to harvest unbiased reactive trajectories without assuming a reaction coordinate. The use of unbiased dynamics sets TPS apart from earlier rare events methods where the sampling and/or dynamics was biased along an assumed coordinate. Chapter 20 will discuss additional methods that use the data from path sampling methods to identify reaction coordinates. This chapter discusses TPS and related interface sampling methods as unbiased ways to harvest reactive trajectories and to compute reaction rates.

19.1 The transition path ensemble

Most readers will be familiar with ensembles of configurations, microstates, and Monte Carlo methods from equilibrium statistical mechanics [11]. Path sampling methods invoke similar ideas for ensembles of dynamical trajectories. In particular, the transition path ensemble includes those paths that would be observed leaving a reactant basin A at equilibrium and then entering a product basin B without first revisiting basin A. In contrast to methods like Nudged

[1] For a discussion of collective variables, order parameters, reaction coordinates, and their differences, see the introduction to Chapter 20.

Elastic Band [12,13] and action-based path optimization methods [14,15] that generate paths between specific configurations, the trajectories from TPS can begin and end at any point within designated *regions* A and B. Figure 19.1.1 shows examples of paths that do and do not belong to the transition path ensemble.

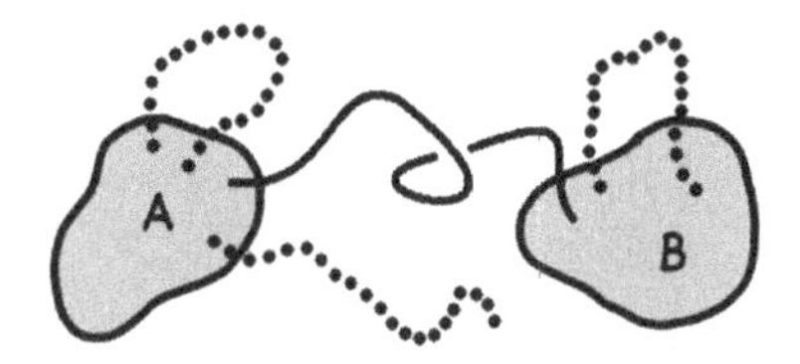

Figure 19.1.1: The dotted curves depict trajectories of duration $\mathscr{T}$ that are not part of the transition path ensemble. These begin and end in A, or begin and end in B, or begin in A but do not end in B, etc. The solid curve depicts a transition path of duration $\mathscr{T}$ that begins in A and ends in B.

To systematically sample these paths, we must formulate a path density for the transition path segments. Consider an ensemble in which all paths have a specific duration $\mathscr{T}$. The probability density for a specific path is, without a normalization factor [16,17],

$$\mathscr{P}[x(t)] = \rho(x_0) \prod_{i=1}^{n} \rho(x_{i\Delta t}|x_{(i-1)\Delta t}) \tag{19.1.1}$$

Here $n = \mathscr{T}/\Delta t$ is the number of time slices on the discretized path, x_t is the full phase space coordinate at time t, and $\rho(x_0)$ is the probability of the initial condition. The coordinates at each time slice depend on those at the previous time slice through the transition probabilities $\rho(x_{i\Delta t}|x_{(i-1)\Delta t})$. The path probability density in equation (19.1.1) applies to all paths of duration $\mathscr{T}$. To focus on the subset of paths that make transitions from A to B, the path density is modified by indicator functions $h_A[x]$ and $h_B[x]$ [16,17].

$$\mathscr{P}_{AB}[x(t)] = \mathscr{Q}_{AB}(\mathscr{T})^{-1} h_A(x_0)\mathscr{P}[x(t)]h_B(x_{\mathscr{T}}) \tag{19.1.2}$$

where as defined in previous chapters

$$h_A(x) = \begin{cases} 1 & x \in A \\ 0 & x \notin A \end{cases} \tag{19.1.3}$$

and with $h_B[x]$ defined similarly. The normalization factor $\mathscr{Q}_{AB}(\mathscr{T})$ is the transition path partition function [16,17]

$$\mathscr{Q}_{AB}(\mathscr{T}) = \int \mathscr{D}x(t) h_A(x_0)\mathscr{P}[x(t)]h_B(x_{\mathscr{T}}) \tag{19.1.4}$$

The symbol $\int \mathscr{D}x(t)$ indicates an integral over all paths – a device that arises in many areas of quantum and statistical mechanics [18]. The discrete representation of the path integral is

$$\int \mathscr{D}x(t) \equiv \int dx_0 \int dx_{\Delta t} \cdots \int dx_{\mathscr{T}}$$

Figure 19.1.2 illustrates the representation of an integral over paths by integration over discrete time slices.

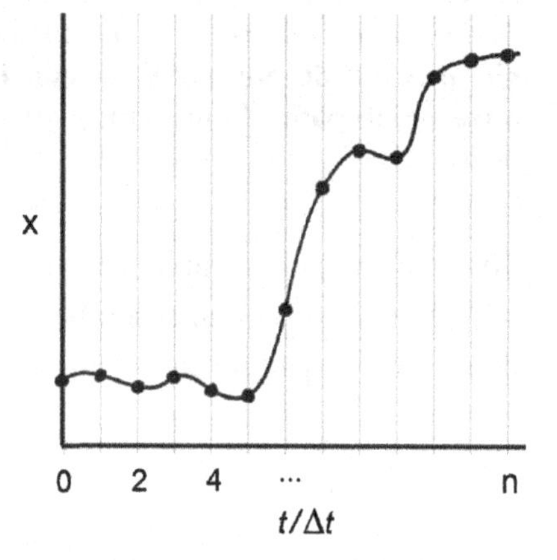

Figure 19.1.2: For sufficiently small Δt, a dynamical path from x_0 to $x_{n\Delta t}$ can be represented as a set of discrete points at times $0, \Delta t, 2\Delta t, ..., n\Delta t$.

Note that the restriction to paths of specific duration $\mathscr{T}$ and to paths that start and end in regions A and B makes the path ensemble much smaller than the complete ensemble of paths. Indeed, the reversible work to convert the complete path ensemble into the more restrictive transition path ensemble is closely related to the reaction rate [19,20]. Bolhuis and coworkers developed methods for sampling the complete path ensemble from which they can simultaneously compute free energies, transmission coefficients, rates, and committors [21,22]. We will consider methods for sampling the transition path ensemble below, but let us begin with two simple limiting cases. Overdamped dynamics and also deterministic Newtonian dynamics, lead to easily computed transition probabilities and path probability densities.

Example: $\mathscr{P}_{AB}[x(t)]$ for overdamped dynamics

Consider overdamped dynamics in one-dimension following the equation of motion $dq/dt = -D\partial\beta F/\partial q + \tilde{R}(t)$ where $\langle\tilde{R}(t)\rangle = 0$ and $\langle\tilde{R}(0)\tilde{R}(t)\rangle = 2D\delta[t]$. The transition probabilities for short times Δt are Gaussian Green's functions that drift and spread with time [23]:

$$\rho(q_{(i+1)\Delta t}|q_{i\Delta t}) = \frac{1}{\sqrt{4\pi Dt}} exp\left[-\frac{1}{4D\Delta t}\left(q_{i\Delta t} - \Delta t\, D\frac{\partial\beta F}{\partial q} - q_{(i+1)\Delta t}\right)^2\right]$$

When the Boltzmann distribution is used for the initial probability $\rho(x_0)$ and Gaussian Green's functions are inserted into equation (19.1.1), it becomes the density of over-damped paths in one dimension. Early work in the direction of TPS [24] proposed to sample paths between specific reactant and product configurations by using these over-damped transition probabilities.

Example: $\mathscr{P}_{AB}[x(t)]$ for Newtonian dynamics

The transition probabilities for deterministic Newtonian dynamics are Dirac delta functions. Let the deterministic time-propagation be accomplished by a function ϕ such that $x_t = \phi_t(x_0)$. The transition probability for any time interval Δt is then given by [17]

$$\rho(x_{(i+1)\Delta t}|x_{i\Delta t}) = \delta\left[x_{(i+1)\Delta t} - \phi_{\Delta t}(x_t)\right]$$

The path density becomes a product of delta functions. In principle, the path partition function is easily integrated because of the delta functions:

$$\mathscr{Q}_{AB}(\mathscr{T}) = \int dx_0 \rho(x_0) h_A(x_0) h_B(\phi_{\mathscr{T}}(x_0))$$

The last integral is entirely determined by $\mathscr{T}$, and by the definitions of A and B. Liouville's theorem [25] says that Newtonian dynamics preserve phase space volumes, so we can interpret $\mathscr{Q}_{AB}(\mathscr{T})$ as the overlap between the time-evolved equilibrium distribution in A with state B after time $\mathscr{T}$.

For intermediate friction, the inertial Green's functions of Adelman can be used as transition probabilities [26]. The 1D calculations are conceptually illuminating, but we know how to compute rates for one-dimensional systems with various models for the bath friction. Transition path sampling became useful with the development of procedures for sampling paths in atomistic simulations.

19.2 Transition path sampling

The ensemble of transition paths could, in principle, be harvested from an infinitely long trajectory by excising the occasional sections that make a transition between states A and B. However, most of the simulation time in this brute force approach would be spent on the uneventful dynamics within states A and B. Transition path sampling (TPS) [8,9,27,28] is a path-space Monte Carlo procedure that directly generates the rare reactive trajectories according to their path probability density, i.e. according to equation (19.1.1). Remarkably, TPS generates the same ensemble of transition paths as would the infinitely long "brute force" trajectory, but TPS wastes no effort in basins A and B and it does not bias the natural dynamics. Like standard Metropolis Monte Carlo, TPS modifies old trajectories to generate new trajectories, and then accepts or rejects the new trajectories according to detailed balance criteria. Procedures for generating new trajectories vary from one TPS implementation to another, and the specific protocol can have a large impact on sampling efficiency.

Convergence to the transition path ensemble is guaranteed if the sampling is ergodic and if the scheme for generating and accepting new paths $x^{(n)}(t)$ from old paths $x^{(o)}(t)$ obeys the detailed balance condition

$$\begin{aligned}&\mathscr{P}_{AB}[x^{(o)}(t)]p_{gen}[x^{(o)}(t) \to x^{(n)}(t)]p_{acc}[x^{(o)}(t) \to x^{(n)}(t)] \\ &\quad = \mathscr{P}_{AB}[x^{(n)}(t)]p_{gen}[x^{(n)}(t) \to x^{(o)}(t)]p_{acc}[x^{(n)}(t) \to x^{(o)}(t)]\end{aligned} \tag{19.2.1}$$

The detailed balance condition can be satisfied by a path-weight version of the Metropolis rule.

$$p_{acc}[x^{(o)}(t) \to x^{(n)}(t)] = min\left\{1, \frac{\mathscr{P}_{AB}[x^{(n)}(t)]p_{gen}[x^{(n)}(t) \to x^{(o)}(t)]}{\mathscr{P}_{AB}[x^{(o)}(t)]p_{gen}[x^{(o)}(t) \to x^{(n)}(t)]}\right\} \tag{19.2.2}$$

Chandler and coworkers investigated several path generation schemes. Their original TPS algorithms employed shooting moves and shifting moves [6,16,17,19]. A shooting move slightly perturbs the momentum at a random time slice t along the old trajectory. A new trajectory is generated by propagating the dynamics forward and backward from the shooting point with perturbed momenta to times 0 and $\mathscr{T}$. A shifting move cuts a length of time δt from one end of an old trajectory and extends the other end of the trajectory by the same time δt to regain a trajectory of duration $\mathscr{T}$. Shifting ensures that TPS will generate trajectories that originate/end at the basin edges as well as trajectories that originate/end deep within the basins [17]. The trajectories from shifting moves may seem redundant, but they are essential for ergodic sampling of paths with fixed duration $\mathscr{T}$. Shooting and shifting moves are illustrated in Figure 19.2.1.

Bolhuis et al. proved that when certain conditions are met, the acceptance rule for shooting and shifting simplifies to [17]

$$p_{acc}[x^{(o)}(t) \to x^{(n)}(t)] = h_A(x_0^{(n)})h_B(x_{\mathscr{T}}^{(n)}) \tag{19.2.3}$$

So new trajectories are accepted if they connect states A and B and rejected otherwise! The simple acceptance rule makes TPS possible and rigorous for complex multidimensional systems where the path density is not easily formulated or computed. The conditions for validity of the simple acceptance rule (19.2.3) are [8,27]

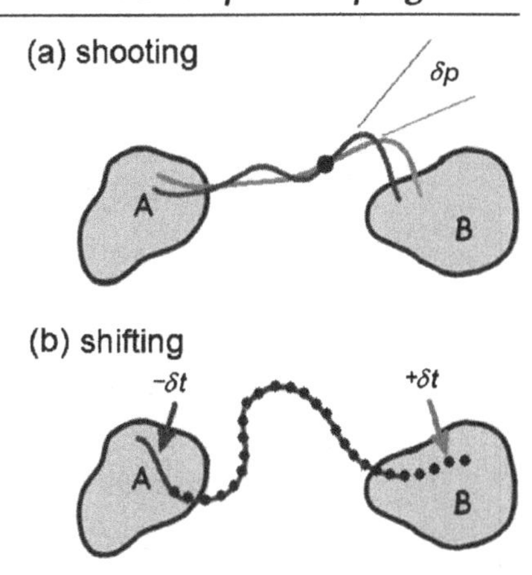

Figure 19.2.1: (a) Shooting begins with the selection of a random time slice (the shooting point) along the old trajectory. The momenta are perturbed at the shooting point and the dynamics are propagated forward and backward in time to $t = 0$ and $t = \mathcal{T}$. The new trajectory is accepted if it connects basins A and B. (b) Shifting moves cut a short time δt from one end of the old trajectory and then extend the dynamics from the other end for the same time δt to regain a trajectory of duration $\mathcal{T}$. Again, the new trajectory is accepted if it connects basins A and B.

1. the dynamics that are used to generate new trajectories, e.g. in a shooting or shifting move, must be the same as the dynamics which give the transition probabilities $\rho(x_{i\Delta t}|x_{(i-1)\Delta t})$,
2. the dynamics that govern the transition probabilities $\rho(x_{i\Delta t}|x_{(i-1)\Delta t})$ must be reversible in the stochastic sense as defined in section (14.1),
3. the distribution $\rho(x)$ for the initial conditions should be that which is automatically generated by the dynamics, e.g. microcanonical for Newtonian dynamics, canonical for Langevin dynamics, etc.

Proofs of these properties for several systems and dynamical equations of motion can be found in the references. Special care must be taken to ensure that the conditions for using acceptance rule (19.2.3) are met. For example, microscopic reversibility with a Nose-Hoover chain requires reversal of the momenta for the real particles and also the momenta of the fictitious Nose-Hoover chain particles. For that reason, TPS is more easily implemented with an Andersen or Langevin thermostat than with a Nose-Hoover thermostat [29]. The basic TPS algorithm with shooting and shifting moves is shown below.

Algorithm: TPS with fixed $\mathcal{T}$, shooting, and shifting

Obtain an initial reactive trajectory, ideally by one of the trajectory annealing methods of Hu et al. [30]. Store the initial trajectory as the "old" trajectory, $x^{(o)}(t)$. Set the fraction of attempts that will be shooting moves to α. Set the standard deviation in shift times to σ. Set parameters related to the size of momentum perturbations.

1. Use a random number to decide whether to perform a shooting move or a shifting move with probability α and $1 - \alpha$, respectively.
2. If a shooting move was chosen in step (1), then:

   (a) Choose a random time slice t_0 along $x^{(o)}$ with uniform probability for each time slice from 0 to $\mathscr{T}$.
   (b) Perturb the momenta at $x^{(o)}(t_0)$. The procedure for generating suitable momentum perturbations should be compatible with the desired ensemble (microcanonical, canonical, zero-angular momentum, etc.) [16,17,27] and with the acceptance criteria (19.2.3). Obtain the new trajectory $x^{(n)}(t)$ by propagating the perturbed momenta from $t = t_0$ backward to $t = 0$ and forward to $t = \mathscr{T}$.
3. If a shifting move was chosen in step (1), then [27]:
   (a) Choose a random shift time δt from the Gaussian distribution with mean zero and standard deviation σ.
   (b) If $\delta t > 0$, propagate the dynamics at the $\mathscr{T}$-end of the trajectory to $t = \mathscr{T} + \delta t$. If $\delta t < 0$, propagate (in reverse) the dynamics at the beginning of the trajectory to $t = \delta t < 0$.
   (c) The trajectory segment of interest from step (3b) goes from $x^{(o)}(\delta t)$ to $x^{(o)}(\mathscr{T} + \delta t)$. Shift its time origin by δt and save it as a new trajectory from $x^{(n)}(0)$ to $x^{(n)}(\mathscr{T})$.
4. If $x^{(n)}(0) \in A$ and $x^{(n)}(\mathscr{T}) \in B$, then accept the new trajectory by overwriting $x^{(o)}$ with $x^{(n)}$. For some applications, it is useful to also accept the reverse transition paths [27, 29]. Accepting paths of type $A \rightarrow B$ and $B \rightarrow A$ doubles the sampling efficiency, but it should only be done when the forward and backward transition paths are indistinguishable. Special care is required for irreversible processes like nucleation. For example, vapor-to-liquid nucleation at undercooled conditions follows very different paths from liquid-to-vapor nucleation at supercooled conditions, and the reverse transitions cannot occur at either non-equilibrium condition.
5. Increment averages based on the properties of the current trajectory (now stored as $x^{(o)}$). Return to step (1).

Shooting and shifting versions of TPS have been applied to many processes: conformational changes in peptides, sugars, nucleotides, and proteins [31–36], micelle fission [37], association of ion pairs [6,38], ligand-exchange processes [39], heterogeneous catalysis [40,41], enzyme catalysis [42,43], hydrophobic polymer collapse [44], and nucleation processes [45–47].

Several early TPS studies revealed that the solvent is more than a dielectric medium. In many cases, changes in solvent density and hydrogen bonding play a central mechanistic role. Three configurations along a transition path for hydrophobic polymer collapse are shown in Figure 19.2.2. The hydrophobic collapse study revealed the mechanistic importance of a solvent bubble that forms around a kink in the polymer as it collapses into a coiled state. Figure 19.2.3 shows reactant, transition state, and product configurations from a transition path in which water evacuates the region around the hydrophobic core of a folding hairpin molecule.

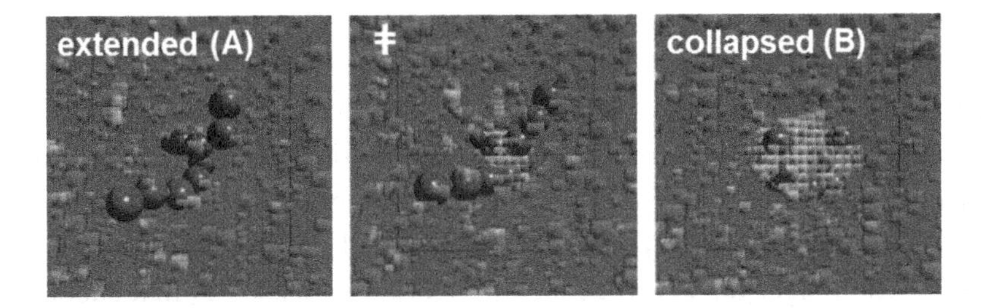

Figure 19.2.2: Showing three configurations from a TPS trajectory for collapse of a hydrophobic polymer in water starting from a coil state (A). The solvent density has been discretized. Light colored cells are low density vapor regions. A vapor bubble forms around the polymer as it collapses. The transition state along the trajectory coincides with formation of the bubble at a polymer kink. [From ten Wolde and Chandler, *Proc. Nat. Acad. Sci. USA* 99, 6539-43 (2004).]

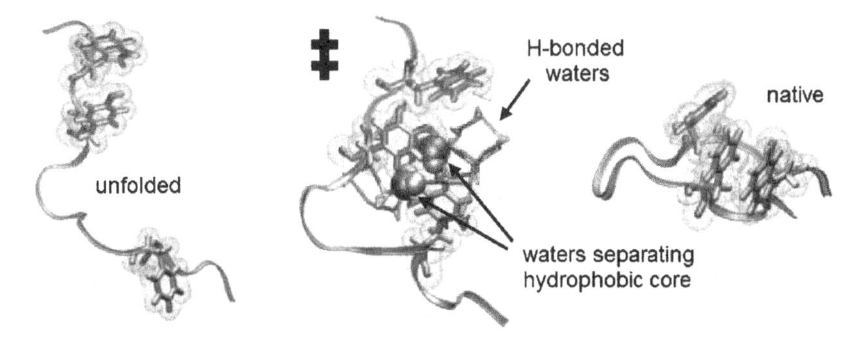

Figure 19.2.3: A TPS study of β-hairpin folding showed that water is partially, but not fully, expelled from the nascent hydrophobic core in the transition state. [From Bolhuis, *Proc. Nat. Acad. Sci. USA* 100, 12129-34 (2003).]

Early applications of TPS illustrated its power, but also revealed some problems with sampling efficiency and ergodicity. Sampling ergodicity is a serious problem when a reaction proceeds via multiple parallel and non-intersecting channels [28]. The shooting and shifting algorithm tends to stay trapped in whichever channel was traversed by the initial transition path. For this reason, the initial path should be prepared carefully. Ideally, TPS should be initiated from a spontaneous barrier crossing trajectory at the actual conditions of interest. Dinner et al. developed path-space annealing strategies which help to obtain an initial transition path in the dominant channel [30]. van Erp [48] and Bolhuis [49] developed replica exchange path sampling algorithms that ensure a representative sample of paths from each separate reaction pathway. Finally, Dellago et al. developed path-space Wang-Landau algorithms to overcome ergodicity limitations [50].

The size of the momentum perturbation, $|\delta \mathbf{p}|$, is an important consideration for the efficiency of the shooting and shifting algorithm. Systems with long transition pathways over rugged energy landscapes have diffusive or chaotic dynamics, i.e. trajectories with very similar initial conditions diverge widely from each other long before reaching the reactant or product state. For many problems with rugged energy landscapes, relaxation from the barrier top takes nanoseconds. Several investigators noted that shooting and shifting becomes very inefficient for transitions that take longer than picoseconds [51–54]. The reason is that shooting from most locations generates a non-transition path, i.e. a path that begins and ends in A or that begins and ends in B. The exceptions are those shooting moves that start from points near the transition state region.

Low sampling efficiency inspired several alternative TPS procedures, especially for processes with rugged PESs. Bolhuis [51] noted that momentum perturbations are not necessary for stochastic simulations with the Andersen or Langevin thermostats. The random forces are sufficient to generate new trajectories during shooting moves. For stochastic equations of motion, Bolhuis et al. developed an easily implemented "one way shooting" algorithm in which only the forward or reverse part of a trajectory, but not both, are regenerated during each shooting move [51]. Grunwald et al. [55] sought to improve sampling efficiency in TPS simulations with deterministic (Newtonian) dynamics. They studied the Lyapunov scaling of the rate at which two trajectories diverge as a function of the initial momentum perturbation size. Their "precision shooting" method enables shooting moves with momentum perturbations $|\delta \mathbf{p}|$ that are far below machine precision limits. Miller and Predescu dealt with low sampling efficiency on rugged energy landscapes by generating new paths that only deviate from old paths in short interior segments [54]. Each of these alternative TPS procedures boosts the acceptance probability by reducing the differences between old and new trajectories.

Another alternative TPS procedure, aimless shooting, maintains high acceptance probability while also generating trajectories that rapidly diverge from the old trajectories [53,56]. Aimless shooting uses a single hybrid shooting/shifting move in which the momenta are drawn fresh from the Boltzmann distribution for each new trajectory. Because trajectories are not "aimed" along the previous trajectory, the sequence of trajectories decorrelates quickly in path space [48,56]. Each aimless shooting move is attempted from one of two shooting points separated by a shift time δt, and each accepted aimless shooting move replaces one of the old shooting points with a new shooting point. The aimless shooting move is illustrated in Figure 19.2.4.

The efficiency of aimless shooting has been demonstrated for several nucleation processes with diffusive barrier crossing dynamics [53,56,57], and also for reactions with inertial barrier crossing dynamics [58–61].

Figure 19.2.4: An old aimless shooting trajectory is shown in black with two candidate shooting points separated by a shift time δt. New momenta are drawn fresh from the Boltzmann distribution to generate the new dashed trajectory. The new trajectory will be accepted if it connects A and B.

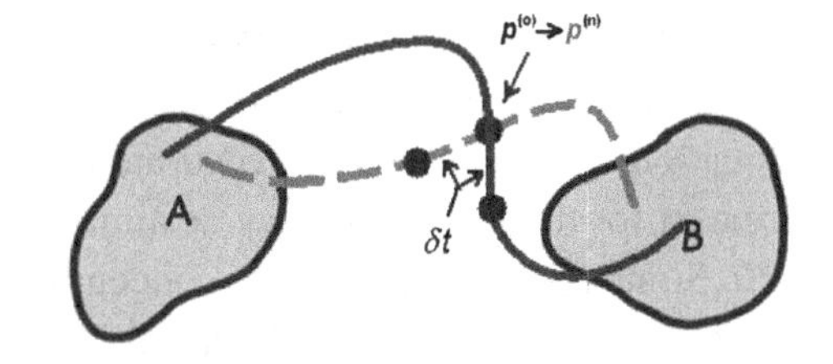

Algorithm: (Canonical TPS) aimless shooting with fixed $\mathcal{T}$

Aimless shooting trajectories are generated in three segments [57]: a "backward trajectory" from $x(0)$ to $x(-\mathcal{T}/2)$, a "connector trajectory" from $x(0)$ to $x(\delta t)$, and a "forward trajectory" from $x(\delta t)$ to $x(\delta t + \mathcal{T}/2)$. The only adjustable parameters in aimless shooting are the time shift δt and the trajectory duration $\mathcal{T}$. Typically, setting $\delta t \sim 0.01\mathcal{T}$ gives favorable acceptance rates [29]. Each aimless shooting move is initiated from the configurations at $x(0)$ or $x(\delta t)$. Obtain an initial reactive trajectory and store it as the "old" trajectory, $x^{(o)}$.

1. Choose $x^{(o)}(0)$ or $x^{(o)}(\delta t)$ as the shooting point with 50% probability for each. Save the shooting point as $x^{(n)}(0)$.
2. Replace the velocities at $x^{(n)}(0)$ with random velocities from the Boltzmann distribution.
3. Propagate the dynamics backward in time from $x^{(n)}(0)$ to $x^{(n)}(-\mathcal{T}/2)$, i.e. reverse the momenta and run a forward trajectory of duration $\mathcal{T}/2$.
4. Propagate the dynamics forward in time from $x^{(n)}(0)$ to $x^{(n)}(\delta t)$. $x^{(n)}(0)$ and $x^{(n)}(\delta t)$ will become the new candidate shooting points if the new trajectory is accepted.
5. Continue the new trajectory in forward time from $x^{(n)}(\delta t)$ to $x^{(n)}(\mathcal{T}/2 + \delta t)$.
6. Accept the new trajectory $x^{(n)}$ if it joins states A and B. Note that reactive trajectories from A to B are not distinguished from reverse trajectories from B to A.

Aimless shooting should be used with dynamics that generate the canonical ensemble, e.g. Langevin dynamics, Andersen thermostat, or similar [29]. The friction/collisions can be weak, but they should be non-zero to ensure detailed balance in the canonical ensemble of transition paths. Nose-Hoover thermostats require extra caution because randomizing the fictitious particle momenta is difficult in some standard MD codes. Finally, note that the initial trajectory was not used except for the two initial shooting points. In some cases, it is useful to initiate the algorithm by preparing approximate transition states for these two configurations.

It can be shown [53] that aimless shooting naturally generates the highest diversity of attempted shooting points near the stochastic separatrix ($p_B = 1/2$) without *a priori* knowledge of its location. The propensity to generate new shooting points near the transition state region is the reason for the high efficiency of aimless shooting [53]. This property is illustrated in Chapter 20. Some have interpreted the concentration of trial shooting points near the transition state as a disadvantage of aimless shooting [28]. Note that the algorithm does generate shooting points across the entire range of committor (p_B) values. Moreover, many aimless shooting moves per shooting point are attempted from those shooting points near $p_B = 0$ and $p_B = 1$. When the shooting point distribution is weighted by the number of shooting attempts per distinct shooting point and plotted vs. p_B, it appears to have support for all values of p_B. These are highly technical issues and we refer interested readers to the literature for details [22,62,63]. See also discussions in Chapter 20.

There is, in principle, a criterion for optimizing the parameter δt. Each new trajectory generates a new shooting point, so the pair of aimless shooting points do a sort of random polka in the transition pathway during the simulation. If δt is too small the acceptance rate will be high, but the shooting points will "diffuse" slowly in the transition pathway. At the other extreme, δt may so large that it exceeds the typical time to descend from the top of the barrier. In that case, newly generated aimless shooting points will already have relaxed to A and B where the probability of generating a new transition path is vanishingly small. An optimum δt value should maximize the rate per shooting move at which the successful shooting points "diffuse" in the transition pathway [29].

Results from fixed-duration TPS algorithms, including aimless shooting and the original shooting/shifting algorithm, are sensitive to the path duration $\mathcal{T}$. The path duration should be long just enough so that, for nearly all trajectories, both endpoints reach either A or B. If $\mathcal{T}$ is too short, the accepted paths will be unusually brief and not representative of the unbiased mechanism. If $\mathcal{T}$ is too long, then computational effort will be wasted on dynamics within the stable basins. The selection of $\mathcal{T}$ can be avoided altogether by using variable-duration TPS algorithms [64].

Instead of terminating trajectories at a specific $\mathcal{T}$, van Erp et al. [64] considered variable-duration TPS algorithms that continue each trajectory forward and backward in time until each end reaches A or B. The variable-duration acceptance probability for a standard shooting move is $p_{acc}[x^{(o)} \to x^{(n)}] = h_A(x_0^{(n)}) h_B(x_{\mathcal{T}}^{(n)}) \min\{1, \mathcal{T}^{(o)}/\mathcal{T}^{(n)}\}$. The acceptance rule for variable-duration shooting involves $\mathcal{T}^{(o)}/\mathcal{T}^{(n)}$ because the old and new trajectories with different durations have different numbers of time slices and therefore different numbers of ways to shoot new trajectories [64]. Mullen et al. showed that aimless shooting can also be performed with variable duration trajectories [65]. Because aimless shooting has two candidate shooting points along each trajectory, variable duration trajectories can be used with no changes to the aimless shooting acceptance rule.

Algorithm: (Canonical TPS) aimless shooting with variable $\mathscr{T}$

Aimless shooting with variable path duration is almost identical to the fixed-$\mathscr{T}$ algorithm [65]. Again shooting moves are always initiated from the configurations $x^{(o)}(0)$ or $x^{(n)}(\delta t)$ along the "old" trajectory.

1. Choose $x^{(o)}(0)$ or $x^{(o)}(\delta t)$ as the shooting point with 50% probability for each. Save the shooting point as $x^{(n)}(0)$.
2. Replace the velocities at $x^{(n)}(0)$ with random velocities from the Boltzmann distribution.
3. Propagate the dynamics backward in time from $x^{(n)}(0)$ until A or B is reached.
4. Propagate the dynamics forward in time from $x^{(n)}(0)$ to $x^{(n)}(\delta t)$.
5. Continue the new trajectory in forward time from $x^{(n)}(\delta t)$ until A or B is reached.
6. Accept the new trajectory $x^{(n)}$ if it goes from A to B or from B to A.

This algorithm should be used with dynamics that generate the canonical ensemble.

Aimless shooting algorithms work well with Langevin dynamics or other NVT simulations, but not for applications that require microcanonical (NVE) trajectories. There are momentum rescaling procedures and precision shooting procedures by which shooting/shifting can generate a microcanonical transition path ensemble, but the implementations are rather complicated. Mullen et al. introduced a simple permutation shooting algorithm that generates the microcanonical transition path ensemble [65,66].

Algorithm: (Deterministic TPS) Permutation shooting with variable $\mathscr{T}$

Obtain an initial "old" transition path $x^{(o)}$. Permutation shooting moves [65] are always initiated from the configurations $x^{(o)}(0)$ or $x^{(n)}(\delta t)$.

1. Choose $x^{(o)}(0)$ or $x^{(o)}(\delta t)$ as the shooting point with 50% probability for each. Save the shooting point as $x^{(n)}(0)$.
2. Randomly select a pair of atoms with identical masses at microstate $x^{(n)}(0)$ and swap their momenta. Repeat m times. m is an adjustable parameter that controls the size of the momentum perturbations.
3. Propagate the dynamics backward in time from $x^{(n)}(0)$ until A or B is reached.
4. Propagate the dynamics forward in time from $x^{(n)}(0)$ to $x^{(n)}(\delta t)$.
5. Continue the new trajectory in forward time from $x^{(n)}(\delta t)$ until A or B is reached.
6. Accept the new trajectory $x^{(n)}$ if it goes from A to B or from B to A.

Permutation shooting trajectories should be generated with Newtonian dynamics and with periodic boundary conditions. If permutation shooting is used for non-periodic NVE systems, it will generate trajectories with an unknown distribution of angular momenta.

Thus far, we have seen TPS algorithms for overdamped dynamics, for Newtonian dynamics, and for molecular dynamics with certain thermostats. There are also important TPS algorithms for systems with discrete states. These do not accelerate the individual hops, but rather they accelerate rarer events like nucleation that occur on even longer time scales. For example, consider coherent precipitate nucleation from a solid solution. Here one encounters the hierarchical time scale separation [67,68]:

$$\text{vibrational relaxation time} \ll \text{vacancy hopping time} \ll \text{nucleation time}$$

kMC offers a dramatic acceleration over molecular dynamics because of the first time scale separation [69,70]. Many investigators have used kMC to simulate solid state nucleation processes, but the second time scale separation still poses a major challenge. kMC trajectories initiated from the metastable alloy can run for a very long time without realizing a nucleation event (basically simulating diffusion in a metastable alloy). In some studies, the nucleation rate has been artificially accelerated by simulating extremely high supersaturations. Hybrid kMC-TPS algorithms can overcome the first time scale separation by using kMC for the underlying dynamics, and they can overcome the second time scale separation by using TPS to focus the kMC effort on nucleation events [67,71]. These path sampling tools have been in the literature for some time, but they are not yet being used in materials science applications.

19.3 Basin definitions and foliations

A key first step in TPS simulations is to define states A and B. These so-called "basin definitions" are usually straightforward, but if defined incorrectly they can badly influence all later results [28,29]. This section appears after the TPS algorithms themselves, because the ideal basin definitions for paths of fixed duration are different from ideal basin definitions for paths of variable duration. Unfortunately there are no universal definitions for either case, but there are some guiding principles that seem to be reliable.

1. Regions defined as A and B absolutely cannot overlap in phase space. If they do overlap, then TPS will eventually find a path from A to a misclassified part of B that never actually leaves A. The subsequent trajectories will just explore excursions from A to its intersection with B.

2. Basins A and B should not encroach on the transition region between the stable states. One of the foremost applications of TPS is in optimizing reaction coordinates. If the basin definitions include transition states, then we cannot use TPS data to learn how transition states differ from reactants and products. This consideration suggests that basin definitions should be as narrow as possible (but see requirement (3)).
3. Especially for paths of fixed duration $\mathscr{T}$, the basin definitions should include the most common equilibrium fluctuations within each basin. If the basins are too small, then even the endpoints of committed trajectories will frequently miss the A, B basins and efficiency will decrease. A key advantage of variable duration TPS algorithms is that they remain efficient even for very small basin definitions.

Basin definitions are usually established by running two separate MD simulations in the reactant and product basins. During these two simulations, the fluctuating values of various order parameters (OPs) are monitored. Basin definitions are established by identifying OPs whose distributions in A and B are clearly separated with no overlap. The OPs do not have to be good reaction coordinates. For example, in simulations of the alanine dipeptide, possible basin definitions include the Ramachandran ϕ and ψ angles, and perhaps other distances and angles.

Later discussions of transition interface sampling methods will require, beyond basin definitions, a special structure called a foliation [20]. A foliation is a set of surfaces that gradually opens around one state (A) and expands until it includes the other state (B). The example below presents a useful way to create a foliation from order parameters.

Example: Conformal foliation from two order parameters

Run two separate unbiased simulations to sample states A and B.

1. Use the two sets of unbiased data to identify two order parameters $q_1(\mathbf{x})$ and $q_2(\mathbf{x})$ whose joint distributions in states A and B are non-overlapping. $q_1(\mathbf{x})$ and $q_2(\mathbf{x})$ should not be redundant descriptions, i.e. they should quantify two unrelated defining characteristics of the A and B ensembles.
2. Using the unbiased data from simulations in A and B, determine the averages $\mathbf{q}_A = \langle q_1, q_2 \rangle_A$ and $\mathbf{q}_B = \langle q_1, q_2 \rangle_B$. Also compute the covariance matrices $\mathbf{S}_A^2 = \langle (\mathbf{q}(\mathbf{x}) - \mathbf{q}_A)\,(\mathbf{q}(\mathbf{x}) - \mathbf{q}_A)^T \rangle_A$ and $\mathbf{S}_B^2 = \langle (\mathbf{q}(\mathbf{x}) - \mathbf{q}_B)\,(\mathbf{q}(\mathbf{x}) - \mathbf{q}_B)^T \rangle_B$.
3. Define quadratic rational function models of the local free energy minima

$$E_A(\mathbf{x}) = \frac{(\mathbf{q}(\mathbf{x}) - \mathbf{q}_A)^T \mathbf{S}_A^{-2} (\mathbf{q}(\mathbf{x}) - \mathbf{q}_A)/2}{1 + (\mathbf{q}(\mathbf{x}) - \mathbf{q}_A)^T (\mathbf{q}(\mathbf{x}) - \mathbf{q}_A)/2}$$

and

$$E_B(\mathbf{x}) = \frac{(\mathbf{q}(\mathbf{x}) - \mathbf{q}_B)^T \mathbf{S}_B^{-2} (\mathbf{q}(\mathbf{x}) - \mathbf{q}_B)/2}{1 + (\mathbf{q}(\mathbf{x}) - \mathbf{q}_B)^T (\mathbf{q}(\mathbf{x}) - \mathbf{q}_B)/2}$$

4. Use $E_A(\mathbf{x})$ and $E_B(\mathbf{x})$ to define

$$b(\mathbf{x}) = ln(E_A(\mathbf{x})/E_B(\mathbf{x}))$$

The isosurfaces of $b(\mathbf{x})$ define a conformal foliation. Isosurfaces with small b ($b(\mathbf{x}) \ll 0$) approximate constant free energy contours in basin A on the (q_1, q_2)-landscape. Isosurfaces with large b ($b(\mathbf{x}) \ll 0$) approximate constant free energy contours in basin B. Isosurfaces of $b(\mathbf{x})$ with intermediate values ($b = ... - 2, -1, 0, 1, 2, 3, ...$) form a conformal foliation that opens around basin A and then close around basin B as b passes from negative to positive values.

See Figure 19.3.1 for an example of this construction on the familiar Muller-Brown potential. This conformal foliation construct is easily generalized to include more than two order parameters. Each additional independent order parameter allows more space to be included within the transition region. To my knowledge this construction has not been used.

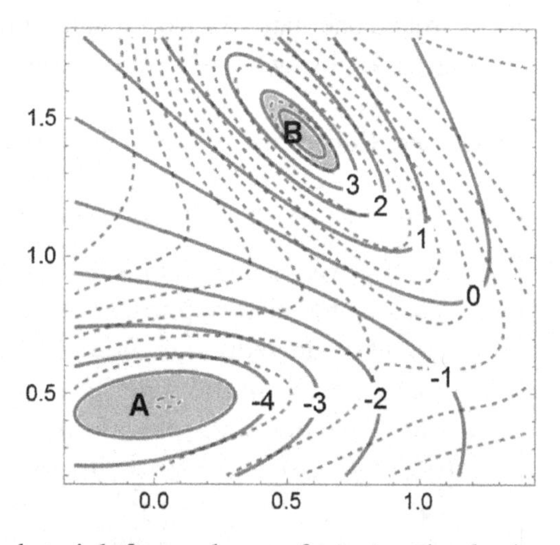

Figure 19.3.1: Dotted lines are contours of the Muller-Brown PES. Solid curves are contours of the conformal foliation $b(\mathbf{x})$, labeled according to the isosurface values $b = ... - 2, -1, 0, 1, 2, ...$. This foliation, based entirely on interpolation between the A and B minima, should facilitate subsequent rate calculations.

The conformal foliation is a general, systematic, and straightforward way of constructing basin definitions and a corresponding foliation of dividing surfaces. Specifically, let A include all configurations with $b(\mathbf{x}) < b_0$, i.e. $A = \{\mathbf{x}|b(\mathbf{x}) < b_0\}$, and let B include all configurations with $b(\mathbf{x}) > b_m$, i.e. $B = \{\mathbf{x}|b(\mathbf{x}) > b_m\}$. With these definitions, one only needs to choose values of b_0 and b_m to define the basins of the appropriate size. The same function $b(\mathbf{x})$ then defines the series of interfaces $b_0, b_1, b_2, ..., b_m$ for computing rates with TPS, transition interface sampling, or forward flux sampling. Following Elber [75], we will refer to the values $b_0, b_1, b_2, ..., b_m$ as *milestones*.

19.4 Rate constants from transition path sampling

Chapter 13 discussed reactive flux methods for obtaining rate constants from the correlation function

$$\bar{h}_B(t) = \frac{\langle h_A(x_0) h_B(x_t) \rangle}{\langle h_A(x) \rangle}$$

The rate constant was obtained from the time derivative of $\bar{h}_B(t)$ after a short molecular relaxation time τ_{mol}. Recall that $d\bar{h}_B/dt$ reaches a plateau value for times $\tau_{mol} < t \ll \tau_{rxn}$ where $\tau_{rxn} = (k_{A\to B} + k_{B\to A})^{-1}$. Reactive flux approaches require a free energy calculation, a TST rate calculation, and short trajectories from the transition state to obtain a transmission coefficient.

The procedure for computing rates with TPS directly uses trajectories without first computing a free energy barrier or a TST rate. In terms of path probability densities, $\bar{h}_B(t)$ is

$$\bar{h}_B(t) = \frac{\int \mathscr{D}x(t) \mathscr{P}[x(t)] h_A(x_0) h_B(x_t)}{\int \mathscr{D}x(t) \mathscr{P}[x(t)] h_A(x_0)} \tag{19.4.1}$$

For times just slightly longer than τ_{mol}, only a tiny fraction of the trajectories initiated in A will have crossed to B. Typically, brute force simulations can only access a few, if any, A to B transitions, which is not enough to accurately compute $d\bar{h}_B/dt$. TPS performs the calculation of $\bar{h}_B(t)$ in stages, somewhat like an umbrella sampling calculation.

Let $b(\mathbf{x})$ be an order parameter, not necessarily a good reaction coordinate. For example, $b(\mathbf{x})$ might be a difference between bond lengths for a chemical reaction, center-to-center distance for association, number of native contacts for protein folding, etc. The foliation can, but does not have to be, a conformal foliation like that shown in Figure 19.3.1. Let $b_0, b_1, b_2, ..., b_m$ be milestones with $b(\mathbf{x}) = b_0$ being easy to reach from A and $b(\mathbf{x}) = b_m$ corresponding to an isosurface around B. If $B = \{\mathbf{x} | b(\mathbf{x}) > b_m\}$ then $\bar{h}_B(t)$ can be written as

$$\bar{h}_B(t) = \int_{b > b_m} db \mathscr{P}_A(b, t) \tag{19.4.2}$$

with

$$\mathscr{P}_A(b, t) \equiv \langle h_A \rangle^{-1} \int \mathscr{D}x(t) \mathscr{P}[x(t)] h_A(x_0) \delta[b(x(t)) - b] \tag{19.4.3}$$

$\mathscr{P}_A(b,t)$ can be computed in stages by defining a series of windows $W(i)$ that cover different regions along the $b(\mathbf{x})$ coordinate. The windows are used to perform a series of TPS simulations to generate path ensembles that gradually convert $\mathscr{P}_A[x(t)]$ to $\mathscr{P}_{AB}[x(t)]$ as shown in Figure 19.4.1.

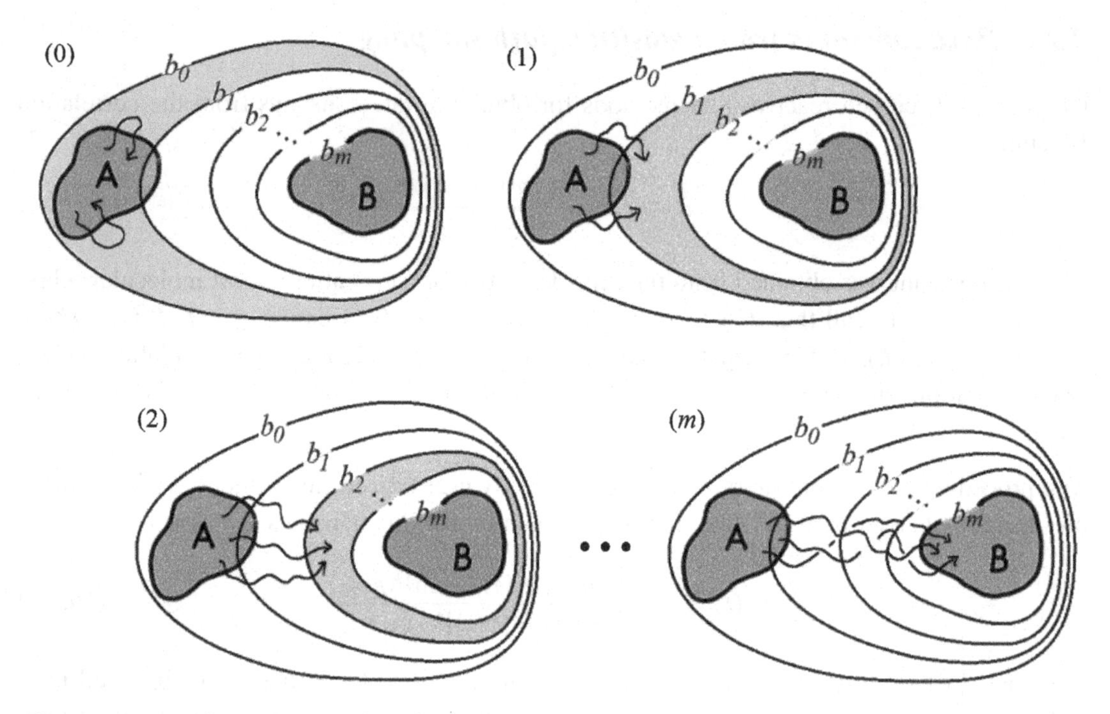

Figure 19.4.1: The path ensemble $\mathscr{P}_A[x(t)]$ can be transformed into the transition path ensemble $\mathscr{P}_{AB}[x(t)]$ by a series of stepwise transformations. An order parameter $b(\mathbf{x})$ with isosurfaces $b(\mathbf{x}) = b_k$ such that A is a subset of $\{\mathbf{x}|b(\mathbf{x}) > b_0\}$ and $B = \{\mathbf{x}|b(\mathbf{x}) > b_m\}$. Trajectories are shown at the first stage which samples $\mathscr{P}_A[x(t)]$, at the last stage which samples $\mathscr{P}_{AB}[x(t)]$, and at two of many intermediate stages.

The windows as drawn in Figure 19.4.1 are mutually exclusive, but in practice each window should overlap with neighboring windows by a small amount Δ. To make the algorithm explicit, define the path destination window at the ith-stage of sampling as

$$W(i) = \{\mathbf{x}\,|(b_i) < b(\mathbf{x}) < (b_{i+1} + \Delta)\}$$

Then a path sampling simulation for each window can be used to compute the family of distributions. Within window $W(i)$, the distribution $\mathscr{P}_{AW(i)}(b,t)$ is proportional to $\mathscr{P}_A(b,t)$:

$$\mathscr{P}_{AW(i)}(b,t) \equiv \frac{\int \mathscr{D}x(t)\mathscr{P}[x(t)]h_A(x_0)h_{W(i)}(x_t)\delta[b(x_t) - b]}{\int \mathscr{D}x(t)\mathscr{P}[x(t)]h_A(x_0)h_{W(i)}(x_t)} \tag{19.4.4}$$

The $\mathcal{P}_{AW(i)}(b,t)$ distributions are matched along the b-axis in the same way that one matches distributions from separate stages of umbrella sampling. The result is the full distribution $\mathcal{P}_A(b,t)$, from which equation (19.4.2) gives $\bar{h}_B(t)$.

Example: Isomerization in two-dimensional model solvent

Dellago et al. studied a two-dimensional model of isomerization in a repulsive WCA fluid (white). In state A the gray atoms share a short bond, and in state B the gray atoms share a longer bond. The dynamical umbrella sampling procedure was implemented using r, the distance between the two atoms, as an order parameter. The figure below shows the system and results of the calculation. [Figure from Dellago et al. *Adv. Chem. Phys.* 123, 1–84 (2003).]

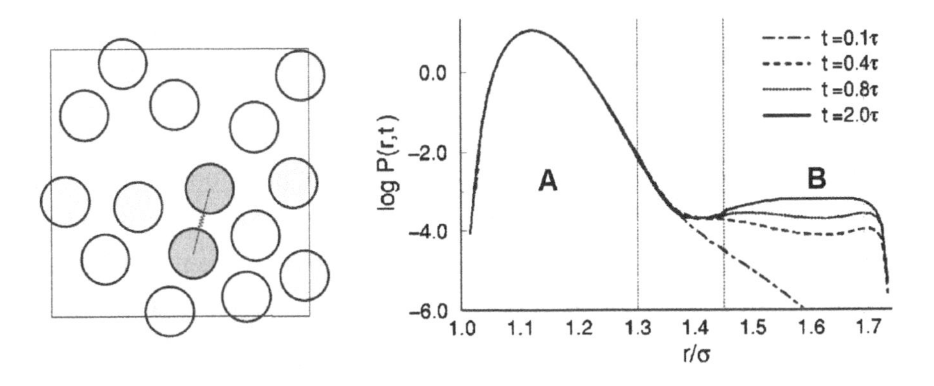

As time increases, the shoulder in state B rises and the peak in state A shrinks to form a bimodal probability density. As t goes to infinity, $\mathcal{P}_A(r,t)$ will become the equilibrium density. One could integrate the total probability to be in state B at each time to obtain $\bar{h}_B(t)$ and the rate according to equation (19.4.2).

In principle, one can compute $\mathcal{P}_A(b,t)$ for many different times to construct the correlation function $\bar{h}_B(t)$ vs. t. However, there is a clever shortcut. Dellago et al. considered trajectories of one long duration $\mathcal{T}$ that visit (but perhaps do not end in) state B. They noted that the average of $h_B(x_t)$ in the ensemble $\mathcal{P}_A[x(t)]$, i.e. $\langle h_B(x_t)\rangle_A$, is extremely small. However, in the ensemble of paths that start in A and *visit* B, the average $\langle h_B(x_t)\rangle_{A,visit B}$ is a dramatically magnified version of $\langle h_B(x_t)\rangle_A$. Sampling paths that visit state B requires a new indicator function

$$h_{\mathcal{B}}[x(t)] \equiv \max_{0<t<\mathcal{T}} h_B(x_t) \tag{19.4.5}$$

The indicator function $\mathcal{h}_{\mathfrak{B}}$ is actually a functional, i.e. its argument is the entire trajectory $x(t)$ and $\mathcal{h}_{\mathfrak{B}} = 1$ if any point along the trajectory visits B. To sample paths that start in A and visit B, the acceptance rule in equation (19.2.3) is modified to $p_{acc} = h_A(x_0^{(n)})\mathcal{h}_{\mathfrak{B}}[x^{(n)}(t)]$. Dellago et al. then derived a convenient expression for the ratio between $\bar{h}_B(t)$ and $\bar{h}_B(t')$:

$$\begin{aligned}
\frac{\bar{h}_B(t)}{\bar{h}_B(t')} &= \frac{\langle h_A(x_0)h_B(x_t)\rangle}{\langle h_A(x_0)h_B(x_{t'})\rangle} \\
&= \frac{\left\langle h_A(x_0)h_B(x_t)\mathcal{h}_{\mathfrak{B}}[x(t)]\right\rangle}{\left\langle h_A(x_0)h_B(x_{t'})\mathcal{h}_{\mathfrak{B}}[x(t)]\right\rangle} \\
&= \frac{\left\langle h_A(x_0)h_B(x_t)\mathcal{h}_{\mathfrak{B}}[x(t)]\right\rangle}{\left\langle h_A(x_0)\mathcal{h}_{\mathfrak{B}}[x(t)]\right\rangle} \cdot \frac{\left\langle h_A(x_0)\mathcal{h}_{\mathfrak{B}}[x(t)]\right\rangle}{\left\langle h_A(x_0)h_B(x_{t'})\mathcal{h}_{\mathfrak{B}}[x(t)]\right\rangle} \\
&= \frac{\langle h_B(x_t)\rangle_{A\mathfrak{B}}}{\langle h_B(x_{t'})\rangle_{A\mathfrak{B}}}
\end{aligned} \tag{19.4.6}$$

The subscript $A\mathfrak{B}$ indicates an average over the ensemble of trajectories which begin in A and visit B between 0 and $\mathcal{T}$. How can we just insert $\mathcal{h}_{\mathfrak{B}}[x(t)]$ inside the averages in step two? It is allowed because $\mathcal{h}_{\mathfrak{B}}[x(t)]h_B(x_t)$ is always equivalent to $h_B(x_t)$. If the trajectory never visits B, then $h_B(x_t)$ is always zero. If $h_B(x_t) = 1$, then $\mathcal{h}_{\mathfrak{B}}[x(t)]$ is also one. The last line defines the ensemble average over paths of duration $\mathcal{T}$ that start in A and visit B. With the new indicators, the new path ensemble, and the new path averages, the ratio $\bar{h}_B(t)/\bar{h}_B(t')$ for any times t, $t' < \mathcal{T}$ can be obtained using data from a single path sampling simulation.

Algorithm: Rates from TPS (abridged)

1. Establish basin definitions for states A and B, and adopt a foliation coordinate $b(\mathbf{x})$.
2. Perform the dynamical umbrella sampling procedure in Figure 19.4.1 for one time t' to obtain $\bar{h}_B(t')$.
3. $\bar{h}_B(t)$ at longer times is computed from $\langle h_B(x_t)\rangle_{A\mathfrak{B}} / \langle h_B(x_{t'})\rangle_{A\mathfrak{B}}$ in an ensemble of paths that start in A at time 0 and visit B in the time interval $[0, \mathcal{T}]$.
4. The plateau in $d\bar{h}_B/dt$ reveals the dynamically correct rate constant, just as in the reactive flux approach.

Although a coordinate $b(\mathbf{x})$ has been used to generate the ensemble of trajectories that visit B, it has no impact on the final results.

Equations (19.4.6) provide an elegant approach to compute rates entirely from dynamical trajectories. Figure 19.4.2 illustrates the time correlation function that emerges from this procedure applied to the two-dimensional isomerization example.

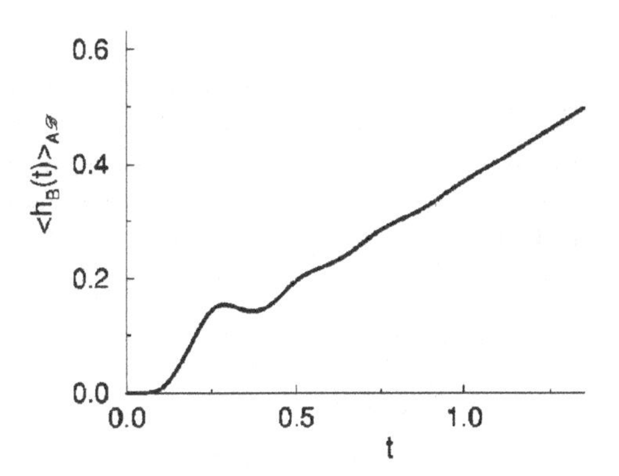

Figure 19.4.2: Instead of computing $\mathscr{P}_A(r,t)$ at several times, one can use a single short-time calculation to get $\bar{h}_B(t')$ and the path ensemble average $\langle h_B(x_t)\rangle_{A\mathfrak{B}}$ to extract $\bar{h}_B(t)$ for all times $t, t' < \mathscr{T}$. The figure shows the correlation $\bar{h}_B(t)$ generated in this way for the two-dimensional isomerization model. [From Dellago et al. *Adv. Chem. Phys.* 123, 1–84 (2003).]

TPS provided a new way of computing rate constants when the reaction coordinate is not known. The shortcut of Dellago et al. improved the efficiency of TPS for computing rates, but the TPS procedure is still more expensive than conventional rare events approaches. For example, Dimelow et al. [34] compared the cost of computing a rate via TPS to the cost of computing a free energy barrier and transmission coefficient. The conventional approach using a reaction coordinate was approximately 20 times faster than the coordinate-free approach. This comparison needs an update to reflect newer and more efficient path sampling algorithms, but it seems likely that umbrella sampling and other procedures still have efficiency advantages.

The key advantage of TPS is that it can be applied even when the reaction coordinate is unknown. Additionally, the order parameter b need not be differentiable, whereas umbrella sampling, blue moon MD, metadynamics, etc. all require forces along the gradient of the order parameter. The gradient of b is also needed to compute the transition state theory rate and the transmission coefficient, whereas TPS obtains a dynamically accurate rate without invoking the gradient of b. The next section shows how ideas from TPS can be used to *efficiently* compute rates that remain independent of the coordinate used for sampling.

19.5 Transition interface sampling

Transition interface sampling (TIS) [64,72] also employs an ensemble of unbiased trajectories to compute rates. TPS employed indicator functions $h_A(\mathbf{x})$ and $h_B(\mathbf{x})$ to detect whether a *configuration* $\mathbf{x}$ is in state A or in state B. The most efficient scheme for computing rates with TPS [19] additionally invoked indicators which tell whether state B was ever visited along a trajectory. TIS invokes a loosely similar indicator which reflects the dynamical history of a trajectory up to the present time [72]. In particular, the point x_t on a trajectory is part of set $\mathscr{A}$ if the trajectory arriving at x_t more recently visited A than B. Similarly, set $\mathscr{B}$ includes the phase space points which emerged from B more recently than A. Figure 19.5.1 shows fragments of a trajectory that belong to sets $\mathscr{A}$ and $\mathscr{B}$.

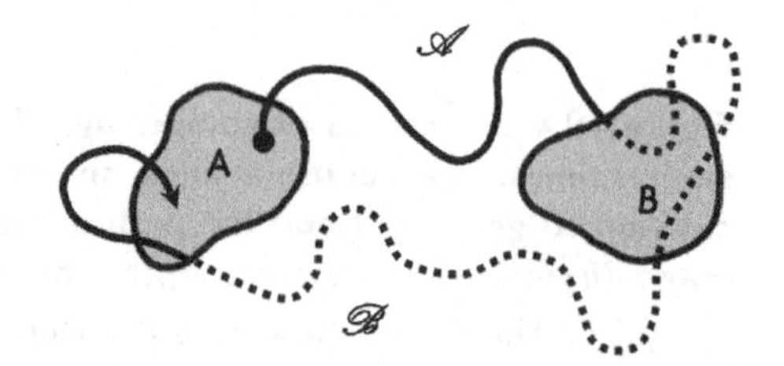

Figure 19.5.1: Phase space points along trajectories that most recently emerged from A belong to $\mathscr{A}$ (solid curves). Phase space points along trajectories that most recently emerged from B belong to $\mathscr{B}$ (dotted curves). Note that brief excursions from A and B do not cause transitions between $\mathscr{A}$ and $\mathscr{B}$.

When a phase space point x belongs to $\mathscr{A}$, the indicator $h_{\mathscr{A}}(x) = 1$. Otherwise $h_{\mathscr{A}}(x) = 0$. Similarly, define an indicator for the set $\mathscr{B}$. Note that the indicators $h_{\mathscr{A}}(x)$ and $h_{\mathscr{B}}(x)$ are strictly 1 or 0 for deterministic dynamics, but for stochastic dynamics the average over noise history makes $h_{\mathscr{A}}(x)$ and $h_{\mathscr{B}}(x)$ smooth functions of x. Throughout this section, averages involving these path history indicators should be taken over the canonical distribution *and* over the stochastic noise [72].

A correlation function analogous to that of the reactive flux method can be defined, but now using $\mathscr{A}$ and $\mathscr{B}$ [10,72].

$$\bar{h}_{\mathscr{B}}(t) = \frac{\langle h_{\mathscr{A}}(x_0) h_{\mathscr{B}}(x_t)\rangle}{\langle h_{\mathscr{A}}(x_0)\rangle} \tag{19.5.1}$$

Unlike the reactive flux correlation function, $\bar{h}_{\mathscr{B}}(t)$ immediately becomes linear [10,72]. There is no transient decay during a short time τ_{mol}. The reason is that trajectories which only temporarily leave A remain in $\mathscr{A}$ unless they have actually reached B. Because $\bar{h}_{\mathscr{B}}(t)$ immediately becomes linear, the rate constant is its zero time derivative. Let the surface $b(x) = b_m$ be a surface enclosing the state B. Then

$$\begin{aligned} k_{A\to B} &= \frac{\langle h_{\mathscr{A}}(x_0)\dot{h}_{\mathscr{B}}(x_t)\rangle}{\langle h_{\mathscr{A}}(x_0)\rangle} \\ &= \frac{\langle h_{\mathscr{A}}(x_0)\dot{b}(x_0)\delta[b(x_0) - b_m]\theta(\dot{b}(x_0))\rangle}{\langle h_{\mathscr{A}}(x_0)\rangle} \end{aligned} \tag{19.5.2}$$

van Erp used the same expression to compute the effective positive flux from A to B [73]. Equation (19.5.2) for $k_{A\to B}$ appears to involve only instantaneous velocities and positions, like the transition state theory, but computing the value of $h_{\mathscr{A}}(x_0)$ requires historical trajectory information. Thus TIS transfers a requirement to follow trajectories into the future, as done in the reactive flux method, to an alternative requirement of retracing their history.

TIS obtains the effective positive flux in stages. Again define a foliation such that its mth surface, $b(x) = b_m$, is the boundary of B. Additionally, the $b(x) = b_0$ surface of the foliation should be chosen such that A and $\{x|b(x) > b_0\}$ are mutually exclusive [72]. An acceptable foliation for TIS is shown in Figure 19.5.2.

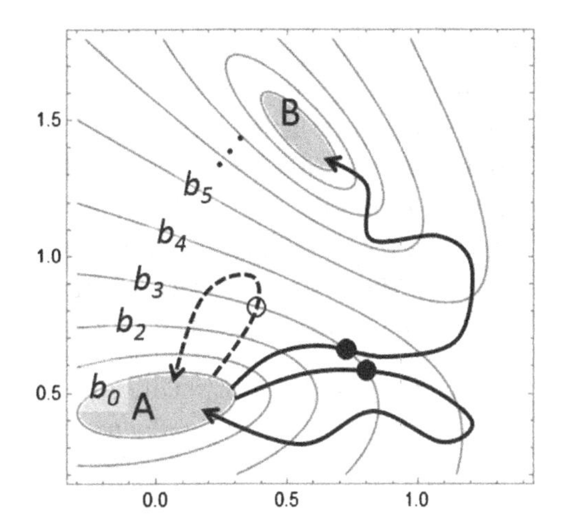

Figure 19.5.2: Foliations $b(x) = b_i$ for $0 \leq i \leq m$. A black dot is shown on trajectories which reach b_4 after reaching b_3. An open circle is shown on the trajectory that reaches b_3 but not b_4 before returning to state A. All trajectories shown emerged from A in backwards time. The fraction of trajectories reaching b_4 after b_3 is used to calculate $p_A(b_4|b_3)$.

van Erp et al. showed that the overall rate $k_{A\to B}$ can be written in terms of two factors. The first is the positive flux $k_{A\to b0}$ through $b(x) > b_0$, i.e. the boundary of state A. $k_{A\to b0}$ is easily computed using a transition state theory formula with standard MD simulation data because transient escapes from A are common [64,72]:

$$k_{A\to b_0} = \frac{\langle \dot{b}(x)\delta[b(x) - b_0]\theta(\dot{b}(x))\rangle}{\langle h_A(x)\rangle} \tag{19.5.3}$$

The second factor is $p_{\mathscr{A}}(b_m|b_0)$, the probability that a trajectory crossing b_0 from A will cross b_m (enter B) without first re-visiting A. The rate is a product of these two factors [64,72]

$$k_{A\to B} = k_{A\to b0} \cdot p_{\mathscr{A}}(b_m|b_0) \tag{19.5.4}$$

where $p_{\mathscr{A}}(b_m|b_0)$ is the probability of crossing b_m after leaving A *without ever returning to* A. Equation (19.5.4) is remarkable because individual trajectories cross the boundary of A with different velocities and at different locations. It would seem important to track their ultimate outcomes individually to know contributions to the overall rate from different parts of the $b(\mathbf{x}) = b_0$ surface. The key to understanding this result is that the definition of $p_{\mathscr{A}}(b_m|b_0)$ discounts that portion of flux from A which is ultimately futile [72]. Equation (19.5.4) is the basis for TIS and also the Forward Flux Sampling algorithm.

Typically, the rate $k_{A\to b0}$ will be much larger than the true rate (hence easily computed by standard MD), and the factor $p_{\mathscr{A}}(b_m|b_0)$ will compensate by being extremely small. To efficiently compute the small probability $p_{\mathscr{A}}(b_m|b_0)$, the calculation is done in stages [64,72].

$$p_{\mathscr{A}}(b_m|b_0) = \prod_{i=0}^{m-1} p_{\mathscr{A}}(b_{i+1}|b_i) \tag{19.5.5}$$

Each factor $p_{\mathscr{A}}(b_{i+1}|b_i)$ counts the probability that a trajectory emerging from A and crossing b_i will cross b_{i+1} before returning to A. The stepwise conditional probabilities can be generated using any variable length TPS algorithm. The acceptance rule for computing $p_{\mathscr{A}}(b_{i+1}|b_i)$ requires trajectories to reach A in backward time, to cross $b(x) = b_i$ at least once, and to reach either A or $b(x) = b_{i+1}$ in forward time. The ensemble of trajectories with this property is called the $[i+]$-ensemble. Note that the conditional probabilities in equation (19.5.5) do *not* assume a Markovian sequence of transitions between milestones. The reason, as emphasized by van Erp [10,72], is that TIS only counts the first crossings.

The accuracy of TIS is formally independent of the coordinate $b(x)$. The coordinate used to construct the foliation does influence the efficiency, but TIS is less sensitive to coordinate error than reactive flux methods [10]. The placement of milestones, i.e. the gaps between b_i and b_{i+1} can also influence the efficiency of TIS. van Erp [10] has noted that by following successful trajectories beyond b_{i+1} at each stage, one can obtain rough estimates for optimal placement of the next b_{i+2} interface.

TIS has been combined with replica exchange ideas in the RETIS algorithm. Here multiple replicas of the TIS simulation are run simultaneously with trajectories being exchanged between the replicas and stages. RETIS allows transition paths to jump between different reaction channels, a capability that other path sampling algorithms do not have [48,49]. Based on the analysis of van Erp, there is a clear efficiency hierarchy: RETIS > TIS > TPS.

TIS and RETIS have been applied to several challenging problems: nucleation, self-assembly, protein folding, biomolecular association, etc. The biomolecular applications are notoriously fraught with long-lived intermediates that lead to very long transition path times. Recall that each TIS trajectory must be followed from intermediate points back to state A at least in the backward direction. If thousands of paths are to be sampled at each stage of TIS, the algorithm can become computationally infeasible.

Milestoning is an alternative method for applications with many stable intermediates [75–77]. Milestoning uses short trajectories and an elegant formalism to estimate the interface-to-interface transition probabilities and the overall reaction rate. The calculations assume that trajectories forget their history between successive interfaces, and therefore Milestoning results depend on the choice of $b(x)$ and the placement of milestones. The partial path TIS (PPTIS) algorithm [10,72,74] can be viewed as a generalization of Milestoning. PPTIS uses short trajectories to estimate conditional interface-to-interface transition probabilities for the next two and most recent two interfaces which have been crossed. The name refers to trajectories that are only 'partially' followed back to state A. The transition paths are then recursively generated using a 'soft-Markovian approximation' in which the trajectories are assumed to forget their history before the most recent two interfaces [10,72]. PPTIS improves on Milestoning because of the soft-Markovian approximation, but its results also depend on the coordinate system $b(x)$ and the interface placement.

19.6 Forward flux sampling

Forward flux sampling (FFS) [52,78,79] is nearly identical to transition interface sampling (TIS) except for a seemingly minor difference in the way the conditional interface-to-interface probabilities are computed. Because of the subtle algorithmic difference, FFS can generate transition paths and rates for irreversible, non-equilibrium, and even non-stationary rare event processes. Like TIS, FFS adopts a coordinate $b(x)$ and a set of milestone surfaces $b(x) = b_0, b(x) = b_1, \ldots,$ and $b(x) = b_m$ to facilitate the sampling. The FFS rate constant is [52,78,79]

$$k_{A\to B} = k_{A\to b0} \cdot p_{\mathscr{A}}(b_m|b_0)$$

with $p_{\mathscr{A}}(b_m|b_0)$ again being computed in stages,

$$p_{\mathscr{A}}(b_m|b_0) = \prod_{i=0}^{m-1} p_{\mathscr{A}}(b_{i+1}|b_i).$$

Again $p_{\mathscr{A}}(b_{i+1}|b_i)$ is the probability that a trajectory emerging from the reactant basin A and crossing milestone b_i will cross b_{i+1} before returning to A. $p_{\mathscr{A}}(b_{i+1}|b_i)$ is computed from the $[i+]$-ensemble in which trajectories must emerge from A, cross b_i, and then may subsequently cross b_{i+1} or else return to state A.

The key difference between FFS and TIS is in the way the $[i+]$-ensembles are generated. TIS generates the $[i+]$-ensemble by shooting a fresh collection of paths that are decorrelated from those sampled in the $[(i-1)+]$-ensemble. FFS generates the $[i+]$-ensemble by splitting [80]. The ith stage of splitting begins with the endpoints of those trajectories from $(i-1)$th stage that reached $b(x) = b_i$. Starting from each of these endpoints, multiple trajectories in the $[i+]$-ensemble are generated by running new trajectories with different realizations of the stochastic noise. Figure 19.6.1 illustrates the splitting procedure and the nature of the final collection of transition paths.

The splitting procedure does not require backward evolution of trajectories. Nor does it require time reversibility (in the stochastic sense) or the existence of an equilibrium probability density. The ability to compute rates for non-equilibrium systems and even for non-stationary driven

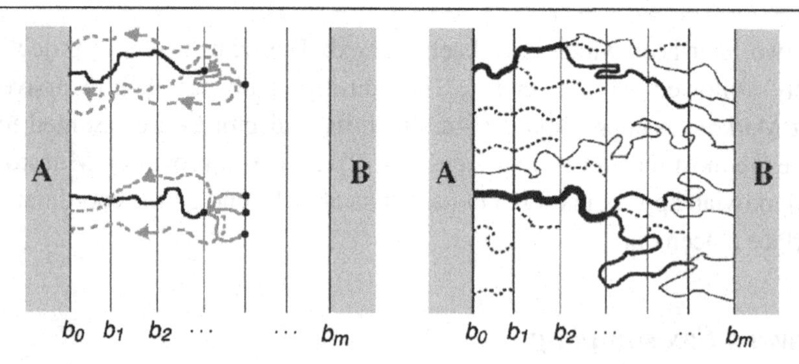

Figure 19.6.1: (Left) In the splitting procedure, those members of the $[(i-1)+]$ ensemble which reach b_i are used to generate the $[i+]$ ensemble. The new trajectories (gray dashed lines) are followed in forward time until they reach b_{i+1} or A. Each of the gray trajectories began from the end of a trajectory (black) that originated in A. Thus there is no need to follow them backward in time. (Right) Successive generations of trajectories in the splitting procedure inherit the characteristics of paths from earlier generations. FFS can generate erroneous results if the last generation of trajectory 'branches' is supported by only a small collection of 'trunks'. [Adapted from Allen et al. *J. Chem. Phys.* 124, 024102 (2006).]

systems [81,82] makes FFS a significant advance. FFS applications include stochastic switching events between stable states in biochemical networks [78], nucleation under intense shear [83], and single molecule pulling experiments with non-stationary pulling forces [81,82]. All of these examples are driven systems with transitions in the forward direction that follow very different paths from trajectories in the reverse direction.

The use of splitting does limit FFS to systems with stochastic dynamics [79,84]. Otherwise, no diversity of trajectories would emerge beyond that collected in the initial crossings of $b(x) = b_0$. The main limitation of FFS is related to the genetic collapse depicted in Figure 19.6.1. Poor sampling of the $[i+]$-ensemble leads to an unusual starting distribution for splitting in the $[(i+1)+]$ ensemble. In this way, sampling errors at one stage are magnified in the later stages [84]. Numerical results confirm that the accuracy of FFS is particularly sensitive to the sampling during the earliest stages [85]. In principle, neither the choice of $b(x)$ or the b_i will influence the results. In practice, this is only true if $k_{A\to b0}$ and the $p_{\mathscr{A}}(b_{i+1}|b_i)$ are perfectly converged [10,84]. A poorly chosen coordinate $b(x)$ or poorly chosen milestone locations $b_0, b_1, ..., b_m$ can increase the risk of a genetic collapse.

Interestingly, FFS has become popular even for studying rare events with stochastically reversible dynamics and in situations where the forward and reverse transition paths should be identical [28]. For example, FFS has been applied to pore translocation [86], alanine dipeptide simulations [87], and protein folding models [88,89]. For those applications where TIS and

FFS are both applicable, TIS is the more robust algorithm. The prevalence of FFS in these applications is likely driven by its ease of implementation [84] – splitting only requires running new trajectories from a "restart file". Computational method developers should take note: there is value in easy implementation!

In addition to FFS, other rare events methods are being developed to compute stationary distributions and fluxes for driven non-equilibrium processes [90]. A particularly important advance is the non-equilibrium umbrella sampling (NEUS) methods by Dinner and coworkers [91,92]. NEUS uses state-to-state fluxes to construct the stationary (non-equilibrium) probability distribution. For example, Dinner and coworkers used NEUS to understand the cyclical fluxes in the biochemical network that controls circadian rythms [93]. Methods that can summarize, understand, and predict the behavior of an ensemble of trajectories evolving according to complex irreversible rules will be important for future work in rare events.

Exercises

1. Give the expression for $\mathscr{P}_{AB}[x(t)]$ in the case of inertial Langevin dynamics on a one-dimensional energy surface.
2. Hummer [94] has shown for overdamped dynamics, that the probability of being on a transition path given a configuration $\mathbf{x}$ is $p(TP|\mathbf{x}) = 2p_B(\mathbf{x})(1 - p_B(\mathbf{x}))$ and that the probability of being at $\mathbf{x}$ within the ensemble of transition paths is $\rho(\mathbf{x}|TP) \propto p_B(\mathbf{x})(1 - p_B(\mathbf{x}))\rho_{eq}(\mathbf{x})$. Sketch these distributions for (i) a parabolic barrier, (ii) the Muller-Brown potential, and (ii) a system that proceeds via the path $A \rightleftarrows I \rightleftarrows B$ with I being a stable intermediate. See Vanden-Eijnden for a generalization of these results to the case of inertial dynamics [95].
3. TPS generates equilibrium paths which are indistinguishable in forward and backward directions owing to time reversal symmetry in the equations of motion. FFS generates only forward transition paths. Comment on applications of TPS and FFS to nucleation. In particular, discuss the quasi-equilibrium assumption in nucleation theory and also the ability (or inability) of TPS and FFS to capture saddle point avoidance phenomena [96].
4. Consider overdamped trajectories on a potential $V(x) = V_{\ddagger} - m\omega^2 x^2/2$ with $A = \{x \,|\, x < -\left(2V_{\ddagger}/m\omega^2\right)^{1/2}\}$ and $B = \{x \,|\, x > \left(2V_{\ddagger}/m\omega^2\right)^{1/2}\}$. The subset of paths that connect A and B are transition paths.
   (a) Compute the probability density for the transition path duration by using the Euler-Maruyama algorithm and shooting trajectories from the barrier top. Hint: Do not investigate every combination of D, $k_B T$, $m\omega^2$, and $V_{\ddagger}$ values. Nondimensionalize the governing equations first to glean the most information from the fewest simulations.
   (b) Calculate the mean transition path duration by using the analytical tools from Chapter 18.

(c) Read Neupane et al. Science, 352, 239–242 (2016). Compare your results from part (a) and (b) to the results in this paper.

5. Recreate the conformal foliation and basin definitions for the Muller-Brown potential [97] as shown in Figure 19.3.1. Use this foliation and overdamped dynamics with the Euler-Maruyama algorithm to compare the efficiency of TPS, TIS, and FFS algorithms for computing the rate of transition from A to B. Set $k_B T = 1$ in the arbitrary units of the Muller-Brown potential. Quantify efficiency in terms of the amount of computation required to establish a 10% confidence interval on the computed estimate of $log\, k$.
6. Permutational isomerization of the 2D LJ_7 cluster is commonly used to test and illustrate computational algorithms [98–100]. If one atom is tagged, and the others are indistinguishable, then the LJ_7 cluster has 20 distinguishable energy minima [101]. The two lowest energy minima (one of which is six-fold degenerate) are shown below.

(a) Let $k_B T/\varepsilon = 0.15$ and define states A and B according to the left and right states in the figure (with B including the degeneracies). Run MD simulations in states A and B to define states A and B. Hint: if configuration $\mathbf{x}$ has any atom with coordination number $CN \geq 6$ then $h_A(\mathbf{x}) + h_B(\mathbf{x}) = 1$, and if the tagged atom has $CN \geq 6$ then $h_A(\mathbf{x}) = 1$.

(b) Run a TPS simulation (you choose the variety) starting from one of the intermediate minima. Compute the overall rate of the transition from A to B. You will need to devise a set of milestones for doing this calculation. How will pathways that are not sampled affect your answer?

(c) Devise a method to classify and catalog the local minima that are encountered along the transition paths. This can be done by doing a local minimization at regular points along the trajectories. Do your transition paths visit all of the minima?

7. Compute the rate of transition from A to B for the LJ_7 cluster again, now using the TIS algorithm. Compare the efficiency of TIS to the standard TPS procedure. Which procedure does a better job of accounting for all transition pathways?
8. Repeat the TPS simulations in problem (6) with a hybrid kMC-TPS algorithm [67]. Use adaptive kMC [102] with event cataloging [103] to parameterize the discrete master equation on the fly.

References

[1] D. Philp, J.F. Stoddart, Angew. Chem., Int. Ed. Engl. 35 (1996) 1154–1196.
[2] J.N. Onuchic, Z. Luthey-Schulten, P.G. Wolynes, Annu. Rev. Phys. Chem. 48 (1997) 545–600.
[3] H. Tributsch, L. Pohlmann, Science 279 (1998) 1891–1895.

[4] V. Agarwal, B. Peters, Adv. Chem. Phys. 155 (2014) 97–160.
[5] R. Zwanzig, Phys. Rev. 124 (1961) 983.
[6] P.L. Geissler, C. Dellago, D. Chandler, J. Phys. Chem. B 103 (1999) 3706–3710.
[7] P.G. Bolhuis, C. Dellago, D. Chandler, Proc. Natl. Acad. Sci. USA 97 (2000) 5877–5882.
[8] P.G. Bolhuis, D. Chandler, C. Dellago, P.L. Geissler, Annu. Rev. Phys. Chem. 53 (2002) 291–318.
[9] P.G. Bolhuis, C. Dellago, Rev. Comput. Chem. 27 (2009) 1–105.
[10] T.S. van Erp, Adv. Chem. Phys. 151 (2012) 27–60.
[11] D. Chandler, Introduction to Modern Statistical Mechanics, Oxford Press, New York, 1997.
[12] H. Jonsson, G. Mills, K.W. Jacobsen, in: B.J. Berne, G. Ciccotti, D.F. Coker (Eds.), Classical and Quantum Dynamics in Condensed Phase Simulations, World Scientific, Singapore, 1998, pp. 385–404.
[13] G. Henkelman, H. Jonsson, J. Chem. Phys. 113 (2000) 9978–9985.
[14] R. Olender, R. Elber, J. Mol. Struct., Theochem 398–399 (1997) 63–71.
[15] R. Elber, A. Ghosh, A. Cardenas, H. Stern, Adv. Chem. Phys. 126 (2003) 93–130.
[16] C. Dellago, P.G. Bolhuis, F.S. Csajka, D. Chandler, J. Chem. Phys. 108 (1998) 1964–1977.
[17] P.G. Bolhuis, C. Dellago, D. Chandler, Faraday Discuss. 110 (1998) 421.
[18] R.P. Feynman, Statistical Mechanics: A Set of Lectures, Westview Press, Boulder, CO, 1998.
[19] C. Dellago, P.G. Bolhuis, D. Chandler, J. Chem. Phys. 110 (1999) 6617–6625.
[20] C. Dellago, P.G. Bolhuis, P.L. Geissler, in: Erice Proceedings, volume 1, 2005, pp. 1–38, chapter 1.
[21] J. Rogal, W. Lechner, J. Juraszek, B. Ensing, P.G. Bolhuis, J. Chem. Phys. 133 (2010) 174109.
[22] P.G. Bolhuis, W. Lechner, J. Stat. Phys. 145 (2011) 841–859.
[23] A. Nitzan, Dynamics in Condensed Phases: Relaxation, Transfer, and Reactions in Condensed Molecular Systems, Oxford University Press, Oxford, 2006.
[24] L.R. Pratt, J. Chem. Phys. 85 (1986) 5045.
[25] M. Tuckerman, Statistical Mechanics: Theory and Molecular Simulation, Oxford University Press, Oxford, UK, 2010.
[26] S.A. Adelman, J. Chem. Phys. 64 (1976) 124–130.
[27] C. Dellago, P.G. Bolhuis, P.L. Geissler, Adv. Chem. Phys. 123 (2001) 1–84.
[28] P.G. Bolhuis, C. Dellago, Eur. Phys. J. 6 (2015) 1–19.
[29] B. Peters, Mol. Simul. 36 (2010) 1265–1281.
[30] J. Hu, A. Ma, A.R. Dinner, J. Chem. Phys. 125 (2006) 114101.
[31] M.F. Hagan, A.R. Dinner, D. Chandler, A.K. Chakraborty, Proc. Natl. Acad. Sci. USA 100 (2003) 13922–13927.
[32] R. Radhakrishnan, T. Schlick, Proc. Natl. Acad. Sci. USA 101 (2004) 5970–5975.
[33] R.B. Best, G. Hummer, Proc. Natl. Acad. Sci. USA 102 (2005) 6732–6737.
[34] R.J. Dimelow, R.A. Bryce, A.J. Masters, I.H. Hillier, N.A. Burton, J. Chem. Phys. 124 (2006) 114113.
[35] C. Dellago, P.G. Bolhuis, Top. Curr. Chem. 268 (2007) 291–317.
[36] J. Hu, A. Ma, A.R. Dinner, Proc. Natl. Acad. Sci. USA 105 (2008) 4615–4620.
[37] R. Pool, P.G. Bolhuis, J. Chem. Phys. 126 (2007) 244703.
[38] J. Marti, F.S. Csajka, D. Chandler, Chem. Phys. Lett. 328 (2000) 169–176.
[39] P.T. Snee, J. Shanoski, C.B. Harris, J. Am. Chem. Soc. 127 (2005) 1286–1290.
[40] C.S. Lo, R. Radhakrishnan, B.L. Trout, Catal. Today 105 (2005) 93–105.
[41] T. Bucko, L. Benco, O. Dubay, C. Dellago, J. Hafner, J. Chem. Phys. 131 (2009) 1.
[42] J.E. Basner, S.D. Schwartz, J. Am. Chem. Soc. 127 (2005) 13822–13831.
[43] R. Crehuet, M. Field, J. Phys. Chem. B 111 (2007) 5708–5718.
[44] P.R. ten Wolde, D. Chandler, Proc. Natl. Acad. Sci. USA 99 (2002) 6539–6543.
[45] D. Zahn, Phys. Rev. Lett. 92 (2004) 40801.
[46] A.C. Pan, D. Chandler, J. Phys. Chem. B 108 (2004) 19681–19686.
[47] M. Grunwald, C. Dellago, Nano Lett. 9 (2009) 2099–2102.
[48] T.S. van Erp, Phys. Rev. Lett. 98 (2007) 268301.

[49] P.G. Bolhuis, J. Chem. Phys. 129 (2008) 114108.
[50] E.E. Borrero, C. Dellago, J. Chem. Phys. 133 (2010) 134112.
[51] P.G. Bolhuis, J. Phys. Condens. Matter 15 (2003) S113–S120.
[52] R.J. Allen, D. Frenkel, P.R. ten Wolde, J. Chem. Phys. 124 (2006) 194111.
[53] B. Peters, B.L. Trout, J. Chem. Phys. 125 (2006) 54108.
[54] T.F. Miller, C. Predescu, J. Chem. Phys. 126 (2007) 144102.
[55] M. Grunwald, C. Dellago, P.L. Geissler, J. Chem. Phys. 129 (2008) 194101.
[56] G.T. Beckham, B. Peters, J. Phys. Chem. Lett. 2 (2011) 1133–1138.
[57] B. Peters, G.T. Beckham, B.L. Trout, J. Chem. Phys. 127 (2007) 34109.
[58] B. Pan, M.S. Ricci, B.L. Trout, J. Phys. Chem. B 114 (2010) 4389–4399.
[59] L. Xi, M. Shah, B.L. Trout, J. Phys. Chem. B 117 (2013) 3634–3647.
[60] R.G. Mullen, J.E. Shea, B. Peters, J. Chem. Theory Comput. 10 (2014) 659.
[61] B.C. Knott, M. Haddad Momeni, M.F. Crowley, L.F. Mackenzie, A.W. Götz, M. Sandgren, S.G. Withers, J. Ståhlberg, G.T. Beckham, J. Am. Chem. Soc. 136 (2014) 321–329.
[62] B. Peters, J. Chem. Phys. 125 (2006) 241101.
[63] B. Peters, Chem. Phys. Lett. 494 (2010) 100–103.
[64] T.S. van Erp, D. Moroni, P.G. Bolhuis, J. Chem. Phys. 118 (2003) 7762.
[65] R.G. Mullen, J.-E. Shea, B. Peters, J. Chem. Theory Comput. 11 (2015) 2421–2428.
[66] R.G. Mullen, J.-E. Shea, B. Peters, J. Chem. Phys. 140 (2014) 41104.
[67] N. Eidelson, B. Peters, J. Chem. Phys. 137 (2012) 94106.
[68] M. Athenes, V.V. Bulatov, Phys. Rev. Lett. 113 (2014) 230601.
[69] A.B. Bortz, M.H. Kalos, J.L. Lebowitz, J. Comput. Phys. 18 (1975) 10–18.
[70] D.T. Gillespie, J. Comput. Phys. 22 (1976) 403–434.
[71] B. Harland, S.X. Sun, J. Chem. Phys. 127 (2007) 104103.
[72] T.S. van Erp, P.G. Bolhuis, J. Comput. Phys. 205 (2005) 157–181.
[73] T.S. van Erp, J. Chem. Phys. 125 (2006) 174106.
[74] D. Moroni, P.G. Bolhuis, T.S. Van Erp, J. Chem. Phys. 120 (2004) 4055–4065.
[75] A.K. Faradjian, R. Elber, J. Chem. Phys. 120 (2004) 10880–10889.
[76] S. Kirmizialtin, R. Elber, J. Phys. Chem. A 115 (2011) 6137–6148.
[77] A.M. West, R. Elber, D. Shalloway, J. Chem. Phys. 126 (2007) 04B608.
[78] R.J. Allen, P.B. Warren, P.R. Ten Wolde, Phys. Rev. Lett. 94 (2005) 18104.
[79] R.J. Allen, C. Valeriani, P.R. ten Wolde, J. Phys. Condens. Matter 21 (2009) 463102.
[80] P. Glasserman, P. Heidelberger, P. Shahabuddin, Oper. Res. 47 (1999) 585–600.
[81] N.B. Becker, R.J. Allen, P.R. ten Wolde, J. Chem. Phys. 136 (2012) 174118.
[82] N.B. Becker, P.R. ten Wolde, J. Chem. Phys. 136 (2012) 174119.
[83] R.J. Allen, C. Valeriani, S. Tanase-Nicola, P.R. Ten Wolde, D. Frenkel, J. Chem. Phys. 129 (2008) 134704.
[84] F.A. Escobedo, E.E. Borrero, J.C. Araque, J. Phys. Condens. Matter 21 (2009) 333101.
[85] Y. Bi, T. Li, J. Phys. Chem. B 118 (2014) 13324–13332.
[86] J.P. Hernandez-Ortiz, M. Chopra, S. Geier, J.J. de Pablo, J. Chem. Phys. 131 (2009) 44904.
[87] C. Velez-Vega, E.E. Borrero, F.A. Escobedo, J. Chem. Phys. 130 (2009) 225101.
[88] E.E. Borrero, F.A. Escobedo, J. Chem. Phys. 127 (2007) 164101.
[89] C. Velez-Vega, E.E. Borrero, F.A. Escobedo, J. Chem. Phys. 133 (2010) 105103.
[90] C. Valeriani, R.J. Allen, M.J. Morelli, D. Frenkel, P.R. ten Wolde, J. Chem. Phys. 127 (2007) 114109.
[91] A. Warmflash, P. Bhimalapuram, A.R. Dinner, J. Chem. Phys. 127 (2007) 154112.
[92] A. Dickson, A.R. Dinner, Annu. Rev. Phys. Chem. 61 (2010) 441–459.
[93] A. Dickson, A. Warmflash, A.R. Dinner, J. Chem. Phys. 131 (2009) 154104.
[94] G. Hummer, J. Chem. Phys. 120 (2004) 516–523.
[95] E. Vanden-Eijnden, in: M. Ferrario, G. Ciccotti, K. Binder (Eds.), Computer Simulations in Condensed Matter: From Materials to Chemical Biology, in: Lecture Notes in Physics, vol. 703, Springer, Berlin, Heidelberg, 2006, pp. 439–478.

[96] A.M. Berezhkovskii, V.Y. Zitserman, Chem. Phys. Lett. 158 (1989) 369.
[97] K. Muller, L.D. Brown, Theor. Chim. Acta 53 (1979) 75–93.
[98] D.J. Wales, Mol. Phys. 100 (2002) 3285.
[99] W. E, W. Ren, E. Vanden-Eijnden, Phys. Rev. B 66 (2002) 52301.
[100] C. Dellago, P.G. Bolhuis, D. Chandler, J. Chem. Phys. 108 (1998) 9236.
[101] D.J. Wales, Energy Landscapes, Cambridge University Press, Cambridge, UK, 2003.
[102] G. Henkelman, H. Jonsson, J. Chem. Phys. 115 (2001) 9657.
[103] J.-F. Joly, L.K. Beland, P. Brommer, N. Mousseau, Phys. Rev. B 87 (2013) 144204.

CHAPTER 20

Reaction coordinates and mechanisms

If we choose well, the results may be useful; if we choose badly, the results (while still formally correct) will probably be useless.

Zwanzig, Noneq. Statistical Mechanics, (2000)

[1]In the chemical sciences and in many areas of physics, activated processes are described and understood in terms of mechanisms. A mechanism summarizes the progression of events during a specific transformation. It should not be a detailed account of each atom's trajectory, nor a universal equation of motion[2] like $f = Ma$ or $i\hbar\partial_t\Psi = H\Psi$. However, mechanistic summaries do share a universal theme across many reactions and other rare events. The system repeatedly makes unsuccessful excursions from the reactant state along the reaction coordinate. After many unsuccessful attempts, the summary climaxes with a large excursion that crosses a transition state and then concludes, usually with a happy ending in the product state. The generic mechanism becomes useful for making specific predictions when we specify the reaction coordinate and the characteristics of the transition state. For example, inserting "nucleus size" and "critical nucleus" transforms the generic summary into the familiar mechanism of classical nucleation theory. Inserting "arclength along the minimum energy path" and "saddle point" transforms the generic summary into the familiar mechanism of a chemical reaction.

Despite its widespread use, the term reaction coordinate means different things to different people. To most chemists, a reaction coordinate refers to the progression of intermediates and transition states along a multistep reaction pathway. To computational chemists, a reaction coordinate refers to arclength along a minimum energy path for a single elementary step. At first, these definitions seem fundamentally different, because they pertain to markedly different resolutions. However, both definitions are types of mechanistic hypotheses, both lead to testable predictions about the kinetics, and (to my knowledge) no inconsistencies arise from the differences in terminology.

---

[1] Much of this chapter can also be found in Peters, *Ann. Rev. Phys. Chem.* 67, 669-90 (2016).

[2] The documented version [1] of Dirac's apocryphal "the rest is chemistry" quip, continues by advocating (in 1929!) the need for practical computational procedures to address important problems in chemistry.

Some misunderstandings do arise from semantic differences between reaction coordinates, order parameters, and collective variables [2–4]. Some use these terms interchangeably, while others give them distinct definitions [5]. To avoid confusion, we adopt the following distinctions.

- A collective variable is any function of the full phase space coordinates.
- An order parameter is a special collective variable that clearly distinguishes reactants from products.
- A reaction coordinate is a special scalar order parameter that quantifies dynamical progress along the pathway from reactants to products.
- All degrees of freedom apart from the reaction coordinate comprise the bath.

The bath is important as a source of thermal activation energy and as a sink that quenches trajectories after activation. However, the entire bath can often be integrated away leaving only a one-dimensional potential of mean force and a simple dynamical model for motion along the reaction coordinate. In many models, the only remnants of the bath are friction and random forces in the dynamics of the reaction coordinate.

One might suspect that coarse graining all the way to one degree of freedom results in an oversimplified and useless model. The previous chapters have shown that inaccurate reaction coordinates lead to overestimated transition state theory rates, overestimated mean first passage times, and slow convergence of algorithms to compute free energies, rates, and transmission coefficients. Even when the free energy profile can be converged with no hysteresis, an inaccurate coordinate may lead to free energy maxima that do not actually correspond to transition states. In contrast, projection onto an accurate reaction coordinate helps to accurately and efficiently estimate rates and to identify the common properties of transition states (see Figure 20.0.1).

In addition to the underlying potential energy surface (PES), the ideal reaction coordinate also depends on the dynamics. In harmonic TST, the transformation to mass weighted coordinates automatically accounts for dynamics, but for processes with rugged energy landscapes the dynamics of collective variables are not easily anticipated. Figure 20.0.2 (adapted from Schenter et al. [6]) shows how dynamics can influence the reaction coordinate and the nature of the activated process.

Reaction coordinates are surprisingly difficult to identify even in systems where the relevant collective variables would seem obvious. For example, the intuitive reaction coordinate for dissociation of an Na^+Cl^- ion pair in aqueous solution is the distance between the two ions [7–10]. However, Geissler et al. showed that distance between ions is not an accurate reaction coordinate [11]. It took another decade and several advances in methods to identify the solvent coordinates for ion pair dissociation [12,13]. Similarly, the conformational transitions

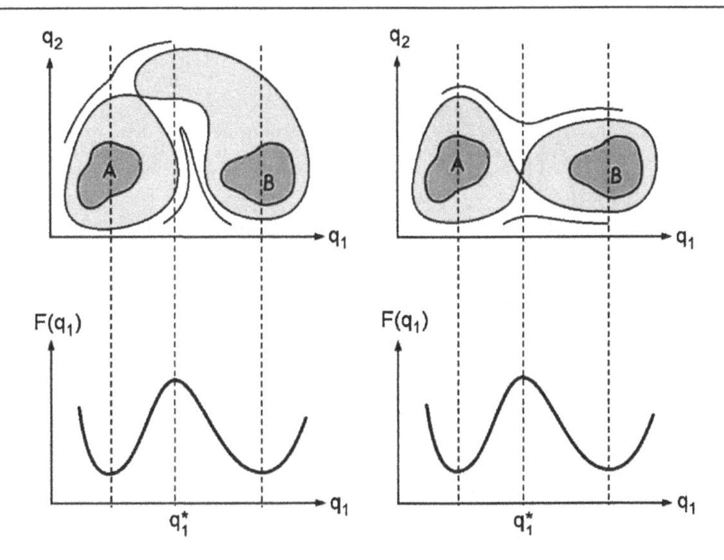

Figure 20.0.1: Important characteristics of the true bottleneck can be obscured when a high dimensional PES (here two-dimensional) is projected to a lower dimensional free energy landscape. Information loss can occur even when advanced sampling methods appear to be working correctly and when the resulting free energy landscape has a barrier in the expected location. For the PES on the right, all states with $q_1 = q_1^*$ are transition states in the multidimensional space. For the PES on the left, states at $q_1 = q_1^*$ are either committed to B (those at large q_2) or committed to A (those at small q_2).

of an alanine dipeptide in aqueous solution were widely studied using Ramachandran angles until Bolhuis et al. demonstrated them to be inaccurate [14]. The work of Bolhuis et al. further illustrated that even choosing the most important *components* of the reaction coordinate is non-trivial. Again it took several years and new methods to identify the solvent reaction coordinate for alanine dipeptide isomerization [15].

Accurate rates can be computed with powerful trajectory-based rare events methods that do not need reaction coordinates. These include the path sampling methods that directly generate rates from trajectories: transition path sampling [16–18], transition interface sampling [19,20], and forward flux sampling [21–23]. Additionally, there are trajectory-based methods for constructing diffusion maps [24–26] and Markov state models [27]. The eigenfunctions from these dynamical models provide natural (but numerical) reaction coordinates, and their eigenvalues provide the spectrum of relaxation times. Trajectory-based methods are invaluable for situations where physically meaningful reaction coordinates cannot be identified. However, they only provide the absolute rate *at the conditions of the dynamical trajectories.* They cannot (yet) predict kinetic trends as functions of temperature, ionic strength, pH, etc. [28] except via multiple simulations at a series of conditions.

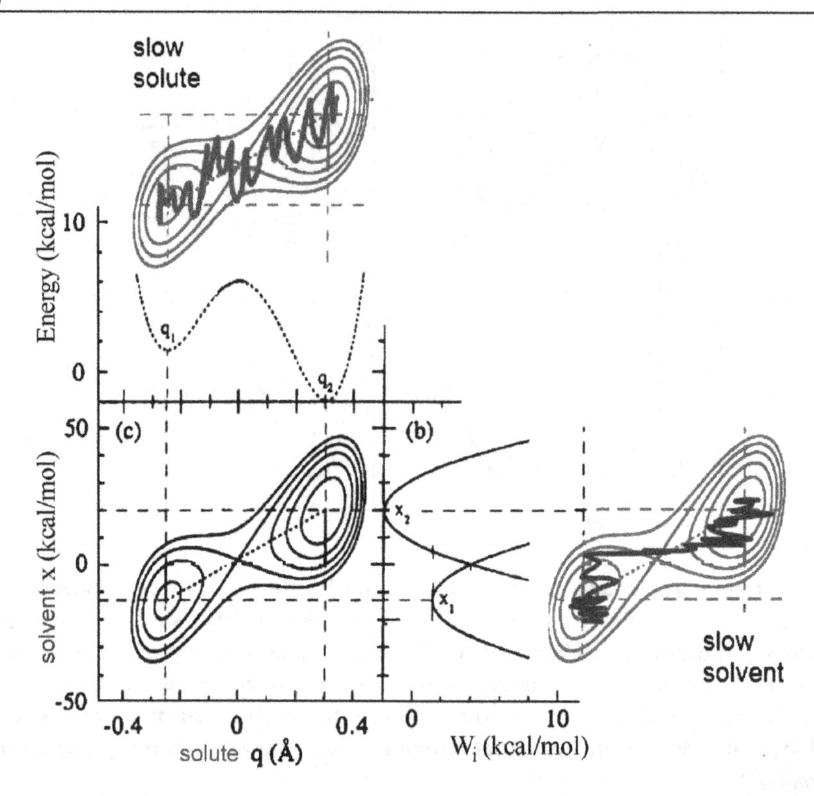

Figure 20.0.2: The lower left panel depicts the free energy landscape as a function of a solute coordinate q and a solvent coordinate x. (In this case, x depicts a vertical energy gap.) The upper left depicts a transition where x is fast relative to q, so that the optimal reaction coordinate is almost perfectly aligned with q. The lower right depicts a transition where x is much slower than q, so that the reaction coordinate is almost perfectly aligned with x. Figure modified from Schenter et al. *J. Phys. Chem.* 105, 9672-85 (2001).

An individual comparison between a measured and predicted rate is rarely sufficient to validate a computational/theoretical model. Predicted rates can be wrong because of minor errors in the force field or for more serious reasons: modeling the wrong mechanism, the wrong active site, ignoring the role of a chemically active solvent, etc. On the other hand, predicted rates can be right for the wrong reasons, e.g. because of a cancellation of errors. Evidence that corroborates both rates *and* trends is far more significant. Simple quasi-equilibrium theories based on a physical reaction coordinate can predict rates *and* trends from a single calculation, and this is their greatest advantage. Of course, the reaction coordinates which lead to these simple but powerful theories are difficult to find, and for some processes they may not exist at all.

20.1 Properties of an ideal reaction coordinate

What characteristics distinguish the reaction coordinate from the bath? Traditionally, investigators studying different phenomena chose reaction coordinates according to very different principles. For example, chemists have largely adopted the variational TST criteria which provides a local approximation to the reaction coordinate near the dividing surface. But the VTST criteria are often impractical and they are not applicable for overdamped barrier crossings. Investigators studying processes with overdamped dynamics (e.g. protein folding, nucleation, etc.) often identify reaction coordinates as the "slow variables". But what does it mean for a variable to be slow? Does it mean that all other variables should adiabatically equilibrate to motion of the slow variable? Or is the slow variable that with the most slowly decaying time correlation function? These two interpretations of "slow" are not always consistent with each other. More recent theoretical frameworks have focused on the committor [29–33], the first non-equilibrium eigenfunction of the master equation [34–37], mean first passage times to the product state [38], and phase-space manifolds or time-dependent structures that eliminate recrossing [39–41]. We encountered some of these objects in Chapters 10, 14, 15, 18, and 19.

So what is the ideal reaction coordinate? Here we attempt to focus on the general characteristics that lead to an accurate and useful reaction coordinate. First and foremost, it should provide an accurate minimalist description of the mechanism. Dellago gave the example shown in Figure 20.1.1.

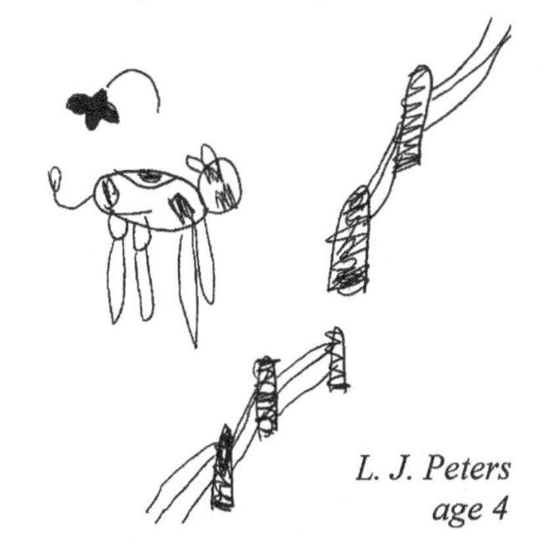

Figure 20.1.1: Consider the cow and the fly as a dynamical system which can cross through a gate (a transition state) in the fence. Common sense says the cow is the reaction coordinate and the fly is an irrelevant detail. But the cow and the fly typically cross the gate together, so *why* do we focus on the cow? The cow's position is a decisive variable. In contrast, if the fly moves across the gate without the cow, he just returns to the cow.

Throughout the discussion below we adopt the following notation: $(\mathbf{x}, \mathbf{p})$ is a point in the complete phase space, $V(\mathbf{x})$ is the potential energy surface, $q(\mathbf{x})$ is a putative reaction coordinate, and $F(q)$ is the free energy as a function of q. We start with some requirements that a good reaction coordinate must satisfy.

Requirement 1. Purely configurational

The reaction coordinate $q(\mathbf{x})$ should depend only on the *instantaneous* point $(\mathbf{x}, \mathbf{p})$ in phase space, and moreover it *should not depend on* $\mathbf{p}$. Inertial dynamics of a configurational variable can still be included via $\dot{q} = \dot{\mathbf{x}} \cdot \nabla q$, as done in the Kramers and Grote-Hynes theories, rather than by explicitly building $\mathbf{p}$-dependence into q. Definitions using both $\mathbf{x}$ and $\mathbf{p}$ have been proposed, but velocity components in q lead to impractical rate theories with prefactors that depend on forces and/or accelerations.

Requirement 2. Foliation

The reaction coordinate $q(\mathbf{x})$ should be a scalar function that monotonically increases as one moves from reactants to products along the reaction pathway. Its isosurfaces should create a foliation, i.e. a series of non-intersecting dividing surfaces as described in Chapter 19 [42].

Requirement 3. Sufficient to predict committor[3]

The committor $p_B(\mathbf{x})$ is the probability to reach the product state B before reaching the reactant state A when trajectories are launched from $\mathbf{x}$ with an equilibrium distribution of momenta [7, 11,43–46]. The procedure for estimating $p_B(\mathbf{x})$ in a high dimensional molecular system is depicted in Figure 20.1.2. The reactants are the ensemble of configurations with $p_B(\mathbf{x}) = 0$, and the products are the ensemble of configurations with $p_B(\mathbf{x}) = 1$. Intermediates take a range of committor values between 0 and 1. Thus, by definition, the committor is a scalar coordinate that quantifies dynamical progress from state A to B for any intermediate configuration $\mathbf{x}$.

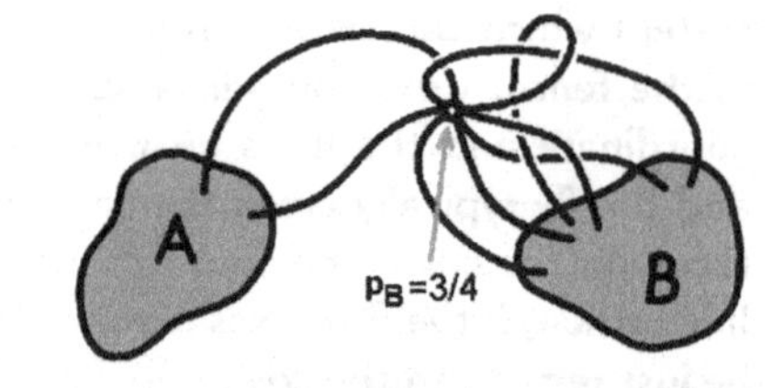

Figure 20.1.2: The committor is the fraction of trajectories that reach state B before reaching state A when initiated with Boltzmann distributed momenta. In the schematic, the estimated committor from eight trajectories would be $\hat{p}_B(\mathbf{x}) = 3/4$.

Purely configurational OPs that create a foliation are common, but coordinates that suffice to predict the committor are extremely special. The value of the reaction coordinate q alone, as opposed to the fully detailed configuration $\mathbf{x}$, should be sufficient to predict the committor [15, 29,47]. Simple physical coordinates that can collapse the high-dimensional $\mathbf{x}$ dependence of

[3] The committor has been called the splitting probability in early chemical physics literature, p-fold in the protein folding literature, and the capacitance in the mathematics literature.

p_B onto a simple q-dependence are difficult to find, and in some cases there may be no such coordinate other than $p_B(\mathbf{x})$ itself.

Analyses of the committor and its relationship to rates and other kinetic properties have dramatically advanced our fundamental understanding of rare events [44,47–51]. We discussed some properties of the committor in Chapter 18. Recall that the committor satisfies a steady-state backward Kolmogorov equation with boundary conditions 0 and 1 at the edges of states A and B, respectively [30,48]. When A and B are cores of the two most stable states, the committor is essentially a shifted and scaled version of the most slowly relaxing eigenfunction. Several additional properties of the committor make it an ideal reaction coordinate.

- The committor arose from a variational optimization of the reaction coordinate for overdamped barrier crossings (see Chapter 18) [47].
- On isosurfaces of the committor, the distribution of hitting points from the transition path ensemble is the same as that from the equilibrium distribution [29].
- The free energy surface and dynamics, when projected onto the committor, result in a one-dimensional model that perfectly preserves the reaction rate (see Chapter 18) [50,51].

If the ideal coordinate $q(\mathbf{x})$ should be sufficient to predict $p_B(\mathbf{x})$, then why not just use $p_B(\mathbf{x})$ itself? Do these requirements say that the committor is perfect, with all other coordinate being suboptimal?

The committor itself is a wonderful reaction coordinate, but some coordinates are *better* in subtle but important ways. Let us take nucleation as an example (see Chapter 18). Let basin A include metastable solutions with only small precritical nuclei. Let B correspond to post-critical irreversibly growing nuclei. Several studies have shown that nucleus size (n) is sufficient to predict p_B [52–54]. However, at a high supersaturation the critical nuclei (those with $p_B = 1/2$) are small, while at a low supersaturation the critical nuclei (also at $p_B = 1/2$) are larger. (See Figure 20.1.3.)

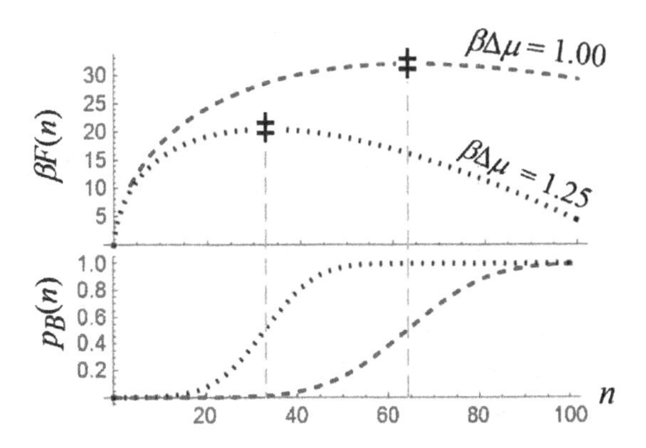

Figure 20.1.3: Classical nucleation theory correctly predicts that nucleus size n is a good reaction coordinate, but p_B depends on n and also on supersaturation, i.e. on the driving force $\beta\Delta\mu$. The curves show $\beta F(n)$ and $p_B(n)$ with $\beta\phi\gamma = 6.0$ for two different values of the driving force $\beta\Delta\mu$.

Clearly, $p_B(\mathbf{x})$ and $n(\mathbf{x})$ are somehow different from each other even though isosurfaces of n are isocommittor surfaces. The difference is that $p_B(\mathbf{x})$ carries a hidden dependence on supersaturation, while $n(\mathbf{x})$ is a true purely configurational coordinate. This difference makes $n(\mathbf{x})$ a better coordinate than $p_B(\mathbf{x})$ itself. In fact, there are three practical problems with directly using the committor as a reaction coordinate.

- $p_B(\mathbf{x})$ is not mechanistically enlightening [55]; it does not identify the molecular features or properties that distinguish transition states from reactants and products. Additionally, rate expressions based on gradients of the committor [30,48,50] do not identify activation parameters with a clear physical interpretation.
- In complex molecular systems, the procedures for evaluating $p_B(\mathbf{x})$ include running hundreds of costly trajectories to estimate the fraction that reach B [11,46], or using simulation data to solve the backward equation [49,56]. Both procedures are non-trivial and both result in estimates of $p_B(\mathbf{x})$ rather than precise values.
- Although $p_B(\mathbf{x})$ can provide rates, it is not easily used to predict kinetic trends [28]. The reason is that $p_B(\mathbf{x})$ carries a hidden dependence on process condition variables that influence the rate: temperature, pressure, ionic strength, supersaturation, etc. Even computing the activation energy would require separate dynamical simulations at several temperatures to determine $p_B(\mathbf{x}|T_1)$, $p_B(\mathbf{x}|T_2)$, $p_B(\mathbf{x}|T_3)$, etc.

Thus the committor satisfies most requirements of an ideal reaction coordinate, but it is too abstruse for mechanistic interpretation, too inconvenient to drive biased sampling methods, and too entangled with process condition variables to yield simple theories and trends. To facilitate applications, we *hope* to find reaction coordinates with one additional property.

Desideratum: A clear mechanistic interpretation

Reaction coordinates with a simple physical interpretation provide mechanistic insight and facilitate the prediction of rates, activation parameters, and trends. For example, consider the reaction coordinates for three classic theories: nucleus size for the classical nucleation theory [57,58], an unstable vibrational mode for harmonic transition state theory [59,60], and an energy gap for the theory of electron transfer [61]. Figure 20.1.4 shows how these physically meaningful reaction coordinates lead to theories with extraordinary capabilities [28].

1. **Harmonic transition state theory (hTST) [59,62]:** Theorists use hTST to predict rates and activation parameters from *ab initio* calculations. Experimentalists use hTST to interpret kinetic measurements [63]. At the foundation of hTST, an unstable mode reaction coordinate quantifies progress through a saddle on the energy landscape. The reaction coordinate naturally emerges from a normal mode analysis of the dynamics in the vicinity of the saddle point. The recipe is applicable to any reaction whose bottleneck is a high saddle on the PES.

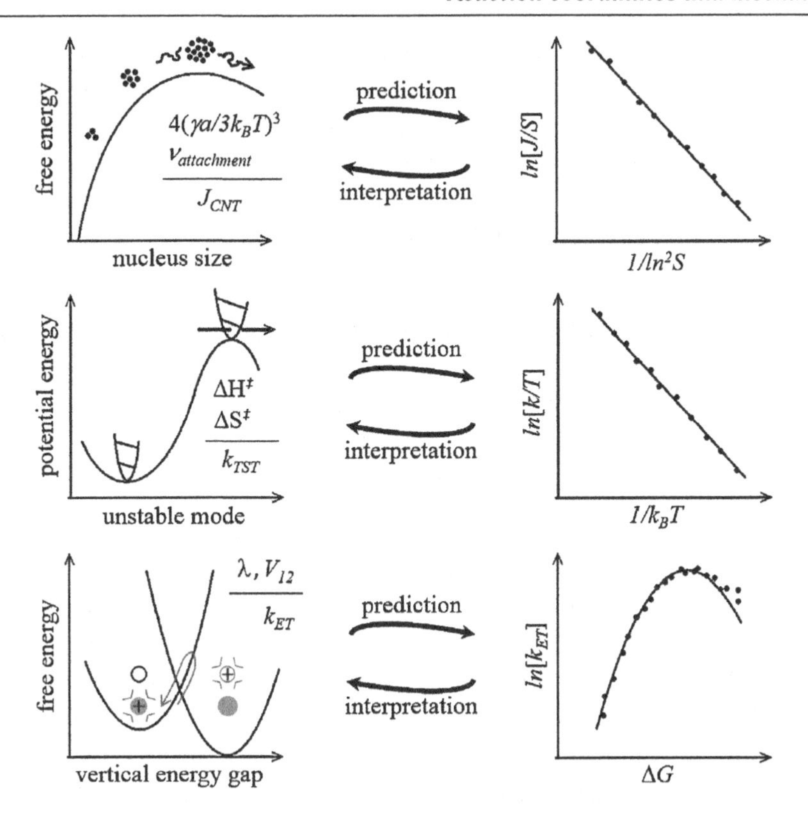

Figure 20.1.4: (Schematic) Classical nucleation theory (top), harmonic transition state theory (middle), and Marcus theory (bottom) are each built upon a reaction coordinate with a clear physical interpretation. These reaction coordinates lead to (i) models of the activation free energy in terms of familiar thermodynamic properties, and (ii) models of the barrier crossing dynamics that involve physical quantities like attachment frequencies and electronic coupling. Theories constructed in this way naturally predict rates, trends, and corollary theories. They also provide an important tool for interpreting the results of experiments. [From Peters et al. *J. Phys. Chem. B* 119, 6349-56 (2015).]

2. **Classical nucleation theory (CNT)** [57,58,64,65]: The absolute nucleation rates predicted by CNT are highly inaccurate, but it remains the most important theory of nucleation for its ability to predict and interpret kinetic trends. CNT reduces the highly complex dynamics and thermodynamics of nucleation to a random walk along the nucleus size coordinate. CNT has been used in many capacities: by theorists for rate predictions, by experimentalists to interpret rates, and by engineers who use it to construct population balance models.
3. **Marcus theory:** Chapter 21 will discuss the vertical energy gap coordinate in the Marcus theory of electron transfer reactions [61,66]. The vertical energy gap quantifies the complex pre-organization of many solvent degrees of freedom into a state where the reactants ($D^- +$

A) have the same energy as the charge transfer products ($D + A^-$). At these configurations the energy gap is zero and the rapid electron transfer event can occur with conservation of energy. The vertical energy gap is an ingenious collective variable that depends on a plethora of solvent degrees of freedom. It would be difficult (perhaps impossible) to divine the energy gap by inspecting electron transfer trajectories or minimum free energy paths. Instead, it was a stroke of genius [67] about dynamics and energy conservation that led to the energy gap and the diabatic states picture of electron transfer.

The activation free energies in these classic theories are not just numerical estimates from a free energy calculation. Instead, the physical nature of the reaction coordinate identifies the activation free energy as a specific type of work, amenable to the formulation of physical free energy models. Likewise, the nature of the reaction coordinate in each theory suggests a model for its dynamics at the bottleneck, and that dynamical model yields the prefactor expression.

Note that each of the three classic theories was developed using a "reaction-coordinate-first" strategy. First, dynamical considerations helped to identify an accurate and physically meaningful reaction coordinate. Then, models for the activation parameters and prefactors were developed based on the nature of the reaction coordinate and its dynamics. The three theories that emerged from this strategy have truly extraordinary capabilities [28]. They (i) predict rates, (ii) predict kinetic trends, (iii) extract activation parameters directly from experimental data, (iv) help construct estimates for the kinetics of one reaction from the kinetics of related reactions (see Chapter 22), and (v) guide the development of "corollary theories". Corollary theories go beyond the explicit predictions of the rate expression [28], e.g. by predicting how solvents influence reaction kinetics [68], or how a surfactant influences the nucleation kinetics [69].

Of course, there are many important activated processes for which the classic theories and their prescribed reaction coordinates are not applicable, cf. conformational transitions of biomolecules, ligand exchange processes, association of complex molecules and ions, pore translocations, non-classical nucleation, or self-assembly processes. The mechanisms of these processes remain poorly understood and accordingly their reaction coordinates remain elusive. Future efforts to understand these and other rare events should strive for simple, broadly applicable, and physically meaningful coordinates like those of the three classic rate theories.

20.2 Variational theories and eigenfunctions

In previous chapters we encountered two important variational theories that provide approximate reaction coordinates. We also encountered spectral theories that provide reaction coordinates that are, in many ways, perfect. Here we briefly recount some advantages and disadvantages of these theories and methods.

Variational TST

Variational transition state theory (VTST) [70–72] optimizes the dividing surface to minimize the TST reaction rate. See Chapter 10 for details on the variational optimization. The reaction coordinate is locally orthogonal to the variationally optimized dividing surface, so VTST effectively identifies the reaction coordinate at least in the vicinity of the transition states. The reaction coordinate from VTST may not remain accurate at early stages and late stages, but for many applications, a local approximation to the reaction coordinate near the transition state is sufficient. Exceptional cases that require a globally accurate reaction coordinate include (i) processes with stable intermediates along the pathway and (ii) reactions that occur in a non-concerted fashion such that the pathway changes direction at the transition state.

For many years, VTST was the only systematic framework for optimizing dividing surfaces and reaction coordinates. VTST contributed to many fundamental advances (including harmonic TST), but for truly complex systems VTST is not a practical optimization criteria. The reason is that variational optimizations with rugged high dimensional PESs are intractable, and variational optimizations with an incomplete set of coordinates may not find the true optimum dividing surface. Thus, in practice, VTST requires *a priori* knowledge of the few most important components of the reaction coordinate to ensure success. Even when the most important variables are known, it should be remembered that VTST is based upon instantaneous *inertial* fluxes at equilibrium. For processes with overdamped dynamics, the optimal reaction coordinate depends on relative *diffusivities* (mobilities) along the important order parameters.

KLBS theory

Kramers-Langer-Berezhkovskii-Szabo (KLBS) theory [47,73,74] (Chapter 18) essentially does for overdamped barrier crossings what VTST did for inertial barrier crossing dynamics. Accordingly, the methods have similar capabilities and limitations. KLBS theory is a powerful tool for theoretical analyses, but like VTST it does not necessarily provide a globally accurate reaction coordinate. The most important limitation of KLBS theory is that it again requires *a priori* knowledge of the most important components of the reaction coordinate. When the few most important variables are not known, KLBS theory requires one to guess (and potentially omit) the most important variables. Alternatively, one can include all potentially important variables, but then the free energy calculations become costly. Thus for processes with poorly understood mechanisms, KLBS theory has severe practical limitations.

Note that processes with rugged free energy landscapes and overdamped dynamics are often studied with "toy models", e.g. with models based on two-dimensional free energy surfaces and isotropic friction/diffusion. These toy models are important pedagogical tools, but they represent highly idealized cases where the dynamics and the free energy landscape are consistent with one another by construction and where all but a few variables have been eliminated from consideration. In real applications, the dynamics and the free energy are consistent with each other only when the chosen coordinates include the correct reaction coordinate and other slow variables [50,75].

Eigenfunctions of the master equation

The conceptual foundations of spectral rate theories were outlined in Chapters 14 and 15. Examples in Chapter 14 were developed from a discrete Markov state model, while Chapter 15 focused on eigenfunctions of master equations [76], Fokker-Planck equations [77,78], and diffusion maps [24,26,36]. For all of these frameworks, we saw that left eigenfunctions are useful coordinate systems for visualizing metastable states and the reaction pathways between them. In fact, the first non-equilibrium "left" eigenfunction ψ_2^L is a natural and essentially perfect reaction coordinate. It increases monotonically as one moves through the high dimensional space from reactants to products, and like the committor it solves the backward Kolmogorov equation. In fact, there are only subtle differences between the committor and ψ_2^L, and these are only evident at the boundaries of the reactant and product states [78]. Thus eigenfunction-based reaction coordinates have all of the properties that make the committor a good reaction coordinate, and eigenfunctions circumvent the need to *a priori* define reactant and product basins in configuration space. However, there are some practical disadvantages to using eigenfunctions of the master equation as reaction coordinates.

1. The proper spectrum of time scales and eigenfunctions only exists for systems that have a well-defined reactant-product equilibrium. For example, they do not exist for irreversible processes like nucleation [79].
2. There is (as yet) no established way to relate the eigenfunctions to a few physical variables with a clear mechanistic interpretation.
3. The eigenfunctions depend not just on $\mathbf{x}$, but also on the conditions of the simulation. The reason, of course, is that the transition rates in the master equation depend on the conditions of the simulation. Because of their dependence on conditions, the eigenfunctions are not easily used to construct simple theories that can predict kinetic trends.
4. The eigenfunctions are constructed based on observed transitions in long unbiased trajectories. To obtain an ergodic master equation, the trajectory must spontaneously cross over all significant barriers. Applications to processes with barriers much larger than $k_B T$ require additional tools to enhance sampling [80,81].

Methods based on the splitting probability (committor) require *a priori* definition of the reactant and product states (A and B). However, *a priori* reactant and product state definitions enable efficient path sampling methods that can "learn" how to predict the committor and/or the forward committor in terms of just a few physically meaningful variables. The rest of this chapter focuses on committors, forward committors, and methods to find simple models for these objects.

20.3 Committor analysis

Can we identify one collective variable $q(\mathbf{x})$ that suffices to predict the committor? Can we also require a concrete and simple physical definition for q so that the reaction coordinate has a clear mechanistic interpretation? The answers to both questions are yes, but the task is not easy. Figure 20.0.1 showed how complications can emerge in a projection from two dimensions to one. Each additional dimension presents another potentially important component of the reaction coordinate. To further complicate matters, anisotropies in the mobility and/or velocity along different coordinates also influence the reaction coordinate as shown in Figure 20.0.2.

Let us first learn to examine the accuracy of a trial coordinate using a procedure called committor analysis [18], a p_{fold} test [46], or a histogram test [11,82]. For an accurate coordinate q, any two configurations $\mathbf{x}$ and $\mathbf{x}'$ for which $q(\mathbf{x}) = q(\mathbf{x}')$ must also have the same committor, i.e. $p_B(\mathbf{x}) = p_B(\mathbf{x}')$. In terms of coordinate isosurfaces, each isosurface of an accurate coordinate q should closely approximate an isosurface of the committor. Thus the ensemble of configurations on isosurface $q(\mathbf{x}) = q_0$ should be characterized by a narrow distribution of committors. Committor analysis quantifies the width of the committor distribution for a trial surface.

A committor analysis can be performed for any isosurface of $q(\mathbf{x})$, i.e. for any value q and corresponding surface $q(\mathbf{x}) = q$. The committor test is often conducted only for the putative transition state isosurface. For processes with a single dominant barrier, the transition state configurations have $p_B(\mathbf{x}) = 1/2$.

Algorithm: Committor analysis

1. Compute the free energy $F(q)$ along trial coordinate $q(\mathbf{x})$.
2. Select a putative transition state location $q*$ along q. Usually, one selects the location of the free energy maximum in $F(q)$ [11].
3. Sample the equilibrium ensemble of configurations on the surface $q(\mathbf{x}) = q*$, e.g. by umbrella sampling [83] with a bias potential $k\,(q(\mathbf{x}) - q*)^2/2$ for a large spring constant k. Approximately 1000 decorrelated configurations are usually sufficient.
4. At each configuration obtained in step (3), estimate the committor by launching many trajectories with Boltzmann distributed momenta. Equations (20.3.1) to (20.3.4) will show how binomial sampling errors can be removed with a deconvolution trick [82], so a crude estimate of the committor at each configuration is acceptable, e.g. from the fraction of 20 trajectories that commit to B.

5. Prepare a histogram of the committor estimates. If q is an accurate reaction coordinate, the histogram will be unimodal and peaked around some characteristic value near $p_B = 1/2$.
6. Insert the mean and standard deviation of the committor histogram into the binomial deconvolution formulas (equations (20.3.1) to (20.3.4)) to obtain the mean and standard deviation of the actual committor distribution.

If the p_B-histogram is bimodal or scattered from 0 to 1, then alternative reaction coordinates should be investigated. If the histogram shows a single peak near $p_B = 1/2$, and if there are no stable intermediates between A and B, then the mechanistic hypothesis embodied by the proposed reaction coordinate is (locally) confirmed. These cases are illustrated in Figure 20.3.1.

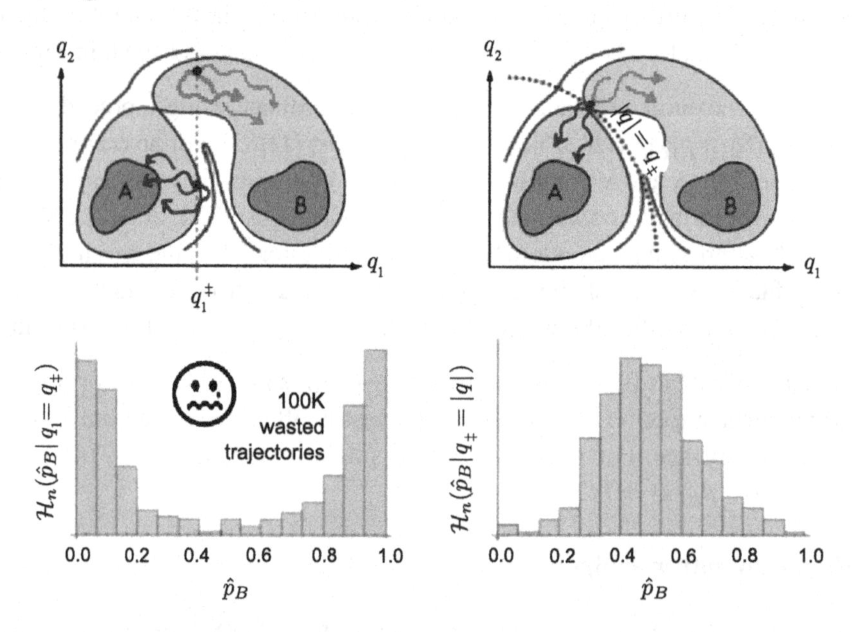

Figure 20.3.1: Schematic depicting committor histogram tests for dividing surfaces obtained from an inaccurate coordinate (left) and from an accurate coordinate (right).

For pathways with metastable intermediates, a sharply peaked histogram with $p_B = 1/2$ is not sufficient evidence for reaction coordinate accuracy [84]. Recall that the committor is approximately constant for all points within a stable intermediate basin, and consider a stable intermediate in which all configurations have $p_B \approx 1/2$. Nearly any dividing surface that cuts through the stable intermediate will generate a committor histogram that is peaked at $p_B \approx 1/2$, but the true reaction coordinate cannot be orthogonal to many different dividing surfaces. Thus, when stable intermediates are a possibility, it is important to test multiple isosurfaces (not just the putative $p_B = 1/2$ isosurface) to ensure reaction coordinate accuracy at all stages of the reaction pathway. It is rather baffling that the most rigorous multi-isosurface tests have been performed in studies of nucleation where metastable intermediates are a rarity [53,85], and not in studies of biomolecular conformational transitions where metastable intermediates are the norm.

Early implementations of the committor test often used 100 trajectories per p_B-estimate and thousands of p_B-estimates [11,14,46]. These early tests were expensive, so they were not widely implemented despite their widely recognized importance. Fortunately, there is an inexpensive, and still rigorous, way to perform the committor test. The binomial distribution $B_n(\hat{p}_B|p_B)$ can be used to deconvolute binomial estimation errors from the actual committor distribution [82]. The histogram is formally a conditional discrete distribution $\mathcal{H}_n(\hat{p}_B|q = q*)$ while the true underlying committor distribution is a continuous conditional density $\rho(p_B|q = q*)$. These two objects are related by the convolution formula

$$\mathcal{H}_n(\hat{p}_B|q = q*) = \int_0^1 dp_B\, \rho(p_B|q = q*) B_n(\hat{p}_B|p_B) \tag{20.3.1}$$

$\mathcal{H}_n(\hat{p}_B|q = q*)$ depends on the number of trajectories per estimate used in the committor analysis procedure. In contrast, $\rho(p_B|q = q*)$ is the protocol independent committor distribution. As the number of trajectories per estimate increases, $\mathcal{H}_n(\hat{p}_B|q = q*)$ approaches $\rho(p_B|q = q*)$. This intuitive limit is why early committor tests used 100s of trajectories per estimate. Instead, the mean and variance of $\rho(p_B|q = q*)$ can be derived from the mean and variance of $\mathcal{H}_n(\hat{p}_B|q = q*)$. These moments are related by

$$\mu = \mu_H \tag{20.3.2}$$

and, to a good approximation [82],

$$\sigma = \sqrt{\sigma_H^2 - \mu_H(1 - \mu_H)/N} \tag{20.3.3}$$

where μ and σ are the mean and standard deviation of the actual committor distribution and μ_H and σ_H are those of the histogram. The quantitative and protocol independent range of committor probability values for a trial dividing surface can then be compactly summarized as [82]

$$p_B = \mu \pm \sigma \tag{20.3.4}$$

Equations (20.3.1) to (20.3.4) show how crude committor estimates can yield accurate values of μ and σ. The error deconvolution procedure reduces the cost of committor analysis by at least a factor of ten, so that the cost is no longer prohibitive even for truly complex systems. An alternative to the binomial deconvolution is to visually compare the perfect binomial distribution to the actual committor histogram [86]. The visual comparison is less quantitative, but it too enables tests with inexpensive p_B-estimates.

Figure 20.3.2 shows an example of coordinate identification and validation for the chemical step in cellulose hydrolysis by cellulase (cel7A) [87]. The cellulase study used QM/MM with an explicit solvent, aimless shooting simulations, likelihood maximization, free energy calculations, transition state theory, and reactive flux calculations. The work of Knott et al. leaves no doubt that state-of-the-art rare events methods, including coordinate identification and validation, are applicable to the most challenging of activated processes.

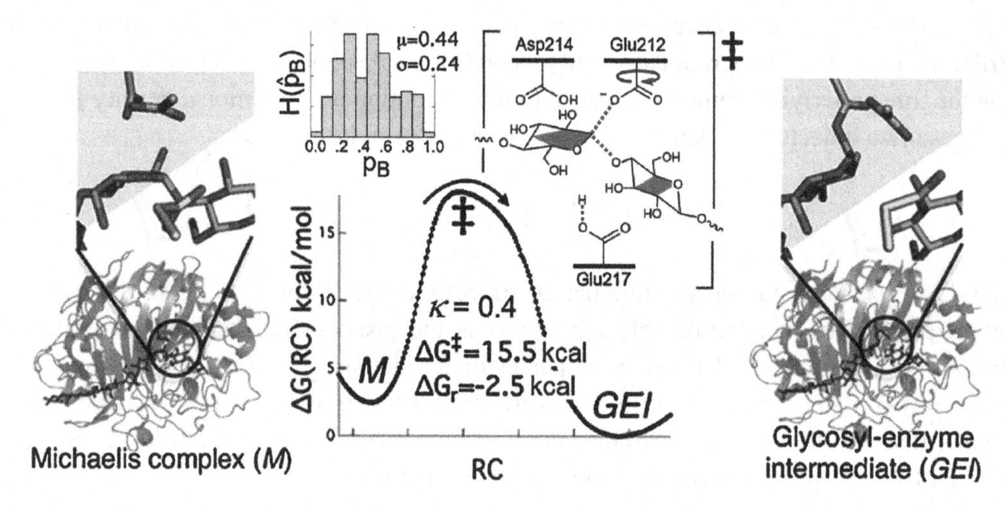

Figure 20.3.2: The center inset shows a histogram test for the reaction coordinate in the hydrolysis of cellulose by a cellulase enzyme. The optimal coordinate combines the dashed bond lengths and a dihedral angle as shown in the upper center panel. Using the accurate reaction coordinate, the authors computed the free energy barrier, transition state theory rate constants, and dynamical corrections. The results show that, when the reaction coordinate is correctly identified, transition state theory can describe enzyme kinetics with only minor dynamical corrections. Interestingly, their reaction coordinate involves no bonds to the transferring hydrogen, consistent with non-adiabatic models of proton transfer. Adapted from Knott et al. *J. Am. Chem. Soc.* 136, 321-9 (2013).

Step 3 in the committor analysis procedure samples configurations from a single trial dividing surface $q(\mathbf{x}) = q*$. The single constraint corresponds to an isosurface of a single scalar reaction coordinate. In contrast, the literature often refers to systems having multiple reaction coordinates. Sometimes, this is only a semantic issue. Certainly, the phrase "reaction coordinates" is less cumbersome than "components of the reaction coordinate". However, some studies obtain peaked histograms by adding constraints on multiple coordinates in step (3) of a committor analysis. These "overconstrained" committor tests have at times been misinterpreted as confirmation of the trial reaction coordinates [88].

Overconstrained tests can be useful, but the results must be interpreted with special care. An overconstrained committor analysis investigates the intersection between multiple surfaces. The constraint intersections have smaller dimensionality than a true dividing surface, and thus an overconstrained committor test has no bearing on the accuracy of a rate calculation [3,89]. First note that some artificial narrowing will accompany each additional constraint. Consider for example, a system with two degrees of freedom x and y. The constraint $x = x*$ divides the space. The constraint $y = y*$ divides the space differently. Now suppose that x and y are both poor coordinates when considered individually, i.e. that $x = x*$ or $y = y*$ each give bimodal committor distributions. Now suppose that we simultaneously apply the constraints $x = x*$ and $y = y*$. The constrained space contains only the single configuration $(x*,\ y*)$ and therefore it has a single committor value [89]. The overconstrained committor test shows that the reaction coordinate involves both x and y, but that was obvious from the beginning. The additional constraint sharpened the committor distribution, but it did not reveal the reaction coordinate. In a multidimensional system, e.g. with a third variable z, sharpness of the doubly-constrained histogram for $(x = x*, y = y*)$ does not even eliminate the importance of z! To eliminate z, we would need to show that z is unimportant for simultaneous constraints at all x and y, not just at $x*$ and $y*$. Thus overconstrained tests *cannot verify* the reaction coordinate. An overconstrained test *can refute* certain combinations of variables which are not sufficient to describe the reaction coordinate. For example, a bimodal histogram at $(x*, y*)$ immediately shows that some variables beyond x and y are important.

20.4 Square error minimization

We have seen how to create detailed maps of abstract reaction coordinates like the committor or the most slowly relaxing eigenfunction. We have also seen how committor analysis can test whether a trial coordinate is sufficient to predict the committor. This section (and those which follow) discuss methods for discovering accurate coordinates among countless candidate variables.

Ma and Dinner developed the first systematic method for generating reaction coordinates in complex systems [15]. Their genetic neural network (GNN) method begins with a collection of transition path sampling trajectories. A training set of committor estimates is then created by shooting additional trajectories from selected locations along the transition paths. The locations are selected so that the training set is evenly distributed with $\hat{p}_B$-values between zero and one.

Finally, a multilevel neural network is optimized to take a list of trial collective variables as inputs and to output a model of the reaction coordinate (see Figure 20.4.1). The optimization is based on a least square error (LSE) prediction of the committor training data. A genetic algorithm accelerates the neural network optimizations and helps to find the best combination of input collective variables [15].

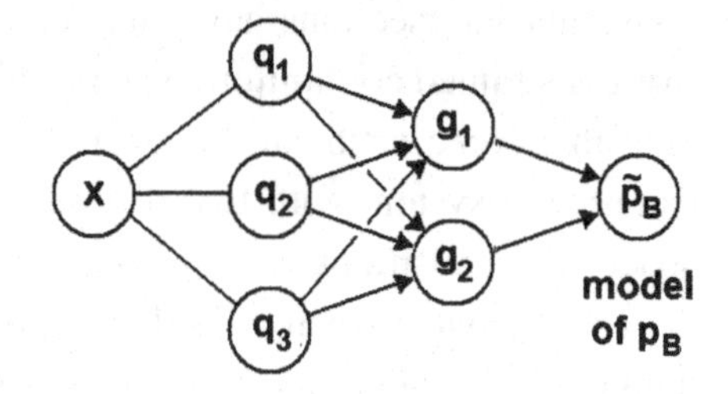

Figure 20.4.1: The Genetic Neural Network (GNN) method uses neural nets to construct models of the committor ($\widetilde{p}_B$) from a list of candidate collective variables. The training data for neural net optimization are p_B-estimates computed at points drawn from the transition path ensemble. Each level in the neural net allows it to model more complex relationships between p_B and the collective variables.

In an impressive first demonstration, GNN identified the missing solvent coordinate for alanine dipeptide isomerization in an explicit water solvent. Specifically, GNN found a solvent induced torque that facilitates the change in Ramachandran angles. GNN used a training set of approximately 400,000 trajectories in the study of alanine dipeptide conformational transitions [15]. Recall that committor analyses (at the time of the GNN study) required approximately 100,000 trajectories per test, and many trial coordinates typically failed these tests before an accurate coordinate could be found. Thus the systematic approach of GNN was a major advance. However, the transition state ensemble for the alanine dipeptide, as predicted by GNN, was also contaminated with reactants and products [15]. Later comparisons to likelihood maximization pointed to the LSE minimization as the source of contamination [90]. Because of the uniform square error penalty, deviations between the model $\widetilde{p}_B$ and the $\hat{p}_B$-estimates are penalized at the same level regardless of the target $\hat{p}_B$-estimate. LSE minimizations might be improved if the squared residuals were first scaled by an estimate for the binomial variance, i.e.

$$S = \sum_i \{\widetilde{p}_B(\mathbf{x}_k) - \hat{p}_B(\mathbf{x}_i)\}^2 / \{n\, \hat{p}_B(\mathbf{x}_i)(1 - \hat{p}_B(\mathbf{x}_i))\}$$

where $\widetilde{p}_B(\mathbf{x}_k)$ is the model of the committor. To my knowledge, this modification has not been attempted.

Borrero and Escobedo developed FFS-LSE [23,91], a least squares minimization approach for extracting reaction coordinates from forward flux sampling data. Because it uses forward flux sampling data, FFS-LSE can identify reaction coordinates for intrinsically non-equilibrium driven processes. The training data for FFS-LSE comes from the A or B outcomes at branching points in the forward flux sampling simulation. Borrero and Escobedo also used ANOVA techniques to test the statistical significance of parameters and components of their reaction coordinates [23]. Escobedo and coworkers applied the FFS-LSE approach to genetic switches [91] and to conformational transitions of biomolecules [92].

20.5 Likelihood maximization

Bayesian inference techniques provide a natural framework for identifying reaction coordinates from the complex multidimensional dynamics. In contrast to methods that generate data according to *a priori* chosen coordinates, constraints, or models (i.e. $p(data|model)$), likelihood analyses assess the plausibility of different models given the data (i.e. $p(model|data)$). Bayes' rule formally relates these two types of probabilities: $p(model|data) = p(data|model)p(model)/p(data)$.[4] $p(data)$ is model independent, i.e. it will be generated independent of the models to be tested. A non-uniform $p(model)$ term allows prior beliefs, as well as new evidence, to influence the probability of a model [93,94]. In the discussion below, our prior belief is that all models are equally plausible. Therefore, the model which maximizes $p(model|data)$ also maximizes $p(data|model)$.

All that remains is to quantitatively relate reaction coordinates, mechanistic models, and data. The outcome at each shooting point in a TPS simulation is a trajectory that commits to either A or B, i.e. a binary realization of the committor. Without computing any $\hat{p}_B$-estimates, the binary shooting data can be used to optimize accurate physical reaction coordinates. Ideally, each shooting point outcome should be independent of the others. Aimless shooting and permutation shooting versions of TPS [90,95,96] (see Chapter 19) were specially designed to generate decorrelated training data with shooting points that are automatically scattered between $p_B = 0$ and $p_B = 1$. Alternatively, the training data can be generated with forward flux sampling [22], transition interface sampling (TIS) [20], replica exchange TIS [97,98], or specialized multiple channel TIS methods [99,100].

The models to be optimized consist of different trial reaction coordinates and their mappings to the committor. Let $\widetilde{p}_B(\mathbf{x}_k)$ be a trial model of the committor. Details on its construction will be given below. The likelihood of the model can be constructed from the training data (the shooting move outcomes) as [95]

$$\mathscr{L}[\widetilde{p}_B(\mathbf{x})] = \prod_{\mathbf{x}_k}^{\to B} \widetilde{p}_B(\mathbf{x}_k) \prod_{\mathbf{x}_{k'}}^{\to A} (1 - \widetilde{p}_B(\mathbf{x}_{k'})) \tag{20.5.1}$$

[4] Informatics methods impacted other research areas long before they impacted rare events methods in chemical physics. I was thinking about these topics in 2005 because of rewarding interactions with my brother-in-law, a biomedical engineer, and because of less rewarding efforts to secure NIH funding for development of methods based on likelihood maximization.

Here $\mathbf{x}_k$ is the kth shooting outcome for which the resulting trajectory reached B, and $\mathbf{x}_{k'}$ in the second product is the k'^{th} shooting point for which the resulting trajectory reached A. Note that $\mathscr{L}[\widetilde{p}_B(\mathbf{x})]$ includes shooting points from both accepted and rejected TPS trajectories. $\mathscr{L}[\widetilde{p}_B(\mathbf{x})]$ is the likelihood of observing the shooting data if the model $\widetilde{p}_B(\mathbf{x})$ is exact. The likelihood becomes large when the model committor $\widetilde{p}_B(\mathbf{x})$ closely matches the real committor $p_B(\mathbf{x})$ over the entire collection of training data [95].

The next step is to propose a relationship between the model committor $\widetilde{p}_B(\mathbf{x})$ and a yet-to-be-specified reaction coordinate $q(\mathbf{x})$. This mapping can be chosen in several different ways. The one constraint is that the committor should become zero in the reactant state and one in the product state. For inertial or overdamped Langevin dynamics on a parabolic barrier [44,47,101] with reaction coordinate $q(\mathbf{x})$, the committor would be

$$\widetilde{p}_B(\mathbf{x}) = \frac{1}{2}\text{erfc}\,[-(q(\mathbf{x}) - q_\ddagger)/\delta q]\} \tag{20.5.2}$$

where $q_\ddagger$ and δq are adjustable parameters. $q_\ddagger$ is the estimated transition state location and $\delta q = (2/|\beta F''(q_\ddagger)|)^{1/2}$ is the estimated width of the region k_BT from the barrier top [55]. The erfc function provides a theoretically motivated form for the mapping, but in practice the barrier is not perfectly parabolic and the dynamics may deviate from the Langevin model. Accordingly, $q_\ddagger$ and δq are best viewed as adjustable parameters to be identified by likelihood maximization.

Now the remaining – and most difficult – task is to construct trial coordinates $q(\mathbf{x})$. Let us start from a long list of potentially important collective variables $q_1(\mathbf{x}_1),\ q_2(\mathbf{x}_1),\ \cdots\ q_M(\mathbf{x}_1)$. The collective variables are evaluated at each shooting point to create a data file with n rows, one per shooting point, as shown below:

Shooting point	Trial coordinate values	Outcome
$\mathbf{x}_1$	$q_1(\mathbf{x}_1)\, q_2(\mathbf{x}_1) \cdots q_M(\mathbf{x}_1)$	B
$\mathbf{x}_2$	$q_1(\mathbf{x}_2)\, q_2(\mathbf{x}_2) \cdots q_M(\mathbf{x}_2)$	A
$\mathbf{x}_3$	$q_1(\mathbf{x}_3)\, q_2(\mathbf{x}_3) \cdots q_M(\mathbf{x}_3)$	A
$\vdots$	$\vdots$	$\vdots$
$\mathbf{x}_n$	$q_1(\mathbf{x}_n)\, q_2(\mathbf{x}_n) \cdots q_M(\mathbf{x}_n)$	B

Now the component variables are used to construct trial coordinates. In the simplest case, a trial coordinate $q(\mathbf{x})$ might be one of the individual $q_1(\mathbf{x}),\ q_2(\mathbf{x}),\ \cdots,\ q_M(\mathbf{x})$ coordinates. If none of the individual $q_1,\ q_2,\ \cdots,\ q_M$ are satisfactory, then they can be mixed in pairs. For example, the linear combination $q_i + \alpha \cdot q_j$ becomes a new coordinate q with an adjustable parameter α to be determined by likelihood maximization [95]. Non-linear functions of the coordinates have also been used [52], e.g. $q = q_i^\alpha q_j^\beta$ where α and β are the adjustable parameters. Beyond pairs, one can try linear or non-linear combinations that mix the $q_1,\ q_2,\ \cdots,\ q_M$ coordinates in triplets. Early implementations of likelihood maximization considered a systematic progression of linear combinations mixing $m = 1, 2$, or 3 collective variables from the table [95,102]. The central idea is to build upwards in complexity, starting from the simplest coordinates and trying all possibilities until a satisfactory model is found.

Combining all of the proposed relationships gives a complete model of the committor:

$$(x_1, x_2, \ldots x_{3N}) \rightarrow (q_1, q_2, \cdots, q_M) \rightsquigarrow \tilde{p}_B(q)$$

where the strange arrow symbolizes mappings that are parsimonious in the sense that q should only depend on a few of the $q_1, q_2, \cdots, q_M$. Each trial reaction coordinate constructed in this manner is a hypothesis that $q(\mathbf{x})$ adequately summarizes the mechanism. By constructing $q(\mathbf{x})$ from only a few collective variables, we can (hopefully) represent the committor in terms of physical variables that reveal a clear and simple mechanistic interpretation.

Of course, most trial reaction coordinates and committor models are terribly inaccurate. How do we identify the most accurate models? $\mathscr{L}[\widetilde{p}_B(\mathbf{x})]$ is a functional of the entire model $\widetilde{p}_B(\mathbf{x})$. It depends on the selection of $q_i(\mathbf{x})$ components within $q(\mathbf{x})$, on free parameters within the parsimonious mapping $q_1, q_2, \cdots, q_M \rightsquigarrow q$, and on the $q_\ddagger$ and δq parameters in the $q \rightarrow \tilde{p}_B$ mapping. Often there are additional adjustable parameters within the $q_1(\mathbf{x}), q_2(\mathbf{x}), \cdots, q_M(\mathbf{x})$ variables, e.g. the cutoff radii, critical bond orders, and k-space cutoffs that commonly enter order parameter definitions. Likelihood maximization can also optimize these parameters. For example, Jungblut et al. [53] optimized the clustering criteria in an algorithm for computing nucleus size, Mullen et al. [13] optimized the distance cutoffs within a family of coordination numbers, and Best and Hummer [103] (with different methods) optimized a weighted sum of native contacts for protein folding. After likelihood maximization has optimized all of the parameters within each trial $\widetilde{p}_B(\mathbf{x})$ model, the optimized trial coordinates assembled from different component variables will have different optimized likelihood scores. Thus likelihood maximization also systematically identifies the most important component variables.

Figure 20.5.1 shows aimless shooting points naturally clustered near the separatrix and also within the transition pathway on the Muller-Brown [104] PES. The shooting points are black if the forward trajectory committed to B, and gray if the forward trajectory committed to A. The visibly discernible gradient in shooting point coloration is the reaction coordinate. A large likelihood score results from a coordinate $q(x, y)$ whose gradient points in the direction of the arrow.

The Muller-Brown PES provides an example to illustrate the method, but as noted earlier these two dimensional models obscure the most important challenge. In a truly high dimensional system, identifying the few most important variables is more difficult than combining the important variables into an accurate scalar coordinate. Likelihood maximization can rapidly screen thousands of trial coordinates and their combinations using a single set of shooting point data. The

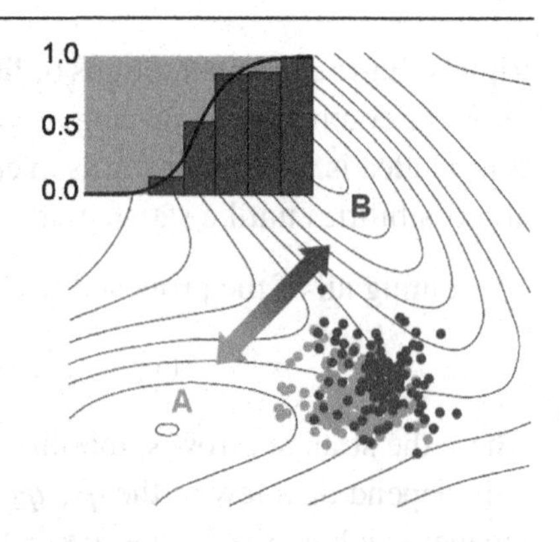

Figure 20.5.1: Aimless Shooting points on the Muller-Brown PES with Langevin dynamics. The points are colored black if the resulting trajectory went to B and gray if the resulting trajectory went to A. The shooting data reveals the direction, location, and lengthscale on which p_B changes from zero to one (inset). Note that the aimless shooting points are naturally clustered within the transition pathway and within about $4k_BT$ (1 contour) from the saddle point. [From Peters et al. *J. Chem. Phys.* 127, 034109 (2007).]

approach has been successful across a diverse range of chemical reactions [87,102,105], nucleation processes [52,53,106], biomolecular conformational transitions [107–109], and other processes [96,110–112]. A few thousand shooting point outcomes are usually sufficient to identify an accurate model.[5] For an illustrative application of likelihood maximization to enzymatic cellulose hydrolysis [87], see Figure 20.3.2.

Some limitations of the likelihood maximization machinery should be noted. First, the erfc model cannot describe pathways with stable intermediates near the barrier top. As shown in Figure 18.4.2, the committor-coordinate mapping has several inflection points when there are multiple intermediates along the reaction path, whereas equation (20.5.2) only has one inflection point. To model more complicated relationships between p_B and q, one could use a sum of two (scaled) erfc functions with different $q_\ddagger$ and δq parameters, or one could optimize a p_B vs. q spline function. Additionally, equation (20.5.2) fails to describe pathways where the reaction coordinate changes between early and late stages. More versatile mappings can be constructed by interpolation along the arclength of a string [114] or by multi-level neural networks [15].

In Figure 20.5.1, the shooting points are clustered near the saddle point. The focused shooting points enhance the sampling efficiency of aimless shooting, but the training data should cover the entire range of p_B values to ensure a globally accurate reaction from $p_B = 0$ to $p_B = 1$ [5, 115]. Note that the contour spacing in Figure 20.5.1 is $4k_BT$, and that shooting points are scattered beyond $4k_BT$ down from the barrier on both sides. Equation (20.5.2) shows that these shooting points are indeed very near $p_B = 0$ and $p_B = 1$. Additionally, the shooting points near A and B are resampled many times in the aimless shooting algorithm because the trajectories at those locations tend to be rejected.

The shooting point distribution can become overly focused in systems with stable intermediates along the reaction pathway. In fact, all TPS algorithms have difficulties with stable interme-

[5] Note that the GNN method is conceptually similar [113], but it required ca. 100,000 trajectories beyond the TPS simulation to build the training data from committor estimates [15].

diates because the shooting points become highly concentrated near local minima. For the purposes of reaction coordinate identification, shooting points near local minima are largely uninformative because the committor essentially remains constant throughout each metastable basin. For systems with local minima, a more efficient approach is to separately analyze each step in the overall transformation. Alternatively, likelihood maximization can use shooting data from more robust algorithms like transition interface sampling.

Occam's razor

Models with more adjustable parameters and more component variables will generally give higher likelihoods. The gain in likelihood upon adding new component variables might reflect physical improvements in the p_B-model, or it may simply reflect overfitting. How can we quantitatively invoke Occam's razor to detect meaningful coordinate improvements? Peters and Trout used the Bayesian Information Criteria (BIC) to identify the point of diminishing returns in the progression of trial coordinate complexity [95]. They constructed reaction coordinates from linear combinations of other component variables, first with $m = 1$ component models, then $m = 2$, then $m = 3$, etc. At the mth stage, likelihood maximization is performed for a total of $M!/(m!(M-m)!)$ models with m-components. The best m-component model was deemed a significant improvement over the best $(m-1)$ component model if their likelihoods differ by more than the BIC. For a data set of n-shooting move outcomes, the BIC is $(ln[n])/2$ [95]. In practice, it may be useful to place an even greater emphasis on parsimony, prioritizing simple mechanistic explanations over full convergence according to the BIC. Alternatively, new L_1-norm optimization methods like LASSO [116] or compressive sensing [117] might be used to obtain simple models from likelihood maximization.

Need for an independent test

Two potential problems require that coordinates obtained by likelihood maximization be subjected to at least one final histogram test [55]:

1. The optimal coordinate obtained by likelihood maximization is only the best among the trial coordinates. Omission of any critically important coordinate will result in an inaccurate reaction coordinate.
2. The data used by likelihood maximization is from the transition path ensemble, but the reaction coordinate should also describe the committor in the equilibrium ensemble. The optimal coordinate from likelihood maximization may perfectly describe the transition pathway, but fail a committor test.

Both pitfalls indicate coordinates that do not adequately describe the mechanism. Coordinates that describe the transition path ensemble, but fail the committor test are particularly tricky. They arise when the transition state isosurface of a seemingly good coordinate cuts through an off-pathway part of the reactant or product basin. The ensemble of configurations on the isosurface will then include legitimate transition states and also committed states within the stable basin. Examples can be seen in models that exhibit catch-bond behavior [118,119]. Even for the catch-bond scenario, there are always coordinates in which a highly curved pathway will appear straight. The challenge is being sufficiently inventive to find them.

20.6 Inertial likelihood maximization

The original likelihood maximization (oLMax) algorithm identified reaction coordinates that approximately predict the committor, but it never finds an exact model for the committor. How do the errors in a model of p_B translate to errors in the computed rate? This is not an easy question to answer. The relationship between rate errors and committor errors is especially complicated for chemical reactions because these tend to have inertial barrier crossing dynamics. oLMax was successful in certain applications with inertial dynamics (cf. Figure 20.3.2), but Mullen et al. [13] found that some coordinates obtained from oLMax *improved* the predicted committors and simultaneously *reduced* the computed transmission coefficient. These results, which at first seem contradictory, illustrate the lack of a simple relationship between the committor distribution and the transmission coefficient.

Recall the third requirement for an accurate reaction coordinate: the reaction coordinate $q(\mathbf{x})$ should be sufficient to predict the committor $p_B(\mathbf{x})$. This requirement says nothing about velocities, but maybe it should. For inertial barrier crossings, both the value of the reaction coordinate *and its velocity* influence the probability to reach state B [101,120]. We can amend the third requirement: the reaction coordinate *and its velocity* should be sufficient to predict the reaction probability. Recall from Chapter 17 that the reaction probability $p_{RX}(q, \dot{q})$ is the probability to reach state B before A as a function of position and velocity along the reaction coordinate q. Integrating over the velocity information recovers the original committor requirement:

$$p_B(q) = \int p_{RX}(q, \dot{q})\rho_{eq}(\dot{q})d\dot{q}$$

so the revised requirement (3) is entirely consistent with the original requirement. For important practical reasons, the reaction probability requirement is *superior* to the committor requirement as shown below.

The reaction probability is closely related to the forward committor $p_B(\mathbf{x}, \dot{\mathbf{x}})$ [30], i.e. to the probability that phase space point $(\mathbf{x}, \dot{\mathbf{x}})$ evolves to state B before A. For stochastic dynamics, the dimensionality reduction from $p_B(\mathbf{x}, \dot{\mathbf{x}})$ to $p_{RX}(q, \dot{q})$ is the inertial analogue of the dimensionality reduction in going from $p_B(\mathbf{x})$ to $p_B(q)$. We could replace the committor with the forward committor throughout the theory of transition paths, but the forward committor is either zero or one at each phase space point for systems with deterministic dynamics. In contrast, the reaction probability varies continuously with $(q, \dot{q})$ for all dynamics in a multidimensional system.

Inertial Likelihood maximization (iLMax) exploits the additional velocity dependence in $p_{RX}(q, \dot{q})$ to facilitate the reaction coordinate optimization. For Langevin (Markovian) dynamics or generalized Langevin (non-Markovian) dynamics on a parabolic barrier [101,120], the reaction probability for initial condition $(q_0, \dot{q}_0)$ has the form

$$p_{RX}(q_0, \dot{q}_0) \cong \frac{1}{2}\text{erfc}\left[-a \cdot \{(q_0 - q_{\ddagger}) + b\dot{q}_0\}\right] \tag{20.6.1}$$

Here $q_{\ddagger}$ is the transition state along coordinate q, and parameters a and b depend on the barrier frequency, the friction, and the reduced mass. Equation (20.6.1) is a theoretically motivated mapping from the position and velocity along a trial coordinate to the reaction probability [121]. (See Chapters 16 and 17.)

A transition path sampling simulation provides shooting point positions, the shooting point velocities, and the shooting move outcomes. If $\mathbf{x}$ is the atomistically detailed shooting point position, then $q = q(\mathbf{x})$ is the trial coordinate value at the shooting point and $\dot{q} = \nabla q \cdot \dot{\mathbf{x}}$ is the velocity along the trial coordinate at the shooting point. iLMax uses these quantities at each of the shooting points to optimize a model of the reaction probability instead of the committor. The inertial likelihood is [121]:

$$\mathscr{L}[q(\mathbf{x})] = \prod_{\mathbf{x}^{(k)}, \dot{\mathbf{x}}^{(k)}}^{\rightarrow B} \widetilde{p}_{RX}(q^{(k)}, \dot{q}^{(k)}) \prod_{\mathbf{x}^{(k)}, \dot{\mathbf{x}}^{(k)}}^{\rightarrow A} \left\{1 - \widetilde{p}_{RX}(q^{(k)}, \dot{q}^{(k)})\right\} \tag{20.6.2}$$

where $q^{(k)} = q(\mathbf{x}^{(k)})$ and $\dot{q}^{(k)} = \nabla q|_{\mathbf{x}^{(k)}} \cdot \dot{\mathbf{x}}^{(k)}$. The first summation includes all shooting points for which the resulting trajectory reached B. The second includes all shooting points that reached A. For iLMax, the forward and reverse outcomes at each shooting point are both used in the likelihood. The forward and time-reversed "backward" trajectories provide separate reaction probability outcomes from two separate points in phase space, $(\mathbf{x}, \dot{\mathbf{x}})$ and $(\mathbf{x}, -\dot{\mathbf{x}})$. In most other respects, iLMax is similar to oLMax. See the algorithm below for a table showing the required input data for iLMax: shooting point configurations, shooting point velocities, and A, B outcomes.

Note that velocity data is used by iLMax, but that $\mathscr{L}[q(\mathbf{x})]$ depends only on the definition of the configurational coordinate $q(\mathbf{x})$ [121]. Velocities always appear as projections onto trial configurational coordinates. Therefore, the iLMax procedure always yields a purely configurational coordinate that is easily used with transition state theory or quasi-equilibrium rate theories.

iLMax consistently finds coordinates with higher transmission coefficients than does oLMax [121]. iLMax also tends to find coordinates which more accurately predict committor. Figure 20.6.1 illustrates the advantages of iLMax over oLMax for an anharmonic double well potential coupled to a bath of eleven harmonic oscillators [121].

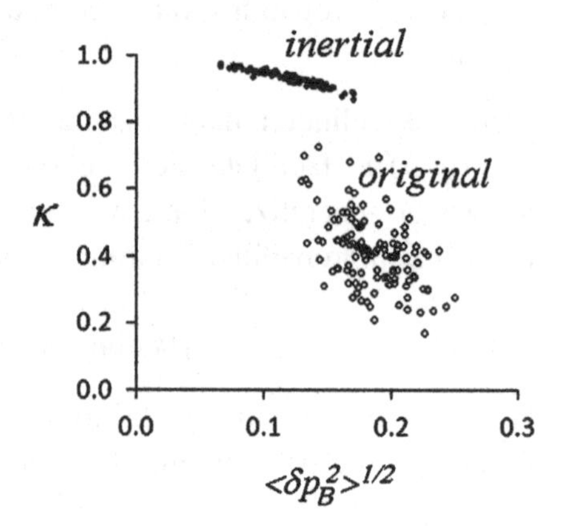

Figure 20.6.1: Transmission coefficients and committor distribution variances from repeated applications of oLMax and iLMax. The system is an anharmonic double well coupled to a harmonic oscillator bath. iLMax consistently finds dividing surfaces with higher transmission coefficients and usually with narrower p_B-distributions. [From Peters, *Chem. Phys. Lett.* 554, 248-53 (2012).]

Why do coordinates from iLMax lead to larger transmission coefficients than those from oLMax? Why do coordinates from iLMax more accurately describe the committor than coordinates from oLMax? There are two key reasons:

1. iLMax extracts more training data from the same collection of TPS trajectories by using both the forward and backward trajectory outcomes [121].
2. The transmission coefficient correlates initial velocity at the transition state to the long time probability of reaching B. By construction, iLMax also identifies coordinates whose initial velocity is strongly correlated to the probability of reaching state B [121].

In a sense, iLMax is a generalization of the VTST approach. VTST optimizes only the dividing surface whereas iLMax uses a similar inputs to optimize the reaction coordinate at all stages along the reaction pathway. Moreover, iLMax requires only a single set of unbiased TPS trajectory data whereas VTST requires free energy barriers and prefactors for each iteration of the variational dividing surface optimization.

For overdamped barrier crossings, where velocities are unimportant, iLMax confirms their unimportance by finding $b \approx 0$ in equation (20.6.1). Thus, for processes with overdamped dynamics, iLMax naturally reverts to the committor optimization approach of oLMax. Because of this property, iLMax should be superior or equal to oLMax for all activated processes [121]. Furthermore, the optimized b and a parameters in equation (20.6.1) can be used to estimate the friction and imaginary frequency for a Langevin model of the barrier crossing dynamics. Thus the reaction coordinate and a simple model of the barrier crossing dynamics can be constructed all starting from a few thousand TPS trajectories and an iLMax analysis [121].

The algorithm below maximizes the inertial log-likelihood for a set of trial reaction coordinates. The likelihood maximum always coincides with the log-likelihood maximum, and the latter varies more slowly making it amenable to numerical optimization.

Algorithm: Inertial likelihood maximization (iLMax) [121]

Use the variable path length version of aimless shooting (or other algorithms) to generate n shooting trajectories. The collection of n-trajectories should include members of the transition path ensemble and also rejected trajectories (non-transition paths). Save the shooting point, the random velocity, and the forward and reverse trajectory outcomes. Compile a list of trial coordinates and velocities along the trial coordinate at each shooting point. The data can be stored as shown in the table below:

Shooting point	Trial coordinate values and velocities		Outcome
$\mathbf{x}_1 \quad \dot{\mathbf{x}}_1$	$q_1^{(1)} \cdots q_M^{(1)}$	$\dot{q}_1^{(1)} \cdots \dot{q}_M^{(1)}$	B
$\mathbf{x}_1 \quad -\dot{\mathbf{x}}_1$	$q_1^{(1)} \cdots q_M^{(1)}$	$-\dot{q}_1^{(1)} \cdots -\dot{q}_M^{(1)}$	A
$\mathbf{x}_2 \quad \dot{\mathbf{x}}_2$	$q_1^{(2)} \cdots q_M^{(2)}$	$\dot{q}_1^{(2)} \cdots \dot{q}_M^{(2)}$	A
$\mathbf{x}_2 \quad -\dot{\mathbf{x}}_2$	$q_1^{(2)} \cdots q_M^{(2)}$	$-\dot{q}_1^{(2)} \cdots -\dot{q}_M^{(2)}$	A
$\vdots$	$\vdots$		$\vdots$
$\mathbf{x}_n \quad \dot{\mathbf{x}}_n$	$q_1^{(n)} \cdots q_M^{(n)}$	$\dot{q}_1^{(n)} \cdots \dot{q}_M^{(n)}$	B
$\mathbf{x}_n \quad -\dot{\mathbf{x}}_n$	$q_1^{(n)} \cdots q_M^{(n)}$	$-\dot{q}_1^{(n)} \cdots -\dot{q}_M^{(n)}$	B

1. Set $m = 1$ and $BIC = [ln(2n)]/2$, because for iLMax, there are two observations per shooting point.
2. Optimize and compare all trial reaction coordinates constructed from m component variables. This algorithm considers only linear combinations of the component variables.
   (a) Construct a trial coordinate q from each m-variable combination among the q_1, q_2, ..., q_M variables. Let the coefficients be c_1, c_2, ..., c_m with $c_1 = 1$ for each combination.
   (b) For each trial coordinate, maximize the inertial log-likelihood (20.6.2) to find the optimal a, b, and $q_{\ddagger}$ parameters in equation (20.6.1) and the c_2, ..., c_m parameters within q. The optimization can be done using BFGS, conjugate gradient, or similar methods [122].
   (c) Identify the optimal combination of m components from step 1b. If the highest log-likelihood with m components exceeds the highest log-likelihood from $m - 1$

components by less than the BIC, the $m-1$ component reaction coordinate was already converged.

3. If the best m-component coordinate is significantly better than the best $(m-1)$-component coordinate, then try coordinates with $(m+1)$-components, i.e. increment m and return to step 2. As noted for oLMax, simple models can be obtained by stopping before full convergence with respect to the BIC.

Once the optimal reaction coordinate has been identified, it should be subjected to a committor test to ensure that the coordinate basis $q_1, q_2, \ldots, q_M$ was sufficient. For inertial reactions (indicated by nonzero optimal b parameter) the transmission coefficient provides an alternative indicator of reaction coordinate accuracy. Note that to recover the oLMax algorithm, simply fix $b=0$ throughout this iLMax algorithm.

Figure 20.6.2 shows results from an application of iLMax to ion-pair dissociation by Mullen et al. [13].

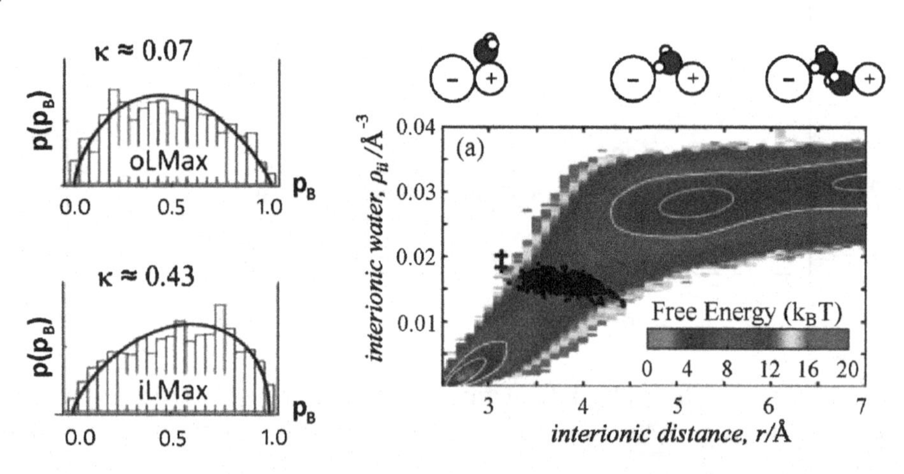

Figure 20.6.2: Mullen et al. used iLMax and oLMax to study NaCl dissociation in water. The reaction coordinates identified by iLMax and oLMax gave similar p_B-histograms. However, the oLMax dividing surface gave a small transmission coefficient, while the iLMax dividing surface gave a large transmission coefficient. The coordinates identified by iLMax quantify the local density and orientation of water molecules between the two ions. Adapted from Mullen et al. *J. Chem. Theory and Comp.* 10, 659-67 (2014).

The main disadvantage of likelihood maximization (and all other methods) is that they still rely on human intuition to propose trial coordinates and mechanistic hypotheses. As yet, there is no procedure for having an epiphany [123], but each successful application provides new insights that (hopefully) can be transferred to similar types of activated processes. Some systematic approaches have emerged, e.g. varying cutoff parameters in coordination numbers, varying bond-order parameters for computing nucleus size, and varying weights in native contact coordinates. It seems likely that progress on previously intractable problems can be made by drawing upon prior families of successful coordinates and by systematic coordinate optimization strategies.

Exercises

1. A flawed (but not uncommon) test of reaction coordinate accuracy presents the average of the committor as a function of a trial reaction coordinate:

$$\overline{p}_B(q) = \langle p_B(\mathbf{x})\delta[q(\mathbf{x}) - q]\rangle \,/\, \langle\delta[q(\mathbf{x}) - q]\rangle$$

   vs. q. For a good coordinate, the test results in a sigmoid shaped curve, but explain why the test also confirms any decent order parameter. Hint: A man with his head in an oven and his feet in the icebox, on average, is actually not very comfortable.
2. Use $D = k_B T/m\gamma$ and its multidimensional generalization to argue that the inertial criteria for dividing surface optimization in variational TST does not necessarily find the optimal dividing surface for an activated process with overdamped dynamics. Discuss how $\mathbf{D}(\mathbf{q})$ might be computationally decomposed into separate factors involving the coordinate masses and frictions. Hint: consider equipartition and friction estimates from Chapter 17.
3. Give expressions for $E(q)$ and $S(q)$ in the free energy decomposition $F(q) = E(q) - TS(q)$. Start from

$$F(q) = -kTln \int d\mathbf{x}d\mathbf{p}\, \rho_{eq}(\mathbf{x}, \mathbf{p})\delta[q - q(\mathbf{x})]$$

   First, consider a conventional coordinate $q(\mathbf{x})$ like a bond length, a nucleus size coordinate, or an energy gap. Then consider the case where q is the committor $p_B(\mathbf{x})$. Explain how your results relate to the difficulty of extracting trends like activation energies from rate expressions like equation (18.7.7) that directly use the committor [123]. Hint: use $p_B(\mathbf{x}) = \int d\mathbf{p}\rho_{eq}(\mathbf{p})p_{RX}(\mathbf{x}, \mathbf{p})$ where $p_{RX}(\mathbf{x}, \mathbf{p})$ is the forward committor.
4. Consider a system with inertial Langevin dynamics, friction γ, mass m, and one-dimensional parabolic barrier $V(q) = -m\omega^2(q - q_\ddagger)^2/2$. Relate the optimal a and b parameters in equation (20.6.1) to γ, m, and ω. Hint: Use the Green's function for inertial Langevin dynamics on a parabolic barrier.
5. Configurations with $p_B(\mathbf{x}) = 1/2$ are called transition states, but they do not always coincide with dividing surfaces that minimize the transmission coefficient [124,125]. Variational transition state theory (VTST) says that the reaction coordinate, more precisely the dividing surface $q(\mathbf{x}) = q_\ddagger$, should minimize the reactive flux, i.e. minimize the effects of recrossing.

(a) Show that a perfect dividing surface with no recrossing must consist entirely of states with $p_B(q) = 1/2$. Also give an example to show that the converse is not true, i.e. a surface with $p_B(q) = 1/2$ may have many recrossings.

(b) In general, the surface with fewest recrossings may be entirely different from the $p_B(q) = 1/2$ surface. Consider a stable intermediate (I) symmetrically situated between symmetrical states (A) and (B) as shown below.

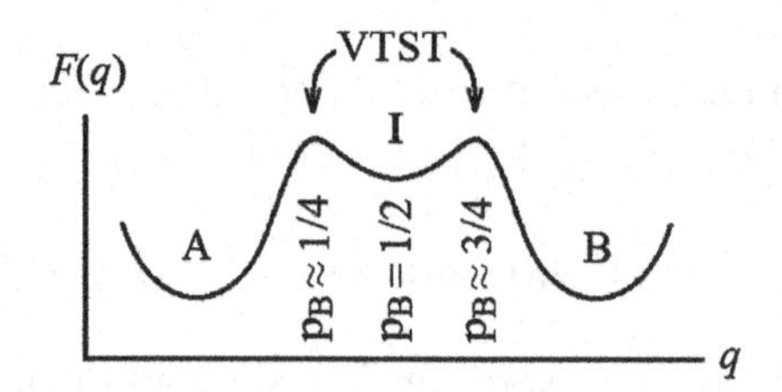

The $p_B(q) = 1/2$ surface, by symmetry, must cut the intermediate state into two halves. However, the (two) dividing surfaces with minimal recrossing are those which separate I from A and B. If I is highly metastable, show that the dividing surfaces which minimize recrossing are located where $p_B(q) = 1/4$ and $p_B(q) = 3/4$. Such inconsistencies between the $p_B(q) = 1/2$ separatrix and the reactive flux/minimal recrossing definitions of the transition states are easily rectified by treating the reaction as a series of two separate elementary steps.

(c) Construct a model for $p_B(q)$ in the situation of part (b) by adding two rescaled erfc functions with different inflection points.

6. Let the p_B-distribution for a trial dividing surface be $\rho(p_B)$. Additionally, suppose that the exact (and unknown) coordinate $q(\mathbf{x})$ has coordinate independent diffusivity D and parabolic free energy profile $F(q) \approx F(q_\ddagger) - |F''(q_\ddagger)|(q - q_\ddagger)^2/2$ near the transition state location. How are the approximate transition states from the trial dividing surface distributed along the true reaction coordinate q? Hint: use equation (18.4.5).

References

[1] P.A.M. Dirac, Proc. R. Soc. A 123 (1929) 714–733.
[2] M.H.M. Olsson, W.W. Parson, A. Warshel, Chem. Rev. 106 (2006) 1737–1756.
[3] B. Peters, J. Chem. Theory Comput. 6 (2010) 1447–1454.
[4] A. Kohen, Acc. Chem. Res. 48 (2015) 466–473.
[5] P.G. Bolhuis, C. Dellago, Eur. Phys. J. 6 (2015) 1–19.
[6] G.K. Schenter, B.C. Garrett, D.G. Truhlar, J. Phys. Chem. B 105 (2001) 9672–9685.
[7] L. Onsager, Phys. Rev. 54 (1938) 554–557.
[8] G. Ciccotti, M. Ferrario, J.T. Hynes, R. Kapral, J. Chem. Phys. 93 (1990) 7137–7147.
[9] R. Rey, E. Guardia, J. Phys. Chem. 96 (1992) 4712–4718.
[10] D.E. Smith, L.X. Dang, J. Chem. Phys. 100 (1994) 3757.
[11] P.L. Geissler, C. Dellago, D. Chandler, J. Phys. Chem. B 103 (1999) 3706–3710.

[12] A.J. Ballard, C. Dellago, J. Phys. Chem. B 116 (2012) 13490–13497.
[13] R.G. Mullen, J.E. Shea, B. Peters, J. Chem. Theory Comput. 10 (2014) 659.
[14] P.G. Bolhuis, C. Dellago, D. Chandler, Proc. Natl. Acad. Sci. USA 97 (2000) 5877–5882.
[15] A. Ma, A.R. Dinner, J. Phys. Chem. B 109 (2005) 6769–6779.
[16] P.G. Bolhuis, C. Dellago, D. Chandler, Faraday Discuss. 110 (1998) 421.
[17] C. Dellago, P.G. Bolhuis, D. Chandler, J. Chem. Phys. 110 (1999) 6617–6625.
[18] P.G. Bolhuis, D. Chandler, C. Dellago, P.L. Geissler, Annu. Rev. Phys. Chem. 53 (2002) 291–318.
[19] T.S. van Erp, D. Moroni, P.G. Bolhuis, J. Chem. Phys. 118 (2003) 7762.
[20] T.S. van Erp, P.G. Bolhuis, J. Comput. Phys. 205 (2005) 157–181.
[21] R.J. Allen, P.B. Warren, P.P. Ten Wolde, Phys. Rev. Lett. 94 (2005) 18104.
[22] R.J. Allen, C. Valeriani, P.R. ten Wolde, J. Phys. Condens. Matter 21 (2009) 463102.
[23] F.A. Escobedo, E.E. Borrero, J.C. Araque, J. Phys. Condens. Matter 21 (2009) 333101.
[24] R.R. Coifman, I.G. Kevrekidis, S. Lafon, M. Maggioni, B. Nadler, SIAM J. Multiscale Model. Simul. 7 (2008) 842–864.
[25] A. Singer, R. Erban, I.G. Kevrekidis, R.R. Coifman, Proc. Natl. Acad. Sci. USA 106 (2009) 16090–16095.
[26] M.A. Rohrdanz, W. Zheng, C. Clementi, Annu. Rev. Phys. Chem. 64 (2013) 295–316.
[27] G.R. Bowman, V.S. Pande, F. Noe, in: G.R. Bowman, V.S. Pande, F. Noe (Eds.), An Introduction to Markov State Models and Their Application to Long Timescale Molecular Simulation, Springer, Berlin, Heidelberg, 2013.
[28] B. Peters, J. Phys. Chem. B 119 (2015) 6349–6356.
[29] W. E, W. Ren, E. Vanden-Eijnden, Chem. Phys. Lett. 413 (2005) 242–247.
[30] W. E, E. Vanden-Eijnden, J. Stat. Phys. 123 (2006) 503–523.
[31] P. Metzner, C. Schutte, E. Vanden-Eijnden, SIAM J. Multiscale Model. Simul. 7 (2009) 1192–1219.
[32] D.J. Wales, J. Chem. Phys. 130 (2009) 204111.
[33] S.V. Krivov, J. Phys. Chem. B 115 (2011) 11382–11388.
[34] W. Zheng, B. Qi, M.A. Rohrdanz, A. Caflisch, A.R. Dinner, C. Clementi, J. Phys. Chem. B 115 (2011) 13065–13074.
[35] P.J. Ledbetter, C. Clementi, J. Chem. Phys. 135 (2011) 44116.
[36] A.L. Ferguson, A.Z. Panagiotopoulos, I.G. Kevrekidis, P.G. Debenedetti, Chem. Phys. Lett. 509 (2011) 1–11.
[37] J.-H. Prinz, B. Keller, F. Noe, Phys. Chem. Chem. Phys. 13 (2011) 16912–16927.
[38] S. Muff, A. Caflisch, J. Chem. Phys. 130 (2009) 125104.
[39] T. Uzer, C. Jaffe, J. Palacian, P. Yanguas, S. Wiggins, Nonlinearity 15 (2002) 957–992.
[40] R. Hernandez, T. Uzer, T. Bartsch, Chem. Phys. 370 (2010) 270–276.
[41] S. Kawai, T. Komatsuzaki, Phys. Chem. Chem. Phys. 12 (2010) 15382–15391.
[42] C. Dellago, P.G. Bolhuis, P.L. Geissler, in: Erice Proceedings, Vol. 1, 2005, pp. 1–38, Chap. 1.
[43] M. Tachiya, J. Chem. Phys. 69 (1978) 2375.
[44] K. Schulten, Z. Schulten, A. Szabo, J. Chem. Phys. 74 (1981) 4426–4432.
[45] D. Ryter, J. Stat. Phys. 49 (1987) 751–765.
[46] R. Du, V.S. Pande, A.Y. Grosberg, T. Tanaka, E.S. Shakhnovich, J. Chem. Phys. 108 (1998) 334.
[47] A. Berezhkovskii, A. Szabo, J. Chem. Phys. 122 (2005) 14503.
[48] E. Vanden-Eijnden, in: M. Ferrario, G. Ciccotti, K. Binder (Eds.), Computer Simulations in Condensed Matter: From Materials to Chemical Biology, in: Lecture Notes in Physics, vol. 703, Springer, Berlin, Heidelberg, 2006, pp. 439–478.
[49] A. Berezhkovskii, G. Hummer, A. Szabo, J. Chem. Phys. 130 (2009) 205102.
[50] A.M. Berezhkovskii, A. Szabo, J. Phys. Chem. B 117 (2013) 13115–13119.
[51] S.V. Krivov, J. Chem. Theory Comput. 9 (2013) 135–146.
[52] G.T. Beckham, B. Peters, J. Phys. Chem. Lett. 2 (2011) 1133–1138.
[53] S. Jungblut, A. Singraber, C. Dellago, Mol. Phys. 111 (2013) 3527–3533.
[54] Y. Bi, A. Porras, T. Li, J. Chem. Phys. 145 (2016) 211909.

[55] B. Peters, Mol. Simul. 36 (2010) 1265–1281.
[56] F. Noe, C. Schutte, E. Vanden-Eijnden, L. Reich, T.R. Weikl, Proc. Natl. Acad. Sci. USA 106 (2009) 19011–19016.
[57] J.W. Gibbs, Trans. Connect. Acad. Sci. 3 (1876) 108–248.
[58] J.W. Gibbs, Trans. Connect. Acad. Sci. 16 (1878) 343–524.
[59] E. Wigner, H. Eyring, Sci. Mon. 44 (1937) 564–567.
[60] E. Wigner, Trans. Faraday Soc. 34 (1938) 29–41.
[61] R.A. Marcus, J. Chem. Phys. 24 (1956) 966.
[62] H. Eyring, Chem. Rev. 17 (1935) 65–77.
[63] W.F.K. Wynne-Jones, H. Eyring, J. Chem. Phys. 3 (1935) 492.
[64] M. Volmer, A. Weber, Z. Phys. Chem. 119 (1926) 277–301.
[65] Y.B. Zeldovich, Acta Physicochim. URSS 18 (1943) 1–22.
[66] R.A. Marcus, N. Sutin, Biochim. Biophys. Acta 811 (1985) 265–322.
[67] R.A. Marcus, Phys. Chem. Chem. Phys. 14 (2012) 13729–13730.
[68] K.J. Laidler, A. Landskroener, Trans. Faraday Soc. 52 (1954) 200–210.
[69] G.G. Poon, B. Peters, J. Phys. Chem. B 120 (2016) 1679–1684.
[70] E. Wigner, J. Chem. Phys. 5 (1937) 720–725.
[71] J.C. Keck, J. Chem. Phys. 32 (1960) 1035–1050.
[72] B.C. Garrett, D.G. Truhlar, in: C. Dykstra (Ed.), Theory and Applications of Computational Chemistry: The First Forty Years, Elsevier, Amsterdam, 2005, pp. 84–87.
[73] L.S. Pontryagin, A.A. Andronov, A.A. Vitt, Zh. Eksp. Teor. Fiz. 3 (1933) 165–180.
[74] J.S. Langer, Ann. Phys. 54 (1969) 258–275.
[75] B. Peters, P.G. Bolhuis, R.G. Mullen, J.-E. Shea, J. Chem. Phys. 138 (2013) 54106.
[76] N.-V. Buchete, G. Hummer, J. Phys. Chem. B 112 (2008) 6057.
[77] N.G. van Kampen, J. Stat. Phys. 17 (1977) 71–88.
[78] J.-H. Prinz, H. Wu, M. Sarich, B. Keller, M. Senne, M. Held, J.D. Chodera, C. Schutte, F. Noe, J. Chem. Phys. 134 (2011) 174105.
[79] J.S. Langer, in: Systems far from Equilibrium, Springer, Berlin, Heidelberg, 1980, pp. 12–47.
[80] X. Huang, G.R. Bowman, S. Bacallado, V.S. Pande, Proc. Natl. Acad. Sci. USA 106 (2009) 19765–19769.
[81] W. Zheng, M.A. Rohrdanz, C. Clementi, J. Phys. Chem. B 117 (2013) 12769–12776.
[82] B. Peters, J. Chem. Phys. 125 (2006) 241101.
[83] G.M. Torrie, J.P. Valleau, J. Comput. Phys. 23 (1977) 187–199.
[84] B. Peters, Chem. Phys. Lett. 494 (2010) 100–103.
[85] D. Moroni, P.R. ten Wolde, P.G. Bolhuis, Phys. Rev. Lett. 235703 (2005) 1–4.
[86] A. Waghe, J.C. Rasaiah, G. Hummer, J. Chem. Phys. 117 (2002) 10789.
[87] B.C. Knott, M. Haddad Momeni, M.F. Crowley, L.F. Mackenzie, A.W. Götz, M. Sandgren, S.G. Withers, J. Ståhlberg, G.T. Beckham, J. Am. Chem. Soc. 136 (2014) 321–329.
[88] S.L. Quaytman, S.D. Schwartz, Proc. Natl. Acad. Sci. USA 104 (2007) 12253–12258.
[89] B. Peters, J. Phys. Chem. B 115 (2011) 12671.
[90] B. Peters, G.T. Beckham, B.L. Trout, J. Chem. Phys. 127 (2007) 34109.
[91] E.E. Borrero, F.A. Escobedo, J. Chem. Phys. 127 (2007) 164101.
[92] C. Velez-Vega, E.E. Borrero, F.A. Escobedo, J. Chem. Phys. 133 (2010) 105103.
[93] P.H. Garthwaite, I.T. Jolliffe, B. Jones, Statistical Inference, Oxford, Oxford, UK, 2002.
[94] D. Husmeier, in: D. Husmeier, R. Dybowski, S. Roberts (Eds.), Probabilistic Modeling in Bioinformatics and Medical Informatics, Springer, London, 2004, pp. 17–58.
[95] B. Peters, B.L. Trout, J. Chem. Phys. 125 (2006) 54108.
[96] R.G. Mullen, J.-E. Shea, B. Peters, J. Chem. Theory Comput. 11 (2015) 2421–2428.
[97] T.S. van Erp, Phys. Rev. Lett. 98 (2007) 268301.
[98] P.G. Bolhuis, J. Chem. Phys. 129 (2008) 114108.

[99] J. Rogal, P.G. Bolhuis, J. Chem. Phys. 129 (2008) 224107.
[100] W.-N. Du, K.A. Marino, P.G. Bolhuis, J. Chem. Phys. 135 (2011) 145102.
[101] R.F. Grote, J.T. Hynes, J. Chem. Phys. 73 (1980) 2715–2732.
[102] B. Pan, M.S. Ricci, B.L. Trout, J. Phys. Chem. B 114 (2010) 4389–4399.
[103] R.B. Best, G. Hummer, Proc. Natl. Acad. Sci. USA 102 (2005) 6732–6737.
[104] K. Muller, L.D. Brown, Theor. Chim. Acta 53 (1979) 75–93.
[105] S. Paul, S. Taraphder, J. Phys. Chem. B 119 (2015) 11403–11415.
[106] W. Lechner, C. Dellago, P.G. Bolhuis, J. Chem. Phys. 135 (2011) 1–14.
[107] J. Vreede, J. Juraszek, P.G. Bolhuis, Proc. Natl. Acad. Sci. USA 107 (2010) 2397–2402.
[108] J. Juraszek, J. Vreede, P.G. Bolhuis, Chem. Phys. 396 (2012) 30–44.
[109] M. Schor, J. Vreede, P.G. Bolhuis, Biophys. J. 103 (2012) 1296–1304.
[110] Y. von Hansen, F. Sedlmeier, M. Hinczewski, R.R. Netz, Phys. Rev. E 84 (2011) 51501.
[111] L. Xi, M. Shah, B.L. Trout, J. Phys. Chem. B 117 (2013) 3634–3647.
[112] C. Leitold, C. Dellago, J. Chem. Phys. 141 (2014) 134901.
[113] W. Li, A. Ma, Mol. Simul. 40 (2014) 1–10.
[114] W. Lechner, J. Rogal, J. Juraszek, B. Ensing, P.G. Bolhuis, J. Chem. Phys. 133 (2010) 174110.
[115] P.G. Bolhuis, W. Lechner, J. Stat. Phys. 145 (2011) 841–859.
[116] R. Tibshirani, J. R. Stat. Soc., Ser. B (1996) 267–288.
[117] S. Foucart, H. Rauhut, A Mathematical Introduction to Compressive Sensing, Springer, Berlin, Heidelberg, 2013.
[118] J. Rogal, W. Lechner, J. Juraszek, B. Ensing, P.G. Bolhuis, J. Chem. Phys. 133 (2010) 174109.
[119] D.E. Makarov, in: A.F. Oberhauser (Ed.), Single Molecule Studies of Proteins, Springer, New York, 2013, pp. 235–268.
[120] H.A. Kramers, Physica A 7 (1940) 284–304.
[121] B. Peters, Chem. Phys. Lett. 554 (2012) 248–253.
[122] R. Fletcher, Practical Methods of Optimization, Wiley, Chichester, 1987.
[123] B. Peters, Annu. Rev. Phys. Chem. 67 (2016) 669–690.
[124] E. Vanden-Eijnden, F.A. Tal, J. Chem. Phys. 123 (2005) 184103.
[125] B. Peters, N.E.R. Zimmermann, G.T. Beckham, J.W. Tester, B.L. Trout, J. Am. Chem. Soc. 130 (2008) 17342–17350.

CHAPTER 21

Nonadiabatic reactions

I realized that what troubled me in this picture for reactions occurring in the dark was that energy was not conserved.

Marcus, Rev. Mod. Phys. (1993)

Thus far we have focused on adiabatic reactions for which the dynamics are confined to a single ground state potential energy surface (PES). *Nonadiabatic* reactions include a range of phenomena where electronic energy levels cross or approach one another closely enough that trajectories evolving on one surface can jump to the other [1]. Closely spaced energy levels and nonadiabatic effects are common, e.g. any reaction involving a metallic surface couples to closely spaced energy levels in the metal. This chapter focuses on thermally activated nonadiabatic reactions involving transitions between two specific diabatic states, e.g. solution phase electron transfer reactions and spin-crossing reactions. Nonadiabatic models can also be applied to many reactions beyond those which involve different electronic states. This chapter concludes with a discussion of proton transfer reactions in enzymes where a nonadiabatic theory often provides a more natural starting point than an adiabatic model.

21.1 Diabatic and adiabatic representations

The terms adiabatic and nonadiabatic refer to two alternative models for dynamical systems that exhibit occasional transitions between nearly independent substates. An adiabatic model explicitly models the dynamics throughout the transition region in which the two subsystems interact with each other. A nonadiabatic model is constructed in parts: there are two models for the two substates, and a separate model that governs the state-to-state transitions between them. We have seen nonadiabatic models before, e.g. see the discussion of empirical valence bond potentials in Figure 7.6.1.

The nearly independent substates in a nonadiabatic model are called diabatic states. Typically, a diabatic state corresponds to an electronic configuration that remains fixed even when the nuclei move to positions where the electronic configuration is no longer stable. Specifically, the wavefunction should retain the same symmetry, bonding, and antibonding orbitals even as the geometry of the molecule is distorted [2]. For example, in a nonadiabatic spin-crossing model the diabatic states would correspond to low and high electron spin configurations. In a nonadiabatic model of electron transfer, the diabatic states would correspond to $|\mathbf{DA}\rangle$ and $|\mathbf{D}^{+}\mathbf{A}^{-}\rangle$ electronic states where **D** and **A** represent the donor and acceptor species. Nonadiabatic models of electronic transitions are natural because (i) the actual transition event is effectively instantaneous relative to nuclear motions, (ii) the conditions which allow the transition are infrequently met, and (iii) the conditions which allow the transition are easily phrased in terms of energy gaps.

Suppose that the Hamiltonian can be written in the form

$$H(\mathbf{x}, q) = \mathbf{T}_q + H_x(\mathbf{x}, q) \tag{21.1.1}$$

where q is a reaction coordinate, $\mathbf{x}$ includes other nuclear and electronic degrees of freedom, respectively. $\mathbf{T}_q$ is the kinetic energy operator for the q-variable. All other terms (the kinetic energy for the $\mathbf{x}$-variables and all potential interactions) live within the $H_{\mathbf{x}}$ term. The adiabatic energy surfaces and energy levels would be obtained by solving $H_x(\mathbf{x},q)\Psi_i(\mathbf{x}|q) = E_i(q)\Psi_i(\mathbf{x}|q)$. If q included *all* of the nuclear positions, the adiabatic calculation would be that which gives the Born-Oppenheimer PES. Our goal is (1) to instead find *diabatic* electronic states $\Phi_i(\mathbf{x}|q)$ and (2) to compute the rates and requirements for transitions between the diabatic states.

Consider for example two electronic wavefunctions functions with some fixed characteristics, e.g. one triplet and one singlet, as diabatic states. At any one static configuration, the states may be orthogonal, but they are coupled by their dependence on q. The (first order) coupling between the two diabatic states is

$$d_{ij}(q) = \left\langle \Phi_i | \frac{\partial}{\partial q} \Phi_j \right\rangle \tag{21.1.2}$$

Perfectly diabatic states $\Phi_i(\mathbf{x}|q)$ and $\Phi_j(\mathbf{x}|q)$ would have $d_{ij}(q) = 0$, but these cannot exist [2]. Diabatic states can be *approximated* by several approaches: minimization of the coupling matrix elements [3,4], block diagonalization starting from accurate adiabatic states [5], valence bond [6] and empirical valence bond [7] approaches.

For spin-forbidden reactions, diabatic states are naturally obtained from the Schrodinger equation by doing the usual electronic structure calculations where spin-orbit coupling is omitted [8]. For electron transfer reactions (especially intramolecular electron transfer reactions), the PES for each diabatic state (i.e. the *diabats*) can be obtained from constrained density functional theory (DFT) with excess charges localized to specific molecular fragments [2,9].

By identifying the appropriate donor (**D**) and acceptor (**A**) fragments within a molecule, constrained DFT calculations can be performed for the reactant state $|\mathbf{DA}\rangle$ and the product state $|\mathbf{D}^+\mathbf{A}^-\rangle$ at each point along the reaction coordinate for electron transfer.

Note an interesting contrast between adiabatic and nonadiabatic reaction rate theories. Adiabatic theories require the probability of reaching an artificially defined dividing surface on a well-defined ground state PES. Nonadiabatic theories use somewhat arbitrary definitions for the diabatic states, but the dividing surfaces are unambiguously defined by the diabats. Specifically, the dividing surface will always be the "seam", i.e. the intersection, between the diabatic energy surfaces.

21.2 Spin-forbidden reactions

Spin-forbidden reactions have reactants and products of different spin, so they require a symmetry forbidden crossing from one energy surface to the other. For molecules containing only light atoms, spin-crossing reactions are effectively impossible. In contrast, transition metal compounds often have several low lying excited states with different configurations of unpaired electrons. Numerous excited states create more opportunities for spin-crossings and the high spin-orbit coupling for transition metals [10] makes a crossing more likely at each opportunity.

The rate of a spin crossing reaction depends on both the spin-orbit coupling and the probability that the system will visit the "seam" between the spin states. The seam is an intersection between the diabatic spin states. As depicted in Figure 21.2.1, a seam has the same dimensionality as a dividing surface ($3N - 1$ where N is the number of atoms). For the (adiabatic) transition state theory, the saddle point was a particularly important point on the dividing surface. Similarly for the nonadiabatic theory of spin-crossing, the minimum energy crossing point (MECP) is a particularly important point on the seam. The overall rate of a spin-crossing reaction also depends on the position of the MECP relative to other features of the energy landscape(s).

Seams are often depicted as maxima along the reaction pathway, but they can also be located near the reactant geometry on the two surfaces, near the product geometry, or even outside the region between the reactants and products. As shown in Figure 21.2.1, the routes downhill from an early (or late) MECP on the two spin surfaces may lead to similar species with different spin. Depending on the spin crossing location, conversion from the reactant to the product of another spin may require both a spin-crossing and (adiabatic) passage through a saddle on one of the spin surfaces.

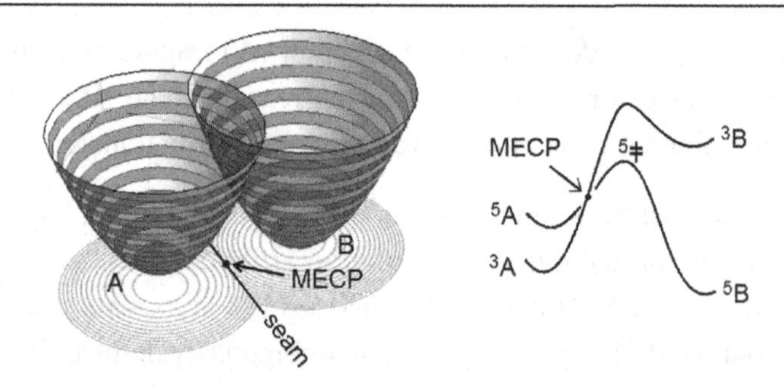

Figure 21.2.1: (Left) Schematic diabats, seam, and MECP in a system with two degrees of freedom. The MECP on the left divides reactants on one spin surface from products on the other spin surface. (Right) Showing a one-dimensional schematic of a spin-crossing en route from a triplet version of species A (^3A) to a quintuplet version of species B (^5B). Because the MECP is essentially part of the (^3A) and (^5A) basins, the spin-crossing does not immediately lead to ^5B. If the transition state on the quintuplet surface is several k_BT above the MECP, the limiting step may still be the conventional rate of crossing ^{5}‡.

When the MECP is a maximum along the path from the reactants to the intended products, the rate of seam crossing determines the rate for the elementary step. Figure 21.2.2 shows the quantities that will be important in the spin-crossing rate calculation.

Figure 21.2.2: Diabatic surfaces for a spin-forbidden reaction are obtained from typical electronic structure calculations without spin-orbit coupling. V_{12} is the spin-orbit coupling strength. Note that the MECP lies just above the saddle point that would emerge on the adiabatic ground state surface if spin-orbit coupling was included. [Adapted from Harvey *Phys. Chem. Chem. Phys.* 9, 331-43 (2007).]

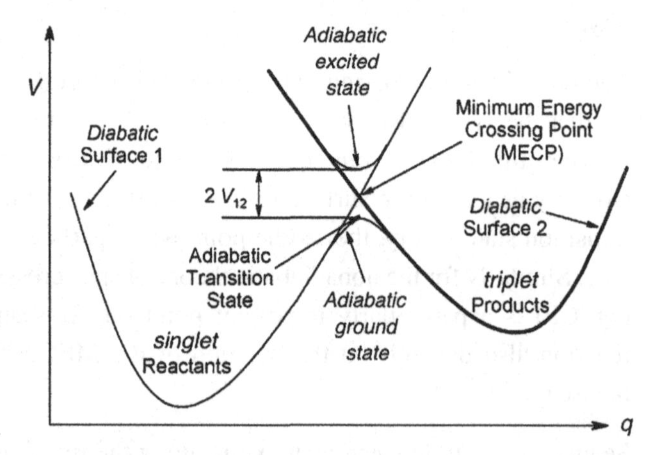

The nonadiabatic rate calculation requires the frequency of reaching the seam as well as the probability of a transition from one surface to the other as a trajectory passes through the seam. Landau [11] and Zener [12] derived the probability of a jump from one *adiabatic* surface to the other. The probability we want is that for a jump from one *diabatic* surface to the other as a trajectory crosses the seam. Thus the probability of a nonadiabatic transition is $P_{NA} = 1 - P_{LZ}$, where P_{LZ} is the Landau-Zener result. Key parameters of the Landau-Zener theory are the absolute velocity $|\dot{q}|$ perpendicular to the seam, the magnitude of the spin-orbit coupling $|V_{12}|$ at the seam, and the difference in gradients on the two diabatic surfaces along the direction perpendicular to the seam, i.e. $\partial E_1/\partial q - \partial E_2/\partial q$. The non-adiabatic transition probability for a single crossing of the seam is

$$P_{NA} = 1 - exp\left[-\theta\left(|\dot{q}|\right)\right] \tag{21.2.1}$$

where the Massey parameter $\theta(|\dot{q}|)$ is

$$\theta = \frac{2\pi\, |V_{12}|^2}{\hbar\, |\dot{q}\partial(E_2 - E_1)/\partial q|} \tag{21.2.2}$$

Equation (21.2.1) is based on several simplifying assumptions: constant velocity through the intersection, constant slopes of the two surfaces, and constant coupling. Nevertheless, it provides a semi-quantitative estimate for the nonadiabatic transition probability. The quantities in the Massey parameter reveal that the transition is more likely for trajectories that pass through the seam slowly, for large spin orbit coupling, and for surfaces that intersect with nearly tangent slopes. Slow crossing velocity and nearly tangent surfaces preserve a situation where the energies of the two spin states are nearly the same for a long time interval along a seam crossing trajectory. The prolonged energy degeneracy creates a large window of opportunity for the electronic degrees of freedom to make an energy-conserving transition. Note that modest changes in the Massey parameter are exponentially magnified to give potentially enormous changes in the non-adiabatic crossing probability. This is why spin-forbidden reactions can occur when transition metals or nearly parallel intersections are involved [13].

At this point, the usual derivation of the overall seam crossing rate assumes that a seam-crossing excursion from the reactant state will involve two correlated seam crossings. The probability of transition on the first pass through the seam is P_{NA}. The probability of hopping on the second pass is $(1 - P_{NA})P_{NA}$, i.e. the probability of no transition on the first pass multiplied by the probability for the subsequent pass en route back to the minimum in state 1 [8]. The overall probability is then assumed to be $P_{NA} + (1 - P_{NA})P_{NA}$ or equivalently, $P_{NA}(2 - P_{NA})$. Key assumptions in this calculation are that the two crossings occur at locations with the same $|\partial E_2/\partial q - \partial E_1/\partial q|$ and at the same absolute velocities along q. It is possible to avoid those assumptions as shown in the derivation below.

The frequency of crossing from state 1 to state 2 at the intersection is

$$\begin{aligned} k_{NA} &= \left\langle \delta[q - q_\times]|\dot{q}|P_{NA}(|\dot{q}|)\right\rangle_1 \\ &= \frac{\int dq \int d\dot{q}\, e^{-\beta(F_1(q)+m\dot{q}^2/2)}\, \delta[q - q_\times]\, |\dot{q}|\, P_{NA}(|\dot{q}|)}{\int dq \int d\dot{q}\, e^{-\beta(F_1(q)+m\dot{q}^2/2)}} \end{aligned} \tag{21.2.3}$$

where $\times$ indicates the crossing point and the subscript 1 indicates that the average is over diabatic state 1. Equation (21.2.3) accurately quantifies the probability of reaching any point along the seam at $q(\mathbf{x}) = q_\times$, but the local force difference $|\partial E_2/\partial q - \partial E_1/\partial q|$ may vary from one point on the seam to another. It is not clear how important the resulting variations in the Massey parameter are, but they are usually ignored. Apart from this approximation, equation (21.2.3) accounts for (1) the probability of reaching the seam, (2) the absolute velocity of the trajectory passing through the seam, and (3) the probability of a transition from one surface to the other. In this formula, transitions can occur from trajectories crossing in either direction.

For sufficiently weak coupling $|V_{12}|$, the Massey parameter will be small for all typical velocities at the seam. To first order in θ, the curve crossing probability for a single pass through the seam is

$$P_{NA}(|\dot{q}|) \approx \theta(|\dot{q}|) \tag{21.2.4}$$

Also assume that the free energy profile is locally parabolic near the reactant state, i.e. near the minimum in diabat 1:

$$F_1(q) \approx m_1\omega_1^2(q - q_{10})^2/2 \tag{21.2.5}$$

Inserting approximations (21.2.4) and (21.2.5) into equation (21.2.3) and then completing the integrals gives

$$k_{NA} = \frac{|V_{12}|^2\sqrt{2\pi m\omega_1^2/k_BT}}{\hbar|\partial E_2/\partial q - \partial E_1/\partial q|} exp\left[-\frac{F_1(q_\times)}{k_BT}\right] \tag{21.2.6}$$

The rate now depends exponentially on $F_1(q_\times)$, the free energy at the intersection $q_\times$ relative to the free energy minimum in the reactant state. Note that the harmonic approximation was only used at the reactant minimum. The free energy at the seam can be separately computed, e.g. with standard partition functions in which the mutual gradient direction has been removed from the vibrational analysis at the MECP.

Formally, computing $F_1(q_\times)$ requires a reaction coordinate q for which $q = q_\times$ coincides with the seam. For intersecting surfaces in high dimensions, one can immediately see that q must be an energy gap and that $q_\times = 0$. The seam is the set of all configurations for which $q = E_2(\mathbf{x}) - E_1(\mathbf{x}) = 0$. The energy gap was extensively used in Marcus' theoretical work and also in the computational work by Warshel – the energy gap coordinate is thus partially responsible for *two* Nobel prizes!

Example: Hemoglobin and triplet O_2

Oxygen binding to hemoglobin involves a triplet to singlet transition. Free oxygen is a triplet, and therefore should be chemically inert and inaccessible for metabolism. To chemisorb oxygen in the singlet state, a spin-state transition must occur. Jensen and Ryde

[14] showed that the hemoglobin has triplet and singlet surfaces with nearly equal energy over a broad spatial region of the entrance channel. They estimate that this feature of the diabatic energy landscapes enhances the rate at which triplet oxygen binds to heme by 10^{11}-fold relative to similar non-enzymatic FeO complexes [14]. Thus, our metabolism is a sort of controlled fire - we make just enough hemoglobin to keep it burning at the optimal rate.

The spin-orbit coupling in transition metal complexes is often between 10 and $100cm^{-1}$ [10]. Nonadiabatic spin-forbidden reactions are most important for 3d and 4d transition metals [13]. For 4f, 5d, and 5f metals the spin orbit coupling is so large that reactions involving a change in spin can essentially occur adiabatically [13].

Several algorithms have been developed to find the MECP [15–17]. The most straightforward is to minimize $\mathcal{L} = E_1 - \lambda(E_2 - E_1)$ with respect to atom positions and Lagrange multiplier λ [15]. Computational studies of spin-forbidden reactions have mostly focused on gas phase reactions, but there have also been analyses of spin-forbidden reactions in enzymes, homogeneous catalysis, and heterogeneous catalysis. See Harvey and coworkers for excellent reviews on the theory, computational methods, and applications of spin crossing calculations [8,18].

21.3 Electron transfer

Many reactions involve the transfer of electrons rather than atoms. Electron transfer, as opposed to ionization, refers to situations where the initial (donor) and final (acceptor) states of the electron are both bound states. Electron transfer is a key elementary step in photosynthesis [19], metabolism [19,20], electrochemistry [21], corrosion [22], vision [23], and molecular electronics applications [24]. Electron transfer can occur within molecules, between metal centers in enzymes, between solvated ions, from electrodes to molecules or ions, and through nanometers of protein residues in biological electron transfer. Often electron transfer precedes, follows, or occurs in concert with reactions that break and make chemical bonds [25,26]. We point the reader elsewhere for more on these reactions and their associated theories [27,28].

This chapter focuses on the most basic electron transfer processes that occur between ions in solution:

$$\mathbf{D} + \mathbf{A} \rightarrow \mathbf{D}^+ + \mathbf{A}^- \tag{21.3.1}$$

where **D** is the electron donor and **A** is the electron acceptor. The reactant state, i.e. the entire left side of the equation, is denoted $|\mathbf{DA}\rangle$. The product state, the entire right side of the equation, is denoted $|\mathbf{D}^+\mathbf{A}^-\rangle$. Typical ionization potentials for isolated molecules are hundreds of kJ/mol making electron transfer impossible in the gas phase. Electron transfer is only possible because polar condensed phase environments stabilize charged species. A typical electron transfer reaction between solvated ions is

$$Fe^{+2} + Ce^{+4} \rightarrow Fe^{+3} + Ce^{+3}$$

In this reaction, the donor (**D**) is an Fe^{+2} ion and the acceptor (**A**) is a Ce^{+4} ion. The reactants are different from the products, which makes this an example of *cross electron transfer*. We will see that the rates of cross electron transfer reactions can be predicted from the rates of *self-exchange electron transfer* reactions, e.g. from

$$Fe^{+2} + Fe^{+3} \rightarrow Fe^{+3} + Fe^{+2} \tag{21.3.2}$$

and

$$Ce^{+3} + Ce^{+4} \rightarrow Ce^{+4} + Ce^{+3} \tag{21.3.3}$$

The relationship between the rates of cross electron transfer and the corresponding self-exchange electron transfer rates is an important mechanistic clue. The rate must have little to do with specific interactions between the donor and acceptor ions. Before diving into the theory, let us consider some complicated aspects of the electron transfer process:

- The transferring electron lacks sufficient energy to escape either the donor or the acceptor state, so the electron must tunnel through the medium between the ions [27].
- The electron transfer event changes the electrostatic potential, perturbs all molecular orbitals, and alters the solvent interactions at the donor and acceptor site [27,29,30].
- It is not sufficient to find the rate of electron transfer between ions at a fixed separation distance, i.e. at a fixed electronic coupling. The distance dependent rate enters a diffusion-reaction model as discussed in Chapter 5 to give the overall rate [31,32].

Despite these complications, some highly simplified models of electron transfer have been remarkably successful. As for spin-forbidden reactions, the atom positions and electronic states provide a natural separation between fast and slow variables. The time scale separation will help to construct a simple nonadiabatic theory.

Marcus' parabolas

Electron transfer reactions differ from other chemical reactions in that the primary change is one of electron density. All atoms are essentially frozen during the electron transfer step, but the moments before and after electron transfer are marked by substantial changes in bonding and solvation. Marcus recognized that electron transfer requires the slow nuclear degrees of freedom to preorganize such that the energy of the reactant and product states are degenerate. Otherwise, energy would not be conserved at the moment of electron transfer [33]. Figure 21.3.1 shows schematic solvent and ion configurations. The process begins with a fluctuation from equilibrium solvation in the reactant state. The fluctuation creates a preorganized transition state with both ions in a non-equilibrium state of solvation. From the preorganized transition state, the electron transfer process occurs with no change in energy. Finally, the solvent reorganizes around the product ions.

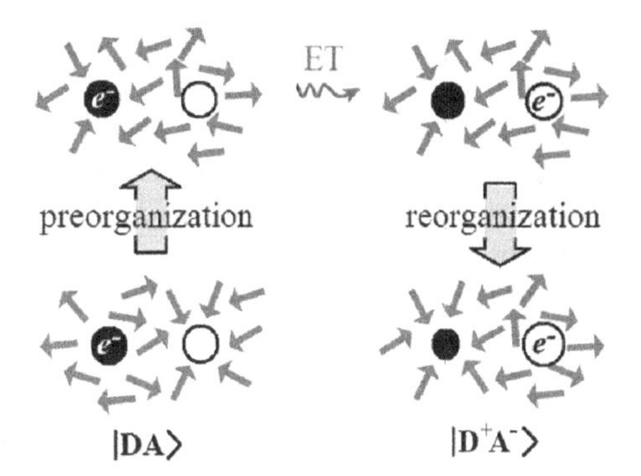

Figure 21.3.1: Preorganization creates a transition state in which the reactant and product electronic states have the same energy because neither the donor nor the acceptor are ideally solvated. After electron transfer occurs, reorganization restores equilibrium solvation for the product ions. [Adapted from Van Voorhis et al. *Ann. Rev. Phys. Chem.* 61, 149-70 (2010).]

Electron transfer and the associated solvent relaxation processes involve many time scales. Some are as fast as the electron transfer event itself, e.g. induced dipole creation. Others are much slower, e.g. the reorientation and motion of solvent molecules into and out of solvation shells. The theory developed by Marcus separates fast processes (electron density changes) from all slower processes (the reorientation and motion of atoms in response to the electron density). His strategy yields a nonadiabatic model based on a continuum description of the solvent and a few basic properties of the ions. The fast electronic degrees of freedom couple to the optical dielectric, i.e. $\epsilon_e = n^2$ where n is the index of refraction. These can be probed spectroscopically. The static dielectric constant ϵ_s includes *all* electronic responses including the slower rotation and translation of solvent molecules. Contributions of fast and slow degrees of freedom to the free energy can be separated by considering solvent polarization as the system moves along a fictitious charge density interpolation path.

Let the electron density of the donor-acceptor system in the reactant state be ρ_0 and let the electron density in the product state be ρ_1. Also define the interpolating electron density

$$\rho_\xi = \rho_0 + \xi \cdot (\rho_1 - \rho_0) \tag{21.3.4}$$

If the density ρ_ξ could be prepared, it would induce an equilibrium nuclear polarization field P_ξ in the solvent. The reversible work required to create the nuclear polarization field *without actually changing the charge distribution* is a diabatic free energy curve, i.e. a diabat. There are two diabats, one starting from the charge distribution in the reactant state and one starting from the product charge distribution. These two free energy diabats are depicted as heavy black curves in Figure 21.3.2. See also Figure 21.3.1 for corresponding schematic molecular configurations.

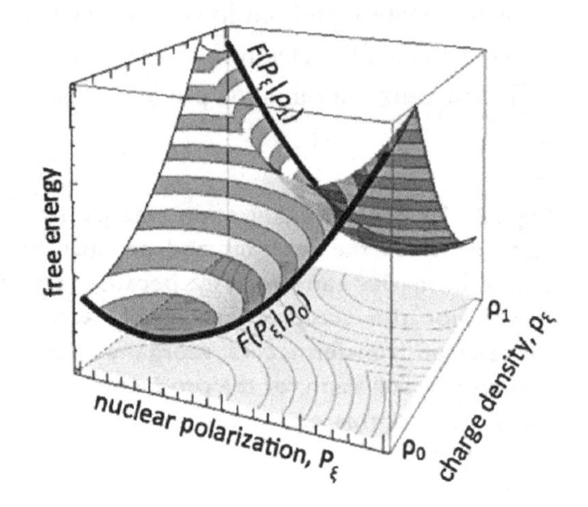

Figure 21.3.2: The diabatic free energies in Marcus theory give the reversible work to create a nuclear polarization P_ξ that would be in equilibrium with an interpolated charge density ρ_ξ. The familiar Marcus curves (usually shown without the charge density axis) are the heavy black curves.

To compute the free energy diabats in the continuum model, define the following three states:

- equilibrium state 0: charge distribution ρ_0 and nuclear polarization P_0
- equilibrium (fictitious) state ξ: charge distribution ρ_ξ and nuclear polarization P_ξ
- non-equilibrium state x: charge distribution ρ_0 and nuclear polarization P_ξ

Let us start with the reversible work to create nonequilibrium state x for the acceptor ion in isolation, $F_{0\to x}$. The overall diabatic free energy curve will then be constructed from similar free energy changes from the donor and the donor-acceptor interactions going from state 0 to state x. For the acceptor ion in isolation, the free energy $F_{0\to x}$ can be computed in two stages:

$$F_{0\to x} = F_{0\to\xi} + F_{\xi\to x} \tag{21.3.5}$$

$F_{0\to\xi}$ is the free energy to transfer a fractional charge while maintaining a state of equilibrium solvation. For an ion of charge z and radius a, in a static dielectric ϵ_S the potential at the surface of the ion is $\Phi = z/\epsilon_S a$. Charging the ion from $z = z_0$ to $z_\xi = z_0 + \xi \cdot (z_1 - z_0)$ changes the free energy by $\int_{z_0}^{z_\xi} \Phi dz$. The result in terms of the reactant (z_0) and product (z_1) ion charges is

$$F_{0\to\xi} = \frac{z_0(z_1 - z_0)\xi}{\epsilon_S a} + \frac{(z_1 - z_0)^2\xi^2}{2\epsilon_S a} \tag{21.3.6}$$

For the $F_{\xi\to x}$ calculation, imagine the ion in state ξ with surface potential $\Phi = z_\xi/\epsilon_S a$. Add a charge δz to the ion but only allow the fast electronic degrees of freedom (with the smaller dielectric ϵ_e) to respond. The surface potential becomes $\Phi(\delta z) = z_\xi/\epsilon_S a + \delta z/\epsilon_e a$. The electronic polarization can be dialed back to that of state 0 by letting $\delta z = z_0 - z_1$. The work done along the way, $\int_0^{z0-z1} \Phi(\delta z) d(\delta z)$, gives the other part of the required free energy

$$F_{\xi \to x} = \frac{z_0(z_0 - z_1)\xi}{\epsilon_S a} + \left(\frac{1}{2\epsilon_e} - \frac{1}{\epsilon_S}\right) \frac{(z_1 - z_0)^2 \xi^2}{a} \tag{21.3.7}$$

Adding the two components of the free energy together gives the quadratic model of Marcus (for a single ion):

$$F_{0 \to x} = \frac{(z_1 - z_0)^2}{2a} \left(\frac{1}{\epsilon_e} - \frac{1}{\epsilon_S}\right) \xi^2 \tag{21.3.8}$$

There is still a need to account for the free energy of polarization at the second ionic center, and for the donor-acceptor interactions during the $0 \to \xi$ and $\xi \to x$ processes. Moreover, adding those two contributions together (both quadratic) only gives the free energy diabat for the reactant state. Two diabatic free energy curves are needed. The second diabat gives the reversible work required to create a similar non-equilibrium polarization around the product ions.

Including all of the pieces above gives the free energy diabats $F_0(\xi) \equiv F(P_\xi|\rho_0)$ and $F_1(\xi) \equiv F(P_\xi|\rho_1)$. The results are

$$F_0(\xi) = \lambda \cdot \xi^2 \tag{21.3.9}$$

and

$$F_1(\xi) = \Delta F + \lambda \cdot (1 - \xi)^2 \tag{21.3.10}$$

where the parameter λ is the reorganization energy:

$$\lambda = \left(\frac{1}{\epsilon_e} - \frac{1}{\epsilon_S}\right)\left(\frac{1}{2R_{\mathbf{A}}} + \frac{1}{2R_{\mathbf{D}}} - \frac{1}{R_{\mathbf{AD}}}\right) \Delta z^2 \tag{21.3.11}$$

Here Δz is the charge transferred, ϵ_e is the dielectric constant at optical frequencies, and $R_{\mathbf{A}}$, $R_{\mathbf{D}}$ are radii for the acceptor and donor, respectively. $R_{\mathbf{AD}}$ is the distance between the donor and acceptor centers.

The reference free energy for both equations (21.3.9) and (21.3.10) is the equilibrium free energy of the reactants. Recall that $F_0(\xi)$ is the free energy to create the equilibrium nuclear polarization corresponding to a degree ξ of charge transfer when the charges have not actually transferred. $F_1(\xi)$ is the free energy to create the analogous non-equilibrium polarization state starting from the product charge distribution.

The transition state and free energy barrier for thermally activated electron transfer can be computed by solving $F_0(\xi) = F_1(\xi)$. The transition state location is

$$\xi_{\ddagger} = (\lambda - \Delta F)/(2\lambda) \tag{21.3.12}$$

Inserting $\xi_{\ddagger}$ into $F_0(\xi)$ gives the free energy barrier

$$\Delta F_{\ddagger} = \frac{(\Delta F + \lambda)^2}{4\lambda} \tag{21.3.13}$$

Several simplifying assumptions in the free energy model limit the accuracy of Marcus theory. First, it does not account for changes in inner sphere coordination in the reorganization energy. Inner-sphere contributions are important when there are significant changes in the distance between the ions and strongly bound ligands or first-shell solvent molecules. The free energy contributions that we have considered are only for "outer sphere" degrees of freedom, e.g. for reorientation of solvent molecules beyond the first solvation shell. More complex models [27,28] that account for both inner and outer sphere reorganization are beyond the scope of this book.

Even for outer sphere electron transfer, the simple model for the free energy diabats makes some problematic assumptions. By using spherically symmetric formulas for the electrostatics, we have implicitly assumed $R_{\mathbf{AD}} \gg R_{\mathbf{A}} + R_{\mathbf{D}}$. Additionally, differences in size between $\mathbf{D}, \mathbf{D}^+$ and $\mathbf{A}, \mathbf{A}^-$ have been ignored to obtain parabolas with the same curvature. Some errors in the model of outer sphere electron transfer can be partially fixed by appealing to experiment. For example, the reorganization energy λ is the sum of ΔF and the optical excitation energy which can be probed experimentally (see Figure 21.3.3) The thermodynamic driving force for electron transfer, ΔF, can also be obtained experimentally from redox potentials.

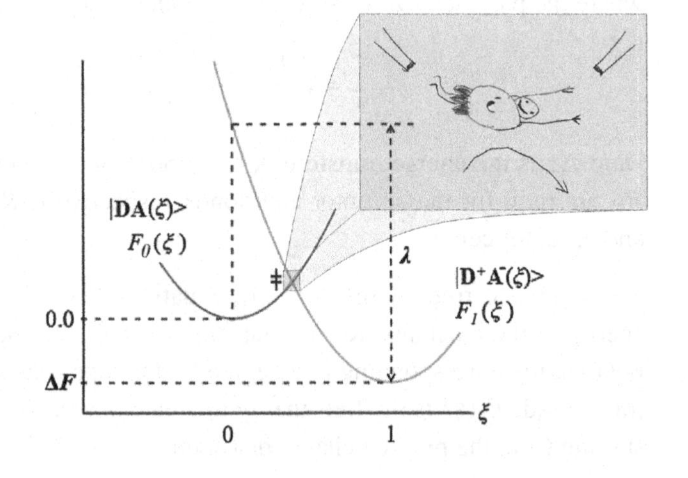

Figure 21.3.3: Diabatic free energy curves from Marcus theory. If the curvatures of the free energy diabats in the reactant and product basins are identical, then the diabats are entirely defined by the reorganization energy λ and the reaction free energy ΔF. Inset: electron as trapeze artist, by Aven Peters, age 5.

Rate calculation

Thus far, the model of electron transfer describes a two-state system in which the only important variables are ξ and the donor-acceptor distance $R_{\mathbf{AD}}$. Computing the electron transfer rate, requires additional information about the electronic coupling between the two diabatic states. A simple two-level Hamiltonian can be constructed as

$$H = \begin{bmatrix} F_0(\xi) & V_{01}(R_{\mathbf{AD}}) \\ V_{10}(R_{\mathbf{AD}}) & F_1(\xi) \end{bmatrix} \tag{21.3.14}$$

Here, the diabatic states are the diagonal elements, and the off-diagonal elements are the electronic coupling terms. The diagonal elements of a proper Hamiltonian would be energies that depend on the full atomistic configuration. In equation (21.3.14), the diagonal terms are the diabatic *free* energy surfaces that were developed in the previous section. They depend only on the variable ξ and not the full nuclear configuration.

Additionally, the off-diagonal coupling terms are assumed to depend only on donor-acceptor distance $R_{\mathbf{AD}}$ and not on the full configuration. Electron transfer rates are observed to decay exponentially with distance between the donor and acceptor centers.

$$|V_{01}|^2 = |V_{01}|_0^2 \, exp[-R/\ell] \tag{21.3.15}$$

where ℓ is a decay constant that depends on the medium between **D** and **A**. The exponential decay of coupling strength with distance is known as Dutton's rule [19]. The theoretical justification comes from the exponential decay of tunneling tail overlaps with donor-acceptor distance [29]. The donor and acceptor states can be envisioned as two potential holes separated by a medium in which the potential of the electron is high. The electron is confined to the holes but its wavefunction has tunneling tails that extend into the medium and overlap with tunneling tails from the neighboring hole state. Electron transfer rates have been measured as a function of distance in various media to ascertain the decay constants. Figure 21.3.4 shows decay rates for various media as measured and compiled by Gray and coworkers [34–37].

A successful electron transfer trajectory must jump from one diabatic surface to the other at $\xi_{\ddagger}$. The nonadiabatic rate including the hopping probability can be obtained from Landau-Zener theory. The derivation closely parallels that for spin-forbidden reaction rates. In the weak coupling limit, the transition probability for a single seam crossing is

$$P_{NA} = 1 - exp[-\theta(|\dot{\xi}|)] \approx \theta(|\dot{\xi}|) \tag{21.3.16}$$

with

$$\theta = \frac{2\pi\, |V_{01}|^2}{\hbar \left|\dot{\xi} d(F_1 - F_0)/d\xi\right|_{\ddagger}} \tag{21.3.17}$$

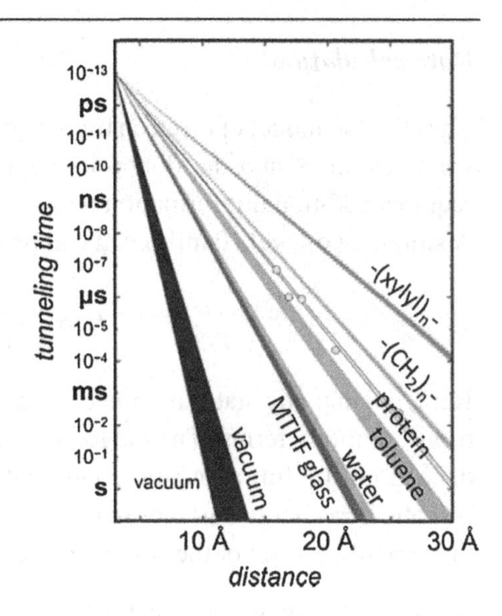

Figure 21.3.4: Decay constants for electron transfer over large distances have been measured by Gray and coworkers. To fix the distance, these studies immobilize the electron transfer center in a protein or in a solid matrix. For alkane bridges, the decay length is approximately $\ell = 1.0$ Angstrom. [Gray, Winkler, *Proc. Nat. Acad. Sci. USA*, 102, 3534-39 (2005)]

Note that we are directly using the free energy surfaces to estimate the slopes at an ensemble of crossing points. This approximation will underestimate the rate if there are strong contributions from rare conformations where the surfaces locally intersect with nearly parallel slopes. The denominator in the Massey parameter can be written in terms of the reorganization energy: $d(F_0 - F_1)/d\xi = 2\lambda$. Therefore in the weak coupling limit, the probability of a transition for a single pass through the transition state is

$$P_{NA} \approx \frac{\pi\,|V_{01}|^2}{\hbar|\dot{\xi}|\lambda}$$

The electron transfer rate is

$$\begin{aligned} k_{ET} &= \left\langle \delta[\xi - \xi_{\ddagger}]|\dot{\xi}|P_{NA}\right\rangle_0 \\ &\approx \frac{\int d\xi \int d\dot{\xi} exp[-\beta(\lambda\xi^2 + m\dot{\xi}^2/2)]\,\delta[\xi - \xi_{\ddagger}]\pi\,|V_{01}|^2/\hbar\lambda}{\int d\xi \int d\dot{\xi} exp[-\beta(\lambda\xi^2 + m\dot{\xi}^2/2)]} \end{aligned} \tag{21.3.18}$$

The subscript 0 indicates that the average is over diabatic state 0. The factor $\delta[\xi - \xi_{\ddagger}]|\dot{\xi}|$ counts the flux of trajectories arriving at the transition state. In contrast to the assumptions of transition state theory, reactive trajectories in this model can arrive at the transition state from either direction. The formulation in equation (21.3.18) avoids the need to estimate the overall transmission probability as a function of the single pass probability [38].

Inserting approximation (21.3.16) into equation (21.3.18) and then completing the integrals gives the famous result by Marcus [28],

$$k_{ET} = \frac{2\pi}{\hbar} \frac{|V_{01}|^2}{\sqrt{4\pi\lambda k_B T}} exp\left[-\frac{(\Delta F + \lambda)^2}{4\lambda k_B T}\right]. \tag{21.3.19}$$

The model is extraordinarily simple with only three parameters λ, ΔF, and $|V_{01}|$. Two of these, λ and ΔF, can be determined in separate experiments. More accurate results can be obtained from more elaborate theories that account for quantized inner sphere vibrations, changes in the ion sizes, better models of the relaxation dynamics, and more precise models of the electrostatic interactions. [27,33,39–42]. But despite its simplifications, equation (21.3.19) already captures the main characteristics of electron transfer between solvated ions. In particular, equation (21.3.19) suggests a way to predict the rates of cross electron transfer from self-exchange rates, and also predicts the remarkable and counterintuitive inverted regime.

Inverted regime

As the driving force increases (i.e. as ΔF becomes more negative), equation (21.3.12) predicts that the transition state shifts to the left along the reaction coordinate ξ. For driving forces $\Delta F > -\lambda$, everything is as expected from the Hammond postulate [43]: larger driving forces lead to earlier transition states and smaller barriers. However, equation (21.3.12) is unusual in predicting an *inverted regime*. For very large driving forces ($\Delta F < -\lambda$), the transition state location actually moves to the left of the reactant minimum and the free energy barrier begins to rise again! Thus Marcus theory predicts a counterintuitive maximum in the rate when $\Delta F = -\lambda$. This was a true prediction because the inverted regime behavior had not been observed.

For the next 25 years, there were just a few hints that the inverted regime was real. As the product diabat shifts downward, so too do excited product states. Electron transfer can occur to these states before crossing the inverted regime barrier, and indeed chemiluminescence was reported for some electron transfer processes at high driving forces. The first strong evidence for the inverted regime came from biphenyl anion (donor) transfer through a rigid hydrocarbon bridge molecule to a series of acceptor molecules [44]. The acceptors were chosen to span a 2eV range of driving forces and, as predicted by Marcus theory, a maximum in the rate was observed for this carefully designed system (see Figure 21.3.5). Also see the photochemical analysis of inverted behavior by Gould and Farid [45].

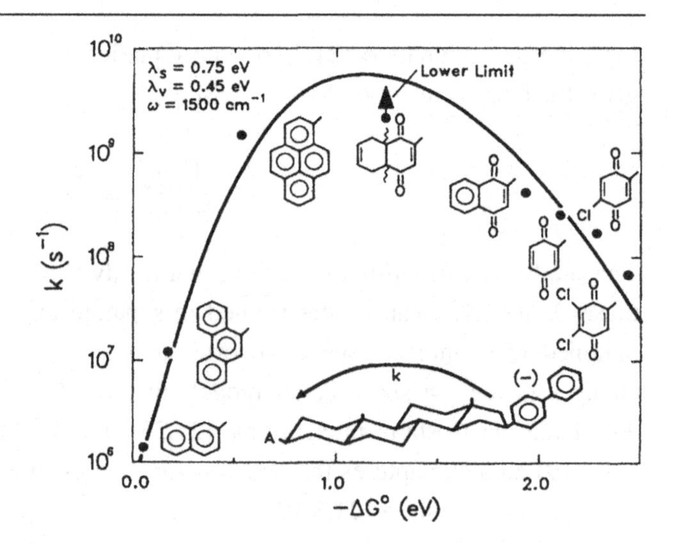

Figure 21.3.5: Marcus theory predicts an inverted regime in which the rate reaches a maximum and then decreases again as a function of electron transfer driving force. The inverted regime was confirmed by Miller et al. for electron transfer to a series of acceptors from one donor through a rigid spacer molecule. [Miller et al., *J. Am. Chem. Soc.* 106, 3047-49 (1984)]

Predictions for cross electron transfer

The diabats in equations (21.3.9) and (21.3.10) are mostly comprised of separate contributions from the solvation of participating ions. According to the continuum model, the donor-acceptor interaction only appears in the $1/R_{\mathrm{DA}}$ term in the reorganization energy. Apart from this one term, the free energy model could be broken into contributions from half reactions: $\mathbf{D} \rightarrow \mathbf{D}^+ + e^-$ and $\mathbf{A} + e^- \rightarrow \mathbf{A}^-$. As an alternative to half-reactions, we could start with the rates and reorganization energies for the self-exchange reactions: $\mathbf{D} + \mathbf{D}^+ \rightarrow \mathbf{D}^+ + \mathbf{D}$ and $\mathbf{A} + \mathbf{A}^- \rightarrow \mathbf{A}^- + \mathbf{A}$. Marcus theory predicts that the reorganization energy for the cross electron transfer reaction is

$$\lambda_{\mathrm{XY}} = \frac{1}{2}(\lambda_{\mathrm{XX}} + \lambda_{\mathrm{YY}}) \tag{21.3.20}$$

where λ_{XX} and λ_{YY} are reorganization energies for the self-exchange electron transfers. It also predicts that

$$k_{\mathrm{XY}} = \sqrt{k_{\mathrm{XX}} k_{\mathrm{YY}} K_{\mathrm{XY}} f_{\mathrm{XY}}} \tag{21.3.21}$$

where k_{XY} is the rate of transfer from the donor form of **X** to the acceptor form of **Y** and where K_{XY} is the equilibrium constant for transfer from the donor form of **X** to the acceptor form of **Y** [46]. The self-exchange definitions k_{XX} and k_{YY} are implicit in the definition of k_{XY}. Details on the function f_{XY} can be found in the review by Marcus and Sutin [46], but in most cases its value is near unity. More accurate (and more complicated) cross relations have been developed, but the simple cross relations above are sufficient to demonstrate another prediction of the Marcus theory. Figure 21.3.6 shows data assembled by Bennett [47] comparing measured and computed cross electron transfer rates for a suite of reactions. Marcus and Sutin discuss deviations from the predicted results in terms of several effects that are not captured in the continuum model [46].

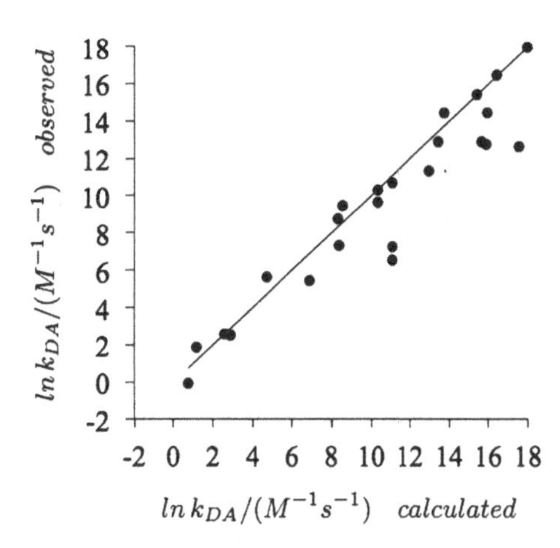

Figure 21.3.6: Cross electron transfer rates can be approximated using the corresponding self-exchange electron transfer rates and equation (21.3.21). **[Data from Bennett, *Prog. Inorg. Chem.* 18, 1 (1973).] Three data points with rates near the diffusion control limit are not shown.**

The cross relations and elements of Marcus theory have also been used to describe proton-coupled electron transfer (PCET) [25] and other chemical reactions [48]. Both proton and electronic transfer components of PCET have non-adiabatic character [26], and the Marcus cross relations have been empirically effective as models of the overall PCET rate.

21.4 Classical MD methods for electron transfer

Somewhat counterintuitively, classical MD simulations and rare events methods can be used to compute free energy barriers for electron transfer. In fact, rare events simulations can test key assumptions in the continuum model of Marcus like the parabolic nature of the diabats and the one common force constant 2λ for forward electron transfer (from the **DA** state) and backward electron transfer (from the $\mathbf{D}^{+}\mathbf{A}^{-}$ state). The enabling realization was that the vertical energy gap is the electron transfer reaction coordinate [49,50], As shown below, the vertical energy gap and its diabatic free energy surfaces have some remarkable properties. The vertical energy gap is

$$\Delta E(\mathbf{x}) = E_{\mathbf{DA}}(\mathbf{x}) - E_{\mathbf{D}^{+}\mathbf{A}^{-}}(\mathbf{x}) \tag{21.4.1}$$

where $E_{\mathbf{DA}}(\mathbf{x})$ and $E_{\mathbf{D}^{+}\mathbf{A}^{-}}(\mathbf{x})$ are all-atom diabatic PESs. These are computed from two different Hamiltonians, one with charges $z_{\mathbf{D}}$ and $z_{\mathbf{A}}$ on the donor and acceptor ions and the other with charges $z_{\mathbf{D}^{+}} = z_{\mathbf{D}} + 1$ and $z_{\mathbf{A}^{-}} = z_{\mathbf{A}} - 1$. The energy gap $\Delta E(\mathbf{x})$ is "vertical" in the sense that the two energies are evaluated at exactly the same nuclear configuration $\mathbf{x}$. The solvent environment is not given a chance to respond to the altered ion charges. The energy gap is easily computed on the fly during an MD simulation [50,51].

The energy gap exactly probes the instantaneous state of solvent polarization around the ions. All solvent-solvent interactions cancel in the energy gap leaving only the change in electrostatic interactions between the solvent and the donor-acceptor pair. For a one-electron transfer from **D** to **A**, the energy gap is

$$\Delta E(\mathbf{x}) = \frac{e^2}{4\pi\epsilon_0}\left\{\frac{(1+z_{\mathbf{D}}-z_{\mathbf{A}})}{R_{\mathbf{AD}}} + \sum_{i\in solvent}\frac{z_i}{R_{\mathbf{A}i}} - \sum_{i\in solvent}\frac{z_i}{R_{\mathbf{D}i}}\right\} \tag{21.4.2}$$

Each term now has a specific physical interpretation. The $1/R_{\mathbf{DA}}$ term is the particle-hole Coulomb contribution. It vanishes for all self-exchange electron transfers and more generally it vanishes when $z_{\mathbf{D}} = z_{\mathbf{A}} - 1$. The last two terms are the ionization potentials of the donor and the electronic affinity of the acceptor, respectively [52].

Transition states for electron transfer are the ensemble of configurations for which $\Delta E(\mathbf{x}) = 0$. The adiabatic free energy as a function of ΔE could be computed using standard rare events methods for computing Landau free energies. The results would look much like the ground state *adiabatic* PES in Figure 21.2.2, but because of an exact result by Bennett [53] it is easier to compute the diabatic free energy curves $F_0(\Delta E)$ and $F_1(\Delta E)$.

Driving force and Zwanzig-Bennett relation

The free energy driving force for electron transfer is formally

$$\beta\Delta F = -ln\frac{\int d\mathbf{x}\, exp[-\beta E_{\mathbf{D}^+\mathbf{A}^-}(\mathbf{x})]}{\int d\mathbf{x}\, exp[-\beta E_{\mathbf{DA}}(\mathbf{x})]} \tag{21.4.3}$$

The driving force ΔF can be written in terms of the energy gap by using thermodynamic perturbation theory [54]. Replacing $E_{\mathbf{D}^+\mathbf{A}^-}(\mathbf{x})$ with $E_{\mathbf{DA}}(\mathbf{x}) + \Delta E(\mathbf{x})$ in equation (21.4.3) gives

$$\beta\Delta F = -\,ln\,\langle exp[-\beta\Delta E(\mathbf{x})]\rangle_{\mathbf{DA}} \tag{21.4.4}$$

where the subscript **DA** indicates an average over the equilibrium distribution according to the reactants, i.e. over the distribution

$$\rho_{DA}(\mathbf{x}) = \frac{exp[-\beta E_{\mathbf{DA}}(\mathbf{x})]}{\int d\mathbf{x}\, exp[-\beta E_{\mathbf{DA}}(\mathbf{x})]} \tag{21.4.5}$$

Equation (21.4.4) is not practical because $\rho_{\mathbf{DA}}(\mathbf{x})$ has vanishingly small support at the locations where $\rho_{\mathbf{DA}}(\mathbf{x})exp[-\beta\Delta E(\mathbf{x})]$ is peaked. In other words, there is little overlap between the reactant and product regions of phase space. Instead, the free energy perturbation must be done in stages. Let the variable ξ interpolate between the reactant and product potential energy surfaces

$$E_{\xi}(\mathbf{x}) = \xi E_{\mathbf{D^+A^-}}(\mathbf{x}) + (1-\xi) E_{\mathbf{DA}}(\mathbf{x}) \tag{21.4.6}$$

The overall free energy change can then be obtained from a series of smaller perturbations

$$\beta \Delta F = -\sum_{i=0}^{1/\Delta\xi} ln \left\langle exp[-\beta(E_{(i+1)\Delta\xi}(\mathbf{x}) - E_{i\Delta\xi}(\mathbf{x}))] \right\rangle_{i\Delta\xi} \tag{21.4.7}$$

where the subscript $i\Delta\xi$ on the average indicates a Boltzmann average using energy $E_{i\Delta\xi}(\mathbf{x})$. The variable ξ in equation (21.4.6) corresponds to the interpolating variable ξ in Marcus' derivation, but it also represents a fractional energy gap between $E_{\mathbf{D^+A^-}}(\mathbf{x})$ and $E_{\mathbf{DA}}(\mathbf{x})$. Alternatively, the two diabats can be generated in terms of the energy gap itself. The diabat corresponding to state **DA** is

$$\beta F_{\mathbf{DA}}(\Delta E) = -ln\, \lambda_T^{-3N} \int d\mathbf{x}\, e^{-\beta E_{\mathbf{DA}}(\mathbf{x})} \delta[\Delta E(\mathbf{x}) - \Delta E] \tag{21.4.8}$$

and the diabat corresponding to state $\mathbf{D^+A^-}$ is

$$\beta F_{\mathbf{D^+A^-}}(\Delta E) = -ln\, \lambda_T^{-3N} \int d\mathbf{x}\, e^{-\beta E_{\mathbf{D^+A^-}}(\mathbf{x})} \delta[\Delta E(\mathbf{x}) - \Delta E] \tag{21.4.9}$$

Techniques like umbrella sampling can generate these diabats by directly using equations (21.4.8) and (21.4.9). However, there is a convenient and conceptually important short cut. Consider the *free energy gap* between the diabatic curves. Use $E_{\mathbf{DA}}(\mathbf{x}) = E_{\mathbf{D^+A^-}}(\mathbf{x}) + \Delta E(\mathbf{x})$ in equation (21.4.8)

$$\begin{aligned} \beta F_{\mathbf{DA}}(\Delta E) &= -ln\, \lambda_T^{-3N} \int d\mathbf{x}\, e^{-\beta(E_{\mathbf{D^+A^-}}(\mathbf{x}) + \Delta E(\mathbf{x}))} \delta[\Delta E(\mathbf{x}) - \Delta E] \\ &= -ln\, \lambda_T^{-3N} e^{-\beta\Delta E} \int d\mathbf{x}\, e^{-\beta E_{\mathbf{D^+A^-}}(\mathbf{x})} \delta[\Delta E(\mathbf{x}) - \Delta E] \end{aligned}$$

The exponential within the average depends on the same function as the argument of the delta function. Thus it can be removed from the average, giving [30,49]

$$F_{\mathbf{DA}}(\Delta E) = F_{\mathbf{D^+A^-}}(\Delta E) + \Delta E \tag{21.4.10}$$

This remarkable result is the Zwanzig-Bennett relation [53,55], and it has several important consequences. For example, if one diabat is a parabola, then the other diabat must also be a parabola with the same curvature [52].

The Zwanzig-Bennett relation is also useful for calculations. Consider the results of an unbiased simulation in the reactant **DA** state. We can compute energy gaps during the reactant simulation to create a histogram of energy gap values in the reactant state. Taking the (negative) logarithm of the histogram provides $F_{\mathbf{DA}}$, within a constant. Because of the Zwanzig-Bennett relation (equation (21.4.10)) the unbiased simulation results also provide a vertically displaced non-equilibrium portion of the other diabat, $F_{\mathbf{D^+A^-}}$. We can do the same procedure for the product state to generate two additional portions of the diabats. Thus from two relatively easy simulations, four sections of the diabatic free energy profiles can be generated. The typical energy gap values in the reactant state will usually be different from the typical energy gap values in the product state, so there will usually be a gap with no data on the ΔE axis. Now we can interpolate between the two sets of unbiased simulation results to reconstruct the pair of diabats over the whole range of ΔE values. The interpolation may seem risky, but theoretical models for the diabats help to ensure accuracy. Blumberger and Sprik [52] have used the Zwanzig-Bennett relation to study electron transfer. It has also been used in recent work to compute polymorph free energy differences [56].

Warshel generalized the Zwanzig-Bennett ideas to simplify the diabatic free energy surface calculations in his seminal paper [49]. To begin, define

$$\beta F_\xi(\Delta E) \equiv -ln\, \lambda_T^{-3N} \int d\mathbf{x}\, e^{-\beta E_\xi(\mathbf{x})} \delta[\Delta E(\mathbf{x}) - \Delta E] \tag{21.4.11}$$

where $E_\xi(\mathbf{x})$ was defined in equation (21.4.6). The free energy diabats of interest are $F_0(\Delta E)$ and $F_1(\Delta E)$. Starting from the definition of $F_0(\Delta E)$:

$$\begin{aligned}\beta F_0(\Delta E) &= -ln\left\{e^{-\beta F_0}\langle\delta[\Delta E(\mathbf{x}) - \Delta E]\rangle_0\right\}\\ &= \beta F_0 - ln\left\{e^{-\beta\Delta F_\xi}\left\langle e^{+\beta(E_\xi - E_0)}\delta[\Delta E(\mathbf{x}) - \Delta E]\right\rangle_\xi\right\}\end{aligned}$$

Because $E_\xi - E_0 = \xi\Delta E$ for all $\mathbf{x}$, $exp[\beta\xi\Delta E]$ factors out of the average. Finally, writing $\langle\delta[\Delta E(\mathbf{x}) - \Delta E]\rangle_\xi$ in terms of free energies yields the useful identity

$$F_0(\Delta E) = -\xi\Delta E + F_\xi(\Delta E) \tag{21.4.12}$$

The $\xi\Delta E$ term steadily tilts the $F_\xi(\Delta E)$ free energy landscape and changes the corresponding probability distribution much like a conventional umbrella potential [57]. Thus this form of the Zwanzig-Bennett relation suggests a family of bias potentials that is linear in ΔE. The equations above outline the route to $F_0(\Delta E)$. Similar analyses yield a recipe for computing $F_1(\Delta E)$. Several reports use these relations to compute free energy profiles for electron transfer [50–52,55,58]. Figure 21.4.1 shows the diabatic free energy profiles that Hwang and Warshel computed in simulations using equations (21.4.11) and (21.4.12).

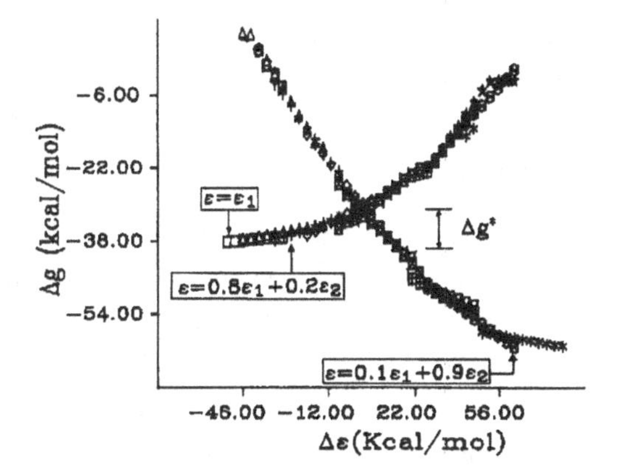

Figure 21.4.1: Hwang and Warshel used the umbrella sampling approach based on equation (21.4.12) to compute diabatic free energy curves for electron transfer between model organic donor and acceptor molecules in solution. [Hwang and Warshel, *J. Am. Chem. Soc.* 109, 715-720 (1987)]

The theories of spin-forbidden reactions and electron transfer processes go far beyond the simple introductions in this chapter. There are a number of potentially important effects that we have not discussed. When do assumptions behind the quadratic free energy surfaces and the Gaussian distribution of energy gaps break down [30,42]? Are continuum models with fast and slow dielectric contributions adequate descriptions of the solvent dynamics [40,41,59]? How can the strong distance dependence of electron transfer be incorporated into stochastic models of the donor-acceptor diffusional encounters [32]? What role do interference effects play in electron transfer over long distances through separate conduits [60]? How can electrode surface chemistry and corrosion be examined with state-of-the-art rare events approaches for electron transfer [22]? Finally, how important are fluctuations in electron transfer bridge conformations and the coupling strength for electron transfer [61–63]. In some cases standard MD simulations have been used to average rates over an ensemble of conformations [64]. Given the extremely strong distance dependence of electron transfer, the ET rate might be dominated by rare strongly coupled conformations that are not seen in a standard MD simulation. Methods to account for these effects have been proposed [65], and it seems like an important direction for the future.

21.5 Nonadiabatic models of enzyme catalysis

Enzyme reactions are far more complex than reactions of small molecules in the gas phase or in solution. The active site of an enzyme is a multifarious solvent with charged residues, non-polar residues, and weakly bound water molecules that are expelled upon substrate binding. The relative importance of the various contributions to the energy landscape for enzyme catalysis has been the focus of many studies and debates [66]. Debates over the importance of dynamical effects in enzymes have been especially vigorous [67–77]. The dynamics of enzymes involve a vast range of different time scales: bond vibrations on the 10–100 fs scale, rotation of side chains on the 10–1000 ps scale, hinge-bending on the ns scale, and helix-coil formation on the

10–100 ns scale [66]. Somewhere in this spectrum of mostly inconsequential motions is the reaction time scale. What does it mean that the reaction time is commensurate with other time scales? Should enzyme reactions be modeled using adiabatic or nonadiabatic theories? How do these choices impact the interpretation of dynamical effects?

Clearly standard harmonic TST as used for gas phase reactions and heterogeneous catalysis is not applicable to enzymes. The transition state ensemble for an enzymatic reaction will include thousands of saddle points that are similar in the active site region, but differ in peripheral regions. We cannot just choose one conformation for the peripheral side chains and apply harmonic transition state theory (TST). For enzymes, a proper TST calculation requires ensemble averaging like the formulation in section 10.5 [78].

Some literature proposals suggest that conformational dynamics couple to the chemical step and accelerate the rate by funneling energy into bond breaking motions [69,79]. Strictly speaking dynamical effects can only *reduce* the reaction rate [76,80]. Moreover, Warshel and coworkers noted that chemical steps (with transition path times of order $100fs$) cannot be dynamically coupled to conformational relaxations with ns and longer time scales [74,81,82].

On the other hand, Karplus noted that conformational dynamics may still be critically important for substrate binding and product release steps [70,72]. Moreover, conformational changes can modulate barriers and tunneling probabilities for the chemical reaction steps. As Warshel noted, the chemical step takes just femtoseconds to traverse the Angstrom-scale width of the barrier top, and on that time scale conformational modes are effectively frozen. However, it may take seconds for the chemical step coordinate to become activated. Thus the chemical step coordinate may be in one sense slower and in another sense faster than the conformational coordinate.

The coupling between processes with disparate time scales can be understood in terms of conformational gating. Figure 21.5.1 depicts two free energy landscapes for conformational gating mechanisms. In both cases, the conformational change dramatically lowers the barrier along the chemical step coordinate. The cases differ in the strength of the coupling between the conformational degrees of freedom and the chemical steps [83,84]. For weak coupling, motion along the conformational and chemical step directions are largely independent. For strong coupling, the conformational fluctuation drives motion along the chemical step toward its transition state. These two cases are naturally described using non-adiabatic and adiabatic models, as shown in Figure 21.5.1. Adiabatic and non-adiabatic models are interconvertible, so both models can, in principle, address both situations. However, choosing the "right" picture affords advantages in both efficiency and mechanistic insight.

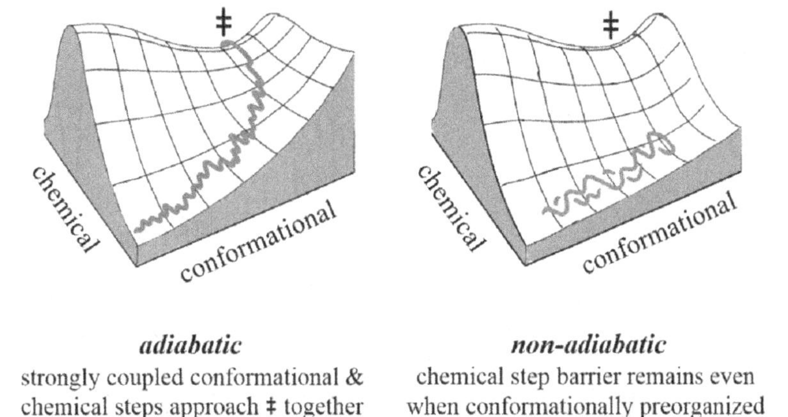

‡
chemical
conformational
‡
chemical
conformational
adiabatic
strongly coupled conformational &
chemical steps approach ‡ together
non-adiabatic
chemical step barrier remains even
when conformationally preorganized

Figure 21.5.1: A conformationally gated enzyme reaction has an insurmountable barrier except when the enzyme adopts certain favorable conformations. When conformational modes are strongly coupled to chemical step, then the conformation and chemical step coordinates cross the transition state in a concerted (adiabatic) manner. When conformational modes are weakly coupled to the chemical step, reactions occur in two distinct (nonadiabatic) stages: a preorganizing conformational fluctuation followed by a chemical step.

In the non-adiabatic picture, the conformational degrees of freedom that gate the chemical reaction enter the rate through a probability to find the system in the conformationally pre-organized state. In the adiabatic picture, changes in the conformational degrees of freedom are part of the optimal reaction coordinate and dividing surface. In both models, the conformational changes that occur *en route* to the transition state contribute to the overall activation free energy.

If conformational gating reduces (without entirely eliminating) the chemical step barrier then nonadiabatic theories provide a natural description. In the nonadiabatic picture, the reaction coordinate is the energy gap between the two diabatic states, e.g. the reactant and product diabats from an empirical valence bond (EVB) model [66,85]. The transition state, instead of a maximum on the adiabatic free energy profile, is the ensemble of configurations where the diabats cross ($\Delta E = 0$) [7,86–89]. Each conformational fluctuation that drives the system through the crossing point creates an opportunity for a transition from one diabatic free energy surface to the other. Figure 21.5.2 depicts the conformational variable (the EVB energy gap) and the chemical step portions of a nonadiabatic enzyme reaction model. In the figure, note how the energy gap modulates the energy landscape along the pure chemical step coordinate.

For certain enzyme reactions, the questions about dynamics can be given a definite answer. For proton transfer steps, dynamical effects are relatively unimportant with tunneling being the more important factor [90]. Garcia-Viloca et al. [80] wrote "The lowering of the activation free energy has been found to accelerate the reaction rate by more than a factor of 10^{11}, whereas

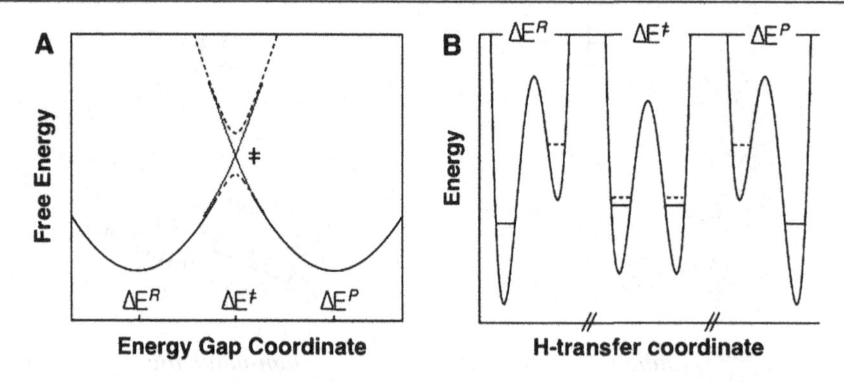

Figure 21.5.2: Nonadiabatic models of enzyme reactions use two free energy surfaces (diabats). The reactant free energy diabat (minimum at ΔE^R) is computed with the proton on the donor, i.e. with the H-transfer chemical step coordinate on the left in figure B. The product free energy diabat (minimum at ΔE^P) is computed with the proton on the acceptor. The reactant and product diabats cross at the non-adiabatic transition state where the energy gap, i.e. the difference in energy along the H-transfer coordinate is zero. The barrier region along the chemical step coordinate makes negligible contributions to the diabats, but it influences the probability of crossing from one diabat to the other at $\Delta E^{\ddagger}$. [Adapted from Benkovic and Hammes-Schiffer, *Science*, 301, 1196–1202 (2003).]

[other effects] contribute no more than a factor of 10^3 to the rate." The transmission coefficient correction to transition state theory is much closer to unity in the vast majority of cases. When the transmission coefficient is as large as 10^3, one can be sure that tunneling is responsible. When the transmission coefficient is as small as 10^{-3}, it is likely that (1) the reaction coordinate has been chosen poorly, or (2) the limiting step involves conformational dynamics and not bond-breaking chemistry. For enzymatic reactions other than proton transfers, there is less evidence, but it again seems unlikely that dynamical effects can exceed a distant secondary importance.

If the conformational fluctuations that modulate the barrier are fast and if the reaction remains slow even from a gated conformation (the ideal non-adiabatic situation), then the considerations above yield a single average rate constant for all enzymes. However, if conformational fluctuations that modulate the barrier are slower than the chemical reaction in the gated conformation, then the enzyme can exhibit *dynamical disorder* [91]. Effectively, the enzyme will have "on" and "off" states resulting from the slow conformational fluctuations. While the enzyme is on, the catalytic cycle may turn over many times before conformational motion turns the enzyme off again. See work by Flomenbom et al. [92] for single molecule experiments and phenomenological models of dynamical disorder in lipase. The discrete on-off case can be generalized to a spectrum of activities with a slow degree of freedom that controls the rate constant.

Exercises

1. For a spin-forbidden reaction, the free energy of the seam can be estimated from a partition function in which the direction of the gradient(s) at the MECP is removed from the Hessian. Outline exactly how this calculation works. Describe the role of masses, zero point energies, projector operations, etc.
2. Explain how the spin magnetic moment and the orbital angular momentum are coupled via, from the electron's point of view, a magnetic field associated with an orbiting proton.
3. The spin orbit coupling for a hydrogenic atom is proportional to $Z^4/(n^3\ell(\ell+\frac{1}{2})(\ell+1))$ where Z is the atomic number, n is the principle quantum number, and ℓ is the orbital angular momentum. Show how this quantity varies for the last valence electrons to fill throughout the periodic table. Plot your results with Z on the horizontal axis and $Z^4/(n^3\ell(\ell+\frac{1}{2})(\ell+1))$ on the vertical axis. How would your plot change if you also included common oxidation states for the transition metals?
4. Use the WKB approximation for tunneling through a square top barrier (below) to justify Dutton's rule for the distance dependence of electron transfer. Give an (approximate) expression for the decay length ℓ in terms of the electron mass and the ionization potential of the donor in the medium between **D** and **A**. Why do electron transfer rates decay most rapidly with donor-acceptor distance in vacuum?

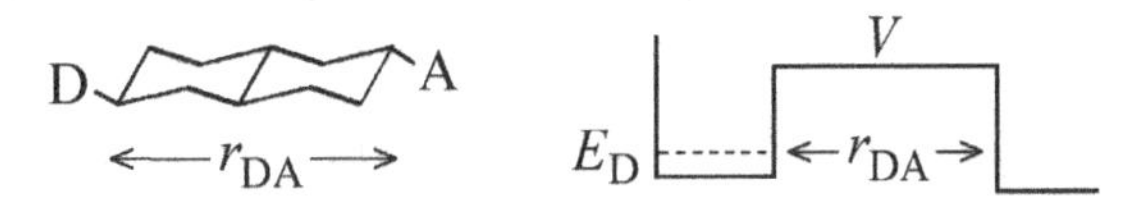

5. The donor charging and solvation contributions to Marcus' free energy diabats were computed in the text. Complete the calculation for the acceptor charging/solvation and donor-acceptor interaction terms to obtain Marcus' expression for the reorganization energy.
6. [Adapted from M. Pilkington] Explain the kinetic trend for the self-exchange electron transfer reactions below:

D, bondlength	A, bondlength	k_{ET}
$[Co(NH_3)_6]^{2+}$ $\ell_{Co-N}=2.11$Å	$[Co(NH_3)_6]^{3+}$ $\ell_{Co-N}=1.94$Å	$10^{-6}M^{-1}s^{-1}$
$[Ru(H_2O)_6]^{2+}$ $\ell_{Ru-O}=2.12$Å	$[Ru(H_2O)_6]^{3+}$ $\ell_{Ru-O}=2.03$Å	$44M^{-1}s^{-1}$
$[Fe(CN)_6]^{4-}$ $\ell_{Fe-C}=1.95$Å	$[Fe(CN)_6]^{3-}$ $\ell_{Fe-C}=1.92$Å	$700M^{-1}s^{-1}$

7. Derive equations (21.3.20) and (21.3.21) for cross electron transfer.

8. Holstein developed a model for electron transfer between H_2 and H_2^+ that (i) includes a quantum description of the bond length motions, and (ii) provides a pedagogical example in which essentially every part of the calculation can be done from first principles. Read the fully quantum mechanical development in Schatz and Ratner [93]. Afterwards, incorporate the derived coupling parameter in a model with purely classical nuclei (like the ones in this chapter) and compare your results.
9. Several predictions from Marcus theory have been used beyond electron transfer to predict the rates of chemical reactions [94]. Consider a series of reactions that includes highly exothermic reactions. Will the most exothermic reactions exhibit an inverted regime kinetics? Why or why not?
10. Consider the two-dimensional PES

$$V(q,r) = V_0(1-q^2)^2 + c(q^2-1)\cdot r + \frac{1}{2}m_r\omega_r^2 r^2$$

with dynamics that include a strong friction on coordinate r. This PES has been used to model an enzyme conformational mode r that brings a proton donor and acceptor closer together, while coordinate q represents the fast chemical step of proton transfer [67,75]. Define a "left" state with

$$F_L(r) = -k_B T \ln \int_{q<0} e^{-\beta V(q,r)} dq$$

and a corresponding "right" state $F_R(r)$. Use transition state theory to construct an expression for the net rate of transfer from L to R at constant r. Incorporate the rate law into a non-adiabatic model comprised of the coupled stochastic equations

$$\frac{\partial \rho_L(r,t)}{\partial t} = D\frac{\partial}{\partial r}\left\{\frac{\partial \beta F_L}{\partial r}\rho_L + \frac{\partial \rho_L}{\partial r}\right\} + \kappa_{R\to L}(r) - \kappa_{L\to R}(r)$$

with a similar equation for $\rho_R(r)$. Hint: both equations are linear. Compute the rate constant from the non-adiabatic model with a steady-state source on the left at $r=0$ and a sink on the right at $r=0$. When does the full two-dimensional transition state theory rate provide a good approximation to the steady-state rate?

References

[1] L.J. Butler, Annu. Rev. Phys. Chem. 49 (1998) 125–171.
[2] T. Van Voorhis, T. Kowalczyk, B. Kaduk, L.-P. Wang, C.-L. Cheng, Q. Wu, Annu. Rev. Phys. Chem. 61 (2010) 149–170.
[3] M. Baer, Chem. Phys. Lett. 35 (1975) 112–118.
[4] M. Baer, Mol. Phys. 40 (1980) 1011–1013.
[5] T. Pacher, L.S. Cederbaum, H. Koppel, Adv. Chem. Phys. 89 (1988) 7367–7381.

[6] L. Pauling, The Nature of the Chemical Bond, Cornell Univ. Press, Ithaca, NY, 1960.
[7] A. Warshel, R.M. Weiss, J. Am. Chem. Soc. 102 (1980) 6218–6226.
[8] J.N. Harvey, Phys. Chem. Chem. Phys. 9 (2007) 331–343.
[9] Q. Wu, T. Van Voorhis, J. Phys. Chem. A 110 (2006) 9212–9218.
[10] A. Earnshaw, Introduction to Magnetochemistry, Academic Press, New York, 1968.
[11] L. Landau, Physik. Z. Soviet Union 2 (1932) 46–51.
[12] C. Zener, Proc. R. Soc. A, Math. Phys. Eng. Sci. 137 (1932) 696–702.
[13] D. Schroder, S. Shaik, H. Schwarz, Acc. Chem. Res. 33 (2000) 139–145.
[14] K.P. Jensen, U. Ryde, J. Biol. Chem. 279 (2004) 14561–14569.
[15] N. Koga, K. Morokuma, Chem. Phys. Lett. 119 (1985) 371–374.
[16] F. Jensen, J. Am. Chem. Soc. 114 (1992) 1596–1603.
[17] M.J. Bearpark, M.A. Robb, H.B. Schlegel, Chem. Phys. Lett. 223 (1994) 269–274.
[18] J.-L. Carreon-Macedo, J.N. Harvey, J. Am. Chem. Soc. 126 (2004) 5789–5797.
[19] C.C. Moser, J.M. Keske, K. Warncke, R.S. Farid, P.L. Dutton, Nature 355 (1992) 796–802.
[20] V.L. Davidson, Biochemistry 41 (2002) 14633–14636.
[21] R.A. Marcus, Annu. Rev. Phys. Chem. 15 (1964) 155–196.
[22] C.D. Taylor, J.D. Gale, H.-H. Strehblow, P. Marcus, in: Molecular Modeling of Corrosion Processes, Wiley, Hoboken, NJ, 2015.
[23] A. Warshel, Nature 260 (1976) 679–683.
[24] A. Nitzan, M.A. Ratner, Science 300 (2003) 1384–1389.
[25] J.M. Mayer, Annu. Rev. Phys. Chem. 55 (2004) 363–390.
[26] S. Hammes-Schiffer, Chem. Rev. 110 (2010) 6937–6938.
[27] A.M. Kuznetsov, J. Ulstrup, Electron Transfer in Chemistry and Biology: An Introduction to the Theory, Wiley, New York, 1999.
[28] V. May, O. Kuhn, Charge and Energy Transfer Dynamics in Molecular Systems, 3rd ed., Wiley-VCH, Weinheim, 2011.
[29] P.F. Barbara, T.J. Meyer, M.A. Ratner, J. Phys. Chem. 100 (1996) 13148–13168.
[30] D.W. Small, D.V. Matyushov, G.A. Voth, J. Am. Chem. Soc. 125 (2003) 7470–7478.
[31] M. Tachiya, J. Chem. Phys. 69 (1978) 2375.
[32] A.V. Baryzkin, P.A. Frantsuzov, K. Seki, M. Tachiya, Adv. Chem. Phys. 123 (2002) 511–616.
[33] R.A. Marcus, Rev. Mod. Phys. 65 (1993) 599–610.
[34] F.A. Tezcan, B.R. Crane, J.R. Winkler, H.B. Gray, Proc. Natl. Acad. Sci. USA 98 (2001) 5002–5006.
[35] H.B. Gray, J.R. Winkler, Proc. Natl. Acad. Sci. USA 102 (2005) 3534–3539.
[36] O.S. Wenger, B.S. Leigh, R.M. Villahermosa, H.B. Gray, J.R. Winkler, Science 307 (2005) 99–102.
[37] H.B. Gray, J. Halpern, Proc. Natl. Acad. Sci. USA 102 (2005) 3533.
[38] M. Newton, N. Sutin, Annu. Rev. Phys. Chem. 35 (1984) 437–480.
[39] N.S. Hush, Electrochim. Acta 13 (1968) 1005–1023.
[40] L.D. Zusman, Chem. Phys. 49 (1980) 295–304.
[41] D.F. Calef, P.G. Wolynes, J. Phys. Chem. 87 (1983) 3387–3400.
[42] M. Tachiya, J. Phys. Chem. 97 (1993) 5911–5916.
[43] J.E. Leffler, Science 117 (1953) 340–341.
[44] J.R. Miller, L.T. Calcaterra, G.L. Closs, J. Am. Chem. Soc. 106 (1984) 3047–3049.
[45] I.R. Gould, S. Farid, Acc. Chem. Res. 29 (1996) 522–528.
[46] R.A. Marcus, N. Sutin, Biochim. Biophys. Acta 811 (1985) 265–322.
[47] L.E. Bennett, Prog. Inorg. Chem. 18 (1973) 1–176.
[48] P. Blowers, R.I. Masel, Theor. Chem. Acc. 105 (2000) 46–54.
[49] A. Warshel, J. Phys. Chem. 86 (1982) 2218–2224.
[50] J.-K. Hwang, A. Warshel, J. Am. Chem. Soc. 109 (1987) 715–720.
[51] R. Kuharski, J.S. Bader, D. Chandler, M. Sprik, M.L. Klein, R.W. Impey, J. Chem. Phys. 89 (1988) 3248.

[52] J. Blumberger, M. Sprik, Theor. Chem. Acc. 115 (2005) 113–126.
[53] C.H. Bennett, J. Comput. Phys. 22 (1976) 245–268.
[54] R. Zwanzig, J. Chem. Phys. 22 (1954) 1420–1426.
[55] J. Blumberger, M. Sprik, J. Phys. Chem. B 109 (2005) 6793–6804.
[56] K. Kamat, B. Peters, J. Phys. Chem. Lett. 8 (2017) 655–660.
[57] G.M. Torrie, J.P. Valleau, J. Comput. Phys. 23 (1977) 187–199.
[58] G. King, A. Warshel, J. Chem. Phys. 93 (1990) 8682.
[59] H. Sumi, R.A. Marcus, J. Chem. Phys. 84 (1986) 4894.
[60] I.A. Balabin, J.N. Onuchic, Science 290 (2000) 114–117.
[61] B.M. Hoffman, M.A. Ratner, J. Am. Chem. Soc. 109 (1987) 6237–6243.
[62] M.D. Newton, Coord. Chem. Rev. 239 (2003) 167–185.
[63] A.C. Benniston, A. Harriman, Chem. Soc. Rev. 35 (2006) 169–179.
[64] S.S. Skourtis, I.A. Balabin, T. Kawatsu, D.N. Beratan, Proc. Natl. Acad. Sci. USA 102 (2005) 3552–3557.
[65] L.W. Ungar, M.D. Newton, G.A. Voth, J. Phys. Chem. B 103 (1999) 7367–7382.
[66] S.J. Benkovic, S. Hammes-Schiffer, Science 301 (2003) 1196–1202.
[67] D. Antoniou, M.R. Abolfath, S.D. Schwartz, J. Chem. Phys. 121 (2004) 6442.
[68] J.R.E.T. Pineda, S.D. Schwartz, Philos. Trans. R. Soc. Lond. B, Biol. Sci. 361 (2006) 1433–1438.
[69] S.L. Quaytman, S.D. Schwartz, Proc. Natl. Acad. Sci. USA 104 (2007) 12253–12258.
[70] K.A. Henzler-Wildman, M. Lei, V. Thai, S.J. Kerns, M. Karplus, D. Kern, Nature 450 (2007) 913–916.
[71] K. Henzler-Wildman, D. Kern, Nature 450 (2007) 964–972.
[72] M. Karplus, Proc. Natl. Acad. Sci. USA 107 (2010) E71.
[73] S.C.L. Kamerlin, A. Warshel, Proc. Natl. Acad. Sci. USA 107 (2010) E72.
[74] S.C.L. Kamerlin, A. Warshel, Proteins 78 (2010) 1339–1375.
[75] B. Peters, J. Chem. Theory Comput. 6 (2010) 1447–1454.
[76] B. Peters, J. Phys. Chem. B 115 (2011) 12671.
[77] L.Y.P. Luk, J. Javier Ruiz-Pernia, W.M. Dawson, M. Roca, E.J. Loveridge, D.R. Glowacki, J.N. Harvey, A.J. Mulholland, I. Tunon, V. Moliner, R.K. Allemann, Proc. Natl. Acad. Sci. USA 110 (2013) 16344–16349.
[78] L. Masgrau, D.G. Truhlar, Acc. Chem. Res. 48 (2015) 431–438.
[79] D. Antoniou, S.D. Schwartz, J. Phys. Chem. B 115 (2011) 2465–2469.
[80] M. Garcia-Viloca, J. Gao, M. Karplus, D.G. Truhlar, Science 303 (2004) 186–195.
[81] M.H.M. Olsson, W.W. Parson, A. Warshel, Chem. Rev. 106 (2006) 1737–1756.
[82] A.V. Pisliakov, J. Cao, S.C.L. Kamerlin, A. Warshel, Proc. Natl. Acad. Sci. USA 106 (2009) 17359–17364.
[83] A. Szabo, D. Shoup, S.H. Northrup, J.A. McCammon, J. Chem. Phys. 77 (1982) 4484–4493.
[84] J.A. McCammon, BMC Biophys. 4 (4) (2011).
[85] A. Warshel, Proc. Natl. Acad. Sci. USA 81 (1984) 444–448.
[86] J. Aqvist, A. Warshel, Chem. Rev. 93 (1993) 2523–2544.
[87] S. Hammes-Schiffer, Biochemistry 41 (2002) 13335–13343.
[88] S.C.L. Kamerlin, A. Warshel, Faraday Discuss. 145 (2010) 71–106.
[89] O. Acevedo, W.L. Jorgensen, Acc. Chem. Res. 43 (2010) 142–151.
[90] S. Hammes-Schiffer, Acc. Chem. Res. 39 (2006) 93–100.
[91] R. Zwanzig, Acc. Chem. Res. 23 (1990) 148–152.
[92] O. Flomenbom, K. Velonia, D. Loos, S. Masuo, M. Cotlet, Y. Engelborghs, J. Hofkens, A.E. Rowan, R.J.M. Nolte, M. Van der Auweraer, F.C. de Schryver, J. Klafter, Proc. Natl. Acad. Sci. USA 102 (2005) 2368–2372.
[93] G.C. Schatz, M.A. Ratner, Quantum Mechanics in Chemistry, Dover, Mineola, NY, 2002.
[94] R.I. Masel, Chemical Kinetics and Catalysis, Wiley, New York, 2001.

CHAPTER 22

Free energy relationships

> *...in a series of homologous reactions the heats of activation decrease with increasing heats of reaction. ... derive the Tafel equation for hydrogen overvoltage and the series of overvoltages at the metals.*
>
> ***Horiuti and Polanyi, (tr. Muller), Acta Physicochem. URSS (1935)***

The introduction to this book stressed that "matching experiments" should not be the final goal of a computational effort. To engineer better catalysts, processes, or materials, it is often more important to identify reliable ways of controlling the kinetics. This chapter explores relationships, some empirical and some derived, between kinetics and driving forces. These relationships are useful for *predicting* trends, e.g. how a change in solvent will change the rate of a chemical reaction, or how a mechanical pulling force will accelerate the unfolding of a protein. The relationships are also useful for *interpreting* kinetic trends in terms of transition state properties. For example, some free energy relationships indicate the location of a transition state, or the size of a critical nucleus, both from measurable quantities.

With some effort we can even go beyond kinetic trends for one reaction, to predict and understand kinetic trends across series of similar reactions. In these efforts, relationships between thermodynamic properties and reactivity have emerged as one of the most fruitful intersections between theory and experiment. For example in organometallic chemistry, we can understand and predict how ligands with different electron withdrawing characteristics influence the rate of a reaction. In catalysis, we can understand and predict how the amount or type of a dopant alters the activity. Modern computing power has made it possible to perform automated and increasingly reliable calculations for hundreds or thousands of slightly different catalysts or substrates. The resulting large data sets have revealed many relationships between molecular structure, electronic structure, and chemical properties including reactivity. Structure-property relationships discovered in this manner often begin as empiricisms and heuristics, but especially in recent catalysis work they have also inspired fundamental advances and *ab initio* tools for new types of analyses.

22.1 BEP relations and the Bronsted catalysis law

One of the first and still most important relationships between thermodynamics and reactivity is that of Bell[1] and Evans and Polanyi (BEP) [2]. BEP postulated that for a series of similar reactions, the more exothermic reactions would have lower activation barriers. BEP relations of the form

$$E_a \approx E_a^o + \alpha \Delta H^o \tag{22.1.1}$$

are widely used [1,3], especially in catalysis [4–6]. Here ΔH^o is the reaction enthalpy and E_a^o is the activation energy for a (hypothetical) isoergic ($\Delta H^o = 0$) reaction in the series. The slope α is related to the position of the transition state (early, midway, or late) along the reaction pathway.

Aqvist and Warshel [4] and van Santen et al. [7] have reviewed the derivations, applications, and recent advances related to BEP relations in catalysis. The BEP relation is a practical computational tool because it is far easier to measure or compute reaction enthalpies than to accurately compute reaction rates. In heterogeneous catalysis, reaction enthalpies themselves can be further correlated to electronic band structures and to adsorbate properties. These correlations have been combined with BEP relations to create "universal BEP relations" that predict activities of many catalysts for many different reactions.

Example: BEP relations in catalysis [8]

Computational work in heterogeneous catalysis shows how ΔH^o can be estimated from even more readily computed quantities, e.g. d-band characteristics of a metal or alloy [5,9]. These studies first correlate d-band energies to atomic adsorbate energies, e.g. $*O$, $*H$, $*N$, and $*C$. The atomic adsorption energies are then correlated to adsorption energies of more complex and partially saturated adsorbates, e.g. $*OH$, $*OCH_3$, $*NH_2$, $*CH$, $*CH_2$, and $*CH_3$. Surface reaction stoichiometries are used to combine correlations for adsorbate enthalpies into correlations for surface reaction enthalpies. Finally, the reaction enthalpies are used in BEP relations for the breaking of different bond types. The result is a modern "aufbau principle" for building correlations that relate d-band characteristics and a small collection of adsorbate properties to activation energies. These correlations-in-series are not extremely accurate, but they are powerful ways of quickly screening the entire periodic table for highly active metals, alloys, dopants, etc.

[1] The BEP acronym sometimes refers to Bell-Evans-Polanyi and sometimes refers to Bronsted-Evans-Polanyi instead. Bronsted most definitely contributed to the development of LFERs with his catalysis law. R.P. Bell, who worked in Bronsted's laboratory, directly used the BEP relation in his own work after it was put forth by Evans and Polanyi. Some sources even point to the much later G.I. Bell. Still others refer to "Bema Hapothle" (Bell-Marcus-Hammond-Polanyi-Thornton-Leffler) principles which describe changes in transition state structure across series of reactants or solvents. Clearly, there were many contributors to the theory of free energy relationships [1].

This approach has now been used to predict the activity of thousands of metals and alloys *in silico*. BEP analyses have led to (1) an improved understanding of chemical bonding and catalysis on metal surfaces, and to (2) the design of now-patented new catalysts [10,8]. Figure 22.1.1 shows the "universal" BEP relation for dissociation of molecules containing single bonds between C, O, and N atoms on transition metal surfaces.

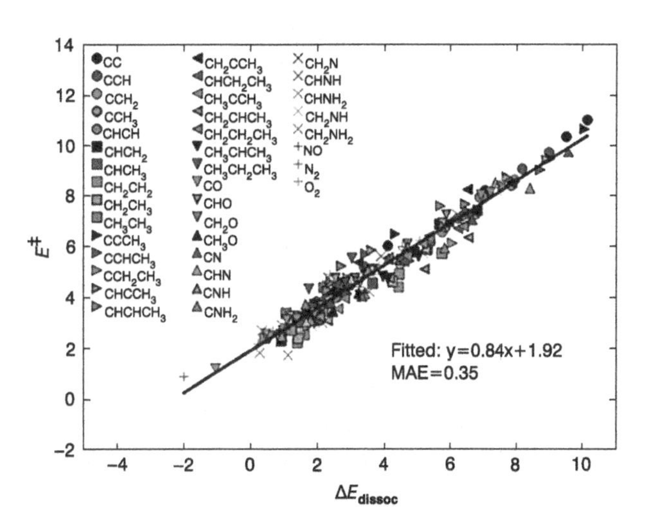

Figure 22.1.1: Wang et al. developed "universal" BEP relation from computed transition states and dissociation energies for C-C, C-O, C-N, N-O, N-N, and O-O bonds on a variety of transition metal surfaces. The typical error is 0.35eV, a result of correlating such different reactions in a single BEP relation. When different reactants are analyzed separately, the MAE is significantly lower. [Wang et al. *Catalysis Lett.* 141, 370-73 (2011)]

Recent progress in catalysis illustrates the accelerated fundamental advances and technological discoveries that are possible with relatively simple calculations that focus on trends. Heuristic derivations of the BEP relation (see section 22.2) suggest general guidelines on the situations where a BEP relation will be valid:

- BEP relations are most reliable when applied to individual elementary steps.
- BEP relations tend to be most accurate for reactions in which the transition state is late, i.e. for cases with α near unity so that transition states resemble the products.
- In practice, there is some ambiguity in what constitutes a related series of reactions. Generally, more closely related groups of reactions will more closely obey the BEP relation.

There are cases where BEP relations fail [11]. In heterogeneous catalysis, BEP relations fail when some members in a series of catalysts are prone to surface reconstructions, preferential segregation of certain alloy components to the surface, or to oxidation.

The Bronsted catalysis law [12] relates the kinetics of proton abstraction to an acid dissociation equilibrium. Specifically, consider a series of reactions

$$HA_i + B^- \xrightarrow{k_i} HB + A_i^- \qquad \Delta G_i^o \tag{22.1.2}$$

with equilibrium constant $K_i = exp\left[-\Delta G_i^o / k_B T\right]$. Bronsted's catalysis law says that, for a series of similar reactions, the equilibrium constants are related to the rate constants by

$$k_i \approx \gamma K_i{}^{\alpha} \tag{22.1.3}$$

where γ and α are constants. If we assume $k_i = k_B T\, h^{-1} \mathcal{V}_o^{\nu-1} exp\left[-\Delta G_i^{\ddagger} / k_B T\right]$ and take logarithms, equation (22.1.3) becomes $\Delta G_i^{\ddagger} \approx \alpha\, \Delta G_i^o - k_B T\, ln\left[k_B T\, \mathcal{V}_o^{\nu-1} / h\gamma\right]$. Now invoke a reference reaction in the series for which $\Delta G_{ref}^o = 0$. This reference reaction has $\Delta G_{ref}^{\ddagger} \approx -k_B T\, ln\left[k_B T\, \mathcal{V}_o^{\nu-1} / h\gamma\right]$. Therefore the Bronsted law (equation (22.1.3)) can also be written as

$$\Delta G_i^{\ddagger} \approx \Delta G_{ref}^{\ddagger} + \alpha\, \Delta G_i^o$$

which is similar to the BEP relation but involves free energies. α is called the Bronsted coefficient or Bronsted slope [13,4],

$$\alpha \equiv \frac{\partial \Delta G^{\ddagger}}{\partial \Delta G^o} \tag{22.1.4}$$

and typically it falls in the range $0 < \alpha < 1$. The Bronsted slope has been interpreted as the position of the transition state along the pathway from reactants to products. The reasons for this interpretation will become clear below when free energy relationships are heuristically derived from a Marcus theory model.

22.2 The Marcus equation

Chapter 21 noted that electron transfer ideas are applicable to many chemical reactions, especially proton transfer reactions [14,15]. Marcus obtained the parabolic free energy diabats as functions of a charge transfer interpolation coordinate ξ. The reactant state at $\xi = 0$ and the product state at $\xi = 1$ are minima for separate free energy parabolas with the same spring constants $m\omega^2$. As depicted in Figure 22.2.1, the free energy landscape in state A is approximately

$$\Delta G_A(\xi) \approx \frac{1}{2} m\omega^2 \xi^2 \tag{22.2.1}$$

In state B, the free energy landscape is approximately,

$$\Delta G_B(\xi) \approx \frac{1}{2} m\omega^2 (\xi - 1)^2 + \Delta G^o \tag{22.2.2}$$

Solving for the point of intersection,

$$\xi^{\ddagger} \approx \frac{1}{2} + \frac{\Delta G^o}{m\omega^2} \tag{22.2.3}$$

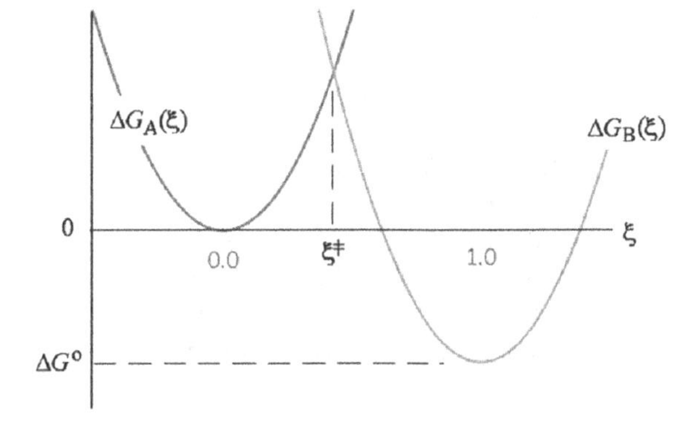

Figure 22.2.1: The Marcus-BEP relation is based on a simple model with two diabatic states.

Inserting $\xi^{\ddagger}$ into $\Delta G_A(\xi)$ gives the activation free energy

$$\Delta G^{\ddagger} = \frac{1}{2} m\omega^2 \left(\frac{1}{2} + \frac{\Delta G^o}{m\omega^2} \right)^2 \tag{22.2.4}$$

For a thermoneutral reference reaction ($\Delta G^o_{ref} = 0$) we can define

$$\Delta G^{\ddagger}_{ref} = \frac{1}{8} m\omega^2 \tag{22.2.5}$$

This allows us to write the activation free energy in terms of ΔG^o and the reference activation free energy. The result is the Marcus equation

$$\Delta G^{\ddagger} = \Delta G^{\ddagger}_{ref} + \frac{\Delta G^o}{2} \left(1 + \frac{\Delta G^o}{8\Delta G^{\ddagger}_0} \right) \tag{22.2.6}$$

The reference activation free energy $\Delta G^{\ddagger}_{ref}$ is essentially a fit parameter. No member of the reaction series actually needs to have $\Delta G^o = 0$ to use the Marcus equation.

Cohen and Marcus tested the Marcus equation against data for several reaction series [16]. Figure 22.2.2 shows their test for a large experimental data set on hydrogen abstraction rates from ketones by carboxylate anions and related bases.

$$CH_3COCHRR' + \mathbf{A}^- \rightarrow CH_3COCRR'^- + H\mathbf{A} \tag{22.2.7}$$

Cohen and Marcus tested some twenty reaction families, with results that generally validated the Marcus equation. The Marcus equation has also been validated for proton coupled electron transfers [17] and proton transfer reactions in enzymes [18].

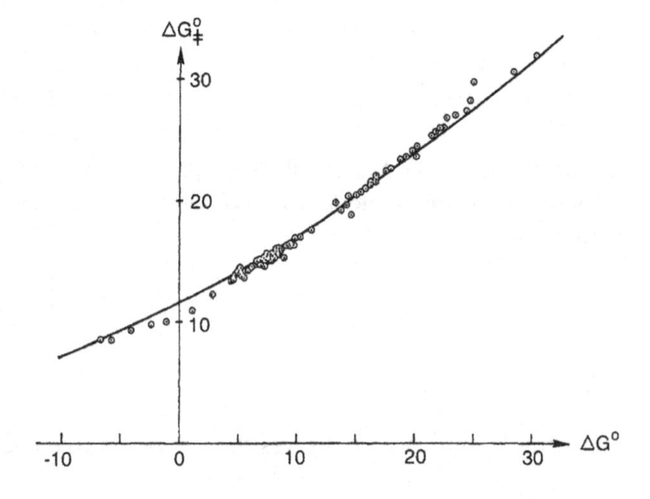

Figure 22.2.2: A test of the Marcus equation for hydrogen abstraction from ketones by a family of carboxylate anions and other bases. Units on both axes are $kcal/mol$**. [From Cohen and Marcus, *J. Phys. Chem.* 78, 4249-56 (1968).]**

Returning to equation (22.2.3), let us write the transition state location in terms of ΔG^o and the activation free energy for the reference reaction. The result is the Marcus-Hammond relation

$$\xi^{\ddagger} = \frac{1}{2} + \frac{\Delta G^o}{8\Delta G^{\ddagger}_{ref}} \tag{22.2.8}$$

Now using equation (22.2.6) to compute the local slope of the free energy relationship gives

$$\alpha \equiv \frac{\partial \Delta G^{\ddagger}}{\partial \Delta G^o} = \xi^{\ddagger}. \tag{22.2.9}$$

This equation explains why the slope in BEP relationships and LFERs has been interpreted as the position of the transition state along the reaction pathway. Equations (22.2.8) and (22.2.9) justify the postulates of Leffler and Hammond [19,20]. Specifically, in a series of similar reactions, more exothermic reactions will have earlier transition states along the reaction coordinate and their transition state structures will more closely resemble the reactants.

Where chemical accuracy is needed, it should be remembered that most FERs have a heuristic basis. Even theoretically derived FERs like the Marcus equation [16] and the nucleation theorem [21] are based on extremely simple models. The true activation energies and free energies are often widely scattered around the values predicted by an FER. Of course, FERs will fail to predict trends when the reaction path or mechanism changes within a seemingly similar series of reactions, or when the limiting step in a multistep mechanism changes to another other step in the sequence.

22.3 Externally controlled driving forces

The previous section discussed free energy relationships that correlate rates across a series of similar reactions. This section discusses similar free energy relations that help understand situations where the driving force for a single reaction is externally controlled. Rate theories that account for externally modulated driving forces have a long history, and similar expressions starting with Eyring [22] have emerged in many different contexts [23–25,21]. The simplest models consider a one dimensional free energy profile as a function of the reaction coordinate with a force applied along the reaction coordinate. The applied force alters the free energy landscape by simply tilting the original landscape. For small forces,[2] the new free energy landscape qualitatively resembles the original free energy landscape, but the stationary points shift to new locations along q and the barrier heights change as shown in Figure 22.3.1.

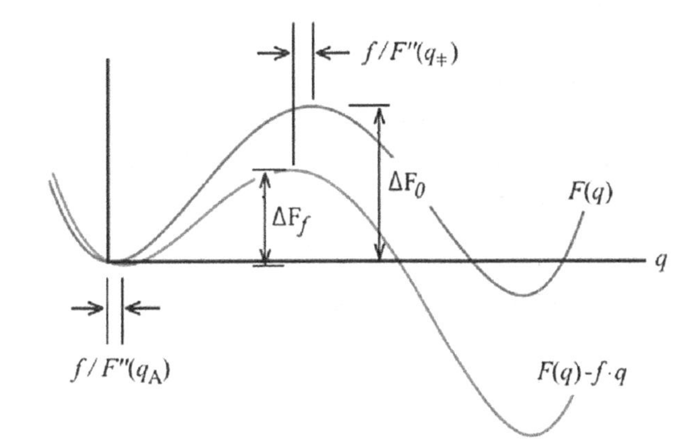

Figure 22.3.1: Pulling along the reaction coordinate q tilts the zero-force free energy landscape and shifts the locations and heights of the free energy barriers.

Changes in the barrier to escape state A include terms of order f which are proportional to the unperturbed distance from reactants to transition states along q. Changes in the barrier also include terms of order f^2 which emerge from force-induced changes in the location of the reactant minimum and transition state. The force-perturbed free energy barrier is

$$\Delta F_f = \Delta F_0 - f \cdot (q_{\ddagger} - q_A) - \frac{1}{2} f^2 \{(1/F''(q_{\ddagger}) - 1/F''(q_A)\} \tag{22.3.1}$$

To first order in the force f, equation (22.3.1) is consistent with the Hammond-Leffler postulate, i.e. a force that pulls along the direction leading to the transition state lowers the barrier and leads to an earlier transition state. Typically, the first order term in the external force dominates

[2] For sufficiently large forces a catastrophe [26] causes the minima lose their mechanical stability. We limit our discussion to the effects of smaller forces.

except where the force is strong.[3] The slope in a free energy relationship has a physical interpretation which is related to the position of the transition state along the reaction coordinate. Some important special cases of equation (22.3.1) include:

- Electrochemistry [29,23]: Suppose that an electrochemical reaction $M^{n+} + n\,e^- \rightarrow M$ involves transfer of a specific charge $n\,e^-$ from the electrode. A Tafel plot shows the log of the electrochemical reaction rate (current) vs. electrical overpotential. Tafel plots are approximately linear with a slope that is proportional to the number of electrons transferred in the limiting step.
- Nucleation [30]: Rates of nucleation are highly sensitive to the chemical potential driving force $\Delta\mu$, i.e. to supersaturation. The nucleation theorem says that $\partial ln J / \partial \Delta\mu \approx n^{\ddagger}$. The nucleation theorem is not a linear (or even quadratic) free energy relationship. Nonetheless, the slope at each driving force $\Delta\mu$ is the distance from reactant to transition state. The nucleation theorem is widely used in experiments to estimate critical nucleus sizes.
- Single molecule pulling experiments [31,32]: New optical tweezers and atomic force microscopy techniques [33,34] can attach tethers to folded proteins or hybridized DNA strands and pull with constant force until the structures "rupture". When the tethers are attached at appropriate locations [35], a constant pulling force[4] reduces the rupture time in a fashion consistent with equation (22.3.1).

The example below considers a model of mean rupture times in constant force pulling experiments. In this case, the slope of the free energy relationship will give the critical extension at the transition state.

Pulling experiments with a constant force

Suppose that the reaction coordinate for spontaneous unfolding of a protein or for unzipping of an RNA hairpin [34] is q. So that we may retain a simple one dimensional model, assume that the direction of pulling coincides with the reaction coordinate q. Finally, assume that the free energy as a function of the reaction coordinate with zero pulling force is $F_0(q)$ and that the mobility along q is coordinate independent (and force independent). For a non-zero force f, the free energy as a function of q becomes $F(q) = F_0(q) - f \cdot q$. We can use the mean first passage time (MFPT) expression from Chapter 18 to compute

[3] D. Makarov has shown that important exceptions arise when the pulling direction is not aligned with the reaction coordinate [27,28].

[4] Pulling experiments can also be done at constant pulling speed. A fascinating new branch of non-equilibrium statistical mechanics concerns relationships between the reversible work (free energy) and the non-equilibrium work distributions along forward and reverse trajectories [36–39].

mean rupture times. The ratio of mean rupture times with and without force is [31,40]

$$\frac{\tau_{MFP}(f)}{\tau_{MFP}(0)} = \frac{\int_{\cap} e^{+(F_0(q)-f\cdot q)/k_BT} dq \int_{\cup} e^{-(F_0(q)-f\cdot q)/k_BT} dq}{\int_{\cap} e^{+F_0(q)/k_BT} dq \int_{\cup} e^{-F_0(q)/k_BT} dq}$$

where subscripts $\cup$ and $\cap$ on the integrals indicate integration over values of q near the well bottom and near the barrier top, respectively.

Dudko et al. [31] examine a model in which the unbiased free energy surface is $F_0(q) = 1.5 \cdot (q/q_0^{\ddagger})\Delta F_0^{\ddagger} - 2.0 \cdot (q/q_0^{\ddagger})^3 \Delta F_0^{\ddagger}$. The integrals that arise from the model of Dudko et al. can be done analytically giving

$$\frac{\tau_{MFP}(f)}{\tau_{MFP}(0)} = (1 - f \cdot q_0^{\ddagger}/2\Delta F_0^{\ddagger})^{-1} exp\left[-\beta\Delta F_0^{\ddagger}\left\{1 - (1 - f \cdot q_0^{\ddagger}/2\Delta F_0^{\ddagger})^2\right\}\right]$$

Instead of MFPTs, define the force dependent rate constant as $k = \tau_{MFP}^{-1}$ and take logarithms

$$lnk(f) - lnk(0) = ln(1 - f \cdot q_0^{\ddagger}/2\Delta F_0^{\ddagger}) + \beta\Delta F_0^{\ddagger}\left\{1 - (1 - f \cdot q_0^{\ddagger}/2\Delta F_0^{\ddagger})^2\right\}$$

which for small forces is

$$lnk(f) = lnk(0) + (\beta\Delta F_0^{\ddagger} - 1/2)\frac{f \cdot q_0^{\ddagger}}{\Delta F_0^{\ddagger}} + \ldots$$

The remaining terms are of order $(f \cdot q_0^{\ddagger}/\Delta F_0^{\ddagger})^2$. The term $-f \cdot q_0^{\ddagger}/2\Delta F_0^{\ddagger}$ is unusual in an LFER. Here, it emerged from a force dependence in the prefactor of the MFPT expression. Presumeably, rupture happens while $\beta\Delta F_0^{\ddagger} \gg 1/2$. Otherwise, the separation of time scales would break down making the simple MFPT formula invalid anyway. Omitting the -1/2 term recovers $\beta\Delta F^{\ddagger} = \beta\Delta F_0^{\ddagger}(1 - f \cdot q_0^{\ddagger}/\Delta F_0^{\ddagger})$ which is the standard $k(f) = k(0)e^{\beta f q_0^{\ddagger}}$ formula of Bell [25]. The key prediction from Bell's formula is that $dlnk/d\beta f$ is $q_0^{\ddagger}$, i.e. the displacement along the reaction coordinate between the transition state and the minimum at zero force. This, of course, is the same interpretation given in the earlier FERs for chemical reactions.

Additional arenas where free energy relationships with externally controlled driving forces include slip-stick friction, electrophoresis, crack-tip initiation, etc. Undoubtedly, some (if not all) of these phenomena have already been studied using free energy relationships.

Closing remarks

My effort to write a text on rate theories and rare events methods started nine years ago. Of course, the task wasn't supposed to take nine years. Like most academic pursuits, I began with unwarranted confidence, and surrendered after being humbled by the many gaps in my knowledge. Nevertheless, the task was a pleasure. Hopefully, each member of my audience has found at least some of these pages enlightening and useful.

Before closing this endeavor, let me briefly comment on the status of the subject and prognosticate on its future. Recent decades saw tremendous advances: degree-of-rate-control analyses, disconnectivity graphs, empirical valence bond, saddle search algorithms, Landau free energy calculations, kinetic Monte Carlo, reactive flux methods, Grote-Hynes theory, Langer-Berezhkovskii-Szabo theory, path sampling, and reaction coordinate optimization methods. Approximately one third of the material in this book did not exist 40 years ago.

The recent flurry of theoretical progress is especially remarkable given that the subject began in the 1800s and that its theoretical foundations were established in a separate flurry during the 1930s. The recent progress has been spurred by organizations like the Centre Europeen de Calcul Atomique et Moleculaire (CECAM), the Telluride Summer Research Conferences (TSRC), and the Institute for Pure and Applied Mathematics (IPAM). These institutions and the agencies that support them have helped chemists, physicists, mathematicians, and engineers discover and draw inspiration from each others' work.

The newest rate theories and rare events methods have transformed tasks that were once impossible into readily executed calculations. Opportunities to use these new tools abound, and there is no end in sight to the current period of rapid fundamental discovery. However, there are also major changes afoot. Many current efforts in the computational sciences are focused on simulations at larger and ever longer time scales and on machine learning tools for data driven models. These new tools have already impacted many areas of research, and they will surely impact the theory of reaction rates and rare events. However, large scale computation should not (and as yet cannot) replace physics-based theoretical models and clever rare events methods. In the words of a recent perspective by Marcus [41], "...as a percentage of total theoretical research effort, [analytical theory] is now small compared with computation. I believe the future can be expected to settle on some insightful combination of both, since both are necessary." Amen

Exercises

1. Following Polanyi, let two potential energy curves $V_A(q)$ and $V_B(q)$ cross at location q_0 with energy E_a as shown in the figure below.

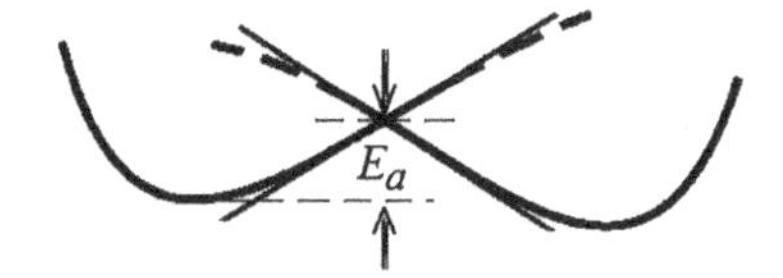

Let $g_A = dV_A/dq$ and $g_B = dV_B/dq$ be their slopes at the crossing point. Use linear approximations at the intersection to show that when $V_B(q)$ is shifted vertically by an amount $\Delta\Delta H$ the new intersection is at energy $E_a' \approx E_a + \gamma \cdot \Delta\Delta H$. Find the parameter γ in terms of g_A and g_B.

2. Summarize the Hammett relations [42] for reaction rates and equilibrium constants of reactions involving substituted benzene rings. Write these as relationships between reaction free energies and activation free energies.
3. [Z. Han] Robinson and Holbrook tabulated the high pressure Arrhenius parameters for a series of ten 1,5-hydrogen transfer reactions in 1,3-dienes. Choose an accurate *ab initio* model chemistry. Compute the energies of the stable reactants and products to develop a BEP correlation for the series of 1,5 hydrogen transfer reactions. Compute the Arrhenius parameters and energies of the reactant and product states for this series. How well does the BEP correlation work? Do you expect a free energy relationship to work better? Why or why not?
4. In the limit of a small force f along the reaction coordinate, what is $dlnK(f)/df$ where $K(f)$ is the force-perturbed equilibrium constant? Express your answer in terms of the product and reactant positions along the reaction coordinate.
5. The nucleation theorem gives the derivative of the reversible work to create the critical nucleus as

$$d\Delta G^{\ddagger}/d\Delta\mu = -n^{\ddagger}$$

See the literature on nucleation [43] for relationships between $\Delta G^{\ddagger}$ and the nucleation rate. At the spinodal, $\Delta\mu = \Delta\mu_S$ and $\Delta G_S^{\ddagger} = 0$. Use the nucleation theorem to write the free energy barrier for $\Delta\mu \neq \Delta\mu_S$ in terms of an integral over estimated nucleus sizes.

6. Let $V_0(x, y) = V_{\ddagger,0} \cdot (1 - 3y^2 - 2y^3) + \frac{1}{2}m\omega_x^2(x - cy/m\omega_x^2)^2$ be the PES in the absence of a pulling force. The reaction coordinate is some linear combination of x and y.
   (a) Consider a force f that pulls along the x-coordinate so that the potential becomes $V_f(x, y) = V_0(x, y) - f \cdot x$. Compute $\lim_{f \to 0} dV_{\ddagger}/df$. Compare your results to the Bell formula.
   (b) Suppose that you were unaware of y, and that you had began your analysis from the free energy at zero force $F_0(x) = -k_BT\, ln \int e^{-\beta V_0(x,y)} dy$. Show that the stationary points in $F_0(x)$ coincide with the x-coordinates of stationary points in $V_0(x, y)$. When a force is applied, the saddle point on $V_f(x, y)$ moves, as does the maximum in the potential $F(x) = F_0(x) - fx$. Do they move in consistent directions?

(c) Makarov examined this situation and concluded that "no one-dimensional model can possibly reproduce the prediction of the two dimensional model..." Clearly, this is true if the one-dimension is x. Assume that x and y are associated with the same mass, perform the usual normal mode analysis to find the reaction coordinate, and repeat the calculations above. Do the transition states and minima from the 1D and 2D models shift in consistent directions? Do the results from the 1D and 2D models become consistent with Bell's formula? Do we always have the ability to pull along the correct reaction coordinate? See Makarov for equations that describe the surprising effects of pulling forces that are not aligned with the reaction coordinate in higher dimensions [27,28].

References

[1] W.P. Jencks, Chem. Rev. 85 (1985) 511–527.
[2] M.G. Evans, M. Polanyi, Trans. Faraday Soc. 34 (1938) 11–24.
[3] J.-K. Hwang, A. Warshel, J. Am. Chem. Soc. 109 (1987) 715–720.
[4] J. Aqvist, A. Warshel, Chem. Rev. 93 (1993) 2523–2544.
[5] B. Hammer, J.K. Norskov, Adv. Catal. 45 (2000).
[6] E. Skulason, G.S. Karlberg, J. Rossmeisl, T. Bligaard, J. Greeley, H. Jonsson, J.K. Norskov, Phys. Chem. Chem. Phys. 9 (2007) 3241–3250.
[7] R.A. van Santen, M. Neurock, S.G. Shetty, Chem. Rev. 110 (2010) 2005–2048.
[8] J.K. Norskov, F. Studt, F. Abild-Pederson, T. Bligaard, Fundamental Concepts in Heterogeneous Catalysis, Wiley, Hoboken, NJ, 2014.
[9] C.H. Christensen, J.K. Norskov, J. Chem. Phys. 128 (2008) 182503.
[10] J.K. Norskov, T. Bligaard, J. Rossmeisl, C.H. Christensen, Nat. Chem. 1 (2009) 37–46.
[11] J.R.B. Gomes, J.M. Bofill, F. Illas, J. Phys. Chem. C 112 (2008) 1072–1080.
[12] J.N. Bronsted, K.J. Pederson, Z. Phys. 108 (1924) 185–235.
[13] R.A. Marcus, J. Phys. Chem. 72 (1968) 891–899.
[14] J.M. Mayer, Acc. Chem. Res. 44 (2011) 36–46.
[15] A. Warshel, J. Phys. Chem. 86 (1982) 2218–2224.
[16] A. Cohen, R.A. Marcus, J. Chem. Phys. 78 (1968) 4249–4256.
[17] J.M. Mayer, Annu. Rev. Phys. Chem. 55 (2004) 363–390.
[18] D.N. Silverman, Biochim. Biophys. Acta, Bioenerg. 1458 (2000) 88–103.
[19] J.E. Leffler, Science 117 (1953) 340–341.
[20] G.S. Hammond, J. Am. Chem. Soc. 77 (1955) 334.
[21] D. Kashchiev, J. Chem. Phys. 76 (1982) 5098.
[22] H. Eyring, J. Chem. Phys. 4 (1936) 283–292.
[23] A.J. Appleby, J.H. Zagal, J. Solid State Electrochem. 15 (2011) 1811–1832.
[24] S.N. Zhurkov, Int. J. Fract. Mech. 1 (1965) 311.
[25] G.I. Bell, Science 200 (1978) 618.
[26] D.J. Wales, Energy Landscapes, Cambridge University Press, Cambridge, UK, 2003.
[27] S.S.M. Konda, J.N. Brandtley, B.T. Varghese, K.M. Wiggins, C.W. Bielawski, D.E. Makarov, J. Am. Chem. Soc. 135 (2013) 12722–12729.
[28] D.E. Makarov, J. Chem. Phys. 144 (2016) 030901.
[29] C. Taylor, S. Wasileski, J.-S. Filhol, M. Neurock, Phys. Rev. B 73 (2006) 165402.
[30] D. Kashchiev, G.M. van Rosmalen, Cryst. Res. Technol. 38 (2003) 555–574.

[31] O. Dudko, G. Hummer, A. Szabo, Phys. Rev. Lett. 96 (2006) 108101.
[32] G. Hummer, A. Szabo, Proc. Natl. Acad. Sci. USA 107 (2010) 10–15.
[33] M.S.Z. Kellermeyer, S.B. Smith, H.L. Granzier, C. Bustamante, Science 276 (1997) 1112–1116.
[34] W.J. Greenleaf, K.L. Frieda, D.A.N. Foster, M.T. Woodside, S.M. Block, Annu. Rev. Biophys. Biomol. Struct. 319 (2008) 630–633.
[35] J.D. Chodera, V.S. Pande, Phys. Rev. Lett. 107 (2011) 98102.
[36] C. Jarzynski, Phys. Rev. Lett. 78 (1997) 2690.
[37] G. Crooks, Phys. Rev. E 61 (2000) 2361–2366.
[38] R.C. Lua, A.Y. Grosberg, J. Phys. Chem. B 109 (2005) 6805–6811.
[39] C. Bustamante, J. Liphardt, F. Ritort, Phys. Today July (2005) 43–48.
[40] O.K. Dudko, G. Hummer, A. Szabo, Proc. Natl. Acad. Sci. USA 105 (2008) 15755–15760.
[41] R.A. Marcus, Phys. Chem. Chem. Phys. 14 (2012) 13729–13730.
[42] L.P. Hammett, J. Am. Chem. Soc. 59 (1937) 96.
[43] D. Kashchiev, Nucleation: Basic Theory with Applications, Butterworth-Heinemann, Oxford, 2000.

Index

M

N

O

P

Q

R

S

T

U

V

W

Z

Glossary of acronyms and constants

DFT density functional theory
FFS forward flux sampling
GH Grote-Hynes
kMC kinetic Monte Carlo
KLBS Kramers-Langer-Berezhkovskii-Szabo
LFER linear free energy relationship
LMax likelihood maximization
LQA local quadratic approximation
MARI most abundant reactive intermediate
MASI most abundant surface intermediate
MC Monte Carlo
MD molecular dynamics
MECP minimum energy crossing point
MEP minimum energy path
MFPT mean first passage time
PES potential energy surface
PMF potential of mean force
PSSA pseudo steady state approximation
QE quasi-equilibrium
QM/MM quantum mechanics/molecular mechanics
RDS rate determining step
RPMD ring-polymer molecular dynamics
RRKM Rice-Ramsberger-Kassels-Marcus
TIS transition interface sampling
TST transition state theory
TPS transition path sampling

Constant	Symbol	Values/Units
Boltzmann's constant	k_B	$8.315\, J/(°C\, mol) = 1.381 \times 10^{-23} J/°C$
Planck's constant	h	$6.626 \times 10^{-34} J \cdot s$
Speed of light	c	$2.998 \times 10^{8} m/s$
Electron charge	e	$1.602 \times 10^{-21} C$
Avogadro's number	$mol,\ N_A$	6.022×10^{23}
Vacuum permittivity	ε_0	$8.854 \times 10^{-12} C^2/(J \cdot m)$
Atomic mass unit	amu	$1.661 \times 10^{-27} kg \approx 1 g/mol$

Printed in the United States
By Bookmasters